Two Manuals Combined into One Book

Combat Tactics

And

Marine Scouting and Patrolling

Department of Defense

Pentagon
Publishing

Field Manual
No. 3-90

Headquarters
Department of the Army
Washington, DC, 4 July 2001

TACTICS

Contents

Figures

Tables

Preface

Doctrine provides a military organization with a common philosophy, a language, a purpose, and unity of effort. Tactics is the art and science of employing all available means to win battles and engagements. Specifically, it comprises the actions taken by a commander to arrange units and activities in relation to each other and the enemy.

PURPOSE

FM 3-90 introduces the basic concepts and control measures associated with the art of tactics. It cannot be read in isolation. To understand FM 3-90, the reader must understand the operational art, principles of war, and links between the operational and tactical levels of war described in FM 3-0, *Operations*. He should understand how the activities described in FM 3-07, *Stability Operations and Support Operations*, carry over and affect offensive and defensive operations and vice versa. He should understand the operations (plan, prepare, execute, and assess) process described in FM 6-0, *Command and Control*, and how that process relates to the military decision making process and troop-leading procedures described in FM 5-0, *Army Planning and Orders Production*. FM 3-90 also refers to joint publications (JPs). Reviewing referenced JPs will also aid in understanding FM 3-90.

FM 3-90 focuses on the tactics used to employ available means to win in combat. Those tactics require judgment in application. The ability to seize and secure terrain, with its populations and productive capacity, distinguishes land forces conducting decisive offensive and defensive operations. FM 3-90 provides a common discussion of how commanders from the battalion task force level through the corps echelon conduct tactical offensive and defensive operations and their supporting tactical enabling operations. It is not prescriptive, but authoritative.

Tactical fundamentals do not change with the fielding of each new piece of equipment. However, the integration of new equipment and organizations usually requires changes in related techniques and procedures. FM 3-90 provides guidance in the form of combat-tested concepts and ideas modified to exploit emerging Army and joint capabilities.

SCOPE

FM 3-90 focuses on the organization of forces, minimum essential control measures, and general planning, preparation, and execution considerations for each type and form of combat operation. It is the common reference for all students of the tactical art, both in the field and the Army school system. A family of subordinate manuals address the techniques and procedures used by tactical combat forces at each echelon. Assessment concepts described in FM 6-0 and FM 6-22, *Army Leadership*, also apply. All operations process considerations are modified as necessary to account for the specific factors of METT-TC existing during each operation.

Tactical enabling operations and special environments discussed in other manuals are not repeated in FM 3-90. That is why FM 3-90 does not discuss why information operations (FM 3-13), river-crossing operations (FM 3-97.13), and combined arms breaching operations (FM 3-34.2).

ADMINISTRATIVE INFORMATION

Readers of FM 3-90 should be aware that the rules regarding the use of land mines are evolving. A commander should seek legal guidance concerning currently applicable rules and policies regarding land mines before directing their employment. The rules of engagement he provides to his subordinates should reflect those guidelines.

The historical examples used in FM 3-90 were deliberately chosen to illustrate the fact that American tacticians should widen their areas of study of the military profession to include other armies and other times.

In FM 3-90, the term *heavy* applies to armored, aviation, cavalry, mechanized infantry and their associated combined arms elements. The term *light* applies to airborne, air assault, and light infantry and their associated combined arms elements. The term *special operations forces* applies to rangers, special forces, special operations aviation, civil affairs, and psychological operations elements. The *initial brigade combat teams* (IBCTs) are the prototypes of a medium force and are included in the discussion of heavy and light forces as appropriate.

In FM 3-90, the term *battalion* applies to a battalion, battalion task force, or a cavalry squadron unless differences in capabilities requires that a specific organization be named. The term *company* refers to a company, company team, battery, or troop size organization.

FM 3-90 uses *planning process* to indicate both the military decision making process and troop leading procedures. Battalion-sized and larger units use the military decision making process. Company-sized and smaller units follow troop leading procedures.

Unless otherwise stated, whenever FM 3-90 uses the singular form of decisive operation or shaping operation; both the singular and plural forms are included.

The glossary lists most terms used in FM 3-90 that have joint or Army definitions. Terms for which FM 3-90 is the proponent manual (the authority) are indicated with an asterisk. Definitions for which FM 3-90 is the proponent manual are printed in boldface in the text. For other definitions the term is italicized and the number of the proponent manual follows the definition.

Cross-references use the new field manual numbering system. The bibliography lists field manuals by new number followed by old number.

Unless stated otherwise, masculine nouns or pronouns do not refer exclusively to men.

Headquarters, US Army Training and Doctrine Command is the proponent for this publication. Send comments and recommended changes and the rational for those changes to: Commander, US Army Combined Arms Center and Fort Leavenworth, ATTN: ATZL-SWW, 1 Reynolds Road, Fort Leavenworth, KS 66027-1352.

PART ONE

Tactical Fundamentals

Chapter 1

The Art of Tactics

War is, above all things, an art, employing science in all its branches as its servant, but depending first and chiefly upon the skill of the artisan. It has its own rules, but not one of them is rigid and invariable. As new implements are devised new methods result in its mechanical execution; but over and above all its mechanical appliances, it rests upon the complex factors of human nature, which cannot be reduced to formulas and rules. The proper use of these thinking and animate parts of the great machine can be divined only by the genius and instinct of the commanders. No books can teach this, and no rules define it.

Captain Francis V. Greene, 1883

Tactics is the employment of units in combat. It includes the ordered arrangement and maneuver of units in relation to each other, the terrain and the enemy to translate potential combat power into victorious battles and engagements (FM 3-0).

CONTENTS

1-1. This is the capstone manual for offensive and defensive operations at the tactical level. This is a manual for professionals and requires dedication and study to master. It is authoritative and provides guidance in the form of combat-tested concepts and ideas modified to take advantage of emerging Army and joint capabilities, focusing on the tactics used to employ available means to win in combat. Those tactics are not prescriptive in nature but require judgment in application.

> ***Tactics*** **is the employment of units in combat.**
>
> ***Techniques*** **are the general and detailed methods used by troops and commanders to perform assigned missions and functions, specifically the methods of using equipment and personnel.**
>
> ***Procedures*** **are standard and detailed courses of action that describe how to perform tasks.**

1-2.The tactics and supporting techniques and procedures described in this manual are only starting points for the tactician, who must understand the difference between tactics and techniques and procedures. Tactics always require judgment and adaptation to the unique circumstances of a specific situation. Techniques and procedures are established patterns that can be applied repeatedly with little or no judgment in a variety of circumstances. Tactics, techniques, and procedures (TTP) provide the tactician with a set of tools to use in developing the solution to a tactical problem. The solution to any specific problem is a unique combination of these TTP or the creation of new ones based on a critical evaluation of the situation. The tactician determines his solution by a thorough mastery of doctrine and existing TTP, tempered and honed by experience gained through training and operations. He uses his creativity to develop solutions for which the enemy is neither prepared, nor able to cope.

THE TACTICAL LEVEL OF WAR

1-3.The levels of war are doctrinal perspectives that clarify the links between strategic objectives and tactical actions. Although there are no finite limits or boundaries between them, the three levels are strategic, operational, and tactical. They apply to all types of military operations.

1-4.The *tactical level of war* is the level of war at which battles and engagements are planned and executed to accomplish military objectives assigned to tactical units or task forces. Activities at this level focus on the ordered arrangement and maneuver of combat elements in relation to each other and to the enemy to achieve combat objectives (JP 1-02). It is important to understand tactics within the context of the levels of war. The strategic and operational levels provide the context for tactical operations. Without this context, tactical operations are reduced to a series of disconnected and unfocused actions. Engagements are linked to battles. One or more battles are linked to winning major operations and campaigns, leading to operational success, which can lead to strategic success. (FM 3-0 discusses major operations and campaigns.)

1-5.A *battle* consists of a set of related engagements that last longer and involve larger forces than an engagement (FM 3-0). Battles can affect the course of the campaign or major operation. A battle occurs when a division, corps, or army commander fights for one or more significant objectives. Battles are usually operationally significant, if not operationally decisive.

1-6.An *engagement* is a small, tactical conflict between opposing maneuver forces, usually conducted at brigade level and below (FM 3-0). An engagement normally lasts only a short time—minutes, hours, or a day. It can result from one side's deliberate offensive movement against an opponent or from a chance encounter between two opponents, such as a meeting engagement. An engagement can be a stand-alone event or one of several related engagements comprising a battle.

1-7. Levels of command, size of units, types of equipment, or types of forces or components are not associated with a particular level of war. National assets, such as intelligence and communications satellites, previously considered principally in a strategic context, are an important adjunct to tactical operations. Actions are strategic, operational, or tactical based on their effect or contribution to achieving strategic, operational, or tactical objectives. Many times the accuracy of these labels can only be determined during historical studies.

1-8. Advances in technology, information-age media reporting, and the compression of time-space relationships contribute to the growing interrelationships between the levels of war. The levels of war help commanders visualize a logical flow of operations, allocate resources, and assign tasks to the appropriate command. However, commanders at every level must be aware that in a world of constant, immediate communications, any single event may cut across the three levels (see FM 3-0).

THE SCIENCE AND ART OF TACTICS

1-9. The tactician must understand and master the science and the art of tactics, two distinctly different yet inseparable concepts. **A *tactician* is an individual devoted to mastering the science and art of tactics.** Commanders and leaders at all echelons and supporting commissioned, warrant, and noncommissioned staff officers must be tacticians to lead their soldiers in the conduct of full spectrum operations.

THE SCIENCE

1-10. **The *science of tactics* encompasses the understanding of those military aspects of tactics—capabilities, techniques, and procedures—that can be measured and codified.** The science of tactics includes the physical capabilities of friendly and enemy organizations and systems, such as determining how long it takes a division to move a certain distance. It also includes techniques and procedures used to accomplish specific tasks, such as the tactical terms and control graphics that comprise the language of tactics. While not easy, the science of tactics is fairly straightforward. Much of what is contained in this manual is the science of tactics—techniques and procedures for employing the various elements of the combined arms team to achieve greater effects.

1-11. Mastery of the science of tactics is necessary for the tactician to understand the physical and procedural constraints under which he must work. These constraints include the effects of terrain, time, space, and weather on friendly and enemy forces. However—because combat is an intensely human activity—the solution to tactical problems cannot be reduced to a formula. This realization necessitates the study of the art of tactics.

THE ART

1-12. **The *art of tactics* consists of three interrelated aspects: the creative and flexible array of means to accomplish assigned missions, decision making under conditions of uncertainty when faced with an intelligent enemy, and understanding the human dimension—the effects of combat on soldiers.** An art, as opposed to a science, requires exercising intuitive faculties that cannot be learned solely by study. The tactician must temper his study and evolve his skill through a variety of relevant, practical experiences. The more experience the tactician gains from practice under a variety of circumstances, the greater his mastery of the art of tactics.

1-13. The tactician invokes the art of tactics to solve tactical problems within his commander's intent by choosing from interrelated options, including—

- Types and forms of operations, forms of maneuver, and tactical mission tasks.
- Task organization of available forces, to include allocating scarce resources.
- Arrangement and choice of control measures.
- Tempo of the operation.
- Risks the commander is willing to take.

1-14. These options represent a starting point for the tactician to create a unique solution to a specific tactical problem. Each decision represents a choice among a range of options; each balances competing demands requiring judgment at every turn. While there may be checklists for techniques and procedures, there are no checklists for solving tactical problems. The commander must not look for a checklist approach to tactics; instead, he must use his experience and creativity to outthink his enemy.

1-15. There are three aspects to the art of tactics that define a competent tactician. The first is the creative and flexible application of the tools available to the commander, such as doctrine, tactics, techniques, procedures, training, organizations, materiel, and soldiers in an attempt to render the enemy's situational tactics ineffective. The tactician must understand how to train and employ his forces in full spectrum operations. The factors of mission, enemy, terrain and weather, troops, time available, and civil considerations (METT-TC) are variables whose infinite mutations always combine to form a new tactical pattern. (FM 6-0 discusses the factors of METT-TC in detail.) They never produce exactly the same situation; thus there can be no checklists that adequately address each unique situation. Because the enemy changes and adapts to friendly moves during the planning, preparation, and execution of an operation, there is no guarantee that a technique which worked in one situation will work again. Each tactical problem is unique and must be solved on its own merits.

1-16. The second aspect of the art of tactics is decision making under conditions of uncertainty in a time-constrained environment and demonstrated by the clash of opposing wills—a violent struggle between two hostile, thinking, and independent opposing commanders with irreconcilable goals. Each commander wants to impose his will on his opponent, defeat his opponent's plans, and destroy his opponent's forces. Combat consists of the interplay between these two opposing commanders, with each commander seeking to accomplish his mission while preventing the other from doing the same. Every commander needs a high degree of creativity and clarity of thought to outwit a willing and able opponent. He must quickly apply his judgment to a less than omniscient common operational picture provided by his command and control (C2) system to understand the implications and opportunities afforded him by the situation. The commander always uses the most current intelligence in order to facilitate his visualization of the enemy and environment. That same C2 system transmits the decisions resulting from his situational understanding to those individuals and units required to engage and destroy the enemy force.

1-17. The third and final aspect of the art of tactics is understanding the human dimension—what differentiates actual combat from the problems encountered during training and in a classroom. Combat is one of the most complex human activities, characterized by violent death, friction, uncertainty, and chance. Success depends at least as much on this human aspect as it does on any numerical and technological superiority.

The Human Dimension—Combat at Han-sur-Nied

When only [300] yards from the bridge, the [1-317th IN] skirmish line was hit by high explosive shells from a detachment of sixteen 40-mm. antiaircraft guns.... The armored infantry froze in their places or tried to reach the shelter of the ditches alongside the road..., while projectiles..., fired with almost sniperlike accuracy, swept...their ranks. The 231st Armored [FA BN] turned its howitzers on the enemy..., but as the German gunners were blasted—arms and legs flying into the air—others ran forward to serve the weapons.

...[1LT Vernon L.] Edwards' [platoon of the 68th Tank BN] started across the bridge. The first tank crossed successfully. The second stalled on the bridge when the platoon commander was hit; for a brief while the tank stood there, [1LT] Edwards' body dangling from the open turret. The third received a direct hit and burst into flame, but was backed off the wooden bridge by its commander after he had ordered his crew to leave the blazing tank. During this effort...[1LT] Daniel Nutter and [CPL] Charles Cunningham, B Company 25th Armored [EN BN], ran forward to cut the wires leading to the demolition charges. [1LT] Nutter, at the enemy end of the bridge, was killed just as he completed his task. [CPL] Cunningham, who had cut the wires at the western end...raced across the bridge, and returned with the body of his commander.

Who [ordered] the final charge probably never will be known. Perhaps it was [LTC Sterling S.] Burnette, who had been standing erect in the open urging his lead company on and...was mortally wounded. [CPT James A.] Craig and a few men rushed the bridge, crossing the 100-foot span "faster than they knew how" amidst a hail of shell fragments and tracer bullets.... [CPT] Craig disposed his little force...and through the afternoon held the approach to the bridge against German tanks and riflemen.

1-18. The tactician cannot ignore the human aspect. He seeks to recognize and exploit indicators of fear and weakness in his enemy, and to defeat the enemy's will, since soldiers remain key to generating combat power. More than any other human activity, continuous combat operations against an intelligent enemy takes a toll on soldiers, severely straining their physical and mental stamina. This creates in soldiers the tangible and intangible effects of courage, fear, combat experience, exhaustion, isolation, confidence, thirst, and anger. If left unchecked these effects can result in decreased vigilance, slowed perception, inability to concentrate, communication difficulties, and an inability to accomplish manual tasks.

1-19. Leaders must be alert to indicators of fatigue, fear, lapses in discipline standards, and reduced morale in friendly and enemy soldiers. They must work to counteract the effects on the friendly force while taking measures to enhance these effects on the enemy. When the friendly force has the initiative, it can force the enemy to conduct continuous operations to react to friendly actions and then exploit the effects of continuous operations on the enemy. These conditions can have a cumulative effect on units that can lead to collapse. The tactician must understand how they affect human endurance and factor them into his plans. He must understand the limits of human endurance in combat. This is the subtle difference between pushing soldiers beyond their limits to exploit success versus resting them to prevent the collapse of unit cohesion. (FM 6-22.5 discusses the effects of continuous combat operations.)

HISTORICAL EXAMPLE

1-20. The following vignette discusses the Battle of Cowpens fought during the American Revolution. It illustrates the need for the tactician to combine the effects of the science of tactics with his application of the tactical art.

Cowpens, 17 January 1781

On 17 January 1781, American BG Daniel Morgan defeated British LTC Tarleton at the Cowpens in South Carolina in a battle that captures the essence of the art of tactics—the use of intuitive faculties that cannot be learned solely by study. Although outnumbered, Morgan's troops fought and won against a previously unbeaten opponent. The battle achieved decisive results with strategic significance.

In December 1780, Morgan was sent with 600 men into the South Carolina highlands under orders to protect Americans, forage, and threaten British control of the highlands. Tarleton's British Legion—numbering nearly 1,000 men and consisting of a combined cavalry-infantry force reinforced with additional infantry and two, three-pounder guns—was sent to pursue Morgan. By 16 January 1781, Tarleton had closed to within six miles of Morgan's force.

Having known of Tarleton's mission since 14 January, Morgan obtained information on Tarleton's tactical style while he began to retreat. With Tarleton so close on 16 January and his own camp nearly six miles from the Broad River, Morgan decided to fight at the Cowpens. While the two forces were now roughly the same size, Morgan had only about one-third the cavalry, one-third the regular line infantry, and no artillery. However, his militia force's rifles had a longer range than the British muskets, and the terrain allowed him to mask his reserves from view. Morgan deployed his forces on the battlefield, confident that Tarleton would not attempt to flank his position. Rather, Morgan believed that Tarleton would initially attack him with part of his cavalry supported by infantry, wait for confusion, and then exploit those vulnerabilities with his cavalry reserve. This had been his pattern of operations in previous engagements with the Americans.

In the main line of battle, Morgan placed his Continentals and Virginia militia, who were former Continentals, on the military crest of the rise under COL Howard's command. They numbered 450. About 150 meters downhill, toward the expected reaction of the enemy, he stationed the bulk of his militia under the command of COL Pickens. Another 150 meters down the hill, he positioned a skirmisher line of 150 militia riflemen. Behind the hill, he placed LTC Washington's 120-strong cavalry force in reserve. Each line was within rifle range but out of musket range of the line behind it. As explained the night before, skirmishers only needed to fire one or two shots and then retire to the second line. In turn, the militia in the second line only needed to fire two volleys. Then they could retire to their left around to the rear of the hill and, protected by the cavalry, reform. The cavalry would counterattack British cavalry as the situation allowed, guard the militia horses, or cover a retreat if necessary. (See Figure 1-1.)

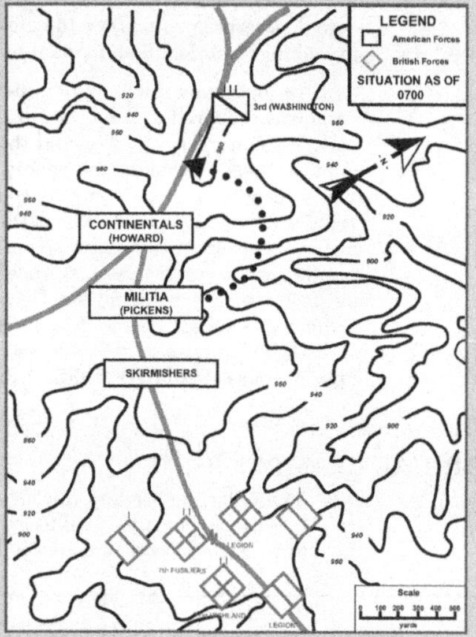

Figure 1-1. Initial Situation

Tarleton deployed his forces from their march formation into a line, with three light infantry companies on the right, the Legion infantry in the center, and one regular British battalion on the left of this main line. He stationed one troop of cavalry on each flank of the main line and one three-pounder gun on either side of the Legion infantry. He kept the other regular British battalion and the remainder of his Legion cavalry in reserve. The British immediately came under fire from the skirmishers. Tarleton sent a troop of dragoons to disperse them while his main forces deployed. The dragoons lost 15 of 50 men. The skirmishers retired to the American second line.

Tarleton then assaulted the second line. His artillery opened fire, but apparently on the third line. The American rifle fire disrupted his formation. When his forces closed to within 50 meters of the second line, they received a volley from the militia that staggered and further disrupted them. The militia's expected second volley was more ragged as it began to withdraw. Seeing this movement, Tarleton ordered his right-hand troop of dragoons to charge the militia as it withdrew. The American cavalry charged this troop and overwhelmed it, driving the dragoons off the field in accordance with Morgan's plan. (See Figure 1-2.)

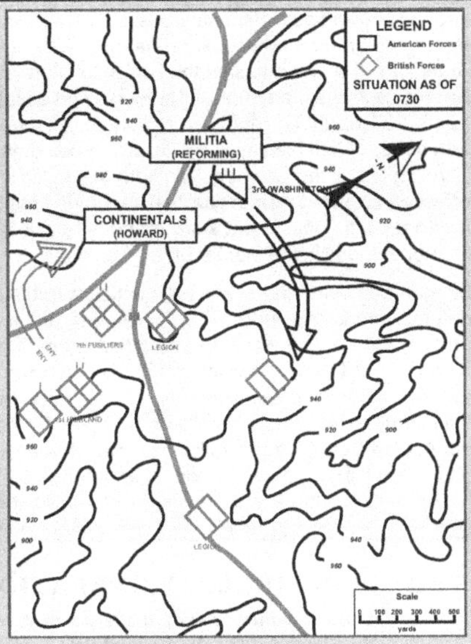

Figure 1-2. British Moves

As the British moved hastily forward to assault the main American position, they further lost their cohesion as a firefight between the two forces ensued. Tarleton ordered his reserve infantry battalion up to the left of his line for this assault, and the cavalry troop on his left to encircle the American line. This move outflanked the American line. Morgan and the third-line commander recognized the danger to their right flank and ordered the right flank units to "refuse" the flank. However, the American units adjoining those right flank units also commenced moving to the rear. This situation could have crumbled the American line except for Morgan's personal order for the rest of the American line to move to the rear with those right flank units.

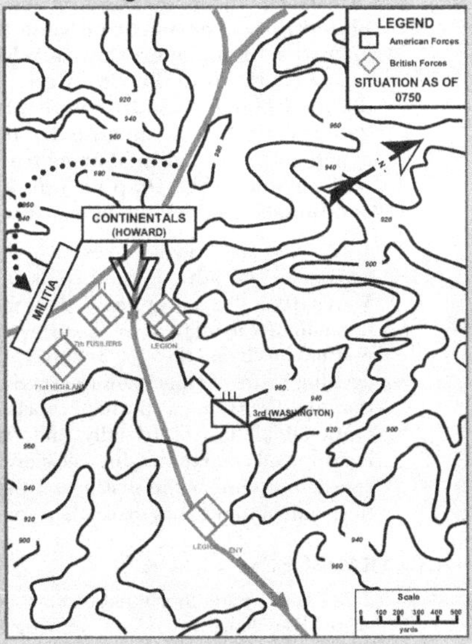

Figure 1-3. Final Battle Stage

Seeing this apparent general withdrawal, Tarleton ordered his forces to close with the Americans. They did, but suffered further disorganization. Just as the British attempted to close, the Americans turned and fired a volley, followed by a bayonet charge into the British lines. Simultaneously, Morgan's cavalry attacked the British right from the rear. Meanwhile, the militia, having reformed, returned to the field on the American right and attacked the British left flank units. (See Figure 1-3.) The battle was over within an hour of Tarleton's first assault. The British losses were 110 killed, 200 wounded, and 700 prisoners, although Tarleton personally escaped with about 140 of his cavalry. The British could not replace the mobile forces that Tarleton lost at the Cowpens. Without a mobile force, the British no longer had an effective counter to American partisans and light forces. The British later won at Guilford Court House, but suffered such heavy losses that they had to abandon their operations in the interior of the Carolinas.

Morgan combined the science of tactics with his application of the tactical art to defeat superior numbers of British forces under Tarleton. Morgan arrived at a unique and creative solution to his tactical problems. Trusted information about Tarleton's style was a crucial part of his deployment plan. Morgan understood the diverse military and social elements of his force (untried militia, Continentals, and volunteers). He asked no more of any element than it could deliver and used the strengths of each to the fullest. Daniel Morgan used tactical art to convert his understanding of American troops, knowledge of human nature, and rapport with his soldiers into the vital components of a brilliant tactical victory.

HASTY VERSUS DELIBERATE OPERATIONS

1-21. **A *hasty operation* is an operation in which a commander directs his immediately available forces, using fragmentary orders (FRAGOs), to perform activities with minimal preparation, trading planning and preparation time for speed of execution. A *deliberate operation* is an operation in which a commander's detailed intelligence concerning the situation allows him to develop and coordinate detailed plans, including multiple branches and sequels. He task organizes his forces specifically for the operation to provide a fully synchronized combined arms team. He conducts extensive rehearsals while conducting shaping operations to set the conditions for the conduct of his decisive operation.**

1-22. Most operations lie somewhere along a continuum between these two extremes. The 9th Armored Division's seizure of the bridge at Remagen in March 1945 illustrates one end, a hasty operation conducted with the forces immediately available. At the other end of the continuum is a deliberate operation, such as the 1st Infantry Division's breach operation during the opening hours of Operation Desert Storm. Ongoing improvements in information and C2 systems continue to assist in the development of a common operational picture of friendly and enemy forces while facilitating decision making and communicating decisions to friendly forces. These improvements can help diminish the distinction between hasty and deliberate operations; they cannot make that distinction irrelevant.

CHOICES AND TRADEOFFS

1-23. The commander must choose the right point along the continuum to operate. His choice involves balancing several competing factors. He bases his decision to conduct a hasty or deliberate operation on his current knowledge of the enemy situation, and his assessment of whether the assets available (to include time), and the means to coordinate and synchronize those assets, are adequate to accomplish the mission. If they are not he takes additional time to plan and prepare for the operation or bring additional forces to bear on the problem. The commander makes that choice in an environment of uncertainty, which always entails some risk.

1-24. The commander may have to act based only on his available combat information in a time-constrained environment. *Combat information* is unevaluated data gathered by or provided to a commander that, due to its highly perishable nature or the critical nature of the tactical situation, cannot be processed into tactical intelligence or other staff products in time to meet the commander's information requirements (FM 6-0). The commander must understand the inherent risk of acting only on combat information since it is vulnerable to enemy deception operations and can be misinterpreted at any stage up through reporting channels. The unit intelligence staff helps the commander assign a level of confidence to combat information he uses in decision making.

1-25. Uncertainty and risk are inherent in tactical operations and cannot be eliminated. A commander cannot be successful without the capability of acting under conditions of uncertainty while balancing various risks and taking advantage of opportunities. Although the commander strives to maximize his knowledge about his forces, the terrain and weather, civil considerations, and the enemy, he cannot let a lack of information paralyze him. The more intelligence on the enemy, the better able the commander is to make his assessment. Less information means that the commander has a greater risk of making a poor decision for the specific situation. A commander never has perfect intelligence, but knowing when he has enough information to make a decision within the higher commander's intent and constraints is part of the art of tactics and is a critical skill for a commander.

1-26. The commander should take the minimum time necessary in planning and preparing to ensure a reasonable chance of success. Reduced coordination at the start of the operation results in less than optimum combat power brought to bear on the enemy, but often allows for increased speed and momentum while possibly achieving surprise. The commander must balance the effects of reduced coordination against the risk that the effects of increased coordination will not match the enemy's improved posture over time. The more time the commander takes to prepare for the operation, including improving his situational understanding, the more time the enemy has to prepare and move additional units within supporting range or distance. Additionally, it reduces the time his subordinates have to conduct their own planning and preparations. If the enemy can improve his disposition faster than the friendly force can, the delays in execution decrease the commander's chances of success.

1-27. It is better to err on the side of speed, audacity, and momentum than on the side of caution when conducting military operations, all else being equal. Bold decisions give the best promise of success; however, one must differentiate between calculated risks and a military gamble. A *calculated risk* is an operation in which success is not a certainty but which, in case of failure, leaves sufficient forces to cope with whatever situations arise (FM 6-0). The willingness to take calculated risks requires military judgment to reduce risk by foresight and careful planning and to determine whether the risk is worth taking to grasp fleeting opportunities. MG Wood's decision to advance east toward the German border with his 4th Armored Division after the breakout from the Normandy beachhead is an example of a justifiable calculated risk. A *military gamble* is an operation that can lead either to victory or to complete destruction of one's force (FM 6-0). Rare situations can arise where even a gamble may be justified; for example, when defeat is merely a matter of time and the only chance lies in an operation of great risk. LTC Chamberlain's decision to conduct a bayonet charge with what was left of the 20th Maine on the second day of the Battle of Gettysburg is an example of a military gamble.

1-28. The commander can be less deliberate in planning and preparing for an operation when facing a clearly less-capable and less-prepared enemy force. In these circumstances, the commander can forego detailed planning, extensive rehearsals, and significant changes in task organization. For example, an attacking battalion task force encountering enemy security outposts just moving into position will conduct actions on contact to immediately destroy the outposts without the loss of momentum. It then follows that against a larger and more prepared enemy, the commander needs more preparation time and a larger force to succeed. If the commander determines that he cannot defeat the enemy with the forces immediately at hand, he must determine what additional measures he must take to be successful. The measures can include any or all of the factors along the continuum.

1-29. This does not imply that a commander conducting a hasty operation foregoes the advantages provided by his combined arms team. A commander who chooses to conduct hasty operations synchronizes the employment of his forces in his head as he issues FRAGOs. He uses tangible and intangible factors, such as the training level and experience of his subordinates, his own experience, perception of how the enemy will react, understanding of time-distance factors, and knowledge of the strengths of each subordinate and supporting unit to achieve the required degree of synchronization.

RISK REDUCTION

1-30. An important factor in reducing a commander's risk is how much intelligence he has about the enemy. As intelligence becomes available, the commander determines where along the continuum of hasty versus deliberate operations he will operate to accomplish his mission. There is no set of rules to determine this point—any choice entails risk. If the commander decides to execute a hasty operation based on limited intelligence, he risks an uncoordinated operation against an enemy about which he knows little. Moreover his forces may not be strong enough to accomplish their mission with minimum casualties. This could lead to piecemeal commitment and potential defeat in detail. He must balance this option against the risk of waiting to attack, which allows the enemy time to reinforce or conduct additional preparation.

1-31. When higher headquarters determines the time to start an operation, or in a defense when the enemy initiates the operation, the commander has little flexibility regarding where to operate along the continuum of hasty versus deliberate operations. In these situations he must use all the time available to conduct planning and preparation. While the military decision making process tasks used in a time-constrained environment are the same as in the full process, many are done mentally by the commander or with less staff involvement. Each commander decides how to shorten the process. A commander may use the complete process to develop the plan, while a subordinate headquarters abbreviates the process. (See FM 5-0 for a discussion of decision making in a time-constrained environment.)

1-32. The commander can reduce the risk associated with any situation by increasing his knowledge of the terrain and friendly, neutral, and enemy forces. He has a greater risk of making a poor decision if his situational understanding is incomplete or faulty. If the commander lacks sufficient information to make an informed choice, his first priority must be to gain the required information to support his decision making while at the same time taking precautions to protect his force from surprise. During an unexpected encounter with the enemy, often an acceptable way to gain that intelligence is to conduct a hasty attack to determine the size and disposition of the enemy force. The commander adapts his reconnaissance and intelligence efforts to the existing situation and picks the appropriate tools to answer his critical information requirements. For example, the commander can retask his reconnaissance assets or increase the size of his reconnaissance effort.

1-33. A commander—supported by a digital C2 system that can access accurate, real-time information—takes advantage of a different operational environment than that facing a commander with an analog C2 system. Greatly improved knowledge of the enemy and friendly situations facilitates his employment of precision fires, his conduct of decisive maneuver at extended ranges, and his provision of responsive and flexible support of his forces. The integration of advanced information technologies, highly capable leaders, and agile organizational systems reduces risk and facilitates the conduct of full spectrum operations.

1-34. Risk reduction does not always mean increasing knowledge of the enemy at the expense of time. A commander can partially compensate for a lack of intelligence by being flexible in his troop dispositions through an increase in the depth of the security area, size and number of security units, and size of the reserve. The commander's choices of combat and movement formations provide the versatility to allow for initial enemy contact with the smallest possible friendly force. This allows the greatest flexibility in meeting unforeseen enemy dispositions. Another way to compensate for increased risk is to allow time and resources for subordinate elements to develop the situation.

1-35. Because uncertainty exists in all military operations, every military decision contains risk. The commander exercises tactical art when he decides how much risk to accept. As shown in Figure 1-4, the commander has several techniques available to reduce the risk associated in a specific operation. Some of these techniques for reducing risk take resources from the decisive operation, which reduces the concentration of effects at the decisive point.

1-36. The commander has the option to redirect the efforts of forces previously used to reduce his risk toward strengthening his decisive operation as more information becomes available. In any operation, the relationship between information, uncertainty, risk, size of reserves and security forces, and the disposition of the main body may change frequently. The commander must continually weigh this balance and make adjustments as needed.

1-37. These adjustments can create problems. Too many changes or changes made too rapidly in task organization, mission, and priorities can have negative effects on the operations process. For example, if a commander changes the task organization of his force too frequently, the force fails to develop the flexibility provided by teamwork. On the other hand, if he fails to change the task organization when dictated by circumstances, the force lacks flexibility to adapt to those changing circumstances. It is then unable to react effectively to enemy moves or act with the concentration of effects that lead to mission success.

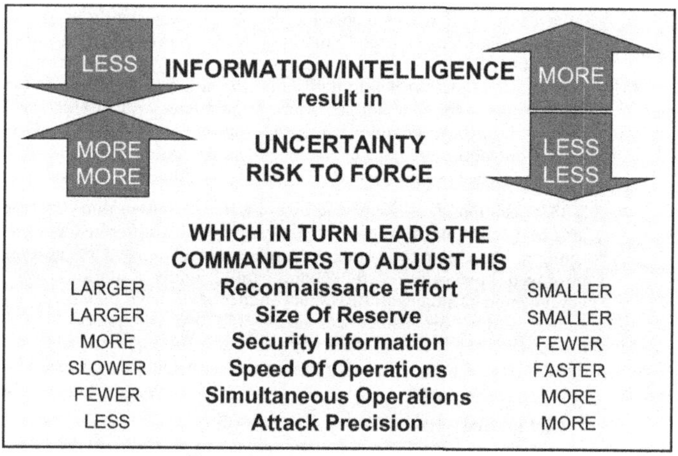

Figure 1-4. Risk Reduction Factors

SOLVING TACTICAL PROBLEMS

1-38. Success in tactical problem solving results from the aggressive, intelligent, and decisive use of combat power in an environment of uncertainty, disorder, violence, and danger. A commander wins by being on the offense, initiating combat on his own terms—at a time and place of his choosing. He never surrenders the initiative once he gains it. He builds momentum quickly to win decisively through the rapid application of available combat power, operating inside the enemy's decision making cycle, and mastering the transitions between the defense to the offensive and vice versa. These rules of thumb allow the commander to maximize friendly and minimize enemy combat power by preventing the enemy from fighting as a combined arms force.

1-39. Offensive action is key to achieving decisive results. Tactical commanders conduct offensive operations to achieve their assigned missions and objectives—destroying enemy forces or seizing terrain—that cumulatively produce the theater-level effects required by the operational commander.

Circumstances may require defending; however, tactical success normally requires shifting to the offense as soon as possible. The offense ends when the forces conducting it accomplish their missions, reach their limits of advance, or approach culmination. Those forces then consolidate, resume the attack, or prepare for other operations.

1-40. A commander wants to initiate combat on his own terms to give himself important advantages. This allows him to mass the effects of his combat power against selected inferior and isolated enemy units in vulnerable locations. Possession of the initiative allows a commander to continually seek vulnerable spots and shift his decisive operation when opportunities occur. A commander seizes, retains, and exploits the initiative by—

- Maneuvering more rapidly than the enemy to gain positional advantage (the place where the effects of fires are most destructive) over the enemy.
- Employing firepower to facilitate and exploit positional advantage.
- Sustaining his forces before, during, and after the engagement with the enemy.
- Achieving and maintaining information superiority.
- Planning beyond the initial operation and anticipating possible events.

A commander never surrenders the initiative once he gains it. He presses the fight tenaciously and aggressively. He accepts risk while leading soldiers and pushing systems to their limits.

1-41. The tactician, notwithstanding his status as a commander or a staff officer, seeks ways to build momentum quickly by seizing the initiative and executing shaping, sustaining, and decisive operations at a high tempo. Momentum helps to retain and complements the initiative. Concentrating the effects of combat power at the decisive place and time overwhelms an enemy and gains control of the situation. Rapid maneuver to place the enemy in a disadvantageous position also builds momentum. Momentum allows the tactician to create opportunities to engage the enemy from unexpected directions with unanticipated capabilities. Having seized the initiative, the tactician continues to control the relative momentum by taking action to maintain focus and pressure, controlling the tempo of operations, and creating and exploiting opportunities, while always assessing the situation and taking calculated risks.

1-42. The commander's C2 system assists the rapid building of momentum by allowing him to see and understand the situation so quickly that his forces can act before the enemy forces can react to the initial situation. His operations process focuses on executing rather than planning. Modern information systems allow compressed planning and effective incremental adjustments to the plan during execution. This allows the commander's forces to adapt more quickly to emerging threats and opportunities as they are identified. Units whose commanders can make and implement decisions faster, even to a small degree, gain an accruing advantage that becomes significant over time; making decisions quickly—even with incomplete information—is crucial.

1-43. The tactician chooses from a number of tactical options to create the solution to the tactical problem facing him. (Chapter 2 lists these options as the types and forms of military operations and forms of maneuver.) Although he solves the specific tactical problem facing him by following the general principles outlined in this manual, there is no single, doctrinally correct, procedurally derived solution to any problem. The tactician who employs the more appropriate tactics given the existing situation has a distinct advantage over his opponent, even if their forces have equal combat power.

1-44. The tactician uses his mastery of the art and science of tactics, his understanding of the situation, and his judgment to create unique solutions appropriate to the mission and the other specific factors of METT-TC. There are usually several solutions that might work, although some will be more effective. He seeks a solution that defeats the enemy in the time available at the least cost in men and materiel. It should be a decisive solution that postures the unit for future missions and provides for the greatest flexibility to account for unexpected enemy actions or reactions. The solution must be in accordance with the higher commander's intent. A thorough understanding of the enemy greatly assists the commander in his development of workable solutions. Commander's visualization is the doctrinal term for this process. (FM 6-0 describes commander's visualization.)

1-45. The tactician learns to cut to the heart of a situation, recognize its important elements, and base his decisions on those important elements as he masters his profession. The ability to do this cannot be acquired overnight. A tactician develops this capability after years of schooling, self-study, and practical training experiences, which eventually develop the intuitive faculties required to solve tactical problems. He rarely gets the opportunity to practice the science and art of tactics under actual combat conditions.

1-46. Doctrine requires human judgment when applied to a specific situation. In choosing a solution to a tactical problem, applicable laws and regulations, the mission, the laws of physics, human behavior, and logistic realities constrain the tactician, not standardized techniques and procedures. The true test of the tactician's solution is not whether it uses the specific techniques or procedures contained in this manual, but whether the techniques and procedures used were appropriate to the situation. Tactical proficiency is not defined by mastery of written doctrine, but by the ability to employ available means to win battles and engagements. A solution may not match any previous doctrinal example; however, the language used to communicate that concept must be technically precise and doctrinally consistent, using commonly understood and accepted doctrinal terms and concepts.

1-47. Transitions between the types and forms of operations are difficult and, during execution, may create unexpected opportunities for Army or enemy forces. The tactician must quickly recognize such opportunities. He develops transitions as branches during the planning process and acts on them immediately as they occur. Transition between one type or form of an operation to another is a complex operational consideration.

1-48. Tactical victory occurs when the opposing enemy force can no longer prevent the friendly force from accomplishing its mission. That is the end goal of all military operations. Decisive tactical victory occurs when the enemy no longer has the means to oppose the friendly force. It also occurs when the enemy admits defeat and agrees to a negotiated end of hostilities. Historically, a rapid tactical victory results in fewer friendly casualties and reduced resource expenditures. However, the tactician avoids gambling his forces and losing his combined arms synchronization in search of rapid victory.

1-49. In closing, solutions to tactical problems are a collective effort. Success results from the commander's plan and the ability of subordinates to execute it. The commander must have full confidence in his subordinates' mastery of the art and science of tactics and in their ability to execute the chosen solution. (See FM 6-0 for a full discussion of this concept.)

Chapter 2

Common Tactical Concepts and Graphic Control Measures

In war, obscurity and confusion are normal. Late, exaggerated or misleading information, surprise situations, and counterorders are to be expected.

Infantry in Battle, 1939

The tactician must understand the common tactical concepts and definitions used by the military profession in the conduct of offensive and defensive operations. This chapter introduces the doctrinal hierarchy that forms the framework by which this manual is organized. The concepts and terms in this chapter are common to most operations. This manual discusses those concepts and terms specific to a type or form of operations in the corresponding chapter. For example, Chapter 4 discusses the objective as a control measure.

DOCTRINAL HIERARCHY

2-1. Figure 2-1 on page 2-2 shows the doctrinal hierarchy and relationship between the types and subordinate forms of operations. While an operation's

CONTENTS

predominant characteristic labels it as an offensive, defensive, stability, or support operation, different units involved in that operation may be conducting different types and subordinate forms of operations, and often transition rapidly from one type or subordinate form to another. The commander rapidly shifts from one type or form of operation to another to continually keep the enemy off balance while positioning his forces for maximum effectiveness. Flexibility in transitioning contributes to a successful operation. A good tactician chooses the right combinations of combined arms to place the enemy at the maximum disadvantage.

Figure 2-1. Doctrinal Hierarchy of Operations

2-2. The commander conducts tactical enabling operations to assist the planning, preparation, and execution of any of the four types of military operations (offense, defense, stability, and support). Tactical enabling operations are never decisive operations in the context of offensive and defensive operations; they are either shaping or sustaining operations. Part IV of this manual discusses tactical enabling operations that are not the subject of a separate field manual. The commander uses tactical enabling operations to help him conduct military actions with minimal risk.

2-3. This hierarchy does not describe discrete, mutually exclusive operations. All tactical missions can contain elements of several different types and subordinate forms. For example, an attacking commander may have one subordinate conducting an envelopment, with another subordinate conducting a frontal attack to fix the enemy. The enveloping force usually attacks once the direct-pressure force makes a movement to contact while repeatedly attacking to keep pressure on the fleeing enemy. The encircling force uses an envelopment to conduct a series of attacks to destroy or clear enemy forces in its path on the way to its blocking position. Once it occupies the blocking position, the unit may transition to a defense as it blocks the retreat of the fleeing enemy force.

THE OPERATIONAL FRAMEWORK

2-4. The *operational framework* consists of the arrangement of friendly forces and resources in time, space, and purpose with respect to each other and the enemy or situation. It consists of the area of operations, battlespace, and the battlefield organization (FM 3-0). The framework establishes an area of geographic and operational responsibility for commanders and provides a way for them to visualize how they will employ forces against the enemy. Army commanders design an operational framework to accomplish their mission by defining and arranging its three components. The commander uses the operational framework to focus combat power. *Combat power* is the total means of destructive and/or disruptive force which a military unit/formation can apply against the opponent at a given time (JP 1-02).

2-5. As part of the military decision making process, the commander visualizes his battlespace and determines how to arrange his forces. The *battlefield organization* is the allocation of forces in the area of operations by purpose. It consists of three all-encompassing categories of operations: decisive, shaping, and sustaining (FM 3-0). Purpose unifies all elements of the battlefield organization by providing the common focus for all actions. The commander organizes his forces according to purpose by determining whether each unit's operation will be decisive, shaping, or sustaining. Those decisions form the basis of his concept of operations. He describes the area of operations (AO) in terms of deep, close, and rear areas when the factors of METT-TC require the use of a spatial reference. FM 3-0 explains the operational framework and battlefield organization.

PRINCIPLES OF WAR

2-6. The nine principles of war defined in FM 3-0 provide general guidance for conducting war and military operations other than war at the strategic, operational, and tactical levels. They are fundamental truths governing combat operations. The principles are the enduring bedrock of Army doctrine. First published in 1923 as general principles in *Field Service Regulations United States Army*, they have stood the tests of analysis, experimentation, and practice. They are not a checklist and their degree of application varies with the situation. Blind adherence to these principles does not guarantee success, but each deviation may increase the risk of failure. The principles of war lend rigor and focus to the purely creative aspects of tactics and provide a crucial link between pure theory and actual application.

TENETS OF ARMY OPERATIONS

2-7. The tenets of Army operations—initiative, agility, depth, synchronization, and versatility—build on the principles of war. They further describe the characteristics of successful operations. While they do not guarantee success, their absence risks failure. FM 3-0 defines the tenets.

Principles of War

- Objective
- Offensive
- Mass
- Economy of force
- Maneuver
- Unity of command
- Security
- Surprise
- Simplicity

Tenets of Army Operations

- Initiative
- Agility
- Depth
- Synchronization
- Versatility

THE FACTORS OF METT-TC

2-8. The six factors of METT-TC—mission, enemy, terrain and weather, troops and support available, time available, and civil considerations—describe the unique situation in which a tactician executes the science and art of tactics. An analysis of the factors of METT-TC is critical during the military decision making process. The METT-TC analytical framework is useful in assessing operations planning, preparing, and executing. The tactician considers these six factors for any type of operation. Their impact on an operation will differ, but each must be considered as factors during the commander's visualization process. That consideration involves both the science and art of tactics. For example, terrain and weather effects on movement rates and fuel consumption are quantifiable and, therefore, part of the science of war. Terrain and weather effects on soldier morale are not totally quantifiable and are part of the art of war. FM 6-0 provides a detailed discussion of the factors of METT-TC.

Factors of METT-TC

- Mission
- Enemy
- Terrain and weather
- Troops and support available
- Time available
- Civil considerations

ELEMENTS OF OPERATIONAL DESIGN

2-9. A major operation begins with a design—an idea that guides the conduct (plan, prepare, execute, and assess) of the operation. The operational design provides a conceptual linkage of ends, ways, and means. The elements of operational design are tools to aid the commander in visualizing major operations and shaping his intent. They help the commander clarify and refine his vision by providing a framework for him to describe the operation in terms of task and purpose. FM 3-0 defines each of these elements. Their utility diminishes with each succeeding tactical echelon. What is envisioned as a decisive point by the land component commander becomes a clearly delineated objective for a battalion task force.

Elements of Operational Design

- End state and military conditions
- Center of gravity
- Decisive points and objectives
- Lines of operation
- Culminating point
- Operational reach, approach, and pauses
- Simultaneous and sequential operations
- Linear and nonlinear operations
- Tempo

BATTLEFIELD OPERATING SYSTEMS

2-10. There are seven battlefield operating systems (BOS). The seven BOS definitions from FM 7-15 are—

> **Battlefield Operating Systems**
> - Intelligence
> - Maneuver
> - Fire support
> - Air defense
> - Mobility/countermobility/survivability
> - Combat service support
> - Command and control

- The *intelligence system* is the activity to generate knowledge of and products portraying the enemy and the environmental features required by a command planning, preparing, and executing operations.
- The *maneuver system* is the movement of forces to achieve a position of advantage with respect to enemy forces. This system includes the employment of forces on the battlefield in combination with direct fire or fire potential. This system also includes the conduct of tactical tasks associated with force projection.
- The *fire support system* encompasses the collective and coordinated use of target-acquisition data, indirect-fire weapons, fixed-wing aircraft, offensive information operations, and other lethal and nonlethal means against targets located throughout an AO.
- The *air defense system* is the employment of all active measures designed to nullify or reduce the effectiveness of attack by hostile aircraft and missiles after they are airborne.
- The *mobility, countermobility, and survivability system*.
  - *Mobility operations* preserve the freedom of maneuver of friendly forces.
  - *Countermobility operations* deny mobility to enemy forces.
  - *Survivability operations* protect friendly forces from the effects of enemy weapon systems.
- The *combat service support system* is the support and services provided to sustain forces during war and military operations other than war.
- The *command and control system* includes all collective tasks associated with supporting the exercise of authority and direction by a properly designated commander over assigned and available forces in the accomplishment of the mission.

The BOS provide the Army a common taxonomy of critical tactical activities. They provide the commander and his staff a means of assessing the planning, preparation, and execution of an operation in discrete subsets.

BASIC TACTICAL CONCEPTS

2-11. The following paragraphs contain basic tactical concepts common to both offensive and defensive operations. They are listed in alphabetical order, not in order of importance. These concepts, along with the principles of war, tenets of Army operations, factors of METT-TC, estimates, input from other commanders, and the commander's experience and judgment allow him to visualize the conduct of operations as he accomplishes his assigned mission.

COMBINED ARMS

2-12. *Combined arms* is the synchronized or simultaneous application of several arms—such as infantry, armor, artillery, engineers, air defense, and aviation—to achieve an effect on the enemy that is greater than if each arm was used against the enemy separately or in sequence (FM 3-0). Weapons and units are more effective when they operate in concert. No single action, weapon, branch, or arm of service generates sufficient power to achieve the effects required to dominate an opponent.

2-13. Combined arms is more than the combat arms working together. Each branch of the Army provides unique capabilities that complement the other branches. A combined arms team consists of two or more arms supporting one another. The commander takes his available combat, combat support (CS), and combat service support (CSS) elements and forms them into a combined arms team focused on mission accomplishment. The proper combination of actions and systems by the combined arms team is the essence of combined arms. The use of combined arms provides complementary and reinforcing effects and may have asymmetrical effects on an enemy force. (See FM 3-0 for more information on symmetrical and asymmetrical effects.)

2-14. Armor, attack helicopter, and infantry units are normally the nucleus of the combined arms team. However, emerging capabilities allow the commander to use any combat arms unit, such as artillery and aviation, to form that nucleus. The commander uses his combat arms forces in different combinations to provide flexibility in conducting different types of operations in varied terrain. For example, a commander may have his infantry lead in mountains and cities when moving dismounted, while his armor leads in open terrain. Attack helicopters can deliver large quantities of precision munitions throughout the AO. A commander can conduct decisive operations using field artillery Multiple Launch Rocket Systems (MLRS) and cannons augmented by the effects of fixed-wing aviation, given the correct conditions. Air defense artillery destroys enemy aerial assets to assist the free movement of the friendly force. Engineers enhance the friendly force's mobility, degrade the enemy's mobility, and assist in providing for survivability of the friendly force. Combat support and CSS members of the combined arms team support the combined arms nucleus by combining capabilities in an appropriate manner to support and sustain the combined arms force.

DECISIVE ENGAGEMENT

2-15. A *decisive engagement* is an engagement in which a unit is considered fully committed and cannot maneuver or extricate itself. In the absence of outside assistance, the action must be fought to a conclusion and either won or lost with the forces at hand (JP 1-02). The unit's mission is what usually results in the acceptance of decisive engagement rather than the unit's physical ability to extricate itself. For example, a unit might become decisively engaged to hold key terrain, defeat a specific enemy force, or secure a specific objective. Less common is a defender's decisive engagement as a result of being placed in a position of disadvantage by an attacker.

DEFEAT IN DETAIL

2-16. ***Defeat in detail*** **is achieved by concentrating overwhelming combat power against separate parts of a force rather than defeating the entire force at once.** A smaller force can use this technique to achieve success against a larger enemy. Defeat in detail can occur sequentially (defeat of separate elements one at a time in succession). For example, a commander can mass overwhelming combat power effects against an enemy element outside the supporting distance of the rest of the enemy force. This allows the commander to destroy the targeted enemy element before it can be effectively reinforced.

FLANKS

2-17. ***Flanks*** **are the right or left limits of a unit.** For a stationary unit, they are designated in terms of an enemy's actual or expected location. (See Figure 2-2.) For a moving unit, they are defined by the direction of movement. (See Figure 2-3.) The commanders tries to deny the enemy opportunities to engage his flanks because a force cannot concentrate as much direct fire on the flanks as it can to the front. Commanders seek to engage enemy flanks for the same reason.

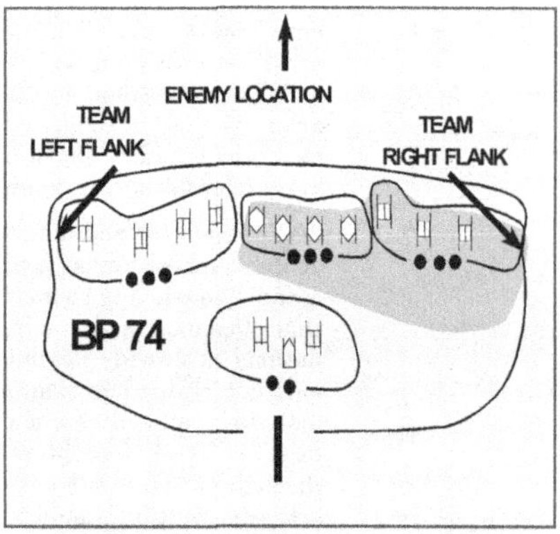

Figure 2-2. Flanks of a Stationary Unit

2-18. **An *assailable flank* is exposed to attack or envelopment.** It usually results from the terrain, the weakness of forces, technical capability of the opponent (vertical envelopment), or a gap between adjacent units. If one flank rests on highly restrictive terrain and the other flank is on open terrain, the latter is immediately recognized as the assailable flank for a heavy ground force. The flank on the restrictive terrain may be assailable for a light force. Sufficient room must exist for the attacking force to maneuver for the flank to be assailable. A unit may not have an assailable flank if both flanks

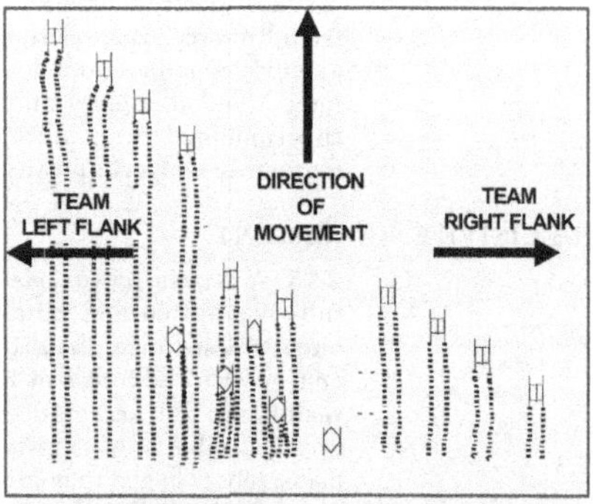

Figure 2-3. Flanks of an Armor-Heavy Team Moving in an Echelon Right Formation

link into other forces. When a commander has an assailable flank, he may attempt to refuse it by using various techniques, such as supplementary positions.

2-19. **A *flanking position* is a geographical location on the flank of a force from which effective fires can be placed on that flank.** An attacking commander maneuvers to occupy flanking positions against a defending force to place destructive fires directly against enemy vulnerabilities. A defending commander maneuvers to occupy flanking positions on the flanks of a hostile route of advance for the same reason. A flanking position that an advancing enemy can readily avoid has little value to the defender unless the enemy does not realize it is occupied.

MANEUVER

2-20. *Maneuver* is the employment of forces on the battlefield through movement in combination with fire, or fire potential, to achieve a position of advantage in respect to the enemy in order to accomplish the mission (JP 1-02). Maneuver creates and exposes enemy vulnerabilities to the massed effects of friendly combat power. A commander employs his elements of combat power in symmetrical and asymmetrical ways so as to attain positional advantage over an enemy and be capable of applying those massed effects.

MUTUAL SUPPORT

2-21. *Mutual support* is that support which units render to each other against an enemy because of their assigned tasks, their position relative to each other and to the enemy, and their inherent capabilities (JP 1-02). Mutual support exists between two or more positions when they support one another by direct or indirect fire, thus preventing the enemy from attacking one position without being fired on from one or more adjacent positions. That same relationship applies to units moving with relation to each other, except they can maneuver to obtain positional advantage to achieve that support. It is normally associated with fire and movement (maneuver), although it can also relate to the provision of CS and CSS.

2-22. In the defense, the commander selects tactical positions to achieve the maximum degree of mutual support. Mutual support increases the strength of defensive positions, prevents the enemy from attempting to defeat the attacking friendly forces in detail, and helps prevent infiltration. In the offense, the commander maneuvers his forces to ensure a similar degree of support between attacking elements.

OPERATION

2-23. An *operation* is a military action or the carrying out of a strategic, tactical, service, training, or administrative military mission (JP 1-02). It includes the process of planning, preparing, executing, and assessing those offensive, defensive, stability, and support operations needed to gain the objectives of any engagement, battle, major action, or campaign. It also includes activities that enable the performance of full spectrum operations, such as security, reconnaissance, and troop movement.

PIECEMEAL COMMITMENT

2-24. ***Piecemeal commitment* is the immediate employment of units in combat as they become available instead of waiting for larger aggregations of units to ensure mass, or the unsynchronized employment of available forces so that their combat power is not employed effectively.** Piecemeal commitment subjects the smaller committed forces to defeat in detail and prevents the massing and synchronizing of combat power with following combat and CS elements. However, piecemeal commitment may be advantageous to maintain momentum and to retain or exploit the initiative. A commander may require piecemeal commitment of a unit to reinforce a faltering operation, especially if the commitment of small units provide all of the combat power needed to avert disaster. The "pile-on" technique associated with search and attack operations employs the piecemeal commitment of troops. (See Chapter 5 for a discussion of search and attack operations.)

RECONSTITUITON

2-25. *Reconstitution* is those actions that commanders plan and implement to restore units to a desired level of combat effectiveness commensurate with mission requirements and available resources. Reconstitution operations include regeneration and reorganization (FM 4-100.9). Reconstitution is a total process. It is not solely a CSS operation, though CSS plays an integral role. The commander conducts reconstitution when one of his subordinate units becomes combat ineffective or when he can raise its combat effectiveness closer to the desired level by shifting available resources. Besides normal support actions, reconstitution may include—

- Removing the unit from combat.
- Assessing it with external assets.
- Reestablishing the chain of command.
- Training the unit for future operations.
- Reestablishing unit cohesion.

2-26. Reconstitution transcends normal day-to-day force sustainment actions. However, it uses existing systems and units to do so. No resources exist solely to perform reconstitution. (See FM 4-100.9.)

RULES OF ENGAGEMENT

2-27. *Rules of engagement* (ROE) are directives issued by competent military authority which delineate the circumstances and limitations under which United States forces will initiate and/or continue combat engagement with other forces encountered (JP 1-02). Operational requirements, policy, and law define the commander's ROE. Rules of engagement impact on how a commander conducts his operations in all four types of military operations by imposing political, practical, operational, and legal limitations on his actions. They may extend to criteria for initiating engagements with certain weapon systems, such as employing unobserved indirect fires within the echelon rear area, or reacting to an attack. They always recognize the right of self-defense and the commander's right and obligation to protect assigned personnel. CJCSI 3121.01A establishes the Joint Chiefs of Staff's standing ROE.

Operational level commanders modify those standing ROE as necessary in response to the factors of METT-TC.

SUPPORTING DISTANCE

2-28. *Supporting distance* is the distance between two units that can be traveled in time for one to come to the aid of the other. For small units, it is the distance between two units that can be covered effectively by their fires (FM 3-0). Supporting distance is a factor of combat power, dispositions, communications capability, and tactical mobility of friendly and enemy forces.

SUPPORTING RANGE

2-29. *Supporting range* is the distance one unit may be geographically separated from a second unit, yet remain within the maximum range of the second unit's indirect fire weapons systems. (FM 3-0). Major factors that affect supporting range are terrain relief, the range of the supporting unit's weapon systems, and their locations in relation to the supported unit's position.

TACTICAL MOBILITY

2-30. ***Tactical mobility* is the ability to move rapidly from one part of the battlefield to another, relative to the enemy.** Tactical mobility is a function of cross-country mobility, firepower, and protection. The terrain, soil conditions, and weather affect cross-country mobility. Heavy ground maneuver units have good tactical mobility—except in restrictive terrain—combined with firepower and protection. They can move on the battlefield against most enemy forces unless faced with an enemy who can defeat their protection and cannot be suppressed by friendly fires. Light ground maneuver units have a tactical mobility advantage against enemy heavy forces in restrictive terrain, but limited firepower and protection. Army aviation maneuver units have good tactical mobility in most types of terrain, good firepower, but limited protection. Extreme weather conditions can restrict the tactical mobility of Army aviation units.

BASIC TACTICAL GRAPHIC CONTROL MEASURES

2-31. This section establishes basic tactical graphic control measures common to offensive and defensive operations. The appropriate chapters of this manual discuss those graphic control measures that apply to only one type of military operation. For example, Chapter 3 discusses the objective as a basic offensive control measure since an objective is a graphic control measure that applies only to offensive operations. These graphics apply to both automated and hand-drawn graphic displays or overlays. This section portrays control measures for use on situation maps, overlays, and annotated aerial photographs. They are also the standard for all simulations, to include those used in live, virtual, and constructive environments.

2-32. Units conducting tactical operations must have clearly defined tasks and responsibilities. The commander uses control measures to establish specific responsibilities that prevent units from impeding one another and impose necessary restrictions. Control measures can be permissive (which allows something to happen) or restrictive (which limits how something is done). Control measures may be graphical, such as boundaries, or procedural, such as target engagement priorities. A commander should establish only the minimum control measures necessary to provide essential coordination and deconfliction between units. Control measures must not unduly restrict subordinates in accomplishing their missions. The commander removes restrictive control measures as soon as possible. FM 1-02 discusses the rules for drawing control measures on overlays and maps.

Basic Tactical Graphic Control Measures

- Air corridor and air control points
- Area of operations and boundaries
- Assembly areas
- Checkpoint
- Contact point
- Critical friendly zones
- Direct fire control measures
- Deep, close, and rear areas
- Engagement area
- Fire support coordination measures
- Fire support targets
- Forward line of own troops
- Line of contact
- Named area of interest
- Obstacle control measures
- Phase lines
- Position areas for artillery
- Routes
- Targeted area of interest

2-33. Well-conceived control measures facilitate the conduct of current and future operations. The commander adjusts his control measures as necessary to maintain synchronization and ensure mission accomplishment as the tactical situation evolves. He balances the risk of introducing additional friction into the operation with the benefits gained by changing them if all of his subordinate elements do not receive the new control measures when contemplating changes to previously established control measures.

2-34. Control measures apply to all forces: combat, CS, and CSS. The commander ensures all higher-echelon control measures, such as phase lines (PLs) and checkpoints, are incorporated into his graphic control measures. When he reports to higher headquarters, he references only the control measures established by that headquarters.

AIR CORRIDOR AND AIR CONTROL POINTS

2-35. An *air corridor* is a restricted air route of travel specified for use by friendly aircraft and established for the purpose of preventing friendly aircraft from being fired on by friendly forces (JP 1-02). It is used to deconflict artillery firing positions with aviation traffic. Low-level transit routes, minimum-risk routes, standard use army aircraft flight routes, and UAV routes are types of air corridors. An air corridor always includes air control points. An *air control point* (ACP) is an easily identifiable point on the terrain or an electronic navigational aid used to provide necessary control during air movement. ACPs are generally designated at each point where the flight route or air corridor makes a definite change in direction and at any other point deemed necessary for timing or control of the operation (FM 3-52). (Figure 2-4 on page 2-12 depicts a generic air corridor and ACPs. See FM 3-52 for more information on aerial control measures.)

AREA OF OPERATIONS AND BOUNDARIES

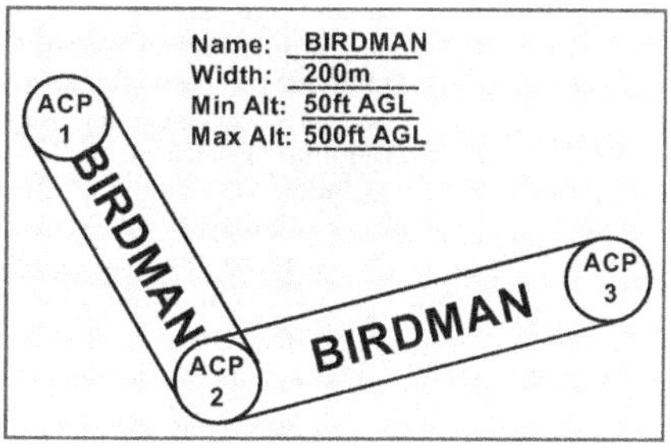

Figure 2-4. Generic Air Corridor

2-36. An AO is the basic control measure for all types of operations. An *area of operations* is an operational area defined by the joint force commander for land and naval forces. Areas of operations do not typically encompass the entire operational area of the joint force commander, but should be large enough for component commanders to accomplish their missions and protect their forces (JP 1-02). The joint force land component commander (JFLCC) or the Army service component command (ASCC) commander will in turn assign their subordinates their own AOs. Those subordinates will further assign their subordinates AOs down to the battalion or company level based on the factors of METT-TC. A unit assigned an AO, the owning unit, may not change control measures imposed by a higher headquarters within their AO. However, it may establish additional control measures to coordinate and synchronize its operations.

> **An *avenue of approach* is the air or ground route leading to an objective (or key terrain in its path) that an attacking force can use.**

2-37. Assigning an AO to a subordinate headquarters maximizes decentralized execution by empowering subordinate commanders to use their own initiative to accomplish their assigned missions. This encourages the use of mission command. (See FM 6-0 for a discussion of mission command.) At the same time it adds the responsibilities listed in Paragraph 2-40 below to the lower headquarters. Conversely, failure to designate subordinate AOs maximizes centralized execution and limits the subordinates' tactical options. The latter choice should be made only when mandated by the factors of METT-TC. For example, a brigade commander responsible for blocking an enemy advance along a single avenue of approach may assign his subordinate battalions battle positions to support a brigade engagement area (EA) instead of subdividing his AO and the avenue of approach into battalion AOs.

2-38. A higher headquarters designates an AO using boundaries. It normally assigns an AO to a maneuver unit, but it may also assign one to CS or CSS units. Having an AO assigned both restricts and facilitates the movement of units and use of fires. It restricts units not assigned responsibility for the AO from moving through the AO. It also restricts outside units from firing into or allowing the effects of its fires to affect the AO. Both of these restrictions can be relaxed through coordination with the owning unit. It facilitates the movement and fires of the unit assigned responsibility for, or owning, the AO. The assigned AO must encompass enough terrain for the commander to accomplish his mission and protect his forces.

2-39. Ideally, the AO is smaller than the commander's area of influence. An *area of influence* is a geographical area wherein a commander is directly capable of influencing operations by maneuver or fire support systems normally under the commander's command or control (JP 1-02). If the commander's area of influence is smaller than his AO, he must consider his options for extending the size of his area of influence. His options include the following techniques:

- Changing the geographical dispositions of his current systems to increase the size of his area of influence and ensure coverage of key areas, installations, and systems.
- Requesting additional assets.
- Requesting boundary adjustments to reduce the size of his AO.
- Accepting the increased risk associated with being unable to provide security throughout the AO.
- Moving his area of influence by phases to sequentially encompass the entire AO.

2-40. All units assigned an AO have the following responsibilities:

- Terrain management.
- Movement control.
- Fires.
- Security.

Selected echelons have an additional responsibility to provide airspace command and control.

Terrain Management

2-41. The commander assigned an AO is responsible for terrain management within its boundaries. A higher headquarters may dictate that another unit position itself within a subordinate unit's AO, but the commander assigned the AO retains final approval authority for the exact placement. This ensures the unit commander controlling the AO knows what units are in his AO and where they are located so that he can deconflict operations, control movement, and prevent fratricide. Only the owning commander assigns subordinate unit boundaries within the AO.

Movement Control

2-42. Units may not move across boundaries into another unit's AO without receiving clearance from the unit owning the AO. Once assigned an AO, the owning unit controls movement throughout the AO. The designation, maintenance, route security, and control of movement along routes within an AO are the responsibility of the owning unit unless the higher echelon's coordinating instructions direct otherwise. The commander may designate movement routes as open, supervised, dispatch, reserved, or prohibited. Each route's designation varies based on the factors of METT-TC. FM 4-01.30 discusses movement planning and control measures.

Fires

2-43. Within its AO, the owning unit may employ any direct or indirect fire system without receiving further clearance from superior headquarters. There are three exceptions: The first and most common is that a unit may not use munitions within its own AO without receiving appropriate clearance if the effects of those munitions extend beyond its AO. For example, if a unit wants to use smoke, its effects cannot cross boundaries into another AO unless cleared with the adjacent owning unit. Second, higher headquarters may explicitly restrict the use of certain munitions within an AO or parts of an AO, such as long-duration scatterable mines. Third, higher headquarters may impose a restrictive fire support coordinating measure (FSCM) within an AO to protect some asset or facility, such as a no-fire area around a camp housing dislocated civilians. These FSCM tend to be linear in nature in a contiguous AO while they are more likely areas in a noncontiguous AO.

2-44. The commander may not employ indirect fires across boundaries without receiving clearance from the unit into whose AO the fires will impact. He may employ direct fires across boundaries without clearance at specific point targets that are clearly and positively identified as enemy.

Security

2-45. The security of all units operating within the AO is the responsibility of the owning commander. This fact does not require that commander to conduct area security operations throughout his AO. (See Chapter 12 for a discussion of area security responsibilities.) He must prevent surprise and provide the amount of time necessary for all units located within the AO to effectively respond to enemy actions by employing security forces around those units. If the commander cannot or chooses not to provide security throughout his AO, he must clearly inform all concerned individuals of when, where, and under what conditions he is not going to exercise this function. The commander generally depicts these locations using permissive FSCM. Each unit commander remains responsible for his unit's local security.

Airspace Command and Control

2-46. *Army airspace command and control* (A2C2) are those actions that ensure the synchronized use of airspace and enhanced command and control of forces using airspace (FM 3-52). The ground maneuver commander manages the airspace below the coordinating altitude, using procedural control measures and positive control measures implemented by his air traffic service organization. Corps and divisions are the echelons that routinely have A2C2 responsibilities, although a commander may provide the resources to accomplish this function to a brigade operating independently.

2-47. Communications, standardized procedures, and liaison normally provide the commander with his required connectivity with the theater airspace control authority. The commander ensures reliable communications through his area communications network. He supervises airspace activities through standardized procedures to prevent real-time conflicts among the various airspace users while achieving the necessary flexibility to ensure the greatest combat effectiveness. The A2C2 section of the battlefield coordination detachment, co-located with the joint air operations center, provides the commander that liaison capability with the airspace control authority (ACA). (See FM 3-52 and JP 3-52 for additional information regarding airspace control doctrine.)

2-48. The vertical dimension, or airspace, of the AO is inherently permissive because all branches and services require the use of airspace. There are procedural and positive airspace control measures (ACM) available to synchronize military operations in the airspace above the AO. Among the procedural ACM is the coordinating altitude, which separates fixed- and rotary-wing aircraft by determining an altitude below which fixed-wing aircraft will normally not fly and above which rotary-wing aircraft will normally not fly. It allows the ground commander to use the airspace above his AO for his organic aviation assets to complement ground maneuver forces, but it is not a boundary for which he has responsibility. The ACA, normally the joint force air component commander, must establish the coordinating altitude, promulgate it through the airspace control plan, address it in the airspace control order, and include a buffer zone for small altitude deviations. Coordinating altitudes are permissive ACM.

Boundaries

2-49. A *boundary* is a line that delineates surface areas for the purpose of facilitating coordination and deconfliction of operations between adjacent units, formations, or areas (JP 0-2). The commander uses graphic control measures to define the limits of an AO and, as such, establish ground forces' responsibilities. He uses ACM to control the vertical dimension. The commander bases his subordinates' boundaries on clearly defined terrain features. This requirement is less important if all units in the AO have precision navigation capabilities. Boundaries should not split roads, rivers, or railways. Responsibility for an avenue of approach and key terrain should belong to only one unit. The commander adjusts his boundaries as necessary in response to the evolving tactical situation. Any areas not delegated to a subordinate remain the responsibility of the commander.

Contiguous and Noncontiguous AOs

2-50. **A commander has a *contiguous AO* when all of his subordinate forces' areas of operations share one or more common boundaries. A commander has a *noncontiguous AO* when one or more of his subordinate forces' areas of operations do not share a common boundary.** The intervening area between noncontiguous AOs remains the responsibility of the higher headquarters. The commander can choose to organize his AO so that his subordinates have contiguous or noncontiguous areas of operations.

2-51. The forward boundary of an echelon is primarily designated to divide responsibilities between it and its next higher echelon. Decisive or shaping operations directed against enemy forces beyond an echelon's forward boundary are the responsibility of the next higher echelon. The higher echelon headquarters normally assigns the lower echelon a forward boundary based on the higher echelon's scheme of maneuver. The ability to acquire and attack targets in the area between the forward boundary of its subordinates and the echelon's forward boundary determines the exact position of that forward boundary. For example, if a division assigns a forward boundary to a brigade, then the division conducts operations beyond the brigade's forward boundary. That area between the brigade's forward boundary and the division's forward boundary is the division's deep area. The rear boundary defines the rearward limits of the unit's area. It usually also defines the start of the next echelon's rear area. Lateral boundaries extend from the rear boundary to the unit's forward boundary. (See Figure 2-5.)

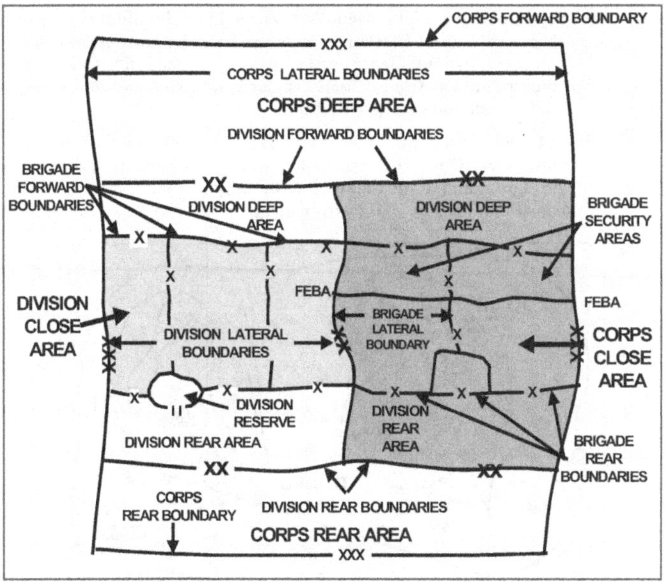

Figure 2-5. Corps with Contiguous Areas of Operations

2-52. The commander bases his decision to establish contiguous AOs on his analysis of the factors of METT-TC. Units with contiguous AOs are normally within supporting distance of one another and may be within supporting range. Other reasons why a commander establishes contiguous AOs include—

- Limited size of the AO in relation to the number of friendly forces.
- Decisive points in close physical proximity to each other.
- Political boundaries or enemy force concentrations require establishing contiguous AOs.
- Reduced risk of being defeated in detail because of an incomplete operational picture or when the friendly force is significantly outnumbered.
- Greater concentration of combat power along a single avenue of approach or movement corridor.

2-53. A noncontiguous AO does not have distinctive forward, rear, and lateral boundaries. It is established by a boundary that encloses the entire area. Subordinate boundaries will be continuous, 360-degree arcs that closely approximate the subordinate unit's area of influence. For example, the commander would normally place a noncontiguous brigade boundary at the limit of observed fires for its security forces. Because noncontiguous boundaries must provide all-around security, they generally allow for less concentration of combat power along a single axis. A brigade, division, or corps commander who establishes noncontiguous AOs for his subordinates still designates an echelon rear area. Battle positions are not AOs since a unit is not restricted from operating outside its battle position. A commander who deploys his subordinates into battle positions is not conducting noncontiguous operations. Chapter 8 defines a battle position.

2-54. Operations directed against enemy forces and systems outside a noncontiguous AO are the responsibility of the organization that owns that location. For example, in Figure 2-6 the middle enemy division is the corps' responsibility since it is not within either of the corps' two divisions' AOs.

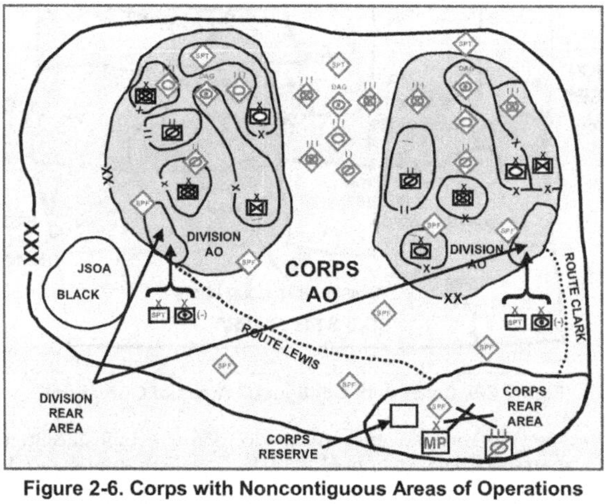

Figure 2-6. Corps with Noncontiguous Areas of Operations

2-55. The commander bases his decision to establish noncontiguous AOs on his analysis of the factors of METT-TC. There is a risk associated with establishing noncontiguous AOs since units with noncontiguous AOs are normally out of supporting range from each other. Overcoming this risk places a premium on situational understanding and tactical mobility. Reasons why a commander establishes noncontiguous AOs include—

- Encompassing key and decisive terrain within his area of influence when he has limited numbers of friendly forces for the size of the AO. ***Key terrain* is any locality, or area, the seizure or retention of which affords a marked advantage to either combatant.**

***Decisive terrain*, when present, is key terrain whose seizure and retention is mandatory for successful mission accomplishment.**

- Comparative weakness of the enemy means that subordinates do not have to remain within supporting range or distance of one another and can take advantage of superior situational understanding and tactical mobility.
- Enemy concentrated in dispersed areas requires a corresponding concentration of friendly forces.
- Existence of large contaminated areas within his AO.

2-56. Using noncontiguous AOs place a premium on the use of innovative means to conduct sustaining operations, including aerial resupply. A commander whose subordinates have noncontiguous AOs has three basic choices for establishing intermittent ground lines of communications (LOCs) with his subordinates:

- Assign a subordinate the mission of providing convoy security for each convoy. For the situation depicted in Figure 2-7, the corps could assign the corps military police brigade the mission of providing convoy security for critical ground convoys traveling between the corps rear area and each division's rear area.
- Assign a subordinate the mission of providing route security for each sustainment route. For the situation depicted in Figure 2-7 the corps could assign the corps armored cavalry regiment an AO that extends four kilometers on either side of LOCs LEWIS and CLARK. The regiment would be assigned the mission of route security within that AO for the period required for sustainment convoys to travel to and from the two subordinate divisions.
- Assume risk by having the corps support command run convoys with only their organic self-defense capabilities, while assigning another unit the mission of responding to enemy contacts beyond a convoy's self-defense capability.

ASSEMBLY AREAS

2-57. **An *assembly area* (AA) is an area a unit occupies to prepare for an operation.** Ideally, an assembly area provides—

- Concealment from air and ground observation.
- Cover from direct fire.
- Space for dispersion; separate each AA by enough distance from other AAs to preclude mutual interference.
- Adequate entrances, exits, and internal routes.
- Good drainage and soil conditions that can sustain the movement of the unit's vehicles and individual soldiers.
- Terrain masking of electromagnetic signatures.
- Terrain allowing observation of ground and air avenues into the AA.
- Sanctuary from enemy medium-range artillery fires because of its location outside the enemy's range.

2-58. The commander assigns each unit its own AA. In Figure 2-7, the example of multiple units occupying one AA is a graphical shortcut taken when the map scale would make depiction of multiple assembly areas unreadable. In reality, the commander would subdivide AA Thomas into two smaller AAs, one for each unit. A unit AA is normally within the AO of another unit. An AA area is usually treated as a noncontiguous AO. This means that a unit has the same responsibilities within its assigned AA as it has for any other AO.

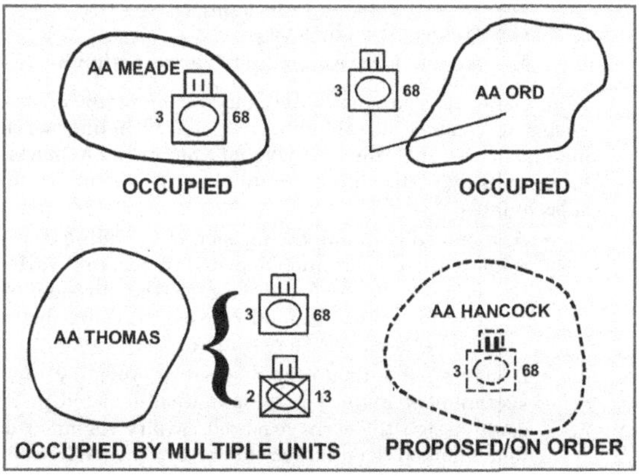

Figure 2-7. Assembly Areas

2-59. The proper location of AAs contributes significantly to both security and flexibility. It should facilitate future operations so movement to subsequent positions can take place smoothly and quickly by concealed routes. Because of their smaller signature, light units can use AAs closer to the enemy than heavy units without excessive risk of enemy detection. The tactical mobility of heavy units allows them to occupy AAs at a greater distance from the line of departure (LD) than light units.

CHECKPOINT

2-60. **A *checkpoint* is a predetermined point on the ground used to control movement, tactical maneuver, and orientation.** Units can also use a checkpoint as a fire control measure in lieu of the preferred control measure, a target reference point. Checkpoints are useful for orientation. Units may use checkpoints to supplement or as substitutes for phase lines (PLs). They are also used in the conduct of CSS. Figure 2-8 depicts Checkpoint 13.

Figure 2-8. Checkpoint 13

CONTACT POINT

2-61. A ***contact point*** **is an easily identifiable point on the terrain where two or more ground units are required to make physical contact.** A commander establishes a contact point where a PL crosses a lateral boundary or on other identifiable terrain as a technique to ensure coordination between two units. He provides a date-time group to indicate when to make that physical contact. Figure 2-9 depicts Contact Point 8.

Figure 2-9. Contact Point 8

2-62. The mutual higher commander of two moving units normally designates the location of contact points and times of contact. When one unit is stationary, its commander normally designates the location of the contact point and the meeting time, and transmits this information to the commander of the moving unit.

CRITICAL FRIENDLY ZONE

2-63. **A *critical friendly zone* (CFZ) is an area, usually a friendly unit or location, that the maneuver commander designates as critical to the protection of an asset whose loss would seriously jeopardize the mission.** The exact size and shape of the CFZ reflects the technical characteristics of the sensor coverage and varies in accordance with the terrain. There is no specific graphic for a CFZ. The designation of a CFZ requires the availability of a target acquisition sensor to cover that area and fire support weapon systems to conduct counterfire. The supporting field artillery unit's automated fire support system is tied to that sensor to place the location of a weapon firing into the CFZ ahead of all other targets in priority for counterfire. This results in an immediate call for fire unless the system operator manually overrides the automated request for fire. (For additional information on the employment of a CFZ, see FM 3-09.12.)

DIRECT FIRE CONTROL MEASURES

2-64. The small unit commander communicates to his subordinates the manner, method, and time to initiate, shift, and mass fires, and when to disengage by using direct fire control measures. The commander should control his unit's fires so he can direct the engagement of enemy systems to gain the greatest effect. The commander uses intelligence preparation of the battlefield (IPB) and reconnaissance to determine the most advantageous way to use direct fire control measures to mass the effects on the enemy and reduce fratricide from direct fire systems. He must understand the characteristics of weapon systems and available munitions (such as the danger to unprotected soldiers when tanks fire discarding sabot ammunition over their heads or near them). Direct fire control measures defined in this manual include engagement criteria, engagement priorities, sectors of fire, and target reference points. Platoon and company maneuver manuals address other direct fire control measures, such as frontal, cross, or depth fire patterns and simultaneous, alternating, or observed techniques of fire.

Engagement Area

2-65. **An *engagement area* (EA) is an area where the commander intends to contain and destroy an enemy force with the massed effects of all available weapons and supporting systems.** This includes organic direct fire systems and supporting systems, such as close air support. Figure 2-10 depicts several EAs used within the context of a battalion defense. The commander determines the size and shape of the EA by the relatively unobstructed intervisibility from the weapon systems in their firing positions and the maximum range of those weapons. The commander designates EAs to cover each enemy avenue of approach into his position. He also can use them to designate known or suspected enemy locations. Once the commander selects his EA, he arrays his forces in positions to concentrate overwhelming effects into these areas. He routinely subdivides his EA into smaller EAs for his subordinates using one or more target reference points or by prominent terrain features. The commander assigns sectors of fires to subordinates to prevent fratricide, but responsibility for an avenue of approach or key terrain is never split. These sectors normally do not affect friendly maneuver. Commanders of units up to battalion task force size normally use this control measure.

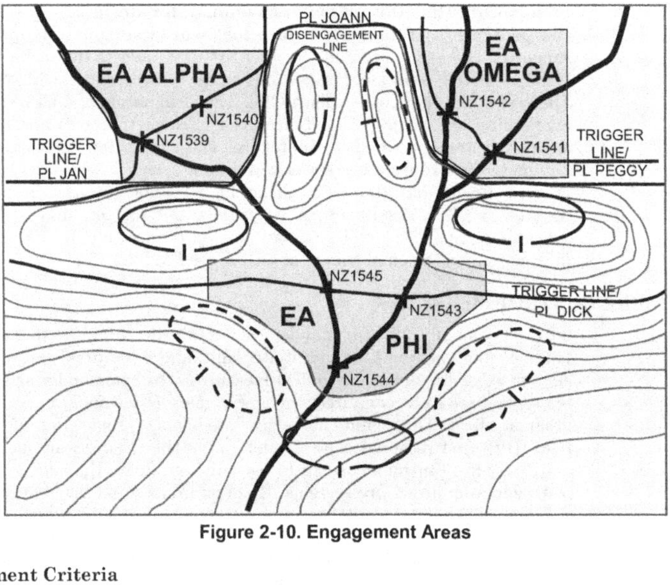

Figure 2-10. Engagement Areas

Engagement Criteria

2-66. ***Engagement criteria* are protocols that specify those circumstances for initiating engagement with an enemy force.** They may be restrictive or permissive. For example, a company commander could tell his 1st Platoon to wait until three enemy tanks reach a target reference point within its EA before initiating fire. Another example is a battalion commander telling one of his company commanders not to engage an approaching enemy unit until it commits itself to an avenue of approach. The commander establishes his engagement criteria in the direct fire plan. Commanders and leaders of small tactical units use engagement criteria in conjunction with engagement priorities and other direct fire control measures to mass fires and control fire distribution.

Engagement Priority

2-67. ***Engagement priority* specifies the order in which the unit engages enemy systems or functions.** The commander assigns engagement priorities based on the type or level of threat at different ranges to match organic weapon systems capabilities against enemy vulnerabilities. Engagement priorities are situationally dependent. The commander uses engagement priorities to distribute fires rapidly and effectively. Subordinate elements can have different engagement priorities. For example, the commander establishes his engagement priorities so that his M2 Bradley fighting vehicles engage enemy infantry fighting vehicles or armored personnel carriers while his M1 Abrams tanks engage enemy tanks. Normally, units engage the most dangerous targets first, followed by targets in depth.

Sector of Fire

2-68. **A *sector of fire* is that area assigned to a unit, a crew-served weapon, or an individual weapon within which it will engage targets as they appear in accordance with established engagement priorities.** (See Figure 2-11.) Battalions and smaller echelons primarily use this direct fire control measure. Each sector of fire can extend from a firing position to the maximum engagement range of the weapon, or it can be an enclosed area at a distance from the firing position. To increase the commander's ability to concentrate fires in a certain area, he should assign each unit or weapon system a primary sector of fire and a secondary sector of fire. The primary sector of fire is that area in which the assigned unit, individual, or crew-served weapon is initially responsible for engaging and defeating the enemy. Fire shifts to the secondary sector, on order, when there are no targets in the primary sector, or when the commander needs to cover the movement of another friendly element. This secondary sector of fire should correspond to another element's primary sector of fire to obtain mutual support. Subordinate commanders may impose additional fire control measures as required.

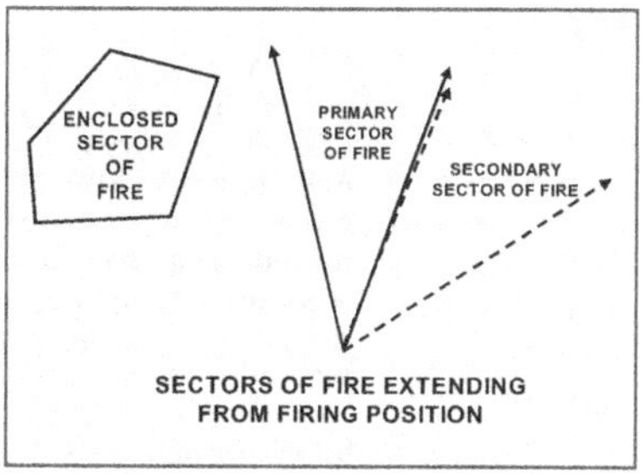

Figure 2-11. Sectors of Fire

Target Reference Point

2-69. **A *target reference point* (TRP) is an easily recognizable point on the ground, such as a building or a road junction, used in conjunction with engagement areas and sectors of fire to initiate, distribute, and control fires.** A TRP may be a natural terrain feature, a manmade artifact, such as a building, or a marker emplaced by the unit. Maneuver leaders at battalion and below designate TRPs to define unit or individual sectors of fire and observation, usually within an EA. A TRP can also designate the center of an area where the commander plans to rapidly distribute or converge fires. A task force commander designates TRPs for his company teams. Company commanders designate TRPs for their platoons, sections, and, in some cases, individual weapons. Platoon leaders or subordinate leaders may designate additional TRPs for their elements as necessary to control direct and indirect fires. The echelon fire support officer can also designate TRPs as indirect fire targets by using the standard target symbol with letters and numbers. Figure 2-12 depicts the symbol for TRP 032.

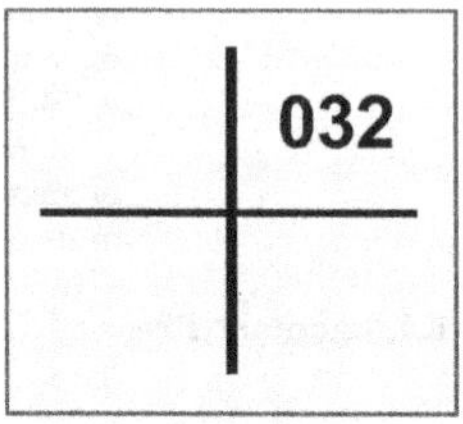

Figure 2-12. Target Reference Point

Trigger Line

2-70. **A *trigger line* is a phase line used to initiate and mass fires into an engagement area or an objective at a predetermined range for all or like weapon systems.** It is located on identifiable terrain—like all phase lines—that crosses an EA, a direction of attack, or an axis of advance. The commander can designate one trigger line for all weapon systems or separate trigger lines for each weapon or type of weapon system. The commander specifies the engagement criteria for this specific situation. The criteria may be either time- or event-driven, such as a certain number or certain types of vehicles to cross the trigger line before initiating engagement. He may reserve to himself the authority to initiate engagement by firing his own weapon or giving the command to fire.

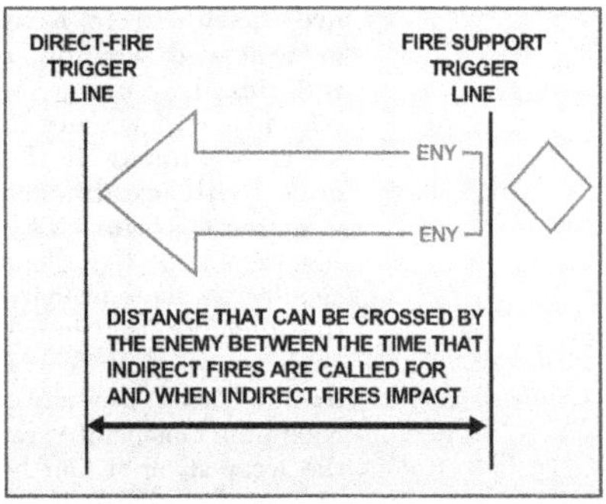

Figure 2-13. Trigger Lines

2-71. The commander designates a PL as the trigger line for his fire support systems. He bases the trigger line's location on the factors of METT-TC, including such variables as the time of flight for artillery shells, positioning of the guns, and the existence of quick-fire links. Its location varies from situation to situation. Its position reflects the distance that the enemy force is likely to traverse in the time it takes from when fires are requested to when artillery rounds impact. (See Figure 2-13.) This gives time for the fire support systems to respond to the initial call for fire. For example, in a desert environment—a battalion task force commander's fire support trigger line is approximately four kilometers beyond the point where he wants to engage the enemy with indirect fires when he has M109A6 howitzers in direct support. It is approximately six kilometers when he has M109A3 howitzers in direct support. The shorter distance reflects the more rapid response capabilities of the M109A6 compared to the M109A3, all other factors being equal.

2-72. The commander can establish another trigger line for his most accurate long-range weapon system in the vicinity of the area where the fire support impacts to capitalize on the asymmetric attack. However, dust and debris resulting from the artillery fire may prevent his direct-fire systems from engaging the enemy. He establishes other trigger lines and TRPs for shorter-range systems. He may give guidance to extremely proficient crews to engage the enemy at longer than normal ranges or give them different engagement priorities than the rest of the force, such as giving priority to engaging air defense or engineer-breaching systems. This could result in losing the effect that the sudden application of massed fires has on an enemy.

2-73. When the enemy reaches these closer trigger lines, the commander establishes a decision point to help him determine if he wants his longer-range systems to continue to fire in depth or to concentrate his fires on a single point. Many factors impact his decision, most of which concern the enemy and how he maneuvers and the effects of the defending force's fires.

DEEP, CLOSE, AND REAR AREAS

2-74. There are times when the factors of METT-TC favor a spatial organization of the entire AO. Korea is an example of large numbers of enemy units concentrated in numerous echelons on a contiguous front across the peninsula. The terrain and the capabilities and doctrine of allied units require the use of a spatial organization. The commander conducts simultaneous decisive, shaping, and sustaining operations within a context of deep, close, and rear areas when this occurs. (See Figure 2-14.)

Deep Area

2-75. When designated, the *deep area* is an area forward of the close area that commanders use to shape enemy forces before they are encountered or engaged in the close area (FM 3-0). It extends from the subordinate's forward boundary to the forward boundary of the controlling echelon. Thus, the deep area relates to the close area not only in terms of geography but also in terms of purpose and time. The extent of the deep area depends on how far out the force can acquire information and strike targets. Commanders may place forces or employ effects in the deep area to conduct shaping operations. Some of these operations may involve close combat. However, most ground maneuver forces stay in the close area.

2-76. New weapon systems and advanced information technology continue to increase the capability of Army forces to engage enemy forces in depth. In the past, deep attacks aimed to slow and disrupt the advance of enemy forces.

Today, Army forces may engage enemy formations with precision fires at substantial distances from the close area. This capability allows the commander to employ greater depth and simultaneity of action than ever before when conducting operations.

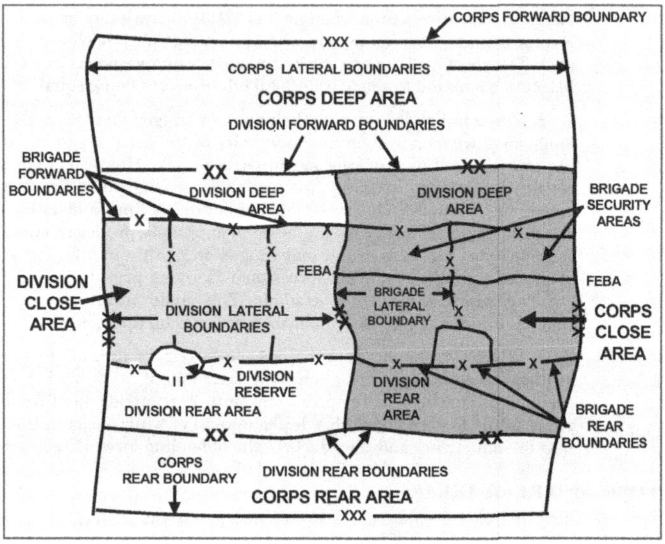

Figure 2-14. Deep, Close, and Rear Areas

Close Area

2-77. When designated, the *close area* is where forces are in immediate contact with the enemy and the fighting between the committed forces and readily available tactical reserves of both combatants is occurring, or where commanders envision close combat taking place. Typically, the close area assigned to a maneuver force extends from its subordinates' rear boundaries to its own forward boundary (FM 3-0). Typically, the close area is where the majority of close combat occurs. It also includes the activities of forces directly supporting the fighting elements, such as direct support field artillery and logistics elements. The close area is historically the only location where the commander could conduct his decisive operation. One unit may conduct the decisive operation, while another conducts shaping operations within a close area. Division commanders whose AOs constitute the corps close area have the option of designating their own deep, close, and rear areas. Figure 2-14 illustrates this option.

Rear Area

2-78. Unlike close and deep areas, a rear area can be designated by the commander regardless of the organization of his AO into contiguous or noncontiguous subordinate AOs. When designated in the context of contiguous AOs, the *rear area* for any command extends from its rear boundary forward to the rear boundary of the next lower level of command. This area is provided primarily for the performance of support functions and is where the majority of the echelon's sustaining operations occur (FM 3-0). The commander designates an individual responsible for conducting his sustaining operations within the rear area after considering the factors of METT-TC. He provides that individual the necessary command and control resources to direct the echelon's sustaining operations. At the corps and division echelons, this is the rear command post. Those operations include the following functions that assure his command's freedom of action and continuity of operations:

- Combat service support.
- Rear area and base security.
- Movement control throughout the AO.
- Terrain management throughout the AO.
- Infrastructure development.

That individual may be the corps deputy commander, assistant division commander (support), or the forward support battalion commander depending on the echelon.

2-79. Regardless of the specific sustaining operations performed by an organization occupying the rear area, its focus on other-than-combat operations leaves it more vulnerable than combat organizations in close areas. Commanders may protect CS and CSS units and facilities in rear areas with combat forces. Geography or other circumstances may cause the commander to designate a noncontiguous rear area. This increases the challenge associated with providing rear area security due to the physical separation from combat units that would otherwise occupy a contiguous close area.

2-80. Between contiguous and noncontiguous AOs, the commander chooses the battlefield organization best suited to the tactical situation. For example, in an area defense the proximity of the enemy, the array and density of friendly forces, and the requirement to protect sustaining functions allows the commander to visualize decisive, shaping, and sustaining operations in terms of discrete areas. In contrast, within the context of a search and attack operation, the lack of information about the enemy, the need to block enemy escape, the existence of noncontiguous rear areas, and the nature of the AO may preclude organization of the battlefield into discrete close, deep, and rear areas.

FIRE SUPPORT COORDINATING MEASURES

2-81. Commanders assigned an AO employ FSCM to facilitate the rapid engagement of targets and simultaneously provide safeguards for friendly forces. Fire support coordinating measures are either permissive or restrictive. Boundaries are the basic FSCM. The fire support coordinator recommends FSCM to the commander based on the commander's guidance, location of friendly forces, scheme of maneuver, and anticipated enemy actions. Once the commander establishes FSCM, they are entered into or posted on all the command's displays and databases. (FM 3-09 explains the use of all FSCM in more detail).

Permissive FSCM

2-82. The primary purpose of permissive measures is to facilitate the attack of targets. Once they are established, further coordination is not required to engage targets affected by the measures. Permissive FSCM include a coordinated fire line, fire support coordination line, and free-fire area.

2-83. **Coordinated Fire Line.** A *coordinated fire line* (CFL) is a line beyond which conventional, direct, and indirect surface fire support means may fire at any time within the boundaries of the establishing headquarters without additional coordination. The purpose of the CFL is to expedite the surface-to-surface attack of targets beyond the CFL without coordination with the ground commander in whose area the targets are located (JP 3-09). Brigades or divisions usually establish a CFL, although a maneuver battalion may establish one. It is located as close as possible to the establishing unit without interfering with maneuver forces to open up the area beyond to fire support. A higher echelon may consolidate subordinate unit CFLs. If this occurs, any changes to the subordinate CFLs are coordinated with the subordinate headquarters. (See Figure 2-15.)

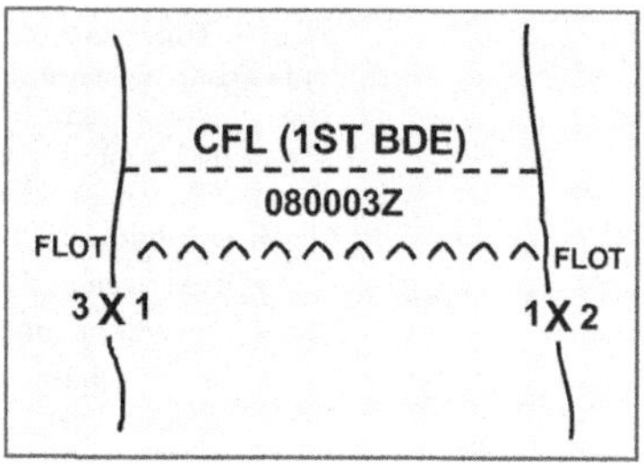

Figure 2-15. Coordinated Fire Line

2-84. **Fire Support Coordination Line.** The *fire support coordination line* (FSCL) is a FSCM that is established and adjusted by appropriate land or amphibious force commanders within their boundaries in consultation with superior, subordinate, supporting, and affected commanders. [See Figure 2-16.] The FSCL facilitates the expeditious attack of surface targets of opportunity beyond the coordinating measure. A FSCL does not divide an area of operations by defining a boundary between close and deep operations or a distinct area [JP 3-09 uses *zone*] for close air support. The FSCL applies to all fires of air, land, and sea-based weapon systems using any type of ammunition. Forces attacking targets beyond an FSCL must inform all affected commanders in sufficient time to allow necessary reaction to avoid fratricide. Supporting elements attacking targets beyond the FSCL must ensure that the attack will not produce adverse effects on, or to the rear of, the line. Short of an FSCL, all air-to-ground and surface-to-surface attack operations are controlled by the appropriate land or amphibious force commander. The FSCL should follow well-defined terrain features. Coordination of attacks beyond the FSCL is especially critical to commanders of air, land, and special operations forces. In exceptional circumstances, the inability to conduct this coordination will not preclude the attack

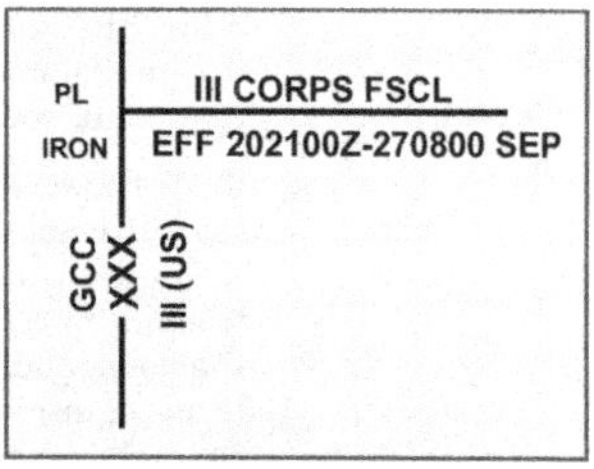

Figure 2-16. Fire Support Coordination Line

of targets beyond the FSCL. However, failure to do so may increase the risk of fratricide and could waste limited resources (JP 3-09).

2-85. The commander designating a FSCL remains responsible for establishing the priority, effects, and timing of fires impacting beyond the FSCL. Coordination for attacks beyond the FSCL is through the air tasking order. The appropriate land or amphibious commander controls attacks short of the FSCL. That commander uses the tactical air control system or the Army air-ground system to control the execution of close air support (CAS).

2-86. **Free-Fire Area.** A *free-fire area* (FFA) is a specific area into which any weapon system may fire without additional coordination with the establishing headquarters (JP 3-09). Normally, division or higher headquarters establish a FFA on identifiable terrain. (See Figure 2-17.)

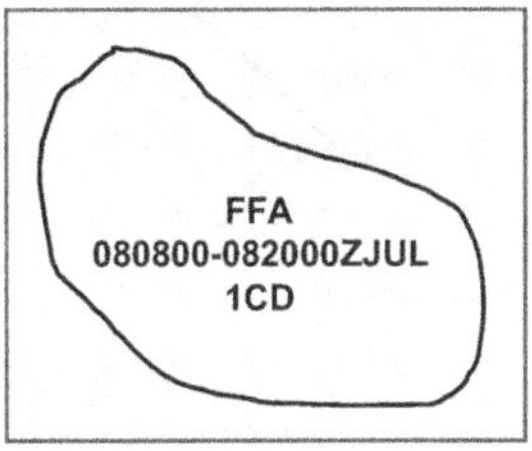

Figure 2-17. Free-Fire Area

Restrictive FSCM

2-87. A restrictive FSCM prevents fires into or beyond the control measure without detailed coordination. The primary purpose of restrictive measures is to provide safeguards for friendly forces. Restrictive FSCM include an airspace coordination area, no-fire area, restrictive fire area, and restrictive fire line. Establishing a restrictive measure imposes certain requirements for specific coordination before the engagement of those targets affected by the measure.

2-88. **Airspace Coordination Area.** The *airspace coordination area* (ACA) is a three-dimensional block of airspace in a target area, established by the appropriate ground commander, in which friendly aircraft are reasonably safe from friendly surface fires. The airspace coordination area may be formal or informal (JP 3-09.3). Time, space, or altitude separates aircraft and indirect fire. The purpose of the ACA is to allow the simultaneous attack of targets near each other by fixed-wing aircraft and other fire support means. Several techniques may be used in this role. The technique selected depends on the time available, tactical situation, unit SOPs, and state of training. (FM 3-52 defines ACM.)

2-89. The airspace control authority establishes formal ACAs at the request of the appropriate ground commander. This is normally a separate brigade or higher-echelon commander. Formal ACAs require detailed planning. The design of the ACA's vertical and lateral limits allows freedom of action for air and surface fire support for the greatest number of foreseeable targets.

2-90. The echelon fire support cell coordinates the location and extent of the ACM with the A2C2) element and the fire direction center. It is located above the target area as recommended by the air liaison element to the fire support cell. The type of aircraft and the ordnance dictate the size of the area. Vital information defining the formal ACA includes minimum and maximum altitudes, a baseline designated by grid coordinates at each end, the width of the ACA from either side of the baseline, and effective times. (See Figure 2-18.)

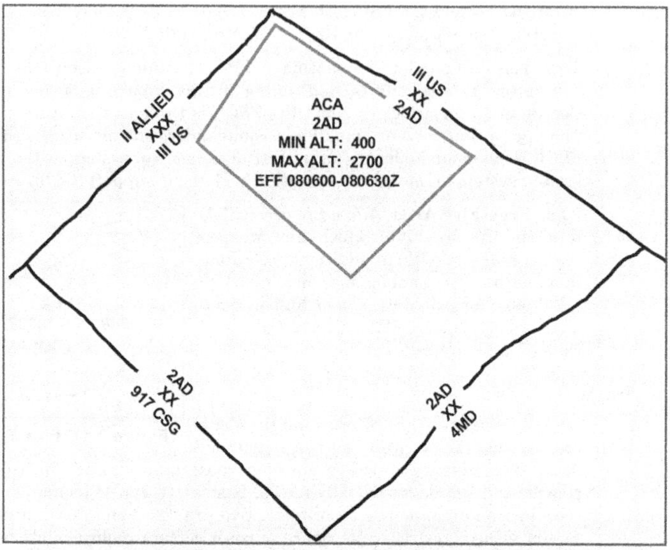

Figure 2-18. A Formal Airspace Coordination Measure

2-91. The maneuver commander may establish informal ACAs. He may separate aircraft and surface fires by distance (lateral, altitude, or a combination) or by time. Lateral separation is effective for coordinating fires against targets that are adequately separated by at least one kilometer from flight routes to ensure aircraft protection from the effects of friendly fires. An example of a lateral separation technique is: "Aircraft stay west of grid line 62." Altitude separation is effective for coordinating fires when aircraft remain above indirect fire trajectories and their effects. This technique is effective when aircrews and firing units engage the same or nearby targets. An example of altitude separation is: "Aircraft remain above 3000 feet mean sea level in quadrant northwest of grid PK7325."

2-92. A combination of lateral and altitude separation is the most restrictive technique for aircraft and may be required when aircraft must cross the gun target line of a firing unit. Time separation requires the most detailed coordination and may be required when aircraft must fly near indirect-fire trajectories or ordnance effects. The commander must coordinate the timing of surface fires with aircraft routing. This ensures that even though aircraft and surface fires may occupy the same space, they do not do so at the same time. Surface and air-to-ground fires should be synchronized. All timing for surface fires will be based on a specific aircraft event time. Fire support personnel and tactical air controllers should select the separation technique that requires the least coordination without adversely affecting timely fires or the aircrew's ability to safely complete the mission.

2-93. **No-Fire Area.** A *no-fire area* (NFA) is a land area designated by the appropriate commander into which fires or their effects are prohibited (JP 3-09). (See Figure 2-19.) A commander uses a NFA to protect independently operating elements, such as forward observers and special operating forces. He can also use it to protect friendly forces in the rear area and for humanitarian reasons, such as preventing the inadvertent engagement of displaced civilian concentrations, or to protect sensitive areas, such as cultural monuments. There are two exceptions to this rule:

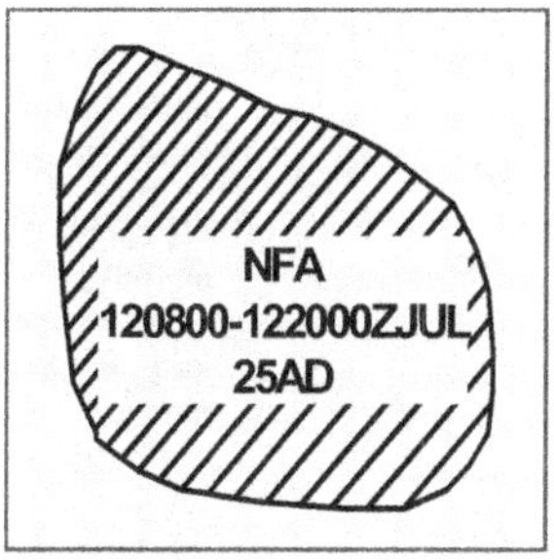

Figure 2-19. No-Fire Area

- The establishing headquarters may approve fires within the NFA on a case-by-case mission basis.
- When an enemy force within a NFA engages a friendly force, the friendly force may engage a positively identified enemy force to defend itself.

2-94. **Restrictive Fire Area.** A *restrictive fire area* (RFA) is an area in which specific restrictions are imposed and into which fires that exceed those restrictions will not be delivered without coordination with the establishing headquarters (JP 3-09). (See Figure 2-20.) The purpose of the RFA is to regulate fires into an area according to the stated restrictions, such as no unguided conventional or dud-producing munitions. Maneuver battalion or larger ground forces normally establish RFAs. On occasion, a company operating independently may establish a RFA. Usually, it is located on identifiable terrain, by grid or by a radius (in meters) from a center point. The restrictions on a RFA may be shown on a map or overlay, or reference can be made to an operation order that contains the restrictions.

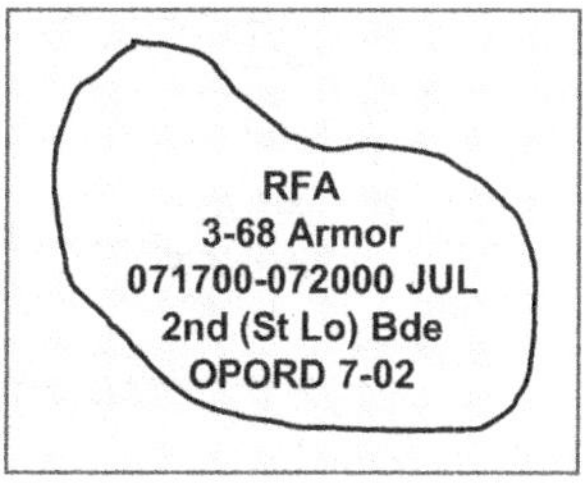

Figure 2-20. Restrictive Fire Area

2-95. **Restrictive Fire Line.** A *restrictive fire line* (RFL) is a line established between converging friendly surface forces that prohibits fires or their effects across that line (JP 3-09). Both or only one of those converging may be moving. Fires and their effects can cross a RFL when the event has been coordinated with the establishing and affected organizations. The purpose of the line is to prevent interference between converging friendly forces, such as what occurs during a linkup operation. The next higher common commander of the converging forces establishes the RFL. Located on identifiable terrain, it is usually located closer to the stationary force—if there is one—than to the moving force. Alternatively, the commander can use a RFL to protect sensitive areas, such as cultural monuments. (See Figure 2-21.)

FIRE SUPPORT TARGETS

2-96. In the fire support context, a *target* is an area designated and numbered for future firing (JP 1-02). There are graphic control measures for point targets, circular targets, rectangular targets, and linear targets. Figure 2-22 depicts these symbols. The commander designates fire support targets using a two-letter and four-digit code established in field artillery doctrine. He may group two or more targets for simultaneous engagement. He may also attack individual targets and groups of targets in series or in a predetermined sequence. When this occurs, it is referred to as a series of targets.

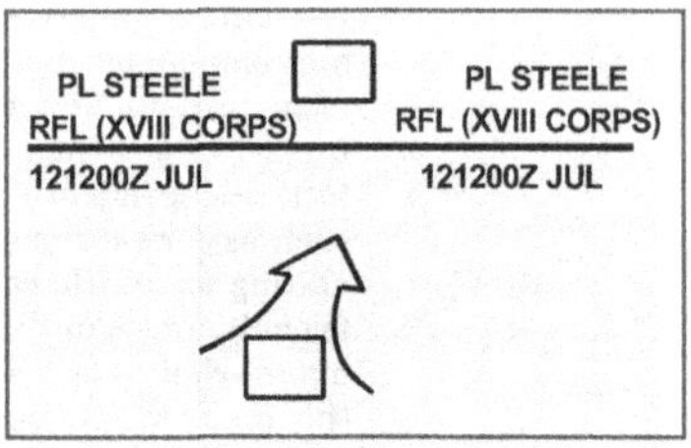

Figure 2-21. Restrictive Fire Line

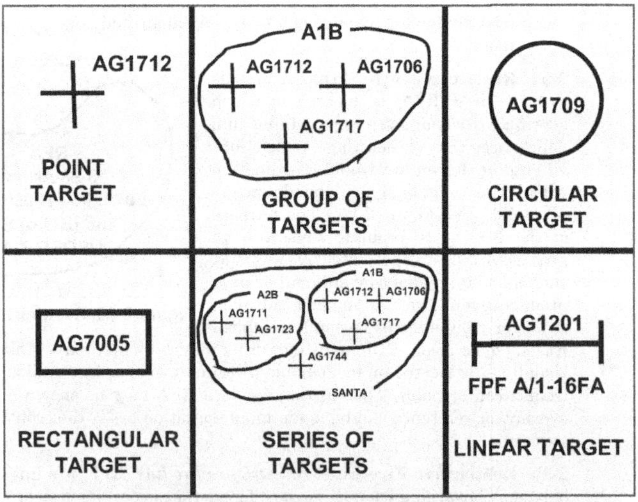

Figure 2-22. Targets

2-97. Doctrine classifies each fire support target as either a planned target or a target of opportunity. Targets of opportunity are not planned in advance and are engaged as they present themselves in accordance with established engagement criteria and rules of engagement. Planned targets are ones on which fires are prearranged, although the degree of this prearrangement may vary.

2-98. Individually planned fire support targets may be further subdivided into scheduled and on-call fires. Scheduled targets are planned targets on which field artillery and other fire support assets deliver their fires in accordance with a pre-established time schedule and sequence. On-call targets are planned targets engaged in response to a request for fires rather than in accordance with an established time schedule. An on-call target requires less reaction time than a target of opportunity. The degree of prearrangement for the on-call target influences the reaction time from request to execution–the greater the prearrangement, the faster the reaction time. Priority targets are an example of on-call targets that have short reaction times since each priority target has a fire unit placed on it when it is not engaged in other fire missions. The final protective fires (FPF) of A Battery, 1st Battalion 16th Field Artillery in Figure 2-22 is an example of a priority target. (See FM 3-09 for additional information regarding fire support.)

FORWARD LINE OF OWN TROOPS

2-99. The *forward line of own troops* (FLOT) is a line which indicates the most forward positions of friendly forces in any kind of military operation at a specific time. The forward line of own troops normally identifies the forward location of covering and screening forces (JP 1-02). In the defense, it may be beyond, at, or short of the forward edge of the battle area (FEBA), depending on the tactical situation. (Chapter 9 defines the FEBA with other defensive control measures.) It does not include small, long-range reconnaissance assets and similar stay-behind forces. Friendly forces forward of the FLOT may have a restrictive fire coordination measure, such as an RFA, placed around them to preclude fratricide. Figure 2-23 depicts the symbol for the FLOT.

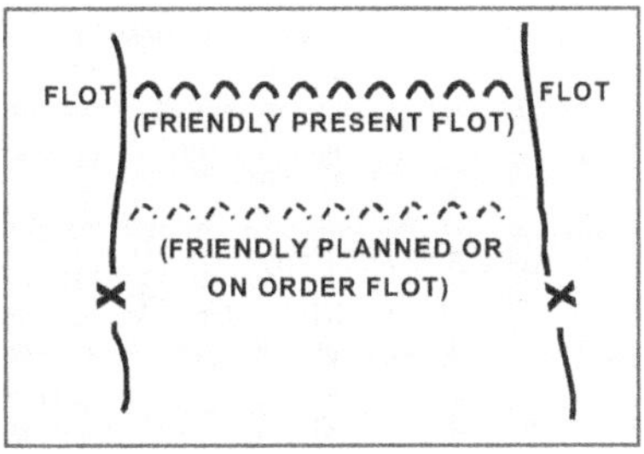

Figure 2-23. Forward Line of Own Troops

LINE OF CONTACT

2-100. **The *line of contact* (LC) is a general trace delineating the location where friendly and enemy forces are engaged.** The commander designates the enemy side of the LC by the abbreviation "ENY." In the defense, a LC is often synonymous with the FLOT. In the offense, a LC is often combined with the LD. Chapter 4 discusses the LD. Figure 2-24 depicts the symbol for the LC.

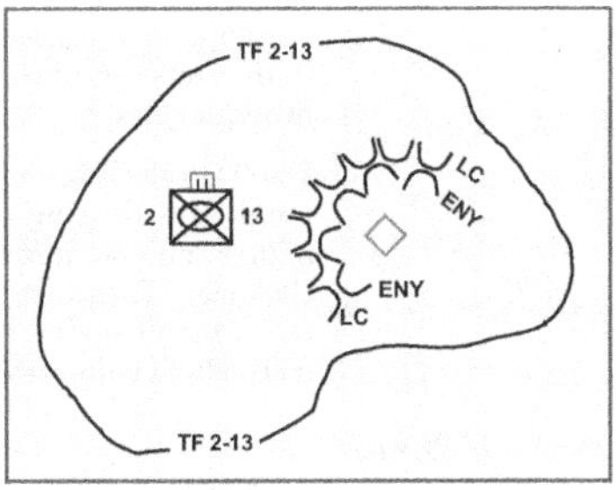

Figure 2-24. Line of Contact

NAMED AREA OF INTEREST

2-101. A ***named area of interest* (NAI) is the geographical area where information that will satisfy a specific information requirement can be collected.** NAIs are usually selected to capture indications of enemy courses of action but also may be related to battlefield and environment conditions. The commander tailors the shape of the NAI symbol to the actual area he wants observed, rather than using a prescribed shape. It is possible to redesignate a NAI as a targeted area of interest on confirmation of enemy activity within the area, allowing a commander to mass the effects of his combat power on that area. Figure 2-25 depicts NAI Augusta.

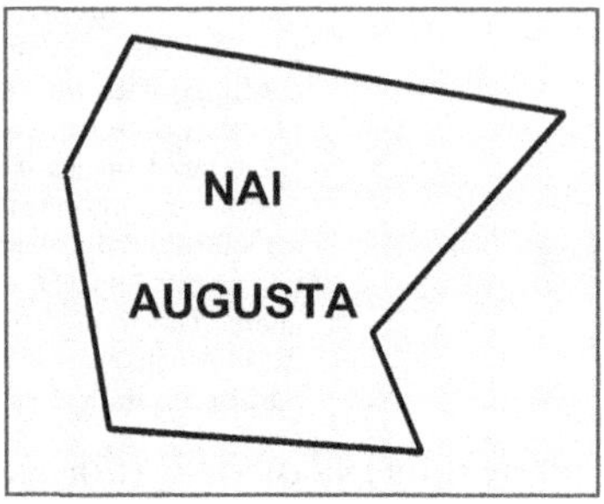

Figure 2-25. Named Area of Interest

OBSTACLE CONTROL MEASURES

2-102. *Obstacles* are any obstruction designed or employed to disrupt, fix, turn, or block the movement of an opposing force, and to impose additional losses in personnel, time, and equipment on the opposing force. Obstacles can exist naturally or can be manmade, or can be a combination of both (JP 1-02). ***Obstacle control measures* are specific measures that simplify the granting of obstacle-emplacing authority while providing obstacle control.** They consist of—

- Zones.
- Belts.
- Groups.
- Restrictions.

Figure 2-26 on page 2-34 summarizes these graphic control measures. Unless he is the senior land component commander within a theater of operations, a commander assigned an AO does not have emplacement authority for other than protective obstacles unless specifically granted the authority by a higher headquarters.

2-103. Procedural measures include restricting specific obstacles by type, munitions, or point locations in a verbal or written order. They also include higher commanders tasking subordinate units to construct or execute specific obstacles. These are known as directed or reserve obstacles and are specified tasks found in the unit order. FM 3-34.1 discusses the detailed procedural methods to both restrict and direct obstacle construction.

Obstacle Zones

2-104. An *obstacle zone* is a division-level command and control measure, normally done graphically, to designate specific land areas where lower echelons are allowed to employ tactical obstacles (JP 1-02). Corps and division commanders use them to grant obstacle-emplacement authority to brigades (including armored cavalry regiments and other major subordinate units). Obstacle zones are permissive, allowing brigades to place reinforcing obstacles to support their scheme of maneuver without interfering with future operations.

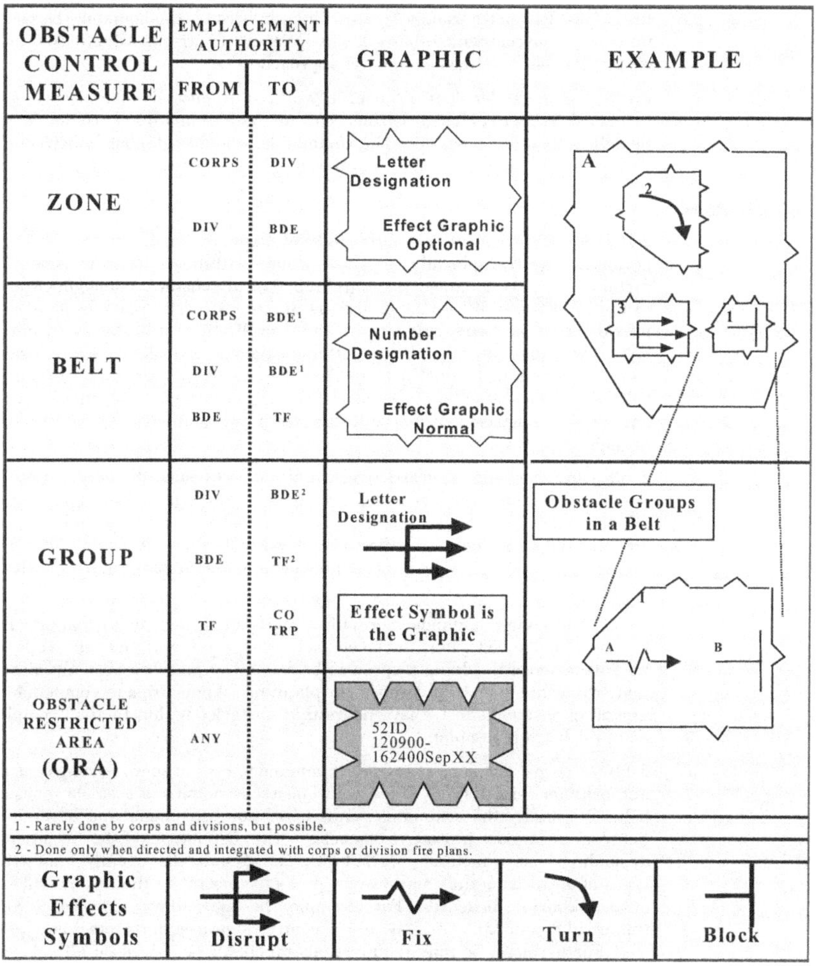

Figure 2-26. Obstacle Control Measure Graphics

2-105. If the obstacle zone encompasses the entire brigade AO, another graphic is unnecessary. Commanders may designate the entire AO as an obstacle zone, with the unit boundaries defining the geographical limits of the zone. Obstacle zones do not cross brigade boundaries. Commanders assign obstacle zones to a single subordinate unit to ensure unity of effort, just as they would when assigning defensive AOs or battle positions. This keeps tactical obstacle responsibility along the same lines as control of direct and indirect fires. This does not normally create vulnerabilities on the boundary between units since the commander bases his assignment of both subordinate AOs and obstacle zones on defined avenues of approach.

2-106. A commander does not normally assign an obstacle effect (block, fix, turn, or disrupt) to an obstacle zone. This allows his subordinate commanders flexibility in using obstacles. The commander should establish construction and resourcing priorities between different obstacle zones.

Obstacle Belts

2-107. An *obstacle belt* is a brigade-level command and control measure, normally given graphically, to show where within an obstacle zone the ground tactical commander plans to limit friendly obstacle employment and focus the defense (JP 1-02). It assigns an intent to the obstacle plan and provides the necessary guidance on the overall effect of obstacles within a belt. They plan obstacle belts within assigned obstacle zones to grant obstacle-emplacement authority to their major subordinate units. Obstacle belts also focus obstacles to support the brigade scheme of maneuver and ensure that obstacles do not interfere with the maneuver of any higher headquarters.

2-108. Obstacle belts are restrictive, but also direct a subordinate unit to construct one or more obstacles to achieve an effect in the area. They do not specify the type or number of obstacles. Obstacle belts do not cross unit boundaries for the same reasons as discussed in obstacle zones. A single unit is responsible for a belt; however, a commander may assign more than one belt to a unit.

2-109. A brigade commander normally assigns an obstacle effect and priority to each obstacle belt. As with the obstacle zone, the target and relative location are apparent. Adding a specific obstacle effect gives purpose and direction to battalion task force obstacle planning. When brigade commanders assign an obstacle effect, they ensure that obstacles within the belt complement the brigade fire plan.

2-110. A corps, division, or brigade commander may authorize emplacement authority for certain types of protective obstacles outside of obstacle zones or belts. Normally, the commander authorizes company team and base commanders to emplace protective obstacles within 500 meters of their positions, depending on the factors of METT-TC. The commander usually limits the types of obstacles a unit may use for protective obstacles that are outside of obstacle-control measures. For example, he may allow only wire- and command-detonated mines outside of control measures for protective obstacles. Furthermore, he may require that minefields be fenced on all sides to prevent fratricide, after obtaining legal guidance concerning current rules and policies.

Obstacle Groups

2-111. ***Obstacle groups* are one or more individual obstacles grouped to provide a specific obstacle effect.** Task forces use obstacle groups to ensure that company teams emplace individual obstacles supporting the task force's scheme of maneuver. In rare cases, brigades, divisions, or even corps may use obstacle groups for specific tactical obstacles. Also, units integrate obstacle groups with their direct- and indirect-fire plans. Brigade and task force commanders can plan their placement anywhere in the obstacle zones or belts, respectively.

2-112. Unlike obstacle zones or belts, obstacle groups are not areas but relative locations for actual obstacles. Commanders normally show obstacle groups using the obstacle-effect graphics. When detailed planning is possible (to include detailed on-the-ground reconnaissance), commanders may show obstacle groups using individual obstacle graphics.

2-113. The company team commander and the engineer can adjust obstacles in the group if the intent and link to the fire plan remain intact. Company team commanders make minor changes to obstacles and fire-control measures based on terrain realities. For example, a commander may move a fixing obstacle group and direct-fire TRPs a hundred meters to avoid having them masked by rolling terrain. A major change to the obstacle group location requires the approval of the commander who ordered the obstacle group emplacement.

Individual Obstacles

2-114. Each type of individual obstacle, such as abatis, antitank ditch, booby traps, mines and minefields, roadblocks, craters, and wire obstacles has its associated graphic. Once a unit constructs an individual obstacle, the obstacle's location is recorded and reported through the chain of command. Commanders must report individual obstacles in sufficient detail so any unit moving through the area can bypass or reduce the obstacle without excessive risk. Each headquarters is responsible to ensure exact obstacle locations are disseminated throughout its organization. Individual obstacle graphics are rarely shown on maps above the battalion echelon and are not depicted in this manual. (FM 3-34.1 defines individual obstacles and establishes the graphics for them.)

Obstacle Restrictions

2-115. Commanders may use obstacle restrictions to provide additional obstacle control and to limit the specific types of obstacles used, such as no buried mines. These restrictions ensure that subordinates do not use obstacles with characteristics that impair future operations. It also allows commanders to focus the use of limited resources for the decisive operation by restricting their use elsewhere. An *obstacle restricted area* (ORA) is a command and control measure used to limit the type or number of obstacles within an area (JP 1-02). The commander with emplacement authority uses ORAs to restrict obstacle placement. The ORA graphic depicts the area effected, the unit imposing the restriction, and the restrictions in effect.

PHASE LINE

2-116. A *phase line* (PL) is a line utilized for control and coordination of military operations, usually a terrain feature extending across the operational area [JP 1-02 uses *zone of action*] (JP 1-02). (See Figure 2-27.) A commander establishes PLs to control the maneuver of his units. Phase lines are not boundaries unless designated as such and do not establish any specific responsibilities between units, unless the operations order so specifies. When possible, the commander places them along easily recognizable terrain features—such as roads, railroad tracks, rivers, and ridgelines—to ensure easy identification. As with boundaries, this is less important if all units are equipped with precision navigation devices. Some PLs have additional designations for specific purposes, such as a LD or a probable line of deployment (PLD). Chapter 5 discusses these specific purposes.

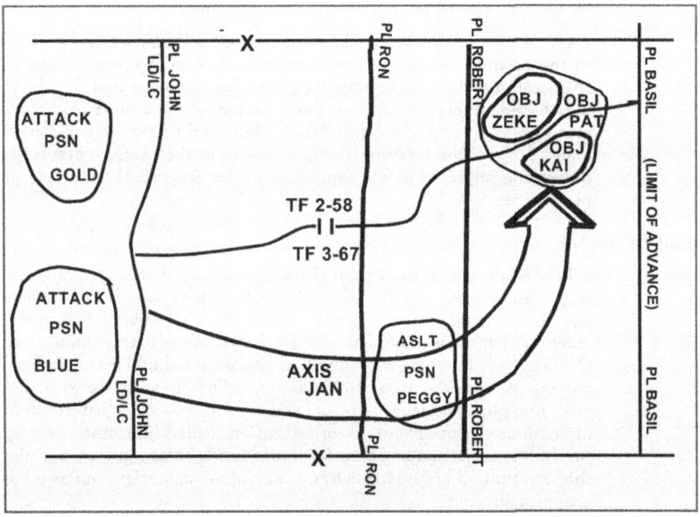

Figure 2-27. Phase Lines Used with Other Control Measures

POSITION AREA FOR ARTILLERY

2-117. **A *position area for artillery* (PAA) is an area assigned to an artillery unit where individual artillery systems can maneuver to increase their survivability. A PAA is not an AO for the artillery unit occupying it.** The commander assigns PAAs for terrain management purposes. Establishing a PAA lets other subordinate units know they should avoid occupying that same terrain, thus avoiding enemy counterfire. While the exact size of a PAA depends on the factors of METT-TC, a Paladin platoon normally requires a PAA encompassing two square kilometers, and a Multiple Launch Rocket System (MLRS) platoon requires nine square kilometers. (See Figure 2-28.)

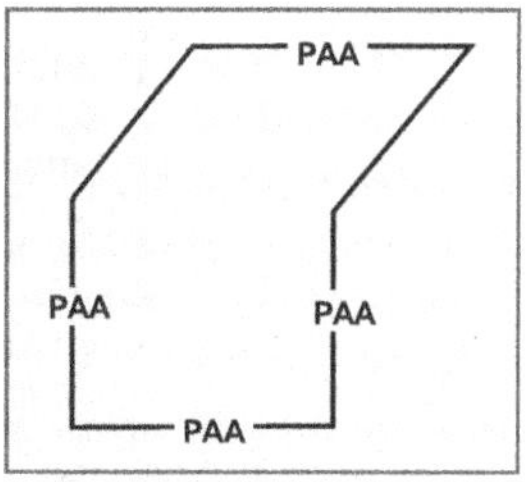

Figure 2-28. Position Area for Artillery

2-118. The maneuver echelon operations officer (G3 or S3) of the unit that owns the terrain establishes the PAA. The occupying artillery unit does not have the same authority and responsibilities toward the PAA that are associated with a unit assigned an AO. For example, other units can move through a PAA without clearing that movement with the artillery unit. The artillery unit occupying a PAA establishes liaison with the unit that owns the AO where the PAA is located. The echelon fire support officer usually conducts this liaison in accordance with standard command and support relationships. (For a discussion of common command and control relationships, see FM 6-0. For a discussion of artillery missions, see FM 3-09.)

2-119. The decision to establish a PAA affects A2C2 for rotary and fixed-wing integration. A PAA is a base upon which to establish future grid-target lines for lateral deconfliction and areas for rotary- and fixed-wing aircraft to avoid depending on high- or low-angle artillery fires.

ROUTE

2-120. A *route* is the prescribed course to be traveled from a specific point of origin to a specific destination (JP 1-02). (See Route Iron in Figure 2-29.) Routes can have different purposes. Those purposes can be added as adjectives to specify different types of routes. Examples of such routes include passing route and main supply route (MSR). The commander can further designate MSRs as open, supervised, dispatch, reserved, or prohibited. The commander can assign names, numbers, or alphanumeric designations to routes within his AO. (See FM 4-01.30 for additional information concerning route classification and marking.)

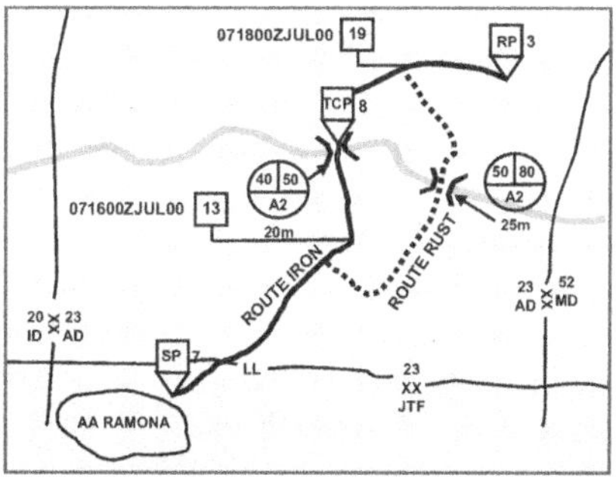

Figure 2-29. Routes

TARGETED AREA OF INTEREST

2-121. **A *targeted area of interest* (TAI) is the geographical area or point along a mobility corridor where successful interdiction will cause the enemy to abandon a particular course of action or requires him to use specialized engineer support to continue. It is where he**

can be acquired and engaged by friendly forces. The commander designates TAIs where he believes his unit can best attack high-payoff targets. The unit staff develops TAIs during the targeting process, based on the currently available products resulting from the IPB process. These TAIs are further refined during wargaming and finally approved by the commander during COA approval. The shape of a TAI reflects the type of target and the weapon system intended to engage that target. They are normally cued by surveillance assets, which include UAVs, combat observation and lasing teams (COLTs), long-range surveillance units (LSUs), fixed-wing reconnaissance aircraft using a variety of sensors, and special operations forces. A commander can designate a TAI for any of his organic or supporting systems, including CAS. Figure 2-30 depicts TAI Whitetail.

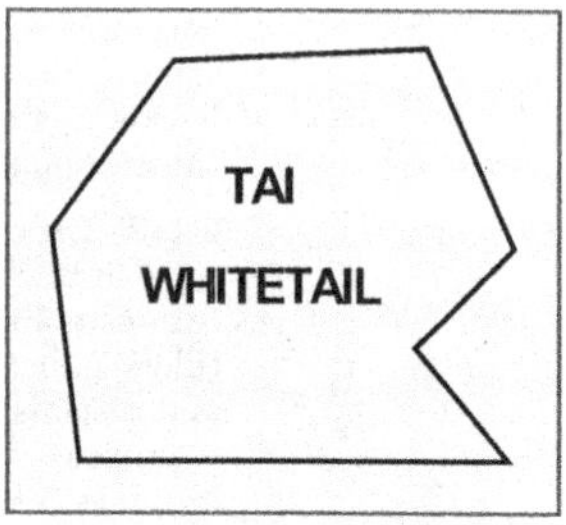

Figure 2-30.
Targeted Area of Interest

PART TWO

Offensive Operations

Chapter 3

The Basics of the Offense

To move swiftly, strike vigorously, and secure all the fruits of victory, is the secret of successful war.

Thomas J. "Stonewall" Jackson, 1863

Offensive operations aim at destroying or defeating an enemy. Their purpose is to impose US will on the enemy and achieve decisive victory (FM 3-0). A commander may also conduct offensive operations to deprive the enemy of resources, seize decisive terrain, deceive or divert the enemy, develop intelligence, or hold an enemy in position. This chapter discusses the basics of the offense. The basics discussed in this chapter apply to all types of offensive operations.

CONTENTS

3-1. The commander seizes, retains, and exploits the initiative when conducting offensive operations. Even in the defense, wresting the initiative from the enemy requires offensive operations.

3-2. Offensive operations are either force- or terrain-oriented. Force-oriented operations focus on the enemy. Terrain-oriented operations focus on seizing and retaining control of terrain and facilities.

CHARACTERISTICS OF OFFENSIVE OPERATIONS

3-3. Surprise, concentration, tempo, and audacity characterize the offense. Effective offensive operations capitalize on accurate intelligence and other relevant information regarding enemy forces, weather, and terrain. The commander maneuvers his forces to advantageous positions before contact. Security operations and defensive information operations keep or inhibit the enemy from acquiring accurate information about friendly forces. Contact with enemy forces before the decisive operation is deliberate, designed to shape the optimum situation for the decisive operation. The decisive operation is a sudden, shattering action that capitalizes on subordinate initiative and a common operational picture (COP) to expand throughout the area of operations (AO). The commander executes violently without hesitation to break the enemy's will or destroy him. FM 3-0 discusses the four characteristics of offensive operations.

Characteristics of Offensive Operations

- Surprise
- Concentration
- Tempo
- Audacity

HISTORICAL EXAMPLE

3-4. The following vignette discusses the offensive operations of the 101st Airborne Division (Air Assault) during OPERATION DESERT STORM. The actions of the 101st Airborne Division between 24 and 28 February 1991 were a shaping operation for the XVIII Airborne Corps.

The 101st Airborne Division (Air Assault), 24–28 February 1991

The 101st attacked on 24 February 1991 to interdict, block, and defeat enemy forces operating in or moving through AO Eagle. On order, they were to attack to the east to assist in defeating the Iraqi Republican Guard Forces Command (RGFC). As part of that attack, the 101st conducted an air assault to establish forward operating base (FOB) Cobra. The FOB was approximately halfway between tactical assembly area (TAA) Campbell and the Euphrates River and Highway 8 (the main road between Basrah and Baghdad). In the FOB, the 101st would build up supplies and forces by both land and air. That FOB would support a further air assault by the division's remaining uncommitted brigade into AO Eagle. This second air assault would cut both Iraqi lines of communication (LOC) and retreat routes of Iraqi forces in Kuwait. From FOB Cobra, the 101st, with two brigades, could launch air assault operations to support other operations.

The operation was scheduled to begin at 0400 on 24 February. The 1st Brigade's ground column departed the TAA at 0700. However, weather delayed the air assault into FOB Cobra, approximately 150 kilometers north of the TAA, until

0730. The assault elements of the brigade cleared the area within the FOB of Iraqi soldiers by 1030, making it secure enough for AH-64 operations. Attack helicopter operations from FOB Cobra into AO Eagle began by 1330 with patrols along Highway 8. The CG, MG J.H. Binford Peay III, decided to bring in 2nd Brigade to FOB Cobra that day while the weather permitted. However, he delayed 3rd Brigade's assault into AO Eagle until 25 February.

That night, reconnaissance assets found an alternative route for the vehicles to use from LZ Sand, in AO Eagle, to Highway 8. The first air assault on 25 February landed three antiarmor companies, two infantry companies, and an artillery battalion into the LZ, but these units were unable to link up with their parent battalions until the next day. The 3rd Brigade's main body cut Highway 8 at 1508 and secured the area by 1848. This later air assault covered 156 miles and cut Highway 8 only 145 miles from Baghdad. Meanwhile, the buildup of FOB Cobra continued, with 3rd Brigade's second lift spending the night there due to weather. The 101st also established contact with 24th Infantry Division (ID) to its east. Farther north in AO Eagle, 3rd Brigade made sporadic contact with Iraqi forces along Highway 8 during the night. With this, the 101st Airborne achieved all of its initial objectives and began planning for subsequent missions. It had pressed the fight, never let the enemy recover from the initial blow, and exploited success at FOB Cobra and AO Eagle. (See Figure 3-1.)

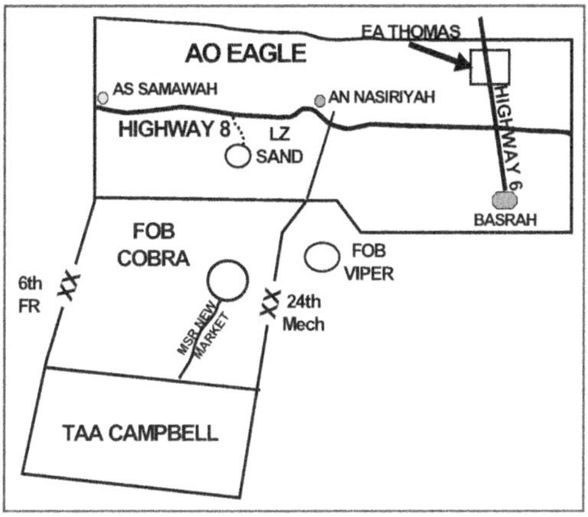

Figure 3-1. Initial Objectives

A continuing sandstorm intensified on 26 February, limiting aerial operations. However, the remaining 3rd Brigade lifts were ferried from FOB Cobra to AO Eagle, and 2nd Brigade closed into FOB Cobra and began planning for subsequent operations. Although the sandstorm curtailed aerial operations, the division continued ground operations. The 3rd Brigade kept Highway 8 closed from ground battle positions supported by DS artillery despite Iraqi attempts to bypass. The 1st Brigade continued the buildup of FOB Cobra by ground transport and established contact with the 82nd Airborne Division. The 24th ID reached the Euphrates that night and cut Highway 8 farther to the east with heavy forces. The tempo was such that planned contingencies were overtaken by events; XVIII Corps had accomplished its initial objectives and issued new orders orienting the corps' main effort eastward. As part of this reorientation, MG Peay decided to establish FOB Viper, 150 kilometers to the east of Cobra, for aerial operations against EA Thomas, 200 kilometers northeast of Viper, to destroy Iraqi forces fleeing north from Basrah. Corps assets, including the 12th Aviation Brigade and the 5th Battalion, 8th Artillery (155, T), reinforced the 101st Division.

The 101st launched its third major air assault at 0830, 27 February. The 2nd Brigade assaulted into FOB Viper with an infantry battalion and a reinforced artillery battalion augmented by engineers and air defense forces. By 1400, four attack helicopter battalions began operations against EA Thomas. The 2nd Brigade continued to close into FOB Viper. The division's attack helicopters cut the last escape route north out of Kuwait and destroyed 14 APCs, eight BM-21 MRLs, four MI-6 helicopters, and two SA-6 units. The Iraqi forces, unable to escape north, lay in the paths of VII Corps and 24th ID. The 1st Brigade had orders to air assault into EA Thomas on 28 February, but the cease-fire precluded this operation. Although the actions of the 101st during these 100 hours were a shaping operation, they illustrate the tactical application of all of the characteristics of offensive operations.

TYPES OF OFFENSIVE OPERATIONS

Types of Offensive Operations

- Movement to contact
- Attack
- Exploitation
- Pursuit

3-5. The four types of offensive operations are movement to contact, attack, exploitation, and pursuit. Entry operations, while offensive in nature, are one of the six subordinate forms of force-projection operations. Force-projection operations are operational level tasks and fall outside the scope of this manual. Joint doctrine addresses force-projection operations. Appendix C addresses tactical airborne and air assault operations as vertical envelopments.

MOVEMENT TO CONTACT

3-6. *Movement to contact* is a type of offensive operation designed to develop the situation and establish or regain contact (FM 3-0). The commander conducts a movement to contact (MTC) when the enemy situation is vague or not specific enough to conduct an attack. A search and attack is a specialized technique of conducting a movement to contact in an environment of noncontiguous AOs. Chapter 4 discusses MTC.

ATTACK

Forms of the Attack

- Ambush
- Spoiling attack
- Counterattack
- Raid
- Feint
- Demonstration

3-7. An *attack* is an offensive operation that destroys or defeats enemy forces, seizes and secures terrain, or both (FM 3-0). Movement, supported by fires, characterizes the conduct of an attack. However, based on his analysis of the factors of METT-TC, the commander may decide to conduct an attack using only fires. An attack differs from a MTC because enemy main body dispositions are at least partially known, which allows the commander to achieve greater synchronization. This enables him to mass the effects of the attacking force's combat power more effectively in an attack than in a MTC.

3-8. Special purpose attacks are ambush, spoiling attack, counterattack, raid, feint, and demonstration. The commander's intent and the factors of METT-TC determine which of these forms of attack are employed. He can conduct each of these forms of attack, except for a raid, as either a hasty or a deliberate operation. Chapter 6 discusses the attack and its subordinate forms.

EXPLOITATION

3-9. *Exploitation* is a type of offensive operation that rapidly follows a successful attack and is designed to disorganize the enemy in depth (FM 3-0). The objective of an exploitation is to complete the enemy's disintegration. Chapter 6 discusses exploitation.

PURSUIT

3-10. A *pursuit* is an offensive operation designed to catch or cut off a hostile force attempting to escape, with the aim of destroying it (JP 1-02). A pursuit normally follows a successful exploitation. However, if it is apparent that enemy resistance has broken down entirely and the enemy is fleeing the battlefield, any other type or subordinate form of offensive operation can transition into a pursuit. Chapter 7 discusses the pursuit.

COMMON OFFENSIVE CONTROL MEASURES

3-11. This section defines in alphabetical order those common offensive control measures that a commander uses to synchronize the effects of his combat power. The commander uses the minimum control measures required to successfully complete the mission while providing the flexibility needed to respond to changes in the situation.

ASSAULT POSITION

3-12. **An *assault position* is a covered and concealed position short of the objective from which final preparations are made to assault the objective.** Ideally, it offers both cover and concealment. These final preparations can involve tactical considerations, such as a short halt to coordinate the final assault, reorganize to adjust to combat losses, or make necessary adjustments in the attacking force's dispositions. These preparations can also involve technical items, such as engineers conducting their final prepare-to-fire checks on obstacle clearing systems and the crews of plow- and roller-equipped tanks removing their locking pins. It may be located near to either a final coordination line (FCL) or a probable line of deployment (PLD). (Paragraphs 3-18 and 3-23 define a FCL and a PLD respectively.)

ASSAULT TIME

3-13. **The *assault time* establishes the moment to attack the initial objectives throughout the geographical scope of the operation.** It is imposed by the higher headquarters in operations to achieve simultaneous results by several different units. It synchronizes the moment the enemy feels the effects of friendly combat power. It is similar to the time-on-target control method for fire mission processing used by the field artillery. A commander uses it instead of a time of attack (defined in paragraph 3-26) because of the different distances that elements of his force must traverse, known obstacles, and differences in each unit's tactical mobility.

ATTACK-BY-FIRE POSITION

3-14. **An *attack-by-fire position* designates the general position from which a unit conducts the tactical task of attack by fire.** (Appendix B defines the tactical mission task of attack by fire.) The purpose of these positions is to mass the effects of direct fire systems for one or multiple locations toward the enemy. An attack-by-fire position does not indicate the specific site. Attack-by-fire positions are rarely applicable to units larger than company size. Figure 3-2 depicts attack-by-fire position BRANDON.

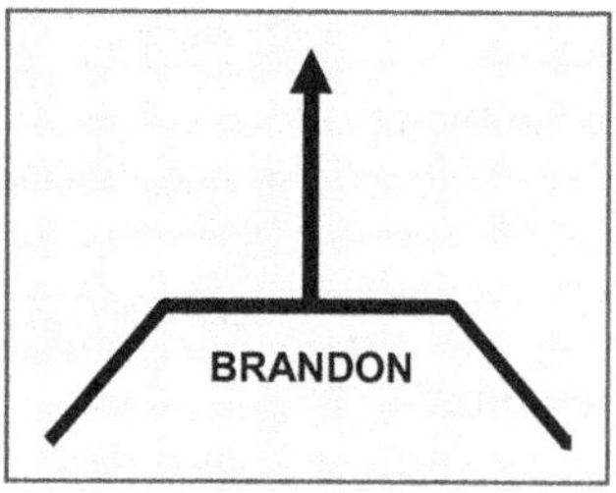

Figure 3-2. Attack-by-Fire Position BRANDON

ATTACK POSITION

3-15. **The *attack position* is the last position an attacking force occupies or passes through before crossing the line of departure.** An attack position facilitates the deployment and last-minute coordination of the attacking force before it crosses the LD. It is located on the friendly side of the LD and offers cover and concealment for the attacking force. It is used primarily at battalion level and below. Whenever possible, units move through the attack position without stopping. An attacking unit occupies an attack position for a variety of reasons; for example, when the unit is waiting for specific results from preparatory fires or when it is necessary to conduct additional coordination, such as a forward passage of lines. If the attacking unit occupies the attack position, it stays there for the shortest amount of time possible to avoid offering the enemy a lucrative target. (Figure 3-3 on page 3-6 shows attack positions BLUE and GOLD used in conjunction with other common offensive control measures.)

AXIS OF ADVANCE

3-16. **An *axis of advance* designates the general area through which the bulk of a unit's combat power must move.** There are three primary reasons why a commander uses an axis of advance. First, to direct the bypass of locations that could delay the progress of the advancing force, such as known contaminated areas. Second, to indicate that he does not require the force to clear the AO as it advances. His force will be required to clear the axis in accordance with specified bypass criteria. The third primary reason is to indicate to a unit involved in offensive encirclement, exploitation, or

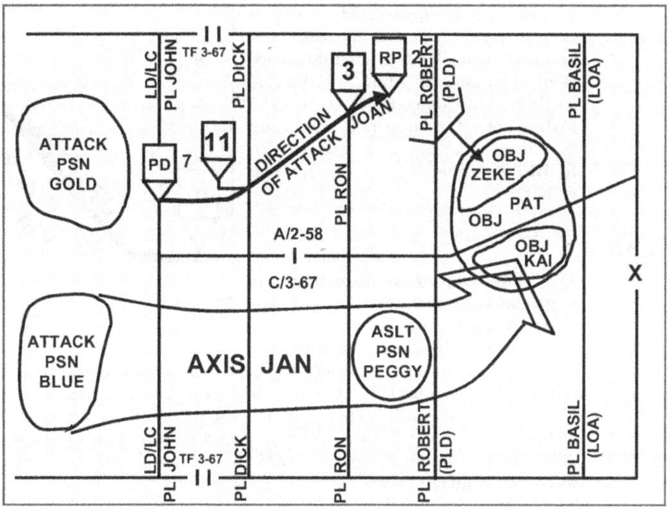

Figure 3-3. Attack Positions Used with Other Common Offensive Control Measures

pursuit operations the need to move rapidly toward an objective. Figure 3-4 depicts axis of advance DEBRA.

DIRECTION OF ATTACK

3-17. **The *direction of attack* is a specific direction or assigned route a force uses and does not deviate from when attacking.** It is a restrictive control measure. The commander's use of a direction of attack maximizes his control over the movement of his unit, and he often uses it during night attacks, infiltrations, and when attacking through smoke. The commander establishes a direction of attack through a variety of means, such as target reference points, checkpoints, global positioning system (GPS) way points, using ground surveillance radar to track the attack force, and the impact of artillery shells. Target reference points placed on recognizable terrain provide the commander with the capability to rapidly shift fires and reorient his maneuver forces. When using a direction of attack, the commander designates a point of departure (PD). (Figure 3-5 depicts direction of attack JOAN and PD 6.)

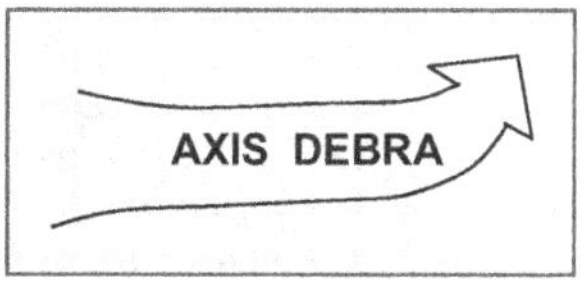

Figure 3-4. Axis of Advance

FINAL COORDINATION LINE

3-18. **The *final coordination line* (FCL) is a phase line (PL) close to the enemy position used to coordinate the lifting or shifting of supporting fires with the final deployment of maneuver elements.** Final adjustments to supporting fires necessary to reflect the actual versus the planned tactical situation take place prior to crossing this line. It should be easily recognizable on the ground. The FCL is not a fire support coordinating measure. (Figure 3-6 shows PL ROBERT as the FCL for the 4th Brigade.)

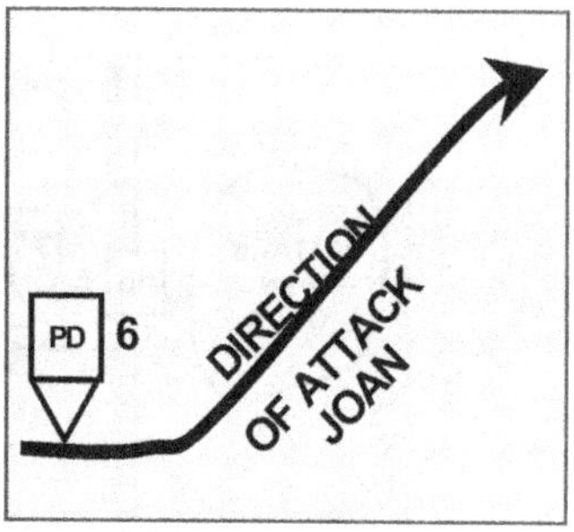

Figure 3-5. Direction of Attack JOAN

LIMIT OF ADVANCE

3-19. **The *limit of advance* (LOA) is a phase line used to control forward progress of the attack. The attacking unit does not advance any of its elements or assets beyond the LOA, but the attacking unit can push its security forces to that limit.** A commander usually selects a linear terrain feature, perpendicular to the direction of attack, on the far side of the objective as the LOA because such a terrain feature is easily identifiable. The commander employs a LOA to prevent overextending the attacking force and reduce the possibility of fratricide by fires supporting the attack. The commander positions a LOA far enough beyond the objective to allow the unit to defend the objective. An LOA prevents units from exploiting success and launching a pursuit; therefore, a commander should only use it if he does not want the unit to conduct an exploitation or pursuit. A forward boundary is always a LOA, but a LOA is not necessarily a forward boundary. In fact, a

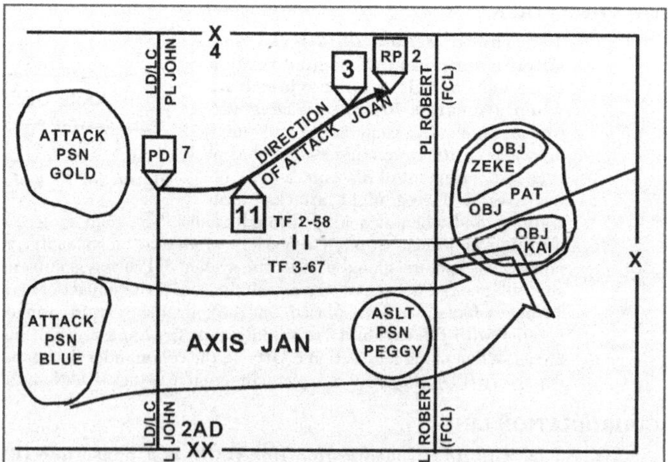

Figure 3-6. Final Coordination Line (FCL) ROBERT Used in Conjunction with Other Offensive Control Measures

LOA and the unit's forward boundary should rarely coincide because of the resulting limitations that a forward boundary places on supporting fires beyond the forward boundary. Figure 3-7 shows PL BASIL used as 4th Brigade's LOA.

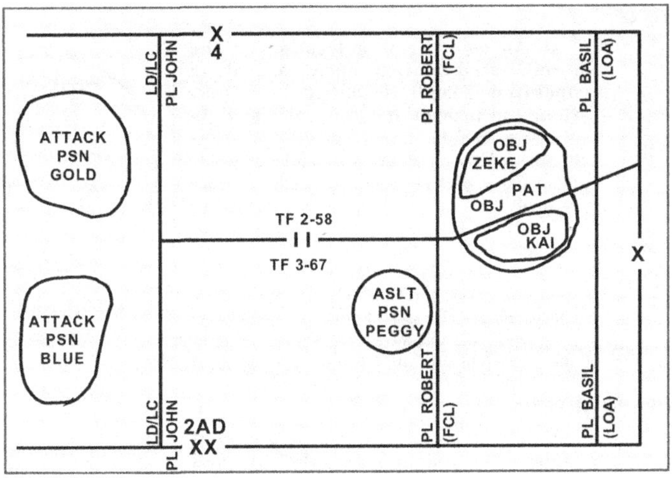

Figure 3-7. Limit of Advance Used with Other Common Control Measures

LINE OF DEPARTURE

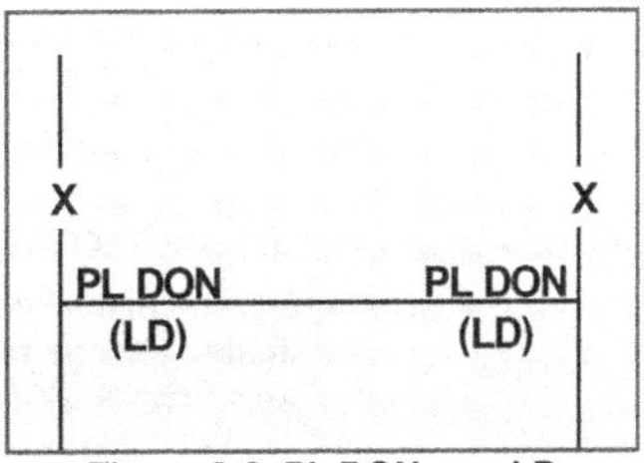

Figure 3-8. PL DON as a LD

3-20. **The *line of departure* is a phase line crossed at a prescribed time by troops initiating an offensive operation.** The purpose of the LD is to coordinate the advance of the attacking force so that its elements strike the enemy in the order and at the time desired. The LD also marks where the unit transitions from movement to maneuver. The commander can also use it to facilitate the coordination of fires. Generally, it should be perpendicular to the direction the attacking force will take on its way to the objective. Friendly forces should control the LD. The commander analyzes the terrain before designating his LD. Different units have different movement rates on leaving their assembly areas (AAs) based on their inherent mobility characteristics and the terrain being crossed. The commander considers these different characteristics when establishing the LD to prevent these differences from affecting the synchronization of the operation. When possible, the commander selects the LD so that the terrain the attack unit traverses before crossing the LD provides sufficient cover for the attacking unit's final deployment into a combat formation before crossing the LD. In many cases the LD is also the line of contact because the unit in contact is conducting the attack from its current positions. Figure 3-8 depicts PL DON as the LD. (Chapter 2 contains a definition for a line of contact.)

OBJECTIVE

3-21. **An *objective* is a location on the ground used to orient operations, phase operations, facilitate changes of direction, and provide for unity of effort.** An objective can be either terrain- or force-oriented. Terrain objectives should be easily identifiable on the ground to facilitate their recognition. The commander determines his force-oriented objectives based on known enemy positions. The commander normally assigns his subordinate commanders only their final objectives, but can assign intermediate objectives as necessary. Figure 3-9 depicts objective STEVE. Objective STEVE is further broken down into two subordinate objectives, objective JOHN and objective HARRY.

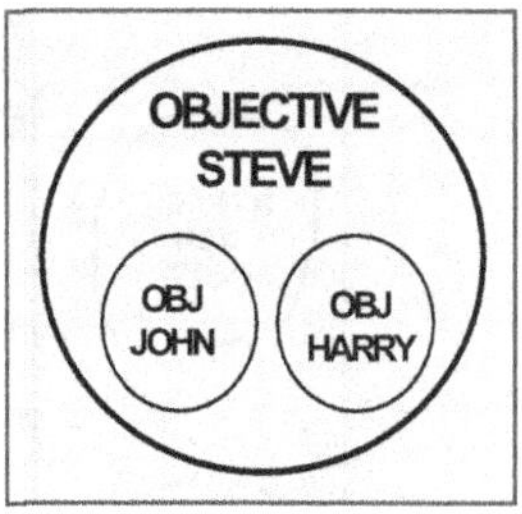

Figure 3-9. Objective STEVE

POINT OF DEPARTURE

3-22. **The *point of departure* is the point where the unit crosses the LD and begins moving along a direction of attack.** Units conducting reconnaissance and security patrols and other operations in a low-visibility environment commonly use a PD as a control measure. Like a LD, it marks the point where the unit transitions from movement to maneuver under conditions of limited visibility. Figure 3-10 depicts PD 7.

Figure 3-10. Point of Departure 7

PROBABLE LINE OF DEPLOYMENT

3-23. **A *probable line of deployment* is a phase line that a commander designates as the location where he intends to completely deploy his unit into assault formation before beginning the assault.** The PLD is used primarily at battalion level and below when the unit does not cross the LD in its assault formation. It is usually a linear terrain feature perpendicular to the direction of attack and recognizable under conditions of limited visibility. The PLD should be located outside the range where the enemy can place the attacking force under effective direct fire. It has no use except as it relates to the enemy. In Figure 3-11, page 3-10, PL ROBERT is also the PLD.

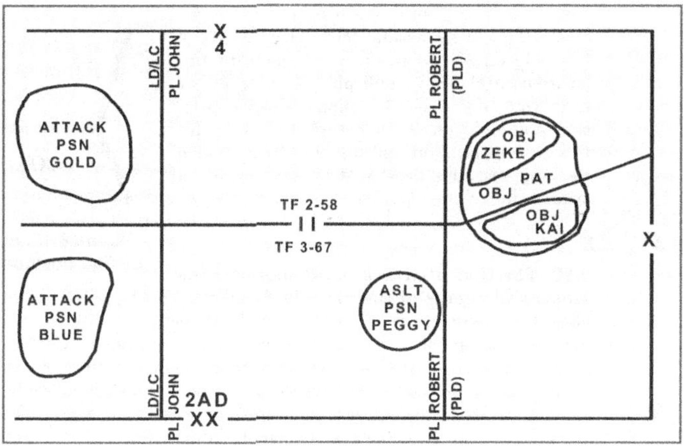

Figure 3-11. Probable Line of Deployment ROBERT Used with Other Control Measures

RALLY POINT

Figure 3-12. Rally Point 14

3-24. **A *rally point* is an easily identifiable point on the ground at which units can reassemble and reorganize if they become dispersed. Alternatively it is an easily identifiable point on the ground at which aircrews and passengers can assemble and reorganize following an incident requiring a forced landing.** Forces conducting a patrol or an infiltration commonly use this control measure. **The *objective rally point* (ORP) is a rally point established on an easily identifiable point on the ground where all elements of the infiltrating unit assemble and prepare to attack the objective.** It is typically near the infiltrating unit's objective; however, there is no standard distance from the objective to the ORP. It should be far enough away from the objective so that the enemy will not detect the infiltrating unit's attack preparations. Figure 3-12 depicts Rally Point 14.

SUPPORT-BY-FIRE POSITION

3-25. **A *support-by-fire position* designates the general position from which a unit conducts the tactical mission task of support by fire.** (Appendix B defines the tactical mission task of support by fire.) The purpose of these positions is to increase the supported force's freedom of maneuver by placing direct fires on an objective that is going to be assaulted by a friendly force. Support-by-fire positions are located within the maximum friendly direct-fire range of the enemy positions. The commander selects them so that the moving assault force does not mask its supporting fires. For this reason, support-by-fire positions are normally located on the flank of the assault force, elevated above the objective if possible. Support-by-fire positions are rarely applicable to units larger than company size. The support-by-fire position graphic depicted in Figure 3-13 indicates the general location and direction from which the unit provides fires; it does not indicate a specific site.

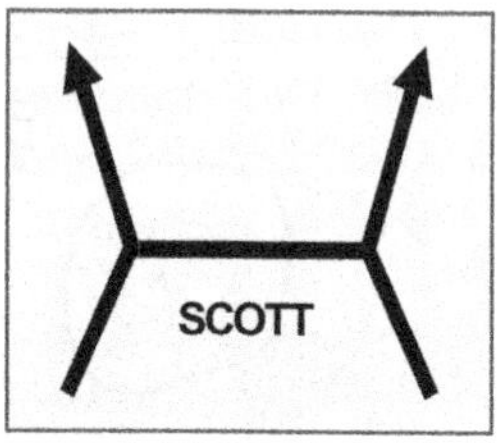

Figure 3-13. Support-by-Fire Position SCOTT

TIME OF ATTACK

3-26. **The *time of attack* is the moment the leading elements of the main body cross the LD, or in a night attack, the PD.** A commander uses it when conducting simultaneous operation where a shaping operation must accomplish its mission to set the conditions for the success of the decisive operation. When determining the time of attack, the commander considers the time required for his subordinate to—

- Conduct necessary reconnaissance, prepare plans, and issue orders.
- Synchronize plans between all subordinate units.
- Complete attack preparations, such as precombat inspections.
- Move to the LD or PD.

3-27. Orders normally designate the time of attack as H-hour. This is normally when the main body crosses the LD. However, H-hour can also designate the time to implement a phase of an operation, such as an airborne or air assault. The headquarters planning the offensive operation specifies the term's exact meaning. This is usually a part of the unit's standing operating procedures (SOP).

FORMS OF MANEUVER

Forms of Maneuver

- Envelopment
- Turning movement
- Frontal attack
- Penetration
- Infiltration

3-28. The forms of maneuver are envelopment, turning movement, frontal attack, penetration, and infiltration. Combined arms organizations seeking to accomplish their assigned mission synchronize the contributions of all battlefield operating systems (BOS) to execute these forms of maneuver. The commander generally chooses one form on which he builds a course of action (COA). The higher commander rarely specifies the specific form of offensive maneuver. However, his guidance and intent, along with the mission that includes implied tasks, may impose constraints such as time, security, and direction of attack that narrow the forms of offensive maneuver to one alternative. Additionally, the AO's characteristics and the enemy's dispositions also help determine the form of offensive maneuver selected. A single operation may contain several forms of offensive maneuver, such as a frontal attack to clear a security area followed by a penetration to create a gap in enemy defenses. An envelopment would follow to destroy the enemy's first line of defense.

ENVELOPMENT

3-29. An *envelopment* is a form of maneuver in which an attacking force seeks to avoid the principal enemy defenses by seizing objectives to the enemy rear to destroy the enemy in his current positions. At the tactical level, envelopments focus on seizing terrain, destroying specific enemy forces, and interdicting enemy withdrawal routes (FM 3-0). Also, at the tactical level, airborne and air assault operations are vertical envelopments. (See Appendix C.) The commander's decisive operation focuses on attacking an assailable flank. It avoids the enemy's strength—his front—where the effects of his fires and obstacles are the greatest. Generally, a commander prefers to conduct an envelopment instead of a penetration or a frontal attack because the attacking force tends to suffer fewer casualties while having the most opportunities to destroy the enemy. An envelopment also produces great psychological shock on the enemy. If no assailable flank is available, the attacking force creates one. The four varieties of envelopment are single envelopment, double envelopment, encirclement, and vertical envelopment. (See Figures 3-14 and 3-15.) For a discussion of the use of vertical envelopments, see Appendix C. For a discussion of encirclement operations, see Appendix D.

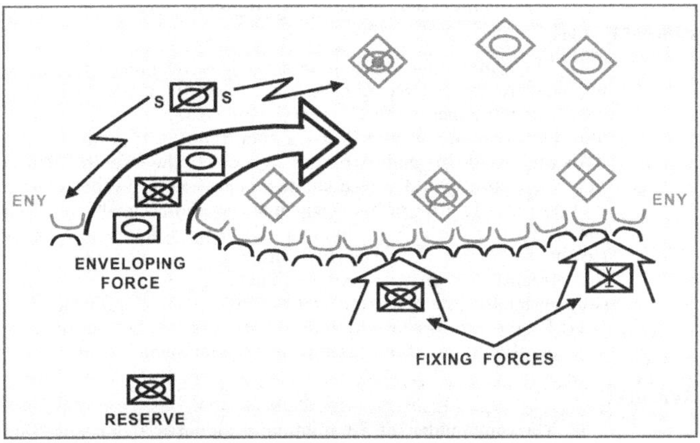

Figure 3-14. Single Envelopment

3-30. Single and double envelopments force the enemy to fight in two or more directions simultaneously to meet the converging efforts of the attack. A double envelopment generally requires a preponderance of force and can be difficult to control. A force seeking to execute a double envelopment must also have a substantial mobility advantage over the defender. A unit performs a double envelopment by conducting a frontal attack as a shaping operation in the center to fix the enemy in place while enveloping both hostile flanks. Because of the forces required, normally only divisions and larger organizations have the resources to execute a double envelopment.

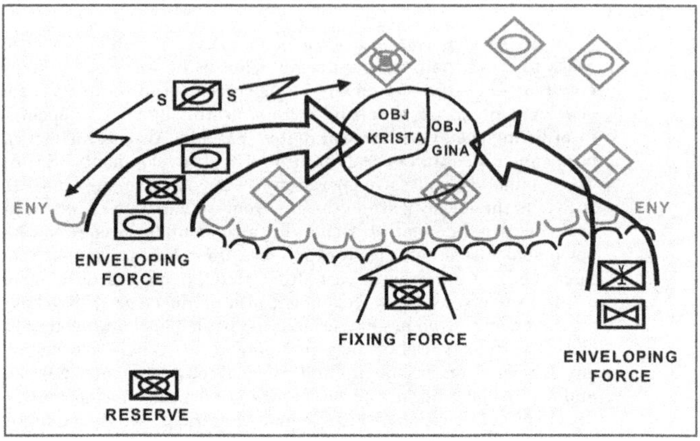

Figure 3-15. Double Envelopment

Organization of Forces

3-31. The commander envisioning a single envelopment organizes his forces into the enveloping force and the fixing force. He also allocates forces to conduct reconnaissance, security, reserve, and sustaining operations. The enveloping force, conducting the decisive operation, attacks an assailable enemy flank and avoids his main strength en route to the objective. The fixing force conducts a frontal attack as a shaping operation to fix the enemy in his current positions to prevent his escape and reduce his capability to react against the enveloping force. A commander executing a double envelopment organizes his forces into two enveloping forces and a fixing force in addition to reconnaissance, security, reserve, and sustaining forces. The commander typically designates the more important of the two enveloping forces as the main effort for resources. That enveloping force will also be the commander's decisive operation if its action accomplishes the mission.

Control Measures

3-32. The commander, at a minimum, designates AOs for each unit participating in the envelopment by using boundaries. He also designates PLs, support-by-fire and attack-by-fire positions, contact points, and appropriate fire coordination measures, such as a restricted fire line or boundary between converging forces, and any other control measures he feels are necessary to control the envelopment. Figure 3-16 on page 3-14 is an example of control measures used when conducting a single envelopment.

Planning an Envelopment

3-33. Successful planning for an envelopment depends on knowing and understanding the enemy and his capabilities. The commander wants to maneuver his enveloping force around or over the enemy's main defenses to secure objectives on the enemy's flank or rear. From those objectives the enveloping force can use its positional advantage to employ superior combat power against a defending enemy oriented in the wrong direction. The commander uses his intelligence assets and personnel to determine the disposition and capabilities of enemy forces to detect and react to their operations.

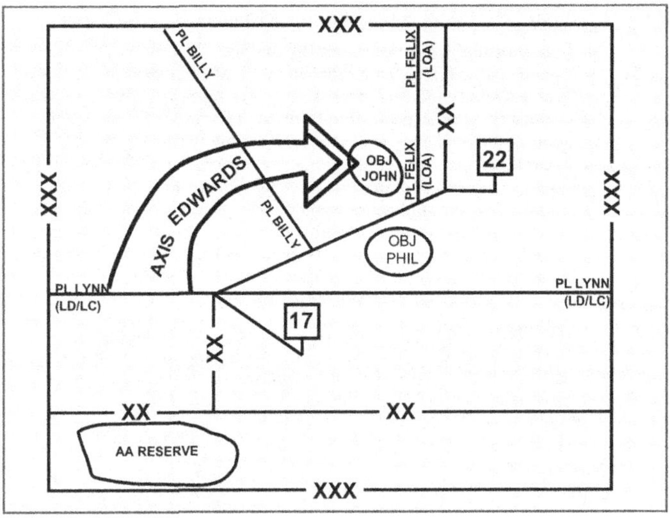

Figure 3-16. Control Measures for Conducting an Envelopment

3-34. The commander plans for the force conducting the envelopment to remain within supporting distance of the fixing force. (If the enveloping force is going outside of supporting distance, it is probably conducting a turning movement, not an envelopment.)

3-35. Sustaining the enveloping force requires deliberate planning because only intermittent ground lines of communication (LOCs) between the rear area and the enveloping force may exist. A *line of communication* is a route, either land, water, and/or air that connects an operating military force with a base of operations and along which supplies and military forces move (JP 1-02).

3-36. The commander plans how he will exploit the success of his envelopment as he encircles the enemy or transitions to a pursuit to complete the destruction of the enemy force. These plans are developed as branches and sequels to the envelopment operation.

Executing an Envelopment

3-37. A successful envelopment depends largely on the degree of surprise the commander achieves against his opponent or the presence of overwhelming combat power. The envelopment's probability of success also increases when the commander's forces have superior tactical mobility, possess air and information superiority, and his shaping operations fix the bulk of the enemy's forces in their current positions. The commander uses his intelligence, surveillance, and reconnaissance (ISR) systems to provide continuous intelligence and combat information to identify changes in enemy COAs throughout the execution of the envelopment.

3-38. Normally, a unit orients the majority of its combat power toward where it expects to engage enemy forces, while placing less combat power on its own flanks. Thus the flanks of most units are more vulnerable to attack. The commander creates an assailable flank using whatever means necessary. The enveloping force then moves rapidly to exploit the situation before the enemy strengthens an assailable flank by preparing positions in depth and by holding mobile forces in reserve. When faced with the threat of envelopment, the enemy commander might move his reserves to meet the enveloping force. Thus, rapid movement around the enemy's flank is essential to prevent him from occupying previously prepared positions. Vigorous shaping operations conducted by ground and air assets aim to prevent him from reconstituting reserves from other portions of his front.

3-39. The enemy may attempt to cut off the enveloping force and extend his flank beyond the area that the enveloping force is attempting to attack through. If the encircling force attempts to outflank such hostile extension, it may become overextended by moving outside of supporting distance from the fixing force. Therefore, it is usually better for the encircling force to take advantage of the enemy's extension and subsequent weakness by penetrating a thinly held area of the enemy's front rather than overextending itself in an attempt to completely outflank the enemy's position.

3-40. The enemy may attempt a frontal counterattack in response to an attempted envelopment. In this case, the fixing force defends itself or conducts a delay while the enveloping force continues the envelopment.

3-41. After the initial envelopment of one flank—which places the enemy at a disadvantage—the commander has many options. He may choose to establish favorable conditions for passing to a double envelopment by using reserves or exploit success by generating additional combat power along the same axis. Alternatively, he can destroy or defeat the enveloped enemy force in place, or transition to another type of operation, such as exploitation or pursuit.

TURNING MOVEMENT

3-42. A *turning movement* is a form of maneuver in which the attacking force seeks to avoid the enemy's principle defensive positions by seizing objectives to the enemy rear and causing the enemy to move out of his current positions or divert major forces to meet the threat (FM 3-0). However, a commander can employ a vertical envelopment using airborne or air assault forces to effect a turning movement. (Appendix C discusses airborne and air assault operations.) A commander uses this form of offensive maneuver to seize vital areas in the enemy's rear before the main enemy force can withdraw or receive support or reinforcements. See Figure 3-17 on page 3-16 for a graphic depiction of a turning movement. This form of offensive maneuver frequently transitions from the attack into an exploitation or pursuit. A turning movement differs from an envelopment because the force conducting a turning movement seeks to make the enemy displace from his current locations, whereas an enveloping force seeks to engage the enemy in his current location from an unexpected direction.

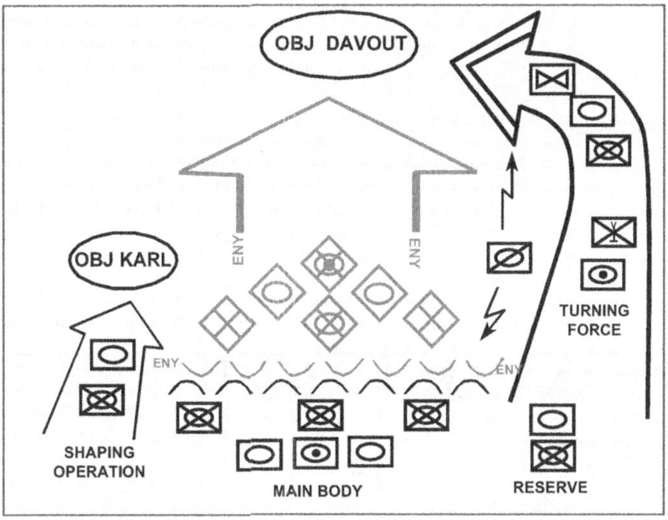

Figure 3-17. Turning Movement—Turning Force Conducting the Decisive Operation

Organization of Forces

3-43. The commander directing a turning movement task organizes his resources into a turning force, a main body, and a reserve. Each of these forces conducts security and reconnaissance operations. Either the turning force or the main body can conduct the echelon's decisive operation given the appropriate factors of METT-TC. A turning movement is particularly suited for division-size or larger forces possessing a high degree of tactical mobility. It is not until a commander has access to the resources of these echelons that he has the combat power to resource a turning force that can operate outside supporting range of his main body to allow the turning force to force enemy units out of their current positions. He bases the exact task organization of these forces on the factors of METT-TC and his concept of operations for the turning movement.

3-44. The maneuver of the turning force is what causes the enemy to leave his position. A turning force normally conducts the majority of its operations outside of the supporting range of the main body and possibly outside its supporting distance. Thus, the turning force must contain sufficient combat, combat support (CS), and combat service support (CSS) capabilities to operate independently of the main body for a specific period of time. This normally requires at least a division-size element.

3-45. The commander task organizes his main body to ensure the success of the turning force. The main body conducts operations, such as attacks designed to divert the enemy's attention away from the area where the turning force conducts its operations. The operations of the main body can be either the echelon's decisive or shaping operations. The commander organizes his reserve to exploit success by either the turning force or the main body. The reserve also provides the commander insurance against unexpected enemy actions.

Control Measures

3-46. The commander designates the AOs for each unit participating in the turning movement by establishing boundaries. He also designates additional control measures as necessary to synchronize the operations of his subordinates. These additional control measures include: phase lines (PLs), contact points, objectives, LOA, and appropriate fire coordination measures. Figure 3-18 depicts these control measures used to synchronize a turning movement that employs an airborne division as the turning force. (Appendix C discusses control measures associated with airborne and air assault operations.)

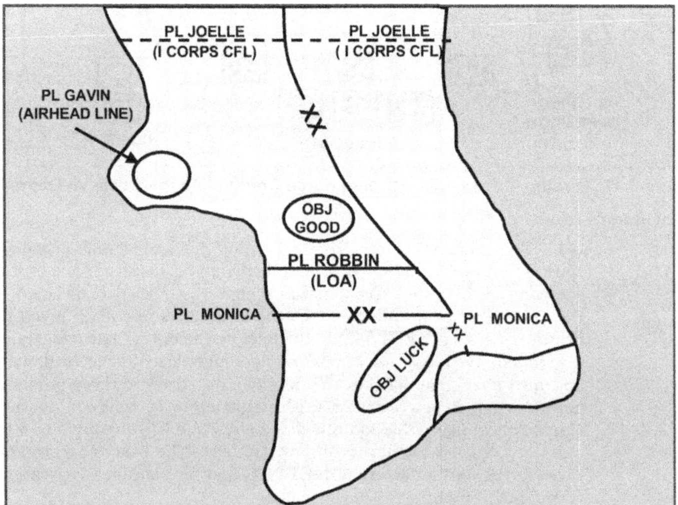

Figure 3-18. Control Measures for a Turning Movement

Planning a Turning Movement

3-47. Selecting the geographic objective of the turning movement is of major importance to the success of the operation. The commander's scheme of maneuver in a turning movement may vary, depending on the specific situation and the factors of METT-TC. In addition to common offensive planning considerations addressed on pages 3-33 through 3-50, the commander conducting a turning movement pays special attention to planning branches and sequels to the turning movement, including—

- Defensive operations by the turning force.
- Link-up operations between the turning force and the main body.
- Retrograde operations for the turning force.

Essential to the planning of the branches and sequels is the linkage between the branch or sequel and specific decision points supported by situation development.

3-48. After developing his tactical plan, the commander plans how the turning force maneuvers to its objective. The commander develops his movement, loading, and staging plans if outside transportation assets are required. He can plan to occupy key terrain that will threaten the enemy's survival or remain mobile and seek ways to exploit the turning force's success. Before initiating the operation, the commander plans how the turning force can exploit success.

3-49. In a turning movement that envisions an early linkup with the main body, the turning force normally plans to defend only that terrain required to protect itself. Once reinforcement or linkup with the main body occurs, the commander plans how to use the turning force to continue the attack or relieve it so it can prepare for subsequent missions.

3-50. The distances between forces and the existence of intermittent LOCs magnify the problems inherent in providing CSS to a combat force during a turning movement. Therefore, in the planning of a turning movement, the commander emphasizes resupply, equipment maintenance, casualty evacuation, graves registration, and prisoner of war handling to deal with these likely problems. Prepackaging company- and battalion-size resupply sets can ease the execution of sustaining operations during periods when CSS units must push supplies to the combat units.

3-51. Planners must consider the provision of all supplies and equipment required for mission accomplishment as an integral part of tactical planning. The commander plans and organizes his CSS operations to support a rapid tempo of highly mobile and widely dispersed operations. Traditional doctrinal supporting distances and responsibilities do not always apply to turning movements. CSS planners recognize this and adjust their plans using available resources. Only supplies required to meet the force's immediate needs are carried into the operation. Excess supplies and equipment can burden the force. Staffs establish and maintain required supply levels in the objective area by phasing supplies into the objective area on an accompanying, follow-up (automatic and on-call), and routine basis. Medical evacuation, resupply, and reinforcement airlifts may be necessary to sustain the force's combat operations. Ammunition and petroleum, oils, and lubricants (POL) normally constitute the major tonnage items. Lift restrictions affect what can be supplied using helicopters and fixed-wing aircraft.

Executing a Turning Movement

3-52. The primary prerequisites of a successful turning movement are moving the turning force to the objective area without incurring unacceptable losses and providing the force with the required combat power and sustainment. A commander can reduce his losses by operating under conditions of friendly air and information superiority, suppressing enemy fires, and having a mobility advantage over the enemy.

3-53. Major sources of firepower to suppress enemy fires are fixed-wing aircraft, attack helicopters, jammers, and Multiple Launch Rocket Systems that cover the entire route taken by the turning force. Other sources include naval surface fire support and artillery units accompanying the turning force.

3-54. When threatened with a turning movement, the enemy commander is in a dilemma. His original defense is misplaced. He must move forces from their original position in meeting the new threat. Often he must commit his available reserves against the new threat. He exposes those forces to friendly fires as he weakens his defense and moves his forces. He must now engage friendly forces on ground he has not chosen or prepared. Whenever possible, the commander tries to reach the decisive location without encountering the enemy. Techniques to accomplish this include outflanking the enemy or using airborne, air assault, and amphibious means to avoid his prepared positions. Once friendly forces find a way deep into the enemy's rear area, the turning force moves rapidly to exploit the situation. It seeks to achieve its mission before the enemy can reposition his committed or uncommitted forces to react. Rapid movement is essential to prevent the enemy from occupying previously prepared positions in his rear. Vigorous shaping operations prevent the enemy from reconstituting reserves from other portions of the enemy front.

3-55. The enemy may counterattack in an attempt to cut off and destroy the turning force and prevent the successful completion of the turning movement. In this case, the turning force's security elements conduct an area or mobile defense or engage in delaying actions while the rest of the turning force continues its mission. Alternatively, the enemy may try to withdraw his forces to a position where his LOCs are not threatened.

INFILTRATION

3-56. An *infiltration* is a form of maneuver in which an attacking force conducts undetected movement through or into an area occupied by enemy forces to occupy a position of advantage in the enemy rear while exposing only small elements to enemy defensive fires (FM 3-0). Infiltration is also a march technique used within friendly territory to move forces in small groups at extended or irregular intervals. (See Chapter 14 for a discussion of infiltration as a movement technique.)

3-57. Infiltration occurs by land, water, air, or a combination of means. Moving and assembling forces covertly through enemy positions takes a considerable amount of time. To successfully infiltrate, the force must avoid detection and engagement. Since this requirement limits the size and strength of the infiltrating force—and infiltrated forces alone can rarely defeat an enemy force—infiltration is normally used in conjunction with and in support of the other forms of offensive maneuver. Historically, the scope of the mission for the infiltrating force has been limited.

3-58. The commander orders an infiltration to move all or a portion of a unit through gaps in the enemy's defenses to—

- Reconnoiter known or templated enemy positions and conduct surveillance of named areas of interest and targeted areas of interest.
- Attack enemy-held positions from an unexpected direction.
- Occupy a support-by-fire position to support the decisive operation.
- Secure key terrain.
- Conduct ambushes and raids to destroy vital facilities and disrupt the enemy's defensive structure by attacking his reserves, fire support and air defense systems, communication nodes, and logistic support.
- Conduct a covert breach of an obstacle or obstacle complex.

3-59. Special operations forces and light infantry units up to brigade size are best suited to conduct an infiltration. In some circumstances, heavy forces operating in small units can conduct an infiltration. However, as the proliferation of technology leads to increased situational understanding, this should increase the ability of heavy forces to avoid enemy contact and move undetected through enemy positions. In the future a commander may conduct an infiltration with heavy forces in coordination with precision fires as a prelude to an attack.

Organization of Forces

3-60. Normally, to be successful, the infiltrating force must avoid detection at least until it reaches its ORP. Thus, the infiltrating force's size, strength, and composition is usually limited. The infiltrating unit commander organizes his main body into one or more infiltrating elements. The largest size element possible, compatible with the requirement for stealth and ease of movement, conducts the infiltration. This increases the commander's control, speeds the execution of the infiltration, and provides responsive combat power. The exact size and number of infiltrating elements are situationally dependent.

3-61. The commander considers the following factors when determining how to organize his forces. Smaller infiltrating elements are not as easy to detect and can get through smaller defensive gaps. Even the detection of one or two small elements by the enemy will not prevent the unit from accomplishing its mission in most cases. Larger infiltrating elements are easier to detect and their discovery is more apt to endanger the success of the mission. Also, they require larger gaps to move through. A unit with many smaller infiltrating elements requires more time to complete the infiltration and needs more link-up points than a similar size unit with only a few infiltrating elements. Many infiltrating elements are also harder to control than fewer, larger elements.

3-62. The commander resources a security force that moves ahead of, to the flanks of, and to the rear of each infiltrating element's main body. These security forces can be given either a screen or a guard mission. (Chapter 12 discusses screen and guard missions.) The sizes and orientations of security elements are also situationally dependent. Each infiltrating element is responsible for its own reconnaissance effort.

3-63. Sustainment of an infiltrating force normally depends on the force's basic load of supplies and those medical and maintenance assets accompanying the infiltrating force. After completing the mission, the commander reopens LOCs to conduct normal sustaining operations.

Control Measures

3-64. Control measures for an infiltration include, as a minimum—

- An AO for the infiltrating unit.
- One or more infiltration lanes.
- A LD or point of departure.
- Movement routes with their associated start and release points, or a direction or axis of attack.
- Linkup or rally points, including ORPs.
- Assault positions.
- One or more objectives.
- A LOA.

The commander can impose other measures to control the infiltration including checkpoints, PLs, and assault positions on the flank or rear of enemy positions. If it is not necessary for the entire infiltrating unit to reassemble to accomplish its mission, the objective may be broken into smaller objectives. Each infiltrating element would then move directly to its objective to conduct operations. (Most of these control measures have been previously described.) The following paragraphs describe using an infiltration lane and a linkup point.

3-65. **An *infiltration lane* is a control measure that coordinates forward and lateral movement of infiltrating units and fixes fire planning responsibilities.** The commander selects infiltration lanes that avoid the enemy, provide cover and concealment, and facilitate navigation. Figure 3-19 depicts the graphic for an infiltration lane. Each unit assigned an infiltration lane picks its own routes within the lane and switches routes as necessary. The left and right limits of the infiltration lane act as lateral boundaries for the unit conducting the infiltration. Attacks by rotary- or fixed-wing aircraft, indirect fires, or munitions effects that impact the lane must be coordinated with the infiltrating unit. Units leaving their assigned lane run the risk of being hit by friendly fires. Company-size units are normally assigned a single infiltration lane, although they can use more than one lane. Larger organizations, battalion and above, are always assigned more than one infiltration lane.

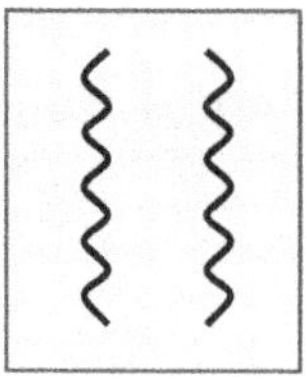

Figure 3-19. Infiltration Lane

3-66. **A *linkup point* is where two infiltrating elements in the same or different infiltration lanes are scheduled to meet to consolidate before proceeding on with their missions.** Figure 3-20 depicts Linkup Point 8. A linkup point is normally positioned in the enemy's rear or along one of his flanks. It should be large enough for all infiltrating elements to assemble and should offer cover and concealment for these elements. It should be an easily identifiable point on the ground. The commander should position his linkup points on defensible terrain located away from normal enemy troop movement routes.

Figure 3-20. Linkup Point 8

Planning an Infiltration

3-67. The activities and functions associated with the process of planning an infiltration are the same as with any other combat operation. That planning takes advantage of that unit's stealth capabilities to surprise the enemy. The planning process synchronizes the BOS that support the infiltrating unit, especially precise, high-resolution intelligence. Without precise, detailed intelligence, infiltration maneuvers become high-risk probing operations, that can be costly and time-consuming. Careful planning, full ISR integration, detailed analysis, and aggressive operations security can permit an infiltrating force to avoid an enemy force, minimize direct contact, and maximize surprise according to the commander's intent.

3-68. After identifying gaps or weaknesses in the enemy's defensive positions, the commander assigns infiltration lanes, contact points, and objectives to subordinate units. These objectives afford the infiltrating force positions of greatest advantage over the enemy and are not required to be to the geographic rear of the targeted enemy force. Each subordinate unit commander picks one or more routes within his assigned lane and establishes additional contact points, rally points, assault points, and other control measures as required. The commander wants each of the routes within an infiltration lane to be far enough apart to prevent an infiltrating element on one route from seeing other infiltrating elements, but close enough so that an infiltrating element could switch quickly to another route if required by the situation. The commander wants each route to provide his infiltrating elements cover and concealment while avoiding known enemy and civilian locations and movement routes to the maximum extent possible. If possible, the subordinate unit commander selects his exact routes during the preparation phase after reconnoitering each infiltration lane. He decides whether his unit will infiltrate as a unit, in smaller elements, or even as two-man buddy teams, depending on the density and strength of the enemy.

3-69. The commander may use single or multiple infiltration lanes depending on the infiltrating force's size, the amount of detailed information on enemy dispositions and terrain accessible, time allowed, and number of lanes available. A single infiltration lane—

- Facilitates navigation, control, and reassembly.
- Requires the existence or creation of only one gap in the enemy's position.
- Reduces the area for which detailed intelligence is required.

3-70. Multiple infiltration lanes—

- Require the existence or creation of more gaps in the enemy's security area.
- Reduce the possibility of compromising the entire force.
- Increase difficulty with maintaining control.

3-71. The sizes and numbers of infiltrating elements are major considerations for the commander when he is deciding whether to use a single lane or multiple infiltration lanes. If the infiltration takes place using multiple elements, contingency plans must address the following situations:

- A lead element, possibly the advance guard, makes contact, but the trail elements have not started infiltrating.
- A lead element infiltrates successfully, but compromises one or more trailing elements.
- A compromised linkup point.

3-72. The commander uses available technology to assist in planning the infiltration and avoiding unintended enemy and civilian contact during the infiltration. This can be as simple as all units using the same infiltrating lane being on the same frequency to facilitate the avoidance of enemy contact. An accurate depiction of enemy systems and locations, tied to rapid terrain analysis, can graphically depict dead spots in the enemy's battlefield surveillance. The commander can then plan how to expand those existing dead spots into infiltration lanes through a precision attack of selected enemy elements and systems.

3-73. The plan also addresses the following considerations:

- Availability of supporting fires, including rotary- and fixed-wing aircraft and offensive information operations—especially electronic attack, throughout the operation, during infiltration and the attack on the objective.
- Linkup or extraction of the infiltrating unit after mission completion.
- Sustainment of the infiltrating force during the operation, to include casualty evacuation.
- Deception operations, such as actions by other units designed to divert enemy attention from the area selected for the infiltration.
- Linkup of the various infiltrating elements.
- Command and control (C2), to include recognition signals.
- Positioning of combat vehicles to support the infiltrating elements.
- Using limited visibility and rough terrain to mask movement and reduce the chance of detection.
- Infiltration of the largest elements possible to maintain control.
- Rehearsals.
- Specially required preparations, such as modifying the unit's SOP regarding the soldier's combat load for the mission. When infiltrating on foot, units carry only required equipment. For example, in close terrain and in the absence of an armor threat, heavy antiarmor missile systems may be a liability.
- Abort criteria.
- Critical friendly zones.

3-74. Planned recognition signals and linkup procedures for the infiltration should be simple and quick. If there has not been any firing or any other noises, signals should not violate noise and light discipline. However, if there have already been assaults, artillery, and small-arms fire, signals, such as whistles and flares, can be used as linkup aids. A lack of time and the short distance involved in many infiltration operations may make conducting formal linkup procedures unnecessary.

Preparing an Infiltration

3-75. Once the commander selects the objective, infiltration lanes, and linkup or rally points, he directs ISR operations to update and confirm the details on which he bases his plan. He identifies enemy sensors and surveillance systems. He then revises the plan to reflect current conditions within the AO.

Executing an Infiltration

3-76. Moving undetected during an infiltration requires a considerable amount of time. The infiltrating unit moves from its AA or current position through the start point and then continues moving along the infiltration route to a release point. If buddy teams or small elements are conducting the infiltration, the unit uses a series of linkup points to reassemble into a coherent unit. Units can use a variety of navigation aids, such as GPS, to remain within the planned infiltration lane, which minimizes their chances of detection by the enemy. At the same time, they report their progress and status using communication systems that provide this information automatically to all command nodes which require this information.

3-77. If the complete unit is conducting the infiltration, the forward security force begins its movement first, followed by the main body. The distance between the forward security force and the main body depends on the factors of METT-TC. The advance guard must be far enough ahead of the main body so that it can either deploy or move to another route if the forward security force discovers the enemy. The forward security force in an infiltration must have enough time to move in a stealthy and secure manner. Enemy units should not be able to move undetected in the gap between the forward security force and the main body.

3-78. As the infiltrating unit moves, the advance guard reports to the commander regarding the cover and concealment of each route, enemy activity, location of danger areas and linkup points, enemy activity on the objective, and other combat information. The unit attempts to avoid enemy and civilian contact; however, contact does not always mean the mission is compromised. The infiltrating unit engages targets first with indirect fires to avoid revealing its presence and exact location. These fires include the conduct of offensive information operations designed to blind enemy ISR assets and prevent the enemy from coordinating an effective response to the infiltration.

3-79. If necessary, the forward security force conducts actions on contact while the main body moves to another route, reconstitutes a forward security force, and continues the mission. If the main body makes contact unexpectedly, it either overruns the enemy force, if the enemy has little combat power, or bypasses him and continues the mission. During the infiltration, the unit ignores ineffective enemy fire and continues to move. The commander may use suppressive fires against the enemy to cover the sounds of the infiltration or to divert the enemy's attention to areas other than where the infiltration lanes are located.

3-80. The infiltrating unit's elements move to an AA or an ORP to consolidate its combat power, refine the plan, and conduct any last-minute coordination prior to continuing the mission. The unit then conducts those tasks needed to accomplish its assigned mission, which could be an attack, raid, ambush, seizing key terrain, capturing prisoners, or collecting specific combat information.

3-81. A commander may need to abort an infiltration operation if the factors of METT-TC change so drastically during the infiltration that the infiltrating force is no longer capable of accomplishing its mission. Examples of changes that might trigger such an action include—

- Significant portions of the infiltrating force's combat power are lost through navigation errors, enemy action, accidents, or maintenance failures.
- Movement or significant reinforcement of a force-oriented objective.
- Detection of the infiltration by the enemy.
- Changes in the tactical situation that make the mission no longer appropriate, such as the initiation of an enemy attack.

The criteria for aborting the operation are developed in the planning process. The decision to abort the infiltration is transmitted to all appropriate headquarters for their action and information.

PENETRATION

3-82. A *penetration* a form of maneuver in which an attacking force seeks to rupture enemy defenses on a narrow front to disrupt the defensive system (FM 3-0). Destroying the continuity of that defense allows the enemy's subsequent isolation and defeat in detail by exploiting friendly forces. The penetration extends from the enemy's security area through his main defensive positions into his rear area. A commander employs a penetration when there is no assailable flank, enemy defenses are overextended and weak spots are detected in the enemy's positions, or time pressures do not permit an envelopment.

Organization of Forces

3-83. Penetrating a well-organized position requires overwhelming combat power in the area of penetration and combat superiority to continue the momentum of the attack. (See Figure 3-21, page 3-26.) The commander designates a breach, support, and assault force. He can designate these elements for each defensive position that he is required to penetrate. He should not withhold combat power from the initial penetration to conduct additional penetration unless he has so much combat power that the success of the initial penetration is assured.

3-84. The commander resources a reserve to deal with expected or unexpected contingencies, such as an enemy counterattack, to avoid diverting the assault element from attacking the final objective of the penetration. He designates additional units follow-and-support or follow-and-assume missions to ensure rapid exploitation of initial success. He designates forces to fix enemy reserves in their current locations and isolate enemy forces within the area selected for penetration.

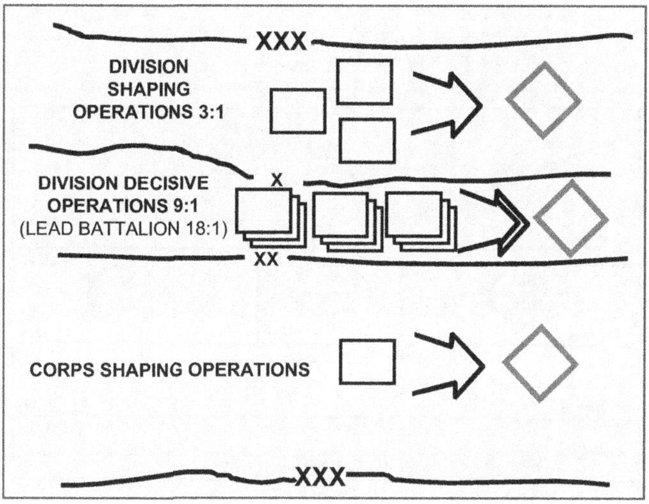

Figure 3-21. Penetration: Relative Combat Power

Control Measures

3-85. A commander assigns, as a minimum, an AO to every maneuver unit, a LD or LC; time of the attack or time of assault; phase lines; objective; and a LOA to control and synchronize the attack. (A commander can use a battle handover line instead of a LOA if he knows where he would like to commit a follow-and-assume force.) The lateral boundaries of the unit making the decisive operation are narrowly drawn to help establish the overwhelming combat power necessary at the area of penetration. The commander locates the LOA beyond the enemy's main defensive position to ensure completing the breach. If the operation results in opportunities to exploit success and pursue a beaten enemy, the commander adjusts existing boundaries to accommodate the new situation. (See Figure 3-22.)

3-86. A commander uses the graphics associated with a breach site, such as gaps and lanes, on the small-scale maps used to control the maneuver of his forces at each point where he penetrates the enemy's defenses. FM 3-34.2 defines the graphics.

3-87. Other control measures available to the commander include checkpoints, support-by-fire and attack-by-fire positions, probable line of deployment, fire support coordinating measures, attack position, assault position, and time of assault. Within the unit's AO, a commander can use either an axis of advance or a direction of attack to further control maneuver.

Planning a Penetration

3-88. The success of the penetration depends primarily on a coordinated and synchronized plan—violently executed at a high tempo to achieve surprise—

against comparatively weak enemy defenses. However, the terrain behind the area selected to penetrate must allow the penetration to proceed from the breach to a decisive objective.

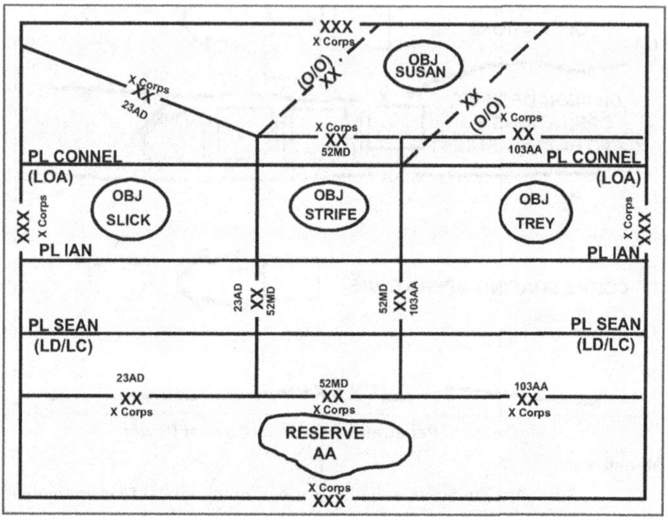

Figure 3-22. Penetration Graphic Control Measures

3-89. The depth of the enemy position and the relative strength of attacking echelons determine the width of the penetration. The availability of artillery, air support, and other combat multipliers for the attacking force helps the commander determine relative combat power. A wider gap allows friendly forces to drive deeper, making it more difficult for the enemy to close the gap. The deeper the penetration, the easier it is for a unit to seize its objective and roll up the enemy's flanks exposed by the breach and the less likely it is that the enemy will be in a position to restore his front by falling back.

3-90. Plans for penetrating a defensive position include isolating, suppressing, and destroying by fire—to include offensive information operations—enemy forces in the area selected for the penetration. These plans should also address how to isolate the area of penetration from support or reinforcement by enemy forces located outside the area. This consideration includes how to fix enemy reserves and long-range weapons in their current locations. Positioning friendly assets so that the commander can mass the effects of their combat power to accomplish these results without giving away the location of the penetration is also a critical part of the plan.

3-91. The commander plans to place the majority of his forces and assets in positions where the effects of their combat power can be felt in the area selected for penetration. The commander's plan for the penetration normally has three phases:

- Breaching the enemy's main defensive positions.
- Widening the gap created to secure the flanks by enveloping one or both of the newly exposed flanks.
- Seizing the objective with its associated subsequent exploitation.

3-92. Planning the sequence of these phases depends on the specific situation. In some situations, if there are weaknesses or gaps in the enemy's front, it is possible for heavy forces to breach the enemy's defenses and proceed straight to the objective. Simultaneously, light units could conduct local envelopment and exploitation operations. In other situations, the commander uses his light forces to create the breach, holding his heavy forces initially in reserve to exploit gaps in the enemy's defenses created by light forces.

3-93. The commander plans shaping operations outside the area of penetration to contain the enemy on the flanks of the penetration and fix his reserves in their current locations. Synchronizing the effects of rotary- and fixed-wing aircraft, artillery fires, and obscuration smoke to delay or disrupt repositioning forces is an example of such shaping operations. These shaping operations will involve the maintenance of operations security and the conduct of deception operations. The commander usually attempts to penetrate the enemy's defensive positions along unit boundaries because defenses tend to be less effective along a boundary.

3-94. The commander plans for the penetration to break through the enemy's defenses so he is unable to reestablish his defense on more rearward positions. Until this event takes place, the commander does not want to divert the strength of his attacking units to widening the gap to secure the flanks of the penetration. However, he must develop plans that address contingencies, such as hostile counterattacks against the flanks of the penetration. The plan should provide assistance to attacking elements as they close with the enemy and support the attack until the enemy's power of resistance is broken.

Executing a Penetration

3-95. After the initial breach of the enemy's main line of resistance, the sequence of the remaining two phases is determined by the factors of METT-TC. If the enemy is in a weak defensive position, it may be possible for the lead attacking force to seize the penetration's final objective while simultaneously widening the initial breach. In other situations, the commander must wait to seize the final objective until the breach is wide enough for other forces, such as reserves and follow-and-assume forces, to be committed. Commanders at all levels must take advantage of success within the commander's intent throughout the penetration.

3-96. **Breaching the Enemy's Main Defensive Positions.** The commander launches the actual penetration on a relatively narrow front. (See Figure 3-23.) He narrows the AO of the unit or units conducting his decisive operation—the penetration—by adjusting unit lateral boundaries to the exact point or points where he wants to penetrate the enemy's defenses. This allows the force conducting the penetration to focus overwhelming combat power. The commander assigns his assault force a close-in objective. His support force locates where it can support by fire both the breach and the assault forces. Local reserves are held in readiness to conduct a forward passage through or around units whose attacks have slowed or stopped.

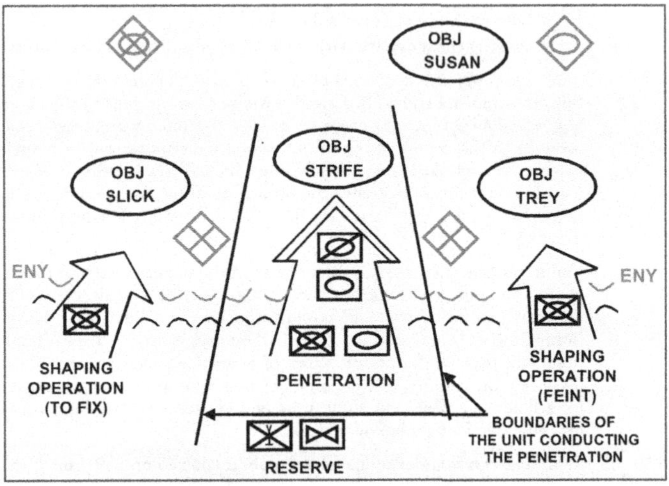

Figure 3-23. Penetration: The Breach

3-97. Shaping operations on the remainder of the hostile front fix the enemy in his current positions and prevent him from disengaging to reinforce enemy units opposing the decisive operation. The commander tracks the battle's progress to ensure that his forces penetrate entirely through the enemy's main defensive positions and not just the enemy's security area.

3-98. The enemy normally tries to slow down or stop the breach to gain time to react to the changing situation. Therefore, the attacking commander rapidly exploits and reinforces success. He piles on resources and additional units as necessary to ensure completing the penetration through the enemy's defensive positions. He also conducts offensive information operations to desynchronize the enemy's reaction.

3-99. **Widening the Breach to Secure the Flanks.** Once the attacking force penetrates the main defenses, it begins to widen the penetration of the enemy's defensive positions by conducting a series of shallow envelopments to roll back its shoulders. (See Figure 3-24, page 3-30.) The task of widening the initial gap of the penetration is normally assigned to a follow-and-support force. That task can also be assigned to the reserve as a contingency mission. If the commander commits his reserve to accomplish that task, he must reconstitute his reserve from another part of his force. Alternatively, he may assume the risk of not having a reserve for the time necessary to accomplish this task. The commander makes plans to meet enemy counterattacks by shifting fires or committing his reserves or follow-and-assume forces. Units can use obstacles on the flanks of the penetration as a combat multiplier to assist in defeating any local enemy counterattack and to provide additional security for the force.

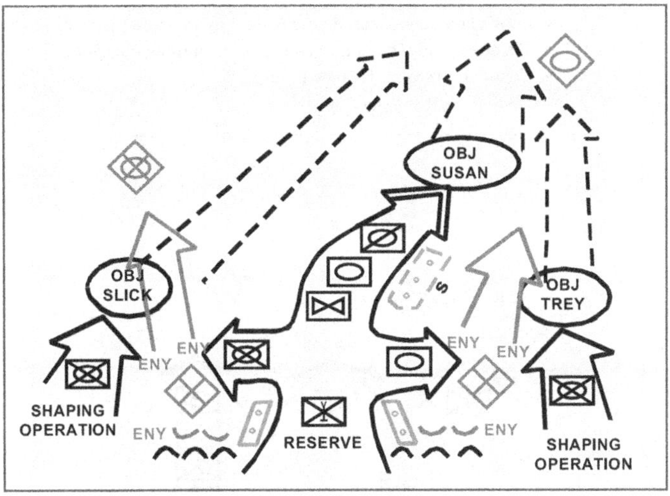

Figure 3-24. Expanding the Penetration

3-100. **Seizing the Objective and Subsequent Exploitation.** The mission of seizing the objective—which may be a specific enemy force—to destroy the continuity of the enemy's defensive position is normally the decisive operation after completing the penetration. Frequently that objective is so far from the area of penetration that the unit or units initially conducting the penetration cannot seize it without a pause. In that case, the commander plans to pass his reserve or follow and assume forces through the initial attacking force early, leaving exploitation beyond the objective to higher echelons. While the exact force mix is METT-TC-dependent, armored, mechanized, and aviation forces are generally suited for subsequent exploitation.

3-101. In large commands, forces may initiate an attack by simultaneously launching two or more convergent penetrations against weak localities on the hostile front. Often this method of attack helps isolate an extremely strong, hostile defense. The commander assigns shaping operations to initially contain any strong localities. When the multiple attacks have advanced sufficiently, the force reduces bypassed enemy forces and unites the penetrating attacks into a single decisive operation.

FRONTAL ATTACK

3-102. A *frontal attack* is a form of maneuver in which an attacking force seeks to destroy a weaker enemy force or fix a larger enemy force in place over a broad front (FM 3-0). At the tactical level, an attacking force can use a frontal attack to rapidly overrun a weak enemy force. A commander commonly uses a frontal attack as a shaping operation in conjunction with other forms of maneuver. He normally employs a frontal attack to—

- Clear enemy security forces.
- Overwhelm a shattered enemy during an exploitation or pursuit.
- Fix enemy forces in place as part of a shaping operation.
- Conduct a reconnaissance in force.

Figure 3-25 depicts a frontal attack.

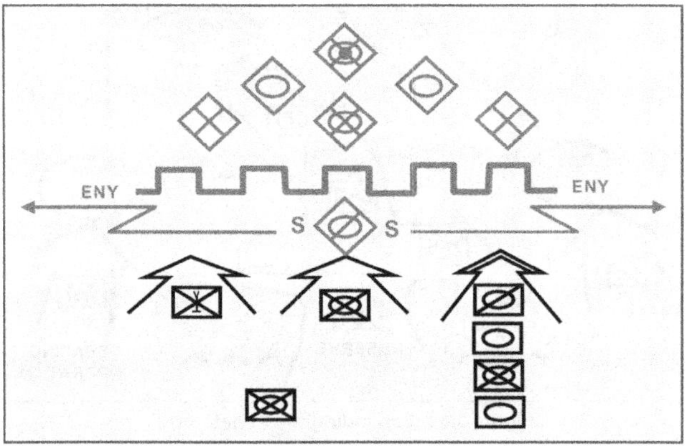

Figure 3-25. Frontal Attack

3-103. It is also necessary to conduct a frontal attack when assailable flanks do not exist. Where a penetration is a sharp attack designed to rupture the enemy position, the commander designs a frontal attack to maintain continuous pressure along the entire front until either a breach occurs or the attacking forces succeed in pushing the enemy back. Frontal attacks conducted without overwhelming combat power are seldom decisive. Consequently, the commander's choice to conduct a frontal attack in situations where he does not have overwhelming combat power is rarely justified unless the time gained is vital to the operation's success.

Organization of Forces

3-104. There is no unique organization of forces associated with this form of maneuver. A commander conducting a frontal attack organizes his unit into an element to conduct reconnaissance and security operations, a main body, and a reserve. The factors of METT-TC dictate the specific task organization of the unit.

Control Measures

3-105. A commander conducting a frontal attack may not require any additional control measures beyond those established to control the overall mission. This includes an AO, defined by unit boundaries, and an objective at a minimum. The commander can also use any other control measure he feels is necessary to control the attack, including—

- Attack positions.
- Line of departure.
- Phase lines.
- Assault positions.
- Limit of advance.
- Direction of attack or axis of advance for every maneuver unit.

A unit conducting a frontal attack normally has a wider AO than a unit conducting a penetration.

Planning a Frontal Attack

3-106. It is seldom possible for a commander to exert sufficient pressure to overwhelm an enemy using a frontal attack, since it strikes the enemy along a significant portion of his front. The force's primary objective is to maintain pressure and help fix the enemy force. The commander's planning effort should reflect these two considerations. When considering employing a frontal attack in a shaping operation, the commander should also consider other means for holding the enemy in position, such as feints and demonstrations employing indirect fires to preclude excessive losses.

Executing a Frontal Attack

3-107. The unit conducting a frontal attack advances on a broad front, normally with its subordinate ground maneuver elements abreast (except for the reserve). This clears the enemy's security area of his security forces and reconnaissance, intelligence, surveillance, and target acquisition assets while advancing the friendly force into the enemy's main defenses. Once the unit makes contact with the enemy, the attacking force's subordinate elements rapidly develop the situation and report enemy dispositions immediately to the commander so he can exploit enemy weaknesses. The attacking force fixes enemy forces in their current locations and seeks to gain a position of advantage to destroy them using fire and movement.

3-108. If the attacking unit discovers a gap in the enemy's defenses, the commander seeks to exploit that weakness and disrupt the integrity of the enemy's defense. After assessing the situation to make sure that it is not a trap, the commander can employ his reserve to exploit the opportunity. He synchronizes the exploitation with the actions of his other combat, CS, and CSS units to prevent counterattacking enemy forces from isolating and destroying successful subordinate elements of his force.

3-109. When the unit conducting the frontal attack can no longer advance, it adopts a defensive posture. The commander may require it to assist the forward passage of lines of other units. It continues to perform reconnaissance of enemy positions to locate gaps or assailable flanks.

COMMON OFFENSIVE PLANNING CONSIDERATIONS

3-110. Visualizing, describing, and directing are aspects of leadership common to all commanders. The tactical commander begins with a designated AO, identified mission, and assigned forces. The commander develops and issues planning guidance based on his visualization in terms of the physical means to accomplish the mission.

3-111. The offense is basic to combat operations. Only by a resolute offense, conducted at a high tempo and to great depth, is total destruction of the enemy attained. The offense has a number of indisputable advantages. The principal advantage enjoyed is its possession of the initiative. Having the initiative allows a commander to select the time, place, and specific tactics, techniques, and procedures used by the attacking force. The attacker has the time and opportunity to develop a plan and to concentrate the effects of his forces and thoroughly prepare conditions for success when he has the initiative. The commander strikes the enemy in unexpected ways at unexpected times and places. He focuses on attacking the right combination of targets, not necessarily the biggest or the closest. These attacks are rapidly executed, violently executed, unpredictable in nature, and disorient the enemy. They enhance the commander's capability to impose his will on his enemy and thus to achieve decisive victory.

3-112. The commander maintains momentum by rapidly following up his attacks to prevent enemy recovery. He denies his enemy any opportunity to adjust to his action in spite of the enemy's desperate attempts to do so. He changes his means and methods before the enemy can adapt to those in current use. The tempo of his operations must be fast enough to prevent effective counteraction. He orchestrates unrelenting pressure by adjusting combinations to meet the offensive's ever-changing demands. He maintains relentless pressure and exploits his gains to make permanent any temporary battlefield success.

3-113. Each battle or engagement, even those occurring simultaneously as a part of the same campaign, has its own unique peculiarities, determined by the actual conditions of the situation. The widespread application of highly accurate and lethal weapons, high degree of tactical mobility, dynamic nature, rapid situational changes, and the noncontiguous and large spatial scope of unit AOs all characterize contemporary combined arms warfare. The commander first able to see the battlefield, understand the implications of existing friendly and enemy operational picture, and take effective action to impose his will on the situation will enjoy tactical success.

3-114. The following discussion uses those physical means—soldiers, organizations, and equipment—that constitute the seven BOS defined in FM 7-15 as the framework for discussing planning considerations that apply to all types and forms of tactical offensive operations. The commander synchronizes the effects of all BOS as part of the visualize, describe, direct, and assess process.

INTELLIGENCE

3-115. A commander uses the products of the intelligence preparation of the battlefield (IPB) process to identify any aspect within his AO or area of interest that will affect how he accomplishes his mission. An *area of interest* is that area of concern to the commander, including the area of influence, areas adjacent thereto, and extending into enemy territory to the objectives of current or planned operations. This area also includes areas occupied by enemy forces who could jeopardize the accomplishment of the mission (JP 2-03).

3-116. The entire staff, led by the echelon intelligence staff, uses the IPB process to identify any aspects of the area of operations or area of interest that will affect enemy, friendly, and third party operations. The IPB process is collaborative in nature and requires information from all staff elements and some subordinate units. All staff and subordinate elements use the results and products of the IPB process for planning. FM 2-01.3 describes the IPB process.

3-117. The commander uses his ISR assets to study the terrain and confirm or deny the enemy's strengths, dispositions, and likely intentions, especially where and in what strength the enemy will defend. These assets also gather information concerning the civilian population within the AO to confirm or deny their numbers, locations, and likely intentions, especially with regard to staying put in shelters or fleeing from combat operations.

3-118. By studying the terrain, the commander tries to determine the principal heavy and light avenues of approach to his objective. He also tries to determine the most advantageous area for the enemy's main defense to occupy, routes that the enemy may use to conduct counterattacks, and other factors, such as observation and fields of fire, avenues of approach, key terrain, obstacles, and cover and concealment (OAKOC). (See FM 6-0 for a discussion of the components of OAKOC.) It is unlikely that the commander has complete knowledge of the enemy's intentions; therefore, he must conduct ISR collection continuously during the battle.

3-119. The echelon intelligence and operations officers, in coordination with the rest of the staff, develop an integrated ISR plan that satisfies the commander's maneuver, targeting, and information requirements. A commander's information requirements are dictated by the factors of METT-TC, but commonly include—

- Locations, composition, equipment, strengths, and weaknesses of the defending enemy force, to include high-priority targets and enemy ISR capabilities.
- Locations of possible enemy assembly areas.
- Location of enemy indirect-fire weapon systems and units.
- Location of gaps and assailable flanks.
- Location of areas for friendly and enemy air assaults.
- Location of enemy air defense gun and missile units.
- Location of enemy electronic warfare units.
- Effects of weather and terrain on current and projected operations.
- Numbers, routes, and direction of movement of dislocated civilians.
- Withdrawal routes for enemy forces.
- Anticipated timetable schedules for the enemy's most likely COA and other probable COAs.
- Locations of enemy C2 and ISR systems and the frequencies used by the information systems linking these systems.

If friendly ISR systems cannot answer the commander's information requirements, his intelligence staff can send a request for information to higher and adjacent units, he can commit additional resources, or he can decide to execute his offensive operation with the current information.

MANEUVER

3-120. The commander conducts maneuver to avoid enemy strengths and to create opportunities to increase the effects his fires. He secures surprise by making unexpected maneuvers, rapidly changing the tempo of ongoing operations, avoiding observation, and using deceptive techniques and procedures. He seeks to overwhelm the enemy with one or more unexpected blows before the enemy has time to react in an organized fashion. This occurs when he is able to engage the defending enemy force from positions that place the attacking force in a position of advantage with respect to the defending enemy force, such as engaging the enemy from a flanking position. His security forces prevent the enemy from discovering friendly dispositions, capabilities, and intentions, or interfering with the preparations for the attack. Finally, he maneuvers to close with and destroy the enemy by close combat and shock effect. *Close combat* is combat carried out with direct fire weapons, supported by indirect fire, air-delivered fires, and nonlethal engagement means. Close combat defeats or destroys enemy forces, or seizes and retains ground (FM 3-0). Close combat encompasses all actions that place friendly forces in immediate contact with the enemy where the commander uses direct fire and movement in combination to defeat or destroy enemy forces or seize and retain ground.

3-121. A commander can overwhelm an enemy by the early seizing and retaining of key and decisive terrain that provides dominating observation, cover and concealment, and better fields of fire to facilitate the maneuver of his forces. If decisive terrain is present, the commander designates it to communicate its importance in his concept of operations, first to his staff and later to subordinate commanders. The friendly force must control decisive terrain to successfully accomplish its mission.

Combat Formations

3-122. **A *combat formation* is an ordered arrangement of forces for a specific purpose and describes the general configuration of a unit on the ground.** A commander can use seven different combat formations depending on the factors of METT-TC:

- Column.
- Line.
- Echelon (left or right).
- Box.
- Diamond.
- Wedge.
- Vee.

Terrain characteristics and visibility determine the actual arrangement and location of the unit's personnel and vehicles within a given formation.

3-123. Combat formations allow a unit to move on the battlefield in a posture suited to the senior commander's intent and mission. A unit may employ a series of combat formations during the course of an attack; each has its advantages and disadvantages. Subordinate units within a combat formation can also employ their own combat formations, consistent with their particular situation. The commander considers the advantages and disadvantages of each formation in the areas of C2 maintenance, firepower orientation, ability to mass fires, and flexibility when determining the appropriate formation for a given situation. All combat formations use one or more of the three movement techniques: traveling, traveling overwatch, and bounding overwatch. (Chapter 13 describes these three movement techniques.)

3-124. The commander's use of standard formations allows him to rapidly shift his unit from one formation to another, giving him additional flexibility when adjusting to changes in the factors of METT-TC. (This results from a commander rehearsing his unit so that it can change formations using standard responses to changing situations, such as actions on contact.) By designating the combat formation he plans to use, the commander—

- Establishes the geographic relationship between units.
- Indicates how he plans to react once the enemy makes contact with the formation.
- Indicates the level of security desired.
- Establishes the preponderant orientation of his weapon systems.
- Postures his forces for the attack.

The number of maneuver units available makes some formations, such as the box and the diamond, impractical for modernized organizations, including the initial brigade combat team and the limited conversion division, which only have three subordinate maneuver units at the brigade and battalion echelons unless task organization occurs.

3-125. **Column Formation.** The unit moves in column formation when the commander does not anticipate early contact, the objective is distant, and speed and control are critical. (See Figure 3-26.) The location of fire support units within the column reflects the column's length and the range fans of those fire support systems. Normally, the lead element uses a traveling overwatch technique while the following units are in traveling formation. Employing a column formation—

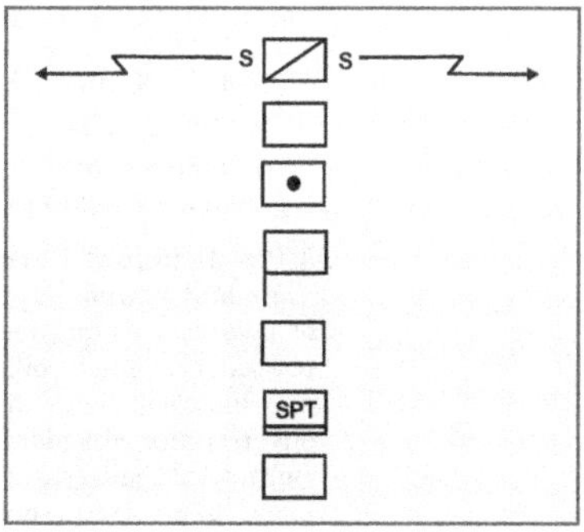

Figure 3-26. Column Formation

- Provides the best formation to move large forces quickly, especially with limited routes and limited visibility.
- Makes enemy contact with a small part of the total force while facilitating control and allowing the commander to quickly generate mass.

- Provides a base for easy transition to other formations.
- Works in restricted terrain.

3-126. A disadvantage of using the column formation is that the majority of the column's firepower can only be immediately applied on the column's flanks. The length of the column impacts movement and terrain management. Additionally, there are the possibilities of inadvertently bypassing enemy units or positions and exposing the unit's flanks or running head on into an enemy deployed perpendicular to the column's direction of movement.

3-127. **Line Formation.** In a line formation, the unit's subordinate ground maneuver elements move abreast of each other. (See Figure 3-27.) A commander employs this formation when he assaults an objective because it concentrates firepower to the front in the direction of movement. A line formation also—

- Facilitates speed and shock in closing with an enemy.
- Allows the coverage of wide frontages.
- Facilitates the occupation of attack-by-fire or support-by-fire positions.

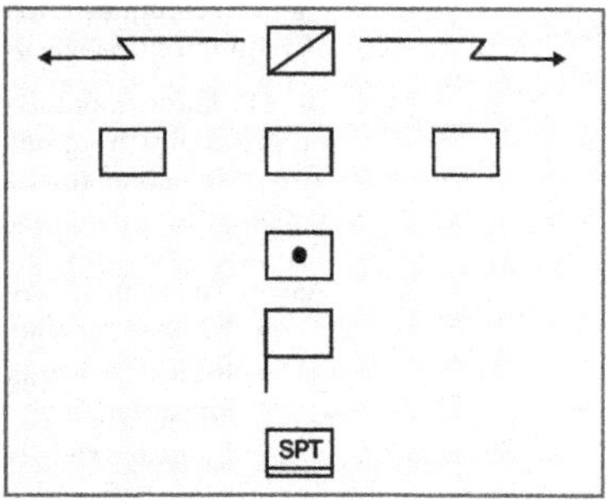

Figure 3-27. Line Formation

3-128. There are also disadvantages of a line formation:

- Provides less flexibility of maneuver than other formations since it does not distribute units in depth.
- Linear deployment allows a unit deployed on line to bring only limited firepower to bear on either flank.
- Provides limited or no reserve.
- Limits overwatch forces.
- Limits control of a unit using a line formation in restricted terrain or under conditions of limited visibility.

3-129. **Echelon Formation.** A commander who has knowledge of potential enemy locations can use an echelon formation to deploy his subordinate ground maneuver units diagonally left or right. (See Figures 3-28 and 3-29, page 3-38.) Units operating on the flank of a larger formation commonly use this formation. Using an echelon formation—

- Facilitates control in open terrain.
- Allows the concentration of

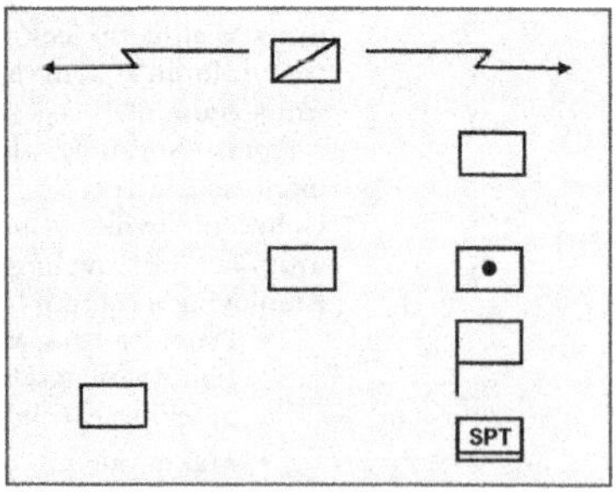

Figure 3-28. Echelon Left Formation

the unit's firepower forward and to the flank in the direction of echelon.

- Allows forces not in contact to maneuver against a known enemy, because all elements probably will not come into contact at the same time.

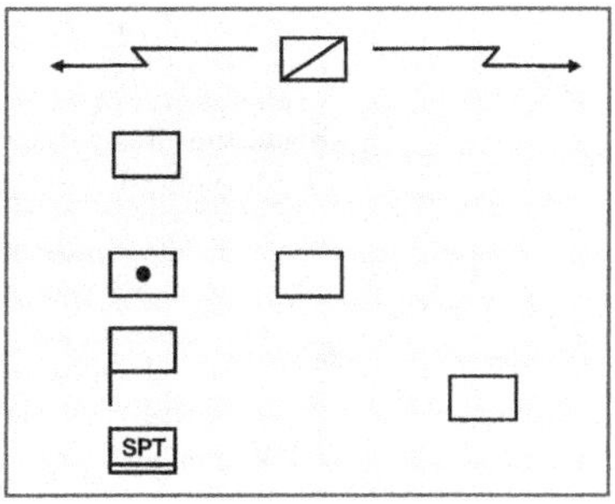

Figure 3-29. Echelon Right Formation

3-130. The primary disadvantages of the echelon formation are that it is more difficult to maintain control over the unit in restricted terrain than a column formation and the lack of security or firepower available on the opposite side of the echelon.

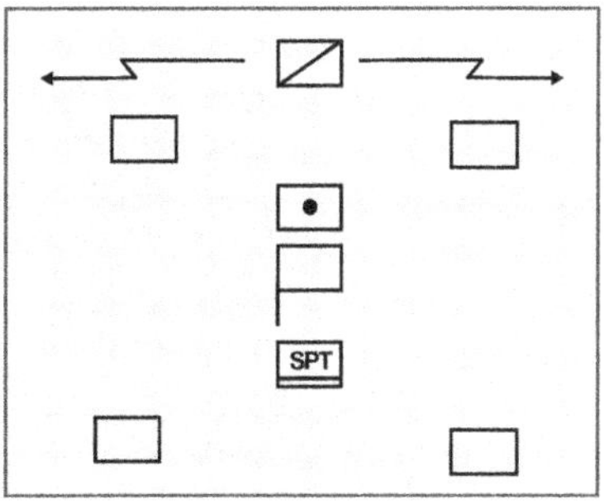

Figure 3-30. Box Formation

3-131. **Box Formation.** The box formation arranges the unit with two forward and two trail maneuver elements. (See Figure 3-30.) A unit with only three maneuver elements cannot adopt the box formation. The subordinate elements of the box usually move in a column formation with flank security. It is often used when executing an approach march, an exploitation, or a pursuit when the commander has only general knowledge about the enemy. Employing a box formation—

- Allows the unit to change quickly and easily to any other formation.
- Facilitates rapid movement, yet still provides all-around security.
- Provides firepower to the front and flanks.
- Maintains control more easily when compared to a line formation.

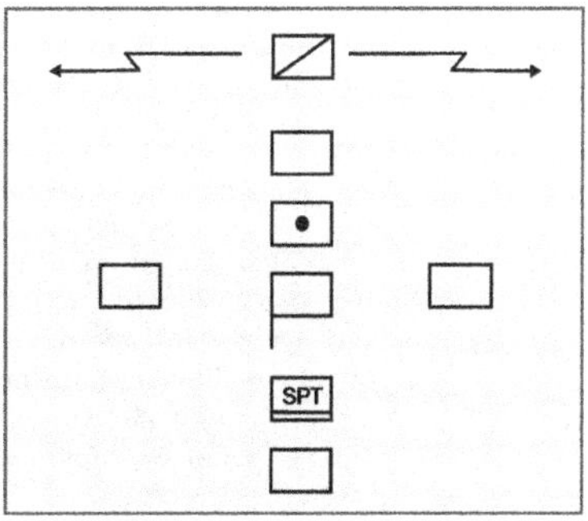

Figure 3-31. Diamond Formation

The primary disadvantages of a box formation are that it requires sufficient maneuver space for dispersion and the availability of multiple routes.

3-132. **Diamond Formation.** The diamond formation arranges the unit with one forward and one trail unit and a unit on each flank. (See Figure 3-31.)

The subordinate elements of the diamond usually move in a column formation with flank security. It is most effective during approach marches, exploitations, or pursuits when the commander has only general knowledge about the enemy. Employing a diamond formation—

- Allows the commander to maneuver either left or right immediately, without first repositioning, regardless of which subordinate element makes contact with the enemy. (This is the chief advantage of and the difference between a diamond and a box formation.)
- Facilitates making enemy contact with the smallest possible force, yet provides all-around security.
- Provides firepower to the front and flanks.
- Changes easily and quickly to another formation.
- Facilitates speed of movement while remaining easy to control.
- Provides an uncommitted force for use as a reserve.

3-133. The primary disadvantages of this formation are that it—

- Requires sufficient space for dispersion laterally and in depth.
- Requires four subordinate maneuver elements.
- Requires the availability of multiple routes.

3-134. **Wedge Formation.** The wedge formation arranges forces to attack an enemy appearing to the front and flanks. (See Figure 3-32.) A unit with only three subordinate maneuver elements can adopt the wedge formation. The commander uses the wedge when contact with the enemy is possible or expected, but his location and dispositions are vague. It is the preferred formation for a movement to contact in an organization with three subordinate maneuver units because it initiates contact with one unit while retaining two other subordinate uncommitted units positioned to maneuver and further develop the situation. Within the wedge, subordinate units employ the formation best suited to the terrain, visibility, and likelihood of contact. Employing a wedge formation—

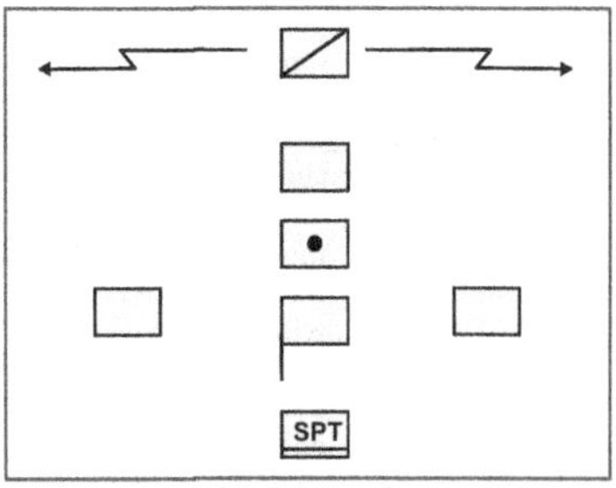

Figure 3-32. Wedge Formation

- Provides maximum firepower forward and allows a large portion of the unit's firepower to be used on the flanks.
- Allows rapid crossing of open terrain when enemy contact is not expected.
- Facilitates control.
- Allows for rapid changes in the orientation of the force.
- Facilitates the rapid change to a line, vee, echelon, or column formation.

3-135. The primary disadvantages to the wedge formation are that it—

- Requires sufficient space for dispersion laterally and in depth.

- Requires the availability of multiple routes.
- Lacks ease of control in restricted terrain or poor visibility.

3-136. **Vee Formation.** The vee formation disposes the unit with two maneuver elements abreast and one or more units trailing. (See Figure 3-33.) This arrangement is well suited for an advance against a known threat to the front. The commander may use this formation when he expects enemy contact and knows the location and disposition of the enemy. Employing a vee formation—

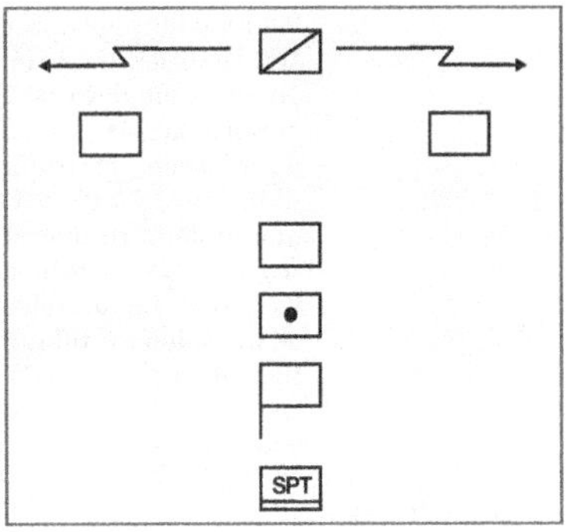

Figure 3-33. Vee Formation

- Provides maximum firepower forward and good firepower to the flanks, but the firepower on the flanks is less than that provided by the wedge.
- Facilitates a continued maneuver after contact is made and a rapid transition to the assault.
- Allows the unit to change quickly to a line, wedge, or column formation.

3-137. The primary disadvantages to this formation are that it—

- Makes reorientation of the direction of movement, such as a 90-degree turn, more difficult than using a wedge.
- Makes control in restricted terrain and under limited-visibility conditions difficult.
- Requires sufficient space for dispersion laterally and in depth.

Limited-Visibility Conditions

3-138. The capability to fight at night and under limited-visibility conditions is an important aspect of conducting maneuver. The commander conducts field training exercises under limited-visibility conditions to ensure that his unit has this capability as part of his standard training program. A commander conducts offensive operations at night or under limited-visibility conditions when a daylight operation continues into the night or when an operation could result in heavy losses if conducted in daylight. Offensive operations conducted in these conditions can achieve surprise, gain terrain required for further operations, and negate enemy visual target acquisition capabilities while taking advantage of the friendly force's night-fighting capabilities.

3-139. All operations conducted in limited visibility or adverse weather require more planning and preparation time than normal. They require designating reference points and establishing navigation aids, such as GPS waypoints. The commander ensures that the night-vision and navigation systems required to maneuver under these conditions are available and functional. The commander rehearses these operations before execution to ensure complete integration and synchronization of the plan. Rehearsals also ensure that his soldiers have the necessary skills to accomplish the mission. Any problem areas require resolution before beginning the operation.

3-140. Night operations degrade the capabilities of soldiers and units. Cognitive abilities degrade more rapidly than physical strength and endurance. Night-vision devices degrade the user's depth perception. This degradation in performance occurs after as little as 18 hours of sustained work. (Additional information concerning the impact of extended operations on soldiers and unit can be found in FM 6-22.5.) The plan should allow time for both soldiers and units to recuperate after conducting a night attack before being committed to other operations. The weight that soldiers must carry also directly affects their endurance. The commander carefully determines the fighting load of his soldiers, taking into account the factors of METT-TC and limits the fighting load of his soldiers conducting night operations. The equipment carried by a soldier for an extended period should never exceed one-third of his body weight.

FIRE SUPPORT

3-141. Fire superiority must be gained and maintained throughout all offensive operations. Fire support uses a variety of methods and assets that attrit, delay, and disrupt enemy forces and enable friendly maneuver. Using preparatory, counterfire, suppression, and nonlethal fires provides the commander with numerous options for gaining and maintaining fire superiority. The commander uses his long-range artillery systems and air support to engage the enemy throughout the depth of his positions.

3-142. Along with the reserve, attacks by indirect-fire systems and close air support are some of the commander's principal means for influencing ongoing actions. Such support helps establish the conditions required for successful mission accomplishment and is key to the commander's ability to react to unexpected situations. Decentralized execution characterizes the employment of fires when conducting offensive operations. Subordinates must have direct access to sufficient firepower to adequately support their maneuvering elements. Simultaneously, the commander retains control over sufficient fire support assets to enable him to mass their effects at critical times and places to support his decisive operation or to respond to the enemy's counteraction.

3-143. Fire support planning is the continuing process of analyzing, allocating, and scheduling fires. It determines how fires are used, what types of targets to attack, what collection assets are used to acquire and track those targets, what assets are used to attack the target, and what assets verify effects on the target. This planning does not stop at the objective or LOA. The commander gives attention to flanks and potential enemy hide positions. Coordination among echelon fire support elements (FSEs) and the proper use of fire support coordinating measures are critical to prevent fratricide.

3-144. The fire support coordinator (FSCOORD) integrates fire support into the unit's maneuver scheme for the commander. The FSCOORD supports the unit's maneuver by planning preparatory fires, harassing fires, interdiction fires, suppressive fires, and deception fires. These fires can be time- or event-driven. The FSCOORD plans fires on known and likely enemy positions, which may include templated enemy positions. Successful massing of indirect fires and fixed-wing attacks requires a FSE that is proficient in the tracking of all battery positions and movements and knows the maximum ordinate requirements. It also requires a tactical air control party proficient in the timely execution of close air support. Fire planning reconciles top-down planning and bottom-up refinement.

3-145. As the attacking force moves forward, preparatory fires sequentially neutralize, suppress, or destroy enemy positions. The commander must weigh its probable effects against achieving a greater degree of surprise against the enemy, especially under conditions of limited visibility, in determining whether to fire an artillery preparation. He may decide to employ smart or brilliant munitions to destroy select high-payoff targets or use these munitions in mass against part of the enemy defense to facilitate a breach and negate the requirement for long-duration preparation fires using conventional munitions.

3-146. The commander may choose to make his initial assault without using preparatory fires to achieve tactical surprise. However, fires are always planned to support each unit's operations so that they are available if needed. Preparatory fires are normally high-volume fires delivered over a short period of time to maximize surprise and shock effect. These preparatory fires also include the conduct of offensive information operations. They can continue while ground combat elements are maneuvering. This consideration applies to all types and forms of offensive operations.

3-147. Artillery and mortars must occupy positions that are well forward and still within supporting range of the flanks or maneuver forces to provide responsive indirect fires. The commander considers the effect that movement by echelon or battery has on the amount of fire support provided. The commander should support his unit's decisive operation with priority of fires. His main effort prior to the initiation of the decisive operation will have priority of fires if the operation contains phases. He places coordinated fire lines (CFLs) as close as possible to friendly maneuver forces and plans on-order CFLs on phase lines so that they can be quickly shifted as the force moves. This allows the expeditious engagement of targets beyond the CFL by the maximum number of available systems.

3-148. The effective assignment of forward observers and target acquisition assets to quick-fire or exclusive nets also provides responsive fires. Quick-fire nets allow the lead observers to associate directly with specific field artillery or mortar fire units. These kinds of communication arrangements enhance fire support responsiveness through streamlined net structures and focused priorities. Communications planning should also include the need for communication nets for the clearing of targets for rotary- and fixed-wing attacks.

3-149. Offensive information operations perform several functions to support the offense. As the friendly force moves through the enemy's security area and closes into his main defensive positions, jamming resources concentrate on neutralizing enemy fire control, target acquisition, and intelligence-gathering systems. The commander uses deception to prevent the enemy from determining the location and objective of his decisive operation. In addition, electronic warfare resources continue to provide intelligence and guidance to both friendly jammers and lethal indirect fire weapon systems so attacking units can destroy enemy C2, ISR, fire support, and other high-value targets. The commander synchronizes the timing and conduct of his offensive operations so they achieve maximum effectiveness.

AIR DEFENSE

3-150. Air defense operations are performed by all members of the combined arms team; however, ground-based air defense artillery units execute the bulk of the Army's air defense tasks. Allocation of air defense artillery assets within a unit depends on the factors of METT-TC. The commander at each echelon establishes his air defense priorities based on his concept of operations, scheme of maneuver, air situation, and the air defense priorities established by higher headquarters. He generally will weight his air defense coverage toward his decisive operation and establish a protective corridor over the terrain traversed by the unit conducting that decisive operation. The commander has the option of retaining all assets under his direct control or allocating assets to subordinate units. Command and control of all air defense assets requires complete and timely communications to ensure proper weapon status for the protection of friendly air support assets.

3-151. Passive air defense measures are an essential part of air and missile defense planning at all levels. All units conduct passive actions in conjunction with their assigned missions. Passive actions reduce the effectiveness of the enemy air threat.

3-152. Targets selected to support echelon tactical air defense efforts include the following—

- Unmanned aerial vehicles with their launchers and control nodes.
- Rotary- and fixed-wing aircraft.
- Facilities supporting enemy air operations, such as airfields, launch sites, logistics support facilities, technical support facilities, forward arming and refueling points, navigation aids, and C2 sites.

These facilities are normally engaged by maneuver and fire support elements and not air defense artillery units. (See FM 3-01 for additional information on using active and passive air defense measures in the offense.)

MOBILITY/COUNTERMOBILITY/SURVIVABILITY

3-153. This BOS has three basic purposes. It preserves the freedom of maneuver of friendly forces. It obstructs the maneuver of the enemy in areas where fire and movement can destroy him. Finally, it enhances the survivability of friendly forces. Time, equipment, and materials may restrict the amount of work accomplished before, during, and after conducting an offensive operation. The commander's plan must realistically reflect these limitations. The plan must provide the desired balance among these three basic purposes and assign support priorities among subordinate units. Normally, priority of support should be concentrated on the unit or units conducting the decisive operation rather than distributed evenly throughout the force. That support will focus on maintaining the mobility of the force conducting the decisive operation while ensuring the survivability of the C2 system.

Mobility

3-154. Mobility is key to successful offensive operations. Its major focus is to enable friendly forces to maneuver freely on the battlefield. The commander wants the capability to move, exploit, and pursue the enemy across a wide front. When attacking, he wants to concentrate the effects of combat power at selected locations. This may require him to improve or construct combat trails through areas where routes do not exist. The surprise achieved by attacking through an area believed to be impassable may justify the effort expended in constructing these trails. The force bypasses existing obstacles and minefields identified before starting the offensive operation instead of breaching them whenever possible. Units mark bypassed minefields whenever the factors of METT-TC allow.

3-155. Maintaining the momentum of an offensive operation requires the force to quickly pass through obstacles as it encounters them. This translates to a deliberate effort to capture bridges and other enemy reserved obstacles intact. Using air assault and airborne forces is an effective technique to accomplish this goal. The preferred method of fighting through a defended obstacle is employing an in-stride breach, because it avoids the loss of time and momentum associated with conducting a deliberate breach. The commander plans how and where his forces conduct breaching operations. He plans his breaching operations using a reverse planning sequence from the objective back to the assembly area. FM 3-34.2 addresses breaching operations in more detail.

3-156. Rivers remain major obstacles despite advances in high-mobility weapon systems and extensive aviation support. River crossings are among the most critical, complex, and vulnerable combined arms operations. Rivers are crossed in-stride as a continuation of the attack whenever possible. The size of the river, as well as the enemy and friendly situations, will dictate the specific tactics, techniques, and procedures used in conducting the crossing. Corps engineer brigades contain the majority of tactical bridging assets. (See FM 3-97.13 for additional information on conducting hasty and deliberate river crossings.)

Countermobility

3-157. Countermobility operations are vital to help isolate the battlefield and protect the attacking force from enemy counterattack, even though force mobility in offensive operations normally has first priority. Obstacles provide security for friendly forces as the fight progresses into the depth of the enemy's defenses. They provide flank protection and deny the enemy counterattack routes. They assist friendly forces in defeating the enemy in detail and can be vital in reducing the amount of forces required to secure a given area. Further, they can permit the concentration of forces for offensive operations in the first place by allowing a relatively small force to defend a large AO. The commander ensures the use of obstacles is fully synchronized with his concept of operations and does not hinder the mobility of the attacking force.

3-158. During visualization, the commander identifies avenues of approach that offer natural flank protection to an attacking force, such as rivers or ridgelines. Staff estimates support this process. Flanks are protected by destroying bridges, emplacing minefields, and by using scatterable mines to interdict roads and trails. Swamps, canals, lakes, forests, and escarpments are natural terrain features that can be quickly reinforced for flank security.

3-159. Offensive countermobility plans must stress rapid emplacement and flexibility. Engineer support must keep pace with advancing maneuver forces and be prepared to emplace obstacles alongside them. Time and resources will not permit developing the terrain's full defensive potential. The commander first considers likely enemy reactions, then plans how to block enemy avenues of approach or withdrawal with obstacles. He also plans the use of obstacles to contain bypassed enemy elements and prevent the enemy from withdrawing. The plan includes obstacles to use on identification of the enemy's counterattack. Speed and interdiction capabilities are vital characteristics of the obstacles employed. The commander directs the planning for aircraft- and artillery-delivered mines on enemy counterattack routes. The fire support system delivers these munitions in front of or on top of enemy lead elements once they commit to one of the routes. Rapid cratering devices and surface minefields provide other excellent capabilities.

3-160. Control of mines and obstacles, and accurate reporting to all units are vital. Obstacles will hinder both friendly and enemy maneuver. Positive C2 is necessary to prevent the premature activation of minefields and obstacles. (See FM 3-34.1 for additional information on using countermobility obstacles in the offense.)

Survivability

3-161. Denying the enemy a chance to plan, prepare, and execute an effective response to friendly offensive operations through maintaining a high operational tempo is a key means a commander employs to ensure the survivability of his force. Using multiple routes, dispersion, highly mobile forces, piecemeal destruction of isolated enemy forces, scheduled rotation and relief of forces before they culminate, and wise use of terrain are techniques for maintaining a high tempo of offensive operations. The exact techniques employed in a specific situation must reflect the factors of METT-TC.

3-162. The commander protects his force to deny the enemy the capability to interfere with ongoing operations. That protection also meets his legal and moral obligations to his soldiers. To protect his force, the commander—

- Maintains a high tempo of operations.
- Conducts area security operations.
- Employs operations security (OPSEC) procedures.
- Executes deception operations.
- Conducts defensive information operations.
- Employs camouflage, cover, and concealment.
- Constructs survivability positions for nondisplacing systems and supplies.
- Conducts operations to defend against enemy use of nuclear, biological, and chemical weapons.

Although this list is not all-inclusive, it typifies the measures a commander takes to secure his force during offensive operations.

3-163. The echelon's OPSEC program and any deception or survivability efforts should conceal from the enemy or mislead him regarding the location of the friendly objective, decisive operation, the disposition of forces, and the timing of the offensive operation. This tends to prevent the enemy from launching effective spoiling attacks. (See FM 3-13 for additional information on OPSEC, deception, and defensive information operations.)

3-164. The commander normally considers the impact of directing the construction of protective emplacements for artillery, air defense units, and logistics concentrations as part of his planning process. This occurs although units do not employ protective positions in the offense as extensively as they do in the defense. The commander may require the hardening of key C2 facilities, especially those with detectable electronic signatures. Maneuver units construct as many fighting positions as possible whenever they halt or pause during offensive operations. They improve existing terrain by cutting reverse-slope firing shelves or slots when possible. (See FM 3-34.112 for more information on constructing protective positions.) Forces conducting offensive operations will continue to use camouflage, cover, and concealment. (See FM 3-24.3 for additional information on those topics.)

3-165. The IPB process contributes to survivability by developing products that help the commander protect his forces, including intervisibility overlays and situation templates. Intervisibility overlays help protect the force. If an enemy cannot observe the friendly force, he cannot engage the friendly force with direct-fire weapons. Situation templates also help protect the force. If a commander knows how fast an enemy force can respond to his offensive operations, he can sequence his operations at times and places where the enemy can respond least effectively. This occurs through determining enemy artillery range fans, movement times between enemy reserve assembly area locations and advancing friendly forces, and other related intelligence items.

3-166. The commander integrates NBC defensive considerations into his offensive plans. Implementing many NBC defensive measures slows the tempo, degrades combat power, and may also increase logistics requirements. NBC reconnaissance consumes resources, especially time. Personnel in protective gear find it more difficult to work or fight. The key fundamental of all NBC defense activities is to avoid NBC attacks and their effect whenever possible. Avoidance includes passive and active avoidance measures. The other fundamentals of NBC defense are protection and decontamination. Avoidance and protection are closely linked. (See FM 3-11 for additional information on NBC defensive considerations.)

COMBAT SERVICE SUPPORT

3-167. The objective of CSS in offensive operations is to assist the tactical commander in maintaining the momentum. The commander wants to take advantage of windows of opportunity and launch offensive operations with minimum advance warning time. Therefore, logistics and personnel planners and operators must anticipate these events and maintain the flexibility to support the offensive plan accordingly. A key to successful offensive operations is the ability to anticipate the requirement to push support forward, specifically in regard to ammunition, fuel, and water. Combat service support commanders must act, rather than react, to support requirements. The existence of habitual support relationships facilitates the ability to anticipate.

3-168. Combat service support maintains momentum of the attack by delivering supplies as far forward as possible. The commander can use throughput distribution and preplanned and preconfigured packages of essential items to help maintain his momentum and tempo. The commander examines his unit's basic load to determine its adequacy to support the operation. He determines his combat load, the supplies carried by his individual soldiers and combat vehicles. His sustainment load consists of what remains of his basic load once his combat load is subtracted. His tactical vehicles carry the sustainment load. The commander also determines the supplies required for likely contingencies. He determines the amount of cross-loading of supplies required by the situation to prevent all of one type of supply from being destroyed by the loss of a single system.

3-169. CSS units and material remain close to the maneuver force to ensure short turnaround time for supplies and services. This includes uploading as much critical materiel—such as POL and ammunition—as possible and coordinating to preclude attempted occupation of a piece of terrain by more than one unit. The commander makes a decision regarding the possibility that CSS preparation for the attack will be detected by enemy forces and give away his tactical plans.

3-170. The availability of adequate supplies and transportation to sustain the operation becomes more critical as it progresses. Supply LOCs are strained, and requirements for repair and replacement of weapon systems mount. Requirements for POL increase because of the distance the combat vehicles of the maneuver force are likely to travel. CSS units in direct support of maneuver units must be as mobile as the forces they support. One way to provide continuous support is to task organize elements of CSS units or complete CSS units with their supported maneuver formations as required by the factors of METT-TC.

3-171. The variety and complexity of offensive operations requires the Army to establish a flexible and tailorable transportation system. There may be a wide dispersion of forces and lengthening of LOCs. Required capabilities include movement control, in-transit visibility of supplies being carried, terminal operations, and mode operations.

3-172. Maintenance assets move as far forward as consistent with the tactical situation to repair inoperable and damaged equipment and to return it to battle as quickly as possible. Crews continue to perform their preventive maintenance checks and services as modified for the climate and terrain in which they find themselves. Battle damage assessment and repair (BDAR) may be critical to sustaining offensive operations. Crews as well as maintenance and recovery teams conduct BDAR to rapidly return disabled equipment for battlefield service by expediently fixing, bypassing, or jury-rigging components. It restores the minimum essential combat capabilities necessary to support a specific combat mission or to enable the equipment to self-recover.

3-173. The burden on medical resources increases due to the intensity of operations and the increased distances over which support is required as the force advances. The commander reallocates medical resources as the tactical situation changes. Medical units can anticipate large numbers of casualties in a short period of time due to the capabilities of modern conventional weapons and the employment of weapons of mass destruction. These mass casualty situations will probably exceed the capabilities of local medical units and require them to alter their normal scope of operations to provide the greatest good for the greatest number. Key factors for effective mass casualty management are on-site triage, emergency resuscitative care, early surgical intervention, reliable communications, and skillful evacuation by air and ground resources.

3-174. Establishing aerial resupply and forward logistics bases may be necessary to sustain maneuver operations such as exploitation and pursuit conducted at great distance from the unit's sustaining base. The unit or support activity at the airlift's point of origin is responsible for obtaining the required packing, shipping, and sling-load equipment. It prepares the load for aerial transport, prepares the pickup zone, and conducts air-loading operations. The unit located at the airlift destination is responsible for preparing the landing zone to accommodate aerial resupply and for receiving the load.

3-175. Raids conducted by ground maneuver forces within the depths of the enemy's rear area tend to be audacious, high-speed, and of short duration. Logistics support is minimal; units carry as much POL and ammunition as possible, taking advantage of any captured enemy supplies. Once the raiding force crosses its LD, only limited, emergency aerial resupply of critical supplies and medical evacuation are feasible because of the absence of a secure LOC. The commander must thoroughly plan for aerial resupply of the raiding force since it entails greater risk than normal operations. Under these conditions, units destroy damaged equipment that is unable to maintain the pace of the operation.

COMMAND AND CONTROL

3-176. The commander's mission and intent determine the scheme of maneuver and the allocation of available resources. The commander reduces the scope of the initial mission if only a few resources are available. For example, a commander could tell his subordinates to clear their AOs of all enemy platoon-size and larger forces instead of clearing their areas of operation of all enemy forces if he lacks the time or forces needed to accomplish the latter task.

3-177. All planning for offensive operations address the factors of METT-TC, with special emphasis on—

- Enemy positions, strengths, and capabilities.
- Missions and objectives for each subordinate element and task and purpose for each BOS manager.
- Commander's intent.
- AOs for the use of each subordinate element with associated control graphics.
- Time the operation is to begin.
- Scheme of maneuver.
- Special tasks required to accomplish the mission.

- Risk.
- Options for accomplishing the mission.

The commander and his staff translate the unit's assigned mission into specific objectives for all subordinates, to include the reserve. These objectives can involve any type or form of operations. If the type of operation assigned has associated forms, the commander may specify which form to use, but should minimize restrictions on his subordinates' freedom of action. FM 5-0 addresses the military decision making process and the format for plans and orders.

3-178. Prior planning and preparations that result in synchronizing the seven BOS increase a unit's effectiveness when executing operations. However, the fluid nature of combat requires the commander to guide the actions of his subordinates during the execution phase. The commander locates himself where he can best sense the flow of the operation and influence its critical points by redirecting the effects of committed forces or employing his reserve. This normally means that he is well forward in the combat formation, usually with the force designated to conduct the decisive operation. Once he makes contact with the enemy, he quickly moves to the area of contact, assesses the situation, and takes appropriate aggressive actions to direct the continuation of the offensive operation.

3-179. In addition to assigning objectives, commanders at all echelons consider how to exploit advantages that arise during operations and the seizure of intermediate and final objectives. The commander exploits success by aggressively executing the plan, taking advantage of junior leader initiative, and employing trained units capable of rapidly executing standard drills. His reserve also provides a flexible capability to exploit unforeseen advantages.

3-180. The commander always seeks to surprise his opponent throughout the operation. Information operations, such as deception, and the choice of an unexpected direction or time for conducting an offensive operation can result in the enemy being surprised. Surprise delays enemy reactions, overloads and confuses enemy C2, induces psychological shock, and reduces the coherence of his defenses. Tactical surprise is more difficult to achieve once hostilities begin, but it is still possible. The commander achieves tactical surprise by attacking in bad weather and over seemingly impassible terrain, conducting feints and demonstrations, making rapid changes in tempo, and employing sound OPSEC measures.

3-181. The commander should anticipate any requirements to shift his main effort during the offensive to press the fight and keep the enemy off balance. The commander develops decision points to support these changes using both human and technical means to validate his decision points.

3-182. The commander retains the capability to rapidly concentrate force effects, such as fires, throughout the extent of his AO during offensive operations. This capability is also critical to the commander when his force crosses linear obstacles. Lanes and gaps resulting from combined arms breaching operations or occurring naturally typically are choke points. There is a tendency for each subordinate element to move out independently as it completes its passage through the choke point. This independent movement detracts from the ability of the whole force to rapidly generate combat power on the far side of an obstacle.

3-183. The commander briefs his plan and the plans of adjacent units and higher echelons to his unit's leaders and soldiers. This helps units and soldiers moving into unexpected locations to direct their efforts toward accomplishing the mission. This exchange of information occurs in all operations.

3-184. The free flow of information between all force elements must be maintained throughout the offensive operation. This requires the commander to maintain communications with all elements of his unit. He plans how to position and reposition his information systems to maintain his common operational picture throughout the operation. The commander plans how to expand his communications coverage to accommodate increased distances as his force advances. Accordingly, he provides for redundant communication means—including wire, radio, visible and ultraviolet light, heat, smoke, audible sound, messengers, and event-oriented communications, such as the casualty-producing device that initiates an ambush.

3-185. A unit with advanced information systems and automated decision aids enjoys reduced engagement times and an enhanced planning process. This assists the unit commander's ability to control the operational tempo of the battle and stay within the enemy's decision making cycle. Greatly improved knowledge of the enemy and friendly situations facilitates the tactical employment of precision fires and decisive maneuver at extended ranges. These digital systems also enhance the commander's freedom to move to those battlefield locations where he can best influence the battle at the critical time and place.

TRANSITION

3-186. A transition occurs when the commander makes the assessment that he must change his focus from one type of military operation to another. The following paragraphs explain why a commander primarily conducting offensive operations would transition to the defense and describe techniques that a commander can use to ease the transition.

3-187. A commander halts an offensive operation only when it results in complete victory and the end of hostilities, reaches a culminating point, or the commander receives a change in mission from his higher commander. This change in mission may be a result of the interrelationship of the other elements of national power, such as a political decision.

3-188. All offensive operations that do not achieve complete victory reach a culminating point when the balance of strength shifts from the attacking force to its opponent. Usually, offensive operations lose momentum when friendly forces encounter heavily defended areas that cannot be bypassed. They also reach a culminating point when the resupply of fuel, ammunition, and other supplies fails to keep up with expenditures, soldiers become physically exhausted, casualties and equipment losses mount, and repairs and replacements do not keep pace. Because of enemy surprise movements, offensive operations also stall when reserves are not available to continue the advance, the defender receives reinforcements, or he counterattacks with fresh troops. Several of these causes may combine to halt an offense. In some cases, the unit can regain its momentum, but this only happens after difficult fighting or after an operational pause.

3-189. If the attacker cannot anticipate securing decisive objectives before his force reaches its culminating point, he plans a pause to replenish his combat power and phases his operation accordingly. Simultaneously, he prevents the enemy from knowing when the friendly forces become overextended.

3-190. Once offensive operations begin, the attacking commander tries to sense when he reaches, or is about to reach, his culminating point. Before reaching this point, he must transition to some other type of military operation. The commander has more freedom to choose where and when he wants to halt the attack if he can sense that his forces are approaching culmination. He can plan his future activities to aid the defense, minimize vulnerability to attack, and facilitate renewal of the offense as he transitions to branches or sequels of the operation. For example, to prevent overburdening the extended LOCs resulting from the advances away from his sustaining base, some of the commander's subordinate units may move into assembly areas before he terminates the offense to start preparing for the ensuing defensive operation.

3-191. A lull in combat operations often accompanies transition. Civilians may present themselves to friendly forces during this period. The commander must consider how he will minimize the interference of these civilians with his military operations while protecting the civilians from future hostile actions. He must also consider the threat they pose to his force and its operations if enemy intelligence agents or saboteurs constitute a portion of the civilian population encountered.

3-192. A commander anticipating the termination of his offensive operation prepares orders that include the time or circumstances under which the offense transitions to another type of military operation, such as the defense, the missions and locations of subordinate units, and C2 measures. As he transitions from the offense to the defense, a commander takes the following actions:

- Maintains contact and surveillance of the enemy, using a combination of reconnaissance units and surveillance assets to develop the information required to plan future actions.
- Establishes a security area and local security measures.
- Redeploys fire support assets to ensure the support of security forces.
- Redeploys forces based on probable future employment.
- Maintains or regains contact with adjacent units in a contiguous AO and ensures that his units remain capable of mutual support in a noncontiguous AO.
- Transitions the engineer effort by shifting the emphasis from mobility to countermobility and survivability.
- Consolidates and reorganizes.
- Explains the rationale for transitioning from the offense to his soldiers.

3-193. The commander redeploys his air defense assets to cover the force's defensive position. A transition to the defense may require the commander to change his air defense priorities. For example, his top priority may have been coverage of maneuver units in the offense. This may shift to coverage of his long-range sensors and weapons in the defense.

3-194. The commander conducts any required reorganization and resupply concurrently with the above activities. This requires a transition in the logistics effort, with a shift in emphasis from ensuring the force's ability to move forward (POL and forward repair of maintenance and combat losses) to ensuring the force's ability to defend on its chosen location (forward stockage construction, barrier, and obstacle material, and ammunition). Transition is often a time in which equipment maintenance can be performed. Additional assets may also be available for casualty evacuation and medical treatment because of a reduction in the tempo.

3-195. The commander should not wait too long to transition from the offense to the defense as his forces approach their culminating point. Without prior planning, transitioning to defensive actions after reaching a culminating point is extremely difficult for several reasons. Defensive preparations are hasty, and forces are not adequately disposed for defense. Defensive reorganization requires more time than the enemy will probably allow. Usually, attacking forces are dispersed, extended in depth, and weakened in condition. Moreover, the shift to defense requires a psychological adjustment. Soldiers who have become accustomed to advancing must now halt and fight defensively—sometimes desperately—on new and often unfavorable terms.

3-196. A commander can use two basic techniques when he transitions to the defense. The first technique is for the leading elements to commit forces and push forward to claim enough ground to establish a security area anchored on defensible terrain. The main force moves forward or rearward as necessary to occupy key terrain and institutes a hasty defense that progresses into a deliberate defense as time and resources allow. The second technique is to establish a security area generally along the unit's final positions, moving the main body rearward to defensible terrain. The security force thins out and the remaining force deploys to organize the defense. In both methods, the security area should be deep enough to keep the main force out of the range of enemy medium artillery and rocket systems.

3-197. In the first technique, the security area often lacks depth because the force lacks sufficient combat power to seize required terrain. In the second technique, the enemy force will probably accurately template the forward trace of friendly units and engage with artillery and other fire support systems. These actions often result in the loss of additional friendly soldiers and equipment and the expenditure of more resources.

3-198. If a commander determines that it is necessary to break off an offensive operation and conduct a retrograde, he typically has his units conduct an area defense from their current locations until he can synchronize the retrograde operation. The amount of effort expended in establishing the area defense depends on the specific factors of METT-TC currently prevailing.

Chapter 4

Movement to Contact

When armies approach each other, it makes all the difference which owns only the ground on which it stands or sleeps and which one owns all the rest.

Winston Churchill: *Their Finest Hour*, 1949

Movement to contact is a type of offensive operation designed to develop the situation and establish or regain contact (FM 3-0). A commander conducts this type of offensive operation when the tactical situation is not clear or when the enemy has broken contact. A properly executed movement to contact develops the combat situation and maintains the commander's freedom of action after contact is gained. This flexibility is essential in maintaining the initiative. All of the tactical concepts, control measures, and planning considerations introduced in Chapters 2 and 3 apply to the conduct of a movement to contact. Many of the attack preparation consideration introduced in Chapter 5 also apply.

CONTENTS

4-1. Purposeful and aggressive movement, decentralized control, and the hasty deployment of combined arms formations from the march to attack or defend characterize the movement to contact. The fundamentals of a movement to contact are—

- Focus all efforts on finding the enemy.
- Make initial contact with the smallest force possible, consistent with protecting the force.
- Make initial contact with small, mobile, self-contained forces to avoid decisive engagement of the main body on ground chosen by the enemy. This allows the commander maximum flexibility to develop the situation.
- Task-organize the force and use movement formations to deploy and attack rapidly in any direction.

- Keep forces within supporting distances to facilitate a flexible response.
- Maintain contact regardless of the course of action (COA) adopted once contact is gained.

Close air support, air interdiction, and counterair operations are essential to the success of large-scale movements to contact (MTCs). Local air superiority or, as a minimum, air parity is vital to the operation's success.

4-2. The Army's improved intelligence, surveillance, and reconnaissance (ISR) capabilities reduce the need for corps and divisions to conduct an MTC since fully modernized units normally have a general idea of the location of significant enemy forces. However, enemy use of complex terrain and offensive information operations designed to degrade the accuracy of the friendly common operational picture will continue to require small tactical units to conduct an MTC.

4-3. A *meeting engagement* is a combat action that occurs when a moving force engages an enemy at an unexpected time and place (JP 3-0). Conducting an MTC results in a meeting engagement. The enemy force may be either stationary or moving. Such encounters often occur in small-unit operations when reconnaissance has been ineffective. The force that reacts first to the unexpected contact generally gains an advantage over its opponent. However, a meeting engagement may also occur when the opponents are aware of each other and both decide to attack immediately to obtain a tactical advantage or seize key or decisive terrain. A meeting engagement may also occur when one force attempts to deploy into a hasty defense while the other force attacks before its opponent can organize an effective defense. Acquisition systems may discover the enemy before the security force can gain contact. No matter how the force makes contact, seizing the initiative is the overriding imperative. Prompt execution of battle drills at platoon level and below, and standard actions on contact for larger units can give that initiative to the friendly force.

HISTORICAL EXAMPLE

4-4. The following vignette discusses the Soviet Manchurian Campaign in the closing days of World War II and illustrates the idea that tacticians can profit from the lessons learned by other armies in other times. It demonstrates the conduct of an MTC at the operational level.

The Soviet Manchurian Campaign, August 1945

Shortly after midnight on 9 August 1945, Soviet assault troops crossed the Soviet-Manchurian border and attacked Japanese positions. This was the vanguard of a force of more than 1.5 million men that was to advance along multiple axes on a frontage of more than 4,400 kilometers. Soviet offensive tactics were shaped by several factors:

- The necessity for speed to increase the effectiveness of maneuver, thus increasing surprise, overcoming initial defenses, and preempting the establishment of subsequent effective defenses.
- The vast expanse of the area of operations (AO).
- The diversity of the terrain, giving rise to large-scale force tailoring.
- The nature of the opposition.

The Soviets conducted their movement to contact operation at the last possible moment. This reinforced strategic surprise and yielded tactical surprise as well. Units deployed for attack from assembly areas 20 to 80 kilometers behind the border and entered from the march, attacking along every possible axis using small, task-organized assault groups with heavy engineer and firepower support. Conduct of operations under adverse weather conditions and at night went contrary to Japanese expectations. The Soviet tendency to bypass fortified positions confused Japanese commanders.

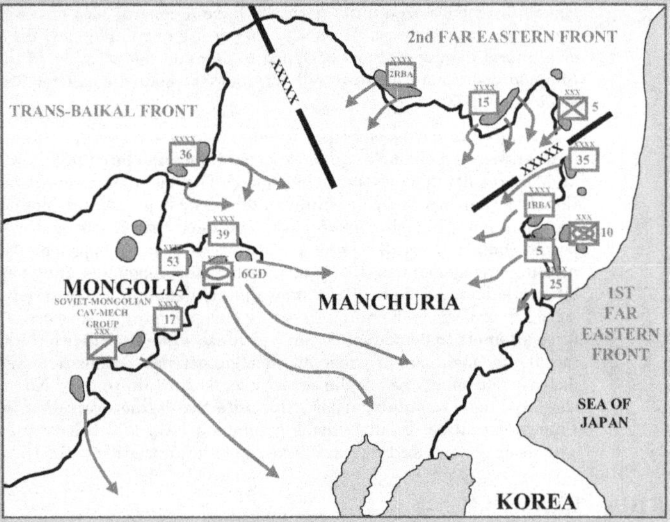

Figure 4-1. Soviet Moves

The Soviets carefully timed the application of their offensive power by first attacking with forward detachments and advance guards in the first echelon, and then with the main force. See Figure 4-1. This perpetuated the momentum of initial assaults and created a momentum that was imparted to army and front-level operations. Often enemy resistance was eliminated before the main columns had to deploy. Forces massed at the critical point on each axis and maneuvered over what was considered to be impassable terrain. All this resulted in a loss of defense coherence that the defending Japanese *Kwantung* Army was never able to regain. In seven days Soviet forces penetrated between 500 and 950 kilometers into Manchuria from their starting points, securing all the objectives necessary for a complete victory over the Japanese.

ORGANIZATION OF FORCES

4-5. A movement to contact is organized with an offensive covering force or an advance guard as a forward security element and a main body as a minimum. A portion of the main body composes the commander's sustaining base. Based on the factors of METT-TC, the commander may increase his security forces by having an offensive covering force and an advance guard for each column, as well as flank and rear security (normally a screen or guard). (See Figure 4-2.) Chapter 12 discusses security operations.

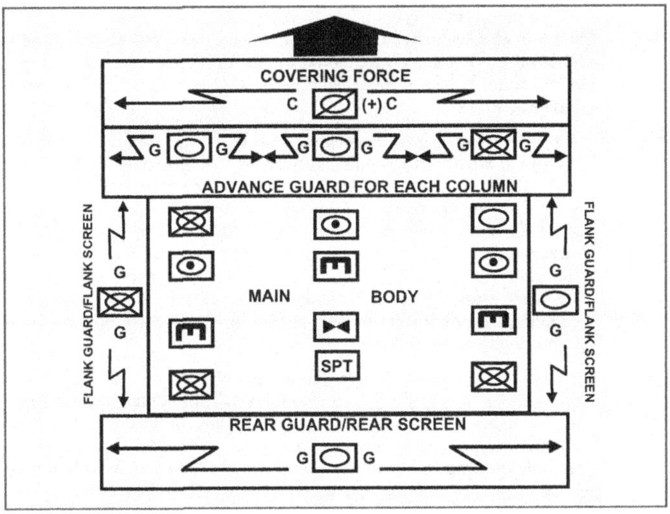

Figure 4-2. Force Organized for a Movement to Contact

4-6. A movement to contact mission requires the commander not to have contact with the enemy main body. However, the commander may still know the location of at least some enemy reserve and follow-on forces. If the corps or division commander has enough intelligence information to target enemy uncommitted forces, reserves, or sustaining operations activities, he normally designates forces, such as long-range artillery systems and attack helicopters, to engage known enemy elements regardless of their geographical location within his AO. At all times the forward security element and the main body perform reconnaissance.

SECURITY FORCES

4-7. A corps or division commander conducting an MTC typically organizes his security element into a covering force to protect the movement of the main body and to develop the situation before committing the main body. A covering force is task-organized to accomplish specific tasks independent of the main body in accordance with the factors of METT-TC, such as conduct mobility and breach operations. This covering force reports directly to the establishing commander.

4-8. If a force conducting an MTC is unable to resource a covering force for independent security operations, it may use an advance guard in the place of a covering force. An advance guard is a task-organized combined arms unit or detachment that precedes a column formation to protect the main body from ground observation or surprise by the enemy. This typically occurs when a brigade or battalion conducts an MTC. In cases where the higher echelon (corps or division) creates a covering force, subordinate elements can establish an advance guard behind the covering force and ahead of the main body. This normally occurs when subordinate units are advancing in multiple parallel columns. In this case, each main body column usually organizes its own advance guard.

4-9. The advance guard operates forward of the main body to ensure its uninterrupted advance by reducing obstacles to create passage lanes, repair roads and bridges, or locate bypasses. The advance guard also protects the main body from surprise attack and fixes the enemy to protect the deployment of the main body when it is committed to action. The elements composing the advance guard should have equal or preferably superior mobility to that of the main body. For this reason, mechanized infantry, cavalry, and armored units are most suitable for use in an advance guard. Engineer mobility assets should also constitute a portion of the advance guard, but the main body can also provide other support.

4-10. The advance guard moves as quickly and as aggressively as possible, but, unlike the covering force, remains within supporting range of the main body's weapon systems. It forces the enemy to withdraw or destroys small enemy groups before they can disrupt the advance of the main body. When the advance guard encounters large enemy forces or heavily defended areas, it takes prompt and aggressive action to develop the situation and, within its capability, defeat the enemy. Its commander reports the location, strength, disposition, and composition of the enemy and tries to find the enemy's flanks and gaps or other weaknesses in his position. The main body may then join the attack. The force commander usually specifies how far in front of his force the advance guard is to operate. He reduces those distances in close terrain and under low-visibility conditions.

4-11. When the command's rear or flanks are not protected by adjacent or following units, it must provide its own flank and rear security. The command can accomplish this by establishing a screen or a guard on its flanks or to its rear. The flank columns of the main body normally provide these flank security elements; for example, the left flank brigade would provide the left flank screen for a division MTC. The rear guard normally comes from one of the subordinate elements of the corps or division and reports directly to the corps or division headquarters. A corps may conduct a flank cover if there is a clearly identified, significant threat from the flank. A flank cover requires significant resources that are unavailable to the main body. Aviation units or intelligence systems may establish a flank screen if the factors of METT-TC allow; however, this increases the risk to the main body. While aviation units can use their combat power to delay enemy forces, intelligence systems can only provide early warning, they cannot trade space for time to "buy" time for the main body to react. (For more specific information concerning reconnaissance operations see FM 3-55. See Chapter 12 for more detailed information concerning security operations.)

MAIN BODY

4-12. The main body consists of forces not detailed to security duties. The combat elements of the main body prepare to respond to enemy contact with the unit's security forces. Attack helicopter units normally remain under division and corps control until contact is made. If the situation allows, the commander can assign a follow and support mission to one of his subordinate units. This allows that subordinate unit to relieve his security forces from such tasks as observing bypassed enemy forces, handling displaced civilians, and clearing routes. This prevents his security forces from being diverted from their primary mission.

4-13. The commander designates a portion of the main body for use as his reserve. The size of the reserve is based upon the factors of METT-TC and the amount of uncertainty concerning the enemy. The more vague the enemy situation, the larger the size of the reserve. The reserve typically constitutes approximately one-fourth to one-third of the force. On contact with the enemy, the reserve provides the commander flexibility to react to unforeseen circumstances and allows the unit to quickly resume its movement.

4-14. The commander tailors his sustainment assets to the mission. He decentralizes the execution of the sustainment support, but that support must be continuously available to the main body. This includes using preplanned logistics packages (LOGPACs). **A *logistics package* is a grouping of multiple classes of supply and supply vehicles under the control of a single convoy commander.** Daily LOGPACs contain a standardized allocation of supplies. Special LOGPACs can also be dispatched as needed.

4-15. The commander frequently finds that his main supply routes become extended as the operation proceeds. Aerial resupply may also be necessary to support large-scale MTCs or to maintain the momentum of the main body. Combat trains containing fuel, ammunition, medical, and maintenance assets move with their parent battalion or company team. Fuel and ammunition stocks remain loaded on tactical vehicles in the combat trains so they can instantly move when necessary. Battalion field trains move in more depth, with the forward support battalion in the main body of each brigade. Aviation units use forward arming and refuel points (FARPs) to reduce aircraft turnaround time.

CONTROL MEASURES

4-16. A commander uses the minimal number and type of control measures possible in an MTC because of the uncertain enemy situation. These measures include designation of an AO with left, right, front, and rear boundaries, or a separate AO bounded by a continuous boundary (noncontiguous operations). The commander further divides the AO into subordinate unit AOs to facilitate subordinate unit actions.

4-17. The operation usually starts from a line of departure (LD) at the time specified in the operations order (OPORD). The commander controls the MTC by using phase lines, contact points, and checkpoints as required. (See

Figure 4-3.) He controls the depth of the movement to contact by using a limit of advance (LOA) or a forward boundary. Figure 4-3 shows an LOA and not a forward boundary. The commander could designate one or more objectives to limit the extent of the MTC and orient the force. However, these are often terrain-oriented and used only to guide movement. Although an MTC may result in taking a terrain objective, the primary focus should be on the enemy force. If the commander has enough information to locate significant enemy forces, then he should plan some other type of offensive action.

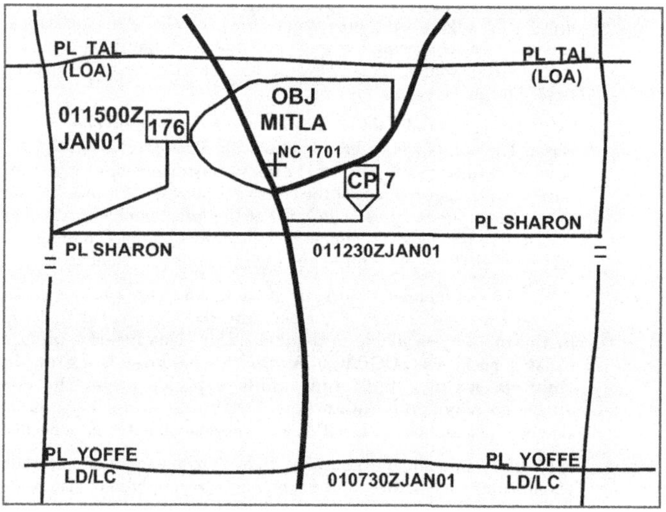

Figure 4-3. Movement to Contact Control Measures

4-18. Corps, division, or brigade commanders use boundaries to separate the various organizational elements of a movement to contact and clearly establish responsibilities between different organizations. Battalion task force commanders use positive control over maneuver units, coupled with battle drills and formation discipline. Company teams are not normally assigned their own areas of operation during the conduct of a movement to contact.

4-19. The commander can designate a series of phase lines that can successively become the new rear boundary of the forward security elements as that force advances. Each rear boundary becomes the forward boundary of the main body and shifts as the security force moves forward. The rear boundary of the main body designates the limit of responsibility of the rear security element. This line also shifts as the main body moves forward. (See Chapter 12 for a discussion of boundaries for a security force.)

4-20. Commanders may use an axis of advance in limited visibility. However, there is the risk of enemy forces outside the axis not being detected, and thus being inadvertently bypassed.

PLANNING A MOVEMENT TO CONTACT

4-21. The commander conducts intelligence, surveillance, and reconnaissance operations to determine the enemy's location and intent while conducting security operations to protect the main body. This includes the use of available fixed-wing aircraft. This allows the main body to focus its planning and preparation, to include rehearsals, on the conduct of hasty attacks, bypass maneuvers, and hasty defenses. The plan addresses not only actions anticipated by the commander based on available intelligence information but also the conduct of meeting engagements at anticipated times and locations where they might occur.

4-22. The commander wants to gain contact by using the smallest elements possible. These elements are normally ground scouts or aeroscouts performing reconnaissance, but may also be unmanned aerial vehicles (UAVs) or other intelligence systems. He may task organize his scouts to provide them with additional combat power to allow them to develop the situation. The unit's planned movement formation should contribute to the goal of making initial contact with the smallest force possible. It should also provide for efficient movement of the force and adequate reserves. The commander can choose to have all or part of his force conduct an approach march as part of the movement to contact to provide that efficient movement. An approach march can facilitate the commander's decisions by allowing freedom of action and movement of the main body. (See Chapter 14 for a discussion of an approach march.)

4-23. The frontage assigned to a unit in a movement to contact must allow it to generate sufficient combat power to maintain the momentum of the operation. Reducing the frontage covered normally gives the unit adequate combat power to develop the situation on contact while maintaining the required momentum. Both the covering force and advance guard commanders should have uncommitted forces available to develop the situation without requiring the deployment of the main body. The commander relies primarily on fire support assets to weight the lead element's combat power but provides it with the additional combat multipliers it needs to accomplish the mission. The fire support system helps develop fire superiority when organized correctly to fire immediate suppression missions to help maneuver forces get within direct-fire range of the enemy.

4-24. The reconnaissance effort may proceed faster in a movement to contact than in a zone reconnaissance because the emphasis is on making contact with the enemy. However, the commander must recognize that by increasing the speed of the reconnaissance effort, he increases the risk associated with the operation.

4-25. Bypass criteria should be clearly stated and depend on the factors of METT-TC. For example, a brigade commander in an open desert environment could declare that no mounted enemy force larger than a platoon can be bypassed. All other forces will be cleared from the brigade's axis of advance. Any force that bypasses an enemy unit must maintain contact with it until handing it off to another friendly element, usually a force assigned a follow and support mission. The commander tasks his forward security force with conducting route reconnaissance of routes the main body will traverse.

4-26. The echelon intelligence officer (G2 or S2), assisted by the engineer and air defense staff representatives, must carefully analyze the terrain to include air avenues of approach. He identifies the enemy's most dangerous COA in the war gaming portion of the military decision making process. Because of the force's vulnerability, the G2 must not underestimate the enemy during a movement to contact. A thorough intelligence preparation of the battlefield (IPB)—by developing the modified combined obstacle overlay to include intervisibility overlays and other products, such as the event templates—enhances the force's security by indicating danger areas where the force is most likely to make contact with the enemy. It also helps to determine movement times between phase lines and other locations. Potential danger areas are likely enemy defensive locations, engagement areas, observation posts (OPs), and obstacles. The fire support system targets these areas and they become on-order priority targets placed into effect and cancelled as the lead element can confirm or deny enemy presence. The reconnaissance and surveillance plan supporting the movement to contact must provide coverage of these danger areas. If reconnaissance and surveillance forces cannot clear these areas, more deliberate movement techniques are required.

4-27. The commander develops decision points to support changes in the force's movement formation or change from an approach march to a combat formation. Uses both human and technical means to validate his decision points, the commander must determine the degree of risk he is willing to accept based on his mission. The commander's confidence in the products of the IPB process and the risk he is willing to accept determine his combat formation and maneuver scheme. In a high-risk environment, it is usually better to increase the distance between forward elements and the main body than to slow the speed of advance.

4-28. Corps and divisions can execute shaping operations as part of a movement to contact although, by definition, a force conducts a movement to contact when the enemy situation is vague or totally unknown. This occurs when the necessary information regarding enemy reserves and follow-on forces is available, but information regarding those enemy forces in close proximity to the friendly force is not available. As in any other type of operation, the commander plans to focus his operations on finding the enemy and then delaying, disrupting, and destroying each enemy force element as much as possible before it arrives onto the direct-fire battlefield. This allows close combat forces to prepare to engage enemy units on their arrival.

4-29. In a movement to contact, the commander can opt not to designate his decisive operation until his forces make contact with the enemy, unless there is a specific reason to designate it. In this case, he retains resources under his direct control to reinforce his decisive operation. He may designate his decisive operation during the initial stages of a movement to contact because of the presence of a key piece of terrain or avenue of approach.

EXECUTING A MOVEMENT TO CONTACT

4-30. Each element of the force synchronizes its actions with adjacent and supporting units, maintaining contact and coordination as prescribed in orders and unit standing operating procedures (SOP). The advance guard maintains contact with the covering force. The lead elements of the main body maintain contact with the advance guard. The rear guard and flank security elements maintain contact with and orient on the main body's movement. These security forces prevent unnecessary delay of the main body and defer the deployment of the main body as long as possible. Reconnaissance elements operate to the front and flanks of each column's advance guard and maintain contact with the covering force. The commander may instruct each column's advance guard to eliminate small pockets of resistance bypassed by forward security force. (See Figure 4-4.)

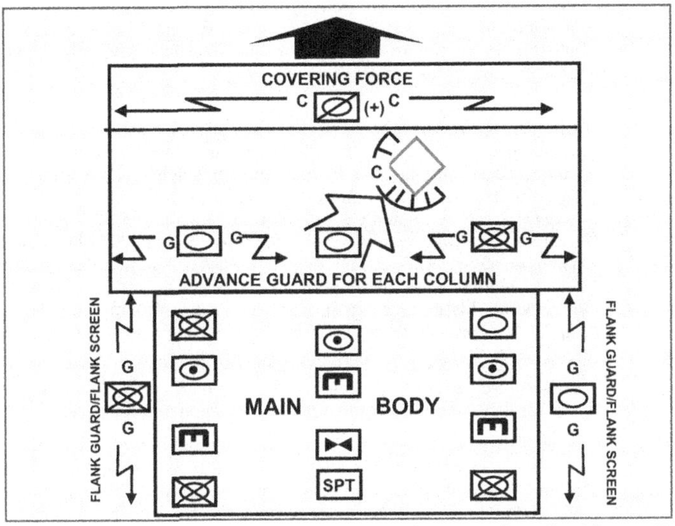

Figure 4-4. A Column Advance Guard Attacking to Destroy a Contained Enemy Force

4-31. The commander of the advance guard chooses a combat formation, based on the factors of METT-TC, to make contact with the smallest possible force while providing flexibility for maneuver. Whatever combat formation is chosen, it must be able to deploy appropriately once the commander becomes aware of the enemy's location. He ensures that the route or axis of advance traveled by the main body is free of enemy forces. It may move continuously (traveling and traveling overwatch) or by bounds (bounding overwatch). It moves by bounds when contact with the enemy is imminent and the terrain is favorable. Some indirect-fire assets, such as a mortar platoon or artillery battery and combat observation and lasing teams (COLTs), may be positioned with the formation. The COLTs can help overwatch the advance guard movement, and indirect fires focus on suppressing enemy weapons, obscuring enemy observation posts, and screening friendly movement.

4-32. The main body keeps enough distance between itself and its forward security elements to maintain flexibility for maneuver. This distance varies with the level of command, the terrain, and the availability of information about the enemy. The main body may execute an approach march for all or part of the movement to contact to efficiently use the available road network or reduce the time needed to move from one location to another. Command posts and trains travel along high-mobility routes within the AO and occupy hasty positions as necessary.

4-33. Behind these forward security elements, the main body advances over multiple parallel routes with numerous lateral branches to remain flexible and reduce the time needed to initiate maneuver. (While it is preferred for a battalion to use multiple routes, battalions and smaller units can move on just one route.) In a movement to contact, the main body's march dispositions must allow maximum flexibility for maneuvering during movement and when establishing contact with the enemy force.

4-34. The commander's fire support systems tend to focus on suppression missions to disrupt enemy forces as they are encountered and smoke missions to obscure or screen exposed friendly forces when conducting a movement to contact. The commander schedules the movements of fire support systems in synchronization with the movement of the rest of the force. Fire support systems that cannot match the cross-country mobility of the combat units cause them to slow their rate of advance. If these units do not slow down, they run the risk of outrunning their fire support. The commander synchronizes the employment of close air support to prevent the enemy from regaining his balance while his ground fire support assets are repositioning. The main body updates its priority target list during a movement to contact operation.

4-35. The same considerations apply to air defense when the enemy possesses aerial capabilities. The commander ensures that his force stays within the air defense umbrella provided by corps and division assets.

4-36. The enemy has a difficult time detecting and targeting the main body when executing a movement to contact until contact is made because of its tempo, momentum, and dispersal; and the attention the commander pays to electromagnetic emission control. Once the force makes contact and concentrates its effects against detected enemy forces, it becomes vulnerable to strikes by enemy conventional weapons and weapons of mass destruction. It must concentrate its combat effects rapidly and disperse again as soon as it overcomes resistance to avoid enemy counteractions.

4-37. Movement should be as rapid as the terrain, the mobility of the force, and the enemy situation permit. Open terrain provides maneuver space on either side of the line of march and facilitates high-speed movement. It also allows for greater dispersal and usually permits more separation between forward security elements and the main body than restricted terrain allows. The commander should never commit his main body to canalizing terrain before these forward security elements have advanced far enough to ensure that the main body will not become fixed within that terrain. The enemy may have also established fire support control measures that allow him to employ nonobserved harassing and interdiction fires to interdict friendly forces traversing these choke points. As the enemy situation becomes known, the commander may shorten the distance between elements of the force to decrease reaction time or he may deploy to prepare for contact.

4-38. At the battalion and company levels, a moving force should move along covered or concealed routes from one covered or concealed position to another, using terrain to minimize its vulnerability to enemy weapons. Further, an overwatching force should cover the moving force. (Chapter 14 describes movement techniques, such as traveling overwatch.) Regardless of the specific movement technique employed, both forces need to provide mutual support and be knowledgeable about their counterpart's sectors of fire.

4-39. The force must attempt to cross any obstacles it encounters without loss of momentum by conducting in-stride breaches. The commander uses his forward security forces in an attempt to seize intact bridges whenever possible. Lead security elements bypass or breach obstacles as quickly as possible to maintain the momentum of the movement. If these lead elements cannot overcome obstacles, the commander directs subsequent elements of the main body to bypass the obstacle site and take the lead. Following forces can also reduce obstacles that hinder the unit's sustainment flow.

4-40. The commander locates himself well forward in the movement formation. Once the formation makes contact with the enemy, he can move quickly to the area of contact, analyze the situation, and direct aggressively. The commander's security elements conduct actions on contact to develop the situation once they find the enemy. Once they make contact with the enemy, a number of actions occur that have been divided into the following sequence. (Units equipped with a full set of digital command and control systems may be able to combine or skip one or more of the steps in that sequence. Those units will conduct maneuver and remain within supporting distance of each other with a significantly larger AOs than units equipped with analog systems.)

GAIN AND MAINTAIN ENEMY CONTACT

4-41. All ISR assets focus on determining the enemy's dispositions and providing the commander with current intelligence and relevant combat information; this ensures that he can commit friendly forces under optimal conditions. The commander uses all available sources of combat information to find the enemy's location and dispositions in addition to his intelligence systems. Corps and divisions employ long-range surveillance units and detachments in conjunction with data provided by available special operating forces, joint, and multinational assets, in addition to their organic ISR assets. The commander may use his surveillance systems to cue the conduct of aerial and ground reconnaissance.

4-42. The enemy situation becomes clearer as the unit's forward security elements conduct actions on contact to rapidly develop the situation in accordance with the commander's plan and intent. By determining the strength, location, and disposition of enemy forces, these security elements allow the commander to focus the effects of the main body's combat power against the enemy main body. The overall force must remain flexible to exploit both intelligence and combat information. The security force should not allow the enemy force to break contact unless it receives an order from the commander. When a strong covering force has not preceded the advance guard, it should seize terrain that offers essential observation.

4-43. ***Actions on contact* are a series of combat actions often conducted simultaneously taken on contact with the enemy to develop the situation.** Actions on contact are:

- Deploy and report.
- Evaluate and develop the situation.
- Choose a course of action.
- Execute selected course of action
- Recommend a course of action to the higher commander.

4-44. Once the lead elements of a force conducting a movement to contact encounter the enemy, they conduct actions on contact. The unit treats obstacles like enemy contact, since it assumes that the obstacles are covered by fire. The unit carries out these actions on contact regardless of whether the enemy has detected its presence. The unit's security force often gains a tactical advantage over an enemy force by using tempo and initiative to conduct these actions on contact, allowing it to gain and maintain contact without becoming decisively engaged. How quickly the unit develops the situation is directly related to its security. This tempo is directly related to the unit's use of well-rehearsed SOP and drills.

Deploy and Report

4-45. When a unit's security element encounters an enemy unit or obstacle, it deploys to a covered position that provides observation and fields of fire. If the security element is under enemy fire, it uses direct and indirect fire to suppress the enemy and restore freedom of maneuver. Simultaneously, the commander of the security element reports the contact using a spot report format to provide all available information on the situation to his higher headquarters. This alerts the commander and allows him to begin necessary actions. (FM 6-99.2 provides the format for a spot report.)

Evaluate and Develop the Situation

4-46. The unit's security force develops the situation rapidly within mission constraints by employing techniques ranging from stealthy, foot-mobile reconnaissance to reconnaissance by fire, which uses both direct and indirect weapons. If possible the commander continues the security mission with other elements not currently in contact with the enemy after evaluating the situation. This helps to develop the situation across the front and provides more maneuver space to execute further actions. As the situation develops, the security force submits additional reports.

Choose a Course of Action

4-47. After the security force makes contact, its commander gathers information, makes an assessment, and chooses a course of action (COA) consistent with his higher commander's intent and within the unit's capability. The unit initiates direct and indirect fires to gain the initiative if it is appropriate to engage the enemy. This allows the security force to resume its mission as soon as possible. The commander cannot allow small enemy forces to delay the movement of the security force. Usually, available intelligence and the concept of the operation indicate the COA to follow. For obstacles not covered by fire, the unit can either seek a bypass or create the required number of lanes to support its maneuver or the maneuver of a supported unit. Once enemy contact is made, these COAs are normally to conduct an attack, bypass, defend, delay, or withdrawal. For obstacles covered by fire, the unit can either seek a bypass or conduct breaching operations as part of a hasty attack.

Execute Selected Course of Action

4-48. The security force commander should determine quickly whether to bypass the enemy or attack. The security force attacks (see Chapter 5) if it has sufficient, immediately available combat power to overwhelm the enemy and the attack will not detract from mission accomplishment. Such attacks are usually necessary to overcome enemy attempts to slow the movement of the security force. If this initial attack fails to defeat enemy defenses, the security force commander must consider other options, such as making a more deliberate attack or assuming the defense while continuing to find out as much as possible about the enemy's positions.

4-49. The security force may bypass the enemy if it does not have sufficient combat power or an attack would jeopardize mission accomplishment. It must request permission to bypass an enemy force unless the operations order provides bypass criteria. The security force commander must report bypassed enemy forces to the next higher headquarters, which then assumes responsibility for their destruction or containment. Alternatively, the security force could keep a minimum force in contact with the bypassed enemy so that he cannot move freely around the battlefield. (See Appendix B for a discussion of bypass as a tactical task.)

4-50. If the security force cannot conduct either a hasty attack or a bypass, it attempts to establish a defense (see Chapter 8). In the defense, the security force maintains enemy contact, continues to perform reconnaissance, and prepares to support other forces. When the security force commander decides to defend, responsibility for further action rests with his higher commander. In the event other COAs would lead to decisive engagements or destruction, the security force conducts those activities necessary to assure self-preservation, such as delay or withdrawal (see Chapter 11), but maintains enemy contact unless the higher commander orders otherwise.

Recommend a Course of Action to the Higher Commander

4-51. Once the security force commander selects a COA keeping in mind his commander's intent, he reports it to his higher commander, who has the option of disapproving it based on its impact on his mission. To avoid delay, unit SOP may provide automatic approval of certain actions. If the higher commander assumes responsibility for continuing to develop the situation, the security force supports his actions as ordered. The higher commander must be careful to avoid becoming overly focused on initial security fights to the determent of operations directed against the enemy main body.

DISRUPT THE ENEMY

4-52. Once contact is made, the main body commander brings overwhelming fires onto the enemy to prevent him from conducting either a spoiling attack or organizing a coherent defense. The security force commander maneuvers as quickly as possible to find gaps in the enemy's defenses. The commander uses his ISR assets to gain as much information as possible about the enemy's dispositions, strengths, capabilities, and intentions. As more intelligence becomes available, the main body commander attacks to destroy or disrupt enemy command and control (C2) centers, fire control nodes, and communication nets. The main body commander conducts operations to prevent enemy reserves from moving to counter his actions.

FIX THE ENEMY

4-53. The commander tries to initiate maneuver at a tempo the enemy cannot match, since success in a meeting engagement depends on effective actions on contact. The security force commander does not allow the enemy to maneuver against the main body. The organization, size, and combat power of the security force are the major factors that determine the size of the enemy force it can defeat without deploying the main body.

4-54. The commander uses his aerial maneuver and fire support assets—including offensive information operations—to fix the enemy in his current positions by directly attacking his combat and command systems and emplacing situational obstacles. The priorities are typically to attack enemy forces in contact, C2 and fire control facilities, fire support assets, and moving enemy forces not yet in contact, such as follow-on forces and reserves. These priorities vary with the factors of METT-TC. Attack helicopters and close air support fixed-wing aircraft working in joint air attack teams (JAAT) are ideally suited to engage the enemy throughout the depth of his area of operations.

4-55. The techniques a commander employs to fix the enemy when both forces are moving are different than those employed when the enemy force is stationary during the meeting engagement. In both situations, when the security force cannot overrun the enemy by conducting a hasty frontal attack, he must deploy a portion of the main body. When this occurs the unit is no longer conducting a movement to contact but an attack.

MANEUVER

4-56. If the security force cannot overrun the enemy with a frontal attack, the commander quickly maneuvers his main body to conduct a penetration or an envelopment. (See Chapter 5 for a discussion of attack.) He does this to overwhelm the enemy force before it can react effectively or reinforce. The commander attempts to defeat the enemy in detail while still maintaining the momentum of his advance. After a successful attack, the main body commander resumes the movement to contact. If he did not defeat the enemy he has three main options: bypass, transition to a more deliberate attack, or conduct some type of defense. In all cases, he makes every effort to retain the initiative and prevent the enemy from stabilizing the situation by conducting violent and resolute attacks. Simultaneously he must maintain his momentum by synchronizing the actions of his combat, combat support, and combat service support elements.

4-57. Main body elements deploy rapidly to the vicinity of the contact if the commander initiates a frontal attack. Commanders of maneuvering units coordinate forward passage through friendly forces in contact as required. The intent is to deliver the assault before the enemy can deploy or reinforce his engaged forces. The commander may order an attack from a march column for one of the main body's columns, while the rest of the main body deploys. The commander can also wait to attack until he can bring the bulk of the main body forward. He avoids piecemeal commitment except when rapidity of action is essential and combat superiority at the vital point is present and can be maintained throughout the attack, or when compartmentalized terrain forces such a COA.

4-58. When trying to conduct an envelopment, the commander focuses on attacking the enemy's flanks and rear before he can prepare to counter these actions. The commander uses the security force to fix the enemy while the main body maneuvers to look for an assailable flank or he uses the main body to fix the enemy while the security force finds the assailable flank.

FOLLOW THROUGH

4-59. If the enemy is defeated, the unit transitions back into an MTC and continues to advance. The movement to contact terminates when the unit reaches the final objective or limit of advance, or it must transition to a more deliberate attack, a defense, or retrograde. (For more discussion of these types of operations, see the respective chapters in this manual.)

SEARCH AND ATTACK

4-60. *Search and attack* is a technique for conducting a movement to contact that shares many of the characteristics of an area security mission (FM 3-0). A commander employs this form of a movement to contact, conducted primarily by light forces and often supported by heavy forces, when the enemy is operating as small, dispersed elements, or when the task is to deny the enemy the ability to move within a given area. The battalion is the echelon that normally conducts a search and attack. A brigade will assist its subordinate battalions by ensuring the availability of indirect fires and other support.

ORGANIZATION OF FORCES

4-61. The commander task organizes his unit into reconnaissance, fixing, and finishing forces, each with a specific purpose and task. The size of the reconnaissance force is based on the available intelligence about the size of enemy forces in the AO. The less known about the situation, the larger the reconnaissance force. The reconnaissance force typically consists of scout, infantry, aviation, and electronic warfare assets. The fixing force must have enough combat power to isolate the enemy once the reconnaissance force finds him. The finishing force must have enough combat power to defeat those enemy forces expected to be located within the AO. The commander can direct each subordinate unit to retain a finishing force, or he can retain the finishing force at his echelon. The commander may rotate his subordinate elements through the reconnaissance, fixing, and finishing roles. However, rotating roles may require a change in task organization and additional time for training and rehearsal.

CONTROL MEASURES

4-62. The commander establishes control measures that allow for decentralized actions and small-unit initiative to the greatest extent possible. The minimum control measures for a search and attack are an AO, target reference points (TRPs), objectives, checkpoints, and contact points. (See Figure 4-5.) The use of TRPs facilitates responsive fire support once the reconnaissance force makes contact with the enemy. The commander uses objectives and checkpoints to guide the movement of subordinate elements. Coordination points indicate a specific location for coordinating fires and movement between adjacent units. The commander uses other control measures, such as phase lines, as necessary. (See Chapters 2 and 3 for definitions of available control measures.)

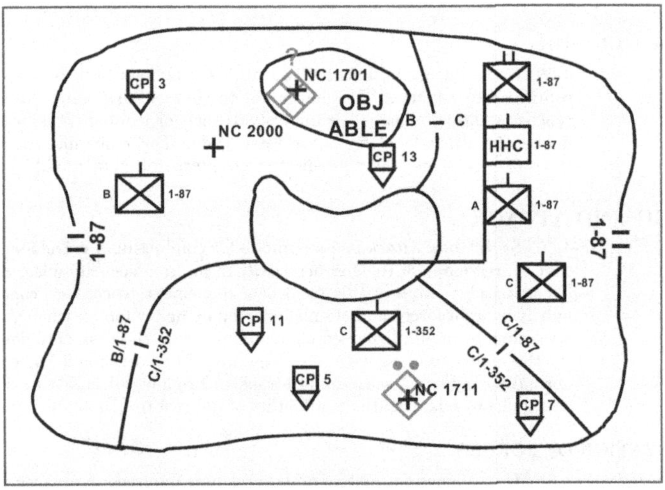

Figure 4-5. Search and Attack Control Measures

PLANNING A SEARCH AND ATTACK

4-63. A commander conducts a search and attack for one or more of the following purposes:

- Destroy the enemy: render enemy units in the AO combat-ineffective.
- Deny the area: prevent the enemy from operating unhindered in a given area; for example, in any area he is using for a base camp or for logistics support.
- Protect the force: prevent the enemy from massing to disrupt or destroy friendly military or civilian operations, equipment, property, and key facilities.
- Collect information: gain information about the enemy and the terrain to confirm the enemy COA predicted as a result of the IPB process.

4-64. The products of the IPB process are critical to conducting a search and attack. They focus the force's reconnaissance efforts on likely enemy locations.

4-65. The search and attack plan places the finishing force, as the decisive operation, where it can best maneuver to destroy enemy forces or essential facilities once located by reconnaissance assets. Typically, the finishing force occupies a central location in the AO. However, the factors of METT-TC may allow the commander to position the finishing force outside the search and attack area. The commander weights this decisive operation by using priority of fires and assigning priorities of support to his other available combat multipliers, such as engineer elements and helicopter lift support. The commander establishes control measures as necessary to consolidate units and concentrate the combat power of the force before the attack. Once the reconnaissance force locates the enemy, the fixing and finishing forces can fix and destroy him. The commander also develops a contingency plan in the event that the reconnaissance force is compromised.

4-66. Fire support plans must provide for flexible and rapidly delivered fires to achieve the commander's desired effects throughout the AO. The commander positions his fire support assets so they can support subordinate elements throughout the AO. The commander must establish procedures for rapidly clearing fires. To clear fires rapidly, command posts and small-unit commanders must track and report the locations of all subordinate elements. Because of the uncertain enemy situation, the commander is careful to assign clear fire-support relationships.

EXECUTING A SEARCH AND ATTACK

4-67. Each subordinate element operating in its own AO is tasked to destroy the enemy within its capability. The commander should have in place previously established control measures and communications means between any closing elements to prevent fratricide. The reconnaissance force conducts a zone reconnaissance to reconnoiter identified named areas of interest (NAIs).

4-68. Once the reconnaissance force finds the enemy force, the fixing force develops the situation, then executes one of two options based on the commander's guidance and the factors of METT-TC. The first option is to block identified routes that the detected enemy can use to escape or rush reinforcement over. The fixing force maintains contact with the enemy and positions its forces to isolate and fix him before the finishing force attacks. The second option is to conduct an attack to fix the enemy in his current positions until the finishing force arrives. The fixing force attacks if that action meets the commander's intent and it can generate sufficient combat power against the detected enemy. Depending on the enemy's mobility and the likelihood of the reconnaissance force being compromised, the commander may need to position his fixing force before his reconnaissance force enters the AO.

4-69. Brigades (and possibly battalions) may establish fire-support bases as part of the operations of their fixing force to provide fire-support coverage throughout the area of operations during search and attack operations conducted in restricted terrain. These positions should be mutually supporting and prepared for all-around defense. They are located in positions that facilitate aerial resupply. The development of these positions depends on the factors of METT-TC because their establishment requires diverting combat power to ensure protecting fire support and other assets located within such bases.

4-70. If conditions are not right to use the finishing force to attack the detected enemy, the reconnaissance or the fixing force can continue to conduct reconnaissance and surveillance activities to further develop the situation. Whenever this occurs, the force maintaining surveillance must be careful to avoid detection and possible enemy ambushes.

4-71. The finishing force may move behind the reconnaissance and fixing forces, or it may locate at a pickup zone and air assault into a landing zone near the enemy once he is located. The finishing force must be responsive enough to engage the enemy before he can break contact with the reconnaissance force or the fixing force. The echelon intelligence officer provides the commander with an estimate of the time it will take the enemy to displace from his detected location. The commander provides additional mobility assets so the finishing force can respond within that timeframe.

4-72. The commander uses his finishing force to destroy the detected and fixed enemy during a search and attack by conducting hasty or deliberate attacks, maneuvering to block enemy escape routes while another unit conducts the attack, or employing indirect fire or close air support to destroy the enemy. The commander may have his finishing force establish an area ambush and use his reconnaissance and fixing forces to drive the enemy into the ambushes.

Chapter 5

Attack

I approve of all methods of attacking provided they are directed at the point where the enemy's army is weakest and where the terrain favors them the least.

Frederick the Great

An *attack* is an offensive operation that destroys or defeats enemy forces, seizes and secures terrain, or both (FM 3-0). When the commander decides to attack or the opportunity to attack occurs during combat operations, the execution of that attack must mass the effects of overwhelming combat power against selected portions of the enemy force with a tempo and intensity that cannot be matched by the enemy. The resulting combat should not be a contest between near equals. The attacker must be determined to seek decision on the ground of his choosing through the deliberate synchronization and employment of his combined arms team.

5-1. Attacks take place along a continuum defined at one end by fragmentary orders that direct the execution of rapidly executed battle drills by forces immediately available. Published, detailed orders with multiple branches and sequels; detailed knowledge of all aspects of enemy dispositions; a force that has been task organized specifically for the operation; and the conduct of extensive rehearsals define the other end of the continuum. Most attacks fall between the ends of the continuum as opposed to either extreme. (Chapter 1 discusses this continuum between hasty and deliberate operations.)

ORGANIZATION OF FORCES

5-2. Once a commander determines his scheme of maneuver, he task organizes his force to give each unit enough combat power to accomplish its

CONTENTS

mission. He normally organizes into a security force, a main body, a reserve, and a sustainment organization. He should complete any changes in task organization in time to allow units to conduct rehearsals with their attached or supported unit. The best place and time for an attacking force to task organize is when it is in an assembly area.

SECURITY FORCES

5-3. Under normal circumstances, a commander resources dedicated security forces during an attack only if the attack uncovers one or more flanks or the rear of the attacking force as it advances. In this case, the commander designates a flank or rear security force and assigns it a guard or screen mission, depending on the factors of METT-TC. Normally an attacking unit does not need extensive forward security forces; most attacks are launched from positions in contact with the enemy, which reduces the usefulness of a separate forward security force. The exception occurs when the attacking unit is transitioning from the defense to an attack and had previously established a security area as part of the defense.

MAIN BODY

5-4. The commander organizes his main body into combined arms formations to conduct his decisive operation and necessary shaping operations. The commander aims his decisive operation toward the immediate and decisive destruction of the enemy force, its will to resist, seizure of a terrain objective, or the defeat of the enemy's plan. His maneuver scheme identifies the focus of the decisive operation. All of the force's available resources operate in concert to assure the success of the decisive operation. The subordinate unit or units designated to conduct the decisive operation can change during the course of the attack. The commander designates an assault, breach, and support force if he expects to conduct a breach operation during his attack.

5-5. If it is impractical to determine initially when or where the echelon's decisive operation will be, such as during a hasty attack, the commander retains flexibility by arranging his forces in depth, holding out strong reserves, and maintaining centralized control of his long-range fire support systems. As soon as the tactical situation clarifies enough to allow the commander to designate his decisive operation, he focuses his resources to support that decisive operation's achievement of its objective. Enemy actions, minor changes in the situation, or the lack of success by other elements cannot be allowed to divert either forces or their effects from the decisive operation.

5-6. The commander may need to designate a unit or units to conduct shaping operations to create windows of opportunity for executing his decisive operation. He allocates the unit or units assigned to conduct shaping operations the minimal combat power necessary to accomplish the missions since he cannot employ overwhelming combat power everywhere. Units conducting shaping operations usually have a wider area of operations (AO) than those conducting a decisive operation. If the commander has sufficient forces as part of his shaping operations, he can assign the tasks of follow and assume or follow and support to subordinate units. (Appendix B defines these two tactical mission tasks.)

RESERVE

5-7. The commander uses his reserve to exploit success, defeat enemy counterattacks, or restore momentum to a stalled attack. Once committed, the reserve's actions normally become or reinforce the echelon's decisive operation, and the commander makes every effort to reconstitute another reserve from units made available by the revised situation. Often a commander's most difficult and important decision concerns the time, place, and circumstances for committing the reserve. The reserve is not a committed force, it is not used as a follow and support force or a follow and assume force.

5-8. In the attack, the combat power allocated to the reserve depends primarily on the level of uncertainty about the enemy, especially the strength of any expected enemy counterattacks. The commander only needs to resource a small reserve to respond to unanticipated enemy reactions when he has detailed information about the enemy. When the situation is relatively clear and enemy capabilities are limited, the reserve may consist of a small fraction of the command. When the situation is vague, the reserve may initially contain the majority of the commander's combat power.

5-9. In addition, the strength and composition of the reserve vary with the reserve's contemplated missions, the forces available, the form of offensive maneuver selected, the terrain, and acceptable risk. For example, in a hasty attack the reserve can contain up to one-third of the force's combat power. Alternatively, in a deliberate attack the reserve is normally sized to defeat the enemy's counterattack forces. The commander should not constitute his reserve by weakening his decisive operation. A reserve must have mobility equal to or greater than the most dangerous enemy ground threat, and be able to fight that threat.

5-10. In an attack the commander generally locates his reserve to the rear of the unit making his decisive operation in a location that provides maximum protection from hostile observation and fire. However, it must be able to move quickly to areas where it is needed in different contingencies. This is most likely to occur if the enemy has strong counterattack forces. For heavy reserve forces, the key factor is cross-country mobility or road networks. For light forces, the key factor is the road network if trucks are available, or the availability of pickup zones (PZs) for air assault forces. The commander prioritizes the positioning of his reserve to reinforce the success of the decisive operation first, then to counter the worst-case enemy counterattack.

SUSTAINMENT ORGANIZATION

5-11. The commander resources his sustaining operations to support the attacking force. A battalion commander organizes his combat service support and other logistics assets into combat and field trains. Higher echelon commanders appoint someone to control sustaining operations within their echelon rear areas. In an attack, the commander tries to position his CSS units well forward. From these forward locations they can sustain the attacking force, providing priority of support to the units conducting the decisive operation. As the attacking force advances, CSS units displace forward as required to shorten the supply lines, using different displacement techniques to ensure uninterrupted support to maneuver units. The size of the force a commander devotes to rear area security depends on the threat in the attacking force's rear area. A significant enemy threat requires the commander to resource a tactical combat force. (Chapter 12 addresses area security operations in more detail.)

CONTROL MEASURES

5-12. Units conducting offensive operations are assigned an AO within which to operate. Within the AO the commander normally designates the following control measures regardless of whether he operates in a contiguous or non-contiguous environment:

- Areas of operations for subordinate units of battalion size or larger.
- Phase line as the line of departure (LD), which may also be the line of contact (LC).
- Time to initiate the operation.
- Objective.

If necessary, a commander can use either an axis of advance or a direction of attack to further control his maneuver forces. (Figure 5-1 depicts the minimum control measures for an attack.)

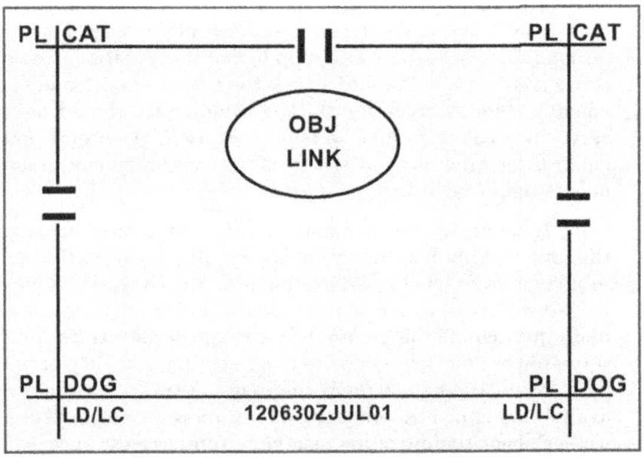

Figure 5-1. Minimum Attack Control Measures

5-13. A commander can use any other control measures necessary to control his attack. Short of the LD/LC, the commander may designate assembly areas and attack positions where the unit prepares for offensive operations or waits for the establishment of the required conditions to initiate the attack. Beyond the LD/LC he may designate checkpoints, phase lines (PLs), probable line of deployment (PLD), assault positions, and direct and indirect fire support coordinating measures. Between the PLD and the objective he can use a final coordination line (FCL), assault positions, support-by-fire and attack-by-fire positions, and time of assault to further control the final stage of the attack. Beyond the objective he can impose a limit of advance (LOA) if he does not want the unit to conduct an exploitation or a pursuit. (Chapters 2 and 3 discuss these control measures. They describe how a commander can use attack positions, axis of advance, combat formations, direction of attack, limit of advance, a LD, objectives, PLD, and a time of attack to help control the operation. Chapter 3 describes the use of AOs, assembly areas, line of contact, phase lines, and common direct and indirect fire coordinating measures.)

5-14. In an attack during limited-visibility conditions, the commander wants positive control over the movement of all attacking elements. He typically imposes additional control measures beyond those he would use in a daylight attack. These additional measures may include using a point of departure (PD) and a direction of attack.

PLANNING AN ATTACK

5-15. In an attack, friendly forces seek to place the enemy in a position where he can easily be defeated or destroyed. The commander seeks to keep the enemy off-balance while continually reducing the enemy's options. In an attack the commander focuses the maneuver effects, supported by the other battlefield operating systems (BOS), on those enemy forces that prevent him from accomplishing his mission and seizing his objective. Planning helps a commander synchronize the effects of his combat power through the military decision making process outlined in FM 5-0 and troop leading procedures outlined in FM 6-0.

5-16. This section addresses those considerations unique to an attack by BOS. There are no unique air defense, mobility/countermobility/survivability, or CSS BOS planning considerations that apply only to the attack. Those BOS planning considerations discussed in Chapter 3 continue to apply.

5-17. ***Fire superiority* is that degree of dominance in the fires of one force over another that permits that force to conduct maneuver at a given time and place without prohibitive interference by the enemy.** The commander plans to focus the effects of friendly systems to achieve fire superiority and allow friendly maneuver forces to breach the enemy's defensive network. The force must gain and maintain fire superiority at critical points during the attack. Having fire superiority allows the commander to maneuver his forces without prohibitive losses. The commander gains fire superiority by using a variety of tactics, techniques, and procedures. This includes using counterfires and precision fires, suppressing enemy positions, and destroying key facilities and assets. Achieving fire superiority requires the commander to take advantage of—

- The range and lethality of available weapon systems.
- Offensive information operations to prevent the enemy commander from synchronizing the effects of his available combat power.
- Movement to place the enemy in a position of disadvantage where his weapons can be destroyed, one or more at a time, with little risk to friendly weapon systems.

INTELLIGENCE

5-18. To employ the proper capabilities and tactics, the commander must have detailed knowledge of the enemy's organization, equipment, and tactics.

He must understand the enemy's strengths and weaknesses. Ideally, this knowledge is available during the military decision making process. The commander and his staff develop enemy situational and weapons templates based on analysis of all available combat information and intelligence data. These templates help to determine the feasibility of available courses of action (COAs) designed to achieve a position of advantage.

5-19. Before the attack a unit conducts ISR operations to ascertain those information requirements addressed in paragraph 3-111. Other ISR requirements can include—

- The location and depth of enemy reserves.
- The location and extent of contaminated areas.
- The location and extent of obstacles, possible breach sites, and enemy engagement areas.
- The location of areas where attacking units could become disoriented, such as rough or restrictive terrain.
- The most favorable routes of approach to the attack objective.
- Areas that the attacker can use for flanking fire and maneuver, such as support-by-fire and attack-by-fire positions.
- Suitability of planned friendly assault, support, artillery, and CSS positions.
- Enemy deception operations.

Commanders and leaders at all echelons personally participate in this process.

5-20. The commander takes every opportunity to gain and refine combat information regarding the enemy. He uses his available ISR assets to gather combat information and process it into intelligence. Information gathered during the planning phase of the plan, prepare, and execute cycle is especially useful in determining the viability of each COA developed. Generally, if a commander does not have good intelligence and, therefore, does not know where the overwhelming majority of the enemy's units and systems are located, he cannot conduct a deliberate attack. He must conduct a movement to contact, conduct a hasty attack, or collect more combat information.

5-21. The two fundamental employment techniques for reconnaissance in the attack are: reconnaissance-pull and reconnaissance-push. In reconnaissance-pull, the reconnaissance objective is to find weaknesses in enemy dispositions that can be exploited by the main force. Reconnaissance is launched over a broad area that allows the reconnaissance elements to identify enemy weaknesses to exploit and enemy strengths to avoid. Once these are identified, the commander exploits the situation by choosing a COA that allows his decisive operation to attack enemy weaknesses and penetrate gaps in the enemy's defense. The commander can then commit forces to widen the gap and envelop the enemy. The reconnaissance elements continue to move, seeking paths of least resistance and pulling the main body deep into the enemy's rear.

5-22. In reconnaissance-push, the reconnaissance objective is to identify the obstacles and enemy forces the attack forces must overcome to assault the objective in a previously chosen location in accordance with a COA selected prior to the reconnaissance. Once friendly reconnaissance elements gain contact with the enemy, they develop the situation within their capabilities. If the objective is an enemy force, the reconnaissance element orients on it to maintain contact and determine as much as possible about its dispositions.

5-23. The commander ensures that reconnaissance and surveillance of the enemy's defensive positions and any terrain critical to the scheme of maneuver continue throughout the attack. If the enemy attempts to modify his defenses, those actions will be detected. In turn, this allows the commander to adjust his scheme of maneuver as the enemy situation becomes clearer. The commander can use human and technological means, acting separately or in combination, to provide the required degree of reconnaissance and surveillance.

5-24. A commander's organic capability to gain information about the enemy and the AO's environment varies by echelon. At the corps echelon these assets include a military intelligence (MI) brigade, an armored cavalry regiment (ACR), and parts of almost every major subordinate command within the corps. Even a company conducts reconnaissance patrols. (For more information on reconnaissance operations, see FM 3-55.)

MANEUVER

5-25. In his plan of attack, the commander seeks to surprise his enemy by choosing an unexpected direction, time, type, or strength for the attack and by exploiting the success of military deception operations. Surprise delays enemy reactions, overloads and confuses enemy C2, induces psychological shock in the enemy, and reduces the coherence of the enemy defense. The commander achieves tactical surprise by attacking in bad weather and over seemingly impassible terrain, conducting feints and demonstrations, maintaining a high tempo, destroying enemy forces, and employing sound operations security (OPSEC). He may plan different attack times for his decisive and shaping operations to mislead the enemy and allow the shifting of supporting fires to successive attacking echelons. However, simultaneous attacks provide a means to maximize the effects of mass in the initial assault. They also prevent the enemy from concentrating his fires against successive attacks.

5-26. In planning the commander and subordinate leaders focus on the routes, formations, and navigational aids they will use to traverse the ground from the LD or PD to the objective. Some terrain locations may require the attacking unit to change its combat formation, direction of movement, or movement technique when it reaches those locations. The unit can post guides at these critical locations to ensure maintaining control over the movement.

5-27. The commander attacks targets throughout the depth of the enemy's defense to keep him off balance and limit his freedom of action. However, at the point of the decisive operation, the commander wants to concentrate the effects of overwhelming combat power against the enemy to shatter the cohesion of his defense. The commander accomplishes this by applying combat power against the enemy at a level of violence and in a manner that he cannot match. For example, the commander could concentrate a tank-heavy battalion task force's shock action and firepower against one enemy rifle platoon's hastily prepared defensive position. (Field Manual 3-0 discusses symmetric and asymmetric attack in more detail.)

5-28. Another aspect of concentration is the ability to rapidly concentrate force effects such as fires and offensive information operations during movement. This is especially critical when crossing linear obstacles. Each subordinate element tends to move out independently when it completes passage through a choke point. This independent movement detracts from the ability of the whole force to rapidly generate combat power on the far side of the obstacle.

Daylight Attacks

5-29. Daylight attacks allow friendly forces to effectively use their equipment while facilitating command and control (C2). They are the least stressful psychologically and physically on the attacking units. One major disadvantage is that the enemy can effectively use his systems to oppose the attack. Another disadvantage is that it does not take advantage of the Army's superior thermal viewer capabilities.

Limited-Visibility Attacks

5-30. The factors of METT-TC normally require an attack conducted during limited visibility to be more deliberate in nature, except when it occurs as part of the follow-up to a daylight attack or as part of an exploitation or pursuit operation. The commander planning a night attack considers how limited visibility complicates controlling units, soldiers, and fires; identifying and engaging targets; navigating and moving without detection; locating, treating, and evacuating casualties; and locating and bypassing or breaching obstacles.

5-31. Commanders attack in limited-visibility conditions to take advantage of American night-vision and navigational superiority against most potential enemy ground forces. Intensively trained forces equipped for such combat have significant advantages over an enemy who is unprepared for limited-visibility combat. When the friendly force's limited-visibility operations capabilities are significantly greater than the enemy's, limited-visibility attacks may be the norm. Table 5-1 on page 5-8 outlines the advantages and disadvantages of conducting limited-visibility attacks.

5-32. Highly trained units equipped with modern night-vision devices conduct limited-visibility attacks in a manner similar to the way they conduct daylight attacks. Units without extensive night-vision devices can use the darkness to their advantage to conceal their movement, allowing them to get as close to the enemy positions as possible if the enemy also does not have extensive night-vision capabilities. Troops that are well trained for limited-visibility operations and take full advantage of the superiority of their night-vision equipment gain significant tactical and psychological advantages when attacking the enemy at night or in other conditions of reduced visibility. The commander should understand the different night-vision capabilities of all elements participating in the attack, to include the enemy's night-vision capabilities, and make any adjustments necessary to his plan based on these differences. The commander should take advantage of his superior night-fighting capabilities whenever possible.

Table 5-1. Advantages and Disadvantages of Limited-Visibility Attacks

ADVANTAGES OF LIMITED-VISIBILITY ATTACKS	DISADVANTAGES OF LIMITED-VISIBILITY ATTACKS
• Defenses are more susceptible to infiltration. • Darkness can conceal the movement of large forces. • Physical and psychological factors favor the attacker, as shock, disorientation, and isolation are easier to achieve. • Air assets can operate more safely because air defenders with only optical sights have greater difficulty acquiring targets at night. • The element of surprise may increase because defenders are more susceptible to deception techniques, such as dummy lights, noise, smoke, and fires. • The defender cannot employ his reserves as quickly at night as he can during daylight conditions.	• Command and control is more difficult. • The defender can react easier to changing situations. • The attacker has difficulty determining the limits of obstacle systems. • Restrictive terrain is more difficult to traverse. • Light, smoke, noise, and fires can deceive the attacker. • The attacker loses momentum because he attacks at a reduced speed to maintain the coherence of his unit. • Land navigation, without GPS, is more difficult at night; units may become separated, cohesion can be lost, and support elements can move to the wrong positions. • The enemy can reposition or emplace obstacles during darkness without being detected by friendly reconnaissance, surveillance, and intelligence assets. • Attacking units are easier to ambush at night. • Adjusting indirect fire is difficult, even with night-vision devices or illumination. • Units require significantly larger quantities of signal ammunition such as smoke, tracers, flares, and illumination rounds. • The task of locating and evacuating casualties is more difficult to execute. • The risk of fratricide may increase.

5-33. The basic organization of forces for a limited-visibility or night attack is the same as for any other attack. However, changing an existing task organization under limited-visibility conditions requires much more time and effort than it does during daylight. Small tactical organizations, such as combat crews and infantry squads, should be resourced as close as possible to full strength even if it means reducing the total number of these small tactical groups.

5-34. The presence or lack of illumination characterizes the conduct of limited-visibility attacks. Nonilluminated attacks offer the best chance of gaining surprise. Illumination, however, is normally planned for every limited-visibility attack so that it can be readily available if required. The commander can choose to conduct a nonilluminated attack until his forces make contact with the enemy. At that point, he can illuminate the objective. The enemy can also choose to employ illumination to increase the effectiveness of his defensive efforts. Units generally conduct nonilluminated attacks although they always plan for illumination. All leaders within the attacking unit must understand the time, conditions, and authority required to employ illumination.

5-35. Illuminated, supported attacks are almost like daylight attacks. They are most effective when speed is essential, time for reconnaissance is limited, or the enemy is weak and disorganized. If the commander employs illumination, it should continue until the force secures the objective. The commander should place the illumination beyond the objective to silhouette objects on the objective. This helps the assaulting force see and fire at withdrawing or counterattacking enemy forces. The commander may also employ illumination in several locations to confuse the enemy about the exact place of attack.

5-36. The commander plans for limited-visibility operations in the same manner that he does for daylight operations, with emphasis on—

- Keeping the plan simple.
- Taking additional time for reconnaissance.
- Taking advantage of easily identifiable terrain features, such as roads and railroad tracks, when establishing control measures.
- Using intermediate objectives as necessary to control and maintain the correct movement direction during the attack.
- Concealing preparations.
- Scheduling initial rehearsals during daylight, with the final rehearsal at night.

5-37. To simplify control problems, the commander may weight his support element over the assault force to reduce the number of friendly soldiers moving on the objective in the darkness. Developing a plan that does not require the unit to change its movement azimuth after it crosses the LD/PD helps to simplify the plan execution.

5-38. The commander must assume that the enemy possesses, in at least limited quantities, the same limited-visibility observation capabilities as his own forces—absent positive information to the contrary—when conducting a limited-visibility attack. Using terrain to mask movement and deployment remains critical because limited visibility may create a false sense of protection from enemy observation. During movement, leaders reduce the distances between vehicles or individual soldiers as necessary to allow one system or soldier to observe the other. This decreases the time necessary to react to enemy contact. The attacking force wants to maintain its momentum; therefore, it does not preserve the alignment of units within the selected combat formation at the expense of additional time. However, it must adhere more closely to the plan of attack than under daylight conditions.

FIRE SUPPORT

5-39. The planning process synchronizes the unit's maneuver with the provision of fire support. It must identify critical times and places where the commander needs the maximum effects from his fire-support assets. The commander combines his maneuver with fires to mass effects, achieve surprise, destroy enemy forces, and obtain decisive results. His guidance gives specified attack criteria for fire support assets, thus focusing the planning and execution efforts on those critical times and events. The specified attack criteria are a compilation of the commander's guidance, desired effects, and high-payoff targets and attack priorities. The amount of time available to plan the operation constrains the commander's ability to synchronize fire-support operations that employ well-matched effects of all available assets against high-payoff targets.

5-40. The goal of the commander's attack criteria is to focus fires on seizing the initiative. The commander emphasizes simple and rapidly integrated fire support plans. This is done using quick-fire planning techniques and good standing operating procedures (SOPs). The commander integrates his fire support assets as far forward as possible in the movement formation to facilitate early emplacement. Fires concentrate (mass) on forward enemy elements to enable maneuver efforts to close with the enemy positions. Fire support isolates forward enemy elements by using long-range fires, air support, and electronic warfare.

5-41. Fires facilitate his unit's maneuver by destroying or neutralizing strong enemy forces and positions. His fire support system must take full advantage of available preparation time to achieve these demanding effects criteria. Fire-support plans feature the following characteristics:

- Targets that are confirmed or denied by ISR efforts.
- Designation of target sensor-to-shooter communication links.
- Possible use of preparation and deception fires to shape the enemy's defense.
- Air support to destroy high-payoff targets on the objective and then shift to reinforcing enemy units, artillery assets, and C2 nodes.
- Proactive suppression of enemy air-defense effort.
- Preparation fires that shift just as the maneuver force arrives on the objective.
- Suppression and obscuration fire plan to support breaching operations.
- Pre-positioned ammunition backed by prepackaged munitions stocks capable of rapid delivery.
- Integration of nonlethal fires, such as electronic attack and PSYOPS, into the attack guidance matrix.
- Integration of primary and backup observers to engage high-priority targets.
- Fire support coordinating measures, accounting for danger close and other technical constraints, to allow maneuver forces to get as close as possible to the objective before lifting fires.
- Signals for lifting and shifting fires on the objective, primarily by combat net radio and by visual signals as a backup means.

These later fire support coordinating measures should also facilitate the massing of fires, including CAS, against high-payoff targets throughout the AO.

COMMAND AND CONTROL

5-42. The commander states the desired effect of fires on the enemy weapon systems, such as suppression or destruction, as part of his planning process. He assigns subordinate units their missions and imposes those control measures necessary to synchronize and maintain control over the operation.

5-43. Using the enemy situational and weapons templates previously developed, the commander determines his probable line of contact and enemy trigger lines. As he arrays his subordinate elements to shape the battlefield, he matches his weapon systems against the enemy's to determine his PLD. Once he determines his PLD, he establishes how long it takes him to move from the LD to the PLD and any support-by-fire positions the attack requires. He establishes when and where his force must maneuver into enemy direct-fire range.

5-44. In addition to accomplishing the mission, every attack plan must contain provisions for exploiting success or any advantages that may arise during the operation. The commander exploits success by aggressively executing the plan, promoting subordinate leader initiative, and using units that can rapidly execute battle drills.

PREPARING AN ATTACK

5-45. Even in fluid situations, attacks are best organized and coordinated in assembly areas. If the commander decides that rapid action is essential to retain a tactical advantage, he may opt not to use an assembly area. Detailed advance planning—combined with digital communications, SOP, and battle drills—may reduce negative impacts of such a decision.

5-46. Unless already in an assembly area, the attacking unit moves into one during the preparation phase. The unit moves with as much secrecy as possible, normally at night and along routes that prevent or degrade the enemy's capabilities to visually observe or otherwise detect the movement. It avoids congesting its assembly area and occupies it for the minimum possible time. While in the assembly area, each unit provides its own local ground security and air defense.

5-47. Units moving to assembly areas send out their quartering parties and link up with their guides at the designated locations. While subordinate units move to and occupy assembly areas, the commander completes the process of planning and coordinating the attack.

5-48. The attacking unit should continue its troop leading procedures and priorities of work to the extent the situation and mission allow prior to moving to attack positions. These preparations include but are not necessarily limited to—

- Protecting the force.
- Conducting task organization.
- Performing reconnaissance.
- Refining the plan.
- Briefing the troops.
- Conducting rehearsals, to include test firing of weapons.

- Moving logistics support forward.
- Promoting adequate rest for both leaders and soldiers.
- Positioning the force for subsequent action.

As part of troop leading procedures, leaders at all levels should conduct a personal reconnaissance of the actual terrain. If a limited-visibility attack is planned, they should also reconnoiter the terrain at night.

5-49. A thorough reconnaissance of the objective, its foreground, and other enemy positions is a critical part of attack preparations. The commander exploits all available ISR assets to provide the necessary information. This includes requesting JSTARS feeds of enemy movements from higher echelons or imagery of enemy obstacles. Reconnaissance forces infiltrate through the enemy security area to conduct an area reconnaissance. They can employ precision munitions and conventional indirect fires to destroy detected enemy outposts while remaining undetected. They locate and attempt to infiltrate the enemy's main defensive positions to confirm his dispositions. When properly task-organized, forces conducting reconnaissance may also be given a mission to conduct covert breaches in the enemy's obstacle complexes to facilitate rapid movement of the decisive or shaping operation.

5-50. During this phase, the commander positions his artillery target-acquisition radars to provide support throughout the AO. Divisions and corps establish quick-fire channels between sensors, such as counterbattery radars and firing units, to rapidly silence enemy indirect fire systems. These channels do not change command relationships or priority of fires.

5-51. The commander exercises and refines his maneuver and fire plans during rehearsals which are an important part of ensuring the plan's coordination and synchronization. As part of the rehearsal process, the commander and his subordinates review the anticipated battle sequence to ensure all units understand the plan, the relationship between fire and movement, and the synchronization of critical events. These critical events include:

- Moving from the assembly area to the line of departure.
- Maneuvering from the line of departure to the probable line of deployment.
- Occupying support-by-fire positions.
- Conducting the breach.
- Assaulting the objective.
- Consolidating on the objective.
- Exploiting success or pursuing a withdrawing enemy.
- Actions of echelon reserves.

The unit should conduct rehearsals under as many types of adverse conditions as possible with time and other restraints to identify and prepare the unit to cope with problems. At lower tactical echelons, the rehearsal includes battle drills, such as creating lanes through minefields.

5-52. From their assembly areas, attacking units move to their respective LDs. (See Figure 5-2.) Units move from assembly areas to the LD in the same way as for any other tactical movement. (Chapter 14 details troop movements.) The number of columns a unit employs in its movement depends on the availability of suitable routes and the friendly and enemy situation.

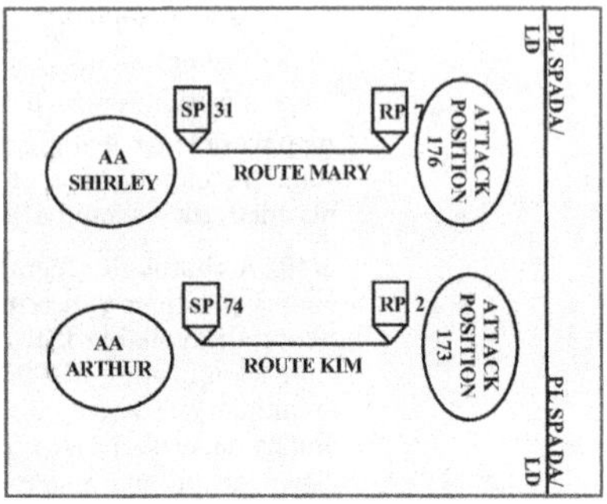

Figure 5-2. Movement from AA to LD

Primarily the tactical situation and the order in which the commander wants his subordinate units to arrive at their attack positions govern the march formation. Using an LD facilitates the simultaneous initiation of the attack at the prescribed time by all attacking units.

5-53. Light infantry units should move by tactical vehicles to the maximum extent possible to avoid prematurely exhausting their soldiers. However, light infantry forces should not travel too far forward in tactical vehicles. The enemy can detect the noise and other battlefield signatures associated with using tactical vehicles at a greater distance than he can detect dismounted infantry soldiers, and will probably respond to the presence of tactical vehicles with direct- and indirect-fire systems. The commander must weigh the need for security against the time required to conduct a foot march and its resulting effects on soldiers.

5-54. Units move rapidly through their attack positions and across the LD, which should be controlled by friendly forces. A unit uses its designated attack position only by exception, such as when it must refuel prior to crossing the LD to ensure sufficient fuel to reach the objective or the conditions required to ensure the success of the planned maneuver are not yet established. A unit does not occupy its attack positions for more than 10 to 15 minutes without initiating actions to protect itself and increase its survivability, such as deploying local security and camouflage nets and starting the construction of fighting and survivability positions. If necessary, a unit can use guides to assist in occupying the attack position. These guides may come from organic resources or from another unit.

5-55. For units attacking on foot using infiltration and stealth, a commander may designate a point of departure for the attacking units instead of an LD. Armor and mechanized infantry units normally use gaps or lanes through the friendly positions to allow them to deploy into combat formations before they cross the LD.

5-56. Preliminary operations for an attack may include using preparatory fires and the relief of units in contact by executing a relief in place or a forward passage of lines. The relief of units may be desirable to continue the momentum of the attack with fresh troops, change the direction of the attack, exploit a weakness in the enemy position with reserve forces, or initiate an offensive on a stabilized front. (Chapter 15 details a relief in place. Chapter 15 details a forward passage of lines.)

5-57. The commander uses available artillery, mortar, CAS, and offensive information operations to conduct preparatory fires. Preparatory fires are developed from the top down, with bottom-up refinement. The subordinate commander most affected by the effects of these preparatory fires must strongly emphasize the bottom-up refinement process. Preparatory fires can accomplish the following functions:

- Destroy the enemy.
- Suppress, neutralize, or disrupt high-value or high-priority targets.
- Gain fire superiority.
- Suppress the enemy in his defensive positions.
- Facilitate the attacking force's maneuver.
- Deceive the enemy.

5-58. If the attacking forces are in contact with the enemy's security zone, preparatory fires may initially destroy or disrupt only the enemy's reconnaissance and security forces and positions. In either case, counterfires and counterbattery fires conducted as part of preparatory fires serve to degrade the enemy's fire-support systems and assist in achieving fire superiority.

5-59. The commander ensures that his maneuver forces have the CS and CSS assets necessary to conduct the operation and continue the momentum of the attack as part of the preparation process. That support effort must anticipate future maneuvers to ensure the uninterrupted sustainment of the maneuver force.

EXECUTING AN ATTACK

5-60. A series of advances and assaults by attacking units until they secure the final objective characterizes the attack. Commanders at all levels must use their initiative to rapidly shift their main effort between units as necessary to take advantage of opportunities and momentum to ensure the enemy's rapid destruction. Attacking units move as quickly as possible, following reconnaissance elements or successful probes through gaps in the enemy's defenses. They shift their strength to reinforce success and carry the battle deep into the enemy's rear. A commander does not delay his attack to preserve the alignment of subordinate units or to adhere closely to the preconceived plan of attack. This manual discusses executing the attack in a five-step sequence:

- Gain and maintain enemy contact.
- Disrupt the enemy.
- Fix the enemy.
- Maneuver.
- Follow through.

This sequence is for discussion purposes only and is not the only way of conducting an attack. The reader should understand that these sequences overlap during the conduct of an attack.

5-61. The commander must avoid becoming so committed to the initial plan that he neglects opportunities. He is prepared to abandon failed attacks and to exploit any unanticipated successes or enemy errors by designating another unit to conduct his decisive operation in response to the changing situation.

5-62. When maneuvering his force, the commander strives to retain freedom of action while protecting his force. Although he may have a detailed plan to defeat the enemy, the commander continually seeks any opportunity to attack to defeat, destroy, or reduce the enemy's combat power or shatter his cohesion and will to fight. The commander avoids dogged adherence to a plan no longer appropriate to current battlefield conditions. The difference between success and failure in combat often depends on the commander's ability to make the plan fit existing circumstances rather than trying to make circumstances fit the plan.

GAIN AND MAINTAIN ENEMY CONTACT

5-63. Gaining and maintaining contact with the enemy when he is determined to break that contact is vital to the success of offensive operations. A defending enemy generally establishes a security area around his forces to make early contact with the attacking forces to determine their capabilities, intent, and chosen COA and to delay their approach. The enemy commander wants to use his security area to strip away friendly reconnaissance forces and hide his dispositions, capabilities, and intent. His goal is to compel the attacking force to conduct a movement to contact against his forces that know the exact location of the attacking forces.

5-64. A commander employs his combat power to overwhelm enemy forces in accordance with his situational understanding. However echelons below division do not normally have the detection, tracking, and weapon systems necessary to conduct decisive or shaping operations directed against enemy forces not currently committed to close combat. The manner in which a unit gains and maintains contact depends on whether the unit is in contact with the enemy's security area or the enemy's main line of resistance and the echelon of the unit in the nested layers of reconnaissance and security. For example, the intent of the corps' reconnaissance effort is to determine the dispositions, composition, direction of movement, and rate of movement of the enemy's significant forces. The corps' armored cavalry regiment, acting as a covering force or advance guard, can fight through a security area, develop the situation, confirm information provided by technical means, and force the enemy to reveal more information than could be acquired solely through using intelligence sensors. This additional information includes locating the enemy's reserve. At a lower level, a battalion constituting the advance guard of the main body can use its scout platoon to conduct a zone reconnaissance that focuses on acquiring updates of enemy positions and obstacles.

5-65. The commander's ability to sense the enemy's actions by gaining and maintaining contact with all significant parts of the enemy force, to include tracking enemy reserves, fire support, and follow-on forces, increases the security of the attacking force. The enemy's attempts to shift major elements of his forces or launch a counterattack will be detected. Additionally, by sending out a force to conduct area reconnaissance with an on-order mission to be prepared to conduct a security mission, the commander can prevent enemy reconnaissance assets from detecting the friendly force's major movements and increase the enemy's risk. The risks to the enemy force increase to the extent friendly forces impede or deny success to enemy ISR assets. Combining these factors results in providing the attacking commander with additional time to take advantage of the changing situation. Moving within the enemy's decision cycle allows the commander to take advantage of his successes by transitioning to the exploitation and pursuit to complete the destruction of the enemy.

5-66. The capabilities of digital C2 systems offer additional techniques a commander can use to gain and maintain enemy contact. The improved common operational picture provided by those systems enhances his situational understanding and ensures rapid, clear communication of orders and intent, thereby reducing the confusion, fog, and friction of battle. Advanced Army and joint intelligence systems feeding those C2 systems enable him to detect and track enemy forces throughout a given AO without having his forces make physical contact with the enemy. The commander's ability to see and understand the situation before the enemy can allows him to act first and maneuver out of contact with the enemy at a high tempo. This allows him to position his forces where they can overwhelm selected elements of the enemy force to disrupt and destroy the enemy's combined arms team. Such attacks—delivered simultaneously with precision by air, ground, and naval systems throughout the width, height, and depth of the battlefield—stun the enemy and rapidly lead to his defeat.

DISRUPT THE ENEMY

5-67. Disrupting one or more parts of the enemy's combined arms team weakens his entire force and allows the friendly commander to attack the remaining enemy force in an asymmetrical manner. The assessment and decisions regarding what to disrupt, when to disrupt, and to what end are critical. For example, the goal of disrupting the enemy's fire-support system is to allow friendly forces to maneuver and mass the effects of their weapon systems against the enemy without being engaged by the enemy's indirect-fire weapons. Attacking forces can accomplish this by attacking enemy forward observers, fire-direction centers, command posts, artillery and rocket systems, or their ammunition supply. Each set of targets requires a different amount of resources. The probability of success, the effectiveness of the attack, and the time necessary to achieve the desired target effects varies with each set of targets.

5-68. Once any type of contact—even sensor contact—is made with the enemy, the commander wants to use the element of surprise to conduct shaping operations that strike at the enemy and disrupt both the enemy's combined arms team and his ability to plan and control his forces. Once this disruption process begins, it continues throughout the attack. The commander uses any existing technological advantage over the enemy in the following areas to aid the disruption process:

- Offensive information operations.
- Lethal firepower effects.
- Range of direct-fire weapons.
- Protection.
- Battlefield mobility.
- Information management.
- C2 systems.

5-69. Whatever form of disruption takes place helps the commander seize, retain, and exploit the initiative; maintain his freedom of action; impose his will on the enemy; set the terms, and select the place for battle. That disruption also allows the commander to exploit enemy vulnerabilities and react to changing situations and unexpected developments more rapidly than the enemy. This disruption effort usually occurs at division level and above because lower echelons lack the necessary reconnaissance, target acquisition, intelligence analysis, and target attack assets to engage forces not committed to close combat.

5-70. The commander plans his shaping operations to occur at the place and time necessary to establish the conditions for his decisive operation. Targets of a shaping operation may include: enemy C2 facilities, ISR assets, fire-support systems, reserves, and logistics support. If a commander executes a shaping operation too early, the enemy has time to recover and respond before friendly forces conducting the decisive operation can complete their maneuver.

5-71. The commander plans to use harassment, suppressive, or interdiction fires against positions likely to contain high-payoff targets to disrupt enemy reactions to the attacking unit's advance. These fires deny the enemy unrestricted use of the terrain and can prevent his reserves from entering the fight before the unit seizes the objective. Additional benefits may result from these fires over time, including increased psychological pressure on enemy forces and a reduction in their mental and physical capabilities by disrupting their sleep and rest patterns.

5-72. Surprise denies the enemy the opportunity to focus and synchronize his combat power against the attacking force. It prevents the enemy from massing his forces or fires at a critical, possibly decisive, place and time. In place of cohesive resistance, surprise can produce confusion, fear, and piecemeal resistance. Factors that contribute to surprise include: the tempo and intensity in executing the attack plan and employing unexpected factors, such as selecting a less than optimal COA, varying operational tactics and methods, conducting deception operations, and ensuring OPSEC.

FIX THE ENEMY

5-73. A primary purpose in fixing the enemy is to isolate the objective of the force conducting the echelon's decisive operation to prevent the enemy from maneuvering to reinforce the unit targeted for destruction. Since war is a contest between thinking opponents, the enemy will oppose the friendly commander's attempts to fix his forces. Every friendly move causes the enemy to attempt to counter that move. The commander does everything in his power to limit the options available to his opponent. Fixing an enemy into a given position or a COA and controlling his movements limit his options and reduce the amount of uncertainty on the battlefield.

5-74. Reducing uncertainty allows the friendly force to use maneuver to mass the effects of overwhelming combat power against a portion of the enemy. It gives the commander more time to modify his plan as necessary and orchestrate the employment of his forces. It allows him to mass forces in one place by using economy of force measures in other areas. The commander may also try to fix an enemy unit, such as the enemy reserve or follow-on force, to prevent it from repositioning or maneuvering against the force conducting his decisive operation.

5-75. Fixing the enemy must be done with the minimum amount of force. The commander normally allocates the bulk of his combat power to the force conducting his decisive operation, so fixing operations are, by necessity, shaping operations that illustrate economy of force as a principle of war. Therefore, the commander must carefully consider which enemy elements to fix and target only those that can significantly affect the outcome of the fight. The longer the requirement to fix these forces, the more resources the commander needs to accomplish the mission. Generally, an enemy force only needs to be fixed until it cannot respond to the actions of the unit conducting the decisive operation in time to affect the outcome. This may require a commander to slow down the rate of march of an enemy unit to prevent it from influencing the outcome of the engagement or battle.

5-76. One method of isolating the objective is to conduct a shaping operation using lethal and nonlethal fires. Lethal fires may range from sniper fire to a joint fire plan designed to totally destroy a selected portion of the enemy force. Nonlethal fires, such as electronic jamming, can prevent the enemy from receiving orders or vital intelligence and combat information.

5-77. Severing enemy lines of communication over prolonged periods of time by using interdiction measures is another way to fix the enemy. These measures can range from air interdiction that destroys bridges and rail switching yards to ambushes conducted by infiltrating combat patrols.

5-78. Another method of fixing the enemy is to tie obstacles into the existing terrain to canalize and slow the movement of enemy reserves. At lower tactical echelons, scatterable minefields employed in accordance with the rules of engagement can seal the objectives from possible enemy reinforcement or counterattacks and neutralize enemy actions to the flanks. Deception operations and activities, such as demonstrations and false preparatory fires, can fix the enemy. Using extensive smoke screens and vehicle mock-ups in a deception effort can also assist in fixing an enemy force.

MANEUVER

5-79. The commander maneuvers his forces to gain positional advantage so he can seize, retain, and exploit the initiative. He avoids the enemy's defensive strength. He employs tactics that defeat the enemy by attacking through a point of relative weakness, such as a flank or the rear.

5-80. Offensive maneuver seeks to achieve a decisive massing of effects at the decisive point, or at several decisive points if adequate combat power is available. The commander exploits maneuver by—

- Taking maximum advantage of dead space and covered and concealed routes to close with the enemy.
- Using his advantages in the effective ranges of weapon systems.
- Repositioning friendly forces rapidly.
- Navigating accurately cross-country.
- Obtaining situational understanding of friendly and enemy locations.
- Taking effective security measures.

- Synchronizing the application of all BOS at a time and place on the battlefield to maximize their effects.

5-81. The key to success is to strike hard and fast, overwhelm a portion of the enemy force, and then quickly transition to the next objective or phase, thus maintaining the momentum of the attack without reducing the pressure. The commander must retain freedom of maneuver with multiple COAs throughout the operation and responsive CSS. Additionally, he must make every effort to locate and track enemy reserve and follow-on forces, which prevents friendly forces from being attacked unexpectedly by significant enemy forces. This allows the commander time to delay, disrupt, or destroy these enemy forces before they can interfere with the attack.

5-82. Depending on the conditions of METT-TC, artillery and mortars may advance with the attacking formation or move forward by bounds. The echelon fire support coordinators (FSCOORDs) position direct support and reinforcing artillery in coordination with their maneuver commanders. The force field artillery headquarters, normally a division or corps artillery headquarters, coordinates position areas for general support and general support-reinforcing artillery units through the fire support officers at corps, division, and brigade. The commander considers the maneuver of fire support assets along with maneuver forces to ensure that proper fire support is available at all times.

5-83. The maneuver process normally follows this sequence:

- Movement from the LD to the PLD.
- Actions at the PLD, assault position, or FCL.
- Breaching operations (discussed in FM 3-34.2).
- Actions on the objective.

The movement from the assembly area to the LD that precedes many attacks is troop movement and is discussed in Chapter 14.

Movement from the LD to the PLD

5-84. The unit transitions from troop movement to maneuver once it crosses the LD. It moves aggressively and as quickly as the terrain and enemy situation allow. It moves forward using appropriate movement techniques assisted by the fires of supporting units. Fire and movement are closely integrated and coordinated. Effective suppressive fires facilitate movement, and movement facilitates more effective fires. Whenever possible, the attacking unit uses avenues of approach that avoid strong enemy defensive positions, takes advantage of all available cover and concealment, and places the unit on the flanks and rear of the defending enemy. Where cover and concealment are not available, the unit uses obscurants to conceal its movement. Any delays in establishing obscuration and suppressive fires prior to crossing the PLD may require the attacking unit to occupy its assault positions.

5-85. Artillery and other fire-support assets move as necessary to ensure that the attacking unit remains within supporting range. Previously conducted analysis of the time it takes the maneuver unit to move from the LD to the PLD and the distances involved ensures that their fire support systems are prepared to provide fire support before maneuver units move inside the effective range of enemy direct-fire weapon systems. The existence of enemy artillery systems that have a longer range than fielded US artillery systems complicates this process. The commander uses fires delivered from fixed- and rotary-wing systems and the autonomous operation capabilities of modernized artillery systems to help counter any enemy range advantage.

5-86. If the commander expects to make enemy contact at or shortly beyond the LD, he deploys his unit so that he can maintain maximum firepower against the enemy's known positions. He chooses the combat formation that best balances firepower, tempo, security, and control in the specific situation. The commander has the option of deploying a security force in front of his attacking unit. He may also employ a flank or rear security force if required by the enemy situation. The commander may not want to change formations during his attack because of the potential loss of momentum resulting from such changes. If the commander finds it necessary to transition from one combat formation to another, he should base the transition on thoroughly trained drills. Once enemy contact is expected, he transitions to the bounding overwatch technique of movement. (Chapter 14 addresses movement techniques.)

5-87. Between the LD and the PLD, the attacking unit secures intermediate objectives only to eliminate enemy positions or bring additional suppressive fires to bear. Fire-support assets engage targets of opportunity. The commander uses CAS and artillery to destroy enemy security forces. As the unit approaches suspected enemy positions or danger areas, the commander directs his forces to occupy predesignated support-by-fire positions. Fire support, suppression, and obscuration are key enablers that allow a force to occupy these positions. Commanders use fires from these positions to suppress enemy forces while the unit continues its advance toward the objective.

5-88. The commander engages known enemy forces with the maximum possible combat power to overwhelm them as quickly as possible. The attacking unit that encounters small enemy units on the way to the objective either quickly overruns or bypasses them if they meet the bypass criteria. The attacking unit then reports the location of bypassed enemy elements to its higher headquarters and maintains contact until they can be handed off to follow and support forces. The commander uses minimal force to maintain that contact to avoid significantly weakening the force conducting his decisive operation.

Actions at the PLD, Assault Position, or FCL

5-89. The attacking unit maintains the pace of its advance as it approaches its PLD. (See Figure 5-3.) The attacking unit splits into one or more assault and support forces once it reaches the PLD if not previously completed. At the PLD infantry soldiers dismount from their infantry fighting vehicles as required by the situation. All forces supporting the assault force should be set in their support-by-fire positions before the assault force crosses the PLD. The commander synchronizes the occupation of these support-by-fire positions with the maneuver of the supported attacking unit to limit the vulnerability of the forces occupying these positions. The commander uses his unit's tactical SOP, prearranged signals, engagement areas (EAs), and target reference points (TRPs) to control the direct fires from these supporting positions. He employs restricted fire lines between converging forces.

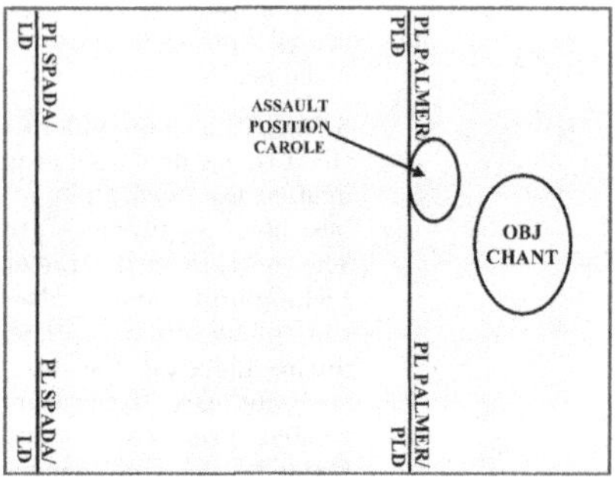

Figure 5-3. PLD and Assault Positions

5-90. The PLD can be co-located with the assault position. (See Figure 5-3.) The commander ensures that the final preparations of his breach force in an assault position do not delay its maneuver to the point of breach as soon as the conditions are set. Whenever possible, the assault force rapidly passes through the assault position. It may have to halt in the assault position while fires are lifted and shifted. In this case, if the enemy anticipates the assault, the assault force deploys into covered positions, screens its positions with smoke, and waits for the order to assault. As long as the assault force remains in the assault position, support forces continue their suppressive fires on the objective.

5-91. Once the support force sets the conditions, the breach force reduces, proofs, and marks the required number of lanes through the enemy's tactical obstacles to support the maneuver of the assault force. The commander must clearly identify the conditions that allow the breach force to proceed to avoid confusion. From the PLD, the assault force maneuvers against or around the enemy to take advantage of the support force's efforts to suppress the targeted enemy positions. The support force employs direct and indirect fires against the selected enemy positions to destroy, suppress, obscure, or neutralize enemy weapons and cover the assault force's movement. The assault force must closely follow these supporting fires to gain ground that offers positional advantage. This COA normally results in the fewest casualties.

5-92. The key to forward movement when the assault force is under enemy direct fire is to return effective fire, which prevents the enemy from firing effectively at the moving assault force. Destructive or suppressive fires are most effective when fired by the stationary support force. These fires prevent the enemy from firing effectively at the moving assault force. Once the support force is in position and the assault force is prepared to move, the support force places a heavy volume of fires on the enemy to destroy, neutralize, or suppress him. The ability of the support force to move to advantageous terrain is critical to accomplishing its purpose of ensuring the assault force's success. Once it suppresses the enemy position, it reduces its rate of fire to sustainable levels to conserve ammunition as the assault force closes on the objective to ensure that it has enough to support the assault. When the assault force nears its objective, the support force increases its rate of fire to ensure the continued suppression of the enemy. This allows the assault force to assault the position before the enemy can react. Either on signal or when the assault begins, the support force ceases fire, shifts its fire to another target area, or walks its fire across the objective in front of the assault force.

5-93. The commander uses smoke to help conceal units and individual weapons. It degrades enemy laser designators, range finders, and directed energy weapons. When planning to employ smoke, the commander remembers that smoke can have the same effects on friendly and enemy forces. If possible during the assault, the commander uses obscuration to blind the enemy and screen friendly movement onto the objective. Obscuration is placed in front of enemy positions, on the far side of obstacles, and in areas that restrict maneuver. The commander may use a smoke haze over rear areas to limit enemy observation. The defeat of enemy thermal viewers requires the use of multispectral smoke.

Assault on the Objective

5-94. The effects of the overwhelming and simultaneous application of fire, movement, and shock action characterize the final assault. This violent assault destroys or drives the enemy from the objective area. Small units conduct the final assault while operating under the control of the appropriate echelon command post. Heavy forces have the option of conducting this final assault in either a mounted or dismounted configuration.

5-95. The commander employs all fire support means to destroy and suppress the enemy and sustain the momentum of the attack. By carefully synchronizing the effects of his indirect-fire systems and available CAS, the commander improves the likelihood of success. He plans fires in series or groups to support maneuver against enemy forces on or near the geographical objective. As the commander shifts artillery fires and obscurants from the objective to other targets, the assault element moves rapidly across the objective. The support element must not allow its suppressive fires to lapse. These fires isolate the objective and prevent the enemy from reinforcing or counterattacking. They also destroy escaping enemy forces and systems. The commander employs offensive information operations, such as electronic warfare, to attack enemy C2 nodes as part of this effort.

5-96. Supporting artillery may need to displace forward during the attack to ensure maximum support is available for the assault. However, changes in position are held to a minimum because they reduce the volume of available fires. The commander balances the need to maintain that amount of fire support against the enemy's counterbattery capabilities with the need to provide continued coverage as the attacking unit continues to move forward. Fire support assets supporting the unit move into their new positions one subordinate unit at a time, by echelon, to maintain fire support to the attack. The commander can use his available CAS to provide supporting fires while his artillery batteries displace.

5-97. Small enemy units moving toward the penetrated area can disrupt the synchronization of this final assault. As small units and weapon systems crews become engaged, they tend to focus on their immediate opponent rather than the overall situation. Loss of situational understanding, combined with the enemy's more detailed knowledge of the terrain, allows small enemy forces to inflict a great deal of damage on the attacking force. The attacking unit's leaders must understand the flow of combat and retain the capability to engage these enemy forces before they can alter the outcome of the assault. The commander can commit his reserve to maintain the attack momentum and keep relentless pressure on the enemy. This action also hinders enemy attempts to stabilize the situation.

5-98. Against a well-prepared, integrated enemy defense, the commander must isolate and destroy portions of the enemy defense in sequence. (See Figures 5-4 and 5-5, page 5-24.) His forces must isolate, suppress, obscure, and bypass selected enemy positions. For example, smoke delivered by field artillery and mortars in front of the objective—between the force and the enemy—screens friendly movement and obscures the enemy's weapon systems. Fires placed on and beyond the flanks of the objective serve to isolate the enemy's position. These fires include: smoke, high explosives, improved conventional munitions, and precision-guided munitions delivered by a mix of field artillery, fixed-wing aviation assets, and attack helicopters. In addition, the commander may employ short-duration scatterable mines in accordance with the rules of engagement in conjunction with terminally guided munitions to help isolate and impair the enemy's ability to counterattack. (Their use must not impede the commander's conduct of exploitation and pursuit operations.) Jamming can be used to cut C2 links between the enemy's maneuver force and its supporting artillery. The commander can also use any available CAS to accomplish the desired effects.

5-99. The commander generates overwhelming combat power in sequence against isolated centers of resistance. The assault element can task organize itself to assault one portion of the objective at a time. For example, within the assault company of a task force attack, two platoons may suppress while one platoon seizes a portion of the company objective. This initial platoon, having seized a foothold, then suppresses to allow a second platoon to continue the assault. The third platoon may have a third portion of the objective assigned to it to seize in turn. The enemy may attempt to reinforce its defending forces or counterattack during the friendly force's attack. Once the attacking force reaches the far side of the objective, selected elements clear remaining pockets of resistance while the bulk of the assault force prepares for a possible enemy counterattack. After the assault force reaches the objective, the support force leaves its support-by-fire position and rejoins the assault force or moves to a blocking position to counter possible enemy counterattacks.

5-100. **Mounted Assault.** In determining whether to conduct a mounted or dismounted attack, the commander considers the primary factors of the terrain, obstacles, and the strength of enemy antiarmor defenses. Mounted assaults accelerate the execution of the operation by allowing the greatest speed and shock action and providing the best protection against small arms and indirect fires while conserving the strength of the infantry soldiers conducting the assault.

5-101. When facing weak, hastily prepared, disorganized resistance, or when attacking with overwhelming combat power in relation to enemy forces on the objective, a heavy force commander can conduct a mounted assault. The commander conducting a mounted assault concentrates all of his supporting fires to destroy and neutralize the enemy and fix local reserves. Tanks and infantry fighting vehicles use cannon and machineguns to engage targets for as long as possible, taking advantage of their accuracy, destructiveness, and small bursting radius of their munitions. As the fires from one type of weapon are lifted or shifted, other weapons increase their rate of fire. The assault force advances close to its objective under the cover of these supporting fires.

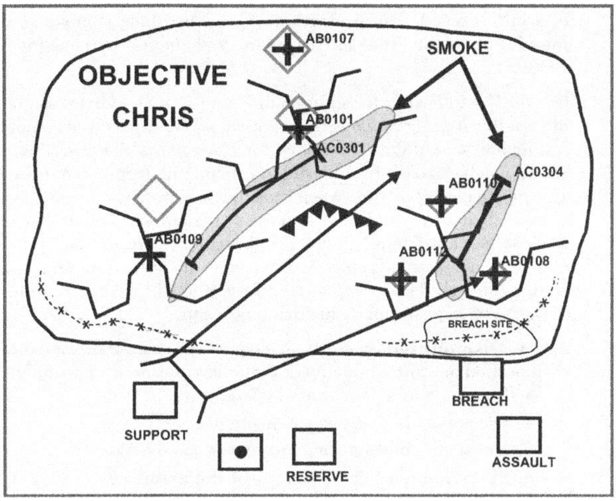

Figure 5-4. Attack of an Objective: The Breach

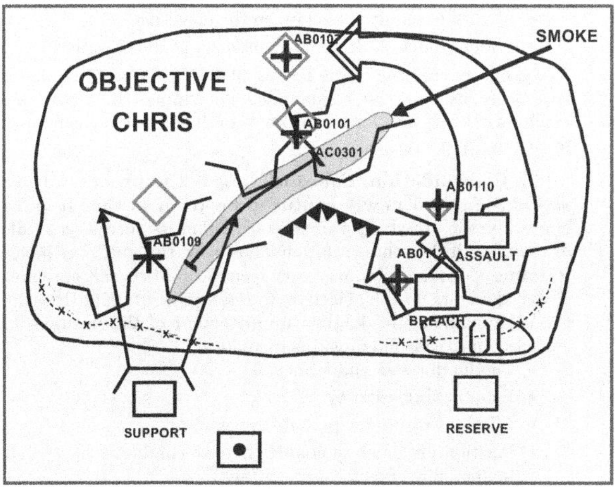

Figure 5-5. Attack of an Objective: The Assault

5-102. The assault force attacks using shock action aided by the firepower of organic systems to rapidly overrun the enemy position as soon as the commander shifts his supporting fires beyond the objective. Mechanized infantry elements move as close as possible to the objective while remaining mounted in their infantry fighting vehicles. When the danger to the mounted infantry elements exceeds the protection offered by their combat vehicle, the commander gives the order for his infantry elements to dismount from their carriers.

5-103. The following technique applies to a heavy force assigned the mission of rapidly clearing an objective against an enemy that does not have a robust antiarmor capability. The heavy force overruns the objective. The accompanying mechanized infantry soldiers dismount from their vehicles on the far side of the objective and sweep the objective from the far side back to the near side to clear any remaining pockets of resistance. The ability of heavy forces to closely follow friendly mortar and artillery fires as they shift across the objective is a major advantage. Any delay in launching the assault after the commander shifts the supporting fires allows the enemy time to move from his protective positions to his firing positions.

5-104. **Dismounted Assault.** A heavy force commander usually conducts a dismounted assault when any of the following conditions apply:

- Terrain favors dismounted operations.
- The enemy is in prepared positions.
- The enemy has a strong antiarmor capability.
- Tanks are not available to lead the assault even though the factors of METT-TC favor their employment.
- Obstacles prevent maneuver across the objective.
- Stealth is required to close on the objective.
- A mounted assault stalls on or short of the objective.

Based on his analysis of the factors of METT-TC and the degree of risk he is willing to accept, the commander determines if, when, and where any mechanized infantry forces in the assault force will dismount from their infantry fighting vehicles.

5-105. **Consolidation. *Consolidation* is the process of organizing and strengthening a newly captured position so that it can be defended.** Normally, the attacking unit tries to exploit its success regardless of the type of the assault. In some situations, however, the unit may have to consolidate its gains. Consolidation may vary from a rapid repositioning of forces and security elements on the objective, to a reorganization of the attacking force, to the organization and detailed improvement of the position for defense. Actions taken to consolidate gains include—

- Conducting reconnaissance.
- Establishing security.
- Eliminating enemy pockets of resistance.
- Positioning forces to enable them to conduct a hasty defense by blocking possible enemy counterattacks.
- Adjusting the fire planning.
- Preparing for potential additional missions.

5-106. Immediately after the assault, the commander must maintain contact with those enemy forces that have abandoned the objective. If he has destroyed all enemy forces on the objective, he takes those actions necessary to regain contact with the enemy. The commander sends out patrols in any direction required to maintain or regain contact with the enemy within his AO. Higher echelons reposition their ISR collection assets and adjust their assigned missions as necessary to maintain that contact.

5-107. The commander also dispatches patrols to ensure contact with any adjacent friendly units. A unit is normally responsible for establishing contact with the units to its front and right as defined by the direction to the enemy. Unless a commander knows that units to his left and rear are preparing to make contact, he takes actions to initiate that contact. Otherwise, a dangerous gap could occur, which the enemy could exploit during a counterattack.

5-108. The task of establishing security is accomplished as soon as the force occupies the objective. Each subordinate element establishes observation posts (OPs) that monitor likely enemy avenues of approach and conduct other security operations. Units must remain aware that the enemy will have defensive fires planned on his former positions, including headquarters bunkers and supply caches.

5-109. Once subordinate units seize the objective, they clear it of enemy forces. They then occupy firing positions to prepare for an enemy counterattack. Normally, an attacking unit does not occupy vacated enemy positions because the enemy is familiar with and normally targets them. Therefore, the attacking unit should position itself away from established enemy positions usually on the next defensible piece of terrain. This positioning is also important because the unit needs to orient on different avenues of approach and in a different direction. The commander positions his armored and antitank systems to cover likely enemy mounted avenues of approach. Mechanized infantry forces normally dismount and orient along likely dismounted and mounted infantry avenues of approach. Overwatching forces, such as antitank systems, orient along likely mounted avenues of approach. Mortars, command posts, and CSS assets move forward to assist in the consolidation.

5-110. The commander should preplan the location and future missions of each element. Artillery and other fire support systems mass fires on enemy assembly areas and troops forming for counterattacks. The commander may alert his reserve to protect the flanks of the attacking units, hold ground seized by them, or counter an enemy counterattack. The commander may use antitank minefields or other obstacles to cover likely enemy avenues of approach. As the unit has time and resources, it improves these obstacles and defensive positions.

5-111. The commander normally designates TRPs, final protective fires, engagement areas, and other direct- and indirect-fire control measures as part of the consolidation process. Once in position, subordinate elements modify preplanned measures and improve the its defensive capabilities as required. As local security is being established, the commander directs subordinate elements to conduct mounted or dismounted patrols along likely enemy avenues of approach. The echelon scout or cavalry unit deploys beyond these local security patrols to conduct its assigned reconnaissance or security mission.

5-112. **Reorganization.** Reorganization includes all measures taken by the commander to maintain the combat effectiveness of his unit or return it to a specified level of combat capability. Commanders of all types of units at each echelon conduct reorganization. Any reorganization actions not completed when conducting the attack are accomplished during consolidation. These actions include—

- Redistributing or cross-leveling supplies, ammunition, and equipment as necessary.
- Matching operational weapon systems with crews.
- Forming composite units by joining two or more attrited units to form a single, mission-capable unit.
- Replacing key personnel lost before or during the battle.
- Reporting unit location and status to keep the next higher commander informed; digitized units can do this automatically.
- Recovering, treating, and evacuating casualties, prisoners of war, and damaged equipment in accordance with its SOP.
- Resupply of its basic loads of ammunition, fuel, and repair parts as time permits.
- Integrating replacement soldiers and systems into the unit.
- Revising communication plans as required. The unit places its C2 facilities in position to conduct further operations and control the consolidation.
- Reestablishing unit cohesion.
- Conducting essential training, such as training replacements on the unit's SOP.

FOLLOW THROUGH

5-113. After seizing the objective, the commander has two alternatives: exploit success and continue the attack or terminate the offensive operation. At brigade echelon and below, the unit maintains contact and attempts to exploit its success. Normally, a division or corps commander makes the decision regarding whether to initiate a general—as opposed to local—exploitation or pursuit or terminate offensive actions.

5-114. After seizing an objective, the most likely on-order mission is to continue the attack. During consolidation, the commander and his staff continue troop leading procedures in preparation for any on-order missions assigned by a higher headquarters. They use available combat information and intelligence products to adjust contingency plans. The commander redirects his ISR collection effort to support his next mission.

5-115. Fire support assets move quickly to take advantage of the natural reduction in support requirements that occur when a position is taken and when the enemy can organize a counterattack to provide depth to a defense. Field artillery units reposition to where they can support a renewed attack when ammunition supply and enemy action permit. Attacks by rotary- and fixed-wing aircraft can provide support while artillery systems reposition. Road conditions, such as destroyed bridges or dislocated civilians, and the unit's existing cross-country mobility abilities given the current environment conditions will also impact on the exact time of the decision to reposition.

5-116. The commander attempts to exploit the deterioration of the enemy position by administering quick and powerful blows before the enemy can reconstitute his defense. Using mass quantities of precision-guided munitions, combined with the action of large, heavy formations and air support, may prove decisive.

5-117. Ordinarily, the enemy attempts to hold his position until nightfall and complete his withdrawal under cover of darkness. The attacking unit maintains relentless pressure, continuing the attack at night. Through these attacks, the unit maintains contact with the enemy, keeps him off balance, and makes his withdrawal from action extremely difficult. If the enemy tries to delay, the unit continues its attack, concentrating its efforts on enveloping or encircling the retrograding enemy force if the enemy is too strong to overrun. An attack aggressively pushed through the hostile front may isolate major elements and force the enemy to evacuate the entire defensive position before he can construct a viable fall-back position.

5-118. When conducting a successful penetration, attacking units penetrate deeper into the hostile position to attack enemy reserves, artillery, C2 centers, and lines of communication. Either the assault or a support unit attacks the enemy's newly exposed flanks to widen the gap. The commander sends forces through the gap that have a high degree of tactical mobility to exploit the advantages gained, attack the enemy from the rear, and prevent his escape. At this time, the commander's force multipliers—such as fixed-wing aviation—concentrate on supporting the ground force exploiting the penetration.

5-119. As part of the follow through to the attack, the commander plans logical sequels to his attack. Attacking forces plan for exploitation. Exploiting forces plan for the pursuit of a defeated enemy. Furthermore, the commander must use his force without overextending its logistics capabilities. The commander must plan to have fresh units pass around or through forward units to sustain the momentum of the attack. He may assign these fresh units the task of follow and support or follow and assume in an effort to maintain the tempo of the attack. (Appendix B discusses both tactical mission tasks in greater detail.) A commander conducting either type of offensive operation envisions how, under what conditions, where, and when he will need to transition to the defense based on possible enemy countermoves and other events.

5-120. If the attacking unit is transitioning to a pursuit or exploitation, it may have to bypass enemy units to maintain the operational tempo. Units bypass enemy forces according to the previously established bypass criteria. As a minimum, the bypassed force is left under observation or fixed in place by other units.

5-121. If the enemy succeeds in withdrawing his major forces from action, the commander intensifies reconnaissance to obtain the information necessary to decide on a COA. Aggressive action may prevent the enemy from reconstituting his defense in a rearward position. The commander may have to delay the renewal of his attack until completing additional reconnaissance so he can formulate a tactically sound plan if the enemy succeeds in occupying new defensive positions.

SPECIAL PURPOSE ATTACKS

5-122. The commander can launch an attack to achieve various results or for special purposes. These subordinate forms of an attack are—

- Ambush.
- Counterattack.
- Demonstration.
- Feint.
- Raid.
- Spoiling attack.

The commander's intent and the factors of METT-TC determine the specific form of attack. As attack forms, they share many of the planning, preparation, and execution considerations of the offense. This section discusses the unique considerations of each form of attack. Demonstrations and feints, while forms of attack, are also associated with military deception operations. (See FM 3-13.)

AMBUSH

5-123. An *ambush* is a form of attack by fire or other destructive means from concealed positions on a moving or temporarily halted enemy (FM 3-0). It may include an assault to close with and destroy the engaged enemy force. In an ambush, ground objectives do not have to be seized and held.

5-124. The two types of ambush are point ambush and area ambush. In a point ambush, a unit deploys to attack a single kill zone. In an area ambush, a unit deploys into two or more related point ambushes. A unit smaller than a platoon does not normally conduct an area ambush.

5-125. Ambushes are categorized as either hasty or deliberate but take place along a continuum. A hasty ambush is an immediate reaction to an unexpected opportunity conducted using SOPs and battle drill. A deliberate ambush is planned as a specific action against a specific target. Detailed information about the target, such as size, organization, and weapons and equipment carried; route and direction of movement; and times the target will reach or pass certain points on its route may be available. Heavy or light forces may conduct an ambush. (Figure 5-6 shows the tactical mission graphic for an ambush.) Doctrine categorizes ambushes as near or far ambushes, based on the proximity of the friendly force to the enemy.

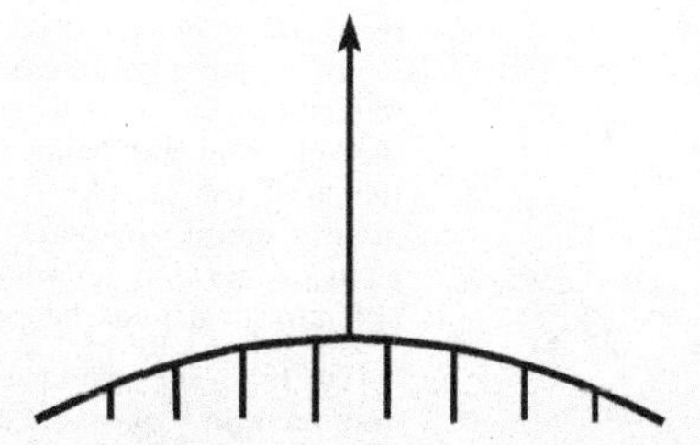

THE ARROW SHAFT'S BASE IS THE CENTER OF MASS OF THE AMBUSH UNIT'S POSITION. THE ARROW POINTS IN THE DIRECTION OF FIRE FROM THE AMBUSH UNIT.

Figure 5-6. Ambush Tactical Mission Graphic

5-126. The typical goal of the ambush force is the death or capture of all enemy personnel located within the kill zone. Another goal could be to destroy certain designated vehicles, such as all missile transporter-erector launchers. Ideally, the ambush force can destroy the ambushed enemy so quickly that he is unable to report the engagement while the ambush force accomplishes its mission.

Organization of Forces

5-127. A typical ambush is organized into three elements: assault, support, and security. The assault element fires into the kill zone. Its goal is to destroy the enemy force. When used, the assault force attacks into and clears the kill zone and may be assigned additional tasks, to include searching for items of intelligence value, capturing prisoners, and completing the destruction of enemy equipment to preclude its immediate reuse. The support element supports the assault element by firing into and around the kill zone, and it provides the ambush's primary killing power. The support element attempts to destroy the majority of enemy combat power before the assault element moves into the objective or kill zone. The security element isolates the kill zone, provides early warning of the arrival of any enemy relief force, and provides security for the remaining ambush force. It secures the objective rally point and blocks enemy avenues of approach into and out of the ambush site, which prevents the enemy from entering or leaving.

Planning an Ambush

5-128. Planning considerations for an ambush include—

- A "no-later-than" time to establish the ambush.
- A tentative ambush formation or, for an area ambush, element locations.
- Insertion and exit routes.
- A forward passage of lines and movement to the ambush site in tactical formation.
- Location of a rally point where the ambush force can reassemble and reorganize if required.
- Actions if the ambush is prematurely detected.
- A scheme of maneuver that maximizes engagement of the enemy's flank or rear, provides early warning of target approach, includes assault element actions in the kill zone, and details how the ambush element displaces from the ambush site.
- Actions at the objective,
- Obstacles to augment the effects of the friendly fire.
- A fire support plan that integrates the direct fire and obstacle plans, which results in the enemy's isolation, inflicts maximum damage, and also supports forces in the rally point.
- The criteria for initiating the ambush; for example, only engage enemy formations of the same or smaller size and withhold fire until the target moves into the kill zone.
- Any required changes to the ambushing unit's fire distribution SOP, based on the factors of METT-TC.

* Rear security measures.

5-129. A point ambush usually employs a line or an L-shaped formation. The names of these formations describe deployment of the support element around the kill zone. **The *kill zone* is that part of an ambush site where fires are concentrated to isolate, fix, and destroy the enemy.** The ambush formation is important because it determines whether a point ambush can deliver the heavy volume of fire necessary to isolate and destroy the target. The commander determines the formation to use based on the advantages and disadvantages of each formation in relation to the factors of METT-TC.

5-130. The assault and support elements generally deploy parallel to the target's route of movement—the long axis of the kill zone—which subjects the target to flanking fire in the line formation. (See Figure 5-7.) The size of the target that can be trapped in the kill zone is limited by the size of the area that can be covered by the support element's weapons. Natural, manmade, and military obstacles—reinforced with tactical obstacles integrated with direct and indirect fires—traps the target in the kill zone. A disadvantage of the line formation is that the target may be so dispersed that it is larger than the kill zone.

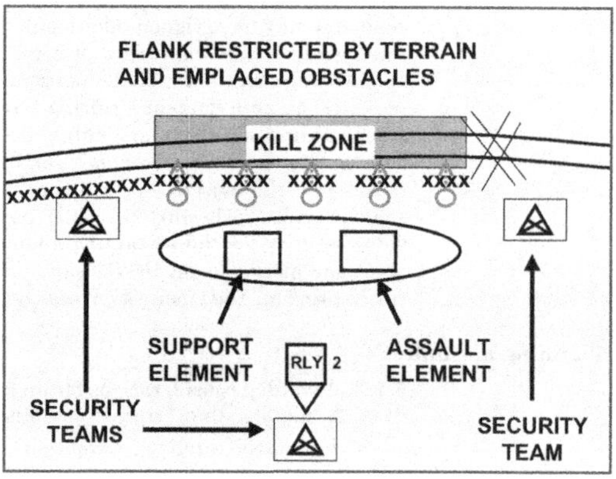

Figure 5-7. Linear Ambush

5-131. The line formation is effective in close terrain, which restricts the target's movement, and in open terrain where one flank is blocked by existing or reinforcing obstacles. The commander may place similar obstacles between the assault and support elements and the kill zone to protect the ambush force from the target's counterambush drills. When the ambush force deploys in a line formation, it leaves access lanes through these protective obstacles so that it can assault the target. An advantage of the line formation is that it is relatively easy to control under all conditions of visibility.

5-132. The L-shaped formation is a variation of the line formation. (See Figure 5-8, page 3-32.) The long leg of the "L" (assault element) is parallel to the kill zone and provides flanking fire. An advantage of the "L" formation is that the short leg (support element) is at the end of the kill zone and at a right angle to it and blocks the enemy's forward movement. It also provides enfilading fire that interlocks with fire from the other leg. The commander can employ an L-shaped formation on a straight stretch of trail, road, stream, or at a sharp bend.

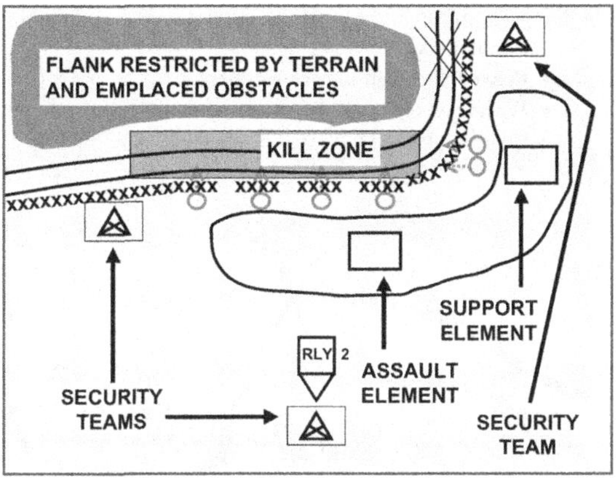

Figure 5-8. L-Shaped Ambush

5-133. An area ambush is most effective when enemy movement is largely restricted to trails or roads. The area should offer several suitable point ambush sites. The commander selects a central ambush site around which he can organize outlying ambushes. Once he selects his site, he must determine the enemy's possible avenues of approach and escape routes. He assigns outlying point ambush sites to his subordinates to cover these avenues. Once they occupy these sites, they report all enemy traffic going toward or away from the central ambush site to the commander. These outlying ambushes allow the enemy to pass through their kill zone until the commander initiates the central ambush. Once the central ambush begins, the outlying ambushes prevent enemy troops from escaping or entering the area. (See Figure 5-9.)

5-134. The ambush unit commander normally specifies the signals required to control the ambush. He changes the meaning of audible and visual signals frequently to avoid setting patterns that the enemy can recognize. Otherwise, the enemy might recognize a signal and react in time to avoid the full effects of the ambush. For example, if a white star cluster is always used to signal withdrawal in a night ambush, an alert enemy might fire one and cause the ambush force to withdraw prematurely. The subordinate elements of the ambush unit must receive communications—in the form of signals—that relay the following information:

- Target approaching, normally given by a member of the security team to warn the ambush commander and the ambush elements of the target's progress.
- Initiate the ambush, given by the ambush unit commander. (This signal should be a mass casualty-producing signal, such as a main gun round from a tank, machine gun fire, the detonation of mines or explosives, or other direct fire crew-served weapons.)

- Lift or shift fire, given when the target is to be assaulted; all fires must stop or be shifted at once so that the assault element can attack before the target can react.
- Assault, given when the assault force is to move into the kill zone and complete its activities.
- Cease fire, given to cease all fires.
- Withdraw from the kill zone or ambush site, given when the ambush is completed or an enemy relief force is approaching.

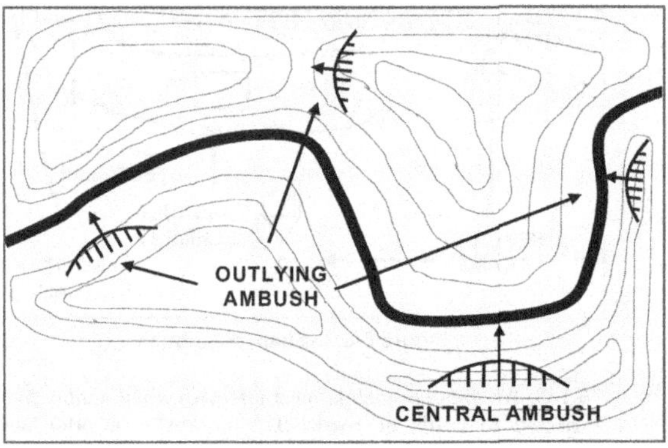

Figure 5-9. Area Ambush

5-135. The commander uses a variety of signals to communicate this information, such as radio transmissions, voice commands, vehicle horns, whistles, or pyrotechnics. All signals must have at least one backup. For example, if the signal to shift fire fails, the assault element should not attack the target unless it receives the backup signal. Signals sent out before initiation of the ambush should not expose the ambush to detection by the enemy. The commander reviews SOP signals to see if they need to be revised or augmented to meet specific situational requirements.

Preparation for an Ambush

5-136. Surprise, coordinated fires, and control are the keys to a successful ambush. Surprise allows the ambush force to seize control of the situation. If total surprise is not possible, it must be so nearly complete that the target does not expect the ambush until it is too late to react effectively. Thorough planning, preparation, and execution help achieve surprise.

5-137. The commander conducts a leader's reconnaissance with key personnel to confirm or modify his plan. This reconnaissance should be undetected by the enemy to preclude alerting him. If necessary, the commander modifies the ambush plan and immediately disseminates those changes to subordinate leaders and other affected organizations. The commander must maintain close control during movement to, occupation of, and withdrawal from the ambush site. Control is most critical when the ambush unit is approaching the target. Leaders enforce camouflage, noise, and light discipline.

5-138. The ambush unit's security element remains at full alert and uses all available observation devices to detect the enemy's approach to the ambush site. Each soldier's duties within each element are rotated as necessary to maintain alertness.

5-139. All elements of the ambush force reconnoiter their routes of withdrawal to the selected rally point. When possible, soldiers or crews reconnoiter the route they will use.

5-140. The commander positions all his weapons, including mines and demolitions authorized by his rules of engagement, to obtain the maximum effectiveness against the target in the kill zone. He coordinates all fires, including those of supporting artillery and mortars. The goals of the support element are to isolate the kill zone, prevent the target's escape or reinforcement, and deliver a large volume of highly concentrated surprise fire into the kill zone. This fire must inflict maximum damage so the assault element can quickly assault and destroy the target.

Execution of an Ambush

5-141. Fire discipline is a key part of any ambush. Fire must be withheld until the ambush commander gives the signal to initiate the ambush. That signal should be fire from the most deadly weapon in the ambush. Once initiated, the ambush unit delivers its fires at the maximum rate possible given the need for accuracy. Otherwise, the assault could be delayed, giving the target time to react and increasing the possibility of fratricide. Accurate fires help achieve surprise as well as destroy the target. When it is necessary to assault the target, the lifting or shifting of fires must be precise. The assault element does not conduct its assault until enemy fires or resistance has been negated or eliminated.

5-142. If the ambush fails and the enemy pursues the ambush force, it may have to withdraw by bounds. It should use smoke to help conceal its withdrawal. Activating limited-duration minefields along the withdrawal routes after the passage of the withdrawing ambush force can help stop or delay enemy pursuit. The commander positions the support element to assist in the withdrawal of the assault element.

5-143. On the commander's order, the ambush force withdraws to the rally point, reorganizes, and starts its return march. At a previously established location, it halts and disseminates any combat information obtained as a result of the ambush to all elements of the ambush force. However, future information systems should be able to disseminate this information without the need to halt a heavy force.

5-144. The commander or his representative debriefs the ambush force to help identify enemy patterns of response, activities, and procedures, both inside and outside the ambush area once the force returns from conducting the ambush. Patterns should be analyzed and reported to all appropriate organizations through intelligence channels. The commander adjusts his tactics, techniques, and procedures to account for these patterns.

COUNTERATTACK

5-145. A *counterattack* is a form of attack by part or all of a defending force against an enemy attacking force, with the general objective of denying the enemy his goal in attacking (FM 3-0). The commander directs a counterattack—normally conducted from a defensive posture—to defeat or destroy enemy forces, exploit an enemy weakness, such as an exposed flank, or to regain control of terrain and facilities after an enemy success. A unit conducts a counterattack to seize the initiative from the enemy through offensive action. A counterattacking force maneuvers to isolate and destroy a designated enemy force. It can attack by fire into an engagement area to defeat or destroy an enemy force, restore the original position, or block an enemy penetration. Once launched, the counterattack normally becomes a decisive operation for the commander conducting the counterattack.

5-146. The commander plans and conducts a counterattack to attack the enemy when and where he is most vulnerable, while he is attempting to overcome friendly defensive positions. Normally, the commander attempts to retain his reserve or striking force to conduct a decisive counterattack once the enemy commits his main force to the attack. The commander assigns objectives to counterattacking forces when he intends for them to assault the enemy. He normally assigns attack-by-fire positions when he intends to counterattack using primarily direct and indirect fires.

5-147. The two levels of counterattacks are major and local counterattacks. In both cases, waiting for the enemy to act first may reveal the enemy's main effort and create an assailable flank to exploit. A defending unit conducts a major counterattack to seize the initiative from the enemy through offensive action after an enemy launches his attack. A commander also conducts major counterattacks to defeat or block an enemy penetration that endangers the integrity of the entire defense, or to attrit the enemy by the defeat or destruction of an isolated portion of the attacking enemy. (See Figure 5-10.)

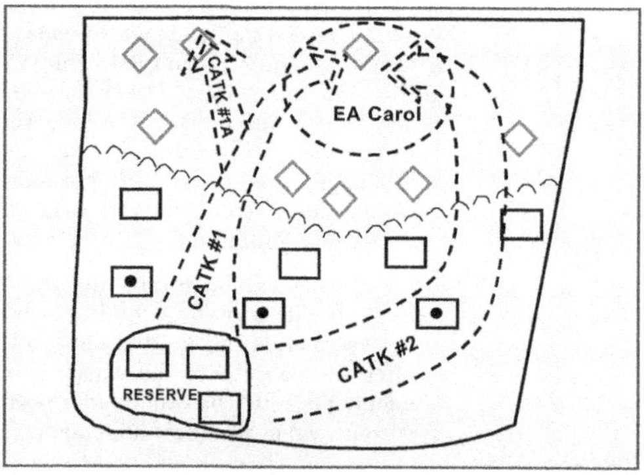

Figure 5-10. Major Counterattack

Organization of Forces

5-148. The commander of a major counterattack force typically organizes his combined arms assets into security, reconnaissance, main body, and reserve forces. He uses those defending forces already in contact with the enemy to fix or contain those same enemy forces. The commander may use a force committed to the counterattack, such as the striking force in a mobile defense, his reserve, another echelon's reserve, or designate any other force he deems appropriate to be the counterattack force. Any changes in task organization should be completed in time to allow units to conduct rehearsals with their attached or supported unit.

5-149. A commander conducts a local counterattack with whatever forces are immediately available to retake positions that have been lost to enemy action or to exploit a target of opportunity. The forces often consist of the reserves of subordinates and defending forces that survive after completing their withdrawal from lost positions. While it is unlikely that the commander changes the task organization of the forces conducting a local counterattack, he organizes the force into a security force and a main body. He may be able to designate an element to conduct reconnaissance.

5-150. The counterattack force is a committed force from the beginning of the defensive operation if the commander's defensive scheme hinges on a counterattack as the defeat mechanism, such as the strike force in a mobile defense. In this case, the commander should designate another force as his reserve.

Planning a Counterattack

5-151. The commander plans the counterattack to strike the enemy when the enemy force is vulnerable. As the enemy force advances, the defense may create gaps between enemy units, exposing the flanks and rear of elements of the attacking force. Immediately after an enemy force occupies a defended position, it is often disorganized and ill prepared to meet a sudden counterattack. Opportunity for effective counterattacks are usually brief; the commander must assess the situation rapidly, and the force must execute the counterattack swiftly. The commander assigns objectives or attack-by-fire positions to counterattacking forces, depending on whether he intends for the counterattacking force to close with and assault the enemy.

5-152. Major counterattack plans are normally developed as a branch or sequel to the main defensive plan. A major counterattack may achieve surprise when it strikes the enemy from an unanticipated direction. For that reason the force directed to conduct a major counterattack, such as the strike force in a mobile defense, should be involved in developing those plans as well as any plans to exploit potential success. Local counterattacks may or may not be the result of previous deliberate planning.

Preparing a Counterattack

5-153. Surprise, coordinated fires, and control are the keys to a successful counterattack. Surprise allows the counterattacking force to seize control of the situation. If total surprise is not possible, it must be so nearly complete that the targeted enemy force does not expect the attack until it is too late to react effectively. Thorough planning and preparation help achieve surprise. The commander adjusts the positioning of his ISR assets and the taskings he gives those assets so he can determine the location and targets for his counterattack.

5-154. The commander conducts a leader's reconnaissance with key personnel to confirm or modify his counterattack plan. If necessary, the commander modifies the plan and disseminates those changes to subordinate leaders and other affected organizations. Each element of the counterattack force reconnoiters its planned axis of advance and routes it will take if possible. The commander maintains close control during movement to and occupation of hide positions and this reconnaissance process so the enemy does not detect the counterattack force prior to initiating the counterattack. Leaders enforce camouflage, noise, and light discipline.

5-155. The commander adjusts the planned positions of his weapon systems to obtain the maximum effectiveness against targets in the planned engagement area. He coordinates all fires, including those of supporting artillery and mortars. He wants his fires to isolate the targeted enemy force in the planned engagement area while preventing the target's escape or reinforcement. These fires must inflict maximum damage quickly before the enemy can respond to the counterattack.

Executing a Counterattack

5-156. A commander should not counterattack unless he has a reasonable chance of success. The commander attempts to retain his reserve for his decisive operation, conducted after the enemy reveals his main effort by committing the majority of his combat power. If the commander orders his reserve to conduct a planned counterattack, the reserve becomes a committed force and the commander should take measures to designate or reconstitute a new reserve.

5-157. The commander conducts the counterattack in the same manner in which he conducts any other attack. He shifts support and priorities of fire, designates targets to be engaged by offensive information operations. The counterattack force also performs those activities discussed in paragraphs 5-61 to 5-122.

5-158. Subordinate commanders initiate local counterattack with the forces on hand when it fits within the higher commander's intent. The conduct of a local counterattack should be swift and violent. It should exploit any disorganization on the part of the enemy, such as the confusion that temporarily exists in an attacking force after it seizes a defended position. A rapidly mounted local counterattack may yield better results than a more deliberate counterattack executed by a higher echelon because of the speed at which it can be launched.

5-159. In the face of a strong enemy penetration, a commander may conduct local counterattacks to retain or seize positions on the shoulders of the enemy's penetration. This prevents the enemy from widening the penetration while forces from other defending units engage the penetrating enemy forces. Holding the shoulders can also prevent the sacrifice of positional depth because the limited gap in the defensive position prevents an attacking enemy from fully exploiting his success.

DEMONSTRATIONS AND FEINTS

5-160. A *demonstration* is a form of attack designed to deceive the enemy as to the location or time of the decisive operation by a display of force. Forces conducting a demonstration do not seek contact with the enemy (FM 3-0). A *feint* is a form of attack used to deceive the enemy as to the location or time of the actual decisive operation. Forces conducting a feint seek direct fire contact with the enemy but avoid decisive engagement (FM 3-0). A commander uses them in conjunction with other military deception activities. They generally attempt to deceive the enemy and induce him to move reserves and shift his fire support to locations where they cannot immediately impact the friendly decisive operation or take other actions not conducive to the enemy's best interests during the defense. Both forms are always shaping operations. The commander must synchronize the conduct of these forms of attack with higher and lower echelon plans and operations to prevent inadvertently placing another unit at risk.

5-161. The principal difference between these forms of attack is that in a feint the commander assigns the force an objective limited in size, scope, or some other measure. Forces conducting a feint make direct fire contact with the enemy but avoid decisive engagement. Forces conducting a demonstration do not seek contact with the enemy. The planning, preparing, and executing considerations for demonstrations and feints are the same as for the other forms of attack.

RAID

5-162. A *raid* is a form of attack, usually small scale, involving a swift entry into hostile territory to secure information, confuse the enemy, or destroy installations. It ends with a planned withdrawal from the objective area on mission completion (FM 3-0). A raid can also be used to support operations designed to rescue and recover individuals and equipment in danger of capture.

5-163. A simplified chain of command is an essential organizational requirement. A raid usually requires a force carefully tailored to neutralize specific enemy forces operating in the vicinity of the objective and to perform whatever additional functions are required to accomplish the objective of the raid. These additional functions can consist of the demolition of bridges over major water obstacles or the recovery of an attack helicopter pilot shot down forward of the forward line of own troops (FLOT). The commander incorporates any necessary support specialists during the initial planning stage of the operation.

5-164. When a commander and his staff plan a raid, they develop COAs that meet ethical, legal, political, and technical feasibility criteria. Planners require precise, time-sensitive, all-source intelligence. The planning process determines how C2, sustainment, target acquisition and target servicing will occur during the raid. Techniques and procedures for conducting operations across the FLOT, given the specific factors of METT-TC expected to exist during the conduct of the raid, are also developed. The commander and his staff develop as many alternative COAs as time and the situation permit. They carefully weigh each alternative. In addition to those planning considerations associated with other offensive operations, they must determine the risks associated with conducting the mission and possible repercussions.

5-165. Time permitting, all elements involved in a raid should be fully rehearsed in their functions. The key elements in determining the level of detail and the opportunities for rehearsal prior to mission execution are time, OPSEC, and deception requirements.

The Raid in Mogadishu, 3–4 October 1993

At 1530 on 3 October 1993, Task Force (TF) RANGER launched another of a series of air assault raids designed to capture key lieutenants of Mohammed Aidid, a clan leader and self-proclaimed general. Aidid and his clan were waging combat operations against UN and US forces and impeding ongoing humanitarian efforts in Somalia. Within 30 minutes TF RANGER had captured several key Aidid lieutenants. The raiding force's HMMWV-equipped ground element was notified to proceed from its assembly area and pick up the prisoners. The ground convoy arrived at the target house, and the prisoners were loaded for the short trip back to the US compound. However, the situation changed drastically during the time it took for the ground convoy to arrive. Two UH-60 Black Hawks had been shot down. In response, TF RANGER attempted to secure the crash sites and was immediately drawn into a series of intense ambushes. The ground convoy with their prisoners could never link up with other task force elements. After several attempts, it arrived with its prisoners back at the US compound at 1818, having suffered almost 70-percent casualties. The commitment of the reaction force was delayed due to the fact that it was not fully briefed (for OPSEC reasons) and was not under direct US control. Failure to maintain a substantial reserve when the situation is not well known and lack of unity of command contributed significantly to TF RANGER's losses that day when the operation quickly transitioned from a raid to a relief of an encircled force.

SPOILING ATTACK

5-166. A *spoiling attack* is a form of attack that preempts or seriously impairs an enemy attack while the enemy is in the process of planning or preparing to attack (FM 3-0). The objective of a spoiling attack is to disrupt the enemy's offensive capabilities and timelines while destroying his personnel and equipment, not to secure terrain and other physical objectives. (See Figure 5-11, page 3-40.) A commander conducts a spoiling attack whenever possible during friendly defensive operations to strike the enemy while he is in assembly areas or attack positions preparing for his own offensive operation or is temporarily stopped. It usually employs heavy, attack helicopter, or fire support elements to attack enemy assembly positions in front of the friendly commander's main line of resistance or battle positions.

5-167. The commander's reasons for conducting a spoiling attack include—

- Disrupt the enemy's offensive preparations.
- Destroy key assets that the enemy requires to attack, such as his fire support systems, fuel and ammunition stocks, and bridging equipment.
- Gain additional time for the defending force to prepare its positions.
- Reduce the enemy's current advantage in the correlation of forces.

The commander synchronizes the conduct of the spoiling attack with his other defensive operations.

5-168. The commander can employ his reserves in a spoiling attack to throw the enemy's offensive preparations off stride. He assumes the risk of not having a reserve or designates another force as his reserve in this case. The following basic considerations affect the spoiling attack:

- The commander may want to limit the size of the force used in executing the spoiling attack.
- Spoiling attacks are not conducted if the loss or destruction of the friendly attacking force would jeopardize the commander's ability to accomplish his defensive mission.
- The mobility of the force available for the spoiling attack should be equal to or greater than that of the targeted enemy force.
- Operations by artillery or aviation systems to prevent enemy elements not in contact from interfering with the spoiling attack are necessary to ensure the success of the operation.

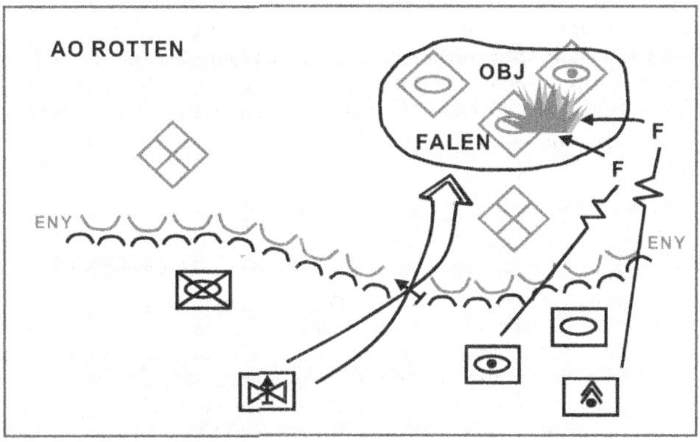

Figure 5-11. Spoiling Attack

5-169. There are two conditions that must be met to conduct a successful and survivable spoiling attack:

- The spoiling attack's objective must be obtainable prior to the enemy being able to respond to the attack in a synchronized and coordinated manner.
- The commander must prevent the force conducting the spoiling attack from becoming overextended.

If the spoiling attack fails to meet both conditions, it will likely fail, with grave consequences to the defense.

Chapter 6

Exploitation

The most important goal of our action is the destruction of the enemy to the last limit of possibility

Field Marshal Prince Mikhail I. Kutuzov

Exploitation is a type of offensive operation that usually follows a successful attack and is designed to disorganize the enemy in depth (FM 3-0). Commanders at all echelons exploit successful offensive actions. Attacks that succeed in annihilating a defending enemy are rare. Failure to aggressively exploit success at every turn may give the enemy time to reconstitute an effective defense by shifting his forces or by regaining the initiative through a counterattack. Therefore, every offensive operation not restricted by higher authority or lack of resources should be followed without delay by bold exploitation. The commander designs his exploitation to maintain pressure on the enemy, compound and take advantage of his disorganization, shatter his will to resist, and seize decisive or key terrain.

CONTENTS

6-1. Exploitation is the primary means of translating tactical success into operational advantage. It reinforces enemy force disorganization and confusion in the enemy's command and control (C2) system caused by tactical defeat. It is an integral part of the concept of the offense. The psychological effect of tactical defeat creates confusion and apprehension throughout the enemy C2 structure and reduces the enemy's ability to react. Exploitation takes advantage of this reduction in enemy capabilities to make permanent what would be only a temporary tactical effect if exploitation were not conducted. Exploitation may be decisive.

6-2. Those plan, prepare, and execute concepts introduced previously continue to apply during an exploitation. Assessment concepts described in FM 6-0 and FM 6-22 also apply. The commander modifies these concepts as necessary to reflect the specific existing factors of METT-TC.

6-3. Local exploitation by the committed force follows a successful attack. A unit conducts a local exploitation when it capitalizes on whatever tactical opportunities it creates in the course of accomplishing its assigned offensive mission. Whenever possible, the lead attacking unit transitions directly to the exploitation after accomplishing its mission in a local exploitation. If this is not feasible, the commander can pass fresh forces (follow and assume) into the lead. The commander acts quickly to capitalize on local successes. Although such local exploitations may appear insignificant, their cumulative effects can be decisive. Subordinate commanders, working within a higher commander's intent, can use their initiative to launch an exploitation. When a commander initiates a local exploitation, he informs his higher headquarters to keep that commander informed of his intentions. This prevents the inadvertent disruption of the higher echelon's battle or campaign and allows the higher headquarters to assess the possibility of general collapse and to direct the initiation of pursuit operations.

6-4. Conduct of a major exploitation is a specific contingency mission assigned to a large unit in anticipation of offensive success by another unit of equivalent size. Divisions and brigades are the echelons that conduct a major exploitation although a corps can conduct a major exploitation as part of a multicorps operation.

ORGANIZATION OF FORCES

6-5. The forces conducting an attack are also the forces that initially exploit that attack's success. Typically, the commander does not assign a subordinate unit the mission of exploitation before starting a movement to contact (MTC) or an attack. The commander reorganizes his unit internally to reflect the existing factors of METT-TC when the opportunity to exploit success occurs. He uses fragmentary orders (FRAGOs) to conduct actions on contact. (See Chapter 4 for a discussion of actions on contact.) If a commander needs additional resources to support the exploitation, he requests them from the appropriate headquarters. The additional resources may include intelligence, surveillance, and reconnaissance (ISR) assets to help identify targets for attack, as well as attack helicopters and controlled munitions, such as the Army tactical missile system, to attack identified targets. Each exploitation force should be large enough to protect itself from those enemy forces it expects to encounter. It should also be a reasonably self-sufficient combined arms force capable of operations beyond the supporting range of the main body.

6-6. The units that create an opportunity to exploit should not be expected to perform the exploitation to an extended depth. If the commander plans to exploit with a specific subordinate unit, he must specify the degree of damage or risk to that force he is willing to accept in the course of the current operation. If the initially attacking units incur significant losses of combat power, the commander should replace them as soon as possible. When the exploiting force's combat power weakens because of fatigue, disorganization, or attrition, or when it must hold ground or resupply, the commander should continue the exploitation with a fresh force. In both cases, the replacement force should have a high degree of tactical mobility so it can conduct the exploitation.

6-7. The exploitation may be more effective if the commander can commit additional forces and assign them the task of either follow and support or follow and assume. The commander assigns follow and support missions to units designated to assist exploiting forces by relieving them of tasks that would slow their advances. The lead unit and any follow and assume or follow and support units exchange liaison teams to facilitate the transfer of responsibilities. Units designated to follow and assume conduct a forward passage of lines and replace the initial exploiting forces when they approach their culminating point. Normally, the next higher commander retains control of the forces performing the tasks of follow and support or follow and assume. (Appendix B expands the discussion of these tasks.) When possible, units assigned these tasks should possess mobility equal to that of the exploiting unit or receive additional engineers and transportation assets to provide the necessary mobility. Once organized, they are committed forces and should receive habitually associated artillery, air defense, engineer, and other combat support (CS) and combat service support (CSS) forces in accordance with the factors of METT-TC. In an exploitation operation projected to cross significant distances, the commander may attach elements of a follow and support unit to the exploiting force to ensure unity of command and effort.

6-8. Since an exploitation operation typically covers a wider front than an attacking force, fire support assets may find their supported elements operating outside normal supporting ranges. They must displace forward to ensure the continued provision of fires on and beyond enemy formations, which may cause some difficulty in supporting the exploiting force's flank elements. To provide the required support, these fire support units, as well as independently operating assets, can be attached to subordinate elements of the exploiting force. Otherwise, the commander can move additional reinforcing fire support units and systems forward to fill the void.

6-9. Responsive air defense coverage provides rapid transition to an exploitation without the loss of momentum. The commander plans on repositioning his air defense artillery assets to ensure this responsiveness. Adequate mobile air defense units should accompany exploiting forces. Air defense arrangements for the initial attack are likely to remain effective throughout the exploitation. However, when the commander extends his formations and assets to cover more area, the air defense coverage becomes less effective. The commander needs to consider the risks associated with moving out from under his air defense artillery umbrella. Alternatively, he can request adjustments in the air defense coverage of higher echelons. Counterair operations by the other services (USAF, USN, and USMC) may provide the desired degree of air defense protection. The commander can use available air interdiction and close air support by fixed-wing aircraft to augment or replace Army fire support assets during an exploitation.

6-10. The exploitation mission demands a force with a significant mobility advantage over the enemy. Attack helicopters and air assault assets may constitute a portion of the exploiting force's combat power. They are extremely useful in seizing defiles, crossing obstacles, and otherwise capitalizing on their mobility to attack and cut off disorganized enemy elements. They can also seize or control key terrain such as important river-crossing sites or vital enemy transportation nodes along the exploiting force's route of advance into and through the enemy's rear area. The commander integrates combat engineers into the exploiting force to help breach obstacles; keep ground forces maneuvering, and provide countermobility protection to the flanks. Typical, problems that degrade an exploiting force's mobility are minefields and other obstacles. The commander also uses engineers to keep his supply routes open.

6-11. The commander retains only those reserves necessary to ensure his flexibility of operation, continued momentum in the advance, and likely enemy responses to the exploitation. (Chapter 5 discusses employment considerations for the reserve.)

RECONNAISSANCE AND SECURITY

6-12. When a commander initiates an exploitation operation, the exact enemy situation may not be clearly known or understood. The commander establishes a reconnaissance force to gain and maintain enemy contact. He complements his reconnaissance effort with sensors and surveillance assets and intelligence products produced by adjacent, higher, and lower echelons to maintain his situational understanding of the strength, dispositions, capabilities, and intentions of all significant enemy elements within his area of interest. The commander normally emphasizes reconnaissance more than security operations when conducting an exploitation. Nevertheless, since the exploiting force moves independently, he addresses the security needs of that force.

6-13. The commander assigns the appropriate security missions to his designated security forces the same way he would for an MTC. (See Chapter 4.) An exploiting corps or division commander typically organizes his forward-most security element into a covering force to protect the main body's movement and develop the situation before he commits his main body. These security elements respond directly to him.

6-14. If an exploiting force is unable to resource a covering force for independent operations, it may use an advance guard in place of a covering force. This is typical for a brigade conducting an exploitation on its own. In some cases when the higher echelon (corps or division) creates a covering force, a brigade may still push out an advance guard behind the covering force. This normally occurs when subordinate units in an exploitation advance in multiple parallel columns.

COMBAT SUPPORT AND COMBAT SERVICE SUPPORT

6-15. Combat support and combat service support arrangements must be extremely flexible during exploitation operations. In the conduct of exploitation operations directed against uncommitted enemy forces or in exploitation operations directed along diverging lines of advance, the commander commonly attaches CS and CSS units to the exploiting maneuver force. Alternatively, the support assets can follow the exploiting force in an echeloned manner along main supply routes (MSRs). Transportation and supplies to sustain the force become increasingly important as the exploitation progresses. As supply lines lengthen, the condition of lines of communications and the conduct of route and convoy security can become problems. The largest possible stocks of fuel, spare parts, and ammunition should accompany the force so that it does not lose momentum because of a lack of support. The exploitation effort may be limited more by vehicle mechanical failures and the need for fuel than by combat losses or a lack of ammunition. Therefore, direct support maintenance support teams accompany the exploiting force to assess the problem and repair disabled vehicles within a limited time period or evacuate them to maintenance collection points for repair by general support maintenance units. The commander should consider using his utility and cargo helicopters to move critical supplies forward during the exploitation.

CONTROL MEASURES

6-16. An exploitation uses fewer control measures than many other operations because of the uncertain enemy situation and the need to provide subordinate commanders with the maximum possible flexibility to take advantage of fleeting opportunities. (See Figure 6-1.) Planners develop graphic control measures as part of the planning process. The commander issues these control measures as part of the attack order to facilitate C2 when the force transitions to an exploitation.

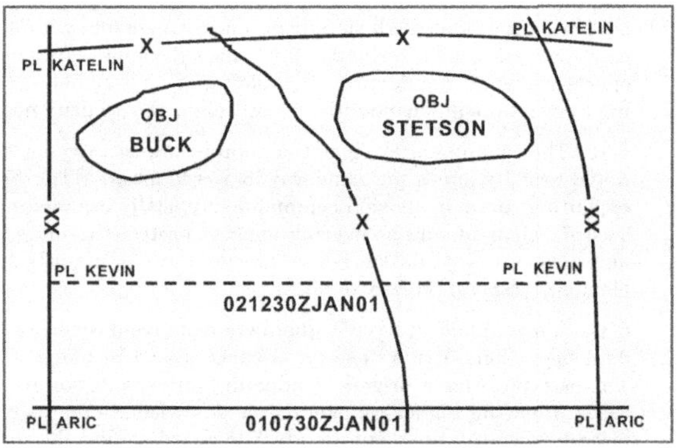

Figure 6-1. Exploitation Control Measures in a Contiguous AO

6-17. A unit conducting an exploitation normally operates in the same area of operations (AO) it was assigned for the attack. The exploiting unit assigns subordinate units their own AOs. Boundaries between subordinate units may change often to take advantage of opportunities. Since an exploiting unit deploys both reconnaissance and security forces, the commander must rapidly adjust his boundaries as the exploiting force advances. The commander designates obstacle-restricted areas to prevent friendly obstacles from hindering the movement of the exploiting force. He designates obstacle zones on the flanks of the exploiting force's movement corridors to enhance his security. He uses phase lines and subsequent objectives to control the conduct of the exploitation. The commander uses objectives to orient the movement of exploiting forces. Although an exploitation may result in taking a terrain objective, the primary focus should be on destroying the enemy force. The commander may establish a limit of advance if he can anticipate a culminating point or some other restriction, such as political considerations regarding an international border, which requires its establishment.

6-18. A commander normally employs permissive fire support control measures during an exploitation. A coordinated fire line (CFL) ensures rapid response. Movement of the CFL is particularly important to provide adequate support as the force continues to advance. Even if the culmination of the exploitation is not anticipated, establishing a forward boundary is important to facilitate operations beyond that boundary by a higher headquarters. The commander can use additional control measures, such as targets and checkpoints, as required.

PLANNING AN EXPLOITATION

6-19. The commander's ability to deny the enemy options by proactive use of his battlefield operating systems is critical to a successful exploitation. He does this by arranging his battlefield operating systems within his opponent's time and space relationship in accordance with the factors of METT-TC.

6-20. The commander must plan for the decentralized execution of an exploitation. His commander's intent is especially important because subordinates must be able to exercise initiative in a rapidly changing, fluid situation. The commander must state the purpose of the exploitation, which may be to force the retrograde of enemy forces from an area, encircle enemy forces so they cannot withdraw, or destroy enemy artillery and other fire support systems. The intent must describe the desired end state. That intent will also determine his decisive and shaping operations and guide the designation of his main effort at any given point.

6-21. A clear commander's intent provides subordinates with guidance on how to integrate their operations into the overall operations of the higher headquarters. Only subordinates who can act quickly can seize all opportunities to damage the enemy or accelerate the tempo of operations. A commander should place minimal restrictions on his subordinates. These may include clear instructions regarding the seizure of key terrain and the size of enemy forces that may be bypassed. Reliable, secure communications between the exploiting force, the follow and support force, and the commander facilitate coordination that can maximize the impact of the exploitation. However, all subordinates should have a clear picture of the desired end state to conduct operations that support it, even if communications are lost.

6-22. Planning for an exploitation begins during the preparation phase of all offensive operations. To avoid losing critical time during the transition from an MTC or an attack to an exploitation, the commander tentatively identifies forces, objectives, and AOs for subordinate units before the offensive operation begins. When the opportunity to exploit occurs, brigade and higher-echelon commanders should initiate the exploitation, either as a branch of or a sequel to the existing operation. The commander's plan should attempt to avoid driving the enemy back in the direction of his own sustaining base.

6-23. During exploitation planning and execution, the commander balances the exploiting force's need for speed and momentum against its need for security as it begins to move beyond supporting range of the rest of the force.

The commander must be careful not to allow an exploiting force to move outside of his main body's supporting distance. Determining the supporting distance requires some knowledge of the enemy's remaining capabilities. Generally, the commander should approach exploitation planning with a sense of guarded optimism. It is an excellent opportunity to shatter enemy cohesion and gain a position of advantage over the enemy. However, the commander cannot allow the exploiting force to fall into an enemy trap where it could be drawn into a salient and destroyed in detail.

6-24. The exploitation may take the form of an MTC with a series of hasty attacks. The commander usually issues a series of FRAGOs that designate—

- Movement formation.
- The position of each major element of the exploiting force within that formation.
- Any required modifications to task organization.
- Bypass criteria.
- Revised or new control measures that assist with the maneuver, such as objectives, boundary changes, a limit of advance (LOA), and FSCM.

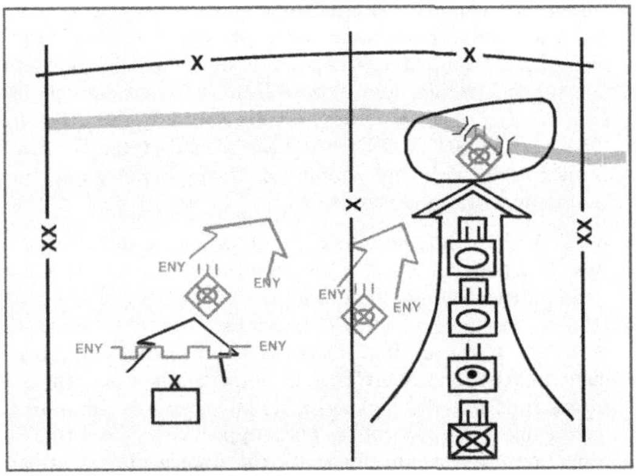

Figure 6-2. Brigade Exploitation: Battalions in Column Formation

6-25. Exploiting forces normally maneuver on a wide front and on at least two axes. The forces on each axis are capable of independent action, depending on the mobility of the force, the road net, and other aspects of the terrain. In some cases, rather than assigning subordinates their own AOs, the commander may designate a movement formation for his entire unit so he can concentrate all his combat power against a specific enemy element. In this case, the commander normally adopts a variation of the column, line, or vee formation. (Chapter 3 discusses combat formations.) (Figure 6-2 shows a brigade conduct an exploitation with its battalions in column.) Movement on parallel routes is preferred; however, the terrain and the enemy situation may force the exploiting force to advance in a column formation. Generally, using a column in the exploitation emphasizes flexibility at the expense of placing maximum firepower forward.

6-26. In exceptional circumstances, when the enemy is clearly incapable of effectively resisting, the commander can choose temporarily not to retain a reserve but commit all his forces to the exploitation. He may employ a line formation with two or more elements abreast without a reserve when the approach to the objective must be made on as wide a front as possible. For example, a commander could use this formation when attempting to secure crossing sites over a major river. (See Figure 6-3.) He could also employ this formation against sporadic and weakening resistance when the enemy lacks a significant counterattack capability or when the counterattack can be blocked by means other than employing the reserve. Despite the lack of a constituted reserve, other actions, such as the effective employment of massed indirect fires, can provide the commander with the flexibility usually provided by the reserve for influencing actions during an exploitation.

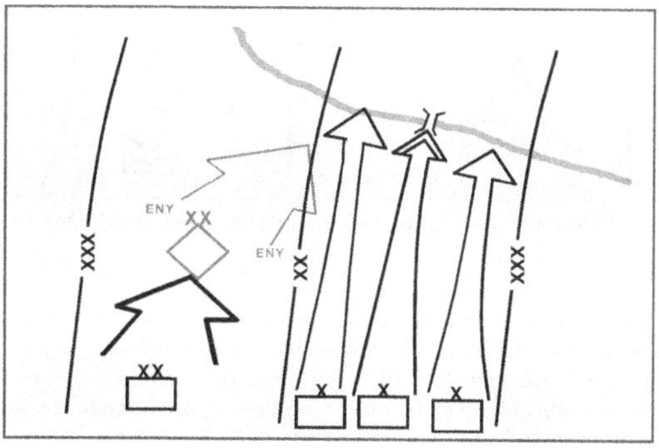

Figure 6-3. Division Exploitation: Brigades Abreast, No Reserve

6-27. A vee formation with two or more elements abreast and a reserve allows the unit to advance on a reasonably wide front with the bulk of the unit's direct firepower oriented forward. This configuration helps when creating gaps in the enemy's defenses. While the bulk of the unit is committed, the reserve is available to exploit the success of the attacking elements, assume the mission of the attacking elements, or counter enemy threats as they develop. (See Figure 6-4.)

6-28. Because of the need to rapidly transition from an attack to an exploitation, fire support planning for the exploitation must take place as part of the planning for the attack. The commander establishes links between his military intelligence, reconnaissance, attack aviation, field artillery, offensive information operations, and supporting fixed-wing assets to expedite the detection and delivery of effects against situationally dependant high-priority targets. He selects those targets regardless of their location within the enemy's defensive area to support the exploitation. During the exploitation, there is little time to revise target lists. Target considerations are similar in nature to those of an MTC. In addition, the exploitation requires a flexible, responsive, and redundant fire control net that must be planned in advance. Coordination with the echelon intelligence officer is critical as the situation develops into exploitation. The exploiting force templates known enemy locations within its AO as danger areas and targets them.

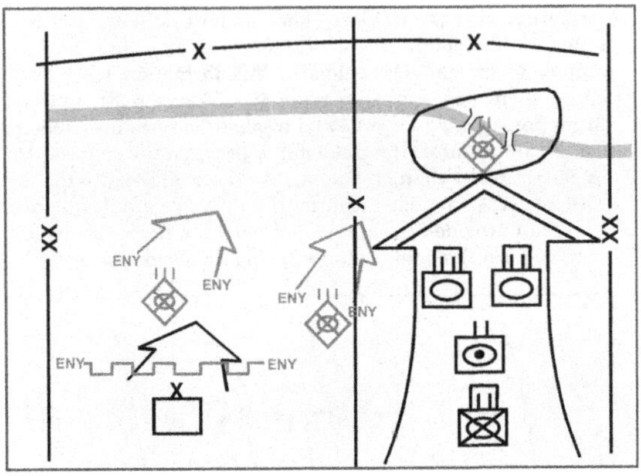

Figure 6-4. Brigade Exploitation: Two Battalions Forward, One in Reserve

6-29. The fire support plan includes allocating support for meeting engagements or hasty attacks that occur during the exploitation. The fire support coordinator plans targets beyond the projected locations of the exploiting maneuver force to shield the force from enemy counterattacks. He then addresses how to provide fire support to the force in its movement to the LOA and targets locations beyond the LOA to interdict the enemy's lines of communication (LOCs).

6-30. The commander plans for artillery displacement as an integral part of the exploitation. Artillery assets must displace at a pace that exceeds normal offensive operations, while maintaining the capability to provide accurate and lethal fires. The commander can normally plan on his forces using less ammunition during an exploitation than in an attack because fleeing enemy forces are normally not in prepared positions, and thus more vulnerable. The commander should also consider using close air support in the exploitation, especially to support those units moving beyond supporting range of the main body. Airborne forward air controllers can help identify and track high-payoff targets forward of the exploiting force.

6-31. The commander plans situational obstacles for each phase of the operation. For example, he plans in accordance with his rules of engagement to emplace scatterable minefields in those areas that could be used by an enemy counterattack force as his forces move forward.

6-32. The enemy may be willing to commit his aircraft against a friendly exploitation that endangers the viability of his defense, buying him time to prepare a defense while weakening the friendly force. The commander plans to move his air defense assets with priority of protection to the decisive operation. He also uses them to protect his lines of communication from enemy air attack, thereby allowing his CS and CSS elements to keep pace with the operation. Planning must address how to rapidly resupply air defense missiles as they are used. It must also allow for adjustments in the priority of protection assigned to different elements during the exploitation.

6-33. The commander must anticipate the exploitation and ensure that his logistics plan supports the force all the way to the LOA. Planning for CSS in the exploitation includes designating future MSRs, logistics release points, unit maintenance collection points, casualty collection points, medical treatment facilities, ambulance exchange points , and the depositing of enemy prisoners of war. In sustaining the exploitation, petroleum, oil, and lubricants (POL) consumption and vehicle maintenance are primary concerns of CSS planners. A significant factor is that an exploiting force tends to travel on a broad front, which may necessitate designating one or more lateral MSRs to handle the dispersion. Logistics operations must be prepared to bound their CSS assets farther forward and move them more often than in an attack.

6-34. Selecting a flexible MSR is critical because it must be able to respond to changes in the direction of the exploitation. Maintaining the MSR is a responsibility of the force engineer. During planning, the commander must specifically address the control of logistics units and convoys. He calls them forward and redirects them as needed. He may have to plan for guides to assist their movement around bypassed enemy positions and obstacles. He may assign some combat elements from the reserve an "on-order" mission to conduct rear area security to help protect CSS elements or secure the MSR. The commander must also ensure adequate plans exist for controlling displaced civilians on the battlefield so that they do not interfere with follow-on maneuver, CS, and CSS assets. This is a critical function of civil-military operations.

EXECUTING AN EXPLOITATION

6-35. Exploitation requires physical and mental aggressiveness to combat the friction of limited visibility, fatigue, bad weather, dangers of fratricide, and extended operations. It requires bold and aggressive reconnaissance, prompt use of firepower, and rapid employment of previously uncommitted units. Exploiting forces maneuver swiftly toward their objectives, sever escape routes, and strike at enemy command posts, communications nodes, reserves, artillery, and CS units to prevent the enemy from reorganizing an effective defense. Well supported by tactical air support, air cavalry, and attack helicopters, exploiting forces should be able to change direction on short notice.

6-36. To maintain sufficient forces to conduct an exploitation, the commander must ensure that his subordinates focus on his intent. They should not dissipate his combat power by seeking minor tactical successes or reducing inconsequential enemy forces. His aim is to reach the final objective with the maximum possible strength as rapidly as possible. The commander must provide his exploiting forces with mobile support, including air resupply, to move emergency lifts of POL and ammunition.

6-37. The transition from attack to exploitation may be so gradual that it is hardly distinguishable; it may also be abrupt. The abrupt transition may occur when a force uses massed quantities of precision munitions, achieves surprise, or overwhelms a much weaker enemy force. Normally, exploitation occurs after the force secures its objective. With adequate support, the commander can launch the exploitation with his initial assault or at any time after that, depending on the effects of the fires and his desires.

6-38. Since the exploitation takes advantage of previous success, forces previously allocated toward attacking enemy forces normally continue their ongoing activities. These activities include—

- Attrition or defeat of enemy reserves prior to their commitment.
- Destruction of enemy countermobility assets prior to their employment on a friendly avenue of advance for the exploiting force.
- Disruption of enemy units attempting to reestablish a coherent defense.
- Disruption of enemy sustaining operations.

This assumes the commander has accurate intelligence data to target these enemy actions.

6-39. Generally, as one part of the attacking force finishes clearing an objective, the commander orders the remaining elements to exploit that success. To accomplish this with minimal confusion, the commander must know where each of his elements is and what combat formation each has adopted. If the commander has previously trained and rehearsed his force to change rapidly from one combat formation to another, to change missions, and to change the direction of advance, he can time the execution of such changes to maintain the initiative over an enemy.

6-40. The commander can also initiate an exploitation when he realizes that the enemy force is having difficulty maintaining its position or cohesion. Updated intelligence is crucial to the commander since it is difficult to accurately predict the exact conditions required to transition from an attack to an exploitation. Therefore, the commander and his subordinates watch the enemy's defenses for indications of disintegration that may signal the opportunity to transition to exploitation. Such indicators include—

- The threat or use of weapons of mass destruction by enemy forces, despite the probable US retaliation, may signal impending enemy collapse.
- Enemy reconnaissance intensifies.
- Rearward movement increases, especially by fire support and reserves.
- The enemy prepares to demolish or destroy his facilities, installations, equipment, and supply stockpiles.
- Various units intermix their vehicles and personnel in combat formations or march columns.
- Number of prisoners captured increases significantly.
- Enemy fire decreases in intensity and effectiveness.
- Fires increase in one or more individual sectors of the front that do not appear to be synchronized with the developing situation and at a time when the amount of defensive fires appears to be decreasing.
- Enemy resistance decreases considerably, or the enemy lacks any type of organized defense.
- Amount of abandoned enemy war materiel encountered increases significantly.

- Reports confirm the capture or absence of enemy leaders.
- Friendly forces overrun enemy artillery, C2 facilities, and supply points.
- Enemy units disintegrate and friendly companies and battalions can defeat enemy battalion- and brigade-size units, respectively.

In any case, the commander ruthlessly exploits vulnerable enemy forces after weighing and accommodating the risks.

GAIN AND MAINTAIN ENEMY CONTACT

6-41. The exploiting force must gain and maintain contact with the enemy. This is a critical aspect of the exploitation since the enemy may be trying to break contact and distance himself from the friendly force to give him time to recover. After a successful attack, the exploiting force must perform aggressive reconnaissance to both its front and flanks. Mission and intent of exploitation determine how much enemy contact is required to maintain pressure on him, compound his disorganization, shatter his will, and seize key or decisive terrain. As discussed in Chapter 5, this reconnaissance effort must start almost immediately after an attacking unit secures its objective. If the commander has dedicated reconnaissance assets, he uses them to maintain enemy contact, observe the enemy's movements, and search for weakly defended enemy positions. If those assets are not available, other maneuver units perform those reconnaissance tasks. While maintaining contact with the enemy, the reconnaissance force tries to locate enemy reserves, uncommitted forces, and blocking positions. This effort helps the exploiting force avoid being led into ambushes as the enemy seeks to recover the initiative by counterattacking.

6-42. When the previously assigned offensive mission is accomplished, units at all echelons push out their reconnaissance and security forces to discover whether the opportunity exists to initiate an exploitation. At brigade and battalion echelons, these reconnaissance and security forces must gain and maintain enemy contact while remaining within the supporting range of their parent brigade or battalion.

6-43. The commander uses air reconnaissance to augment his ground reconnaissance. He can employ aerial sensors, such as JSTARS, air cavalry, and unmanned aerial vehicles in advance of ground maneuver reconnaissance. This allows aerial observation of named and targeted areas of interest that facilitate the unit's movement and cue the attack of high-payoff targets. Scout and attack helicopters can locate enemy positions and engage the enemy to disrupt his movement and preparations. Aviation assets surge to maintain constant contact with the enemy and keep pressure on him.

DISRUPT THE ENEMY

6-44. Exploitation presumes the enemy has already been somewhat disrupted. An exploitation seeks to maintain or increase this disruption by preventing the enemy from effectively reconstituting his defenses. At the division and corps levels, the commander combines the effects of his operations against enemy reserves and uncommitted forces with the rapid maneuver of his close combat forces to maintain this disruption. Attack helicopters can maneuver in front of exploiting ground maneuver forces to destroy high-payoff targets. The commander integrates available fixed-wing aircraft into his plan for attacking these targets. Rapid advances by the exploiting force keep the enemy force off balance and degrade his intelligence and surveillance capabilities, thus providing some security from attack. The commander uses all available resources to maintain pressure on the enemy, using both overwhelming combat power and asymmetric weapon systems. The commander never allows the enemy an opportunity to recover from the initial blow. The exploiting force's fire support system must deliver massed fires quickly to respond to any contingencies that arise during the exploitation.

FIX THE ENEMY

6-45. An exploiting force has three goals in fixing an enemy force. First, it tries to break down the enemy's combined arms organization by fixing enemy units in positions out of supporting distance of each other. This allows the exploiting force to defeat the enemy in detail. Second, the commander attacks out-of-contact enemy forces before they can adversely affect the exploitation. By attacking these enemy forces, the commander seeks to fix them in their current positions or force them to move to locations where they can be harmlessly contained until the exploiting force or a follow and support force can engage and defeat them. Third, it achieves a specific targeting effect—such as causing 15-percent casualties—that disrupts the enemy commander's plan.

MANEUVER

6-46. During an exploitation, the exploiting force maneuvers to maintain pressure on the enemy. The commander can use any heavy and mobile light forces, such as airborne and air assault elements, to secure terrain objectives or choke points critical to the advance and to cut enemy lines of escape. The commander takes advantage of available vertical envelopment capabilities. The exploiting force clears only enough of its AO to permit its advance. It cuts through enemy logistics units and LOCs to seize objectives vital to the enemy's defense. It attacks from the march to overrun weak enemy formations. In accordance with the bypass criteria, the exploiting force can contain and bypass those enemy pockets of resistance too small to jeopardize the mission while its commander reports these enemy forces to adjacent units, following units, and higher headquarters.

6-47. If an enemy unit is too strong for the leading elements of the exploiting force to overrun and destroy, succeeding elements of the force conduct a hasty attack based on the combat information provided by its leading elements. Such enemy forces are rarely attacked frontally. In almost all cases, the commander uses another form of maneuver to produce faster and better results with fewer casualties. While the exploiting force is seeking one or more assailable flanks, available fire support systems continue to engage the enemy to divert attention from the attempted envelopment and destroy as much enemy combat power as possible.

6-48. The exploiting force may face prepared belts of defensive positions in depth when it is exploiting the initial success of the attack. Therefore, the exploiting force must move rapidly to attack and destroy the enemy before he can settle into his subsequent or supplemental positions. The more rapidly this can be done, the less likely it is that succeeding defensive lines will be fully prepared and the less effort it will take to penetrate each successive defensive position. The exploiting force repeats this process as many times as necessary until it breaks completely through the enemy's defenses.

6-49. The commander's primary concern when initiating an exploitation resulting from a successful attack is to shift his force into the appropriate combat formation and task-organize it with additional capabilities and resources to take advantage of a short window of opportunity. Assuming that the force accomplishes this with relative ease, he must control the formation as it moves and prevent its overextension. The commander must anticipate the enemy's reaction to his actions. The real danger to the exploiting force is not the immediate enemy but the enemy not yet engaged. Overextension is a risk inherent in exploitation. While the commander must be concerned with this, he must also guard against being overcautious.

6-50. Surrender appeals and ultimatums are particularly effective when directed against enemy units that have been surrounded, isolated, or bypassed. JP 3-53 and FM 3-05.30 detail the techniques for communicating with the enemy.

6-51. While the exploiting force is conducting its operations, the follow and support force, if available—

- Widens or secures the shoulders of a penetration.
- Destroys bypassed enemy units.
- Relieves supported units that have halted to contain enemy forces.
- Blocks the movement of enemy reinforcements.
- Opens and secures lines of communications.
- Guards prisoners, key areas, installations, and lines of communication.
- Controls dislocated civilians.

FOLLOW THROUGH

6-52. Once the exploitation begins, forces move to attack enemy forces without any operational pauses. Exploitation continues around the clock so the enemy cannot escape the relentless offensive pressure. The exploiting force retains terrain only as necessary to accomplish its mission. The commander must be careful not to dissipate combat power to achieve minor tactical successes or to reduce small enemy forces. Once he reaches the LOA, the commander quickly shifts his attention to survivability and countermobility because of the possibility of an enemy counterattack.

6-53. At some point a unit conducting an exploitation reaches a culminating point or transitions to a pursuit. Culmination can occur for the variety of reasons, such as friendly losses or the enemy's commitment of his reserve. The commander, when he makes an assessment that his force is approaching culmination, should transition to another type of operation. On the other hand, a pursuit enables the commander to complete his destruction of the enemy.

Chapter 7

Pursuit

In pursuit you must always stretch possibilities to the limit. Troops having beaten the enemy will want to rest. They must be given as objectives, not those that you think they will reach, but the farthest they could possibly reach.

Field Marshal Viscount Allenby of Meggido, Order to XXI Corps, 1917

A *pursuit* is an offensive operation designed to catch or cut off a hostile force attempting to escape, with the aim of destroying it (JP 1-02). Pursuit operations begin when an enemy force attempts to conduct retrograde operations. At that point, it becomes most vulnerable to the loss of internal cohesion and complete destruction. A pursuit aggressively executed leaves the enemy trapped, unprepared, and unable to defend, faced with the options of surrendering or complete destruction. The rapid shifting of units, continuous day and night movements, hasty attacks, containment of bypassed enemy forces, large numbers of prisoners, and a willingness to forego some synchronization to maintain contact with and pressure on a fleeing enemy characterize this type of offensive operation. Pursuit requires swift maneuver and attacks by forces to strike the enemy's most vulnerable areas. A successful pursuit requires flexible forces, initiative by commanders at all levels, and the maintenance of a high operational tempo during execution.

CONTENTS

7-1. The enemy may conduct a retrograde when successful friendly offensive operations have shattered his defense. In addition, the enemy may deliberately conduct a retrograde when—

- He is reacting to a threat of envelopment.
- He is adjusting his battlefield dispositions to meet changing situations.
- He is attempting to draw the friendly force into fire sacks, kill zones, or engagement areas.
- He is planning to employ weapons of mass destruction.

Therefore, the friendly force must always consider the enemy's actions whenever it sees an opportunity to conduct a pursuit.

7-2. Division is the lowest echelon equipped with the intelligence assets to determine if the enemy is conducting a retrograde under Army of Excellence tables of organization and equipment. When faced with enemy attempts to break contact, lower echelons act to maintain contact until a division or corps commander directs them to initiate a pursuit operation.

7-3. Unlike an exploitation, which may focus on seizing key or decisive terrain instead of the enemy force, the pursuit always focuses on destroying the fleeing enemy force. This is seldom accomplished by directly pushing back the hostile forces on their lines of communication (LOCs). The commander in a pursuit tries to combine direct pressure against the retreating forces with an enveloping or encircling maneuver to place friendly troops across the enemy's lines of retreat. This fixes the enemy in positions where he can be defeated in detail. If it becomes apparent that enemy resistance has broken down entirely and the enemy is fleeing the battlefield, any type of offensive operation can transition to a pursuit.

7-4. Conducting a pursuit is a calculated risk. Once the pursuit begins, the commander maintains contact with the enemy and pursues retreating enemy forces without further orders. The commander maintains the pursuit as long as the enemy appears disorganized and friendly forces continue to advance. Like exploitation, pursuit tests the audacity and endurance of soldiers and leaders. In both operations, the attacker risks becoming disorganized. Extraordinary physical and mental effort is necessary to sustain the pursuit, transition to other operations, and translate tactical success into operational or strategic victory.

7-5. The commander must be aware of any approaching culmination point. The enemy is usually falling back on his supply base, and potentially on fresh units, while friendly forces become less effective as they expend resources faster than they can be replaced. Reasons to discontinue the pursuit include the presence of fresh enemy forces, greatly increased resistance, fatigue, dwindling supplies, diversion of friendly units to security missions, and the need to contain bypassed enemy units.

7-6. Those plan, prepare, and execute concepts introduced previously continue to apply during a pursuit. Assessment concepts described in FM 6-0 and FM 6-22 also apply. The commander modifies them as necessary to account for the specific existing factors of METT-TC.

ORGANIZATION OF FORCES

7-7. Normally, the commander does not organize specifically for a pursuit ahead of time, although he may plan for a pursuit as a branch or sequel to his offensive operation. Therefore, he must be flexible to react when the situation presents itself. The commander's maneuver and sustainment forces continue their ongoing activities while he readjusts their priorities to better support the pursuit. He acquires additional support from his higher headquarters in accordance with the factors of METT-TC. For most pursuits, the commander organizes his forces into security, direct-pressure, encircling, follow and support, and reserve forces. The commander can employ available airborne and air assault units as part of his encircling force because of their ability to conduct vertical envelopments. Given sufficient resources, there can be more than one encircling force. The follow and support force polices the battlefield to prevent the dissipation of the direct-pressure force's combat power. Appendix B addresses the duties of a follow and support force. The reserve allows the commander to take advantage of unforeseen opportunities or respond to enemy counterattacks.

7-8. There are two basic organizational options in conducting a pursuit; each involves a direct-pressure force. The first is a frontal pursuit that employs only a direct-pressure force. The second is a combination that uses a direct-pressure force and an encircling force. The combination pursuit is generally more effective. Either the direct-pressure force or the encircling force can conduct the decisive operation in a combination pursuit.

FRONTAL

7-9. In a frontal pursuit, the commander employs only a direct-pressure force to conduct operations along the same retrograde routes used by the enemy. (See Figure 7-1.) The commander chooses this option in two situations. The first is when he cannot create an encircling force with enough mobility to get behind the enemy force. The second is when he cannot create an encircling force capable of sustaining itself until it links up with the direct-pressure force. Either situation can occur because of restrictive terrain or because an enemy withdraws in a disciplined, cohesive formation and still has significant available combat power.

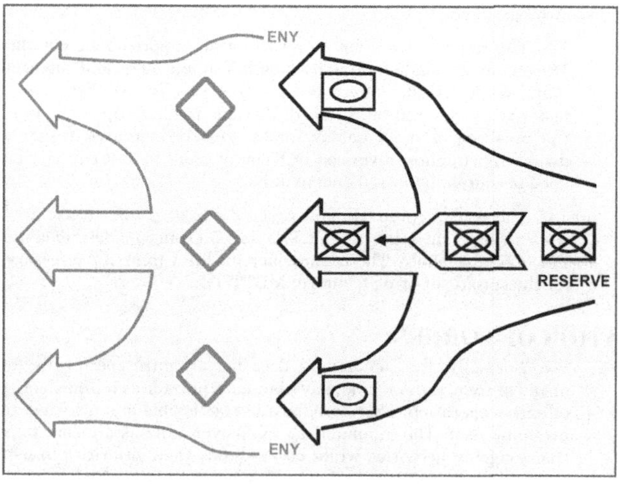

Figure 7-1. Frontal Pursuit

COMBINATION

7-10. In the pursuit, the most decisive effects result from combining the frontal pursuit with encirclement. (See Figure 7-2.) In the combination pursuit, the direct-pressure force initiates a frontal pursuit immediately on discovering the enemy's initiation of a retrograde operation. This slows the tempo of the enemy's withdrawal (or fixes him in his current position if possible), and may destroy his rear security force. The direct-pressure force's actions help to set the conditions necessary for the success of the encircling force's operation by maintaining constant pressure. The encircling force conducts an envelopment or a turning movement to position itself where it can block the enemy's escape and trap him between the two forces, which leads to complete annihilation.

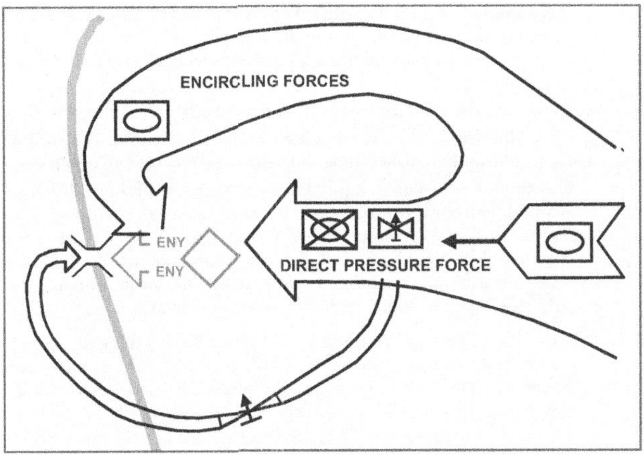

Figure 7-2. Combination Pursuit

7-11. The direct-pressure force conducts hasty attacks to maintain contact and apply unrelenting pressure until it destroys the enemy force. The direct-pressure force prevents enemy disengagement and subsequent reconstitution of the defense and inflicts maximum casualties. It forces the enemy to deploy frequently to delay the direct-pressure force and restricts his ability to disengage and rapidly move away. The direct-pressure force must be at least as mobile as the enemy. Heavy forces are ideally suited to this role, but the commander can employ light forces if the enemy is also foot-mobile. The direct-pressure force organizes to conduct a movement to contact and must be able to conduct a series of hasty attacks. It must be powerful enough to defeat enemy rear guard actions and maintain pressure on the enemy's main body.

7-12. The mobility of the encircling force must be equal—preferably superior—to the withdrawing enemy. If there is no inherent mobility differential, the commander must create one. This differential can also result from the direct-pressure force forcing the enemy to deploy. The commander can enhance, and sometimes create, this mobility advantage by conducting countermobility operations against the enemy, specifically targeting locations such as choke points or bridges that will hinder the fleeing enemy's withdrawal. Heavy, air assault, and airborne forces are well suited for this mission. Attack helicopters are also effective when used as part of the encircling force. The encircling force must be strong enough to protect itself from the enemy's main body and slow or stop it until the friendly direct-pressure force can combine with the encircling force to destroy the enemy. It must be capable of mounting a hasty defense without placing itself at risk of annihilation. The encircling force must be self-contained since it normally operates out of supporting range of friendly indirect-fire systems. Therefore, it frequently has its supporting artillery attached. The primary mission of the encircling force is to prevent the enemy's escape by trapping him between the encircling force and the direct-pressure force. The commander can assign other missions to the encircling force, such as—

- Destroying the enemy's weapons of mass destruction and their delivery means.
- Linking up with airborne or air assault forces in their airheads.
- Reporting terrain conditions and other combat information beyond that normally addressed in the unit standing operating procedures.

The commander can assign the encirclement mission, wholly or in part, to available airborne or air assault units because their vertical envelopment capabilities allow friendly forces to be inserted deeper into enemy-controlled territory than would be possible with ground operations. The time required to plan airborne operations and stage airlift platforms impacts on the utility of airborne forces in small-scale pursuit operations.

7-13. The direct-pressure and encircling forces require engineer support to create lanes through obstacles, which enables them to move rapidly and continuously. The commander should place his engineers well forward in his movement formations to quickly breach any obstacles that cannot be bypassed. Engineers accompanying the encircling force must also be prepared to conduct countermobility and survivability tasks.

CONTROL MEASURES

7-14. The commander uses control measures to retain his tactical options to converge on the most important axis or to redirect his pursuit effort on a new axis. These control measures should be flexible and capable of rapid adjustments to reflect changing conditions. This flexibility is also necessary when engaging advancing enemy reserves or counterattack forces.

7-15. Centralized planning and decentralized execution characterize the pursuit. The commander balances the need to prevent fratricide with the need to allow subordinates to take advantage of fleeting opportunities in a pursuit with rapidly moving forces and a rapidly changing situation. The commander designates an area of operations (AO) for each maneuver unit involved in the pursuit. He establishes few control measures for the direct-pressure force other than phase lines and checkpoints because of the pursuit's nature. He uses these phase lines to designate a forward and rearward boundary for the direct-pressure force. The forward boundary relieves the direct-pressure force of any responsibility beyond the forward boundary. It also gives the higher headquarters flexibility to deal with the encircling force and enemy elements located beyond that forward boundary. The rear boundary becomes the boundary between the direct-pressure force and the follow and support force.

7-16. If the encircling force is a ground element, the control measures are almost identical to those of an envelopment. The commander must designate a route, an axis of advance, or an AO adjacent to that of the direct-pressure force to allow the encircling force to move parallel to and eventually get ahead of the fleeing enemy force. He designates a terrain objective as a guide for the encircling force. (See Objective HAWKE in Figure 7-3.) However, he may change this objective rapidly and frequently, based on the progress of the encircling force and the enemy. The objective should be a piece of ground that provides the encircling force good, defensible terrain that the enemy cannot easily bypass. The commander often selects choke points, such as defiles and bridges, as objectives for his encircling force.

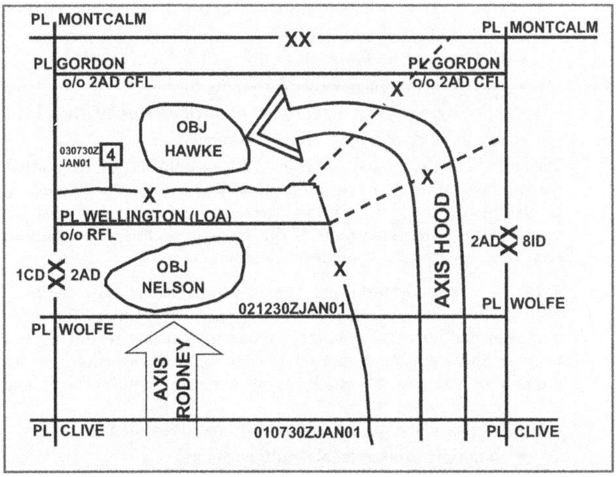

Figure 7-3. Pursuit Control Measures

7-17. The commander establishes a boundary or a restricted fire line between the encircling force and the direct-pressure force before the encircling force reaches its objective. He establishes other fire support coordinating measures (FSCM) around the area currently occupied by the encircling force to relieve it of unnecessary fire support coordination responsibilities. He directs security operations beyond the encircling force, allowing it to engage the withdrawing enemy without devoting resources to flank and rear security. The commander establishes additional control measures to control the convergence of both elements of the friendly force, such as phase lines and contact points.

PLANNING A PURSUIT

7-18. The commander anticipates an enemy retrograde operation as either a branch or a sequel to the plan. The plan should identify possible direct-pressure, encircling, follow and support, and reserve forces and issue on-order or be-prepared missions to these forces. The commander should employ the maximum number of available combat troops in the pursuit. He bases the details of his plan on the enemy's anticipated actions, the combat formation of the attacking troops, and the amount of planning time available. The commander also considers—

- Possible routes the enemy might use to conduct his retrograde operations.
- Availability of his intelligence, surveillance, and reconnaissance assets to detect enemy forces and acquire targets in depth.
- Scheme of maneuver.
- Availability and condition of pursuit routes.
- Availability of forces to keep the pressure on the enemy until his destruction is complete.
- Critical terrain features.
- Use of reconnaissance and security forces.
- Allocation of precision-guided munitions and aviation support.
- Availability of CS and CSS resources.

Pursuit planning must address the possibility of defending temporarily during operational pauses while making preparations to continue the pursuit or to consolidate gains. However, the use of an operational pause generally results in the abandonment of the pursuit because the enemy is able to use that time to organize a coherent defense.

7-19. The commander must specifically address how to detect the enemy retrograde operations; otherwise, the enemy may succeed in breaking contact. The commander relies on active reconnaissance, an understanding of enemy tactics, and knowledge of the current tactical situation. He must watch for signs that indicate the enemy is preparing to conduct a retrograde, such as when the enemy—

- Lacks the capability to maintain his position or cohesion.
- Conducts limited local counterattacks.
- Intensifies his reconnaissance and intelligence efforts.
- Increases the amount of rearward movements and changes the type of elements conducting them, especially by fire support and reserves.
- Prepares his facilities, installations, equipment, and supply stock-piles for demolition and destruction.
- Decreases fire in intensity and effectiveness through the AO.
- Increases his fires in one or more individual sectors of the front, which does not appear to be in accordance with the developing situation, and at a time when the amount of defensive fires seems to be decreasing.

The presence or absence of any of the above signs may not necessarily indicate the start of a retrograde operation. The enemy could be attempting to draw friendly forces into an ambush or setting up a counterattack as part of his defense. The decision of when to start a pursuit is part of the art of tactics.

7-20. When the commander initiates a pursuit, he often creates the encircling force from uncommitted or reserve elements. Normally, these forces do not have fire support assets allocated to them. The commander must plan how to redistribute his fire support assets to properly support the encircling force. Attack helicopters and close air support are well suited to support the encircling force.

7-21. Engineer mobility and countermobility assets are instrumental in sustaining the rate of advance and hindering the enemy's withdrawal. Engineers prepare the route of advance and support the lateral dispersion of units transitioning to the pursuit and the movement of the reserve. During the pursuit, the commander must plan for his engineers to provide assault bridging and emergency road repairs to sustain the tempo of the pursuit. The commander also plans to use his engineer assets to block any bypassed enemy's withdrawal routes by using antitank and command-operated mines, demolitions, and obstacles.

7-22. Logistics units should plan for increases in the demand for fuel and maintenance as the tempo of operations increases. In the pursuit, priority of logistics normally goes to units having the greatest success. Logistics planners need to anticipate success since the depth of the pursuit depends on the capability of logistics assets to support the operation. The logistics elements supporting the pursuing force should be as mobile as possible. Logistics planners are particularly concerned with supporting the encircling force, such as casualty evacuation over possibly unsecured LOCs. The commander may need aerial resupply or heavily guarded convoys to support this force. Security for logistics convoys and LOCs are major planning considerations.

7-23. The commander uses all available logistics assets to provide essential support to the force pursuing the enemy. His pursuit plan must result in a force prepared to conduct wide-ranging operations using all available maneuver assets throughout his AO to complete the destruction and morale collapse of the enemy force.

EXECUTING A PURSUIT

7-24. The decisive operation in a pursuit destroys the withdrawing enemy. This generally occurs as a result of encircling the enemy between the direct-pressure and the encircling forces or a major geographic barrier—such as an unfordable river—and his defeat in detail. The timely and correct decision to initiate a pursuit is critical to its success. If the enemy begins a retrograde undetected, he avoids the constant pressure that results in disrupting that operation. The commander expects the enemy forces to conduct retrograde operations at times advantageous to them—usually at night or during bad weather.

7-25. A pursuit is often conducted as a series of encirclements in which successive portions of the fleeing enemy are intercepted, cut off from outside support, and captured or destroyed. (Appendix D discusses encirclement operations.) The direct-pressure force conducts a series of hasty attacks to destroy the enemy's rear security force, maintain constant pressure on the enemy's main body, and slow the enemy's withdrawal. At every opportunity, the direct-pressure force fixes, slows down, and destroys enemy elements, provided such actions do not interfere with its primary mission of maintaining constant pressure on the enemy's main body. The direct-pressure force can bypass large enemy forces if it can hand them off to follow and support units, or if they do not pose a risk to the direct-pressure force.

7-26. As soon as the commander designates a unit as the encircling force and directs its actions, the force moves as swiftly as possible by the most advantageous routes to cut off the enemy's retreat. If the encircling force cannot move farther and faster than the enemy, it attacks the enemy's main body from the flank. When this occurs, the commander should constitute and dispatch a new encircling force.

GAIN AND MAINTAIN ENEMY CONTACT

7-27. At the first indication of an enemy retrograde, the brigade or lower-echelon commander who discovers the enemy's rearward movement acts to maintain contact with the enemy across a wide area without waiting for orders from higher headquarters. This ensures that the enemy does not break contact and conduct an orderly retirement. These forces in contact constitute the nucleus of the direct-pressure force. As the situation permits, they reform into a movement column with reconnaissance and security elements in the lead and, if necessary, to the flank.

7-28. During a pursuit, the reconnaissance effort is intensive. Reconnaissance elements concentrate on all routes the enemy could use when conducting a retrograde operation. These elements provide information on the disposition of retreating enemy formations and on the forward movement of his reserves as the pursuit develops. The tactical situation during a pursuit may become obscure because of its potential depth. Much of the combat information needed during a pursuit is located behind the fleeing enemy force. Therefore, air reconnaissance, backed by technical intelligence systems, is vital to the overall reconnaissance effort. It can determine—

- The beginning of the rearward movement of enemy sustainment forces.
- The composition of retrograding forces and their direction of movement.
- The composition and direction of enemy reserve forces moving forward.
- The nature of obstacles and intermediate defensive positions.

Information about fresh enemy reserves and prepared positions is vital at the stage when a pursuit force may be approaching a culminating point; it may be the basis for terminating the pursuit.

7-29. The primary mission of the encircling force's reconnaissance assets is to find routes for the encircling force to allow it to move behind withdrawing enemy units and establish blocking positions. This mission may force these reconnaissance assets to operate outside the supporting range of the main body as they try to maneuver behind the retrograding enemy force. The encircling force avoids combat when possible until it reaches its assigned objective area. However, en route to its objective, it overruns any small enemy positions while bypassing larger enemy units. Forward security elements of the encircling force conduct activities to prevent the enemy from interfering with the forward movement of the encircling force's main body. These security elements move rapidly along all available roads or routes and overrun or bypass small enemy pockets of resistance. If they encounter strongly held enemy positions, they attempt to find routes around or through these positions. The encircling force can then avoid these enemy positions and occupy blocking positions before withdrawing enemy forces can reach them. If necessary, the encircling force organizes a hasty defense behind the enemy to block his retreat.

DISRUPT THE ENEMY

7-30. Keeping the enemy from reconstituting an effective defense is critical to success. Constant pressure by direct-pressure forces and echelon fire support systems disrupts and weakens the enemy. The commander uses lethal and nonlethal direct and indirect fires to keep pressure on the enemy. The enemy commander must not be allowed to freely adjust his dispositions to counter the actions of the friendly force. Artillery fire and air strikes harass and disrupt the enemy's attempts to move engaged forces to the rear or bring previously uncommitted forces into action. In a pursuit, decisive operations may include the ground maneuver of the direct-pressure or the encircling force. Fire support targets in a pursuit include fires on enemy columns and troop or vehicle concentrations at road junctions, defiles, bridges, and river crossings. They also include the repulsion of enemy counterattacks, destruction or delay of enemy reserves, and destruction of the enemy's fire support means. The commander conducts offensive information operations against the enemy's command and control (C2) system as an integral part of this disruption process, with emphasis on destroying or degrading the enemy's capability to reconstitute and synchronize an effective defense.

FIX THE ENEMY

7-31. Using movement and fire effects or fire potential, the commander fixes a withdrawing enemy. If the direct-pressure force disrupts the enemy's C2 system, his ability to counter friendly efforts is significantly degraded, and the goal of fixing the enemy is much easier to accomplish.

7-32. The enemy attempts to use his reserves to restore the integrity of his defenses or prevent his withdrawing force from being overrun. Fixing enemy reserves is essential to the pursuit's success and is normally the focus of echelon shaping operations. The direct-pressure force fixes enemy reserves in place or slows them down so that they remain outside supporting distance until the withdrawing enemy force is completely annihilated.

MANEUVER

7-33. To execute the pursuit, the commander normally combines a frontal pursuit with an encirclement. The direct-pressure force conducting the frontal pursuit advances in a column formation as quickly as possible. After a penetration, existing gaps between the different units of the direct-pressure force are likely to increase in size. Aware of the vulnerability of his open flanks in this situation, the commander must deploy his reserves where they can respond to dangers on his flanks. He does not expect a uniform rate of advance on all axes. Some columns may move rapidly while others are still engaged in penetrating the enemy's rear guard defensive positions or meeting enemy counterattacks.

7-34. The actions of the direct-pressure force should facilitate the commitment of an encircling force that moves parallel to the rearward-moving enemy. The depth of the pursuit depends on the size of the forces involved. It takes a division-level or higher commander to make the decision to initiate a pursuit because of the resources necessary to conduct a pursuit. The commander directing the initiation of a pursuit informs his higher commander of his intentions. This allows even greater resources to be devoted to the pursuit and avoids desynchronizing the higher headquarters' major operation or campaign.

7-35. The direct-pressure force normally employs an advance guard to prevent the enemy from ambushing the main body of the direct-pressure force and to overrun or bypass small enemy forces. The security element moves on multiple avenues of advance. If it encounters enemy units beyond its capacity to defeat, it conducts actions on contact to develop the situation. The commander uses combat information provided by these actions on contact to guide the main body of the direct-pressure force to destroy withdrawing enemy forces. These actions of the direct-pressure force may or may not be in conjunction with the actions of any encircling force.

7-36. The commander does everything possible to place his encircling force behind the withdrawing enemy and trap the bulk of that enemy force between the encircling force and the direct-pressure force. The direct-pressure force maintains enough pressure on the withdrawing enemy force so the encircling force can envelop it. To perform this task, the direct-pressure force must be strong enough to overcome any enemy rear guard before the enemy's main body can make a successful withdrawal. Once in position, the encircling force defends or attacks as necessary, responding to the enemy's actions and those of the direct-pressure force to complete the enemy's encirclement.

7-37. The pursuing force must not give the enemy time to reorganize for an all-around defense after it is encircled. If the enemy forms a perimeter, the pursuing commander must repeatedly split it into smaller elements until he destroys the encircled enemy force. If time is not critical, the commander can keep the encirclement closed, defeat enemy breakout attempts, and weaken the enemy by fires alone. He can greatly accelerate the collapse of a large, encircled enemy force by using psychological operations, precision-guided weapons, and improved conventional munitions in mass. (Appendix D addresses the reduction of an encircled enemy force.) If the resulting encirclement does not destroy the withdrawing enemy force, the commander conducts additional pursuit operations until the enemy is destroyed.

FOLLOW THROUGH

7-38. Once the commander initiates a pursuit, he continues pursuing the enemy until a higher commander terminates the pursuit. Conditions under which a higher commander may terminate a pursuit include the following—

- The pursuing force annihilates or captures the enemy and resistance ceases.
- The pursuing force fixes the enemy for follow-on forces.
- The high commander makes an assessment that the pursuing force is about to reach a culminating point.

7-39. A pursuit often transitions into other types of offensive and defensive operations. If the enemy attempts to reorganize, forces conducting a pursuit execute hasty attacks. They conduct an exploitation to capitalize on the success of these attacks and then move back into pursuit. Forces conducting a pursuit may also transition into a defensive operation if the pursuing force reaches a culminating point. This usually occurs when the enemy introduces strong reinforcements to prepare for a counteroffensive.

PART THREE

Defensive Operations

Chapter 8

Basics of Defensive Operations

So the defensive form of war is not a simple shield, but a shield made up of well-directed blows.

Carl von Clausewitz, *On War*, 1832

Defensive operations defeat an enemy attack, buy time, economize forces, or develop conditions favorable for offensive operations. Defensive operations alone normally cannot achieve a decision. Their purpose is to create conditions for a counteroffensive that allows Army forces to regain the initiative (FM 3-0). Other reasons for conducting defensive operations include—

- Retaining decisive terrain or denying a vital area to the enemy.
- Attritting or fixing the enemy as a prelude to offensive operations.

CONTENTS

- Surprise action by the enemy.
- Increasing the enemy's vulnerability by forcing him to concentrate his forces.

8-1. While the offense is the most decisive type of combat operation, the defense is the stronger type. The inherent strengths of the defense include the defender's ability to occupy his positions before the attack and use the available time to prepare his defenses. Preparations end only when the defender retrogrades or begins to fight. The defender can study the ground and select defensive positions that mass the effects of his fires on likely approaches. He combines natural and manmade obstacles to canalize the attacking force into his engagement areas (EAs). He can coordinate and rehearse his defensive plan while gaining intimate familiarity with the terrain. The defender does not wait passively to be attacked. He aggressively seeks ways of attriting and weakening attacking enemy forces before the initiation of close combat. He maneuvers to place the enemy in a position of disadvantage and attacks him at every opportunity, using his direct and indirect fires. The fires include the effects of offensive information operations and joint assets, such as close air support. The static and mobile elements of his defense combine to deprive the enemy of the initiative. He contains the enemy while seeking every opportunity to transition to the offense.

HISTORICAL EXAMPLE

8-2. The following historical example illustrates how conducting a defense can attrit and fix an enemy as a prelude to offensive actions.

The Battle of Kursk, July 1943

Using an area defense, the Red Army defeated the German Army's last Eastern Front operational-level attack at Kursk. The Red Army maximized its defensive advantage using mass, security, objective, and offensive as principles of war.

Soviet intelligence discovered the German offensive objective and concept: a double envelopment of the Kursk salient by panzer-heavy forces. The Red Army massed forces in the most threatened areas. The Soviets reinforced the two fronts defending the salient, prepared defenses, and established a strategic reserve behind the salient. They weighted the forward defenses on the northern and southern shoulders within the salient. They developed their defenses in depth, carefully tying them to the terrain and organizing infantry positions for all-around defense. Above all, they organized an antitank defense, with mutually supporting positions and mobile counterattack forces at all levels. Nearly 6,000 antitank (AT) guns and 3,300 tanks packed the defense.

The German attack in the northern part of the salient would fall on the 13th Army. The 13th Army consisted of 12 rifle divisions (RDs) organized into four rifle corps (RCs) supported by 700 guns, separate tank brigades, assault gun regiments, and antitank regiments. Within 30 kilometers of the front, the 13th Army established three fortification belts. Within each belt there were large numbers of mutually supporting antitank positions. Each position consisted of four to six AT guns, with protection provided by infantry, machine guns, and obstacles.

The 29th RC occupied the 13th Army's main defensive position in a sector 19 kilometers wide and 15 kilometers deep, with the 15th RC on its right, the 70th Army on its left, and the 17th Guards Rifle Corps (GRC) rearward in the army second echelon. At the start of the battle, the 29th RC consisted of three rifle divisions (the 15th, 81st, and 307th), with supporting tank and artillery units. It deployed the 15th RD and 81st RD, with 12 to 15 antitank positions each, as the corps' first echelon. The 307th RD was the corps' second echelon. Both first-echelon divisions also deployed in two echelons. Each division established a battalion security force to its front.

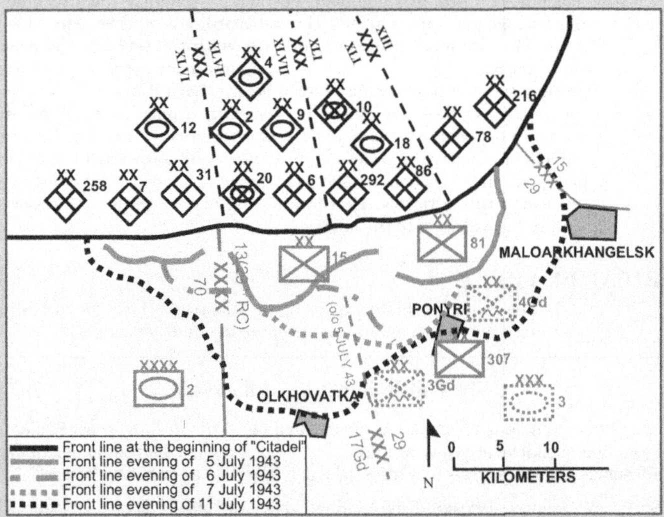

Figure 8-1. Movements of an Area Defense

During a week of intense defensive combat, the 29th RC conducted an area defense, making the Germans pay so dearly for every gain that they reached their culminating point short of Kursk. See Figure 8-1. Tested by nine German divisions, the 29th RC was able to keep German forces from breaking through its area of operations (AO), despite having its initial three divisions rendered combat-ineffective. In the course of the defense, the 29th RC inflicted 10,700 German casualties and destroyed an estimated 220 tanks and 71 guns. Key to the defense was the construction of those mutually supporting antitank positions, organized for all-around defense, with extensive engineer works to enhance the terrain. The 29th RC employed its attached forces aggressively, creating combined arms teams to hold terrain or maneuver against German forces within the defensive belt. It employed counterattacks to retake key terrain or gain time to develop defenses.

TYPES OF DEFENSIVE OPERATIONS

8-3. There are three basic types of defensive operations: the area defense, the mobile defense, and the retrograde. These three types have significantly different concepts and pose significantly different problems. Therefore, each type of defensive operations must be dealt with differently when planning and executing the defense. Although the names of these types of defensive operations convey the overall aim of a selected defensive operation, each typically contains elements of the other and combines static and mobile elements.

8-4. Although on the defense, the commander remains alert for opportunities to attack the enemy whenever resources permit. Within a defensive posture, the defending commander may conduct a spoiling attack or a counterattack, if permitted to do so by the factors of METT-TC. (Chapter 5 discusses these two forms of attack.)

AREA DEFENSE

8-5. The *area defense* a type of defensive operation that concentrates on denying enemy forces access to designated terrain for a specific time rather than destroying the enemy outright (FM 3-0). The focus of the area defense is on retaining terrain where the bulk of the defending force positions itself in mutually supporting, prepared positions. Units maintain their positions and control the terrain between these positions. The decisive operation focuses on fires into EAs possibly supplemented by a counterattack. The reserve may or may not take part in the decisive operation. The commander can use his reserve to reinforce fires; add depth, block, or restore the position by counterattack; seize the initiative; and destroy enemy forces. Units at all echelons can conduct an area defense. (Chapter 9 discusses the area defense.)

MOBILE DEFENSE

8-6. The *mobile defense* is a type of defensive operation that concentrates on the destruction or defeat of the enemy through a decisive attack by a striking force (FM 3-0). The mobile defense focuses on defeating or destroying the enemy by allowing him to advance to a point where he is exposed to a decisive counterattack by the striking force. The decisive operation is a counterattack conducted by the striking force. The striking force is a dedicated counterattack force constituting the bulk of available combat power. A fixing force supplements the striking force. The commander uses his fixing force to hold attacking enemy forces in position, to help channel attacking enemy forces into ambush areas, and to retain areas from which to launch the striking force.

8-7. A mobile defense requires an AO of considerable depth. The commander must be able to shape the battlefield, causing the enemy to overextend his lines of communication (LOCs), expose his flanks, and dissipate his combat power. Likewise, the commander must be able to move around and behind the enemy force he intends to cut off and destroy. Divisions and larger formations normally execute mobile defenses. However, subordinate echelons may participate as part of the fixing force or the striking force. (Chapter 10 discusses the mobile defense.)

RETROGRADE

8-8. The *retrograde* is a type of defensive operation that involves organized movement away from the enemy (FM 3-0). The enemy may force these operations, or a commander may execute them voluntarily. The higher commander of the force executing the retrograde must approve the retrograde operation before its initiation in either case. The retrograde is a transitional operation; it is not conducted in isolation. It is part of a larger scheme of maneuver designed to regain the initiative and defeat the enemy. (Chapter 11 further discusses the retrograde.)

COMMON DEFENSIVE CONTROL MEASURES

8-9. The commander controls the defense by using control measures to provide the flexibility needed to respond to changes in the situation and allow the defending commander to rapidly concentrate combat power at the decisive point. Defensive control measures within a commander's AO include designating his security area, the battle handover line (BHL), and the main battle area (MBA) with its associated forward edge of the battle area (FEBA). (Chapter 12 discusses security operations.) (Paragraph 8-13 defines the FEBA.) The commander can use battle positions and additional direct fire control and fire support coordinating measures (FSCM) in addition to those control measures introduced in Chapter 2 to further synchronize the employment of his combat power. He can designate disengagement lines to trigger the displacement of his forces.

BATTLE HANDOVER LINE

8-10. **The *battle handover line* (BHL) is a designated phase line on the ground where responsibility transitions from the stationary force to the moving force and vice versa.** The common higher commander of the two forces establishes the BHL after consulting with both commanders. The stationary commander determines the location of the line. The BHL is forward of the FEBA in the defense or the forward line of own troops (FLOT) in the offense. The commander draws it where elements of the passing unit can be effectively supported by the direct fires of the forward combat elements of the stationary unit until passage of lines is complete. The area between the BHL and the stationary force belongs to the stationary force commander. He may employ security forces, obstacles, and fires in the area. (Figure 8-2, page 8-6, depicts a BHL used in conjunction with other control measures for a rearward passage of lines.)

MAIN BATTLE AREA

8-11. **The *main battle area* (MBA) is the area where the commander intends to deploy the bulk of his combat power and conduct his decisive operations to defeat an attacking enemy.** In the defense, the commander's major advantage is that he normally selects the ground on which the battle takes place. He positions his forces in mutually supporting positions in depth to absorb enemy penetrations or canalize them into prepared EAs, defeating the enemy's attack by concentrating the effects of overwhelming combat power. The natural defensive strength of the position has a direct bearing on the distribution of forces in relation to both frontage and depth. In addition, defending units typically employ field fortifications and obstacles to improve the terrain's natural defensive strength. The MBA also includes the area where the defending force creates an opportunity to deliver a decisive counterattack to defeat or destroy the enemy.

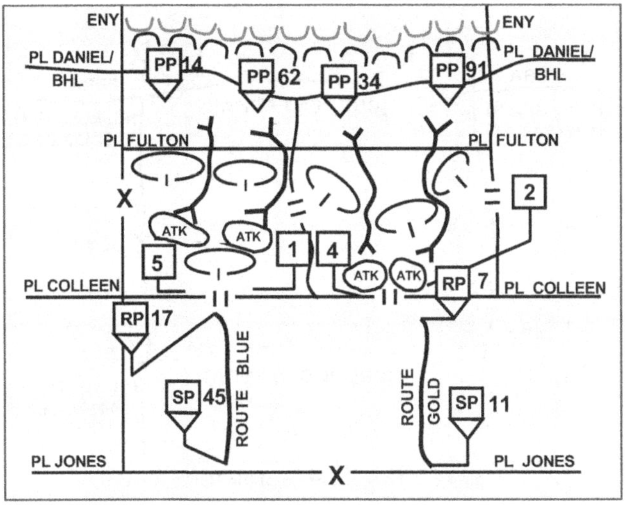

Figure 8-2. Use of a BHL in a Rearward Passage of Lines

8-12. The MBA extends from the FEBA to the unit's rear boundary. The commander locates his subordinate unit boundaries along identifiable terrain features and extends them out beyond the FLOT by establishing forward boundaries. Unit boundaries should not split avenues of approach or key terrain. The commander selects the MBA based on the products of the intelligence preparation of the battlefield (IPB) process and his own analysis using the factors of METT-TC. The IPB process indicates how the enemy will most likely use the available avenues of approach.

FORWARD EDGE OF THE BATTLE AREA

8-13. The *forward edge of the battle area* (FEBA) is the foremost limits of a series of areas in which ground combat units are deployed, excluding the areas in which the covering or screening forces are operating, designated to coordinate fire support, the positioning of forces, or the maneuver of units (JP 1-02). The US Army uses a FEBA only in defensive operations. The FEBA is not a boundary, but conveys the commander's intent. It marks the foremost limits of the areas in which the preponderance of ground combat units deploy, excluding the areas in which security forces are operating. MBA forces can temporarily move forward of the FEBA to expedite the retrograde operations of security forces. The commander designates a FEBA to coordinate fire support and to maneuver his forces. A phase line designating the forwardmost point of the MBA indicates the FEBA. The FEBA shows the senior commander's planned limit for the effects of direct fires by defending forces. Defending units must address this area in their scheme of maneuver and exchange information regarding tactical plans at the coordinating points. (Figure 8-3 graphically depicts the current FEBA and a proposed FEBA.)

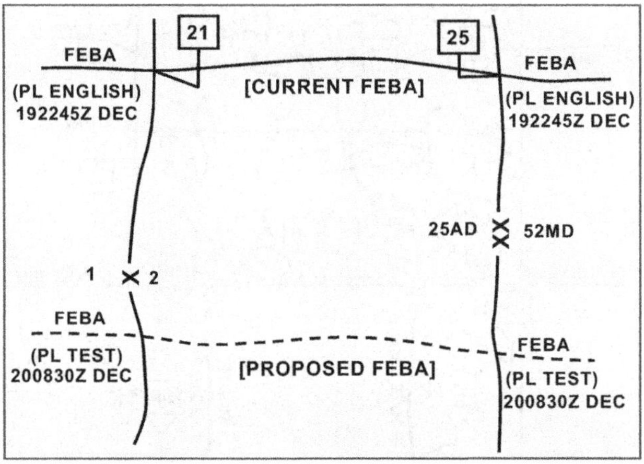

Figure 8-3. Forward Edge of the Battle Area

BATTLE POSITIONS

8-14. **A *battle position* is a defensive location oriented on a likely enemy avenue of approach.** The battle position is an intent graphic that depicts the location and general orientation of the majority of the defending forces. A commander's use of a battle position does not direct the position of the subordinate's entire force within its bounds since it is not an AO. (See Figure 8-4, page 8-8.) Units as large as battalion task forces and as small as squads or sections use battle positions. They may occupy the topographical crest of a hill, a forward slope, a reverse slope, or a combination of these areas. The commander selects his positions based on terrain, enemy capabilities, and friendly capabilities. A commander can assign all or some of his subordinates battle positions within his AO. (See Figure 8-5, page 8-8.)

8-15. The commander may assign his subordinates battle positions in situations when he needs to retain a greater degree of control over the maneuver of his subordinate units than what he has with only an AO, as he controls maneuver outside the general location of the battle position. He may assign multiple battle positions to a single unit, which allows that unit to maneuver between battle positions. The commander specifies mission and engagement criteria to the unit assigned to a battle position. Security, combat support (CS), and combat service support (CSS) forces may operate outside a unit's battle position.

8-16. Battle positions are not normally held at all costs. The commander assigning a unit to a battle position should specify when and under what conditions the unit displaces from the position. If a unit is ordered to defend a battle position, its commander has the option of moving off the battle position. If that unit is directed to retain a battle position, its commander needs to know the specific conditions that must exist before his unit can displace.

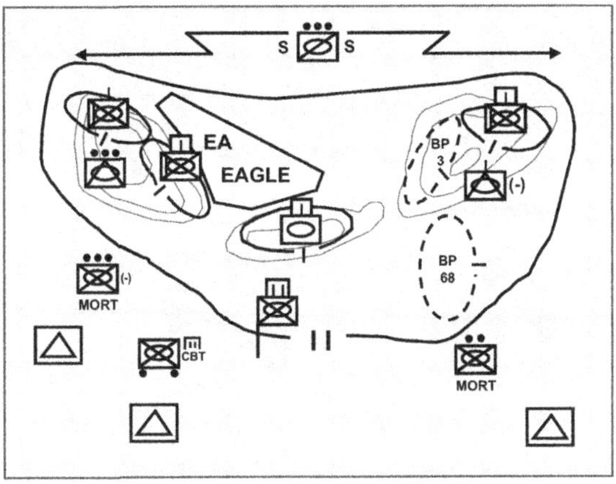

Figure 8-4. Task Force Battle Position

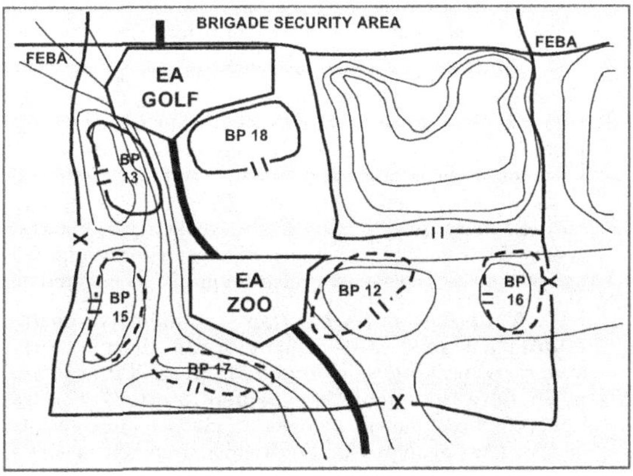

Figure 8-5. AO and Battle Position Control Measures Used in Combination

8-17. There are five kinds of battle positions—primary, alternate, supplementary, subsequent, and strong point. (See Figure 8-6.) When assigning battle positions, the commander always designates the primary battle position.

He designates and prepares alternate, supplementary, and subsequent positions as time and other resources permit and if the situation, especially terrain, requires them.

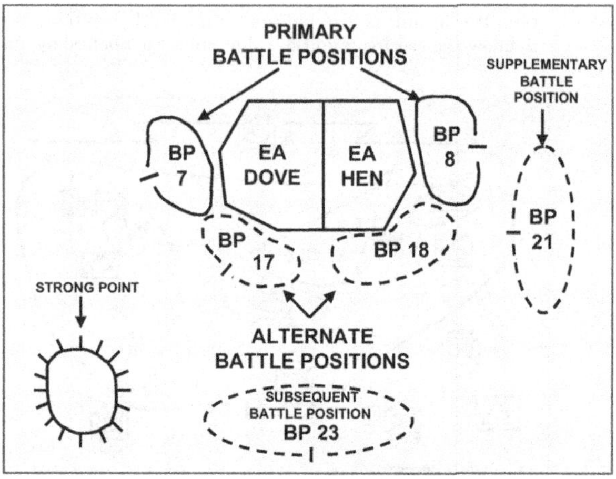

Figure 8-6. Five Kinds of Battle Positions

8-18. **The *primary position* is the position that covers the enemy's most likely avenue of approach into the AO.** It is the best position from which to accomplish the assigned mission, such as cover an EA.

8-19. **An *alternate position* is a defensive position that the commander assigns to a unit or weapon for occupation when the primary position becomes untenable or unsuitable for carrying out the assigned task.** It covers the same area as the primary position. He locates alternate positions so the occupant can continue to fulfill his original task, such as covering the same avenue of approach (AA) or EA as the primary position. These positions increase the defender's survivability by allowing him to engage the enemy from multiple positions. For example, a unit moves to its alternate positions when the enemy brings suppressive fires on the primary position.

8-20. **A *supplementary position* is a defensive position located within a unit's assigned AO that provides the best sectors of fire and defensive terrain along an avenue of approach that is not the primary avenue where the enemy is expected to attack.** For example, an AA into a unit's AO from one of its flanks normally requires establishing supplementary positions to allow a unit or weapon system to engage enemy forces traveling along that avenue.

8-21. **A *subsequent position* is a position that a unit expects to move to during the course of battle.** A defending unit may have a series of subsequent positions. Subsequent positions can also have primary, alternate, and supplementary positions associated with them.

8-22. **A *strong point* is a heavily fortified battle position tied to a natural or reinforcing obstacle to create an anchor for the defense or to deny the enemy decisive or key terrain.** The commander prepares a strong point for all-around defense. (See Figure 8-7.) He positions strong points on key or decisive terrain as necessary. The unit occupying the strong point prepares positions for its weapon systems, vehicles, soldiers, and supplies. The commander also establishes a strong point when he anticipates that enemy actions will isolate a defending force retaining terrain critical to the defense.

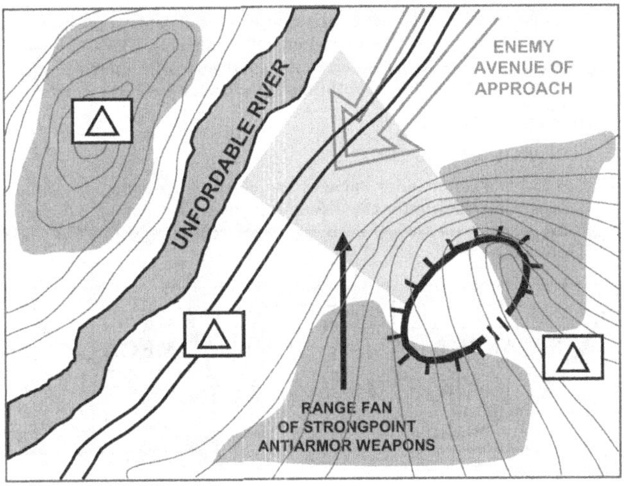

Figure 8-7. Strong Point Defense

8-23. Before assigning a strong point mission, the commander must ensure that the strong point force has sufficient time and resources to construct the position, which requires significant engineer support. A minimally effective strong point typically requires a one-day effort from an engineer unit the same size as the unit defending the strong point. Normally, companies and battalions occupy strong points, although brigades may construct them. The commander does not normally establish strong points for units smaller than company size. This is because a platoon or squad cannot secure a perimeter large enough to encompass all required assets and supplies.

FIRE SUPPORT COORDINATING MEASURES

8-24. The commander tries to engage the enemy at extended ranges and attrit him as his attack advances. To control indirect fires in the defense, the commander uses those common FSCM introduced in Chapter 2. He can also employ final protective fires.

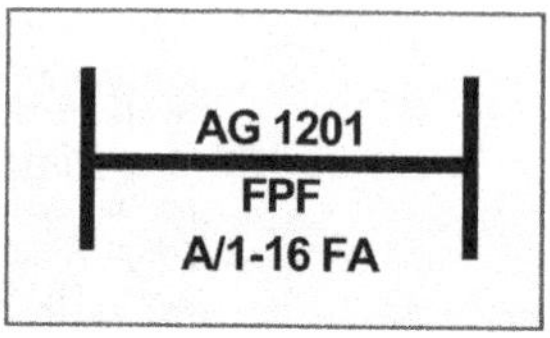

Figure 8-8. Final Protective Fire

8-25. *Final protective fires* (FPFs) are immediately available preplanned barriers of fires designed to impede enemy movement across defensive lines or areas (JP 3-09). Both direct- and indirect- fire weapons can provide FPFs. The commander can only assign each firing battery or platoon a single FPF. A FPF is a priority target for an element or system, and those fire units are laid on that target when they are not engaged in other fire missions. When the enemy initiates his final assault into a defensive position, the defending unit initiates its FPFs to kill enemy infantry soldiers and suppress his armored vehicles. Selected crew-served weapons fire along predesignated final protective lines (FPLs) to break up infantry assaults. (Figure 8-8 depicts a FPF.)

DIRECT FIRE CONTROL MEASURES

8-26. The commander engages the enemy force with all available defensive fires when they enter the defending unit's EA. Chapter 2 defines these direct fire control measures, such as target reference points and EAs. (See Figure 8-9.)

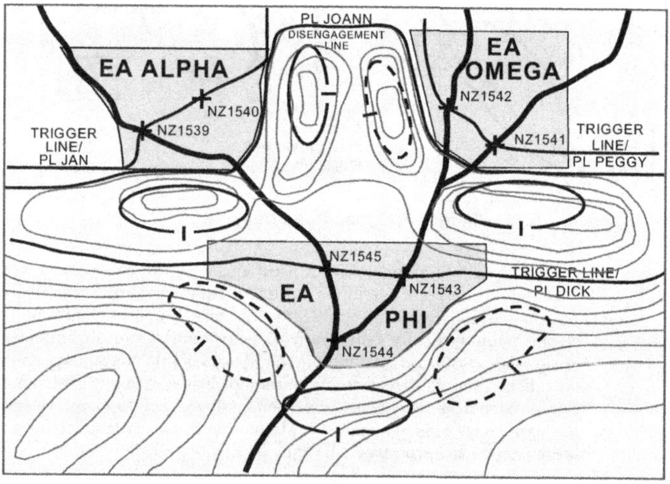

Figure 8-9. Direct Fire Control Measures

DISENGAGEMENT LINE

8-27. **A *disengagement line* is a phase line located on identifiable terrain that, when crossed by the enemy, signals to defending elements that it is time to displace to their next positions.** Phase Line JOANN is a disengagement line in Figure 8-9. The commander uses these lines in the delay and the defense when he does not intend for the defending unit to become decisively engaged. He establishes criteria for the disengagement, such as number of enemy vehicles by type, friendly losses, or enemy movement to flanking locations. Multiple disengagement lines, one for each system in the defense, may exist.

COMMON DEFENSIVE PLANNING CONSIDERATIONS

8-28. At the onset of the attack, the defending commander yields the initiative to the enemy. However, he exploits prepared, mutually supporting positions organized for all-around defense and uses his knowledge of the terrain to slow the enemy's momentum. The defending force maintains its security and disrupts the enemy's attack at every opportunity. The defending commander hinders enemy offensive preparations by using long-range fires and deep maneuver to reduce the force of the enemy's initial blows and start the process of wresting the initiative from the enemy. He draws the enemy into EAs where he can initiate combat on his own terms. The commander surprises the enemy as concentrated and integrated fires violently erupt from concealed and protected positions. He then counterattacks the enemy, repeatedly imposing unexpected blows. The widespread application of highly accurate and lethal weapons, high degree of tactical mobility, dynamic situational changes, and extended spatial scope of unit AOs all characterize contemporary combined arms warfare. The commander exploits small tactical success and opportunities to build momentum rapidly. The commander first able to see the battlefield, understand the common operational picture's implications, and take effective action will defeat his opponent's combined arms team, shatter his cohesion, degrade his strength and ability to concentrate, and destroy his exposed forces.

8-29. The defending force does not have to kill every enemy soldier, squad, or combat system to be successful. It only has to destroy the enemy's ability to synchronize his combined arms team or his will to fight. Those events signal a transition period that affords the defending commander the opportunity to seize the initiative and return to the offense.

8-30. A defense is more effective when there is adequate time to thoroughly plan and prepare defensive positions. Lack of preparation time may cause the commander to maintain a larger-than-normal reserve force or accept greater risks than usual. All units must be capable of mounting a defense with minimal preparation, but a strong defense takes time to organize and prepare. If the enemy attack does not take place at the predicted time, the commander should use the additional time to improve his unit's defensive positions. He can increase the effectiveness of the security area, establish additional alternate and supplementary positions, refine the defensive plan to include branches and sequels, conduct defensive rehearsals, and maintain vehicles and personnel. To gain time to organize a defense, the commander may order his security force to conduct a delay while the main body disengages and moves to more advantageous positions. The security force must know how long it needs to delay the enemy for the main body to prepare its defense and be task organized to conduct a delay. (Chapter 11 discusses the delay.)

8-31. The common defensive planning considerations addressed in the following paragraphs apply to all types of defensive operations. In the defense, synchronizing the effects of his combat and supporting systems allows a commander to apply overwhelming combat power against selected advancing enemy forces to unhinge the enemy commander's plan and destroy his combined arms team. Defensive synchronization is normally the result of detailed planning and preparation among the various units participating in an operation. While these activities may be separated in time and space, they are synchronized if their combined consequences are felt at decisive times and places. All defensive operations are a mix of static and dynamic actions. As an operation evolves, the commander knows that he will probably be required to shift his decisive and shaping operations to press the fight and keep the enemy off balance. Synchronized prior planning and preparation bolster the commander's combat power, increasing the effectiveness of the defense. The commander must remain cognizant of the possibility of dislocated civilians attempting to move through his positions in an effort to escape approaching enemy forces throughout the defense.

INTELLIGENCE

8-32. During the planning process, the commander uses intelligence products to identify probable enemy objectives and various approaches. He studies patterns of enemy operations and the enemy's vulnerability to counterattack, interdiction, electronic warfare, air attacks, and canalization by obstacles. The commander must also examine the enemy's capability to conduct air attacks against his force, insert forces behind friendly units, and employ nuclear, biological, and chemical weapons. He must determine how soon follow-on forces can join the fight against an enemy attacking in echelons.

8-33. The commander uses his intelligence, surveillance, and reconnaissance (ISR), and engineer assets to study the terrain. By studying the terrain, the commander tries to determine the principal enemy and friendly heavy, light, and air avenues of approach. He wants to determine the most advantageous area for the enemy's main attack, as well as other factors of observation and fields of fire, avenues of approach, key terrain, obstacles, and cover and concealment (OAKOC). (See FM 6-0 for a detailed discussion of OAKOC.)

8-34. The commander approves an integrated ISR plan that provides early identification of as many of the following requirements as possible:

- Locations, composition, equipment, strengths, and weaknesses of the advancing enemy force.
- Enemy reconnaissance objectives or goals.
- Locations of possible enemy assembly areas.
- Location of enemy indirect fire weapon systems and units.
- Location of gaps, assailable flanks, and other enemy weaknesses.
- Location of areas for enemy helicopter and parachute assaults.
- Location of artillery and air defense gun and missile units.
- Location of enemy electronic warfare units.
- Location, numbers, and intentions of civilian populations.
- Effects of weather and terrain on current and projected operations.

- Likely withdrawal routes for enemy forces.
- Numbers, routes, and direction of movement of dislocated civilians.
- Anticipated timetable for the enemy's most likely COA.
- Locations of enemy command posts, fire direction control centers, electronic warfare sites, and target acquisition sensor and target fusion sites and the frequencies they are using.

It is unlikely that the commander has complete knowledge of the enemy's intentions; therefore, he must plan to continue his intelligence efforts during the battle. (FM 2-0 provides an overview of the intelligence process and the capabilities of technical surveillance systems. FM 3-55 discusses reconnaissance assets available at each echelon. FMs 3-11.9 and 3-34.170 discuss the specialized tasks associated with NBC and engineer reconnaissance.)

8-35. The commander's ability to see the enemy is critical to the conduct of all defensive operations. Defensive plans must address the sustainment, replacement, and reconstitution of ISR assets throughout the preparation and execution of the defense.

MANEUVER

8-36. The commander's intent is to defeat the enemy force's attack by overwhelming it with repeated, unexpected blows before it conducts its final assault on friendly defensive positions. As the enemy attack fails, the enemy must attempt to withdraw or transition to a defense in the face of friendly counterattacks. If the enemy succeeds in overrunning a key defensive position, the defending force counterattacks to overwhelm the enemy before he can either organize that position for defense or exploit his success.

Exploit the Advantages of Terrain

8-37. The defending commander exploits the defending force's advantages of occupying the terrain where the fight will occur. The defending force engages the attacker from locations that give the defending force an advantage over the attacking enemy. These locations include defiles, rivers, thick woods, swamps, cliffs, canals, built-up areas, and reverse slopes. Defensive positions in the MBA should make use of existing and reinforcing obstacles. The commander may choose to shape the battlefield by defending one area to deny terrain to the enemy while delaying in another area to deceive the enemy commander into believing he has achieved success.

8-38. The defending commander plans how to use key terrain to impede the enemy's movement. He seeks out terrain that allows him to mass the effects of his fires but forces the enemy to commit his force piecemeal into friendly EAs. This exposes portions of the enemy force for destruction without giving up the advantages of fighting from protected positions. Examples of key terrain include terrain that permits the defending force to cover a major obstacle system by fire, and important road junctions and choke points that impact troop movements, such as the movement of reserves and LOCs.

8-39. The commander determines the probable force ratios he will face and arrays his forces accordingly. The terrain impacts how fast the enemy can close on his positions and how much time is available to employ combat multipliers, such as indirect fires. Once the commander arrives at acceptable force ratios—or the degree of risk he must take is clear—he allocates his available forces and begins planning his EAs.

8-40. On each enemy AA, the commander determines where he wants to destroy the enemy. He arrays forces allocated to that AA around this point to establish an EA. He uses obstacles and fires to canalize enemy forces into this EA. The commander takes actions to increase the kill probabilities of his various weapon systems at different ranges. This includes establishing range markers for direct fire weapons, confirming the zero on his weapons, or clearing obstacles that might snag the cables over which the commands of his wire-guided munitions, like the TOW missile, travel.

8-41. Generally, defending forces have the advantage of preparing the terrain by reinforcing natural obstacles, fortifying positions, and rehearsing operations. First, they prepare the ground to force the piecemeal commitment of enemy forces and their subsequent defeat in detail. Second, they prepare the ground to force the enemy to fight where he does not want to fight, such as in open areas dominated by terrain that offers adequate cover and concealment for the occupying friendly forces. The defending force tries to guide or entice the enemy into prepared EAs. Units employ and continuously strengthen obstacles and fortifications to improve the natural defensive strength of the position, which has a direct bearing on the distribution of forces, frontages, and depth of the defense.

8-42. Terrain features that favor defensive operations include—

- A series of parallel ridges across the line of hostile advance.
- Unfordable streams, swamps, lakes, and other obstacles on the front and flanks.
- High ground with good observation and long-range fields of fire.
- Concealed movement routes immediately behind defensive positions.
- Limited road network in front of the line of contact to confine the enemy to predictable avenues of approach.
- Good road network behind the line of contact that allows the commander to reposition his forces as the battle progresses.

The opposite of the terrain conditions listed above degrades a force's ability to conduct defensive operations. For example, terrain with a limited road net that canalizes the defending force allows the enemy to predict its movement and take steps to interdict that movement.

Maintain Security

8-43. Security operations seek to confuse the enemy about the location of the commander's main battle positions, prevent enemy observation of preparations and positions, and keep the enemy from delivering observed fire on the positions. They also try to force the attacking enemy to deploy prematurely. They can offset the attacker's inherent advantage of initiative regarding the time, place, plan, direction, strength, and composition of his attack by forcing him to attack blind into prepared defenses. The commander must not permit enemy reconnaissance and surveillance assets to determine the precise location and strength of defensive positions, obstacles, EAs, and reserves. First, the defending force conducts reconnaissance to gain and maintain contact with the enemy. Second, each echelon normally establishes a security area forward of its MBA. All units conduct aggressive security operations within their AO, including the rear area, to seek out and repel or kill enemy reconnaissance and other forces. Units implement operations security (OPSEC) measures and other defensive information operations to deny the enemy information about friendly dispositions. (See Chapter 12 for more information on security operations.)

Disrupt the Enemy Attack at Every Opportunity

8-44. The defending force conducts operations throughout the depth of the enemy's formation in time and space to destroy his key units and assets, particularly his artillery and reserves, or disrupt their timely introduction into battle at the point of engagement. This allows the defending force to regain the initiative. It conducts spoiling attacks to disrupt the enemy's troop concentrations and attack preparations. The defending force counterattacks enemy successes rapidly with its reserve, the forces at hand, or a striking force before the enemy can exploit success. It conducts offensive information operations to assist this process.

Mass the Effects of Combat Power

8-45. The defending force must mass the effects of its combat power to overwhelm the enemy and regain the initiative. The commander uses economy of force measures in areas that do not involve his decisive operation to mass the effects of his forces in the area where a decision is sought. This decisive point can be a geographical objective or an enemy force. In an area defense, defending units use EAs to concentrate the effects of overwhelming combat power from mutually supporting positions. In a mobile defense, the commander uses the striking force to generate overwhelming combat power at the decisive point. Another way he can generate the effects of mass is through committing his reserve.

Ensure Mutual Support

8-46. Mutual support exists when positions and units support each other by direct, indirect, lethal, and nonlethal fire, thus preventing the enemy from attacking one position without being subjected to fire from one or more adjacent positions. Mutual support increases the strength of all defensive positions, prevents defeat in detail, and helps prevent infiltration between positions. Tactical positions achieve the maximum degree of mutual support between them when they are located to observe or monitor the ground between them or conduct patrols to prevent any enemy infiltration. At night or during periods of limited visibility, the commander may position small tactical units closer together to retain the advantages of mutual support. Unit leaders must coordinate the nature and extent of their mutual support.

Heavy Forces

8-47. When the majority of a defending force consists of mechanized or armored units, the commander can conduct a defense designed to take advantage of the tactical mobility and protection offered by organic combat vehicles. Heavy forces can maneuver to delay the advance of a strong enemy force and then immediately change from a mobile to a static form of defense or counterattack. Such forces are well suited for use as security and MBA forces. They are more suited for operations within an NBC contaminated environment than light forces because of their built-in protection.

Light Forces

8-48. When facing enemy light forces, the commander deploys and uses defending light forces in the same manner as heavy forces are used against other heavy forces. Light forces facing a heavy enemy are primarily used in static roles within the MBA or in security roles within the rear area. When facing heavy enemy forces, light infantry forces are most effective when fighting from prepared defenses or in close terrain, such as swamps, woods, hilly and mountainous areas, and urban areas where they can take advantage of their foot mobility and short-range infantry and anti-armor weapons.

8-49. The commander uses an air assault unit in the same manner as other light forces once it deploys into its landing zones (LZs). (See Appendix C for a discussion of air assault operations.) However, there may be more problems in extracting such a force, particularly if it is in direct contact with the enemy. Because of its mobility and potential reaction speed, an air assault force is often well-suited for a reserve role during defensive operations. Its tasks might include—

- Rapid reinforcement of a threatened position.
- Occupation of a blocking position, possibly in conjunction with existing defensive positions.
- Rear area security operations, such as containment of an enemy airborne or helicopter assault.
- Reinforcement of encircled friendly forces.
- Flank protection.

Rotary- and Fixed-Wing Aviation

8-50. Aviation assets are particularly valuable in the defense because of their speed, mobility, and versatility. Their tasks can include—

- Conducting reconnaissance and security operations.
- Conducting shaping operations to establish the necessary conditions for decisive operations by other forces through attritting, disrupting, and delaying the enemy.
- Conducting counterattacks and spoiling attacks.
- Controlling ground for limited periods where a commander does not wish to irrevocably commit ground forces; for example, forward of an executed obstacle.
- Blocking enemy penetrations.
- Closing gaps in a defense plan before the arrival of ground maneuver forces.
- Facilitating the disengagement of ground forces.
- Countering enemy activities in the rear area, in particular enemy airborne or air assault forces.
- Using available utility and cargo helicopters in their normal roles to support the defensive effort, such as resupplying the defending force with Class IV barrier material or facilitating casualty evacuation.

- Assisting in the countermobility effort.
- Providing long-range biological surveillance.

FIRE SUPPORT

8-51. In the defense, the commander uses his fire support systems to neutralize, suppress, or destroy enemy forces; to delay or disrupt the enemy's ability to execute a given COA; and to enhance the effects of massed direct fires. Thus fire support systems support both the commander's decisive and shaping operations.

8-52. The defending force is more effective if it can locate and attack enemy forces while the enemy is stationary and concentrated in assembly areas or advancing along LOCs, as opposed to when he is deployed in combat formations within the MBA. To accomplish this, the defending force must employ its fire support system throughout its AO. It must be closely linked to target acquisition means, including ISR assets.

8-53. As the commander develops his defensive plans, he must visualize how to synchronize, coordinate, and distribute the effects of indirect and direct fire at the decisive time and place. He places permissive FSCM as close as possible to friendly positions to facilitate the rapid engagement of attacking enemy forces. Prior coordination facilitates the massing of the effects of fires before enemy targets concentrated at obstacles and other choke points can disperse. Proper distribution of fire effects ensures the massing of overwhelming combat power at the decisive point. Proper fire distribution also ensures that high-payoff targets are destroyed without wasting assets through repetitive engagement by multiple friendly systems.

8-54. Indirect fires have the greatest impact on the enemy when they are synchronized with direct fires and the use of obstacles, defensive positions, and counterattack plans. The commander must integrate the defensive fire and obstacle plans from the beginning. Indirect fires complement the effects of obstacles and can disrupt enemy attempts to breach or bypass these obstacles. For the plans to work, all elements in the fire support chain—from forward observers in fire support teams to the fire support coordinator including the supporting tactical air control party—must understand the commander's intent, the scheme of maneuver, and the obstacle plan.

8-55. There are various fire support considerations for each phase of the fight. As part of his shaping operations during defense preparations, a commander tries to disrupt the enemy's attack preparations by—

- Conducting harassing fires on choke points and likely enemy assembly areas.
- Employing air support on known, suspected, and likely enemy locations.
- Attritting his resources by continuously engaging high-payoff targets.
- Conducting offensive information operations to degrade the enemy's ability to command and control his forces.
- Employing counterfires to engage and destroy enemy artillery and mortar systems attempting to deliver suppressive fires.

- Providing fires in support of the unit's security operations, such as a unit conducting the tactical mission task of counterreconnaissance.

In some situations it may be better to wait to execute a counterfire mission until the fighting begins in the MBA. However, when defending forces enjoy qualitative advantages in fire support, the advantages accruing from a counterfire battle usually outweigh the risks to the defending maneuver force. The defender's ability to mass fires quickly and then rapidly reposition its forces is a major factor in disrupting the enemy and establishing the required conditions for successful decisive operations.

8-56. The commander employs fires to support his security forces, using precision and other munitions to destroy enemy reconnaissance and other high-payoff targets. This also helps to deceive the enemy about the location of the MBA. He supports the security force by planning the delivery of the effects of fires at appropriate times and places throughout his area of influence to slow and canalize the enemy forces as they approach the security area. This allows the security force to engage the enemy on more favorable terms. To prevent fratricide, he places no fire areas over his security forces. Finally, he uses fires to support the withdrawal of the security force once its shaping mission is complete and the defending unit is prepared to conduct MBA operations.

8-57. Air support can play an important part in delaying enemy forces following or attempting to bypass rearward-moving defending forces. Air operations contribute to overcoming the enemy's initial advantage of freedom of action. Often, only aircraft are available to initially oppose an enemy penetration until ground forces can redeploy to engage it. Close air support (CAS) can be instrumental in disrupting an enemy advance. It can operate with Army helicopters and artillery assets to form a joint air attack team (JAAT). The commander also incorporates artillery fires with electronic warfare and joint systems to suppress enemy air defenses while CAS hits a target. Air interdiction can delay, destroy, or neutralize enemy follow-on forces, thereby providing the commander with additional time to prepare his defensive positions.

8-58. Once the fight moves into the MBA, fire support assets continue to target enemy combat units to force them to deploy. At the same time, fire support assets inflict casualties, disrupt the cohesion of the enemy's attack and impede his ability to mass combat power. Fire support assets continue to attack enemy follow-on forces before they can be committed to the MBA. This further isolates the attacking enemy force. They attack C2 facilities and logistics sites in depth to contribute to isolating the attacking enemy. The commander takes advantage of the range and flexibility of his fire support weapons to mass fires at critical points, such as obstacles and EAs, to slow and canalize the enemy to provide better targets for direct fire systems. Fire support systems cover barriers, gaps, and open areas within the MBA. Tasks assigned to these fire support systems include closing obstacle gaps or reseeding previously breached obstacles in accordance with the rules of engagement. Other tasks include—

- Massing fires to suppress enemy direct and indirect fire systems to facilitate defensive maneuver, especially the counterattack and disengagement.
- Neutralizing or isolating enemy forces that have penetrated the defensive area and impeding the movement of enemy reserves.

- Attacking enemy artillery and forward air defense elements.
- Using jamming to degrade or destroy the enemy's ability to transmit data and information.
- Reallocating fire support assets, after identifying the enemy's main effort, to reinforce fires in the most vulnerable areas.
- Separating attacking enemy combat vehicles from light infantry, disrupting the enemy's combined arms team.

8-59. In response to shallow enemy penetrations, artillery commanders normally reposition their systems laterally, away from that point. This allows artillery systems to provide fire support throughout the area of penetration.

AIR DEFENSE

8-60. Freedom of movement is essential to successful defensive operations. In a hostile air environment, the defending force must establish air defense in depth around critical points, areas, units, and activities. The dedicated air defense artillery resources probably cannot provide adequate cover completely throughout the AO against all possible threats; therefore, the commander must establish priorities for coverage and assume risk.

Active Air Defense

8-61. Normally, the commander's priorities for air defense protection in the defense begin with his C2 facilities. Because they are generally fixed or semi-fixed sites with high-electronic signatures, they are susceptible to attack by enemy aircraft. Air defense coordinators examine air avenues of approach toward C2 facilities and position guns and missiles to prevent enemy aircraft from reaching their targets.

8-62. Logistics support areas, main supply routes (MSRs), and other logistics sites are also relatively fixed and easily identified from the air. Passive air defense measures help prevent detection. However, once the enemy detects them, he will attempt to attack them. Therefore, route and point security missions require air defense units to locate along the MSR and in positions to protect fixed locations. The commander allocates his air defense assets to protect these locations in accordance with the factors of METT-TC.

8-63. The air defense responsibility may be most critical in forward areas since the commander will task air defense artillery (ADA) units along the FEBA to engage enemy aircraft providing CAS or attempting low-level penetration of friendly air defenses en route to a target in the friendly rear area. Air defense assets protecting combat forces in forward battle positions and strong points are more exposed to destruction by enemy direct and indirect systems than air defense systems located elsewhere on the battlefield. The commander must take steps to ensure their survivability, such as placing man-portable air defense missile gunners inside combat vehicles when not actively engaging enemy aircraft.

8-64. The reserve or striking force is initially a stationary hidden force. However, it is easy to observe from the air as it moves on its commitment by the commander. It is especially vulnerable once discovered. Therefore, the commander positions air defense assets to protect the reserve or striking force, whether it is stationary or moving.

8-65. Air defense systems that protect the reserve and the striking force must be as mobile and protected as the forces they are protecting. The less mobile equipment is usually kept in more static roles. The commander continually coordinates his air defense activities with his air and artillery operations to avoid fratricide. Air defense units and support assets move in support of the defensive effort. If the enemy can disrupt this support from the air, it will affect the defense. Correct assessment of enemy air corridors and tactics is essential to guarantee protection and management of these resources.

8-66. The destruction of key bridges or the closing of choke points interrupts the defender's freedom of movement. The force must protect these positions to sustain the defense and allow the conduct of counterattacks. The commander locates air defense assets to protect these vital locations.

Passive Air Defense

8-67. The commander also uses passive air defense measures to protect his force. *Passive air defense measures* are all measures other than active defense taken to minimize the effects of the hostile air action (FM 3-01.8). Passive defense measures are of two types: attack avoidance and damage-limiting measures. Both include the use of cover, concealment and camouflage, and deception.

8-68. **Attack Avoidance.** Attack avoidance means taking steps to avoid being seen by the enemy. If the force cannot be seen, the probability of it being hit diminishes to near zero. The commander uses the same techniques, procedures, and materials for concealment from aerial observation as for concealment from ground observation. He employs three principles to enhance concealment—

- **Siting.** Siting means selecting the most advantageous position in which to hide a man, an object, or an activity. This is often the shadows provided by woodlines, wadies, and buildings.
- **Discipline.** Success in any concealment effort hinges on strict concealment discipline by units and individual soldiers. The unit should avoid activities that change the appearance of an area or reveal the presence of military equipment. Laxness and carelessness will undoubtedly reveal a position. Tracks, spoil, and debris are the most common signs of military activity, which indicate concealed objects. Ensure that new tracks follow existing paths, roads, fences, or natural lines in the terrain pattern. Do not end exposed routes at a position, but extend them to another logical termination. If practical, the unit should brush out, camouflage, or cover its tracks. It should cover or place spoil and debris to blend with the surroundings. The unit adds artificial camouflage when the terrain and natural vegetation are such that natural concealment is not possible.
- **Construction.** Adding natural materials to blend with the surrounding terrain augments this type of concealment.

8-69. There are three fundamental methods of concealing installations and activities—hiding, blending, and disguising.

- **Hiding.** Hiding is the complete concealment of an object by some form of physical screen. For example, sod placed over mines in a minefield hides the mines; the overhead canopy of trees hides the objects beneath from aerial observation; a net hides objects beneath it; a defilade position hides objects from ground observation. In some cases, the screen may be invisible. In other instances, the screen may be visible, but it hides the activity behind it.
- **Blending.** Blending is arranging or applying camouflage materials on, over, and around the object so that it appears to be part of the background. Examples include applying face paint to the exposed areas of skin, and adding burlap, paint, and live vegetation to helmets and clothing to closely resemble or blend into the background. Units can apply the same technique for equipment or structures.
- **Disguising.** Clever disguises can often mislead the enemy about the friendly force's identity, strength, and intention, and may draw his fire from real assets. Therefore, the simulation of objects, pieces of equipment, or activities may have military significance. Inflatable tanks, tents, and buildings can look like the real thing to an aerial observer.

8-70. In addition to hiding equipment, units can avoid detection by using mud for glassy surfaces and unfilled sandbags over windshields. Camouflage is one of the basic weapons of war. Soldiers must understand the importance, the principles, and the techniques of camouflage. All personnel must ensure the effectiveness of all camouflage measures and maintain strict camouflage discipline.

8-71. **Damage-Limiting Measures.** The other type of passive air defense, damage limiting, is also used for survival. These measures attempt to limit damage if the enemy detects the position. If the enemy is to destroy any equipment, he is forced to do it one piece at a time. Enemy forces should never be able to put a unit out of action with just a single attack. The commander uses the same measures taken to limit damage from field artillery attack—dispersion, protective construction, and cover.

- **Dispersion.** Dispersed troops and vehicles force the attacker to concentrate on a single small target that he will likely miss. The wider the dispersion, the greater the potential for limiting damage.
- **Protective Construction.** Using cover, natural or manmade, acts to reduce damage and casualties. Folds in the earth, natural depressions, trees, buildings, and walls offer damage-limiting cover; individuals and units should seek them out and use them habitually. If deployment is in flat terrain lacking cover, digging in or sandbagging can offer some protection. The unit employs smoke if it is moving and cannot use natural cover or cannot build fortifications. Smoke makes target acquisition much more difficult for the attacker.
- **Cover.** Cover emphasizes the importance of passive defense against an air attack. The unit must do everything it can to avoid an attack in the first place, but if it is attacked, it uses cover and dispersion to limit the amount of damage.

Air Defense Role in Reconnaissance and Surveillance

8-72. A commander can direct his air defense systems to deploy forward with scouts along potential air corridors based on the aerial IPB developed by his intelligence and air defense officers. This provides early warning of enemy air infiltration and allows timely engagement of enemy aerial platforms attempting to insert dismounted reconnaissance, infantry, and antiarmor teams. The air defense systems can report stationary locations of enemy aircraft to assist the supported unit in confirming templated LZs. This allows the unit to quickly react to potential ground threats by calling for indirect fires or employing a quick reaction force to defeat this threat. The commander assigns a clear mission to these systems to ensure that they do not compromise the supported unit's integrated ISR plan by prematurely engaging enemy aerial reconnaissance platforms. He establishes a well-defined trigger event to prevent this from happening. Additionally, he ensures the integration of ADA unique munitions into the supported unit's CSS plan based on the planned time that these assets will be forward.

MOBILITY/COUNTERMOBILITY/SURVIVABILITY

8-73. An attacking enemy has the initiative in terms of where and when he will attack. A defending commander must take a wide range of actions to protect the mobility of his force while degrading the mobility of the enemy. He takes those steps simultaneously to protect his force from losses due to enemy actions.

Mobility

8-74. During the defense, mobility tasks include maintaining routes, coordinating gaps in existing obstacles, and supporting counterattacks. Engineers also open helicopter LZs and tactical landing strips for fixed-wing aircraft. Maintaining and improving routes and creating bypass or alternate routes at critical points are major engineering tasks because movement routes are subjected to fires from enemy artillery and air support systems. These enemy fires may necessitate deploying engineer equipment, such as assault bridging and bulldozers, forward. The commander can also evacuate dislocated civilians or restrict their movements to routes not required by his forces to enhance his mobility. He can do this provided he coordinates the action with the host nation or the appropriate civil military operations agency and fulfills his responsibilities to displaced civilians under international law.

8-75. Priority of mobility support is first to routes used by counterattacking forces, then to routes used by main body forces displacing to subsequent positions. This mainly involves breaching obstacles and improving combat roads and trails to allow tactical support vehicles to accompany moving combat vehicles. Careful coordination ensures leaving required lanes or gaps in obstacles for repositioning main body units and committing the counterattack force during the defense. Chemical reconnaissance systems also contribute to the force's mobility in a contaminated environment.

Countermobility

8-76. In the defense, the commander normally concentrates his engineer efforts on countering the enemy's mobility. A defending force typically requires large quantities of Class IV and V material and specialized equipment to construct fighting and survivability positions and obstacles. With limited assets, the commander must establish priorities among countermobility, mobility, and survivability efforts. He ensures that his staff synchronizes these efforts with the echelon's logistic plans.

8-77. The commander may plan to canalize the enemy force into a salient. In this case, he takes advantage of the enemy force's forward orientation by fixing the enemy and then delivering a blow to the enemy's flank or rear. As the enemy's attacking force assumes a protective posture, the defending commander rapidly coordinates and concentrates all effects of his fires against unprepared and unsupported segments of the enemy force in rapid sequence. The unit may deliver these fires simultaneously or sequentially.

8-78. When planning obstacles, commanders and staffs must consider not only current operations but also future operations. The commander should design obstacles for current operations so they do not hinder future operations. Any commander authorized to employ obstacles can designate certain obstacles that are important to his ability to shape the battlefield as high-priority reserve obstacles. He assigns responsibility for preparation to a subordinate unit but retains authority for ordering their execution or final completion. An example of a reserve obstacle is a highway bridge over a major river. Such obstacles receive the highest priority in preparation and, if ordered, execution by the designated subordinate unit.

8-79. A commander integrates reinforcing obstacles with existing obstacles to improve the natural restrictive nature of the terrain to halt or slow enemy movement, canalize enemy movement into EAs, and protect friendly positions and maneuver. He may choose to employ scatterable mines in accordance with the rules of engagement. Direct and indirect fires must cover obstacles to be effective. This requires the ability to deliver effective fires well beyond the obstacle's location. When possible, units conceal obstacles from hostile observation. They coordinate obstacle plans with adjacent units and conform to the obstacle zone or belts of superior echelons.

8-80. Effective obstacles force the enemy to attempt to breach them if he wants to maintain his momentum and retain the initiative. While the defending force is aware that the enemy is going to breach an obstacle, the enemy tries to conceal exactly where and when he will try to breach. The defending force's plan addresses how to counter such a breach, to include reestablishing the obstacle by using scatterable mines and other techniques.

8-81. Given time and resources, the defending force generally constructs additional obstacle systems to its flanks and rear. These systems can provide additional protection from enemy attacks by forcing the enemy to spend time and resources to breach or bypass the obstacle. This, in turn, gives the defending force more time to engage enemy forces attempting to execute breach operations or bypass these obstacles.

8-82. The commander designates the unit responsible for establishing and securing each obstacle. He may retain execution authority for some obstacles or restrict the use of some types of obstacles to allow other battlefield activities to occur. He allows his subordinate commanders some flexibility in selecting the exact positioning of obstacles. However, all units must know which gaps—

through obstacles and crossing sites—to keep open for the unit's use, as well as the firing and self-destruct times of scatterable mines to prevent delays in movement. The commander must be specific and clear in his orders for firing demolitions, emplacing obstacles, and closing lanes. As each lane closes, the closing unit reports the lane's closure to the higher, subordinate, and adjacent headquarters to preclude displacing units from moving into areas with unmarked or abandoned obstacles.

8-83. Tactical and protective obstacles are constructed primarily at company level and below. Small unit commanders ensure that observation and fires cover all obstacles to hinder breaching. Deliberate protective obstacles are common around fixed sites. Protective obstacles are a key component of survivability operations. They are tied in with FPFs and provide the friendly force with close-in protection. Commanders at all echelons track defensive preparations, such as establishing Class IV and V supply points and start or completion times of obstacle belts and groups. The commander plans how he will restore obstacles the enemy has breached. He uses artillery, air, or ground systems to reseed minefields.

8-84. FM 3-34.1 provides additional information about obstacles and obstacle integration, such as planning factors relating to emplacing obstacles and obstacle function versus lethality. It also describes the methods and essential principles for planning protective obstacles.

Survivability

8-85. Since the attacking enemy force usually has the initiative in terms of where and when it will attack, a defending commander must take a wide range of actions to protect his force from losses due to enemy actions. These steps include ensuring all-around defense, NBC defense, and using smoke.

8-86. The survivability effort for the defense must enable units to concentrate firepower from fixed positions. To avoid detection and destruction by the enemy, units move frequently and establish survivability positions quickly. To provide flexibility, units may need primary, alternate, and supplementary positions. This is particularly true of units defending key or decisive terrain. Units enhance their survivability through concealment, deception, dispersion, and field fortifications.

8-87. Survivability tasks include using engineer equipment to assist in preparing and constructing trenches, command post shelters, and artillery firing, radar, and combat vehicle fighting positions. The commander provides guidance on the level of protection—such as hull defilade or overhead cover, system priorities, and early use of specialized engineer systems that can construct survivability positions. He should protect supply stocks against blast, shrapnel, incendiaries, and NBC contamination. Supplies loaded on tactical vehicles can be protected against almost anything but a direct hit by constructing berms large enough to accommodate the vehicles and deep enough to keep supplies below ground level. The force's engineer officer can advise CSS logistics operators about storage area site selection that reduces the requirements for engineer survivability support without reducing the degree of protection provided. FMs 3-34.1 and 3-34.112 provide additional information concerning the construction and maintenance of survivability positions.

8-88. The commander should avoid predictable defensive preparations because an enemy will tend to attack lightly defended areas. Major positions, facilities, and operational logistics sites may require special camouflage. Camouflage measures that provide this protection include constructing dummy positions and decoys. The commander carefully plans the use of such measures within the framework of real positions and ongoing and future operations. The echelon's OPSEC program and any deception efforts conducted in accordance with guidance from higher echelons should conceal from the enemy or mislead him about the location of the MBA and the disposition of friendly forces.

8-89. **Ensure All-Around Defense.** Units employ all-around security at all times although they deploy the bulk of their combat power against likely enemy avenues of approach. This is because the battlefield offers many opportunities for small enemy elements to move undetected.

8-90. **NBC Defense.** Because defending units are often in fixed positions, they increase their vulnerability to weapons of mass destruction. The commander specifies the degree of risk he is willing to accept and establishes priorities for his NBC defense units. He positions forces and installations to avoid congestion, but he must not disperse to the extent that he risks defeat in detail by an enemy employing conventional munitions.

8-91. The commander determines the mission oriented protective posture (MOPP) level assumed by his force if the MOPP level has not already been established by a higher headquarters. Environmental factors determine where he places his NBC detection devices. He ensures that his unit can conduct hasty and deliberate decontamination of its soldiers and equipment. He drills his unit on measures taken in response to the enemy's use of weapons of mass destruction.

8-92. The commander should employ NBC reconnaissance units along movement routes and at potential choke points. Proper use of these assets enables the commander to reduce casualties and complete his mission. (FMs 3-11 and 3-12 detail NBC defense operations.)

8-93. **Smoke and Obscuration.** The commander uses smoke to disrupt the enemy's assault or movement formations and deny his use of target acquisition optics, visual navigation aids, air avenues of approach, LZs, and drop zones (DZs). Smoke creates gaps in enemy formations, separating or isolating attacking units, and disrupting their planned movement. Bispectral obscuration can blind attackers who lack thermal viewers or other enhanced optical systems. It prevents overwatching enemy elements from observing and engaging the defender, whereas defending forces with advanced optical systems can acquire and engage the enemy within the smoke. The commander can use smoke to facilitate friendly target acquisition by highlighting enemy systems against a light background while degrading the enemy's optics. Smoke used to mask obstacles located in low-level flight corridors and on LZs and DZs can prevent an enemy from using them or greatly increase his risk.

8-94. The commander uses his smoke-generation capabilities to mark targets and screen and obscure friendly positions. Modern bispectral obscurants provide protection from thermal as well as visual viewing devices. This generated capability must be carefully sited with regard to enemy systems and friendly capabilities. Improper use can create an advantage for the enemy. The effectiveness of smoke depends on weather conditions and the quantity of smoke employed. The commander coordinates the use of smoke generators, artillery/mortar smoke, and smoke pot employment. The capabilities of each of these smoke-producing systems are complementary and most effective when used together to achieve synergistic effects. Using smoke can also enhance the effects of deception operations and cover friendly movement to include a river crossing. (FM 3-11.50 provides details on planning, preparing, and executing smoke operations.)

COMBAT SERVICE SUPPORT

8-95. The commander addresses several CSS considerations unique to the defense in his plan. Priorities for replenishment are normally ammunition and materials to construct obstacles and defensive positions. There is normally a reduced need for bulk fuel. There may be an increased demand for decontaminants and chemical protective equipment. The defense should consider stockpiling or caching ammunition and limited amounts of petroleum products in centrally located positions within the main battle area. The commander should plan to destroy those stocks if necessary as part of denial operations. The supply of obstacle materials in a defense can be a significant problem that requires detailed coordination and long lead times. The commander should not overlook the transportation and manpower required in obtaining, moving, and uncrating barrier material and associated obstacle creating munitions, such as demolition charges and mines.

8-96. The logistics officer (G4 or S4) and the commanders of the logistics units supporting the defending force must understand the commander's tactical intent. They can then establish service support priorities in accordance with the commander's intent and plan logistics operations to ensure the supportability of the operations. Logistics plans should address the provision of CSS during branches and sequels to the defense plan, such as a counterattack into the flank of an adjacent unit.

8-97. Combat units top off regularly with supplies in case an enemy breakthrough disrupts the replenishment flow. At the battalion and brigade level the commander ensures that his CSS operators deliver combat-configured loads to his combat units on a scheduled basis. Combat-configured loads are packages of potable and nonpotable water, NBC defense supplies, barrier materials, ammunition, POL, medical supplies, and repair parts tailored to a specific size unit. This eliminates the need to request supplies and reduces the chance that a lapse in communications will interrupt the supply flow and jeopardize the integrity of the defense. The supported combat unit is resupplied using this push system until it issues instructions to the contrary. The commander can use utility and cargo helicopters to deliver supplies directly from the rear area to the defending unit. Advances in information systems should allow these combat-configured push packages to be accurately tailored to the demands of the supported combat units.

8-98. As a technique, the defending force conducts resupply during periods of limited visibility if the commander does not expect the enemy to conduct a limited-visibility attack. This tends to reduce the chance for enemy interference with the resupply process but also tends to lengthen the amount of time it takes to complete the process. Resupply should take place during daylight hours if the commander expects the enemy to conduct a limited visibility attack. The commander may be required to infiltrate resupply vehicles to reduce detection chances when the enemy possesses a significant air, satellite, or unmanned aerial vehicle capability. The commander may also use smoke to help conceal his logistics operations.

8-99. The CSS commander remains responsible for the defense of his unit. Concealment is an important factor in reducing the risk factors of these units. The commander must plan for the reconstitution of CSS capability lost to enemy activities.

8-100. Terrain management is a critical consideration in the rear area. The commander seeks to position each CSS unit where it can best fulfill its support tasks while using minimal resources to maintain security in conjunction with other units located in the rear area. In contiguous operations, the commander positions his CSS facilities farther to the rear in a defense than in the offense to avoid interfering with the movement of units between battle positions or the forward movement of counterattack forces. It also should be located far enough behind friendly lines that likely enemy advances will not compel the relocation of critical CSS at inopportune times. At the same time CSS must be close enough to provide responsive support. In noncontiguous operations, the commander positions his CSS facilities within the perimeters of his combat units to provide security and avoid interrupting support services. The commander distributes his similar functional CSS units throughout his defensive area in both environments. This distribution allows him to designate one support unit to pick up the workload of a displacing second support unit until that unit is operational.

8-101. The defending commander provides maintenance support as far forward as possible to reduce the need to evacuate equipment. The thrust of the maintenance effort is to fix as far forward as possible those systems that can be quickly returned to the unit in combat-ready condition. He must ensure that multifunctional forward logistics elements contain the maximum variety of DS personnel with appropriate equipment, such as repair sets, kits, and outfits to ensure rapid repair of weapon systems.

8-102. The commander must plan to augment his available ambulances if a mass-casualty situation develops. Units should always plan for mass casualties and have an evacuation plan, including air evacuation, that specifies the use of nonstandard air and ground platforms.

8-103. The conduct of troop movements and resupply convoys is critical to a successful defense. Staffs balance terrain management, movement planning, and traffic-circulation control priorities. They plan multiple routes throughout the AO and closely control their use. The commander may allocate mobility resources to maintain MSRs in a functional condition to support units and supplies moving forward and to evacuate personnel and equipment to the rear. Military police ease these movements, prevent congestion, and respond to maneuver plan changes. Civil affairs and host nation agencies are involved as necessary to minimize the impact of displaced civilians on unit and convoy movements. The commander coordinates air and ground movements supporting the commander's maneuver scheme with any other affected services. Commanders also coordinate such movements with any affected organic and external Army aviation, fire support, air defense units, and ground maneuver units.

8-104. During the preparatory phase of the defense, logistics operators normally pre-position supply stocks, particularly ammunition and barrier materials, in the battle positions of defending forces. They also establish maintenance and medical collection points. Logistics operators must address these and other logistics preparations in the planning process to avoid compromising the operation. These logistics preparations can also be included in military deception plans.

COMMAND AND CONTROL

8-105. A defensive mission generally imposes few restrictions on the defending commander. It allows freedom of maneuver within assigned boundaries, but requires him to prevent enemy penetration of the rear boundary. Defending an AO is a typical mission for battalion and higher-echelon units. This mission allows the commander to distribute forces to suit the terrain and plan an engagement that integrates direct and indirect fires. The commander must ensure that subordinate unit defensive plans are compatible and that control measures, such as contact points and phase lines, are sufficient for flank coordination when assigning AOs. The defensive plan must address what happens when it succeeds and the opportunity exists to transition from defense to offense.

8-106. Defensive operations are often difficult to conduct because they may occur against an enemy who has the initiative and usually superior combat power. The commander must have a clear understanding of the battlefield situation to mass the effects of his forces to disengage committed forces. He takes advantage of war gaming that takes place in the military decision making process to derive his decision points. He bases these decision points on enemy and friendly actions, such as shifting fires, moving between battle positions, and rearming part or all of the defending force. He may require additional signal support to sustain communications across wide frontages characteristic of many defensive operations.

8-107. Because the enemy has the initiative, the commander may have to frequently shift his shaping operations to contain the enemy's attack until he can seize the initiative. This may require him to adjust subordinate unit AOs, repeatedly commit and reconstitute his reserve, and modify the original plan.

8-108. The defending commander may change his task organization to respond to the existing or projected situation, such as forming a detachment left in contact prior to conducting a withdraw. Whenever possible the commander ensures that changes in task organization take place between units that have previously trained or operated together to take advantage of established interpersonal relationships. The commanders of such recently reorganized units place special attention on ensuring that each element directs its efforts toward accomplishing the overall unit's mission, thus obtaining the maximum combat capability provided by combined arms. This requires them to ensure synchronizing objectives, control measures, movement routes, defensive positions, and specifically assigned tasks. It also requires using standing operating procedures by each element of the task-organized unit. Failure to synchronize the effects of task-organized elements has often resulted in mission failure in training and actual operations.

8-109. To break through the MBA, the enemy often attacks along the boundaries of defending units when he can identify them. Therefore, it is extremely important for commanders at every echelon to ensure that the plan for their part of the defense is properly coordinated not only within their units but also with flanking and supporting units. This coordination is best done by personal visits to subordinate commanders on the ground. The staff should promptly pass on decisions reached during coordination to all concerned. The following planning aspects require attention in the coordination process:

- Understanding the superior commander's intent and concept of operations.
- Understanding the tactics to be applied by flanking and supporting units.
- Selecting boundary locations that do not increase the coordination problem.
- Planning for mutual support.
- Surveillance and target acquisition plans.
- Location and composition of security forces.
- Obstacles and demolition plans.
- Fire plans, to include employing AT systems, illumination, and smoke.
- Air defense coverage areas.
- Employing the reserve in conjunction with information operations and fire support systems, such as artillery and aviation.
- Boundaries and other control measures.
- Communications.

8-110. Because C2 facilities tend to be more stationary in the defense, the commander should place them in hardened areas or protective terrain and reduce their electronic signature. They must remain capable of rapidly relocating to respond to battlefield developments.

COMMON DEFENSIVE SCENARIOS

8-111. Certain common defensive scenarios have their own unique planning considerations. The following section addresses these scenarios and the unique considerations associated with—

- Defense against airborne and air assault attacks.
- Defense of a linear obstacle.
- Perimeter defense.
- Reverse slope defense.

DEFENSE AGAINST AIRBORNE AND AIR ASSAULT ATTACKS

8-112. Defeating an enemy airborne or air assault begins with a good IPB process to determine the enemy's capabilities to conduct vertical envelopment and identify enemy airfields, pickup zones, DZs, and LZs. Armed with an appreciation of the enemy's capability to conduct vertical envelopment, the commander takes steps to counter the threat before they launch, during their movement to the DZ, or at the LZ. After prioritizing the risk of each potential DZ or LZ to his operation, the commander establishes systematic surveillance of these areas to alert him if the enemy attempts to insert his forces. Units also sight their weapons to cover the most probable DZs and LZs. The fire support plan includes these zones in its target list for conventional munitions and scatterable mines and reflects current rules of engagement and host nation restrictions. Units and engineers emplace obstacles in these locations and block avenues of approach from such areas to critical friendly installations and activities as part of their countermobility and rear area survivability efforts.

8-113. Once enemy forces succeed in landing, the key to a successful defense is speed in containing and counterattacking the inserted enemy force before it becomes organized and reinforced. Field artillery and attack helicopters must commit rapidly to take advantage of the concentration of targets in the insertion area. Affected base and base cluster defense forces and available response forces keep the enemy force under observation at all times, calling in and designating targets for available fire support systems. The commander rapidly musters and commits available heavy units and combat systems to take advantage of enemy light forces' vulnerabilities to attack by armored vehicles while they remain concentrated in the insertion area. If more enemy troops land and succeed in consolidating, local base and base cluster defense forces and the response force try to fix the enemy force in a chosen location to allow a tactical combat force (TCF) to counterattack. If the enemy force is too large for the TCF to reduce, the commander may need to commit his reserve.

DEFENSE OF A LINEAR OBSTACLE

8-114. A commander may conduct either an area or mobile defense along or behind a linear obstacle. An area defense is normally preferred because it accepts less risk by not allowing the enemy to cross the obstacle. Linear obstacles such as mountain ranges or river lines generally favor a forward defense. The defending force seeks to defeat any enemy attempt to secure a bridgehead across the linear obstacle. Local defending units immediately and violently counterattack any enemy bridgeheads established to destroy enemy forces located within the bridgehead, while higher echelons attempt to isolate enemy bridgehead sites. If the enemy secures a bridgehead and strikes out rapidly, it could quickly penetrate the defending force. This requires the commander to conduct retrograde operations, either a delay or a withdrawal.

8-115. It is extremely difficult to deploy in strength along the entire length of a linear obstacle. The defending commander must conduct economy of force measures in some areas. Within an area defense, the commander's use of a defense in depth accepts the possibility that the enemy may force a crossing at a given point. The depth of the defense should prevent the enemy from rapidly exploiting its success. It also defuses the enemy's combat power by forcing him to contain bypassed friendly defensive positions in addition to continuing to attack positions in greater depth. Once the enemy force secures several bridgeheads, the defending force moves to contain them. The defending force commander may choose not to counterattack until he can mass overwhelming combat power. He will probably choose to eliminate the bridgeheads sequentially in this case. However, he risks allowing the enemy to establish and fortify bridgehead crossing sites sufficiently to prevent the counterattack force from eliminating them.

8-116. The mobile defense gives the enemy an opportunity to cross the obstacle with a portion of his force. The commander conducting a mobile defense along a linear obstacle normally employs minimal forces along the obstacle as his fixing force. This generally allows the enemy to cross in at least one location. Once the enemy has partially crossed and the obstacle divides his forces, the commander conducts shaping operations to isolate the enemy bridgehead. Once the bridgehead is isolated, the defending commander launches a decisive attack by the striking force to destroy that isolated enemy bridgehead. He may also choose this technique when the enemy is likely to use weapons of mass destruction.

8-117. Alternatively, in a mobile defense the commander may take advantage of terrain or smoke to hide a striking force until the enemy's forward elements pass this force. Until committed, the striking force maintains a perimeter defense. This technique closely resembles the use of stay-behind forces. Similarly, the commander may order units inadvertently bypassed by the enemy not to break out immediately so that he may capitalize on their position to destroy the enemy.

PERIMETER DEFENSE

8-118. The commander can employ the perimeter defense as an option when conducting an area or mobile defense. The commander uses it in many other circumstances, such as when his unit is bypassed by the enemy or in base and base cluster defense in the rear area.

8-119. A perimeter defense is oriented in all directions. Aggressive patrolling and security operations outside the perimeter are prerequisites for a successful perimeter defense. These activities can be undertaken by the unit within the perimeter or by another force, such as the territorial defense forces of a host nation. The unit can organize a perimeter defense to accomplish a specific mission, such as protecting a fire base, or providing immediate self-protection, such as during resupply operations when all-around security is required. The commander establishes a perimeter when the unit must hold critical terrain, such as a strong point, or when it must defend itself in areas where the defense is not tied in with adjacent units. This occurs when the unit is operating behind enemy lines or when it is securing an isolated objective, such as a bridge, mountain pass, or airfield. A unit may also form a perimeter when it has been bypassed and isolated by the enemy and it must defend in place, or it is located in the friendly rear area within the confines of a base or base cluster. (See Figure 8-10.) However, divisions and corps can also organize a perimeter defense when necessary.

8-120. A major characteristic of a perimeter defense is a secure inner area with most of the combat power located on the perimeter. Another characteristic is the ease of access for resupply operations. The commander coordinates direct and indirect fire plans to prevent accidentally engaging neighboring friendly units and noncombatants. Normally, the reserve centrally locates to react to a penetration of the perimeter at any point.

8-121. Perimeters vary in shape depending on the terrain and situation. If the commander determines the most probable direction of enemy attack, he may weight that part of the perimeter to cover that approach. The perimeter shape conforms to the terrain features that best use friendly observation and fields of fire. The commander can increase the effectiveness of the perimeter by tying it into a natural obstacle, such as a river, which allows him to concentrate his combat power in more threatened sectors.

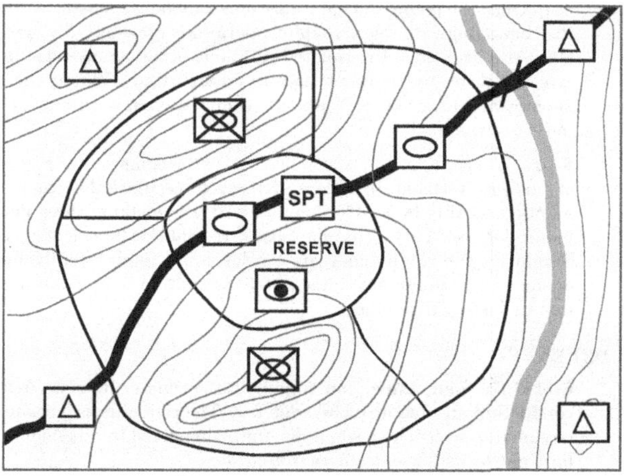

Figure 8-10. Perimeter Defense

Organization of Forces

8-122. The commander may employ all of his forces forward along the perimeter or establish a defense in depth within the perimeter. The commander employs patrols, raids, ambushes, air attacks, and supporting fires to harass and destroy enemy forces before they make contact with the perimeter, thus providing defense in depth with both techniques.

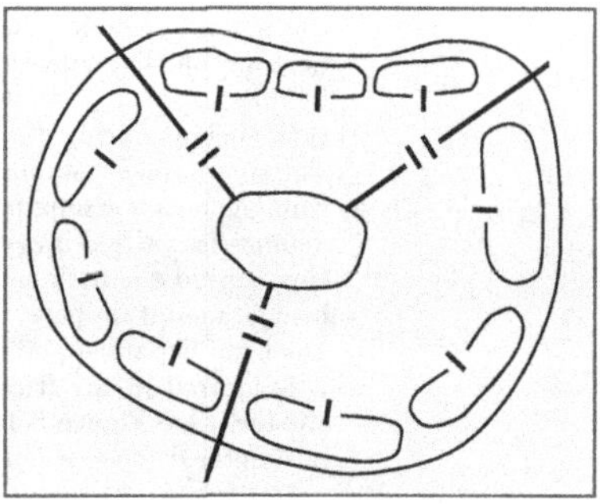

Figure 8-11. All Company Teams on the Perimeter

8-123. In the first technique, he places all of his subordinate units in positions along the perimeter. He divides the perimeter into subordinate unit AOs with boundaries and coordinating points. (See Figure 8-11.) This reduces the possibility of fratricide within the perimeter and maximizes combat power on the perimeter.

8-124. Constructing an outer and inner perimeter creates some depth in the defense in the second technique. Using a brigade assembly area as an example, the commander places two companies in each battalion task force along the outer perimeter and one company in reserve along the inner perimeter. (See Figure 8-12.) This configuration gives depth to the battalion task force's positions and facilitates control. It also gives one company from each battalion task force the mission to support frontline platoons. It enables the company commander to locate any indirect fire systems, such as mortars, near the reserve platoon, enhancing control and security. Alternatively, the commander could elect to assign two battalion task forces to the outer perimeter and a third battalion to an inner perimeter, retaining a larger, more cohesive central reserve. (See Figure 8-13.)

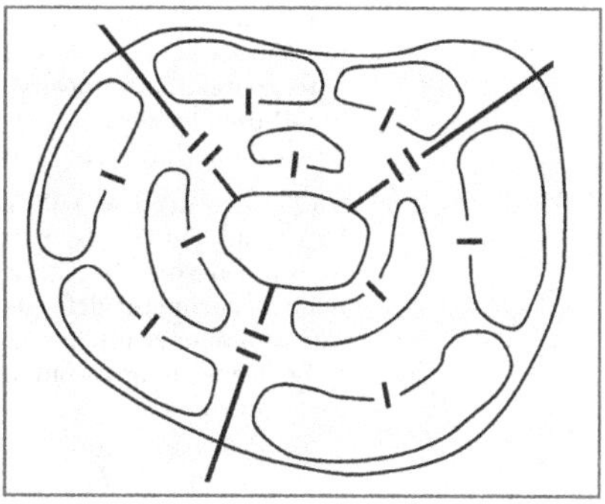

Figure 8-12. Three Battalion TFs on Perimeter, Co/Teams Positioned in Depth

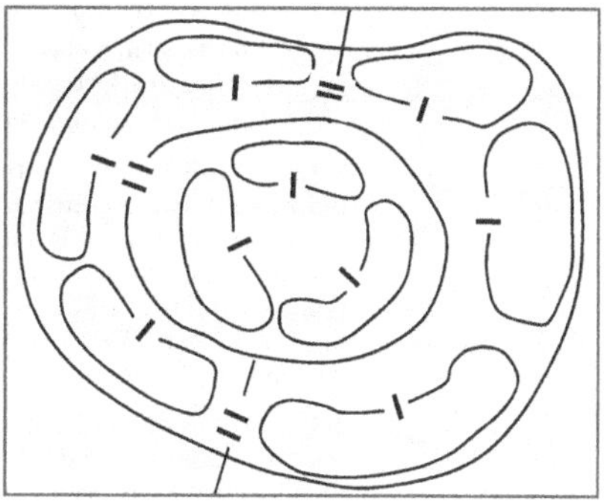

Figure 8-13. Two Battalion TFs on the Perimeter, One in Reserve

8-125. The commander positions his forces within the perimeter to decrease the possibility of an enemy simultaneously suppressing his inner and outer perimeter forces with the same fires regardless of the method used. Friendly forces within the perimeter must be capable of providing mutual support. The commander covers gaps on the outer perimeter between units in open terrain with fires. He should allow no gaps between defensive fighting positions when his unit is in restrictive terrain with restricted fields of fire and observation. This may mean that a unit defends along a narrower frontage than on more open terrain. The commander may also have to employ all of his subordinate units on the line formed by the perimeter. The commander ensures that outer perimeter positions have rearward protection from inner perimeter weapons once he establishes the inner perimeter.

8-126. The commander normally assigns combat vehicles supporting the defense firing positions on the perimeter to cover the most likely mounted avenues of approach. He should select and prepare alternate and supplemental firing positions and routes to and from them. If the perimeter has several mounted avenues of approach leading to it, the commander may elect to hold his combat vehicles in hide positions until the enemy approaches. Units prepare routes, firing positions, and range cards in advance for all positions. Small-unit leaders must ensure that vehicles do not destroy communication wires when they displace from one position to another.

8-127. The need to hold or protect features—such as bridges, airfields, or LZs—from enemy observation and fires may restrict the positioning of units within a perimeter. These factors, as well as the inability to achieve depth, make a perimeter defense vulnerable to penetration by heavy enemy forces. The commander reduces these vulnerabilities by—

- Developing reconnaissance and surveillance plans that provide early warning.
- Positioning antiarmor weapon systems on armor-restrictive terrain to concentrate fires on armor approaches.
- Providing as much depth as the diameter of the perimeter to allow the proper placement of security elements and the reserve and the designation of secondary sectors of fire for antiarmor weapons.
- Constructing obstacles to fix or block enemy forces, so that friendly units can effectively engaged them.
- Using smoke and deception.

8-128. If isolation from other friendly units drives the commander to form a perimeter, such as during rear operations, CS and CSS elements from other units may seek the perimeter's protection. These elements are given defensive missions based on their capabilities. The commander coordinates and integrates any fire support provided from outside the perimeter into the overall defensive plan. This extra fire support conserves the ammunition of units within the perimeter.

8-129. The commander normally employs any reconnaissance assets, such as a scout platoon, outside the perimeter to provide early warning. He may augment security with squad-size or smaller observation posts that are provided and controlled by units on the perimeter. He positions these security elements to observe avenues of approach. Patrols cover areas that cannot be observed by stationary elements. Any security forces operating outside the perimeter must coordinate their passage of lines into and out of the perimeter with the appropriate perimeter units.

8-130. The reserve may be a designated unit or a provisional force organized from available personnel and equipment. The reserve forms a second line of defense behind the perimeter forces. Ideally, the reserve is mobile to react to enemy action along any part of the perimeter. The commander positions the reserve to block the most dangerous AA and assigns on-order positions on other critical avenues. The commander may task available combat vehicles initially occupying firing positions on the perimeter with the mission of reinforcing the reserve.

Control Measures

8-131. The commander in a perimeter defense designates the trace of the perimeter, battle positions, coordinating points, and lateral and forward boundaries. He can use EAs, target reference points, final protective fires, and principal direction of fire as fire control measures. The commander designates checkpoints, contact points, passage points, and passage routes for use by local reconnaissance, surveillance, and security elements operating outside the boundary of the perimeter. (See Figure 8-14.)

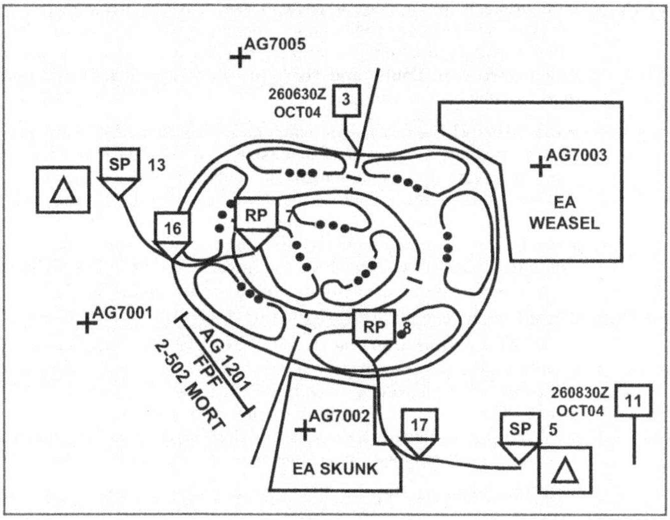

Figure 8-14. Perimeter Defense Control Measures

Planning a Perimeter Defense

8-132. The defending commander positions his forces and plans fire and movement so he can respond to the widest possible range of enemy actions. He prepares plans, including counterattack plans. He rehearses, evaluates, and revises these plans as needed. The availability of LZs and DZs protected from enemy observation and fire is a major consideration when selecting and organizing the perimeter defense. The commander must emphasize supply economy and protect existing supply stocks since aerial resupply is vulnerable to weather and enemy fires. The commander considers the following fundamentals when planning a perimeter defense.

8-133. **Use of Terrain.** Proper evaluation and organization of the area are essential to maximize the effectiveness of a force conducting perimeter defense. Factors considered are—

- Natural defensive characteristics of the terrain.
- Using artificial obstacles to enhance the natural defensive characteristics of the terrain.
- Existing roads, railways, and waterways used for military LOCs and civilian commerce.

- Controlling land areas surrounding the perimeter to a range beyond that of enemy mortars and rockets and also controlling water approaches.

8-134. **Security.** Early warnings of pending enemy actions ensure the commander time to react to any threat. Combat outposts, patrols, sensors, target acquisition radars, and aerial surveillance provide early warning. Civilian informants and actions of indigenous personnel near the position are excellent indicators of pending enemy actions. Security measures vary with the enemy threat, forces available, and the other factors of METT-TC; however, all-round security is essential.

8-135. **Mutual Support.** The commander positions his defending forces to ensure mutual employment of defensive resources, such as crew-served weapons, observation, and maneuver elements. Mutual support between defensive elements requires careful planning, positioning, and coordination because of the circular aspects of the perimeter defense. He uses surveillance, obstacles, prearranged indirect fires, and the provision for maneuver elements to exploit or reinforce fires to control any gaps in the perimeter. Defensive plans provide for using all available support, including field artillery systems firing danger close, attack helicopters, and close air support.

8-136. **All-Around Defense.** In defensive planning, the commander has to be prepared to defend against enemy attack from any direction. His plans are sufficiently flexible, and he positions his reserve to permit reaction to any threat. The commander commits maneuver elements and available supporting weapons to detect, engage, and destroy the attacking enemy force. He assigns all personnel within the perimeter positions and sectors of fire.

8-137. **Defense in Depth.** Alternate and supplementary positions, combat outposts, and mutually supporting strong points forward of the perimeter extend the depth. The commander plans fires throughout the defensive area up to the maximum range of available weapons. He may place portable obstacles around critical locations within the perimeter during periods of reduced visibility to disrupt the enemy's plan based on visual reconnaissance and add depth to the defense.

8-138. **Responsiveness.** Attacks against a perimeter may range from long-range sniper, mortar, or artillery and rocket fire to attacks by demolition teams or major forces. The enemy has the advantage of deciding when, where, and with what force he will attack. The commander prepares plans, to include counterattack plans, and rehearses, assesses, and revises them as necessary. The defensive plan contains procedures for timely response by fire support teams and maneuver forces.

8-139. **Maximum Use of Offensive Action.** Since the objective of the perimeter defense is to maintain a secure position, the commander uses offensive actions to engage enemy forces outside the base. On initial occupation of the perimeter, friendly forces take offensive actions to destroy enemy forces in the immediate area. Once the perimeter area is clear, a relatively smaller force can defend the perimeter, thereby releasing other forces for their primary operations. The commander employs patrols, raids, ambushes, aerial attacks, and supporting fires to harass and destroy enemy forces to prevent their regaining the capability to threaten the perimeter. The commander maintains constant communications with his subordinates within the perimeter and provides them the information necessary to maintain a common operational picture among all units located within the perimeter. He directs them to conduct appropriate actions to remove threats located within their AOs and sectors of fire.

Executing a Perimeter Defense

8-140. Attacks against a perimeter may range from long-range sniper, mortar, or rocket fire; to attacks by suicide demolition squads; to attacks by major enemy ground and air forces. Mortars, artillery, tanks, and antiarmor missile systems from within the perimeter engage the enemy at long ranges. As the enemy comes within small arms range, other weapons on the perimeter engage him. If the assault continues, the force employs its available FPFs. If the enemy penetrates the perimeter, the reserve blocks the penetration or counterattacks to restore the perimeter. After committing the initial reserve, the commander must reconstitute another reserve to meet other threats. This force normally comes from an unengaged unit on another portion of the perimeter. If the commander uses an unengaged force to constitute a new reserve, he must retain sufficient forces to defend the vacated sector, unless he is forced to assume that degree of risk.

8-141. Combat service support elements may provide support from within the perimeter or from another location, depending on the mission and the status of the unit forming the defensive perimeter, type of transport available, weather, and terrain. Units in contested areas without secure ground LOC are often resupplied by air.

REVERSE SLOPE DEFENSE

8-142. The commander organizes a reverse slope defense on the portion of a terrain feature or slope with a topographical crest that masks the main defensive positions from enemy observation and direct fire. All or part of the defending force may employ this technique. It is generally useful at lower tactical levels, such as battalion and below.

8-143. The commander bases a successful reverse slope defense on denying the topographical crest to the enemy. Although the defending unit may not occupy the crest in strength, controlling the crest by fire is essential for success. This defensive situation reduces the effects of massive indirect fire (mortar, artillery, and close-air support) and draws the battle into the small-arms range of infantry weapons. Using the reverse slope defense provides the defending force with an opportunity to gain surprise. Its goal is to make the enemy commit his forces against the forward slope of the defense, causing his forces to attack in an uncoordinated fashion across the exposed topographical crest. Firing from covered and concealed positions throughout the battle area, the defending force maintains a distinct advantage over the exposed enemy forces and canalizes them through unfamiliar terrain into kill zones. (Figure 8-15 shows the terminology associated with the reverse slope defense.)

8-144. The commander chooses to conduct a reverse slope defense when—

- The crest and forward slope are untenable because the enemy enjoys a quantative or qualitative advantage in firepower at that point.

- His weapons cannot depress enough to engage.
- The crest and forward slope offer little or no cover and concealment.
- The forward slope has been lost or has not been seized.
- Units on the flanks can adequately cover the forward slope.
- Variance in the force's tactical pattern is advisable to deceive or surprise the enemy.
- The commander is forced to assume a hasty defense while in contact with or in proximity to the enemy.

8-145. The reverse slope defense may deceive the enemy regarding the true location and organization of the main defensive positions. This defense protects the main defensive positions from preparation fires and causes the enemy to deploy into assault formations prematurely. The forward crest of the main defensive positions limits the enemy's observation. It reduces the effectiveness of enemy indirect fires and close air support and renders his direct fire weapons ineffective. The defending force may bring surprise fires to bear on the enemy as he crests the high ground. Units on the reverse slope have more freedom of movement until the crest is lost.

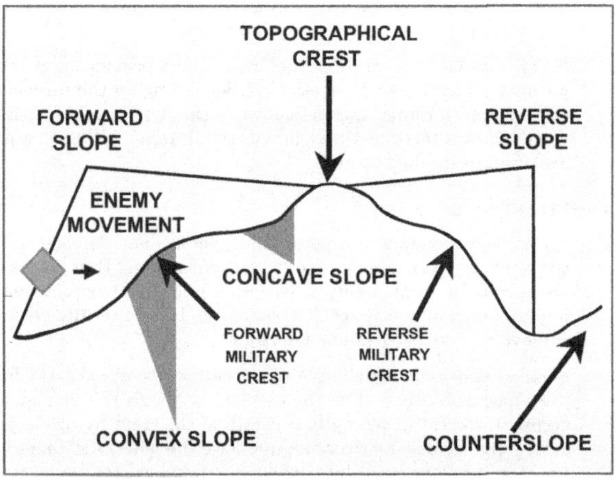

Figure 8-15. A Hill in Cross-Section

8-146. Using the reverse slope defense has several disadvantages:

- The effective range of direct fire weapons may be limited.
- Once security elements withdraw, the enemy can advance largely unimpeded until he has crested the high ground in front of the main defensive positions.
- The enemy has the advantage of attacking downhill.
- Maintaining observation of the enemy is difficult.
- In some cases the best locations for obstacles can only be covered from positions on the forward slope.

Organization of Forces

8-147. The commander places his overwatching elements forward of the topographic crest and on the flanks of the position in a valley or depression. Another variation available to the commander is to organize a system of reverse slope defenses firing to the oblique defilade, each covering the other. A commander uses an oblique defilade to protect his defending systems from enemy frontal and flanking fires and from fires coming from above. For example, in Figure 8-16, the two units defending on the reverse slope cannot engage half of the hill to their direct front because of line of sight restrictions caused by small forests, but they can cover each other using oblique defilade.

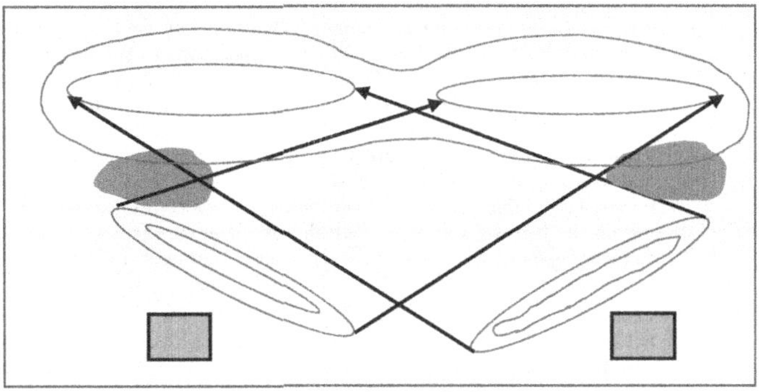

Figure 8-16. Oblique Defilade

8-148. The defending force positions its reconnaissance and security elements where it can observe the forward slope, the terrain forward of it, and other approaches to the defending position. Security elements destroy enemy reconnaissance assets, delay the enemy, disorganize his attack, and deceive him regarding the exact location of the main defense. The commander should position his reconnaissance and surveillance assets in observation posts (OPs) located near or forward of the topographical crest to provide long-range observation of both the enemy's flanks and front. Forces manning these OPs, which can be provided by the commander's reserve, may vary in size from a two-man buddy team to a rifle squad or a multiple combat vehicle section in each position. The commander should employ sufficient forces to provide observation and a security screen for the MBA on ground that should be retained. During darkness and periods of reduced visibility, he should strengthen these detachments in size and numbers to provide security against infiltration or surprise attack. Aggressive night combat patrols and ambushes are an essential part of the security process.

8-149. In order to achieve surprise and limit the enemy's ability to maneuver, the commander organizes the main defensive positions to mass the effects of his concentrated fires on the enemy as he crosses the topographical crest. In a reverse slope defense, the key position denies enemy penetration and supports forward elements by fire. The defending force maintains observation and fires over the entire forward slope as long as possible to destroy enemy forces, thus preventing the enemy from massing for a final assault. From defensive positions on the reverse slope, the close-in battle builds in intensity. The defending force does not fire its direct fire weapons, which are located throughout the MBA (adjacent slope positions, counterslope positions, or reverse slope positions), until suitable targets appear. At the same time, the force shifts the effects of its indirect fires to those areas forward of the crest and forward military slope.

8-150. When possible other units on complementary terrain should support units in reverse slope positions. This is especially desirable when those supporting units can observe and place fires on the crest and forward slope. In a defense on a counterslope (reverse forward slope), fires must cover the area immediately in front of the reverse slope positions to the topographical crest. The commander organizes defensive positions to permit fires on enemy approaches around and over the crest and on the forward slopes of adjacent terrain features if applicable. The key factors that affect the organization of these areas are mutually supporting covered and concealed positions, numerous existing and reinforcing obstacles, the ability to bring devastating fires from all available weapons onto the crest, and a counterattack force. Depending on the terrain, the most desirable location for the reserve may be on the counterslope or the reverse military crest of the counterslope.

Control Measures

8-151. Defensive control measures introduced in previous chapters apply equally to the reverse slope defense. The commander places his EAs and obstacles on the reverse slope. The topographical crest normally marks the far edge of the EA. He must dominate it by fires to prevent the enemy from successfully engaging the defending force.

Executing a Reverse Slope Defense

8-152. When executing a reverse slope defense, the commander places special emphasis on—

- A fire support plan to prevent the enemy's occupation and use of the topographical crest.
- The proper organization of the forward slope to provide observation across the entire front and security to the main battle positions.
- A counterattack plan that specifies measures necessary to clear the creast or regain it from the enemy.
- Fire support to destroy, disrupt, and attrit enemy forces on the forward slope.

8-153. The commander normally places his final protective fires along the topographical crest and employs them as the enemy reaches the first row of defiladed obstacles. He uses his reserve to counterattack and expel the enemy from the topographical crest if massed indirect fires do not defeat the attack. As always, in a reverse slope defense, the commander can employ his designated reserve to conduct rear area security operations, prepare withdrawal routes, provide flank security, and conduct other actions with the understanding that this increases the time required to reassemble the reserve and prepare it to support the defense.

8-154. The reverse slope defense pursues offensive opportunities through surprise and deceptive actions. It is uniquely suited to infantry forces in mountainous terrain. When conducting a reverse slope defense, surprise results from defending in a manner for which the enemy is unprepared. Once this defense is employed successfully to halt an enemy attack, it may have limited further value because the effect of surprise will be difficult to attain. (For additional information on the use of a reverse slope defense, see FM 3-21.30 and other brigade- and lower-echelon field manuals.)

TRANSITION

8-155. If a defense is successful, the commander anticipates and seeks the opportunity to transition to the offense. If the defense is unsuccessful, the commander needs to transition from a defensive posture into retrograde operations. Transition from one type of operation to another requires mental as well as physical agility on the part of the commanders, staffs, and units involved as well as accurate situational assessment capabilities.

8-156. Deliberate contingency planning for either event greatly assists the transition process and allows the commander to set the conditions necessary for a successful transition. Such planning addresses the need to control the tempo of operations, maintain contact with both enemy and friendly forces, and keep the enemy off balance. It establishes the procedures and priorities by which a unit reconstitutes itself for the next mission. In accordance with the factors of METT-TC, it establishes the required organization of forces and control measures necessary for success.

8-157. Such contingency planning decreases the time needed to adjust the tempo of combat operations when a unit transitions from defensive to offensive operations. It does this by allowing subordinate units to simultaneously plan and prepare for subsequent operations. Preparations typically include resupplying unit basic loads and repositioning or reallocating supporting systems. (Chapters 3-7 address the planning, preparation, and execution of all types of offensive operations.)

8-158. Contingency planning also reduces the amount of time and confusion inherent when a unit is unsuccessful in its defensive efforts and must transition to retrograde operations. It does this through designating units to conduct denial operations and early evacuation of casualties and inoperative equipment. The intent of retrograde operations is to preserve the force as a combat-capable formation until the commander can establish those conditions necessary for a successful defense. (Chapter 11 discusses retrograde operations.)

TRANSITION TO THE OFFENSE

8-159. A defending commander transitioning to the offense anticipates when and where the enemy force will reach its culminating point or require an operational pause before it can continue. At those moments, the combat power ratios most favor the defending force. The enemy force will do everything it can to keep the friendly force from knowing when it is becoming overextended. Indicators that the enemy is approaching this point include—

- Enemy forces begin to transition to the defense—this defense may be by forces in or out of contact with friendly forces.
- Enemy forces suffer heavy losses.
- Enemy forces start to deploy before encountering friendly forces.
- Enemy forces are defeated in most engagements.
- Enemy forces are committed piecemeal in continued enemy attacks.
- Enemy reserve forces are identified among attacking forces.
- Examination of captured or killed enemy soldiers and captured or destroyed enemy equipment and supplies shows that the enemy force is unable to adequately sustain itself.
- A noticeable reduction in the tempo of enemy operations.
- Local counterattacks meet with unexpected success.

8-160. The commander must be careful that he is not the target of enemy information operations designed to tempt him to abandon the advantages of fighting from prepared defensive positions. He ensures that his force has the assets necessary to accomplish its assigned offensive mission.

8-161. In a mobile defense, transitioning to the offense generally follows the striking force's attack. In an area defense, the commander designates a portion of his force to conduct the attack, selecting units based on his concept for achieving his mission. However, he allocates available reserves to this effort.

8-162. As the commander transitions his force from the defense to the offense, he takes the following actions—

- Establishes an LD for his offensive operation. This may require him to conduct local, small-scale attacks to secure terrain necessary for the conduct of the offensive operation or destroy enemy forces that could threaten the larger offensive operation.
- Maintains contact with the enemy, using combinations of his available ISR assets to develop the information required to plan future operations and avoid being deceived by enemy information operations.
- Redeploys his combined arms team based on the probable future employment of each element of that team. For example, fire support assets would tend to move forward so that additional enemy forces and terrain would be encompassed within their range fans.
- Maintains or regains contact with adjacent units in a contiguous AO and ensures that his units remain capable of mutual support in a noncontiguous AO.
- Transitions the engineer effort by shifting the emphasis from countermobility and survivability to mobility.
- Provides his intent for transitioning from the defense to the offense to his commanders and soldiers.

8-163. The commander redeploys his air defense assets to provide coverage of combat forces and other assets necessary to conduct offensive operations. This may require him to change or modify his air defense priorities. For example, his top priority in the defense may have been his long-range sensors and weapons. This may shift to providing priority air defense coverage of his ground combat arms units and combat engineers.

8-164. The commander conducts any required reorganization and resupply concurrently with the above activities. This requires a transition in the logistics effort, with a shift in emphasis from ensuring a capability to defend from a chosen location to an emphasis on ensuring the force's ability to advance and maneuver. For example, in the defense, the sustainment effort may have focused on the forward stockage of Class IV and V items and the rapid evacuation of combat-damaged systems. In the offense, the sustainment effort may need to focus on providing POL and forward repair of maintenance and combat losses. Transition is often a time in which deferred equipment maintenance can be performed. Additional assets may also be available on a temporary basis for casualty evacuation and medical treatment because of a reduction in the tempo of operations.

8-165. The commander should not wait too long to transition from the defense to the offense as the enemy force approaches its culminating point. Enemy forces will be dispersed, extended in depth, and weakened in condition. At that time, any enemy defensive preparations will be hasty and enemy forces will not be adequately disposed for defense. The commander wants the enemy in this posture when he transitions to the offense. He does not want to give the enemy force time to prepare for the defense. Additionally, the psychological shock on enemy soldiers will be greater if they suddenly find themselves desperately defending on new and often unfavorable terms while the commander's own soldiers will enjoy a psychological boost by going on the offense.

8-166. A commander can use two basic techniques when he transitions to the offense. The first, and generally preferred, technique is to attack using forces not previously committed to the defense. This is because defending MBA units may still be decisively engaged. These attacking forces may come from his reserve or consist of reinforcements. Since these forces have not recently been actively involved in combat, they are more likely to—

- Be at authorized strength levels.
- Enjoy a higher combat system operationally ready rate.
- Have leaders and soldiers who are more likely to be rested and thus capable of prolonged, continuous operations.
- Have a complete basic load of supplies.
- Have the time and energy to plan and prepare for offensive action.
- Be able to maneuver out of physical contact with the enemy.

8-167. A drawback to the use of this technique is the requirement to conduct a forward passage of lines. Additionally, enemy ISR systems are likely to detect the arrival of significant reinforcements.

8-168. Another consideration of using units not in contact occurs when they are operating in noncontiguous AOs. The commander rapidly masses the effects of overwhelming combat power in his decisive operation. This might require him to adopt economy of force measures in some AOs while temporarily abandoning others in order to generate sufficient combat power. (See Chapters 3 and 5 for offensive planning, preparing, and executing considerations.)

8-169. The second technique is to conduct offensive actions using the currently defending forces. This technique generally has the advantage of being more rapidly executed and thus more likely to catch the enemy by surprise.

Speed of execution in this technique results from not having to conduct an approach or tactical road march from reserve AAs or, in the case of reinforcements, move from other AOs and reception, staging, organization, and integration (RSO&I) locations. Speed also results from not having to conduct a forward passage of lines and perform liaison necessary to establish a common operational picture that includes knowledge of the enemy force's patterns of operation. The primary disadvantage of this technique is that the attacking force generally lacks stamina and must be quickly replaced if friendly offensive operations are not to culminate quickly.

8-170. If units in contact participate in the attack, the commander must retain sufficient forces in contact to fix the enemy. He concentrates the attack by reinforcing select subordinate units so they can execute the attack and, if necessary, maintain the existing defense. He can also adjust the defensive boundaries of subordinate units so entire units can withdraw and concentrate for the attack.

TRANSITION TO THE RETROGRADE

8-171. A defending commander transitions from the defense to the retrograde for those reasons outlined in paragraph 11-1. A retrograde usually involves a combination of delay, withdrawal, and retirement operations. These operations may occur simultaneously or sequentially. As in other operations, the commander's concept of operations and intent drive planning for retrograde operations. Each form of retrograde operation has its unique planning considerations, but considerations common to all retrograde operations are risk, the need for synchronization, and rear operations. The planning, preparing, and executing considerations associated with retrograde operations are found in Chapter 11, but a number of key considerations receive special emphasis during the transition from the defense to the retrograde.

8-172. To accomplish the above purposes, the transition to retrograde operations must be accompanied by efforts designed to—

- Reduce the enemy's strength and combat power.
- Provide friendly reinforcements.
- Concentrate forces elsewhere for the attack.
- Prepare stronger defenses elsewhere within the AO.
- Lure or force part or all of the enemy force into areas where it can be counterattacked.

8-173. The complexity and fluidity of retrograde operations and the absolute need to synchronize the entire operation dictates the need for detailed, centralized planning and decentralized execution. Planning for retrograde operations begins with the preparation of plans for the follow-on mission and is driven by the commander's concept of operation and his intent.

8-174. The nature of retrograde operations involves an inherent risk of degrading the defending force's morale. Therefore, maintaining offensive spirit is essential among subordinate leaders and soldiers. Rearward movements may be seen as a defeat, or as an action that could result in isolation of the force. The commander must be well forward and visible. He must ensure that the leaders and soldiers understand the purpose and intent of the operation and their role in accomplishing the mission. Thorough planning, effective control, and aggressive leadership will minimize risk during the retrograde or enhance the probability of success.

8-175. The commander's ISR requirements dramatically increase as his forces begin their movement to other locations and the combat capabilities of units in contact are subsequently reduced. It is imperative that an integrated ISR collection plan be in place to identify and locate enemy attempts to pursue, outflank, and isolate the defending force as it transitions to the retrograde.

8-176. As the commander transitions to the retrograde, he makes every effort to conserve his combat power. He considers the need to—

- Balance the risk of conserving combat power while remaining disposed to the intent of the defensive mission.
- Disengage and withdraw units with the least tactical mobility and nonessential elements prior to the retrograde of the main body.
- Use mobile forces to cover the retrograde of less mobile forces.
- Use the minimum essential combat power necessary to provide security for the retrograde of the main body.

Chapter 9

The Area Defense

Regardless of the considerations which dictated the adoption of a defensive attitude, the tactics of defensive combat are essentially to develop the maximum firepower against an advancing enemy, to reduce our own losses by a better knowledge and utilization of the terrain, and thereby to stop the enemy's advance or throw him back by counterattack.

FM 100-5, *Field Service Regulations: Operations*, May 1941, Para 652

The *area defense* is a type of defensive operation that concentrates on denying enemy forces access to designated terrain for a specific time rather than destroying the enemy outright (FM 3-0). An area defense capitalizes on the strength inherent in closely integrated defensive organization on the ground. The commander may assign corps, divisions, and separate brigades the task of conducting an area defense as part of their mission. Subordinate echelons defend within their assigned areas of operations (AOs) as part of the larger-echelon's operation.

9-1. A commander should conduct an area defense when the following conditions occur:

- When directed to defend or retain specified terrain.
- When he cannot resource a striking force.
- The forces available have less mobility than the enemy.
- The terrain affords natural lines of resistance and limits the enemy to a few well-defined avenues of approach, thereby restricting the enemy's maneuver.
- There is enough time to organize the position.
- Terrain constraints and lack of friendly air superiority limit the striking force's options in a mobile defense to a few probable employment options.

CONTENTS

9-2. The commander conducting an area defense combines static and mobile actions to accomplish his assigned mission. Static actions usually consist of fires from prepared positions. Mobile actions include using the fires provided by units in prepared positions as a base for counterattacks and repositioning units between defensive positions. The commander can use his reserve and uncommitted forces to conduct counterattacks and spoiling attacks to desynchronize the enemy or prevent him from massing.

ORGANIZATION OF FORCES

9-3. The commander organizes his force to accomplish reconnaissance, security, main battle area (MBA), reserve, and sustaining operations. He has the option of defending forward or defending in depth. When the commander defends forward within an AO, he organizes his force so that he commits most of his combat power early in the defensive effort. To accomplish this he may deploy forces forward or plan counterattacks well forward in the MBA or even beyond of the MBA. If the commander has the option of conducting a defense in depth, he uses his security forces and forward MBA element to identify, define, and control the depth of the enemy's main effort while holding off secondary thrusts. This allows him to conserve his combat power, strengthen his reserve, and better resource the counterattack.

INTELLIGENCE, SURVEILLANCE, AND RECONNAISSANCE OPERATIONS

9-4. The commander directs his intelligence, surveillance, and reconnaissance (ISR) assets to determine the locations, strengths, and probable intentions of the attacking enemy force before and throughout the defensive operation. The commander places a high priority on early identification of the enemy's main effort. He may need to complement surveillance with combat actions that test enemy intentions. Fighting for information can have two benefits—it can force the enemy to reveal his intentions and disrupt his preparations.

9-5. In the defense, ISR operations overlap the unit's planning and preparing phases. Leaders performing reconnaissance tasks must understand that they often deploy before the commander fully develops his plan and they must be responsive to changes in orientation and mission. The commander ensures that his staff fully plans, prepares, and executes reconnaissance missions.

SECURITY OPERATIONS

9-6. The commander balances the need to create a strong security force to shape the battle with the resulting diversion of combat power from his main body's decisive operation. The commander usually allocates security forces to provide early warning and protect those forces, systems, and locations necessary to conduct his decisive operation from unexpected enemy contact. On a battlefield where forces are contiguous with one another, the location of security forces is usually in front of the main defensive positions. On a noncontiguous battlefield they are located on avenues of approach between the protected force and known or suspected enemy locations.

9-7. Battalion and brigade security forces normally conduct screen or guard missions. At division level and above, the commander may use a covering force. A division commander may elect to have his security force conduct a guard mission if a corps covering force exists. Because an area security mission usually ties in closely with flank units, flank security forces are needed if there are gaps on the unit's flanks, which occurs during noncontiguous operations, or if gaps develop during the operation. A flank screen or guard is critical if an enemy avenue of approach into the defended area from the flanks could be uncovered during the defense. A commander does not normally assign a force the mission of conducting rear guard or rear cover during contiguous operations since it is unlikely that his force's rear area will become uncovered during the defense. He resources rear area security forces, to include a tactical combat force (TCF) or accepts the risk to his sustainment effort of not performing this function.

MAIN BATTLE AREA OPERATIONS

9-8. The commander builds his decisive operation around identified decisive points, such as key terrain or high-payoff targets. The commander's decisive operation in an area defense focuses on retaining terrain by using fires from mutually supporting, prepared positions supplemented by one or more counterattacks and the repositioning of forces from one location to another. The commander's decisive operation normally involves close combat since an area defense emphasizes terrain retention.

9-9. The commander normally positions his main body—the bulk of his combat power—within the MBA where he wants to conduct his decisive operation. The commander organizes his main body to halt, defeat, and ultimately destroy attacking enemy forces. The majority of the main body deploys into prepared defensive positions within the MBA. However, mobile elements of the force are ready to deploy where and when needed.

RESERVE OPERATIONS

9-10. The commander's defensive plan should be able to succeed without using his reserve. However, the most likely mission of the reserve is to conduct a counterattack in accordance with previously prepared plans. A lower-echelon commander uses his reserve primarily to conduct local counterattacks to restore his defense's integrity or to exploit an opportunity. A senior commander uses his reserve to seize the initiative from the enemy when the opportunity presents itself. For example, a corps commander may target the effects of his reserve against enemy fire support and follow-on forces to achieve that effect.

9-11. The reserve is not a committed force. The commander can assign it a wide variety of tasks on its commitment, and it must be prepared to perform other missions. In certain situations, it may become necessary to commit the reserve to restore the integrity of the defense by blocking an enemy penetration or reinforcing fires into an engagement area (EA). These secondary tasks include—

- Reinforcing the defense of committed forces.
- Blocking or containing enemy forces that penetrate friendly defensive positions.
- Relieving depleted units and providing for continuous operations.

- Reacting to threats directed against the friendly force's sustainment effort. This includes acting as the echelon TCF when a separate TCF cannot be resourced.
- Extending the flanks of a defending unit to prevent its envelopment.
- Covering a retrograde movement.

9-12. Defending commanders are usually hard-pressed to establish and resource reserve forces because they are normally facing an enemy with superior combat power. Nevertheless, commanders at each echelon down to the battalion task force retain reserves as a means of ensuring mission accomplishment and for exploiting opportunities through offensive action. (Company commanders may retain a reserve based on the factors of METT-TC.) Commanders do not hold artillery and other fire support systems in reserve. (Such systems committed to rear area security operations are not in reserve.) Each echelon's reserve must have the mobility and striking power required to quickly isolate and defeat breakthroughs and flanking attempts. It must be able to seize and exploit fleeting opportunities in a powerful manner to throw the enemy's overall offensive off balance. The commander must resource his reserve so it can repeatedly attack, regroup, move, and attack again.

9-13. The size of the reserve is relative to the commander's uncertainty about the enemy's capabilities and intentions. The more uncertainty that exists, the larger the reserve. The reverse is also true. If the commander knows the size, dispositions, capabilities, and intentions of the enemy, he requires only a comparatively small reserve.

9-14. In some situations, the commander may not be able to resource a separate reserve. Therefore, he may constitute all or a portion of his reserve from his security force after it conducts a rearward passage of lines through MBA units. If the security force is the reserve for an area defense, the commander must withdraw it so it has sufficient time to occupy its reserve position, perform the necessary degree of reconstitution, and prepare plans for its reserve role. However, this is not the preferred option. Before battle handover, the senior commander must state the acceptable risk to the security force or the disengagement criteria in quantifiable terms, such as friendly strength levels, time, or event. In this case, after completing the rearward passage, the security force moves to an assembly area to prepare for its subsequent operations. This area should be free from enemy interference and clear of MBA units, main supply routes (MSRs), and the movements of other portions of the reserve.

9-15. The operations of the reserve usually become the echelon's decisive operation once committed. However, the commander can commit his reserve in a shaping operation to allow his ongoing decisive operation to achieve success. It no longer constitutes the force reserve on its commitment in either case, so the commander should designate another uncommitted force as his reserve. If he does not have that flexibility, he must hold his reserve for commitment at the decisive moment and accept risk.

CONTROL MEASURES

9-16. The commander organizes an area defense by designating his MBA and assigning AOs, battle positions (BPs), or both to subordinate units located within the MBA. He creates a security area in front of the MBA. When possible, the boundaries of the subordinate elements of the security force coincide with those of the major defending units in the MBA. The security area should be deep enough to make the enemy displace as much of his supporting forces as possible, such as cannon artillery, sensors, and air defense artillery gun systems, before carrying his attack into the MBA. The commander also designates his rear area. (See Chapter 12 for a discussion of security operations.)

9-17. Area defense maneuver graphic control measures also include EAs, the forward edge of the battle area (FEBA), battle handover line (BHL), strong points, target reference points (TRPs), named areas of interest (NAIs), targeted areas of interest (TAIs), decision points, and various other fire control and countermobility control measures. (Figure 9-1 depicts the most common control measures. Chapters 2 and 8, and Appendix B define these defensive control measures.)

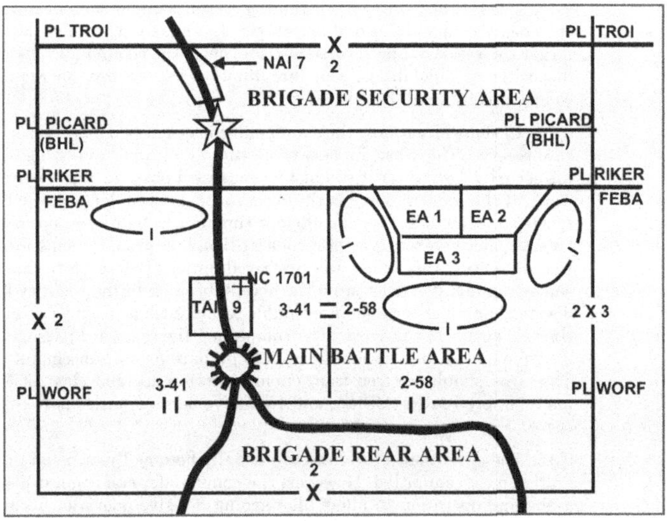

Figure 9-1. Typical Control Measures for an Area Defense

9-18. If the commander assigns a BP and an AO to a subordinate, the subordinate commander has specific guidance on the initial positioning of his forces. The commander ensures the synchronization of each of his subordinate units' defensive plans, and that his control measures, such as contact points and phase lines, are sufficient to ensure the continued control of his subordinates. He is responsible for fire and movement planning between the positions of his subordinate units. If subordinate unit commanders prepare their defensive plans in isolation, one or more assailable flanks between subordinate units could easily develop. (The organization of forces, control measures, planning, preparation, and execution of a passage of lines—a tactical enabling operation—are the subject of Chapter 16.)

PLANNING AN AREA DEFENSE

9-19. The key to a successful area defense is the integration and synchronization of all available assets. The commander achieves this when he can employ the effects of his combined arms team at the decisive time and place. (The general defensive planning considerations addressed in Chapter 8 apply to the area defense.) The commander assigns missions, allocates forces, and apportions combat support (CS) and combat service support (CSS) resources within the battlefield organization of decisive, shaping, and sustaining operations. He decides where to concentrate his effort and where to take risks. The commander can rapidly redirect attack aviation and artillery systems initially allocated to shaping operations to support decisive operations at the appropriate time. (See Figure 9-2 for a graphical depiction of the organization of forces for an area defense in a contiguous AO. See Figure 9-3 for a graphical depiction of the organization of forces for an area defense in a noncontiguous AO.)

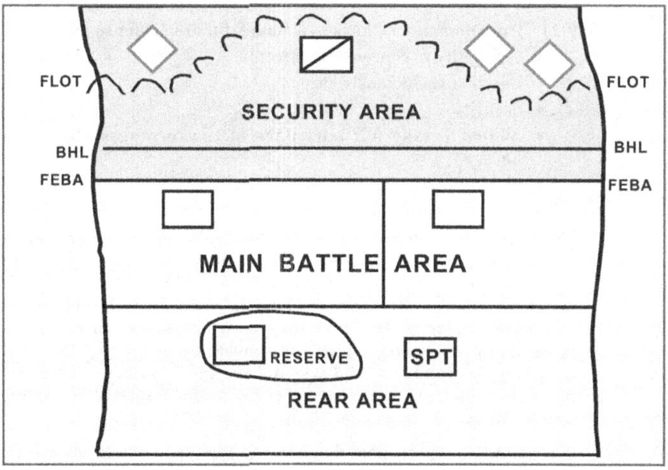

Figure 9-2. Organization of Forces for an Area Defense—Contiguous Area of Operations

9-20. The commander describes his concept of operation in sufficient detail so that his staff and subordinate commanders understand precisely how he intends to fight the battle. He ensures the coordination of maneuver and supporting actions among his subordinates. (FM 5-0 discusses the military decision making process and troop leading procedures.)

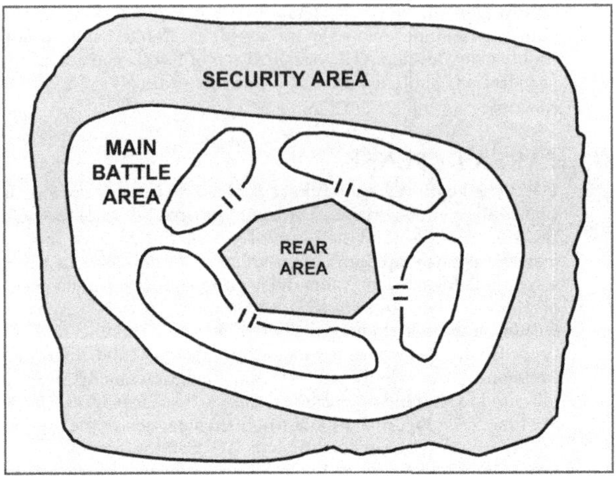

Figure 9-3. Organization of Forces for an Area Defense—Noncontiguous Area of Operations

9-21. The commander's keys to a successful area defense are—

- Capability to concentrate effects.
- Depth of the defensive area.
- Security.
- Ability to take full advantage of the terrain, such as intervisibility lines.
- Flexibility of defensive operations.
- Timely resumption of offensive actions.

The crux of the commander's defensive challenge is to gain time to ensure a synchronized, effective defense. The commander organizes his defensive effort based on an analysis of the factors of METT-TC and the higher commander's concept. He decides where to concentrate his efforts and how to economize his forces. He forces the enemy forces to enter his EAs. To succeed in its area defense mission, the unit must also counteract the enemy's initiative. The commander should take advantage of available offensive opportunities that do not risk the integrity of his defense, such as a spoiling attack or counterattack.

9-22. In planning an area defense, the commander may choose between two forms of defensive maneuver. He can organize either a defense in depth or a forward defense. A higher commander may dictate the form of maneuver or impose restrictions that eliminate a subordinate commander's form of maneuver. These restrictions can include time, security concerns, and directed retention of specific terrain. These two deployment choices are not totally exclusionary. Part of a commander's forces can conduct a forward defense while the other part conducts a defense in depth.

9-23. In determining the form of maneuver, the commander decides where the defensible terrain is located within his assigned AO based on its terrain characteristics and his estimate of the enemy's chosen course of action (COA). Those terrain characteristics include terrain relief patterns, avenues of approach into and within the AO, the location of any key or decisive terrain, existing obstacles and choke points, to include rivers and fording sites. The other factors of METT-TC also influence the commander's decision.

POSITION SELECTION

9-24. Attempting to defend everything defends nothing. Therefore, the commander carefully designs his defense plan to ensure his defending force can halt the enemy attack and develop an opportunity to seize the initiative and undertake offensive operations. The cohesion of the defending force has a significant impact on the overall effectiveness of the defense. The commander must be prepared to adjust the defensive dispositions to meet changes in the enemy's dispositions to maintain that cohesion if the defense is to remain viable.

9-25. The area defense concept requires that defensive positions accomplish their mission independently or in combination by defeating the enemy by fire, absorbing the strength of the attack within the position, or destroying the enemy with a local counterattack. The commander combines the advantages of fighting from prepared positions, obstacles, planned fires, and local counterattacks to isolate and overwhelm selected enemy formations. He must be prepared to rapidly shift the nature and location of his main effort throughout his AO. The commander may have to reposition defending units within their defensive positions or reposition between terrain features as he masses overwhelming effects against the attacking enemy. The defensive plan should designate axes of advance and routes for the commitment or movement of reserves, or the forward or rearward passage of one unit through another. It should identify air axes for aerial maneuver by attack helicopters, air assault units, or fixed-wing aircraft. This capability to reposition is dependent on the defending force having superior tactical mobility. Without tactical mobility, defending forces stay in their prepared positions and accept the possibility of becoming decisively engaged.

9-26. The commander assigning the defensive mission defines the area to defend. A commander defending on a broad front is forced to accept gaps and conduct noncontiguous operations. His forward line of own troops (FLOT) will be discontinuous. Defending shallow areas of operations reduces flexibility and requires the commander to fight well forward. Narrow frontages and deep areas of operations increase the elasticity of an area defense by increasing the commander's maneuver options.

9-27. The ideal area defense is where effective mutual support exists throughout the width and depth of the defender's tactical positions. The commander organizes and occupies these positions based on their natural defensive strength; their retention ensures the integrity of his defense whether he employs a defense in an AO, defends by BP, or employs a combination of both. He maintains tactical integrity within each defensive area. A unit conducting an area defense normally addresses the security requirements of each flank by assigning responsibility to a subordinate element or organizing a security force to specifically execute that mission.

Defense in Depth

9-28. A defense in depth is normally the commander's preferred option. Forces defending in depth absorb the momentum of the enemy's attack by forcing him to attack repeatedly through mutually supporting positions in depth. Depth gives the commander's fire support assets time to generate devastating effects and affords him multiple opportunities to concentrate the effects of overwhelming combat power against the attacking enemy. This also provides more reaction time for the defending force to counter the attack. The commander gathers more information about the attacking enemy's intentions before the enemy commits to a COA. This reduces the risk of the enemy force quickly penetrating the main line of defense.

9-29. The commander also employs a defense in depth when the enemy has the capability to employ large quantities of precision-guided munitions or weapons of mass destruction. Defense in depth results in friendly units and facilities being dispersed throughout the defensive AO. The commander takes area damage-control measures to reduce the effects of weapons of mass destruction on the friendly force and denies the enemy lucrative targets. The degree of dispersal adopted by defending forces is both a function of the enemy's capabilities and the friendly forces' capability to rapidly concentrate overwhelming combat power at decisive points.

9-30. The commander positions his units in successive layers of battle positions along likely enemy avenues of approach when he conducts a defense in depth. (See Figure 9-4, page 9-10.) The commander usually decides to conduct a defense in depth when—

- The mission is not restrictive and allows the commander to fight throughout the depth of the battlefield.
- The terrain does not favor a defense well forward, and there is better defensible terrain deeper within the AO.
- The AO is deep compared to its width, and there is significant depth available.
- The cover and concealment on or near the FEBA is limited.
- The enemy has several times the combat power of the defender.

9-31. Divisions and corps employing a defense in depth can conduct an area defense on a wider frontage than they can if they adopt a forward defense because a forward defense has no time or space to reposition forces. A defense in depth allows the commander to use his security and forward MBA forces to identify the enemy's decisive operation and control the depth of the enemy's penetration into the MBA. By their defensive actions, they provide the commander with time to react to enemy actions and allow him to take offensive steps that eliminate enemy options, such as conducting a counterattack into the flank of an enemy force.

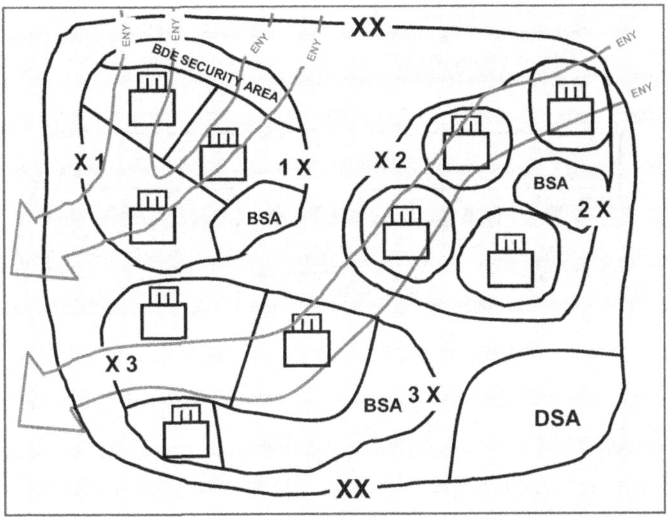

Figure 9-4. Division Conducting a Defense in Depth with Subordinate Brigades Deployed in Noncontiguous Areas of Operations with Enemy Avenues of Approach Depicted

Forward Defense

9-32. The commander conducts his decisive operation from forward defensive positions near the FEBA in a forward defense. (See Figure 9-5.) He concentrates a significant portion of his available combat power into EAs along the FEBA. His intent is to prevent significant enemy penetration into the defensive area. The commander conducting a forward defense fights to retain these positions along the FEBA and violently counterattacks any enemy penetration. However, if the enemy penetrates the main defensive positions, the defender's lack of depth may allow the enemy to rapidly exploit success.

9-33. In general, the commander uses a forward defense when a higher commander directs him to retain forward terrain for political, military, economic, and other reasons. Alternatively, a commander may choose to conduct a forward defense when the terrain in that part of his AO—including natural obstacles—favors the defending force because—

- The best defensive positions are located along the FEBA.
- Strong natural obstacles are located near the FEBA.
- Natural EAs occur near the FEBA.
- Cover and concealment in the rear portion of the AO are limited.

POSITIONING THE RESERVE

9-34. Whatever the commander's choice—forward or in depth—once the enemy commits his forces, the defending commander has the ability to seize the initiative by counterattacking over familiar ground to destroy a halted, disorganized enemy while protected by overwatching fires from friendly positions. Whenever possible, the commander should direct these counterattacks against the enemy's rear or flanks. The commander's reserve is a key component of the counterattack.

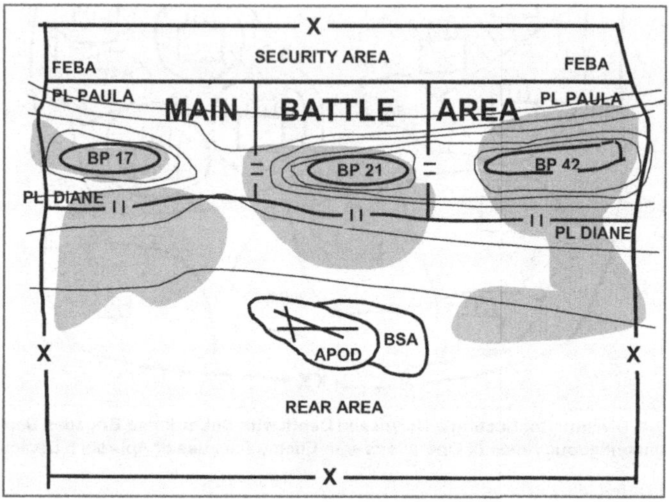

Figure 9-5. Brigade Conducting a Forward Defense in a Contiguous Area of Operations

9-35. When deciding where to place his reserve, the commander decides whether to orient his reserve on its most likely mission or its most important mission. He expends significant effort during the planning process to ensure he can effectively use his reserve when needed. He may locate his reserve within the AO where it can employ the road network to rapidly displace throughout the AO in response to a number of opportunities or contingencies. The commander must consider terrain, MSRs of forward units, enemy avenues of approach, and probable enemy penetrations when determining the exact location for his reserve. He may choose to initially position his reserve in a forward location to deceive the enemy and obscure subordinate unit boundaries, especially those of dissimilar units such as armor and light infantry.

9-36. In restrictive terrain that lacks routes for movement, the commander can task organize his reserve into small elements and position them where they can react quickly to local combat developments. This dispersion improves force protection but reduces the ability of the reserve to mass fires. Covered lateral and forward high-speed deployment routes should be available. The reserve must have movement priority along those routes. He must ensure the maintenance of communication between these dispersed elements. This may require establishing retransmission nodes for combat net radios. In open terrain, the commander maintains a centrally located reserve positioned somewhat farther from the FLOT. He considers the enemy's potential to employ weapons of mass destruction and conduct air interdiction when deciding where to position his reserve.

9-37. Whenever possible, the commander positions his reserve beyond the enemy's direct fire range. This is easier to achieve at higher echelons than at lower echelons. The reserve takes defensive measures to prevent being acquired and attacked by enemy indirect fire systems. These include camouflage, local security, and control of electronic emissions.

9-38. The commander also plans how to reconstitute his reserve once he commits his original reserve. Forces most easily designated are subordinate unit reserves. If his higher headquarters has not committed its reserve, he has more flexibility and can take greater risk in employing his reserve.

SPOILING ATTACKS AND COUNTERATTACKS

9-39. A spoiling attack preempts or seriously impairs the enemy's ability to launch an attack, while a counterattack prevents the enemy from exploiting his successes. The forces conducting either form of attack must be large and strong enough to develop the situation, protect themselves, and force the enemy to react, placing his plan at risk.

9-40. The commander considers the enemy situation and estimates the time and distance factors of any follow-on enemy forces in planning either a spoiling attack or a counterattack by his reserve and other forces. Then he determines which of his units will attack, where they will be after the attack, and what interdiction is necessary to isolate the targeted enemy element. (See Figure 9-6.) His counterattacking forces plan to avoid enemy strength when possible. The most effective attacks seize strong positions that permit the counterattacking force to deliver fire on an exposed enemy unit's flanks and rear. If it is tasked to stay and defend against enemy follow-on forces, the counterattacking force must establish a viable defensive position before any following enemy units can make contact.

9-41. Counterattack plans include assumptions regarding the size and shape of the anticipated penetration or enemy formation; the strength and composition of the enemy force; and the status of the reserve and forces in the MBA. Other factors that affect the counterattack include the capability to contain the enemy, shaping operations to support the attack, and the strength and responsiveness of the reserve at the time of the counterattack.

9-42. The commander's staff prepares counterattack plans and then allocates subordinate headquarters sufficient time to make their plans. The control measures for a counterattack are the same ones discussed in Chapter 5 for the attack. If possible, the commander distributes his counterattack plans along with the basic defense plan. Reserve unit commanders conduct detailed counterattack planning that includes conducting reconnaissance, selecting multiple routes, determining time and space factors, rehearsing, coordinating with appropriate elements of the forward defending force, and fire planning. The commander adjusts his counterattack plans as necessary based on the lessons learned during rehearsals.

9-43. Enemy movement into an NAI helps the commander determine the enemy's scheme of maneuver and possible objectives. He uses decision points and NAIs throughout his AO to trigger his counterattack. The commander identifies TAIs for attack to support his operations.

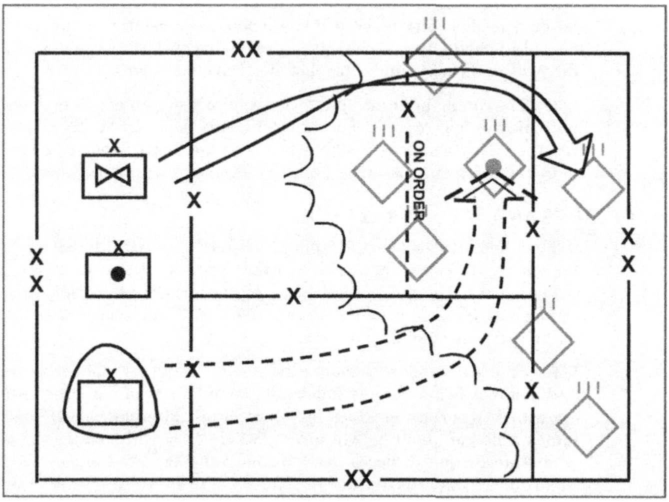

Figure 9-6. Division Counterattack

PREPARING AN AREA DEFENSE

9-44. Preparations focus on planning those additional ISR operations required to answer the commander's critical information requirements, refining the plan, increasing coordination and synchronization, and conducting shaping actions within the force's capability and operations security guidelines. If the commander decides that he must conduct a deliberate defense but knows that the enemy will attack before he is prepared, he may have to commit substantial forces to security operations or conduct a spoiling attack. This buys time and space to prepare for a deliberate defense.

9-45. A unit normally transitions to the defense after it completes the deployment process of force projection, completes offensive operations, or is in an assembly area. The commander issues a warning order stating the mission and identifying any special considerations. His staff conducts detailed planning while the rest of the unit completes its current mission. The staff coordinates for the pre-positioning of ammunition and barrier material in a secure area near the unit's defensive positions before starting the operation.

9-46. Before occupying any position, leaders at all echelons conduct some type of reconnaissance. This reconnaissance effort is as detailed as the factors of METT-TC permit. It may consist of a simple map reconnaissance or a more detailed leaders' reconnaissance and initial layout of the new position.

9-47. The defending unit occupies its defensive positions as soon as practical after receiving the mission. It conducts reconnaissance of the defensive area and establishes a forward security area before occupying the positions The unit may pre-position supplies such as ammunition and barrier materiel once it establishes security. The unit can accomplish many defensive tasks simultaneously; the factors of METT-TC are the deciding consideration in establishing priorities of work. Those priorities may be—

- Establishing local security and deploying a security force.
- Identifying EAs where the commander wants to engage and destroy the enemy.
- Planning fire control measures, such as TRPs, trigger lines, and final protective fires to support the EAs.
- Positioning key weapon systems to engage into the EAs and TRPs and develop range cards and sector sketches.
- Positioning observers who can see both targets and trigger lines.
- Siting obstacle groups to support weapon systems.
- Designating and clearing fields of fire.
- Preparing primary fighting positions based on the anticipated fighting conditions, such as the time of day and weather conditions.
- Emplacing obstacles and surveying indirect fire targets to support these obstacles.
- Providing concealment and camouflage for fighting and survivability positions as they are constructed.
- Positioning any available critical friendly zones over friendly positions by establishing sensor coverage and quickfire links between the sensor and shooter.
- Installing night and limited-visibility aids, such as thermal hot spots and chemical lights on TRPs during daylight.
- Updating range cards and sector sketches as required.
- Preparing alternate fighting positions.
- Designating and preparing supplementary positions.
- Designating hide positions and rehearsing movements to and from fighting positions. (Units may place their combat and tactical vehicles in hide positions at any time while preparing the defensive position.)
- Positioning the reserve.
- Establishing contact points with any adjacent units so that the defensive efforts of both units can be tied together.
- Emplacing wire for communications.
- Improving mobility on counterattack routes.
- Prestocking ammunition in revetments or bunkers where it can survive the enemy's preparatory fires.
- Rehearsing movements under daylight and limited-visibility conditions.
- Continuing to improve the defense.

9-48. Survivability positions enhance the strength of a defensive position by providing soldiers and weapon systems with some degree of cover from enemy fires. Units initiate construction of survivability positions in accordance with their priority of work and continue to build and improve them until the last possible moment. The overhead cover provided varies with the location of the sheltered troops and enemy capabilities. As time and resources allow, the defending unit improves communication routes throughout its defensive positions to ease movement of supplies and forces, particularly the reserve. It quickly establishes wire communications among its various subordinate elements to reduce its electromagnetic signature.

9-49. The defending unit rehearses how to move from its hide positions to its primary positions and how it will occupy alternate and supplementary positions to continue to engage the enemy if he progresses into the unit's defensive positions. These rehearsals establish the time necessary to conduct these movements under different environmental conditions. It modifies existing plans based on the results of rehearsals and changes in the factors of METT-TC. The commander takes steps to ensure that the routes taken during these rehearsals do not show obvious signs of heavy use. These steps can include the conduct of only dismounted rehearsals, only moving one vehicle per platoon, and taking steps to eliminate signs of movement such as sweeping snow back over the tracks made during the rehearsal.

9-50. The commander ensures close coordination among his subordinates. During the preparation phase, he can take his subordinate commanders to a vantage point in the MBA to rehearse the battle and plan coordination among their units if such a site is available. This helps the commander in transmitting his intent and in establishing common control measures for subordinate units.

9-51. The location, composition, and movement of the reserve are essential elements of friendly information. Enemy reconnaissance efforts focus on finding the reserve and reporting when and where it is committed. Avoiding detection by the enemy is vital to the success of the reserve.

9-52. The CSS rehearsal should be integrated into the maneuver rehearsal to verify that routes for support do not cross or conflict with routes used by reserve forces or other maneuver elements. The commander should balance the use of ammunition caches against the defending unit's ability to guard them. The commander should also ensure that alternate MSRs are adequate to accommodate contingency plans and that changing MSRs can be accomplished effectively.

9-53. The commander ensures that his combat multipliers are completely integrated with his intended maneuver. This includes the use of camouflage, deception, and smoke to confuse enemy reconnaissance assets. Having key representatives from each of these multipliers simultaneously rehearse the plan with his subordinate maneuver unit is an effective technique for ensuring integration. After issuing the order and receiving briefbacks from his subordinate commanders and other leaders, the commander verifies that they have a common understanding of the plan and can execute it with minimal guidance.

EXECUTING AN AREA DEFENSE

9-54. A defending unit within the MBA uses a variety of tactics, techniques, and procedures to accomplish the mission. At one end of the defensive continuum is a totally static defense oriented on terrain retention. This defense depends on the use of firepower from fixed positions to deny the enemy terrain. At the other end is a dynamic defense focused on the enemy. That defense depends on maneuver to disrupt and destroy the enemy force.

9-55. A commander combines the static element to control, stop, or canalize the attacking enemy and the dynamic element to strike and defeat him. A successful area defense uses forces in relatively fixed positions to create the opportunity for the reserve to strike at the enemy from an unanticipated direction and strength. (See Figure 9-7.) The defending force repeatedly lures the enemy into EAs where it kills selected portions of the enemy force.

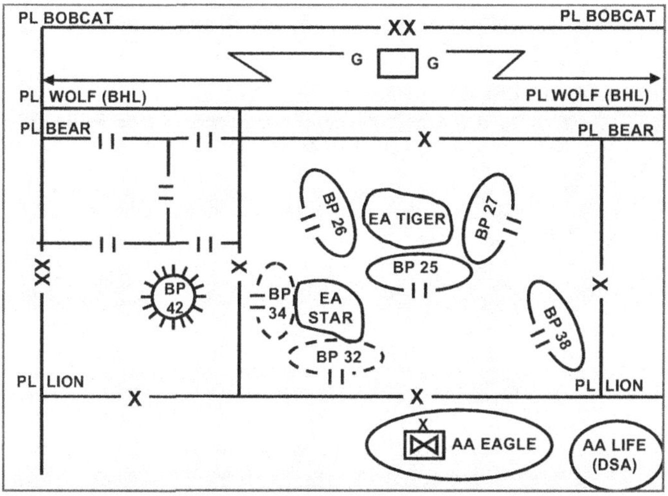

Figure 9-7. Area Defense Using Static and Dynamic Elements

9-56. In an area defense, defending forces fight mainly from prepared, protected positions to concentrate combat power effects against attempted enemy breakthroughs and flanking movements. The commander uses mobile forces to cover gaps between defensive positions, reinforce those positions as necessary, and counterattack to seal penetrations or block enemy attempts at flanking movements.

9-57. Conducting shaping operations in an area defense is similar to shaping operations in the offense. The factors of METT-TC determine how closely the commander synchronizes his shaping operations with his decisive operations. The commander conducts shaping operations designed to regain the initiative by limiting the attacker's options and disrupting the enemy's plan. He conducts shaping operations to prevent enemy forces from massing and creates windows of opportunity for decisive offensive operations, allowing his force to defeat them in detail. The commander also employs shaping operations to disrupt enemy operations by attacking command posts at critical stages in the battle or by striking and eliminating key elements, such as river crossing equipment and supplies in a region that contains numerous unfordable rivers. Reconnaissance and security operations are normally components of the echelon's shaping operations.

9-58. This manual divides execution into five steps:

- Gain and maintain enemy contact.
- Disrupt the enemy.
- Fix the enemy.
- Maneuver.
- Follow through.

This does not imply that these steps occur sequentially; they may occur simultaneously.

GAIN AND MAINTAIN ENEMY CONTACT

9-59. Gaining and maintaining contact with the enemy in the face of his determined efforts to destroy friendly ISR assets is vital to the success of defensive operations. As the enemy's attack begins, the defending unit's first concerns are to identify committed enemy units' positions and capabilities, determine the enemy's intent and direction of attack, and gain time to react. Initially, the commander accomplishes these goals in the security area. The sources of this type of intelligence include reconnaissance and security forces, intelligence units, special operations forces, and aviation elements. The commander ensures the distribution of a common operational picture throughout the force during the battle as a basis for subordinate commanders' actions. (See FM 6-0.) The commander uses the information available to him, in conjunction with his military judgment, to determine the point at which the enemy is committed to a COA.

9-60. The security force seeks to strip enemy reconnaissance forces and hide the defending force's dispositions, capabilities, and intent at the same time as friendly ISR assets help to determine the enemy's chosen COA. Ideally, the fight in the security area should force the enemy to conduct a movement to contact against a prepared defense.

9-61. A single force in the security area can perform both reconnaissance and security functions. The security force uses every opportunity for limited offensive action to delay and harass the enemy and to gain information. As the security element displaces, the commander makes preparations to pass it through or around the MBA force as quickly as possible by using multiple passage points, gaps, or lanes along the FEBA. This usually occurs in one location at a time until the security force has completely withdrawn. However, the security force may pass in sequence based on enemy pressure. Transfer of responsibility occurs forward of the FEBA at the BHL. (See Figure 9-8, page 9-18.) Taking advantage of previous liaison and plans, the security force makes any required last-minute coordination with MBA forces at contact points to ensure its rapid passage through the MBA force.

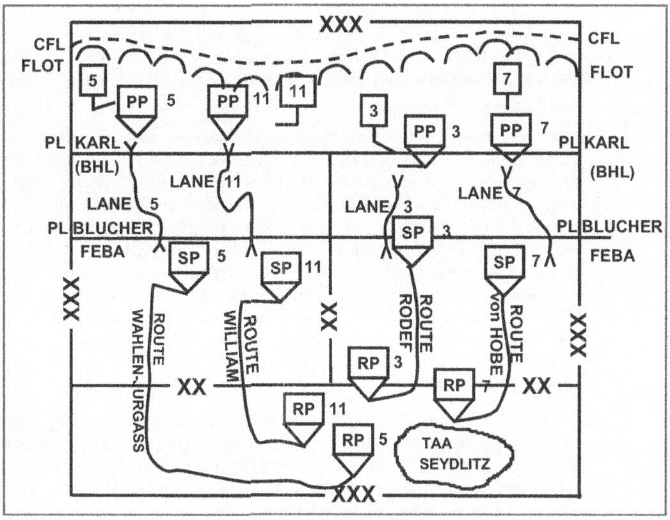

Figure 9-8. Battle Handover Line

9-62. The entire security force should not withdraw automatically as soon as the first enemy units reach the FEBA. The commander can leave in place security elements located in areas where the enemy has not advanced. The security force adjusts to the enemy's advance and continues to conduct security operations as far forward as possible. It continues to resist the enemy's shaping operations, such as the enemy's reconnaissance effort, thereby upsetting his coordination and allowing the MBA commander to fight one engagement or battle at a time. Doing this increases the chances for success even if the enemy attack penetrates into the MBA in some areas. In some cases, the security force can attack the enemy force from its rear, engage high-payoff targets, or drive between echelons to isolate leading enemy units.

9-63. As the enemy force approaches the MBA, the commander may order reconnaissance and surveillance assets within his security force to displace to one or both sides of the enemy penetration and continue to maintain surveillance. By observing and providing access to enemy flanks, reconnaissance and surveillance elements can facilitate the conduct of friendly counterattacks. However, to prevent the encirclement of these assets, the commander may plan to monitor those areas where the enemy has not advanced into the MBA solely by technical means.

9-64. Battle handover between the security force and MBA forces requires close coordination and occurs as quickly and efficiently as possible to minimize their vulnerability to enemy fire. The security force commander must retain freedom to maneuver until he initiates the passage of lines. The commander's fire support assets help cover the withdrawal of security forces.

Combat support and CSS elements of the security force should move to the rear as early as possible to avoid hampering the movement of combat forces. Normally, battalion-size units of the security force hand off the battle to the brigades through which they pass. (See Chapter 16 for a discussion of rearward passage of lines.)

9-65. The commander must consider the security force's next mission prior to battle handover between the security force and the MBA force. Factors that may affect his decision are the status of the security force, its subsequent mission preparation requirements, and the size and nature of the reserve required by the situation. He may decide to employ it immediately as his reserve, which would release his initial reserve for other tasks. Alternately, the commander may decide to use the security force to conduct additional security operations on the flanks of MBA forces as the battle progresses. However, it may be some time before the security force is ready for commitment. Therefore, the commander is more likely to wait until the security force has been reconstituted and the initial reserve committed before designating the former security force as his reserve.

9-66. The commander should base the location of his security force's assembly area on its follow-on mission. The commander wants those assembly areas located to rapidly support ongoing operations yet keep withdrawn security units from interfering with ongoing decisive and shaping operations. After passage, the security force normally moves to these locations to prepare for subsequent operations. At a minimum he must rearm and refuel the security force. Additional CSS concerns include casualty evacuation, maintenance requirements, and resupply of the other classes of supply.

DISRUPT THE ENEMY

9-67. The commander executes his shaping operations to disrupt the enemy regardless of his location within the AO. After making contact with the enemy, the commander seeks to disrupt his plan, his ability to control his forces, and his combined arms team. Ideally, the results of the commander's shaping operations should force a disorganized enemy, whose ability to synchronize its elements has been degraded, to conduct a movement to contact against prepared defenses. Once the process of disrupting the enemy begins, it continues throughout a defensive operation.

9-68. The commander initiates his shaping operations simultaneously with the preparation of his MBA positions. These shaping operations typically focus on high-payoff targets, command and control nodes, engineer, fire support, and air defense assets for destruction or disruption. They can also force the enemy to use avenues of approach covered by friendly EAs. These shaping operations destroy the enemy's cohesion and disrupt the tempo of his approach to the MBA. This, in turn, disrupts the timely introduction of enemy follow-on forces into the engagement. For example, offensive information operations directed against the enemy's C2 nodes and air defense assets increase the enemy's vulnerability to other shaping operations while simultaneously slowing the enemy's reaction to these shaping operations. (FM 3-13 discusses offensive information operations.) Follow-on engagements focus on degrading the enemy's fire support and engineer assets, thereby disrupting the movement of his approaching units.

9-69. Other targets for shaping operations include enemy reconnaissance and intelligence assets. Destroying these assets allows the commander to repeatedly force enemy units to deploy into combat formations on ground of his choosing, thus contributing to the disruption and desynchronization of the enemy's plan. The timing of these shaping operations is important. The enemy cannot be allowed to recover from their effects prior to the decisive operation. The commander may also execute offensive operations to further disrupt the enemy, such as spoiling attacks, raids, ambushes, feints, or demonstrations.

FIX THE ENEMY

9-70. The commander does everything in his power to limit the options available to the enemy when conducting an area defense. In addition to disrupting the enemy, the commander conducts shaping operations to constrain the enemy into a specific COA, control his movements, or fix him in a given location. These actions limit the enemy's options. While executing these operations, the commander continues to find, and delay or attrit enemy follow-on and reserve forces to keep them from entering the MBA.

9-71. The commander has several options to help him fix an attacking force. The commander can design his shaping operations—such as securing the flanks and point of a penetration—to fix the enemy and allow friendly forces to execute decisive maneuver elsewhere. Previously discussed in Chapter 8, combat outposts and strong points can also deny enemy movement to or through a given location. A properly executed military deception operation can constrain the enemy to a given COA.

9-72. The commander uses obstacles covered by fire to fix, turn, block, or disrupt to limit the options available to the enemy. Properly executed obstacles are a result of the synthesis of top-down and bottom-up obstacle planning and emplacement. Blocking forces can also affect enemy movement. A blocking force may achieve its mission from a variety of positions depending on the factors of METT-TC.

MANEUVER

9-73. In an area defense, the decisive operation occurs in the MBA. This is where the effects of shaping operations, coupled with sustaining operations, combine with the decisive operations of the MBA force to defeat the enemy. The commander's goal is to prevent the enemy's further advance through a combination of fires from prepared positions, obstacles, and mobile reserves.

9-74. Generating massed effects is especially critical to the commander conducting the defense of a large area against an enemy with a significant advantage in combat power. The attacker has the ability to select the point and time of the attack. Therefore, the attacking enemy can mass his forces at a specific point, thus dramatically influencing the ratio of forces at the point of attack. An enemy three-to-one advantage in overall combat power can easily turn into a local six-to-one or higher ratio. The defending commander must quickly determine the intent of the enemy commander and the effects of terrain. This allows his units and their weapon systems to use agility and flexibility to generate the effects of combat power against the enemy at those points and restore a more favorable force ratio.

9-75. Forces in the MBA assume responsibility for the battle at the BHL. As the security force approaches the FEBA, it may be necessary to increase the intensity of fire support from the MBA to allow the security force to break contact. Both direct and indirect fire assets from MBA forces provide support to cover the withdrawal of the security force and to close passage lanes through obstacle complexes. The commander may also employ smoke to assist the security force break contact with the enemy. The security force's withdrawal through the forward positions of the MBA must be carefully planned and coordinated. The commander must guard gaps in obstacles left for the withdrawal of the security force and arrange for closing them after the passage of the security force.

9-76. After the enemy reaches the MBA, he tries to find weak points and attempts to force a passage, possibly by a series of probing attacks. As the attack develops, defending units engage the enemy's lead forces. The enemy advance may slow because of canalization and the increased density of forces resulting from limited maneuver space, presenting good targets for defensive fire and air support. The maximum effects of simultaneous and sequential fires are brought to bear at this stage of the battle.

9-77. The commander's subordinate elements maneuver using massed direct and indirect fire and movement to gain positional advantage over the assaulting enemy force. The commander also directs the engineer obstacle and sustainment effort by his assignment of priorities. The commander must reposition his forces to meet the enemy where he is rather than where the commander would like him to be. The commander directs operations and supports his subordinate elements by providing the necessary CS and CSS. He controls the commitment of the reserve and, at division echelon and above, engages enemy follow-on forces. If enemy follow-on forces can be delayed, the enemy's attack may be defeated in detail, one echelon at a time. If the defending unit can force the enemy to commit follow-on forces sooner than planned, it can disrupt the enemy's timetable, which can lead to the creation of exploitable gaps between the committed and subsequent echelons.

9-78. Gaps between defensive positions may be necessary, but they are not left where the commander expects the enemy's probable main effort. They are kept under surveillance, covered by fire or, where possible, blocked by barriers or repositioned friendly forces. The commander clearly defines the responsibility for dealing with each enemy penetration. He leverages the use of choke points and obstacles to prevent enemy penetration. If the enemy succeeds in penetrating the MBA, the commander blocks the penetration immediately and destroys this enemy force as soon as possible; hence, the need for a mobile reserve. He may extend his actions within the depth of his AO to counter enemy penetrations that cannot be stopped farther forward.

9-79. The commander never allows the attacking enemy to consolidate unless it fits his scheme of maneuver. He conducts a local counterattack with all available local resources to prevent the enemy from consolidating his gains. The lowest possible echelon conducts this local counterattack; however, the commander must be aware of the problem of piecemeal commitment. A unit does not abandon a position unless it fits within the higher commander's intent or he grants permission to do so. If the defending force is unable to repulse the enemy, it tries to contain the enemy penetration until it can attack in concert with major counterattacking forces. The commander coordinates his counterattacks with the efforts of his fire support system.

9-80. Although the commander plans for the counterattack in his defensive planning, he is aware that his plan may not correspond exactly with the existing situation when he launches the counterattack. As the situation develops, the commander reassesses his plan based on his revised situational understanding that results from an updated common operational picture as new intelligence and combat information becomes available to answer the following basic questions:

- Is a counterattack feasible, or should the commander use the reserve to contain enemy successes?
- When and where should the defending forces counterattack?
- In the case of enemy penetrations, what should the defending forces counterattack and what should they block or contain?
- Is there enough time to complete the counterattack before the arrival of enemy follow-on forces?
- Can he conduct a counterattack using his fire support systems?

9-81. When counterattacking, the commander employs all available resources necessary to ensure success. The reserve usually becomes the echelon's decisive operation on its commitment, so he avoids its premature or piecemeal commitment. One of the commander's most critical decisions is committing the reserve. He may reinforce his reserve force before committing it to give it greater capability to counter enemy action. The commander does not counterattack as an automatic reaction to an enemy penetration, nor does he commit the reserve solely because the enemy has reached a certain phase line or other location. Fire support assets and local counterattacks by forces already defending could destroy, disrupt, or attrit enemy penetrations, thus relieving the commander of the need to commit his reserve. When possible, the commander launches the counterattack when the enemy presents his flank or rear, overextends himself, or his momentum dissipates. Once the commander identifies the flanks of the enemy's main effort, he can target counterattacks to isolate and destroy enemy forces within the MBA.

9-82. Sometimes the commander may determine that he cannot afford to use his reserve to counterattack. Therefore, he must use his resources to block, contain, or delay the enemy to gain time to employ higher-echelon reserves. In these cases, the commander and his staff must plan how to integrate reinforcing companies and battalions into the defensive scheme, adjust boundaries, and place BPs. He plans the routes these units will use, and what adjustments will be necessary in existing C2 arrangements. He can speed the process of positioning and moving reinforcements or the reserve by designating routes and providing traffic-control personnel and guides at contact points to lead and brief them on the situation. Scouts, military police, and divisional cavalry units can provide traffic control.

FOLLOW THROUGH

9-83. The purpose of defensive operations is to retain terrain and create conditions for a counteroffensive that regains the initiative. The area defense does this by causing the enemy to sustain unacceptable losses short of his decisive objectives. A successful area defense allows the commander transition to an attack. An area defense could also result in a stalemate with both forces left in contact with each other. Finally, it could result in the defender being overcome by the enemy attack and needing to transition to a retrograde operation. Any decision to withdraw must take into account the current situation in adjacent defensive areas. Only the commander who ordered the defense can designate a new FEBA or authorize a retrograde operation.

9-84. During this follow-through period, time is critical. Unless the commander has a large, uncommitted reserve prepared to quickly exploit or reverse the situation, he must reset his defense as well as maintain contact with the enemy. Time is also critical to the enemy, because he will use it to reorganize, establish a security area, and fortify his positions.

9-85. There is a difference between local counterattacks designed to restore the defense and a decisive operation designed to wrest the initiative from the enemy and then defeat him. To conduct a decisive counterattack, the defending force must bring the enemy attack to or past its culminating point before it results in an unacceptable level of degradation to the defending force. To do this, the defending force must disrupt the enemy's ability to mass, causing him to disperse his combat power into small groups or attrit his forces to gain a favorable combat power ratio. The defending force must continue to disrupt the enemy's ability to introduce follow-on forces and to destroy his sustainment system. In the defense, the commander must prepare to quickly take advantage of fleeting opportunities, seize the initiative, and assume the offense. Ideally, he already has a counterattack plan appropriate to the existing situation. He must rapidly reorganize and refit selected units, move them to attack positions, and attack. Alternatively, he must conduct an attack using those units already in contact with the enemy, which is normally the least favorable COA.

9-86. It is extremely difficult for the enemy to fight a defensive battle in response to a friendly counterattack after he reaches a culminating point for the following reasons:

- His defensive preparations are hasty.
- His forces are not adequately organized for defense.
- Reorganizing for a defense requires more time than the friendly commander should allow.
- The enemy force is dispersed, extended in depth, and weakened.
- Enemy attacks rarely culminate on ground ideally suited for defense.
- Physical fatigue.

9-87. The shift to defense requires enemy soldiers to make a psychological adjustment. Soldiers who have become accustomed to advancing, and thus winning, must now halt deep in the opposing force's territory and fight defensively, sometimes desperately, on new and often unfavorable terms. If the enemy commander decides to conduct retrograde operations to more defensible ground, his soldiers find it even harder to adjust psychologically.

9-88. If the defensive battle leads to a stalemate with both forces left in contact with each other, the defending force must seek to retain the initiative and set the conditions for the next encounter. The commander must prepare the defending unit to move rapidly to a subsequent defensive position during a lull in the battle because it is risky to defend from the same position twice. The enemy will know the location of the defending force's position and subject them to his supporting fires unless the defending force moves. Nevertheless the defending unit should normally stay in place and continue to fight unless it can suppress the enemy's approaching forces or take other actions to distract the enemy. This is because of the risk to a unit when it moves out of its prepared positions while still under enemy pressure.

9-89. If the defending unit is unable to maintain the integrity of its defense, it must transition to a retrograde operation or risk destruction. The commander must analyze how to execute this transition and prepare contingency plans. If the situation requires a retrograde movement, the commander conducts the operation according to the retrograde fundamentals and principles addressed in Chapter 11. In the retrograde, if the defending force can trade space for time without sustaining unacceptable losses, the commander can usually reestablish the conditions required for a successful defense.

Chapter 10

The Mobile Defense

A swift and vigorous transition to attack—the flashing sword of vengeance—is the most brilliant point of the defensive.

Carl von Clausewitz, *On War*, 1832

The *mobile defense* is a type of defensive operation that concentrates on the destruction or defeat of the enemy through a decisive attack by a striking force (FM 3-0). It focuses on destroying the attacking force by permitting the enemy to advance into a position that exposes him to counterattack and envelopment. The commander holds the majority of his available combat power in a striking force for his decisive operation, a major counterattack. He commits the minimum possible combat power to his fixing force that conducts shaping operations to control the depth and breadth of the enemy's advance. The fixing force also retains the terrain required to conduct the striking force's decisive counterattack. The area defense, on the other hand, focuses on retaining terrain by absorbing the enemy into an interlocked series of positions, where he can be destroyed largely by fires.

10-1. The factors of METT-TC may dictate that a unit conducts a mobile defense when defending against an enemy force with greater combat power but less mobility. A commander may also employ a mobile defense when defending a large area of operations (AO) without well-defined avenues of approach, such as flat, open terrain. The mobile defense is preferred in an environment where the enemy may employ weapons of mass destruction because this type of defense reduces the vulnerability of the force to attack and preserves its freedom of action. Future technology associated with command and control (C2) should improve the ability of the friendly force to gain and maintain a common operational picture, which reduces the risk associated with this type of defense. Among these risks are—

CONTENTS

- The fixing force may be isolated and defeated in detail because of the need to resource the striking force to the detriment of the fixing force.
- Operations in noncontiguous AOs associated with conducting a mobile defense can lead to defeat in detail.
- Enemy operations may impair the ability of the striking force to react at critical points.
- The enemy may not move into the area intended by the defending commander.
- The attacking enemy retains at least some momentum as he approaches the desired engagement areas (EAs).
- The defending force may not gain an accurate picture of the enemy's locations and dispositions required by the striking force to launch decisive operations in time to react.
- The decentralized operations required by the mobile defense increase the potential for fratricide.

HISTORICAL EXAMPLE

10-2. The concept of a mobile defense did not enter into Army doctrine until it had the chance to review German lessons learned as a result of its World War II experiences in Russia. The following historical example illustrates how conducting a mobile defense can result in recapturing the initiative and accomplishing the mission.

Manstein's Donbas Operation, February 1943

In January 1943, the Soviets launched a number of successful offensives following their Stalingrad counteroffensive. By the end of the month, this culminated in plans to drive German forces back to the Dniepr River. The Soviet high command (STAVKA) approved plans to liberate simultaneously the Donets Basin industrial area, Kharkov, and Kursk, and drive the Germans as far west as possible. The plan required that operations be continued without an operational pause, using forces weakened by previous operations, tenuously sustained by overextended supply lines with virtually no operational reserve.

German Field Marshal Manstein's mission was to preserve the German southern wing in the Donets area. His defensive concept consisted of allowing Soviet forces to advance in some areas, holding tightly to a few critical positions, and deliberately reducing his own forces in other areas to create a striking force capable of mounting a coordinated counterattack. See Figure 10-1. Reinforcements began arriving for his *Army Group Don*. He deployed his *1st Panzer Armee* to defend Voroshilovgrad as a fixing force, *4th Panzer Armee* and *Army Detachment Hollidt* to defend the central and southern parts of Manstein's lines, and the 1st *SS Panzer Korps* to defend Kharkov. The *1st SS Panzer Korps*, consisting of the *1st (Leibstandarte Adolf Hitler)*, *2nd (Das Reich)*, and *3rd (Totenkopf) SS Panzer Divisions) (PzDiv)*, formed his striking force.

STAVKA continued to pursue its offensive plans. However, the farther west the Soviet forces moved, the more overextended their supply lines became. On 20 February, Manstein's plan went into action. The *2nd SS PzDiv* attacked from south of Krasnograd and struck the Russian 6th Army and linked up with the *15th*

Infantry Division at Novo Moskovsk, thereby severing communications between the Soviet 267th Rifle Division (RD) and the 106th Rifle Brigade and the rear. On 21 February, the German units consolidated their positions and prepared to advance on Pavlograd. Meanwhile, *XL Panzer Korps* attacked the Southwestern Front's mobile group, ultimately routing it. Units of the *3rd SS PzDiv* moved into the Krasnograd area to prepare for their advance on Pavlograd. Despite this new situation, the Soviet front did not deviate from its offensive plans.

On 22 February, the *2nd SS PzDiv* drove through to Pavlograd and cut off the 35th Guards RD's communications with its 6th Army headquarters. The *3rd SS PzDiv* advanced, widening the breach between the Soviet 6th Army's main forces and the 267th RD. On 23 February, the *6th* and *17th PzDivs*, previously the 4th *Panzer Armee* (fixing force) reserve, began their offensive, smashing the 6th Army and 1st Guards Army and cutting the supply lines of and virtually encircling the 25th Tank Corps, which had been ordered to continue its advance. The *2nd SS PzDiv* consolidated positions at Pavlograd. The *3rd SS PzDiv* advanced against the 16th Guards Tank Brigade and the 35th Guards RD. Its southern column reached positions just northeast of Pavlograd. The *6th* and *17th PzDivs* advanced northward from the southeast, both divisions ultimately linking up with the *1st SS Panzer Korps* to advance farther north on 24 February.

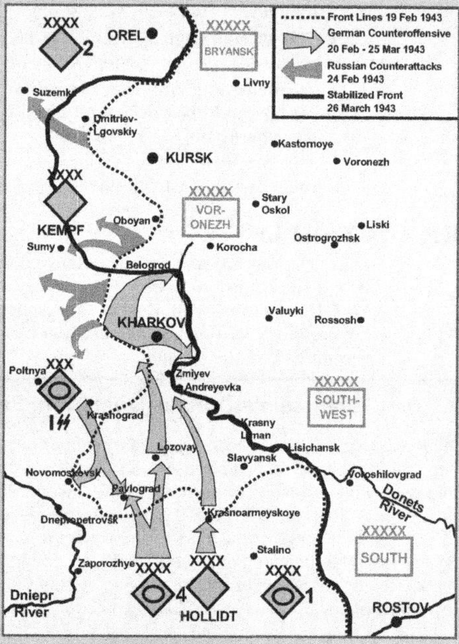

Figure 10-1. Donbas Movements

By the evening of 24 February, Vatutin, the Soviet Southwest Front commander, finally recognized the dangerous situation his forces were facing and ordered what remained of the front's right flank to go over to the defensive. The Germans continued their counteroffensive and ultimately recaptured Kharkov on 14 March.

ORGANIZATION OF FORCES

10-3. Units smaller than a corps do not normally conduct a mobile defense because of their inability to fight multiple engagements throughout the width, depth, and height of the AO, while simultaneously resourcing striking, fixing, and reserve forces. Typically, the striking force in a mobile defense may consist of one-half to two-thirds of the defender's combat power. (See Figure 10-2.) Division and smaller units generally conduct an area defense or a delay as part of the fixing force as the commander shapes the enemy's penetration or they attack as part of the striking force. Alternatively, they can constitute a portion of the reserve.

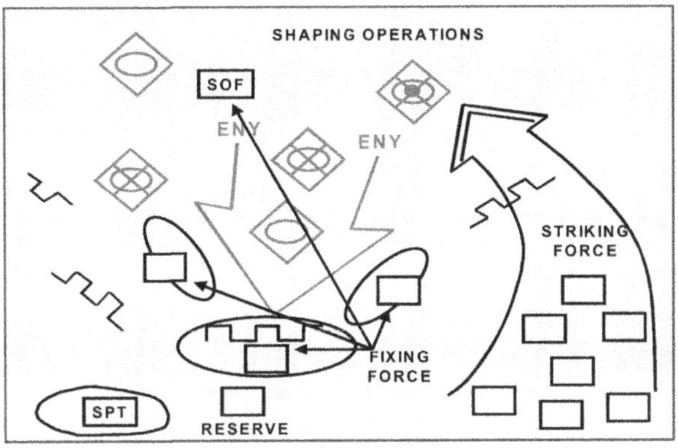

Figure 10-2. Mobile Defense

10-4. The commander organizes his main body into two principal groups—the fixing force and the striking force. In the mobile defense, reconnaissance and security, reserve, and sustaining forces accomplish the same tasks as in an area defense. (See Figure 10-3.) The commander completes any required adjustments in task organization before he commits his units to the fight.

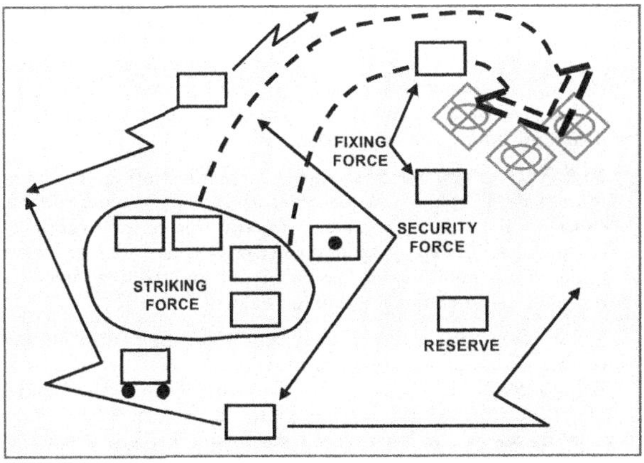

Figure 10-3. Organization of Forces for a Mobile Defense

10-5. Organized by the commander with the minimum combat power needed to accomplish its mission, the fixing force turns, blocks, and delays the attacking enemy force. It tries to shape the enemy penetration or contain his advance. Typically, it has most of the countermobility assets of the defending unit. The fixing force may conduct defensive actions over considerable depth within the main battle area (MBA). However, it must be prepared to stop and hold terrain on short notice to assist the striking force on its commitment. The operations of the fixing force establish the conditions for a decisive attack by the striking force at a favorable tactical location. The fixing force executes its portion of the battle essentially as a combination of an area defense and a delaying action. The actions of the fixing force are shaping operations.

10-6. The striking force decisively engages the enemy as he becomes exposed in his attempts to overcome the fixing force. The term "striking force" is used rather than reserve because the term "reserve" indicates an uncommitted force. The striking force is a committed force and has the resources to conduct a decisive counterattack as part of the mobile defense. It is the commander's decisive operation.

10-7. The striking force contains the maximum combat power available to the commander at the time of its counterattack. The striking force is a combined arms force that has greater combat power and mobility than the force it seeks to defeat or destroy. The commander considers the effects of surprise when determining the relative combat power of the striking force and its targeted enemy unit. The striking force is normally fully task organized with all combat support (CS) and combat service support (CSS) assets before its actual commitment. The commander positions engineer mobility-enhancing assets with the lead elements of the striking force.

10-8. The striking force is the key to a successful mobile defense. All of its contingencies relate to its attack. If the opportunity does not exist to decisively commit the striking force, the defender repositions his forces to establish the conditions for success. The striking force must have mobility equal to or greater than that of its targeted enemy unit. It can obtain this mobility through proper task organization, countermobility operations to slow and disrupt enemy movements, and mobility operations to facilitate the rapid shifting of friendly formations. The striking force requires access to multiple routes because an attacking enemy normally goes to great length to deny the defending force freedom of action.

10-9. The commander responsible for orchestrating the overall mobile defense should retain control of the striking force unless communication difficulties make this impossible. Normally this is the overall defending force commander. The commander's most critical decisions are when, where, and under what conditions he should commit his striking force. The commander normally accompanies the striking force.

10-10. Resourcing a reserve in a mobile defense is difficult and requires the commander to assume risk. He generally uses his reserve to support the fixing force. However, if the reserve is available to the striking force, it exploits the success of the striking force. If the reserve is composed largely of aviation forces and long-range fire support systems, it may have contingencies to support the fixing and striking forces.

CONTROL MEASURES

10-11. A commander conducting a mobile defense uses control measures to synchronize conducting the operation. These control measures include designating the AOs of the fixing and striking forces with their associated boundaries, battle positions, and phase lines. He designates a line of departure or a line of contact as part of the graphic control measures for the striking force. He may designate an axis of advance for the striking force. He can designate attack-by-fire or support-by-fire positions. The commander uses EAs, target reference points, targeted areas of interest, and final protective fires as necessary. He designates named areas of interest to focus the efforts of his intelligence, surveillance, and reconnaissance (ISR) assets. This allows him to determine the enemy's course of action (COA). He designates checkpoints, contact points, passage points, passage routes, and passage lanes for use by reconnaissance and surveillance assets, security units, and the striking force. (See Figure 10-4.)

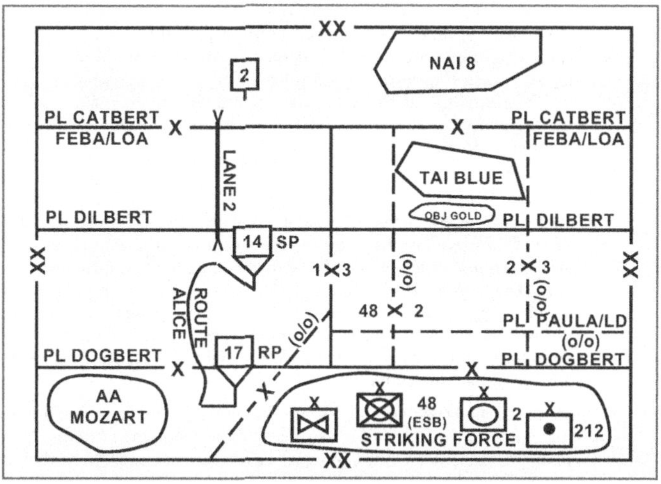

Figure 10-4. Mobile Defense Control Measures

10-12. The commander must provide the striking force commander with control measures to focus his force at the decisive time and place and to deconflict fires with the fixing force. As a minimum, the striking force commander needs to know the anticipated objective decision points that could lead to the commitment of his force, limit of advance, and boundaries of his AO. If the overall commander imposes either an axis of attack or a direction of attack as a control measure, he restricts the striking force commander's freedom of maneuver. However, such restrictions may be necessary to avoid contact with enemy forces that could distract the striking force from accomplishing its primary mission. These control measures may have to be drawn "on the fly" while the commander, his staff, and his subordinates move to take advantage of an opportunity to commit the striking force in a decisive counterattack. They should also help the commander recover the integrity of his defense if the striking force is not successful in its attack. (Chapters 2, 5, and 8 explain these control measures.)

PLANNING A MOBILE DEFENSE

10-13. The key to successful mobile defensive operations is the integration and synchronization of all available assets to maximize the combat power of the defending unit, particularly the striking force. The commander achieves integration and synchronization when he can employ their combined effects at decisive times and places. (The general defensive planning considerations addressed in Chapter 8 apply to the mobile defense.)

MANEUVER

10-14. The commander's ability to maintain the mobility advantage of his forces is an important aspect of the mobile defense. This mobility advantage may result from or be enhanced by countermobility actions directed against the enemy force. In his mobile defense plan, the commander ensures that his forces—including reserves and the striking force—can move freely around the battlefield, while at the same time restricting the enemy's mobility, slowing his momentum, and guiding or forcing him into areas that favor the friendly defensive effort.

FIRE SUPPORT

10-15. The effectiveness of a mobile defense is based on the carefully planned fires of all weapons. The striking force conducts the commander's decisive operation in a mobile defense. It requires continuous and concentrated fire support. The commander weights his decisive operation, in part, by allocating to it field artillery and other fire support weapon systems. He must rapidly shift indirect fire support from the fixing force to the striking force. These fire support systems do not have to move with the striking force if it remains within supporting range.

10-16. If the striking force's planned maneuver places it outside the supporting range of the defending commander's fire support systems, he must either plan the movement of fire support assets to locations where they can support the striking force or incorporate them into the striking force. Fire support assets can partially compensate for a lack of maneuver forces in the striking force. The commander takes precautions to prevent fratricide as the striking force approaches the fixing force's EAs, while supporting air and artillery assets try to interdict enemy movements.

AIR DEFENSE

10-17. In the mobile defense, air defense is normally initially used to cover—

- Security forces and fixing force units in forward areas.
- C2 facilities.
- Critical assets, including fire support systems, reserves, and the striking force.
- Sustainment resources.

- Choke points along movement corridors planned for use by reserves or the striking force.

Once the commander commits the striking force, it receives priority of support as the decisive operation. If the striking force attacks to extended depths, the commander ensures that it and other critical assets remain within the coverage of available air defense systems. This may require him to reposition air defense radars and systems to maintain air defense coverage of the defending force.

MOBILITY/COUNTERMOBILITY/SURVIVABILITY

10-18. The majority of the commander's countermobility and survivability assets support the operations of the fixing force. The majority of the commander's mobility assets support the operations of the striking force. Situational obstacles provide him a tremendous advantage in the mobile defense. These obstacles are a combat multiplier because they enable the commander to use economy of force measures. He uses situational obstacles to exploit enemy vulnerabilities, exploit success, separate enemy follow-on forces, and provide flank protection.

COMBAT SERVICE SUPPORT

10-19. When planning for the mobile defense's sustaining operations, logistics operations planners must look beyond the fixing force's shaping operations to prepare to support the striking force's decisive counterattack. The greater the distance the striking force must cover when moving from its assembly area (AA) to its final objective, the greater the amount of supplies needed to support that move. Once committed, units in the striking force require priority of fuel, ammunition, and maintenance support over comparable units in the fixing force. Casualty evacuation will be a challenge because the fixing force will likely suffer a higher percentage of casualties but the lines of communications to the striking force must also support casualty treatment and evacuation. When the striking force must move a considerable distance from its sustaining base, the commander should consider establishing an intermediate support base (ISB). Before establishing an ISB, he must weigh the benefits of establishing the base against the cost in terms of combat power or effort diverted from the support mission to secure the ISB.

PREPARING A MOBILE DEFENSE

10-20. Preparations for conducting a mobile defense include developing the fixing force's defensive positions and EAs as discussed in Chapter 8. The commander aggressively uses his reconnaissance assets to track enemy units as they approach. Engineers participate in conducting route and area reconnaissance to find and classify existing routes. They improve existing routes and open new routes for use during the battle.

10-21. The striking force assembles in one or more areas depending on the width of the AO, the terrain, enemy capabilities, and the planned manner of employment. Before the enemy attack begins, the striking force may deploy all or some of its elements forward in the MBA to—

- Deceive the enemy regarding the purpose of the force.

- Occupy dummy battle positions.
- Create a false impression of unit boundaries, which is important when operating with a mix of heavy and light forces or multinational forces.
- Conduct reconnaissance of routes between the striking force's AAs and potential EAs.

10-22. The enemy attempts to discover the strength, composition, and location of the units that constitute the fixing force and the striking force. The commander uses security forces and information operations to deny the enemy this information and degrade the collection capabilities of enemy ISR assets. The commander routinely repositions to mislead the enemy and to protect his force. In addition, his plans and preparations incorporate defensive information operations. The commander normally tries to portray an area defense while hiding the existence and location of the striking force.

EXECUTING A MOBILE DEFENSE

10-23. This manual divides the execution of a mobile defense into five phases for discussion purposes. The length and nature of each phase, if it occurs at all, varies from situation to situation according to the factors of METT-TC. The phases of defensive operations are gain and maintain enemy contact, disrupt the enemy, fix the enemy, maneuver, and follow through.

10-24. The commander must have the flexibility to yield terrain and shape the enemy penetration. He may even entice the enemy by appearing to uncover an objective of strategic or operational value to the enemy. The striking force conducts the decisive operation—the attack—once the results of the actions of the fixing force meet the commander's intent.

GAIN AND MAINTAIN ENEMY CONTACT

10-25. The commander conducting a mobile defense focuses on discovering the exact location of the enemy and his strength to facilitate the effectiveness of the striking force. The security force (guard or cover) or the fixing force confirms the enemy's COA and main avenues of approach. The commander normally tasks other ISR assets to determine the location of enemy reserves and follow-on forces. Early detection of the enemy's decisive operation provides the commander with reaction time to adjust the fixing force's positions and shape the enemy penetration, which, in turn, provides the time necessary to commit the striking force. The striking force commander requires as close to real-time updates of the enemy situation as are possible to ensure that the striking force engages the enemy at the right location and time.

10-26. While conducting operations, the security force determines what routes the enemy is using, where the enemy is strong or weak, and where gaps in and between enemy formations exist. This information aids the commander in his attempt to seize the initiative. That information also increases the striking force's agility by identifying opportunities. Further, it helps pull the striking force along the path of least resistance as it maneuvers to employ its combat power at the critical time and place.

DISRUPT THE ENEMY

10-27. In a mobile defense, the commander conducts shaping operations designed to shape the enemy's penetration into the MBA and disrupt the enemy's introduction of fresh forces into the fight. These shaping operations help establish the preconditions for committing the striking force by isolating the object of the striking force and destroying the enemy's key C2 nodes, logistics resupply units, and reserves. Whenever possible the commander sequences these shaping operations, to include offensive information operations, so that the impact of their effects coincides with the commitment of the striking force. To generate a tempo that temporarily paralyzes enemy C2, the intensity of these shaping operations may increase dramatically on the commitment of the striking force. The commander continues to conduct shaping operations once the striking force commits to prevent enemy forces from outside the objective area from interfering with executing the decisive counterattack.

FIX THE ENEMY

10-28. Fixing the enemy is the second half of shaping operations and results in establishing the conditions necessary for decisive operations by the striking force. Typically, the commander of the defending force allows the enemy force to penetrate into the defensive AO before the striking force attacks. (See Figure 10-5.) The fixing force may employ a combination of area defense, delay, and strong point defensive techniques to shape the enemy penetration. The intent of the fixing force is not necessarily to defeat the enemy but to shape the penetration to facilitate a decisive counterattack by the striking force. The commander ensures that the missions and task organization of subordinate units within the fixing force are consistent with his concept for shaping the enemy penetration. Defensive positions within the fixing force may not be contiguous since the fixing force contains only the minimum-essential combat power to accomplish its mission.

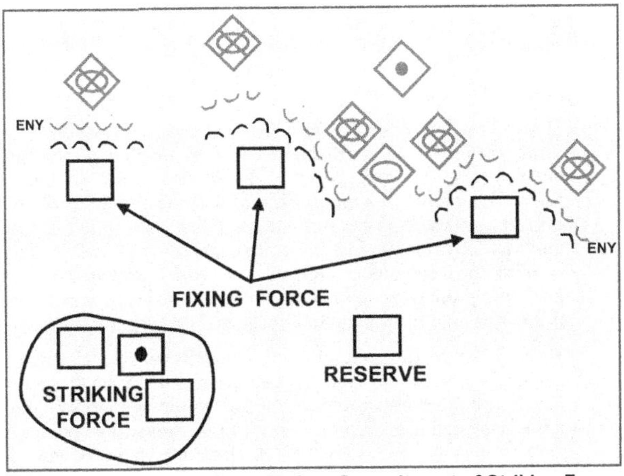

Figure 10-5. Mobile Defense Before Commitment of Striking Force

10-29. The fixing force's extensive use of obstacles supports this shaping effort and helps to gain an overall mobility advantage over the enemy. The commander may want to yield ground quickly to make the enemy think he has been successful or to entice him to a decisive point where the striking force can attack. Normally, in a mobile defense, the commander retains ground only to facilitate the commitment of his striking force.

10-30. When conducting a mobile defense, the commander may need to commit his reserve to reinforce the fixing force and help shape the battlefield. He positions his reserve so it effectively reacts to the most likely contingency and the enemy's most dangerous COA. Without a reserve, the commander assumes significant risk in attempting to shape the enemy penetration. Circumstances may also force the commander to employ elements of the striking force to assist the fixing force. If that occurs, the commander prefers to use his available long-range fire support assets and attack helicopters. They are the best choice because of their ability to rapidly disengage and shift their effects to support the efforts of the rest of the striking force on its commitment.

MANEUVER

10-31. The commander's situational understanding is critical in establishing the conditions that initiate the striking force's movement and in determining the general area that serves as a focus for the counterattack. Situational understanding includes identifying those points in time and space where the counterattack proves decisive. A force-oriented objective or an EA usually indicates the decisive point. The staff synchronizes the unit's activities in time and space to sufficiently mass the effects of the striking force at the right time and place.

10-32. The actions of the striking force are the echelon's decisive operation on its commitment. The commander's ISR systems focus entirely on tracking the enemy's advance. The striking force commander continuously receives intelligence and combat information updates that allow him to adjust his counterattack as necessary to defeat the targeted enemy. Once the enemy starts his attack, any forward-deployed elements of the striking force withdraw to AAs or attack positions and prepare for their commitment in counterattack.

10-33. The defending commander launches his striking force in a counterattack when its offensive power, relative to that of the targeted attacking enemy element, is the greatest. (See Figure 10-6.) Piecemeal commitment of the striking force in support of local objectives jeopardizes the success of the overall operation. The striking force must execute the counterattack rapidly and violently, employing all combat power necessary to ensure success. The striking force may be committed at a time different than anticipated and in an entirely different area than previous contingency plans envisioned. Thus, it must be able to respond to unexpected developments rapidly and decisively.

10-34. Because the striking force normally attacks a moving enemy force, it generally assumes a combat formation with a covering force, an advance guard, a main body, and either a follow-and-support or a follow-and-assume force. The striking force attempts to take advantage of obstacles, such as rivers or obstacle zones that block the enemy's movement. The commander designates flank responsibilities and may even allocate a designated force against a particularly vulnerable flank. However, the striking force moves quickly and takes risk on its flanks, using its speed of movement and superior situational understanding to provide security.

10-35. The striking force attacks in a formation that provides maximum combat power forward to devastate the targeted enemy force and achieve decisive results. The striking force takes advantage of its mobility and fire power to seize the initiative by overwhelming the enemy force with swift, violent blows that cripple the enemy's command and control system, disrupt his formations, and destroy his combat systems. The commander ensures that his fire support and fixing force capture the enemy's attention and posture the enemy for attack by the striking force. During the counterattack, he may have one element of the striking force occupy support-by-fire positions to suppress the enemy, while another striking force element prepares to assault the objective. Either heavy or light forces may make this assault. (Chapter 5 discusses the actual conduct of an assault on an objective.)

10-36. Engineers should be well forward to enhance the mobility of the striking force. These lead engineers search for existing obstacles and clear the route as much as possible within their capabilities. Follow-on engineers expand breaches, improve routes, and replace assault bridges with more permanent structures. Engineers with flank units focus on countermobility to protect the flanks.

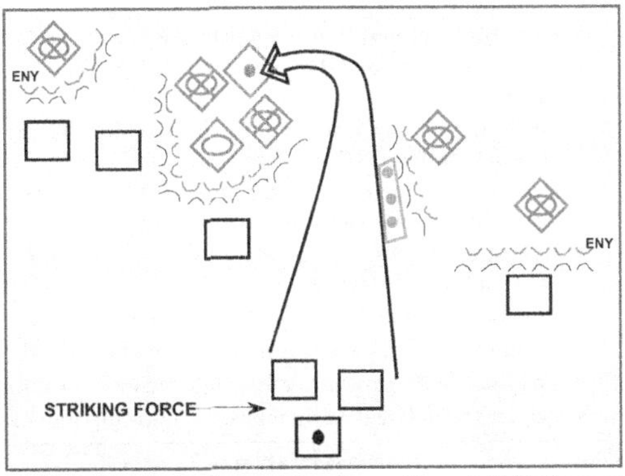

Figure 10-6. Mobile Defense After Commitment of Striking Force

FOLLOW THROUGH

10-37. All defensive operations intend to create the opportunity to transition to the offense. In a mobile defense, that transitional opportunity generally results from the success of the striking force's attack. The commander exploits his success and attempts to establish conditions for a pursuit if his assessment of the striking force's attack is that there are opportunities for future offensive operations. (Chapters 6 and 7 discuss exploitation and pursuit.) If conducting the mobile defense is unsuccessful and the enemy retains the initiative, the commander must either reestablish a viable defense or conduct retrograde operations. (Retrograde operations are the topic of Chapter 11.)

Chapter 11

The Retrograde

The withdrawal should be thought of as an offensive instrument, and exercises framed to teach how the enemy can be lured into a trap, closed by a counter-stroke or a devastating circle of fire.

B.H. Liddell Hart, 1944

The *retrograde* is a type of defensive operation that involves organized movement away from the enemy (FM 3-0). The enemy may force these operations or a commander may execute them voluntarily. In either case, the higher commander of the force executing the operation must approve the retrograde. Retrograde operations are transitional operations; they are not considered in isolation.

11-1. The commander executes retrogrades to—

- Disengage from operations.
- Gain time without fighting a decisive engagement.
- Resist, exhaust, and damage an enemy in situations that do not favor a defense.
- Draw the enemy into an unfavorable situation or extend his lines of communication (LOCs).
- Preserve the force or avoid combat under undesirable conditions, such as continuing an operation that no longer promises success.
- Reposition forces to more favorable locations or conform to movements of other friendly troops.
- Position the force for use elsewhere in other missions.
- Simplify the logistic sustainment of the force by shortening LOCs.
- Position the force where it can safely conduct reconstitution.
- Adjust the defensive scheme, such as secure more favorable terrain.

CONTENTS

- Deceive the enemy.

11-2. The three forms of retrograde are delay, withdrawal, and retirement. In each form, a force moves to the rear, using combinations of combat formations and marches. (Chapter 3 discusses combat formations; Chapter 14 discusses troop movement.) The commander may use all three forms singularly or in combination with other types and forms of offensive or defensive operations.

11-3. Retrogrades can negatively affect the participating soldiers' attitude more than any other type of operation because they may view the retrograde as a defeat. A commander must not allow retrograde operations to reduce or destroy unit morale. Leaders must maintain unit aggressiveness. By planning and efficiently executing the retrograde and ensuring that soldiers understand the purpose and duration of the operation, the commander can counter any negative effects of the operation on unit morale. After completing a retrograde operation, the commander may reconstitute the force. FM 4-100.9 establishes the basic principles of reconstitution.

HISTORICAL EXAMPLE

11-4. The following historical example illustrates how conducting a retrograde operation can preserve an army for future operations.

The Atlanta Campaign, 1864

The first two months of the Atlanta campaign illustrate the successful conduct of a delay in the face of superior forces. Between 5 May and 17 July, Johnston held Sherman to an average gain of one mile a day while preserving his freedom of maneuver and his army for future operations. This part of the campaign contains examples of successful delays, withdrawals, and retirements. Confederate actions at Resaca early in the campaign will be used to illustrate an unassisted withdrawal under enemy pressure. See Figure 11-1, page 11-2.

In May 1864, Confederate GEN Joseph E. Johnston and his 55,000-man *Army of Tennessee* had the mission of defending Atlanta. Johnston faced 110,000 Union soldiers, organized into seven corps under MG William T. Sherman's overall command. Johnston's campaign strategy was to force Sherman to culminate before reaching Atlanta, conserving his army's strength until he crippled the Union army in a defensive battle, and then launch a counteroffensive.

Union forces began the campaign on 5 May with an advance from positions southeast of Chattanooga. Forced to withdraw from his initial positions at Dalton because of a turning movement around his left flank by two Union corps, Johnston raced to position his forces to defend Resaca, Georgia. Johnston intended to hold at Resaca until he could cross his entire force over the Oostanaula River to its southern bank. At Resaca were three bridges that supported the Confederate's line of communication with the logistics base at Atlanta.

Johnston positioned the three corps—then constituting his army—to defend Resaca as they arrived on 13 May. May 14th and 15th saw attacks around Resaca, with neither side gaining a marked advantage. However, the right flank of the Union army moved within cannon range of the bridges. Forces that Sherman sent four miles to the southwest on another turning movement crossed the river on 15 May, although they did not break out from their bridgehead on that day. Sherman intended to follow with his main force and either envelope Johnston or strike his flank during a retreat. Faced with these prospects, Johnston decided to withdraw across the river. A deceptive Confederate attack late on the 15th convinced Sherman that Johnston intended to stay in his current positions. The withdrawal commenced at midnight. Johnston left skirmishers along the line (a detachment left in contact), withdrawing forces in succession from his corps farthest from the bridges, then from his center corps, and finally from the corps closest to the bridges. One division from his right-wing corps constituted the army's rear guard. After the rear guard crossed the bridges, Johnston's engineers commenced their destruction. During the three hours required to withdraw Johnston's army across the river, Union forces did not detect the withdrawal until the Confederates began to destroy the bridges.

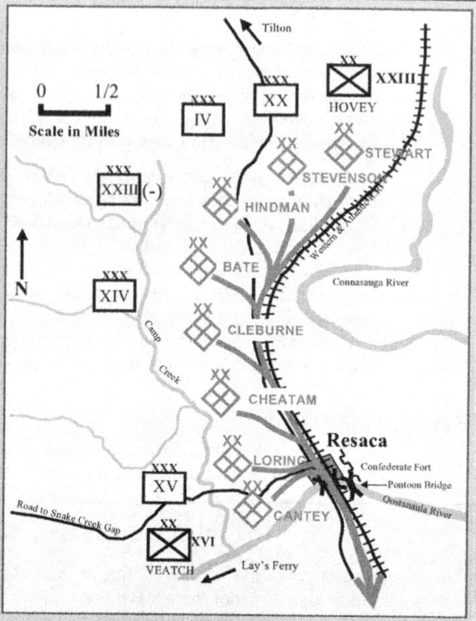

Figure 11-1. Withdrawal from Resaca

Johnston's mix of retrograde and defensive operations preserved his army as a constant threat to Sherman. Nevertheless, his strategy was unacceptable to Confederate President Jefferson Davis. Davis replaced him on 17 July 1864 with GEN John B. Hood. Hood attacked Sherman three time in two weeks, suffering heavy casualties and failing to seize the initiative. On 1 September, Sherman seized Atlanta, achieving the Union's strategic objective in the Western Theater before the national elections.

DELAY

11-5. A *delay* is a form of retrograde [JP 1-02 uses *an operation*] in which a force under pressure trades space for time by slowing down the enemy's momentum and inflicting maximum damage on the enemy without, in principle, becoming decisively engaged (JP 1-02, see delaying operation). The delay is one of the most demanding of all ground combat operations. A delay wears down the enemy so that friendly forces can regain the initiative through offensive action, buy time to establish an effective defense, or determine enemy intentions as part of a security operation. Normally in a delay, inflicting casualties on the enemy is secondary to gaining time. For example, a flank security force conducts a delay operation to provide time for the protected force to establish a viable defense along its threatened flank. Except when directed to prevent enemy penetration of a phase line (PL) for a specific duration, a force conducting a delay normally does not become decisively engaged.

11-6. A delay operation can occur when the commander does not have enough friendly forces to attack or defend. It may also occur, based on a unit's mission, in conjunction with a higher commander's intent. The decision to conduct a delay may not be based on the unit's combat power, but on the other factors of METT-TC. For example, during security operations, the commander may conduct a delay as a shaping operation to draw the enemy into an area where he is vulnerable to a counterattack. Another example would be a delay instituted as an economy of force effort to allow the commander to conduct an offensive operation elsewhere.

11-7. The ability of a force to trade space for time requires depth within the area of operations (AO) assigned to the delaying force. The amount of depth required depends on several factors, including the—

- Amount of time to be gained.
- Relative combat power of friendly and enemy forces.
- Relative mobility of the forces.
- Nature of the terrain.
- Ability to shape the AO with obstacles and fires.
- Degree of acceptable risk.

Ordinarily, the greater the available depth, the lower the risk involved to the delaying force and the greater the chance for success.

11-8. A delay succeeds by forcing the enemy to repeatedly concentrate his forces to fight through a series of defensive positions. A delaying force must offer a continued threat of serious opposition, forcing the enemy to repeatedly deploy and maneuver. Delaying forces displace to subsequent positions before the enemy is able to concentrate sufficient resources to decisively engage and defeat delaying forces in their current position. The length of time a force can remain in a position without facing the danger of becoming decisively engaged is primarily a function of the factors of METT-TC, such as the relative combat power and the terrain and weather.

ORGANIZATION OF FORCES

11-9. The commander normally organizes the delaying force into a main body, a security force, and a reserve. The security force usually conducts a screen forward of the initial delay positions. For a divisional cavalry squadron or a corps cavalry regiment conducting a delay, the security force executing the screen mission may consist of scouts or air cavalry. For a brigade or battalion conducting a delay, the security force may consist of the brigade reconnaissance troop, battalion scouts, or another element tasked to conduct security operations.

11-10. The main body, which contains the majority of the delaying force's combat power, may use alternate or subsequent positions to conduct the delay. The commander usually deploys his main body as a complete unit into a forward position when conducting a delay from subsequent positions. He divides his main body into two parts, roughly equal in combat power, to occupy each set of positions when conducting a delay from alternate positions.

11-11. The commander normally retains a reserve to contain enemy penetrations between positions, to reinforce fires into an engagement area (EA), or to help a unit disengage from the enemy. All of these missions require that the reserve has the mobility and strength to strike with such force that an enemy has no option but to face the immediate threat.

11-12. The extended frontages and ranges common to retrograde operations make the provision of fire support difficult and act to limit the commander's ability to mass fires. Therefore, retrograde forces, especially delay forces, often have more than the normal allocation of fire support assets. The commander's risk of losing artillery systems and their ammunition also increases when he is supporting retrograde operations. Therefore, he balances his decision to commit fire support systems forward against anticipated requirements in subsequent battle stages. In particular, he protects his towed artillery systems from being overrun by a mobile enemy. He can use available rotary- and fixed-wing aircraft to augment or replace his artillery systems.

11-13. Combat support (CS) and combat service support (CSS) assets are widely dispersed and often attached to the units they support because of the width of the AOs normally assigned in a delay. Engineer priorities are normally countermobility first, then mobility. However, restrictive terrain that impedes friendly movement may require the commander to reverse the priorities. Close coordination is necessary so that engineer obstacles are covered by fire and do not impede the planned withdrawal routes of delaying forces or the commitment of a counterattacking reserve force. The delaying force should have a greater-than-normal allocation of fire support systems.

11-14. The requirement to maintain continuous support during the delay requires CSS organizations to echelon their assets. This echelonment, coupled with the wide dispersion of combat forces that is inherent in a delay, complicates conducting the delay.

CONTROL MEASURES

11-15. The delay consists of a series of independent small-unit actions that occur simultaneously across the front. Subordinate commanders must have freedom of action. The tactical mission graphic for the delay appears in Figure 11-2. It is not a control measure. The control measures used in the delay are the same as those introduced in Chapter 8. Common graphics used in a delay include AOs, PLs, battle positions (BPs), coordination points, checkpoints, EAs, trigger lines, target reference points (TRPs), and disengagement lines. (See Figure 11-3.) The commander designates contact points in front of, between, and behind units to assist coordination, ensure continuity of the delay, and draw attention to enemy avenues of approach into unit flanks. (Chapter 16 addresses passage points with the passage of lines discussion.)

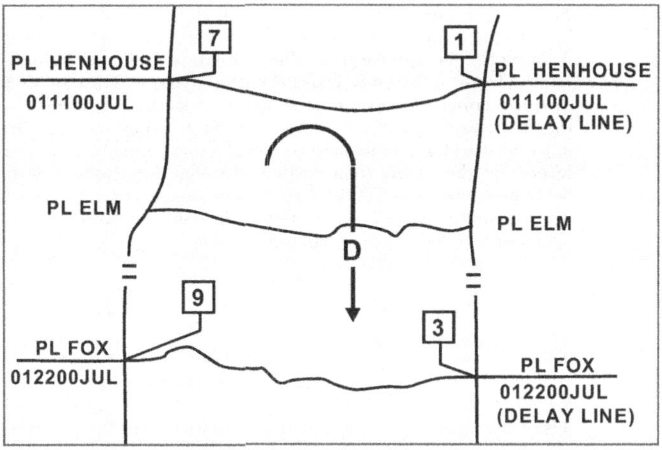

Figure 11-2. Delay Tactical Mission Graphic

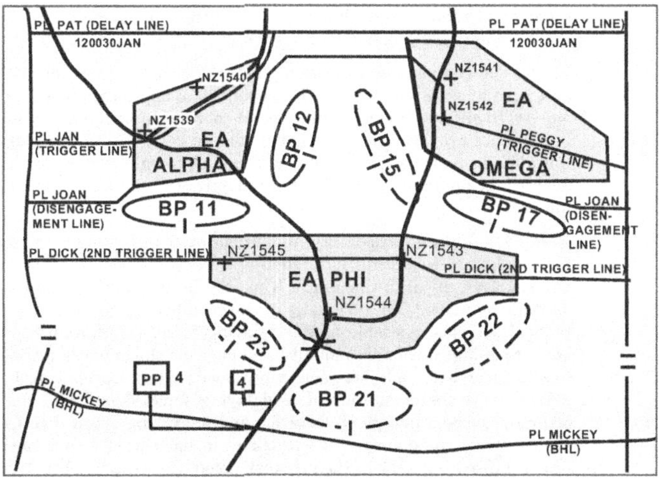

Figure 11-3. Control Measures for a Delay

11-16. In planning for a delaying action, the commander assigns an AO to each committed unit down through the company or troop level. The commander assigns each likely enemy avenue of approach to only one subordinate unit when he designates subordinate units' AOs. When the commander draws the boundaries of these subordinate AOs, he includes the terrain that controls fire and observation into those areas.

11-17. The commander designates additional PLs beyond those established by his higher commander as necessary to control movement during the delay. **A *delay line* is a phase line where the date and time before which the enemy is not allowed to cross the phase line is depicted as part of the graphic control measure.** Designating delay lines is a command decision that imposes a high degree of risk on the delaying unit. The delaying unit must do everything in its power—including accepting decisive engagement—to prevent the enemy from crossing that line before the time indicated. A delay line may also be event-driven. For example, a commander can order a delaying unit to prevent penetration of the delay line until his engineers complete construction of a rearward obstacle belt.

PLAN

11-18. Unit commanders and soldiers must understand and exercise the basics of defensive operations outlined in Chapter 8 to conduct a successful delay. However, these defensive basics have unique considerations, and the significance of these considerations varies depending on the factors of METT-TC. In a delay, units operate on extended frontages at great risk from advancing enemy forces. The tactical situation constantly changes with maneuver opportunities existing for only extremely short periods. Subordinate commanders must have the flexibility to take immediate action to retain the integrity of their forces. This helps retain their freedom of maneuver and inflict maximum destruction on the enemy.

11-19. The commander identifies ground and air avenues for enemy attacks and friendly counterattacks. When avenues of approach diverge or pass from one AO to another, adjacent units must coordinate with each other. Using the intelligence preparation of the battlefield process, the commander designates initial and subsequent delay positions on key terrain that covers likely enemy avenues of approach throughout the depth of the AO allocated to the delay mission.

11-20. Maintaining a mobility advantage over the attacker by the delaying force is key to successfully conducting a delay. Robust engineering and fire support are critical to this effort. The commander plans to maintain this advantage by taking full advantage of the mobility inherent in the combat and tactical systems available to the delaying force. In addition he takes other steps to enhance friendly mobility and degrade the enemy's mobility, such as building combat trails between delay positions and preparing bridges over major rivers for demolition. The delaying force should be capable of constructing large numbers of obstacles and delivering long-range fires. For example, while the enemy seeks to travel in movement formations that allow him to press his attack at the greatest speed, the delaying force's aim is to engage the enemy as early and often as possible. This forces the enemy out of those formations through a multiple series of time-consuming deployments into an assault formation.

11-21. The air defense portion of the plan has three main considerations—the protection of the force while it is in position, the protection of any forces left in contact, and the protection of the force as it moves to the rear. Priority of protection should be toward maintaining the mobility of the force. Air defense assets remain mobile yet able to engage aerial targets with little advance warning. These assets should work in teams, able to move to the rear in alternating bounds. This ensures that dedicated air defense assets will always be in position, with the flexibility to keep pace with the operations. These firing points are not obvious positions that an enemy will probably target as part of his preparatory or support fires. Early warning of enemy air attack is provided over combat net radios using the command net at the brigade echelon and below.

11-22. Flanks and gaps between units are always areas of concern for a commander. In a linear deployment, the enemy can bypass or outflank the delaying force if coordination between adjacent friendly units is weak, or if one unit creates a gap by moving rearward too rapidly. Therefore, the commander normally designates BPs to guard approaches into his AO. Adjacent units of different commands must exchange liaison.

11-23. Displacement criteria should specify at what point—either event, or time-driven—the delaying force should begin its displacement. The commander should calculate enemy closure rates for the terrain and compare them to friendly displacement rates between positions. By comparing time and distance factors, he can calculate his movement window of time. By applying the enemy's probable rates of advance and formations to the avenues of approach, the commander can decide what obstacles to use and where to emplace them (covered by fires). It also helps the commander determine if and where decisive engagement is likely or required to achieve the delay objective. Careful consideration of the factors of METT-TC, especially terrain analysis, is an inherent part of delay planning.

Parameters of the Delay Order

11-24. An order for a delay mission must specify certain parameters. First, it must direct one of two alternatives: delay within the AO or delay forward of a specified line or terrain feature for a specified time. That time is usually based on another unit completing its activities, such as establishing rearward defensive positions. A mission of delay within the AO implies that force integrity is a prime consideration. In this case, the delaying force delays the enemy as long as possible while avoiding decisive engagement. Generally, this force displaces once predetermined criteria have been met, such as when the enemy force reaches a disengagement line. The control measures are the same for both alternatives, except that during a delay forward of a specified line for a specified time, the commander annotates the PL with the specified time. (See Figure 11-4.) If the commander establishes a delay line, mission accomplishment outweighs preservation of the force's integrity. It may require the force hold a given position until ordered to displace.

11-25. The second parameter is that the order must specify the acceptable risk. Acceptable risk ranges from accepting decisive engagement in an attempt to hold terrain for a given time to maintaining the integrity of the delaying force. The depth of the AO available for the delay, the time needed by higher headquarters, and subsequent missions for the delaying force determine the amount of acceptable risk. A delay mission that does not specify times, control of key terrain, or other guidance and control measures implies a lower degree of risk.

11-26. Third, the order must specify whether the delaying force may use the entire AO or must delay from specific BPs. A delay using the entire AO is preferable, but a delay from specific positions may be required to coordinate two or more units in the delay. To enhance command and control and to coordinate the battle across a broad front, the commander assigns units down to platoon-level specific BPs. However, he may assign them missions to delay within their AO if that best supports the scheme of maneuver.

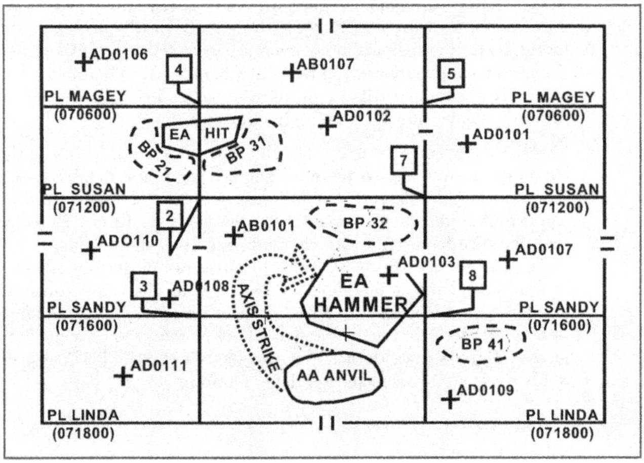

Figure 11-4. Delay Forward of a Specified Line for a Specified Time

Alternate and Subsequent Positions

11-27. The commander normally assigns his subordinate units contiguous AOs that are deeper than they are wide. He uses obstacles, fires, and movement throughout the depth of each assigned AO. He may be forced to fight from a single set of positions if the delay is only planned to last a short time or the AO's depth is limited. If the delay is expected to last for a longer period, or if sufficient depth is available, he may delay from either alternate or successive positions.

11-28. In both techniques, delaying forces normally reconnoiter subsequent positions before occupying them and, if possible, post guides on one or two subsequent positions. Additionally, in executing both techniques, it is critical that the delaying force maintains contact with the enemy between delay positions. (The advantages and disadvantages of the two techniques are summarized in Table 11-1, page 11-10.)

11-29. A commander normally prefers to use alternate positions when he has adequate forces and his AO has sufficient depth. In a delay from alternate positions, two or more units in a single AO occupy delaying positions in depth. (See Figure 11-5, page 11-10.) As the first unit engages the enemy, the second occupies the next position in depth and prepares to assume responsibility for the operation. The first force disengages and passes around or through the second force. It then moves to the next position and prepares to reengage the enemy while the second force takes up the fight. Alternate positions are normally used when the delaying force operates on a narrow front. A delay from alternate positions is particularly useful on the most dangerous avenues of approach because it offers greater security than a delay from successive positions. However, it requires more forces and continuous maneuver coordination. Additionally, the delaying forces risk losing contact with the enemy between delay positions.

Table 11-1. Advantages and Disadvantages of Delay Techniques

METHOD OF DELAY	USE WHEN	ADVANTAGES	DISADVANTAGES
Delay from Subsequent Positions	• AO is wide. • Forces available do not allow themselves to be split.	• Masses fires of all available combat elements.	• Limited depth to the delay positions. • Less available time to prepare each position. • Less flexibility.
Delay from Alternate Positions	• AO is narrow. • Forces are adequate to be split between different positions.	• Allows positioning in depth. • Allows more time for equipment and soldier maintenance. • Increases flexibility.	• Requires continuous coordination. • Requires passage of lines. • Engages only part of the force at one time.

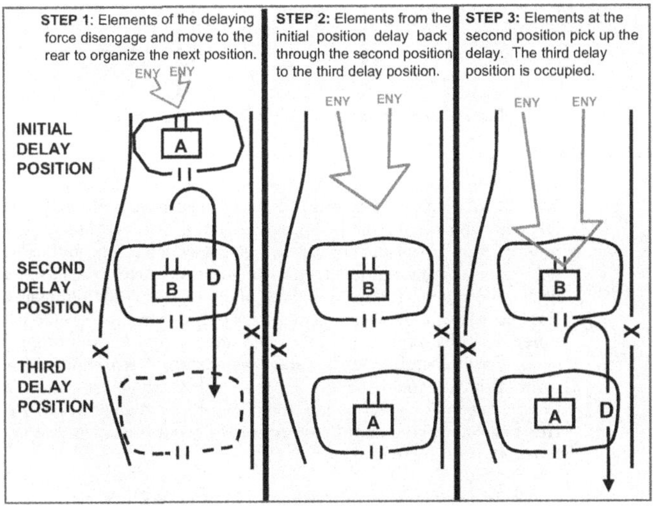

Figure 11-5. Delay from Alternate Positions

11-30. The commander uses a delay from subsequent positions when the assigned AO is so wide that available forces cannot occupy more than a single tier of positions. (See Figure 11-6.) In a delay from subsequent positions, all delaying units are committed to each of the series of BPs or across the AO on the same PL. Most of the delaying force is located well forward. The mission dictates the delay from one BP or PL to the next. The commander staggers the movement of delaying elements so that not all elements are moving at the same time.

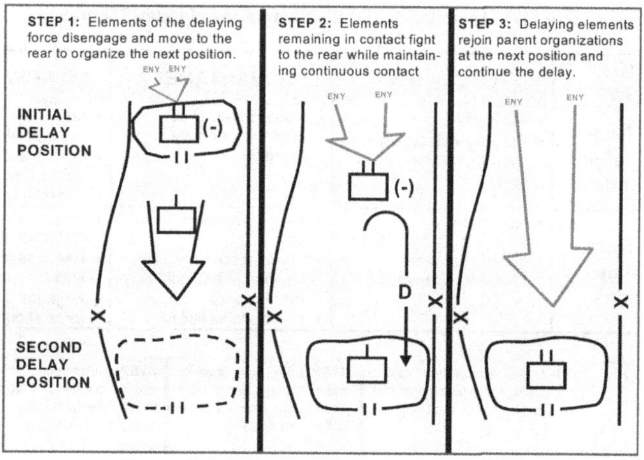

Figure 11-6. Delay from Subsequent Positions

Effects of Terrain

11-31. The commander always takes advantage of the terrain when planning how he positions his forces and conducts operations. He wants the terrain to favor his actions and not be a neutral factor. The terrain dictates where a delaying force can orient on a moving enemy force and ambush it. During a delay, compartmentalized terrain facilitates shorter displacements initiated at closer range to the enemy. The commander conducting operations in such terrain looks for locations that restrict the enemy's movement and prevent him from fully exploiting his combat superiority. On the other hand, flat or open terrain requires earlier displacements at greater distances to stay in front of the advancing enemy. In open terrain, the most important consideration in selecting a position is a good, long-range field of fire. **A *field of fire* is the area that a weapon or group of weapons may cover effectively from a given position.**

11-32. In restricted terrain, where a light force conducts the primary action, positions may be close together, except when conducting a delay using air assault techniques. In open terrain, delay positions are often far apart. In selecting positions, the commander considers natural and artificial obstacles, particularly when the enemy has numerous armored combat systems.

11-33. The commander identifies routes that reinforcements, artillery units, command posts (CPs), and CSS elements will use and keeps them under his control and free of obstacles. Alternate routes should be available so that a friendly force can bypass choke points if they are closed or contaminated.

11-34. Disengaging from the enemy while displacing from one position to the next is difficult. The unit's disengagement plans include the following:

- The maneuver concept of operations for tactical elements after disengagement, which includes the movement routes for each small unit.
- Fires to suppress the enemy and cover the unit's movement.
- Offensive information operations to disrupt enemy C2 at critical moments.
- Screening smoke to conceal the unit's movement, conduct a deception operation, or cover passage points.
- Contact and passage points if moving through friendly lines.
- Disengagement start times.
- The earliest time for CS and CSS elements to move.
- Designating units responsible for closing lanes through obstacles and executing reserve obstacles.

Intelligence

11-35. When conducting a delay operation, the commander may not get the most effective use of his intelligence assets. The commander echelons his organic and supporting ISR systems rearward to maintain at least partial coverage of the AO during the delay. This increases his need to ensure the effective management of collection assets. However, he must rely on a downward flow of intelligence and combat information, such as unmanned aerial vehicle and joint surveillance target attack radar system data, from higher echelons to make up for the degradation in collection capabilities that occurs when systems displace.

11-36. Initially, intelligence assets attempt to determine if the enemy recognizes the delay. Subsequently, the focus is on how the enemy reacts to the delay. Intelligence, surveillance, and reconnaissance systems monitor and predict enemy attempts to envelop the flanks or strike the rear of the rearward-moving friendly force. They also focus on actions of any enemy airborne, air assault, and attack aviation units that may try to interdict the movement of the friendly force. The delaying commander must detect the enemy's advance early to adjust his maneuver scheme and concentrate sufficient combat power to effectively delay the enemy.

Maneuver

11-37. A delay is one of the most difficult forms of defensive operations to execute. The primary reason is that the delaying force must engage the enemy sufficiently to slow his movement, yet not become decisively engaged. Also, the delaying unit must maintain continuous coordination with any flank units as it displaces rearward.

11-38. There are many similarities in the tactics and techniques of a mobile defense and a delay. However, units conducting a delay normally do not become decisively engaged except to prevent the enemy from prematurely crossing a delay line or to risk a part of the force to prevent the whole delaying force from being jeopardized.

11-39. Heavy forces—armor, mechanized infantry, and armored cavalry elements supported by indirect fires—are highly suitable for delay operations in most terrain. Their organic firepower allows them to engage the enemy effectively at long ranges, and their mobility allows them to move quickly between successive positions or to a flank. Their vehicles provide protection that simplifies battlefield movement. These same characteristics also allow heavy reserve forces to rapidly launch counterattacks to extract delaying forces from untenable situations.

11-40. Light forces are especially suited to conduct delays in broken, close, and built-up terrain. They take advantage of such terrain, reinforced by the extensive use of situational obstacles, to hinder the mobility of enemy combat systems and supporting tactical vehicles. They can also participate in stay-behind operations. (See paragraphs 11-118 to 11-122.) This type of terrain offers cover for the movement of light forces and favors using ambushes against the enemy. Because of the restrictions on organic motorized transportation assets and the limited protection available to light infantry units, the commander must specifically plan for their displacement. While all light forces can move rapidly by air, a delay offers little opportunity for airborne forces to use their unique capability.

11-41. The commander may employ air assault forces in a manner similar to that of other light infantry units in a delay. However, they possess additional useful capabilities in a delay operation. They can rapidly deploy, redeploy, and disperse in open terrain if the weather is suitable and the necessary landing zones and pickup zones exist. The combination of light infantry, attack helicopters, and fire support systems found in air assault units allows the delaying commander to rapidly concentrate combat power at key locations to attrit the enemy through repeated ambushes. The combined arms nature of air assault units also makes them extremely useful for conducting security and reserve operations over large geographical areas against heavy and light enemy forces. However, their extraction is a high-risk activity when pressured by a heavy enemy or in the presence of a significant air defense threat.

11-42. The mobility, lethality, and long range of attack aviation firepower make it invaluable to the force conducting a delay. The commander can also use them to conduct counterattacks and spoiling attacks as part of his combined arms team. Other uses of Army aviation in a delay include the rapid rearward movement of CSS assets, the deployment of light forces, and reconnaissance.

Command and Control

11-43. Centralized planning and decentralized execution characterize command and control in a delay operation. Communications are essential to the success of this type of operation, and the commander ensures that redundancy is built into his communications architecture. Digital command and control systems help ensure that redundancy through the promulgation of a common operating picture and a distributed database. This allows one CP to temporarily assume the duties of another CP if the latter is destroyed.

11-44. The echelon rear CP is normally the first CP within an echelon to displace during a delay. It displaces by echelon with other CSS assets. The echelon main CP controls the movement of forces not in contact. It displaces by echelon with the main body. The echelon tactical CP usually remains forward to control and support forces in contact.

MOBILITY/COUNTERMOBILITY/SURVIVABILITY

11-45. Normally, countermobility is the most important engineer task, unless the delaying force must cross one or more major obstacles, in which case the major engineer task is mobility, specifically breach operations. The commander must set realistic and specific priorities for the engineer effort. He monitors its progress to prevent it from dissipating throughout the area. The commander employs his engineers in depth. This is crucial when the commander conducts noncontiguous operations or when the enemy attacks deep into the rear area of a force conducting contiguous operations, or when the enemy has the ability to employ weapons of mass destruction. The maneuver element provides security for the engineers so that they can concentrate their efforts on engineer tasks.

11-46. Because of the importance of mobility and countermobility tasks, a unit conducting a delay probably has few engineer assets to devote to the survivability function. Units should maximize the use of smoke when and where weather conditions allow to provide concealment for movement and assembly. Smoke curtains, blankets, and haze may protect withdrawing columns, critical points, and routes. The commander takes precautions to ensure that the smoke does not provide a screen for the enemy's advance. (See FM 3-11.50.)

PREPARE

11-47. The defensive preparations outlined in Chapter 8 also apply when conducting a delay. As always, resources—including the time available—determine the extent of preparations. The commander assigns a high priority to reconnaissance. Additionally, the preparation of subsequent positions receives a higher priority than it does in either a mobile or an area defense. It is not always possible to complete all preparations before starting the delay operation. Consequently, delaying units continue to prepare and adapt plans as the situation develops.

11-48. In the delay, the commander uses BPs in a manner similar to the defense. However, when organizing his BPs, he places more emphasis on width than depth, as well as reconnaissance and preparing routes for displacement. Within each BP, most of the available firepower is oriented toward the expected enemy avenue of approach. However, the commander must provide adequate flank and rear security since the delaying unit must furnish its own security. Each crew and squad should be familiar with the routes from its primary positions to alternate, supplementary, and sequential positions. In preparing a BP, the commander conducting a delay places less emphasis on installing protective obstacles, final protective fires (FPFs), and ammunition stockpiling than he would in either an area or a mobile defense. In a delay, BPs are sometimes referred to as delay positions.

EXECUTE

11-49. The complex nature of a delay requires the subordinate elements of a delaying force to execute different yet complementary actions. In a single delaying operation, attacks, area defenses, mobile defenses, and other actions may occur in any sequence or simultaneously. For example, the commander may elect to assign one delaying element the task of holding a key road intersection for a period of time so a reserve force can strike the enemy's flank. Therefore, the enemy must deploy into a hasty defense, which delays his attack.

11-50. The commander deploys his security force well forward of his initial delay position to give early warning of any enemy approach. When the security force detects and reports an approach, the commander reconciles these reports against his decision support and event templates to confirm the enemy's probable course of action. Based on his interpretation of how the battle will unfold, the commander can direct one subordinate element to maneuver in a manner designed to draw the advancing enemy into a position of disadvantage.

11-51. The security force fixes, defeats, and destroys the enemy's reconnaissance and security elements without risking decisive engagement. It directs fires at the approaching enemy force as far forward of the delay positions as possible. Engaging a moving enemy at long ranges tends to inflict far more casualties on him than he can inflict on the delaying force; it also slows his tempo of operations. The more a delaying force can blind an enemy and eliminate his reconnaissance assets, the more likely he is to hesitate and move with caution.

11-52. Once the security force makes contact with the enemy, it maintains contact. As the enemy approaches, it moves by bounds back to the flanks of the defending units, keeping the enemy under constant observation. This helps prevent the enemy from finding gaps between delaying units and attacking the exposed flanks of delaying units. The security force uses covered, concealed, and coordinated routes to avoid enemy and friendly fires.

11-53. Recovering security assets may be more difficult if the security force needs to pass through the range fan of friendly tanks and other direct-fire weapons in their movement. Recovery should be to the flanks of delay positions and not through EAs and TRPs unless necessary. Security forces should move so that they do not reveal the locations of other friendly elements.

11-54. The main body uses a variety of tactics to execute the delay. These include ambushes, counterattacks, spoiling attacks, artillery raids, jamming, and close air support. The commander of the delay force preserves his freedom to maneuver by engaging the enemy with sufficient force to temporarily stop his advance. The delay force uses obstacles and defensive positions in depth to slow and canalize the enemy and exploit the mobility of its combat systems to confuse and defeat the enemy. Once a delay starts, units displace rapidly between positions. Whenever possible, the commander grasps any fleeting opportunity to seize the initiative, even if only temporarily. By aggressively contesting the enemy's initiative through offensive action, the delaying force avoids passive patterns that favor the attacking enemy. The delaying force may conduct strong counterattacks from unexpected directions to temporarily confuse the enemy commander. Attacking an enemy throws him off stride, disorganizes his forces, confuses his picture of the fight, and helps prolong the delay. In turn, this confusion may affect the enemy's tempo and momentum. It also affects the movement of enemy reserves and other follow-on forces. However, the delaying force seeks to avoid decisive engagement.

11-55. In a delay, the commander uses his fire support assets to delay enemy forces, inflict casualties on them, and assist the friendly force to gain a mobility advantage over them. Indirect fires continue throughout the delay. The effects of the commander's fire support assets can disrupt the enemy's follow-on forces and restrict the immediate battle to his committed forces. Close air support and attack helicopters can engage enemy forces before they come within range of the supporting field artillery systems. The commander should weigh the effects required, however, since attack aviation is a limited resource and CAS aircraft are a fleeting resource. Massing of fires, to include the killing power of the unit in contact, should be the objective. However, this should not delay integration of CAS aircraft, given limited loiter times.

11-56. Artillery and mortar systems support the direct-fire fight to prevent the enemy from conducting a combined arms attack on the delay position. As the enemy encounters each situational obstacle, he is engaged by these fire support systems. These fires should cause enemy armored forces to button up and slow down. Artillery and mortar systems can use fires to separate enemy formations by striking the enemy when he concentrates near choke points and in EAs. Integrating fires and obstacles makes it difficult for the enemy to traverse EAs. The delaying force breaks the enemy's momentum by forcing him to deploy and by inflicting casualties. Fires assist delaying forces by—

- Assisting in disengaging maneuver forces.
- Suppressing the enemy.
- Degrading the enemy's ability to move and communicate.
- Obscuring the enemy's overwatching support by fire positions and degrading his ISR and target acquisition systems.
- Reinforcing or closing breaches or lanes in obstacles.
- Executing FPFs.
- Screening friendly displacements and disengagements by using smoke. (This also degrades the enemy's terminal guidance of his precision-guided munitions.)
- Destroying high-payoff targets.
- Supporting limited counterattacks.

11-57. As the enemy approaches the delay position, he crosses one or more trigger lines and moves into EAs within the range of the delaying force's anti-armor missiles, tank cannons, and small arms. The commander holds his direct fire until the enemy is positioned where the fire plan and scheme of maneuver require their use. He controls these fires from the delaying force in the same manner as in any defensive operation. The more damage the delaying force can inflict on the enemy, the longer it can stay in position.

11-58. As the enemy presses his attack and attempts to maneuver against the delaying force, the commander constantly assesses the action to guide the displacements of the delay force to anticipate possible decisive engagement while accomplishing the delay mission. When the enemy begins to think he is successfully maneuvering against a friendly position, he is engaged by indirect fires while the delaying force disappears behind a cloud of smoke, dust, and exploding munitions. Intense FPFs and fires aimed at and behind recently evacuated friendly delay positions allow the delaying force to disengage from an attacking enemy.

11-59. Division and brigade commanders generally decentralize execution of a delay to battalion and lower levels. Those senior commanders must rely on their subordinates to execute the mission and request help when needed. The commander establishes the acceptable risk and displacement criteria. Subordinates displace once they meet the previously established delay criteria. This displacement may be a preplanned event or time dependent. The senior commander monitors the delay and intervenes when the displacement of one unit threatens the survival of another.

11-60. The delaying force relies heavily on artillery fires and air support to suppress the enemy so the force can disengage, move, and occupy new positions. If a subordinate element cannot maintain separation from the enemy, the commander can shift additional combat multipliers and other resources to that particular AO to counter the enemy's unplanned success. As one subordinate element displaces, the delaying commander may order other subordinate elements to change their orientation to cover the move. Each displacing element travels along its designated route, using reserve demolitions as required and requesting additional fire support if the enemy is able to maintain contact.

11-61. Passing through obstacle lanes during displacement between positions poses significant risks to the delaying force. The unit passing through a linear obstacle becomes more vulnerable to enemy attack because of the danger of the delaying force becoming congested on the far side of the obstacle. Obstacle lanes also increase the amount of time required for a passing unit to transit through a given area. The commander must attempt to prevent the enemy from engaging the passing unit until it can redeploy into a tactical formation.

11-62. The commander retains his reserve for the decisive moment. As with aviation, the reserve should not be committed early in the delay unless its integrity is threatened. Typically, the commander commits his reserve to help a unit disengage and regain its ability to maneuver or to prevent the enemy from exploiting an advantage. The reserve normally uses a support-by-fire position for this task. If the reserve is committed early, the commander's ability to influence the battle is greatly reduced unless he can reconstitute a new reserve. It is possible to commit the reserve several times throughout the battle, but only when it can be extracted, redesignated, or otherwise reconstituted quickly.

11-63. In the delay, the force's CSS elements should be located outside of enemy artillery range but be able to provide adequate support. Artillery ammunition stocks must be capable of sustaining the quantity of fire support required in the delay. Maintenance operations focus on evacuating rather than returning damaged vehicles to combat. Unless vehicles can be fixed quickly on the spot, the unit should evacuate them to the rear area because vehicles left behind must be destroyed to prevent their capture.

TERMINATION OF A DELAY

11-64. A delay operation terminates when the delaying force conducts a rearward passage of lines through a defending force, the delaying force reaches defensible terrain and transitions to the defense, the advancing enemy force reaches a culminating point, or the delaying force goes on the offense after being reinforced. If the advancing enemy force reaches a culmination point, the delaying force may maintain contact in current positions, withdraw to perform another mission, or transition to the offense. In all cases, the senior commander must plan for the expected outcome of the delay executed by a subordinate. If he expects a friendly counterattack, he plans for the forward passage of the counterattack force, husbands resources to ensure relative combat superiority, and provides for the smooth handoff of appropriate AOs.

WITHDRAWAL

11-65. A *withdrawal*, a form of retrograde, is a planned operation in which a force in contact disengages from an enemy force (FM 3-0). The commander may or may not conduct a withdrawal under enemy pressure. Subordinate units may withdraw without the entire force withdrawing. A unit conducts a withdrawal for a variety of reasons, which are listed at the beginning of this chapter. In addition, a withdrawal may precede a retirement operation.

11-66. Although the commander avoids withdrawing from action under enemy pressure, it is not always possible. He may conduct a withdrawal when the situation requires rapid action to save the command from disaster. This usually occurs after a tactical reverse or after a unit reaches its culminating point. When an aggressive enemy becomes aware of a friendly force's withdrawal or its intention to withdraw, he attempts to exploit the withdrawal, using all his capabilities to try to turn the withdrawal into a rout. He may have ground and air superiority and continuously attempt to pursue, encircle, and destroy the withdrawing force. He will try to use a combination of direct pressure and enveloping forces and fires to isolate the withdrawing friendly force for later destruction.

11-67. Withdrawals are inherently dangerous because they involve moving units to the rear and away from what is usually a stronger enemy force. The heavier the previous fighting and the closer the contact with the enemy, the more difficult the withdrawal. Operations security (OPSEC) is extremely important. A unit usually confines its rearward movement to times and conditions when the enemy cannot observe the activity, so that he cannot easily detect the operation. To help preserve secrecy and freedom of action, for example, the commander must consider visibility conditions and times when enemy reconnaissance satellites can observe friendly movements. Operations security is especially critical during the initial stages of a delay when the majority of CS and CSS elements displace.

11-68. A unit withdraws to an assembly area or a new defensive position. Alternatively, it can withdraw indirectly to either area through one or more intermediate positions. When preparing the new position, the commander balances the need for security with the need to get an early start on the defensive effort.

ORGANIZATION OF FORCES

11-69. The commander typically organizes his withdrawing unit into a security force, a main body, and a reserve. He also organizes a detachment left in contact (DLIC) and stay-behind forces if required by his scheme of maneuver. He avoids changing task organization unless his subordinates have sufficient planning time. However, circumstances may dictate rapid task organization changes immediately before the withdrawal, such as when the unit must conduct an immediate withdrawal to prevent encirclement.

11-70. The security force maintains contact with the enemy until ordered to disengage or until another force takes over. It simulates the continued presence of the main body, which requires additional allocation of combat multipliers beyond those normally allocated to a force of its size. The greater its mobility and range advantages over the enemy, the easier for the security force to successfully cover the main body's withdrawal. The commander organizes the majority of available combat power to the security force as a rear guard or a rear-covering force; the most probable threat to a withdrawing force is a pursuing enemy. However, the commander must maintain all-around security of the withdrawing force. When the enemy can infiltrate or insert forces ahead of the withdrawing force, the commander may establish an advance guard to clear the route or AO. He designates a flank guard or screen, if required.

11-71. When a security zone exists between the two main opposing forces, the existing security force can transition on order to a rear guard or rear-covering force. It then conducts delay operations until ordered to disengage and break contact with the enemy. When the withdrawing force is in close contact with the enemy, a security zone does not normally exist. Withdrawals under these conditions require that security forces adopt different techniques. One technique is to establish a DLIC to provide a way to sequentially break contact with the enemy.

11-72. **A *detachment left in contact* is an element left in contact as part of the previously designated (usually rear) security force while the main body conducts its withdrawal.** Its primary purpose is to remain behind to deceive the enemy into believing the parent unit is still in position while the majority of the unit withdraws. It simulates—as nearly as possible—the continued presence of the main body until it is too late for the enemy to react by conducting activities, such as electronic transmissions or attacks. The DLIC must have specific instructions about what to do when the enemy attacks and when and under what circumstances to delay or withdraw. If the DLIC must disengage from the enemy, it uses the same techniques as in the delay. If required, this detachment receives additional recovery, evacuation, and transportation assets to use after disengagement to speed its rearward movement.

11-73. Two methods to resource the DLIC exist. The first is for each major subordinate element of the withdrawing force to leave a subelement in place.

For example, in a brigade withdrawal, each task force leaves a company team in contact. Typically, these teams fall under a senior DLIC commander designated by the brigade commander. Alternatively, one major subordinate command of the withdrawing force can stay behind as the DLIC. For example, a brigade could leave one battalion task force as the DLIC, which then expands its security responsibilities to cover the width of the AO. (See Figure 11-7.)

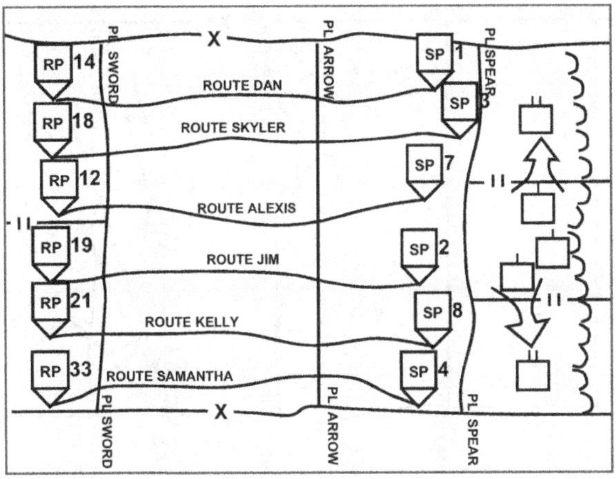

Figure 11-7. Detachment Left in Contact

11-74. Often when a DLIC is used, the commander creates an additional security force behind the existing main defensive positions to assist in the withdrawal process. The commander can create an additional force from the withdrawing unit or from an assisting unit. The DLIC can delay to this additional security force and join it, or delay back, conduct battle handover, and then conduct a rearward passage of lines. In either case, the additional security force becomes the rear guard.

11-75. The main body of the withdrawing force consists of all elements remaining after the commander resources his security force and his reserve. He generally finds it difficult to resource a reserve, but he makes every attempt to do so. When the complete formation withdraws under pressure, the reserve may take limited offensive action, such as spoiling attacks, to disorganize, disrupt, and delay the enemy. It can counter penetrations between positions, reinforce threatened areas, and protect withdrawal routes. Reserves may also extricate encircled or heavily engaged forces.

CONTROL MEASURES

11-76. Withdrawing forces must apply combat power to protect themselves while simultaneously moving combat power away from the enemy. This requires careful coordination among all forces. Throughout the operation, the commander must tightly control rearward movement and maintain the ability to generate decisive combat power at key times and places. As shown in Figure 11-8, the control measures used in the withdrawal are the same as those in a delay or a defense. The routes used by each unit in the withdrawal and the block movement times are also withdrawal control measures.

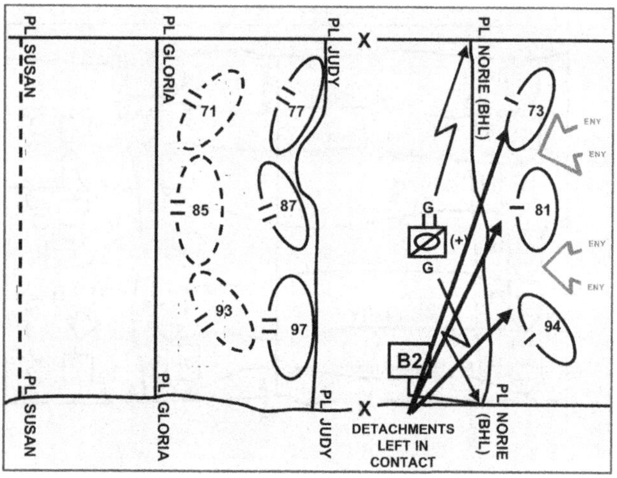

Figure 11-8. Withdrawal Control Measures

PLANNING A WITHDRAWAL

11-77. The commander plans and coordinates a withdrawal in the same manner as a delay. Some factors of METT-TC apply differently because of the differences between a delay and a withdrawal. A withdrawal always begins under the threat of enemy interference. Because the force is most vulnerable if the enemy attacks, the commander always plans for a withdrawal under pressure. He then develops contingencies for a withdrawal without pressure. In both cases, the commander's main considerations are to—

- Plan a deliberate break from the enemy.
- Displace the main body rapidly, free of enemy interference.
- Safeguard the withdrawal routes.
- Retain sufficient combat, CS, and CSS capabilities throughout the operation to support forces in contact with the enemy.

11-78. A withdrawal may be assisted or unassisted. It may or may not take place under enemy pressure. These two factors combined produces the four variations shown in Figure 11-9, page 11-22. That figure also depicts the tactical mission graphic for a withdrawal and a withdrawal under enemy pressure. The withdrawal plan considers which variation the force currently faces. Each variation requires a different blend of the three retrograde options.

11-79. A commander prefers to conduct a withdrawal while not under pressure and without assistance. Actions by the enemy, as well as the additional coordination needed because of the presence of an assisting unit, complicate the operation.

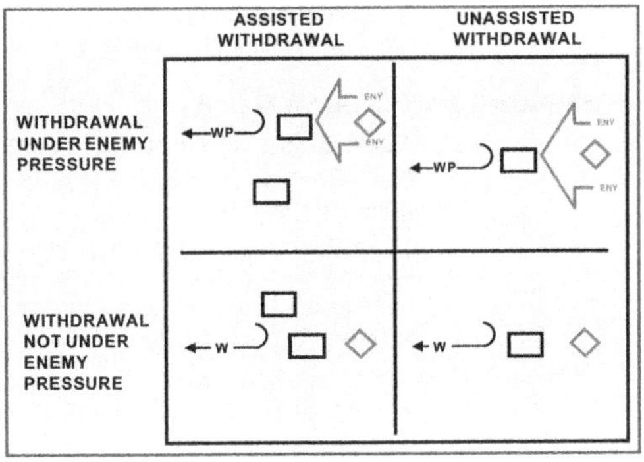

Figure 11-9. Types of Withdrawals

11-80. A withdrawing force can receive assistance from another force in the form of—

- Additional security for the area through which the withdrawing force will pass.
- Information concerning withdrawal routes.
- Forces to secure choke points or key terrain along withdrawal routes.
- Elements to assist in movement control, such as traffic control points.
- Required combat, CS, and CSS, which can involve conducting a counterattack to assist the withdrawing unit in disengaging from the enemy.

11-81. In a withdrawal under enemy pressure, all units withdraw simultaneously when available routes allow, using delaying tactics to fight their way to the rear. In the usual case, when simultaneous withdrawal of all forces is not practical, the commander decides the order of withdrawal. Several factors influence his decision:

- Subsequent missions.
- Availability of transportation assets and routes.
- Disposition of friendly and enemy forces.
- Level and nature of enemy pressure.
- Degree of urgency associated with the withdrawal.

The commander must make three interrelated key decisions: when to start the movement of selected CS and CSS elements, when forward elements should start thinning out, and when the security force should start its disengagement operations. The commander avoids premature actions that lead the enemy to believe a withdrawal is being contemplated. Commanders must anticipate enemy means of interference and plan for employing security forces, attack helicopters, and close air support.

11-82. The commander conducting a withdrawal without enemy pressure can plan when to begin the withdrawal. He has the option of taking calculated risks to increase the displacement capabilities of his force. For example, he may order the main body to conduct a tactical road march instead of moving in tactical formations. The commander can plan for stay-behind forces as part of the operation. (The stay-behind operations starts on page 11-30.)

PREPARING A WITHDRAWAL

11-83. Before withdrawing, the main body dispatches quartering parties to help it occupy the new position. (Chapter 14 details the responsibilities of a quartering party.)

11-84. In an unassisted withdrawal, the withdrawing unit establishes its own security force and reserve. It reconnoiters and secures the routes it will use in its rearward movement while sustaining itself during the withdrawal. The withdrawing unit must disengage from the enemy.

11-85. By concealing supplies along movement routes, CSS operators can simplify support requirements and reduce the enemy's ability to interfere with logistics operations. This allows CSS units to withdraw earlier than they otherwise could. The commander carefully considers whether to place his supplies in caches. Once cached, supplies are difficult to recover if the operation does not go as planned. Other than medical items, the unit evacuates or destroys all supplies to prevent their capture. The commander establishes his destruction criteria, which is time- or event-driven for each class of supply.

EXECUTING A WITHDRAWAL

11-86. Typically, when under enemy pressure, the less heavily engaged elements of the withdrawing force withdraw first. The more heavily engaged units generally withdraw under the cover of a security force using support provided by available fire support and electronic warfare assets. They take advantage of obstacles to assist in breaking contact with the enemy. The commander conducts night movements and uses obscuration smoke to screen friendly movement while reducing both the accuracy of enemy direct-fire systems and his ability to visually observe friendly movements. The security force continues to use alternate and successive positions until the entire force breaks contact with the enemy.

11-87. The security force may remain in position and maintain a deception. The main body moves rearward to intermediate or final positions as rapidly as possible. After the main body withdraws a safe distance, the security force begins its rearward movement. Once the security force begins moving, it assumes the duties of a rear guard. Even if the enemy does not pursue the withdrawing force, the security force continues to act as the rear guard unless the commander assigns that mission to another element. However, if not pursued by the enemy, the security force may remain in a march column. (Chapter 14 provides a definition of a march column.)

11-88. On order, the main body moves rapidly on multiple routes to reconnoitered positions. It may occupy a series of intermediate positions before completing the withdrawal. Usually CS and CSS units, along with their convoy escorts, move first and precede combat units in the withdrawal movement formation. The commander needs to maintain the disciplined use of routes during a withdrawal. Despite confusion and enemy pressure, subordinate units must follow specified routes and movement times.

11-89. When the main body withdraws, its reserve remains well forward to assist the security force and other units by fire and counterattack. The reserve can launch spoiling attacks to disorganize and delay the enemy and extricate encircled or heavily engaged forces.

11-90. If the security force and the reserve cannot prevent the enemy from closing on the main body, the commander must commit some or all of the main body to prevent the enemy from further interfering with the withdrawal. The main body delays or defends if the security force fails to slow the enemy. In this event, the withdrawal resumes at the earliest possible time. If the enemy blocks movement to the rear, the commander shifts to alternate routes to bypass the interdicted area. Alternatively, he can attack through the enemy.

TERMINATING A WITHDRAWAL

11-91. Once the withdrawing force successfully disengages from the enemy, it has two options. It can rejoin the overall defense under more favorable conditions or transition into a retirement and continue its movement away from the enemy and toward its next mission.

RETIREMENT

11-92. A *retirement* is a form of retrograde in which a force out of contact with the enemy moves away from the enemy (JP 1-02). Figure 11-10, page 11-24 shows the tactical mission graphic for a retirement. A retiring unit organizes for combat but does not anticipate interference by enemy ground forces. Typically, another unit's security force covers the movement of one formation as the unit conducts a retirement. However, mobile enemy forces, unconventional forces, air strikes, air assaults, or long-range fires may attempt to interdict the retiring unit. The commander must plan for enemy actions and organize the unit to fight in self-defense. The commander usually

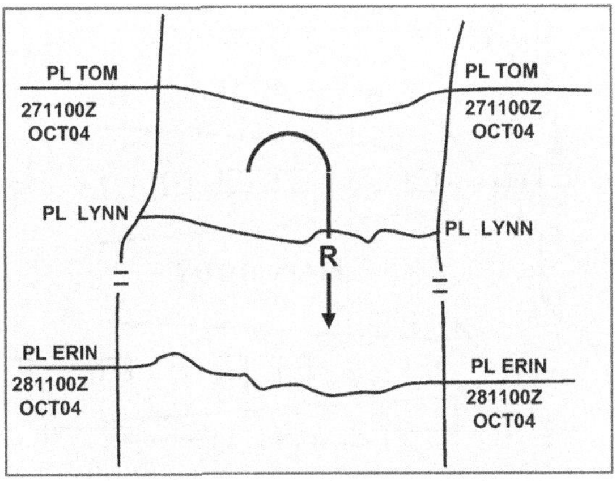

Figure 11-10. Retirement Tactical Mission Graphic

conducts retirement operations to reposition his forces for future operations or to accommodate the current concept of the operation.

11-93. When a withdrawal from action precedes a retirement, the actual retirement begins after the unit breaks contact and organizes into its march formation organization. (While a force withdrawing without enemy pressure can also use march columns, the difference between the two situations is the probability of enemy interference.) Units conduct retirements as tactical road marches where security and speed are the most important considerations.

11-94. The retiring unit generally moves toward an assembly area, which should support the preparations for the unit's next mission. When determining the routes the retiring force takes to the assembly area, the commander considers the unit's capability to support defensive actions if combat occurs during the retirement.

11-95. The initial action in a retirement is to move CSS units and supplies to the rear. At the designated time, the retiring unit executes a withdrawal from action and forms into a march formation. The unit can first move into an assembly area if this step is necessary before moving into a march formation to reestablish command and control or resupply. Once it forms a march formation, the force is prepared to initiate the retirement. During the initial phase, the force retires in multiple small columns. As the distance from the enemy increases, smaller columns can consolidate into larger ones for ease of movement control. Road nets and the potential for hostile interference influence how and when this consolidation occurs.

ORGANIZATION OF FORCES

11-96. The commander normally designates security elements and a main body in a retirement. (See Figure 11-11.) The formation and number of columns employed during a retirement depend on the number of available routes and the potential for enemy interference. The commander typically wants to move his major elements to the rear simultaneously. However, a

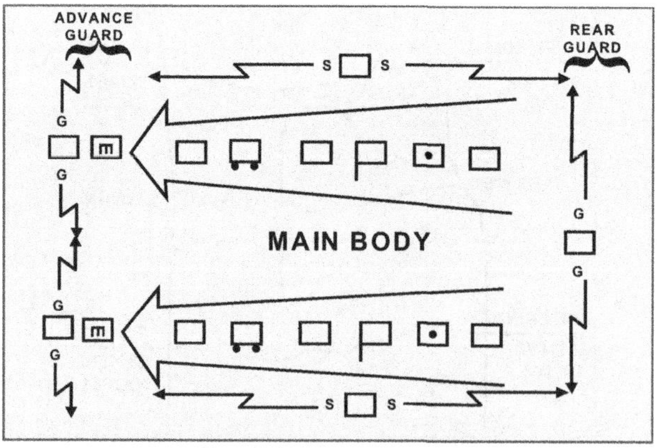

Figure 11-11. Organization of Forces for a Retirement Operation

limited road net or a flank threat may require echelonment of the movement in terms of time and ground locations.

11-97. The terrain and the enemy threat dictate whether the retiring force establishes a single rear security force, which is usually a rear guard, or whether each column forms a separate rear security force. These security forces protect the rearward moving columns from surprise, harassment, and attack by any pursuing enemy force. Their size and composition depend on the strength and imminence of the enemy threat. These security elements generally remain in march columns unless there is a potential for enemy interference. If the enemy establishes contact, the rear security element conducts a delay.

11-98. The retiring march columns normally require an advance guard augmented by engineers. The commander assigns a flank security element to prevent potential enemy interference with the retiring force's extended columns. The commander may designate flank security responsibilities to subordinate march units.

11-99. The main body organizes in a manner opposite that of an approach march. (Chapter 14 explains the approach march.) The movement of CS and CSS units should precede the movement of combat forces. When necessary, elements of the main body can reinforce the rear guard or any other security element. Because fire support elements and attack helicopter elements of the main body can respond most rapidly, they are usually the first elements tasked for this mission.

CONTROL MEASURES

11-100. The control measures used in a retirement are the same as those in a delay and a withdrawal. As in a withdrawal, thorough planning and strict adherence to routes and movement times facilitate an orderly retirement. Typically, the commander controls movement using movement times, routes, and checkpoints. (Chapter 14 discusses movement control measures.)

COMBAT SERVICE SUPPORT

11-101. During retrograde operations, CSS units echelon their movements to maintain adequate support to the committed force. They maintain maximum dispersion consistent with control and local security. Their goal is to provide uninterrupted support and maximum protection during the time it takes to conduct the retrograde operation. By echeloning support, the commander reduces the amount of time each CSS unit spends moving, preventing it from performing its primary support tasks. High-priority assets may require added protection to prevent their loss or capture. To reduce congestion and interference with the operations of combat and combat support units, the commander should displace his CSS assets as early as possible, normally in an environment that provides the enemy limited visibility. The early displacement of CSS units can also prevent revealing friendly future operations to the enemy.

11-102. The commander anticipates the effects of retrograde movements on logistics support to ensure adequate support for the operation and the prompt evacuation of casualties. Retrograde movements generally result in increased distances between CSS and combat units, which makes providing this support more difficult. Executing retrograde operations generally requires more Class III, and possibly more Class V, supplies than other types of defensive operations. These supplies must be available for emergency issue. These two factors combine to increase the demand for transportation assets and the allocation of space on main supply routes. This, in turn, increases the need for movement management and pre-positioned services and supplies. Combat service support units carry and cache necessary fuel and ammunition stocks as required by the specific situation.

11-103. The logistics support provided must be mobile to cope with demands of the fluid tactical situation that typically occurs during a retrograde operation. The commander prevents unnecessary supplies from accumulating in areas that will be abandoned. Only essential medical and logistics support should be located in the area involved in the retrograde operation.

11-104. The commander establishes his maintenance, recovery, and evacuation priorities and his destruction criteria for inoperable equipment in paragraph 4 of the operations order. Maintenance requirements generally overwhelm the organic capabilities of forward units during a retrograde operation. Forward units place as much maintenance, recovery, and evacuation assets forward as possible to augment or relieve combat elements of the burden of repairing unserviceable equipment. Recovery and evacuation vehicles position themselves at critical locations to keep disabled vehicles from blocking movement routes. Forward units evacuate systems that cannot be repaired within established timelines, using all available means, such as equipment transporters and armored vehicles with inoperative weapon systems. When recovery and evacuation are impossible, units destroy inoperable equipment to prevent capture. When possible, units destroy the same vital components in each type of system to prevent the enemy from rapidly exploiting captured friendly systems through battlefield cannibalization.

11-105. The commander assigns transportation priorities to the movement of combat troops and their supplies, the movement of obstacle materials to impede the enemy, and the evacuation of casualties and repairable equipment. He keeps his main supply routes open and decontaminated as necessary. Units control the back-haul of transportation assets before the retrograde begins, reducing the amount of transportation needed to support the operation.

11-106. Generally, the commander prefers to use many separate supply routes rather than just a few main supply routes. Some routes remain open for traffic moving to the front while the bulk of CS and CSS units displace farther rearward. Routes reserved for evacuating displaced civilians avoid crossing or otherwise interfering with the unit's main supply routes to the maximum extent possible.

11-107. The commander bases his medical evacuation priorities on the availability of transportation assets and the results of casualty triage by medical personnel. Medical elements supporting the retrograding force must provide rapid evacuation of casualties to medical facilities. Medical evacuation requirements are especially demanding in the large AOs common to the retrograde. The commander should consider augmenting the ground ambulance capabilities of his forward medical units.

11-108. Military police elements of the retrograde force are involved primarily in battlefield circulation control to ensure smooth traffic flow. The commander may augment his military police force to establish traffic control points and route and convoy security. They also help control dislocated civilians and enemy prisoners of war.

UNIQUE RETROGRADE SITUATIONS

11-109. Conditions that require conducting denial and stay-behind operations can arise during retrograde operations. These two operations have their own unique planning and execution considerations.

DENIAL OPERATIONS

11-110. ***Denial operations*** **are actions to hinder or deny the enemy the use of space, personnel, supplies, or facilities.** It may include destroying, removing, and contaminating those supplies and facilities or erecting obstacles. It is inevitable that, on occasion, an enemy will be in a position to capture friendly equipment and supplies. This situation often occurs during retrograde or defensive operations. As a result, the defending commander may be required to conduct denial operations. The principles of denial are:

- The commander should deny his enemy the use of military equipment and supplies.
- Steps taken to deny equipment and supplies to the enemy should, if possible, not preclude their later use by friendly forces.
- The commander orders the destruction of military equipment and supplies only when friendly forces cannot prevent them from falling into enemy hands.
- The user is responsible for denying the enemy the use of its military equipment and supplies by means of its destruction, removal, or contamination.
- Deliberately destroying medical equipment and supplies and making food and water unfit for consumption is unlawful under the terms of the Geneva Conventions.

In denial operations, the definition of a unit's military equipment and supplies could expand to include military installations and any civilian equipment and supplies used by the friendly force. Under the law of war the destruction of civilian property is only permitted where required by immediate military necessity. The determination of whether there is sufficient necessity to justify destruction is a complex analysis that requires consideration of moral, political, and legal considerations.

11-111. The commander who orders the denial operation must consider the potential value of the military equipment and supplies to an enemy when determining the priorities and the extent of the denial operation. Examples of high priorities for denial include—

- Classified equipment, material, and documents.
- POL.
- Sophisticated weapon systems or electronic equipment.
- Heavy weapons and associated ammunition.

- Communications equipment.
- Ferrying and bridging equipment.
- Air, sea, and land transport systems.

Of lesser priority for denial would be any other military supplies, equipment, or facilities that may be of use to an enemy.

11-112. The commander must issue detailed instructions to deny military equipment and supplies to prevent the enemy from directly using such assets. Denial must also prevent an enemy from repairing a system through the cannibalization of several systems. The unit must destroy the same parts in each type of system.

11-113. Denial differs from countermobility operations because the commander designs denial operations to deprive the enemy of some or all of the short-term benefits of capturing a geographical region. The impact of denial operations on civilian inhabitants and the environment of the region act as a moral and a legal restraint on their use and scope by US forces. The commander should involve his staff judge advocate and civil-military operations officer in planning denial operations.

11-114. The commander ensures that executing the denial plan does not adversely affect his future operations. This includes carefully considering the force's demolition policy in relation to the purpose of the rearward movement and the contemplated subsequent actions of the force. Widespread demolitions during a retrograde may become a greater hindrance to a friendly force moving back into the area than to the enemy during the friendly retrograde. For example, destroying the transportation infrastructure increases friendly logistics difficulties once the area is recaptured. Removing or destroying militarily significant supplies and equipment, such as fuel, obstacle materials, and rail cars, from an area requires the friendly force to bring similar assets with them when they reoccupy the area.

11-115. The commander can expand a denial operation to prevent the enemy from exploiting resources, such as fuel, minerals, and the indigenous population; routes of communication, such as river locks, railroad switching yards, road interchanges, and bridges; and facilities, such as telephone exchanges, radio and television stations, and the industrial plants of a region. The defending force can assist civil authorities in evacuating the civilian population. The defending force either removes the resources, supplies, and facilities from the geographical area being abandoned to the enemy or destroys them in place. Such denial operations may be either total or limited in nature.

11-116. Total denial operations can produce long-term political, economic, military, and environmental effects. Total denial operations have operational-level, and possibly strategic-level, impact. Total denial operations consume large quantities of transportation and engineer resources and require considerable time to plan and execute.

11-117. Limited or partial denial operations are particularly suitable if the defending force expects to regain control of the geographical area within a short time. The removal or destruction of only a few key components can reduce a facility to limited utility, yet it allows for the facility's quick restoration of all functions once it is returned to friendly control. American forces only destroy discrete targets of significant military value. Limited denial operations normally do not affect the advance of properly supported enemy combat formations possessing cross-country mobility. However, they can seriously impede an enemy's road-bound and rail-bound logistics support if executed with skill and imagination according to an overall plan.

STAY-BEHIND OPERATIONS

11-118. **A *stay-behind operation* is an operation in which the commander leaves a unit in position to conduct a specified mission while the remainder of his forces withdraw or retire from an area.** The force should consist of enough combat, CS, and CSS elements to protect and sustain its fighting capability for the duration of the mission. A stay-behind force may also result from enemy actions that bypass friendly forces.

11-119. The main purpose of a stay-behind force is to destroy, disrupt, and deceive the enemy. This force has a high-risk mission because of the danger that it will be located, encircled, and destroyed by the enemy. Resupply and casualty evacuation are also extremely difficult. A commander considers assigning this mission only after a thorough METT-TC analysis. The stay-behind force attacks enemy combat forces and C2, CS, and CSS elements from unexpected directions. (See Figure 11-12.) These attacks may cause enemy follow-on forces to be more cautious and to slow down to clear possible attack and ambush sites. The stay-behind force may be required to conduct a breakout from encirclement and linkup operations after it completes its mission. (Appendix D discusses the conduct of a breakout from encirclement.)

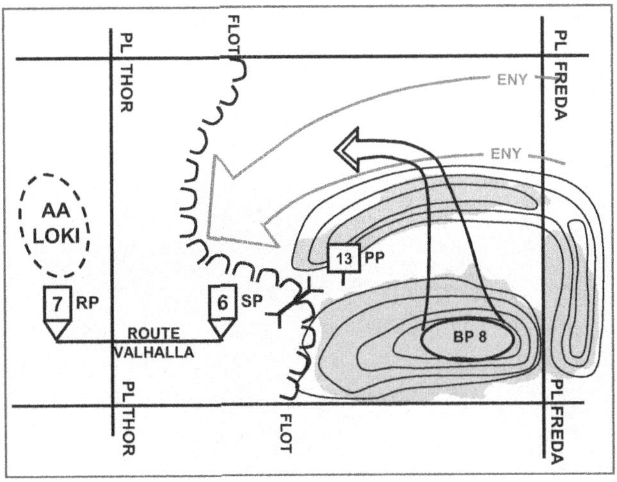

Figure 11-12. Stay-Behind Force

11-120. A light infantry, stay-behind force surprises an enemy by conducting a series of raids and ambushes. The light infantry force can be inserted via infiltration, air assault, or parachute; it can also be a bypassed force. Attacks in the enemy rear area by heavy forces can cover a larger area than attacks by light infantry forces.

11-121. Stay-behind operations eventually require the force to reenter friendly lines or link up with other elements, often in more than one location. The commander must carefully coordinate this reentry to prevent fratricide. The return routes for the stay-behind force must be the best-covered and concealed routes available. Obstacles along these routes that cannot be bypassed should have guarded lanes or gaps.

11-122. A stay-behind operation is not a suicide mission. The commander conducts this operation only when he has confidence that the stay-behind force will rejoin the main body, extract itself in alternative ways, or the main body will fight its way forward to link up with the stay-behind force.

PART FOUR

Tactical Enabling Operations

Chapter 12

Security Operations

The officers and men who permit themselves to be surprised deserve to die, and the commanding general will spare no efforts to secure them their desserts.

D. H. Hill, 1863

***Security operations* are those operations undertaken by a commander to provide early and accurate warning of enemy operations, to provide the force being protected with time and maneuver space within which to react to the enemy, and to develop the situation to allow the commander to effectively use the protected force.** The ultimate goal of security operations is to protect the force from

CONTENTS

surprise and reduce the unknowns in any situation. A commander may conduct security operations to the front, flanks, or rear of his force. The main difference between security operations and reconnaissance operations is that security operations orient on the force or facility being protected, while reconnaissance is enemy and terrain oriented. Security operations are shaping operations.

12-1. There are five forms of security operations—screen, guard, cover, area security, and local security.

- ***Screen* is a form of security operations that primarily provides early warning to the protected force.**
- ***Guard* is a form of security operations whose primary task is to protect the main body by fighting to gain time while also observing and reporting information and preventing enemy ground observation of and direct fire against the main body. Units conducting a guard mission cannot operate independently because they rely upon fires and combat support assets of the main body.**
- ***Cover* is a form of security operations whose primary task is to protect the main body by fighting to gain time while also observing and reporting information and preventing enemy ground observation of and direct fire against the main body.**
- ***Area security* is a form of security operations conducted to protect friendly forces, installations, routes, and actions within a specific area.**
- ***Local security* consists of low-level security operations conducted near a unit to prevent surprise by the enemy.**

The screen, guard, and cover, respectively, contain increasing levels of combat power and provide increasing levels of security for the main body. However, more combat power in the security force means less for the main body. Area security preserves the commander's freedom to move his reserves, position fire support means, provide for command and control, and conduct sustaining operations. Local security provides immediate protection to his force.

12-2. All maneuver forces are capable of conducting security operations. Ground and air cavalry units and scout platoons are specially trained, equipped, and organized to conduct security missions; however, there are rarely enough of them to meet all the security needs of a force. A commander should ensure that subordinate maneuver units train to perform specific security missions. This allows subordinate units to add these missions to their mission-essential task list (METL) and train for them. Habitual relationships with attachments are required to obtain proficiency in these missions.

12-3. A maneuver force commander normally designates the security area within which his security force operates. In this chapter, the force (or facility) being secured is called the main body. When discussing the forms of security operations, the terms *stationary* and *moving* describe the actions of the main body, not the security force.

12-4. All forces, regardless of whether they are combat, combat support (CS), or combat service support (CSS), have an inherent responsibility to provide for their own local security. Local security consists of observation posts (OPs), local security patrols, perimeter security, and other measures to provide close-in security of a force. This chapter focuses on security operations conducted by one force or a subordinate element of a force that provides security for the larger force. Echelon-specific manuals discuss local security.

FUNDAMENTALS OF SECURITY OPERATIONS

12-5. Successful security operations depend on properly applying five fundamentals:

- Provide early and accurate warning.
- Provide reaction time and maneuver space.
- Orient on the force or facility to be secured.
- Perform continuous reconnaissance.
- Maintain enemy contact.

PROVIDE EARLY AND ACCURATE WARNING

12-6. The security force provides early warning by detecting the enemy force quickly and reporting information accurately to the main body commander. The security force operates at varying distances from the main body based on the factors of METT-TC. As a minimum, it should operate far enough from the main body to prevent enemy ground forces from observing or engaging the main body with direct fires. The earlier the security force detects the enemy, the more time the main body has to assess the changing situation and react. The commander positions ground security and aeroscouts to provide long-range observation of expected enemy avenues of approach, and he reinforces and integrates them with available intelligence collection systems to maximize warning time.

PROVIDE REACTION TIME AND MANEUVER SPACE

12-7. The security force provides the main body with enough reaction time and maneuver space to effectively respond to likely enemy actions by operating at a distance from the main body and by offering resistance to enemy forces. The commander determines the amount of time and space required to effectively respond from information provided by the intelligence preparation of the battlefield (IPB) process and the main body commander's guidance regarding time to react to enemy courses of action (COA) based on the factors of METT-TC. The security force that operates farthest from the main body and offers more resistance provides more time and space to the main body. It attempts to hinder the enemy's advance by acting within its capabilities and mission constraints.

ORIENT ON THE FORCE OR FACILITY TO BE SECURED

12-8. The security force focuses all its actions on protecting and providing early warning to the secured force or facility. It operates between the main body and known or suspected enemy units. The security force must move as the main body moves and orient on its movement. The security force commander must know the main body's scheme of maneuver to maneuver his force to remain between the main body and the enemy. The value of terrain occupied by the security force hinges on the protection it provides to the main body commander.

PERFORM CONTINUOUS RECONNAISSANCE

12-9. The security force aggressively and continuously seeks the enemy and reconnoiters key terrain. It conducts active area or zone reconnaissance to detect enemy movement or enemy preparations for action and to learn as much as possible about the terrain. The ultimate goal is to determine the enemy's COA and assist the main body in countering it. Terrain information focuses on its possible use by the enemy or the friendly force, either for offensive or defensive operations. Stationary security forces use combinations of OPs, aviation, patrols, intelligence collection assets, and battle positions (BPs) to perform reconnaissance. Moving security forces perform zone, area, or route reconnaissance along with using OPs and BPs, to accomplish this fundamental.

MAINTAIN ENEMY CONTACT

12-10. Once the security force makes enemy contact, it does not break contact unless specifically directed by the main force commander. The security asset that first makes contact does not have to maintain that contact if the entire security force maintains contact with the enemy. The security force commander ensures that his subordinate security assets hand off contact with the enemy from one security asset to another in this case. The security force must continuously collect information on the enemy's activities to assist the main body in determining potential and actual enemy COAs and to prevent the enemy from surprising the main body. This requires continuous visual contact, the ability to use direct and indirect fires, freedom to maneuver, and depth in space and time.

HISTORICAL EXAMPLE

12-11. Military history contains numerous examples of the importance of security operations. The following historical example illustrates the major role of security operations in ensuring the success of an operation. This non-US example illustrates that the study of other armies and other times can contribute toward helping to understand the art and science of tactics.

Operation Bagration, 1944

During Operation Bagration, 22 June to 29 August 1944, the Red Army destroyed the German *Army Group Center* and recaptured the last significant part of the Soviet Union remaining under German control. Soviet security operations played a major role in this operation's success. Soviet field regulations of 1944 specified the purposes of security operations: prevent surprise attack of the main body by enemy ground or air forces, prevent enemy reconnaissance, and give friendly forces time and conditions for deployment against the enemy.

From April through June, the Red Army conducted security operations against German reconnaissance and intelligence activities. During this period, Soviet operations directed against German sustaining operations and facilities, conducted by partisans, kept the Germans so busy conducting area security operations that they had few resources to devote to ground reconnaissance. The Red Army Air Force kept German aerial reconnaissance from looking deep into the Red Army's rear to operational depths. All the Soviet fronts (army groups) preparing for the summer offensive established a 25-kilometer-deep security area against German ground reconnaissance. Frontline divisions conducted numerous and frequent patrols to counter German reconnaissance efforts and maintain regular physical contact with adjoining divisions.

The Soviets used all available assets to maintain the security of forces involved in the operation. Throughout the operation, the Red Army Air Force provided aerial cover, especially for mobile groups and forward detachments. Because the marshy nature of much of the terrain prevented using large mechanized formations, the Soviets used horse cavalry corps, augmented with tanks, to cover ground unsuitable for heavier forces and maintain contact between separated elements of their forward mobile forces.

On the flanks of Operation Bagration, the Soviet 1st Guards Tank Corps served as a covering force against reinforcements or relief efforts from the German *Army Group North*. A combined horse-cavalry mechanized group served the same role in the south for the 1st Byelorussian Front against the German *Army Group North Ukraine*.

GENERAL CONSIDERATIONS FOR SECURITY OPERATIONS

12-12. There are a number of general considerations when conducting security operations. These apply to all forms of security operations but are most applicable to screen, guard, and cover missions.

COMMON SECURITY CONTROL MEASURES

12-13. Security operations are depicted on overlays using a lightning bolt on either side of the symbol representing the unit conducting the security operation and are labeled with the letter S, G, or C to denote screen, guard, or cover. The end of the lightning bolt has arrowheads that touch the designated operational graphics, which define the left and right limits of the security operation. (See Figure 12-1.)

12-14. The screen, guard, and cover have many common control measures, starting with boundaries defining the security area. The main body commander establishes the security area. For a security force operating to the front of the main body, the lateral boundaries of the security area are normally an extension of the lateral boundaries of the main body. The security force's rear boundary is normally the battle handover line (BHL).

12-15. To establish a screen to the rear of a force, the lateral boundaries are also an extension of the boundaries of the main body, with the screening force's rear boundary being the rear boundary of the entire force. For a flank screen, the lateral boundaries of the security area are an extension of the rear boundary of the main body and its forward edges of the battle area (FEBA) or forward line of own troops (FLOT). The rear boundary of a flank screen is the lateral boundary of the main body. The rear boundary or another phase line (PL) may serve as a BHL between the security force and the main body to control the passing of responsibility for the enemy to the main body. Normally, the responsibility of the flank security force begins at the trail element of the advance security force or the lead combat element in the main body. It ends at the rear of the main body or the lead element of the rear security force. The main body commander clarifies responsibilities as necessary.

12-16. Either the main body or the security force commander designates additional PLs to control the operation. These PLs may serve as subsequent screen or delay lines. Each element of the security force must report when crossing or occupying them. Displacement to these subsequent PLs is event-driven. The approach of an enemy force, relief of a friendly unit, or movement of the protected force dictates the movement of the security force. The security force commander normally assigns additional lateral boundaries within the security area to delineate the areas of operations (AOs) for subordinate units.

12-17. The commander uses checkpoints and named areas of interest (NAIs) to indicate specific areas of interest and to coordinate movement and surveillance. He uses contact points to facilitate coordination with flank units during front and rear security missions or between elements of a security force within the security area. Units conducting flank security for a moving force physically contact the main body at contact points. If the security force commander wants to ensure coverage of a specific NAI or avenue of approach, he establishes OPs.

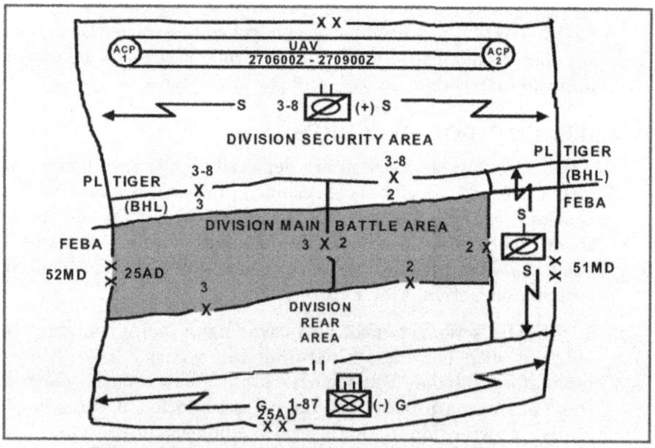

Figure 12-1. Common Security Control Measures

PLANNING CONSIDERATIONS FOR SECURITY MISSIONS

12-18. In addition to the planning considerations applicable to other types of operations discussed in this manual, such as control of key terrain and avenues of approach, the commander assigning a security mission and the security force commander must address special considerations:

- Force to be secured.
- Location and orientation of the security area.

- Initial observation post locations.
- Types of OPs.
- Time to establish the security force.
- Criteria for ending the security mission.
- Augmentation of security forces.
- Intelligence support to security operations.
- Special requirements or constraints.
- Fire planning.
- Integration of ground and air operations.
- Planning the engineer effort.
- Reporting.
- Positioning of command and control (C2) and CSS assets.
- Combat service support.

Force to Be Secured

12-19. The main body commander must designate the exact force to secure. This designation determines the limits of the security force's responsibilities. The security force must orient on the force it is securing. If the main body moves, the security force also moves to maintain its position in relation to the main body. Table 12-1 shows the typical size of security forces for a given echelon. The limited capabilities of most maneuver platoons prohibit them from having a mission separate from their parent company. Scout platoons are the exception to this rule.

Table 12-1. Typical Size of Security Forces for a Given Mission and Echelon

	SECURITY MISSION			
ECHELON	Screen	Advance Guard	Flank/Rear Guard	Cover
Battalion/Task Force	PLATOON	CO/TM		
Brigade	CO/TM	BN TF	CO/TM	BN TF(+)
Division	DIV CAV BN TF	DIV CAV (+) BDE	DIV CAV BN TF	DIV CAV (+) BDE
Corps	AR CAV SQD BN TF BDE	ACR	AR CAV SQD(+) or BN TF	ACR (+) or Division
Echelons Above Corps (JTF/Numbered Army)	ACR (+)	DIV (+) or Corps	ACR or BDE	DIV (+) or Corps

Location and Orientation of the Security Area

12-20. The main body commander determines the location, orientation, and depth of the security area in which he wants his security force to operate. He identifies specific avenues of approach and NAIs he wants covered. Depth in the security area provides the main body with more time to react to approaching enemy ground units. Occupying a deep security area allows the security force to destroy enemy reconnaissance assets without compromising critical OPs or positions. It also prevents the enemy from penetrating the security area too easily and prevents gaps from occurring when OPs or units displace or are lost. The wider the area to secure, the less the security force can take advantage of the increased depth because it will have fewer forces to position in depth. A very shallow security area may require a guard to provide needed reaction time.

12-21. The security force commander conducts a detailed analysis of the terrain in the security area. He establishes his initial dispositions (usually a screen line) as far forward as possible on terrain to allow good observation of avenues of approach. Next, he assigns clear responsibility for identified avenues of approach and designated NAIs. For a screen or guard, the initial screen line must be within supporting range of the main body, yet provide the desired amount of early warning.

Initial Observation Post Locations

12-22. The security force commander determines tentative initial OP locations along or behind the screen line to ensure effective surveillance of the sector and designated NAIs. The unit or asset that occupies each OP may shift its exact location to achieve the commander's intent. A commander may place more than one OP along a high-speed avenue of approach to allow an enemy contact to be tracked from one OP to another, thus maintaining enemy contact without requiring security forces to displace. The security force commander tasks subordinate units to perform reconnaissance and combat patrols to cover gaps between OPs. To prevent fratricide, the commander places a restrictive fire support coordinating measure around OP locations.

Types of Observation Posts

12-23. Observation posts may be either mounted or dismounted. Mounted OPs can use their vehicular optics, weapon systems, and speed of displacement. However, an enemy can detect them more readily than dismounted OPs. Dismounted OPs provide maximum stealth but lack the speed of displacement, optics, and weapons of mounted OPs. It takes a minimum of two soldiers to man an OP, and then for no more than 12 hours. Observation posts manned for more than 12 hours require, as a minimum, an infantry squad or scout section to ensure continuous operation. The screening force patrols dead space and the area between OPs, conducts resupply operations, and rests or sustains its personnel.

Time the Security Force Must Be Established

12-24. The main body commander must determine when to establish the security force. He decides this based on the activity of the main body and expected enemy activity. He must allow enough time for the security force to move into and occupy the security area to prevent enemy forces from penetrating the security area undetected. The factors of METT-TC influence how the security force deploys to and occupies the screen line. If the security mission is the result of a current reconnaissance mission, the security force is already positioned to begin its mission. This occurs frequently when a reconnaissance mission halts at a designated PL. Analyzing the factors of METT-TC determines which deployment technique meets mission requirements.

Criteria for Ending the Security Mission

12-25. Security missions are usually time- or event-driven. The criteria for ending a security mission can be an action by the main body (such as completing a specific mission), a fixed-time period (for example, not allowing enemy penetration of a PL for two hours), or criteria based on the enemy force (such as its size). To terminate its security mission, the security force commander normally requires the permission of the main body commander to withdraw behind the rear boundary.

Augmentation of Security Forces

12-26. The main body commander is responsible for reinforcing the security force. When the security area is large, additional combat and CS assets may reinforce the security force's organic combat power. Any unique requirement posed by the mission may require assets not organic to the security force. Ground surveillance radars, engineers, and chemical reconnaissance elements are common attachments at the company or troop level.

Intelligence Support to Security Operations

12-27. Intelligence assets can greatly enhance security operations. These assets can conduct rapid surveillance of large areas to detect enemy presence. Remote sensors, unmanned aerial vehicles, battlefield surveillance radars, signal intelligence systems, and downlinks from theater and national assets can expand the area under surveillance and cue the security force. Advanced aircraft, such as the OH-58D Kiowa Warrior and the AH-64D Longbow, can detect and report enemy forces at extended ranges with thermal imaging and other advanced detection equipment. This permits a commander to concentrate his security force on likely enemy avenues of approach, NAIs, targeted areas of interest (TAIs), and restrictive terrain that degrades sensor performance. The commander can use his intelligence assets to detect enemy movements. This gains time to reposition his security force and mass other assets to counter enemy actions. The commander increases the size of his security force to reduce his risk if he cannot anticipate sufficient advance warning from his intelligence assets.

Special Requirements or Constraints

12-28. The main body commander may impose special requirements or constraints, including engagement, disengagement, and bypass criteria. He may order the security force not to become decisively engaged or fall below a certain combat strength. He may be willing to accept a lesser degree of security, which results from either the loss of more terrain or reduced preparation time by the main body, to preserve his security force for later use.

Fire Planning

12-29. The main body commander positions his fire support assets to support his screen and guard forces. He allocates additional artillery to support a covering force. If the security force is assigned a wide AO, the commander may have to position his fire support assets to provide effective coverage of only the most likely enemy avenues of approach. This is particularly important for a screen because often the screen force can rely only on indirect fire to delay or disrupt the enemy. Providing adequate indirect fire support to the security force may require the main body to position its artillery well forward in the formation of the main body.

Integration of Ground and Air Operations

12-30. Integrating ground and air operations is critical to the success of many security missions. Aviation units, especially air cavalry, assist in reconnaissance of the security area as the ground element of the security force moves forward. They can perform the following tasks:

- Extend the screen in front of the flank security element's screen line.
- Screen forward of the ground security force.
- Conduct reconnaissance of areas between ground maneuver units.
- Assist in maintaining contact between the security force and the main body.
- Assist in clearing the area between the flank security element and the main body during moving flank security missions.
- Assist in disengaging ground units, which is especially valuable when conducting battle handover and passage of lines with the main body.
- Monitor terrain that is hard to reach or would require too much time to cover with ground reconnaissance assets.

Planning the Engineer Effort

12-31. Countermobility plays a critical role in the security area. With properly integrated obstacles, the security force can maintain a mobility advantage over the enemy. The commander may mass engineer support in the security area initially and then shift support to the main battle area (MBA) once those units are prepared to begin developing engagement areas. They also enhance the mobility of the security force by identifying repositioning routes and task organizing engineers to provide breaching capability. However, the senior commander must consider the impact of prioritizing the countermobility effort in the security area rather than in the MBA or at the decisive point. In the offense, a commander can employ situational obstacles, covered by fire, on the flanks of an advancing force to provide additional security.

Reporting

12-32. The security force reports enemy activities to the main body. The main body headquarters is responsible for disseminating that information to other affected friendly forces. The main body commander ensures that the security force has access to all pertinent intelligence and combat information obtained by the main body. This supplements the security force's capabilities. By continuously exchanging information, both the security force commander and the main body commander have time to choose a suitable COA. Force digitization greatly assists commanders in maintaining a common operational picture.

Combat Service Support

12-33. The unit logistics staff ensures that security element sustainment requirements are embedded in the unit's logistic chain and CSS orders and annexes. A key component in security element sustainment is developing, maintaining, and using standing operating procedures (SOP). Logisticians and operators must use the SOP and should be involved in their development. The SOP should be exercised and tested during training and changes made as needed. The commander ensures that his staff includes the sustainment of these security elements in logistics rehearsal.

12-34. The security element commander designates the individual within the security element who is responsible for sustaining the element. This is normally the senior noncommissioned officer within platoons and companies assigned security tasks. For example the platoon sergeant of a scout platoon assigned the mission of establishing a flank screen for a battalion task force would coordinate with the first sergeant of the adjacent company team to include the scout platoons sustainment requirements in the company team's logistics package. Likewise the first sergeant of a brigade reconnaissance troop assigned a screening mission coordinates directly with the brigade logistics and forward support battalion staffs for resupply and medical treatment. This individual must have access to the appropriate nets to coordinate logistic support and casualty evacuation.

12-35. That individual coordinates with the appropriate supporting logistics point of contact as soon as possible after receiving the security mission warning order. Coordination includes such items as the mission of the security element, the AO assigned to the security element, the routes it will take to that area from its current location, and movement times. He gives the exact sustainment requirements for the security element—including any specialized items of supply required by the mission, such as cratering charges—to the supporting logistics element. He ensures the support element establishes communication links with the security element and receives a copy of the support element's CSS overlay.

12-36. The commander must place special attention on treating and evacuating casualties for security elements operating removed from normal medical support because of time, terrain, or distance factors, or a need for the security element to remain undetected by the enemy. For this reason, the security element should include as many soldiers trained as combat lifesavers as is possible. The more combat lifesavers within the security element, the more prepared it is for casualties.

Positioning of C2 and CSS Assets

12-37. The security force commander positions himself where he can best control the operation. This is often where he can observe the most dangerous enemy avenue of approach. He positions his command post to provide continuous control and reporting during initial movements. His combat trains position behind masking terrain but remain close enough to the combat elements of the security force to provide rapid response. They are best sited along routes that provide good mobility laterally and in-depth.

MOVEMENT INTO SECURITY AREAS FOR STATIONARY SECURITY MISSIONS

12-38. All stationary security missions are established in a similar manner. In deploying into the security area, the security force must deal with competing requirements: to establish the security area quickly to meet mission requirements, and to provide the necessary level of security for itself. The security force moves into the security area using one of three basic methods: tactical road march, movement to contact, or zone reconnaissance.

12-39. The fastest but least secure method of deploying is a tactical road march from the rear boundary of the security area to the initial positions. The security force moves to a release point on the rear boundary. From the release point, subordinate elements deploy to occupy initial positions, moving by the quickest means possible. This method is appropriate when enemy contact is not expected, time is critical, or an aviation unit is conducting a zone reconnaissance forward of the ground element and has found no enemy in the security area.

12-40. In the second method, the security force conducts a movement to contact from a line of departure (usually the rear boundary of the security area) to the initial positions. This method is slower than a tactical road march but more secure. It is appropriate when enemy contact is likely, time is limited, terrain reconnaissance is not needed, or an aviation unit is conducting zone reconnaissance forward of the ground element and enemy forces have been detected in the security area.

12-41. The most secure method for moving to the initial positions is for the security force to conduct a zone reconnaissance from the security area rear boundary to its initial security line positions or the forward limit of the security area. Given adequate time, this method is preferred because it allows the security force to clear the security area and become familiar with the terrain that it may have to defend. The security force can reconnoiter potential subsequent positions and fire support system firing positions as it moves to its initial positions. A zone reconnaissance is appropriate when time is available and information about the enemy or terrain is unknown. While this technique provides information of tactical value on the enemy and terrain in the area, it may also be time consuming. Using air reconnaissance forward of the ground units increases the speed and security of the movement.

MOVEMENT DURING MOVING FLANK SECURITY MISSIONS

12-42. There are three techniques of occupying and moving in a flank security area for moving security missions based on how the security force crosses the line of departure:

- Security force crosses the line of departure (LD) separately from the main body and deploys to perform the mission.
- Security force crosses the LD separately from main body; lead elements conduct a movement to contact.
- Security force crosses the LD with the main body and conducts a zone reconnaissance out to the limit of the security area.

12-43. The security force should not be required to make its own penetration when it faces prepared enemy defenses. This may prevent or significantly delay the security force from assuming its duties. These three techniques are often combined.

12-44. In the first technique, illustrated in Figures 12-2 and 12-3, page 12-12, the security force crosses the LD separately from the main body and deploys to perform the mission. The security force then conducts a tactical road march, an approach march, or tactical movements parallel to the main body and drops off OPs or occupies BPs along the flank of the main body. This technique keeps the two forces from interfering with each other during deployment. It is appropriate when another force penetrates the line of contact, the main body is not in contact with the enemy and is moving quickly, the LD is uncontested, and the IPB process indicates that enemy contact is not likely in the area through which the security force is moving. It is the fastest but least secure technique.

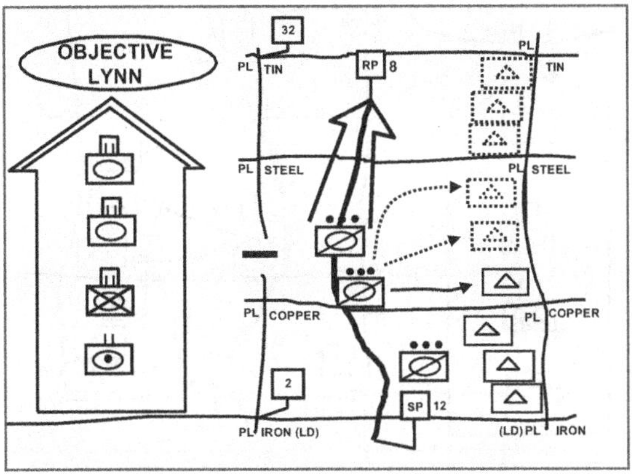

Figure 12-2. Security Force Crossing the LD Separately from the Main Body to Establish a Flank Screen

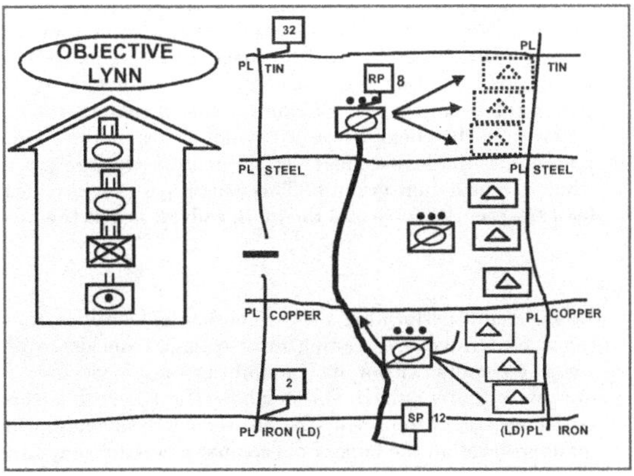

Figure 12-3. Security Force Continuing to Cross the LD Separately from the Main Body to Establish a Flank Screen

12-45. In the second technique, the security force crosses the LD separately from the main body, and its lead elements conduct a movement to contact. Follow-on elements occupy positions as they are reached. (See Figure 12-4.) This technique is appropriate to use when the main body is moving slower than in the first method, the LD is uncontested, and the IPB process indicates possible enemy contact. It is slower than the previous technique but provides better security.

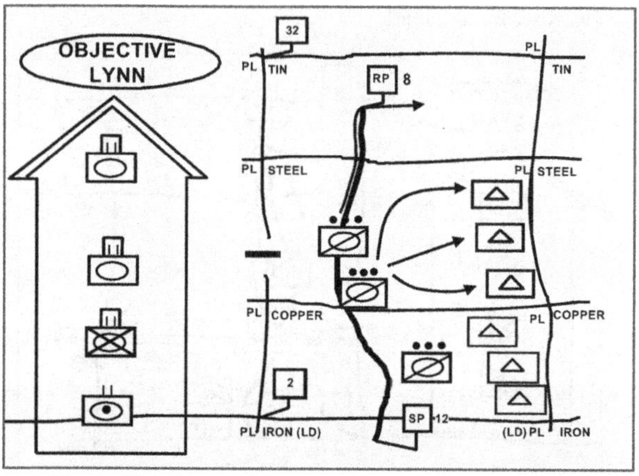

Figure 12-4. Second Technique Used by a Moving Flank Security Force to Establish a Moving Flank Screen

12-46. Finally, in the third technique, the security force crosses the LD with the main body and conducts a zone reconnaissance out to the far limit of the security area. (See Figure 12-5, page 12-14.) This technique is appropriate when the LD is also the line of contact, the main body makes its own penetration of the enemy defenses along the line of contact, the main body is moving slowly, and the enemy situation is not clearly understood. The security force may follow the lead element of the main body through the gap and deploy when the situation permits. This technique provides increased security for both the security force and the main body; it is also the most time consuming.

SCREEN

12-47. A unit performing a screen observes, identifies, and reports enemy actions. Generally, a screening force engages and destroys enemy reconnaissance elements within its capabilities—augmented by indirect fires—but otherwise fights only in self-defense. The screen has the minimum combat power necessary to provide the desired early warning, which allows the commander to retain the bulk of his combat power for commitment at the decisive place and time. A screen provides the least amount of protection of any security mission; it does not have the combat power to develop the situation.

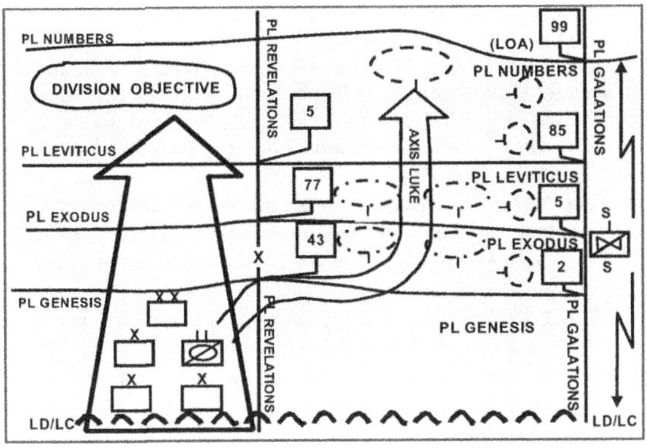

Figure 12-5. Third Technique Used by a Moving Flank Security Force to Establish a Flank Guard or Cover

12-48. A screen is appropriate to cover gaps between forces, exposed flanks, or the rear of stationary and moving forces. The commander can place a screen in front of a stationary formation when the likelihood of enemy action is small, the expected enemy force is small, or the main body needs only limited time, once it is warned, to react effectively. Designed to provide minimum security with minimum forces, a screen is usually an economy-of-force operation based on calculated risk. If a significant enemy force is expected or a significant amount of time and space is needed to provide the required degree of protection, the commander should assign and resource a guard or cover mission instead of a screen. The security element forward of a moving force must conduct a guard or cover because a screen lacks the combat power to defeat or contain the lead elements of an enemy force.

12-49. A security force normally conducts a screen by establishing a series of OPs and patrols to ensure adequate surveillance of the assigned area. The commander uses reconnaissance patrols (mounted, dismounted, and aerial), relocates OPs, and employs technical assets to ensure continuous and overlapping surveillance. The commander also employs terrain data base analytical support systems to ensure the integration of his reconnaissance and surveillance assets to provide that necessary coverage.

CRITICAL TASKS FOR A SCREEN

12-50. Unless the commander orders otherwise, a security force conducting a screen performs certain tasks within the limits of its capabilities. A unit can normally screen an avenue of approach two echelons larger than itself, such as a battalion scout platoon screening a battalion-size avenue of approach or a cavalry troop screening a regimental or brigade-size avenue of approach. If a security force does not have the time or other resources to complete all of these tasks, the security force commander must inform the commander assigning the mission of the shortfall and request guidance on which tasks must be completed and their priority. After starting the screen, if the security unit commander determines that he cannot complete an assigned task, such as maintain continuous surveillance on all avenues of approach into an AO, he reports and awaits further instructions. Normally, the main force commander does not place a time limit on the duration of the screen, as doing so may force the screening force to accept decisive engagement. Screen tasks are to—

- Allow no enemy ground element to pass through the screen undetected and unreported.
- Maintain continuous surveillance of all avenues of approach larger than a designated size into the area under all visibility conditions.
- Destroy or repel all enemy reconnaissance patrols within its capabilities.
- Locate the lead elements of each enemy advance guard and determine its direction of movement in a defensive screen.
- Maintain contact with enemy forces and report any activity in the AO.
- Maintain contact with the main body and any security forces operating on its flanks.
- Impede and harass the enemy within its capabilities while displacing.

ORGANIZATION OF FORCES

12-51. A screen normally requires the subordinate elements of the security force to deploy abreast. A screen force normally organizes itself into a number of OPs determined by the number of avenues of approach into the main force and any additional NAIs it must cover, as specified by the main force commander. The screening force may retain a small reaction force or reserve to extract endangered OPs.

12-52. The size of the avenue of approach kept under surveillance varies by echelon. Normally, a unit maintains observation over avenues of approach used by operationally significant enemy forces. These are normally avenues of approach used by enemy forces one echelon smaller than the friendly unit. For example, a battalion maintains surveillance over enemy company-size avenues of approach, while the corps maintains surveillance over division-size avenues of approach. The situation may require the unit to maintain surveillance over mobility corridors that can be used by enemy units two echelons smaller than the friendly force.

SCREEN CONTROL MEASURES

12-53. The control measures necessary to conduct a screen were previously discussed in this chapter under common security control measures in paragraphs 12-13 to 12-17. (Figure 12-6, page 12-16 displays examples of control measures associated with a screen.)

EXECUTING A STATIONARY SCREEN

12-54. In setting up the screen, the screening force establishes OPs with overlapping fields of observation. The screen commander adjusts the location of his screening elements to take advantage of established links with higher-echelon sensors and collection assets. Patrols reconnoiter areas that cannot be observed from an OP. The force retains a small reserve if possible. If forces are available and the depth of the security area allows, the screening force establishes OPs in-depth on high-speed avenues of approach. The commander plans routes between the initial and subsequent screen lines to facilitate rapid occupation of subsequent screen lines. The screening force reserve deploys in-depth and positions itself to react to contingencies that develop during the screen. The screening force takes advantage of its surveillance, target acquisition, and night-observation equipment and information provided by higher-level systems to expand the area and quality of security provided.

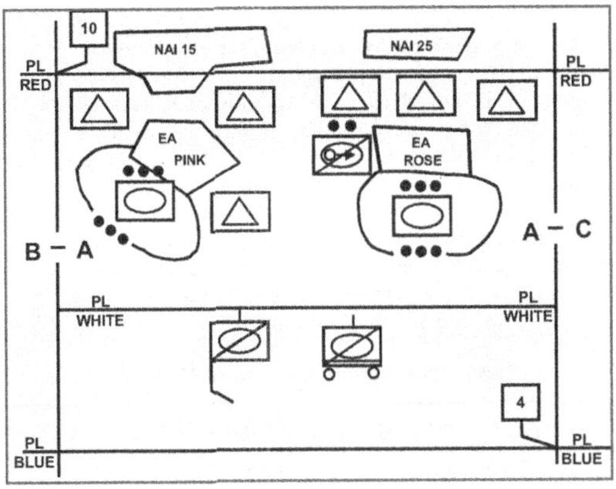

Figure 12-6. Control Measures Used in a Screen Mission

12-55. Observation posts should remain undetected while those manning them report the presence of enemy elements. Prompt, accurate reporting is essential to keep the assets constituting the screen from being overrun or unknowingly bypassed. Once the enemy is detected, the OP uses fire support channels to direct engagement of the enemy at maximum range. This helps the OP avoid detection by the enemy and prevents the enemy from penetrating the screen line. The screening force may destroy enemy reconnaissance assets with direct fire if indirect fire cannot accomplish this task. It also attempts to slow the movement of other enemy elements, primarily using indirect fires and close air support.

12-56. As enemy pressure threatens the security of the OP, the unit reports and requests to move to the next screen line. The commander may have previously established criteria that allow the screening force to displace to subsequent screen lines, based on certain enemy or friendly actions. These criteria should allow subordinates to use their initiative when conducting operations. When displacing from one screen line to another, the screening force emphasizes rapid movement while maintaining contact with the enemy. This ensures that any gaps that occur during movement are quickly closed. The screen's C2 elements displace as required to maintain control and keep from being overrun. The force repeats this procedure as often as necessary.

12-57. The screening force commander decides when to move from one screen line to another. However, the main body commander decides when the screening force can move behind the PL that designates the rear boundary of the security area and hand off the battle to the main body.

EXECUTING A MOVING SCREEN

12-58. The screening force may use several methods to move the screen as the protected force moves. Table 12-2 summarizes each method's advantages and disadvantages.

12-59. A force maintains a moving screen along the flanks and rear of the protected force. The screen movement is keyed to time and distance factors associated with the main body's movement. (See Figure 12-7.) Responsibilities for a moving flank screen begin at the front of the main body's lead combat element and end at the rear of the protected force. They do not include front and rear security forces. A force executes a moving screen in the same way it conducts a stationary screen, except for the movement techniques.

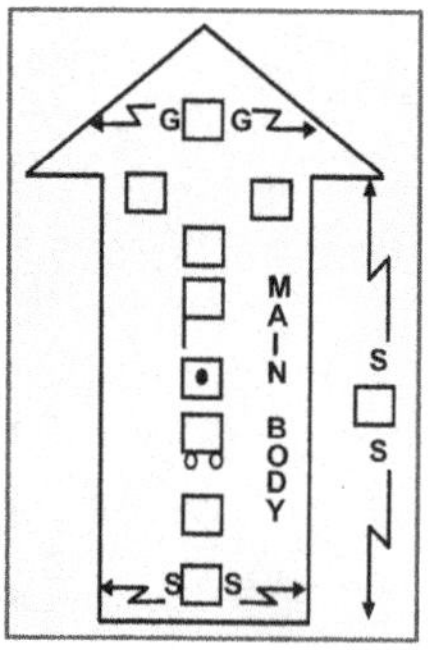

Figure 12-7. Moving Flank Screen

Table 12-2. Screen Movement Methods

METHOD	CHARACTERISTICS	ADVANTAGES	DISADVANTAGES
Alternate Bounds by OPs	• Main body moves faster • Conducted by platoon or company/ troop • Contact is possible • Conducted from rear to front	• Very secure method • Maintains maximum surveillance over the security area	• Execution takes time • Disrupts unit integrity
Alternate Bounds by Units	• Main body moves faster • Conducted by platoon or company/ troop • Contact is possible • Conducted from rear to front	• Execution does not take a great deal of time • Maintains good surveillance over the security area • Maintains unit integrity	• May leave temporary gaps in coverage
Successive Bounds	• Main body is moving slowly • Conducted by platoon or company/troop • Contact is possible • Conducted simultaneously or in succession • Unit should maintain an air screen during ground movement	• Most secure method • Maintains maximum surveillance • Maintains unit integrity	• Execution takes the most time • Unit is less secure when all elements are moving simultaneously • Simultaneous movement may leave temporary gaps
Continuous Marching	• Main body is moving relatively quickly • Performed as a route reconnaissance • Enemy contact is not likely • Unit should maintain an air screen on the flank	• OPs displace quickly • Maintains unit integrity	• Least secure method

12-60. The commander considers the factors of METT-TC in his decision regarding the movement method employed. Figures 12-8 and 12-9, page 12-18, illustrate four methods of controlling movement along a screen line:

- Alternate bounds by individual OPs from the rear to the front. (This method is usually employed at the company/troop level and below.)
- Alternate bounds by subordinate units from the rear to the front.
- Successive bounds by units along the screen line.
- Continuous marching along the route of advance.

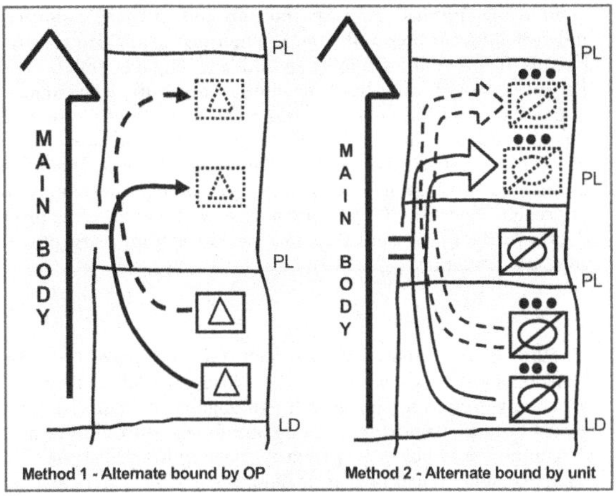

Figure 12-8. Displacement Methods for a Flank Screen

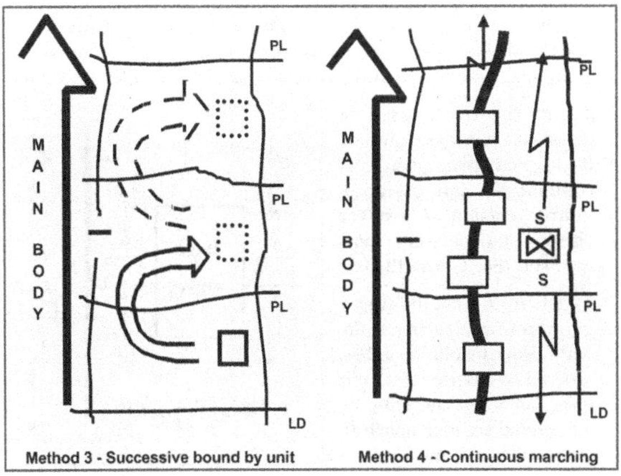

Figure 12-9. More Displacement Methods for a Flank Screen

SCREENING OPERATIONS DURING LIMITED VISIBILITY

12-61. Limited visibility often affects the screening force's ground and air observation capabilities. During limited visibility, the screening force uses all available night and thermal observation devices and depends more on electronic surveillance devices. Although the screening force can use technical reconnaissance and surveillance assets to offset limited visibility, it should also adjust its techniques and procedures to the conditions. For example, the commander of a screening force may need to adjust the number and location of his OPs in limited-visibility conditions. He can establish more OPs to cover avenues of approach that become masked in these conditions. He plans for indirect illumination and uses it when necessary. He closely coordinates his patrols to prevent misidentification and engagement by friendly elements. Rigorous sound and light discipline prevents compromise and potential bypass of OPs by enemy reconnaissance forces. Near OPs and along dismounted avenues of approach, the screening force can use trip flares, protective minefields, and mechanical devices, such as noisemakers integrated into tanglefoot obstacles, to detect the enemy and warn of his approach. Additional OPs along enemy avenues of approach can provide depth to facilitate detecting enemy forces that may have eluded forward security elements.

GUARD

12-62. A guard differs from a screen in that a guard force contains sufficient combat power to defeat, cause the withdrawal of, or fix the lead elements of an enemy ground force before it can engage the main body with direct fire. A guard force routinely engages enemy forces with direct and indirect fires. A screening force, however, primarily uses indirect fires or close air support to destroy enemy reconnaissance elements and slow the movement of other enemy forces. A guard force uses all means at its disposal, including decisive engagement, to prevent the enemy from penetrating to a position were it could observe and engage the main body. It operates within the range of the main body's fire support weapons, deploying over a narrower front than a comparable-size screening force to permit concentrating combat power.

12-63. The three types of guard operations are advance, flank, and rear guard. A commander can assign a guard mission to protect either a stationary or a moving force. (See Figure 12-10.)

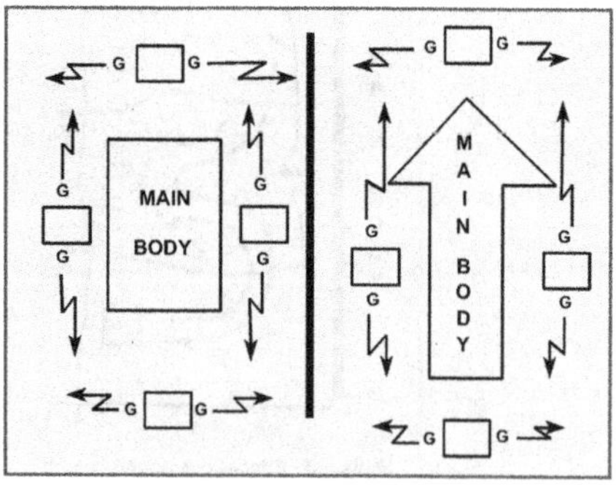

Figure 12-10. Guard Locations

12-64. A unit conducting a guard performs certain tasks within its capabilities unless ordered otherwise. If a unit does not have the time or other resources to complete all of these tasks, it must inform the commander assigning the mission of the shortfall and request guidance on which tasks to complete or the priority of tasks. After starting the guard, if the unit determines that it cannot complete an assigned task, such as cause deployment of the enemy advance guard, it must report this to the commander and await further instructions. Guard tasks—

- Destroy the enemy advance guard.
- Maintain contact with enemy forces and report activity in the AO.
- Maintain continuous surveillance of avenues of approach into the AO under all visibility conditions.
- Impede and harass the enemy within its capabilities while displacing.
- Cause the enemy main body to deploy, and then report its direction of travel.
- Allow no enemy ground element to pass through the security area undetected and unreported.
- Destroy or cause the withdrawal of all enemy reconnaissance patrols.
- Maintain contact with its main body and any other security forces operating on its flanks.

12-65. A commander employs a guard when he expects enemy contact and requires additional security beyond that provided by a screen. The multiple requirements of the guard mission are often performed simultaneously over relatively large areas. While the guard force's exact size is determined by prevailing METT-TC conditions, Table 12-1 provides general guidance on the size of an echelon's guard force.

ORGANIZATION OF A GUARD FORCE

12-66. Whether the guard is for a stationary (defending) or moving (attacking) force, the various types of guard missions and knowledge of the terrain and enemy dictate the specific task organization of the guard force. The guard force commander normally plans to conduct the guard mission as an area defense (Chapter 9), a delay (Chapter 11), a zone reconnaissance (see FM 3-55), or a movement to contact (Chapter 4) mission within the security area.

CONTROL MEASURES

12-67. The commander uses graphic control measures to control the operations of his guard force within the security area. The assigned mission also influences the size of the AO given to subordinate elements. For example, a movement to contact normally takes place across a narrower frontage than if the same unit makes a zone reconnaissance to allow adequate concentration of combat power.

12-68. The guard force may task its subordinate elements to conduct screen missions to the front and flanks of the guard force. This provides early warning of enemy forces and helps maintain contact with flank forces and any higher-echelon security force. An example of the latter would be a corps covering force operating in front of a division advance guard. The presence of a higher-echelon security force also influences how the guard force commander organizes his force and conducts operations. It specifically impacts the areas of fire support and CSS.

ADVANCE GUARD

12-69. An advance guard for a stationary force is defensive in nature. It defends or delays in accordance with the main body commander's intent. An advance guard for a moving force is offensive in nature. (See Figure 12-11.) The advance guard develops the situation so the main body can use its combat power to the greatest effect. The main body's combat power must not be dissipated through piecemeal commitment. The full combat power of the main body must be available immediately to defeat the main enemy force.

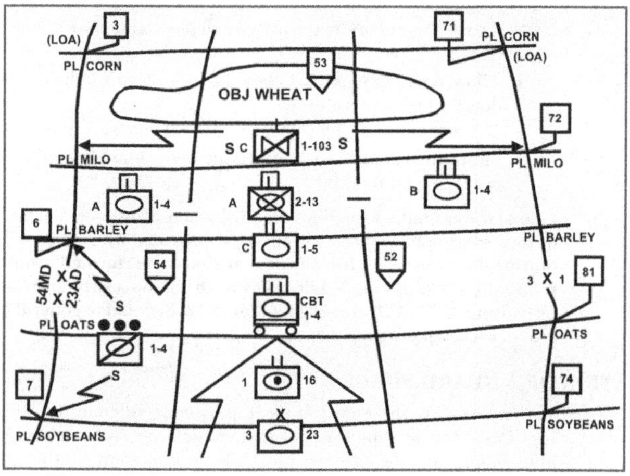

Figure 12-11. Advance Guard for a Division Shaping Attack

12-70. An advance guard for a moving force normally conducts a movement to contact. It organizes and uses the graphics of a movement to contact. (See Chapter 4.) Ground subordinate elements of a guard are normally deployed abreast to cover the axis of advance or the main body's AO.

12-71. The advance guard is responsible for clearing the axis of advance or designated portions of the AO of enemy elements. This allows the main body to move unimpeded, prevents the unnecessary delay of the main body, and defers the deployment of the main body for as long as possible.

12-72. The advance guard may operate behind the security force of a higher echelon. For example, a division may use its cavalry squadron (reinforced) as an offensive covering force, while each subordinate brigade column organizes one of its battalion task forces into an advance guard. (See Figure 12-12, page 12-22.) In these situations, the higher-echelon security force will initially develop the situation. A commander may task the advance guard to—

- Coordinate and conduct the rearward passage of lines of the covering force.
- Reduce obstacles to create lanes or improve existing lanes as required to support the maneuver of the main body.

- Eliminate enemy forces bypassed by the covering force.
- Coordinate and conduct a forward passage of lines through the covering force and fix enemy forces in the enemy's main defensive positions to allow the friendly main body to maneuver.

12-73. The movement of multiple security forces and the handoff of a detected enemy force from the higher-echelon security force to the lower-echelon security force is controlled using PLs, checkpoints, contact points, BHLs, and disengagement criteria, in addition to other graphic control measures. As a minimum, the covering force has a rear boundary that is also the forward boundary of the advance guard.

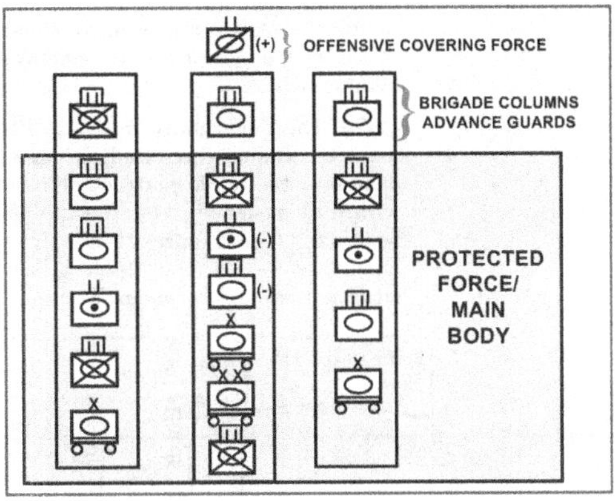

Figure 12-12. Multiple Security Forces

12-74. The advance guard engages in offensive operations when necessary to accomplish the mission. After the guard makes enemy contact, the commander determines whether the guard mission requires an attack, a defense, or a delay based on the factors of METT-TC. For example, if the guard force has sufficient combat power to defeat an enemy, it may conduct a hasty attack or defend from its current location. The guard force will not assault strong enemy positions from the front if this can be avoided. The advance guard then destroys the withdrawing enemy force as it exposes itself by moving to other positions. If the advance guard encounters an enemy force that it cannot stop from interfering with the movement of the main body, the security force reports its presence to the main body. It then establishes a defense, continues reconnaissance operations, and prepares to pass elements of the main body forward while facilitating the deployment of the main body.

12-75. If the guard force does not have enough combat power to defeat an approaching enemy, and the depth of the security area permits, the commander can delay back one or more positions before becoming decisively engaged. This reduces the enemy's combat power. Unless the security force is relieved of the guard mission, it must accept decisive engagement to prevent enemy ground forces from using direct fires to engage the main body.

FLANK GUARD

12-76. A flank guard protects an exposed flank of the main body. A flank guard is similar to a flank screen except that the commander plans defensive positions in addition to OPs.

12-77. The commander of the main body designates the general location of the flank guard's positions. Areas of operation assigned to the flank guard should be sufficiently deep to provide early warning and reaction time. However, flank guards must remain within supporting range of the main body. To determine the guard force's exact initial positions, the flank guard commander considers the front and rear of the flank of the main body, the axis taken by the main body, the enemy's capabilities, and the available avenues of approach.

12-78. The flank guard moves to its initial positions using one of the movement techniques discussed on pages 12-11 to 12-13. On reaching the initial positions, the flank guard establishes defensive positions in assigned BPs or within its assigned AO and establishes a screening element forward of these positions. (See Figure 12-13.) In situations where knowledge about the enemy is vague, the flank guard maintains a larger reserve than in situations where the enemy's actions are more predictable.

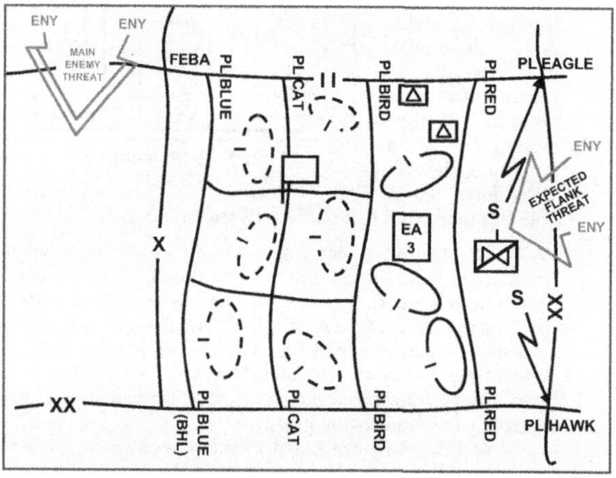

Figure 12-13. Stationary Flank Guard

12-79. Once the flank guard makes contact with the enemy, it can attack; defend to defeat or fix enemy ground forces in their current positions before they can engage the main body; or conduct a delay as required by the situation.

12-80. When conducting a moving flank guard, a commander must address additional considerations beyond those applying to conducting a moving flank screen. Instead of occupying a series of OPs, the security force plans a series of BPs. The tasks associated with a guard mission apply to a moving flank guard. However, the number and location of echelon-specific avenues of approach over which the security force maintains continuous surveillance change as the main body moves. The security force monitors potential enemy avenues of approach for as long as they threaten the main body.

12-81. The lead element of a moving flank guard must accomplish three tasks. It must maintain contact with the protected force's main body, reconnoiter the area between that main body and the flank guard's routes of advance, and reconnoiter the flank guard's route. It accomplishes these tasks by conducting a zone reconnaissance. The speed of the main body determines how thoroughly it can carry out the reconnaissance. The exact size of the AO for any given unit conducting a guard is METT-TC dependent. For example, on typical central European terrain, an AO wider than 10 kilometers from the guard line to the boundary of the main body should not be assigned to a company or troop. An organization of this size quickly finds itself unable to match the movement of the main body. When the distance from the guard line to the main body boundary exceeds 10 kilometers, the commander of the flank security element should use two or more company-size elements abreast. This ensures that the element making contact with the main body is not overtasked and can match the tempo of the main body. An air cavalry troop may maintain contact with the main body, or a following ground element may perform route reconnaissance along the flank guard's route of advance. Under these conditions, the lead security element does not reconnoiter BPs or occupy them unless required when making contact.

12-82. The rest of the flank guard marches along the route of advance and occupies BPs as necessary. Criteria for the route are the same as in a moving flank screen. The commander designates company-size BPs parallel to the axis of the main body. He places these BPs outside the flank guard's route of advance and along avenues of approach into the flank guard. The flank guard occupies OPs along a screen line forward of these BPs.

12-83. Since the flank guard is moving in one direction and orienting on providing protection to the secured force in another direction, the flank guard commander plans control measures to facilitate this dual orientation. These control measures are normally associated with the moving screen, as well as PLs that run parallel to the direction of movement of the main body. The commander uses these PLs to control the delay or defense if the enemy attacks from the flank being protected. (See Figure 12-14.) He may also assign the flank guard an objective that secures the flank for the main body's objective or otherwise serves to orient its security efforts.

12-84. The flank guard regulates its movement along the route of advance by the pace of the main body, the distance to the objective, and the enemy situation. The three methods of movement are successive bounds, alternate bounds, or continuous marching. (See Chapter 14.) If the main body stops, the flank guard occupies blocking positions. As the speed of the main body changes, the flank guard changes its movement methods. The guard commander must not allow the force to fall behind the main body or present a lucrative target by remaining stationary along the route.

12-85. If the flank guard becomes overextended, the guard commander informs the main body commander and recommends one of the following COAs:

- Reinforce the flank guard.
- Reduce the size of the flank guard's AO.
- Screen a portion of the area and guard the rest.

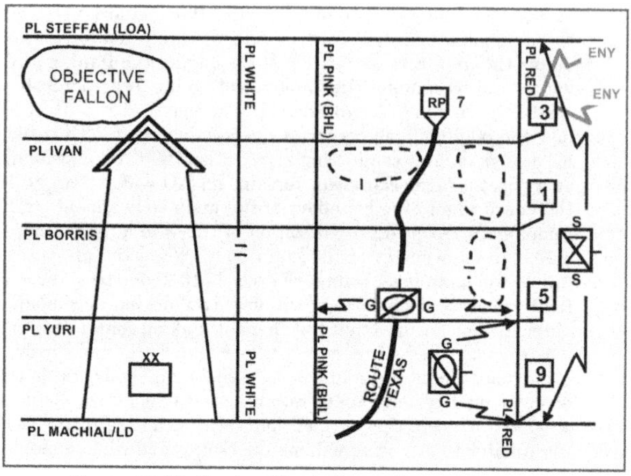

Figure 12-14. Moving Flank Guard Control Measures

REAR GUARD

12-86. The rear guard protects the exposed rear of the main body. This occurs during offensive operations when the main body breaks contact with flanking forces or during a retrograde. The commander may deploy a rear guard behind both moving and stationary main bodies. The rear guard for a moving force displaces to successive BPs along PLs or delay lines in depth as the main body moves. The nature of enemy contact determines the exact movement method or combination of methods used in the displacement (successive bounds, alternate bounds, and continuous marching).

12-87. During a retrograde, the rear guard normally deploys its ground maneuver elements abreast, behind the main body's forward maneuver units, generally across the entire AO. After the main body conducts a rearward passage of lines, the rear guard accepts battle handover and then defends or delays. Alternatively, the rear guard may conduct a relief in place as part of a deception plan or to take advantage of the best defensive terrain. In both cases, the rear guard establishes passage points and assists the rearward passage of the main body, if necessary. The rear guard accomplishes its defensive mission in the same way as any other guard operation after the main body clears the security area. As the main body moves, the rear guard moves to subsequent PLs in depth. Contact with the enemy force may eventually be lost if it does not follow the retrograding friendly force. Fighting a defense or a delay is necessary if the enemy detects the movement and attacks. (Chapter 11 discusses retrograde operations.)

COVER

12-88. The covering force's distance forward of the main body depends on the intentions and instructions of the main body commander, the terrain, the location and strength of the enemy, and the rates of march of both the main body and the covering force. The width of the covering force area is the same as the AO of the main body.

12-89. **A *covering force* is a self-contained force capable of operating independently of the main body, unlike a screening or guard force.** A covering force, or portions of it, often becomes decisively engaged with enemy forces. Therefore, the covering force must have substantial combat power to engage the enemy and accomplish its mission. A covering force develops the situation earlier than a screen or a guard force. It fights longer and more often and defeats larger enemy forces.

12-90. While a covering force provides more security than a screen or guard force, it also requires more resources. Before assigning a cover mission, the main body commander must ensure that he has sufficient combat power to resource a covering force and the decisive operation. When the commander lacks the resources to support both, he must assign his security force a less resource-intensive security mission, either a screen or a guard.

12-91. A covering force accomplishes all the tasks of screening and guard forces. A covering force for a stationary force performs a defensive mission, while a covering force for a moving force generally conducts offensive actions. A covering force normally operates forward of the main body in the offense or defense, or to the rear for a retrograde operation. Unusual circumstances could dictate a flank covering force, but this is normally a screen or guard mission.

ORGANIZATION OF A COVERING FORCE

12-92. Whether the cover is for a stationary (defending) or moving (attacking) force, the various types of cover missions, as well as knowledge of the terrain and enemy, dictate the specific task organization of the covering force. The covering force commander normally plans to conduct the cover mission as an area defense (Chapter 9), a delay (Chapter 11), a zone reconnaissance (see FM 3-55), or a movement to contact (Chapter 4) mission within the security area.

12-93. The commander normally assigns subordinate units one of these missions or the mission of screen or guard. The covering force uses those organizations and control measures associated with these missions. In addition, the commander establishes those control measures necessary for conducting the covering force's passage of lines (forward and rearward). (See Chapter 16.)

12-94. Although the commander can deploy any mobile force as a covering force, the corps covering force is normally built around the armored cavalry regiment or a division. Both have the C2 structures necessary for the forces involved and the capability to cover the geographical area typically required in a cover security mission. The corps commander tailors this unit to be self-contained by reinforcing it with assets such as attack helicopters, field artillery, engineers, air defense, tank, and infantry units with appropriate CSS to sustain the resulting force. A covering force is usually allocated additional artillery and engineer support beyond that normally given to a force of its size because it is operating beyond the main body's supporting range. The covering force commander normally maintains a sizable reserve to conduct counterattacks in the defense and to defeat enemy counterattacks in the offense.

12-95. A division covering force is normally a reinforced brigade, often with the divisional cavalry squadron as part of the covering force, to perform reconnaissance or other security missions. If the division AO is narrow enough, an adequately reinforced cavalry squadron may perform a cover mission. At both corps and division echelons, the amount of reinforcement provided to the covering force determines the distance and time it can operate away from the main body. These reinforcements typically revert to their parent organizations on passage of the covering force. Brigades and battalions typically organize a guard force instead of a covering force because their resources are limited.

12-96. Since one task of the covering force is to deceive the enemy into thinking he has found the main body, the commander should supply the covering force with combat systems that are representative of the main body. For example, if the main body has organic or reinforcing systems, such as MLRS, available to it, the commander should organize the covering force with the same systems.

OFFENSIVE COVER

12-97. An offensive covering force seizes the initiative early for the main body commander, allowing him to attack decisively. Figure 12-15 shows an attacking main body with an advance covering force and a flank guard.

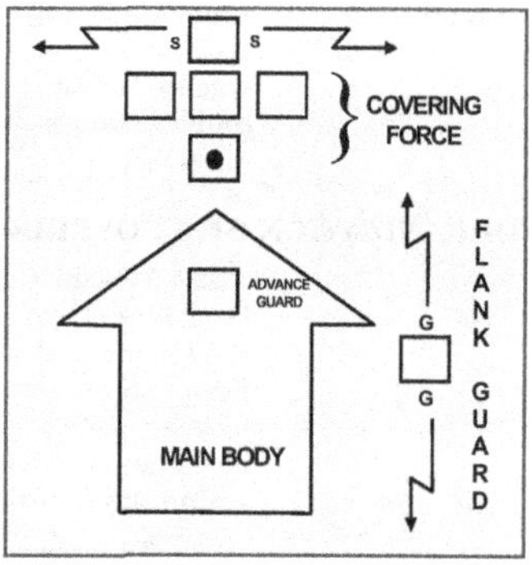

Figure 12-15. Attack Using a Covering Force

12-98. Unless the commander orders otherwise, an offensive covering force performs specific tasks within its capabilities. If a unit does not have the time or other resources to complete all of these tasks, it must inform the commander assigning the mission of the shortfall and request guidance on which tasks to complete or the priority of tasks. After starting the mission, if the unit determines that it cannot complete an assigned task, such as destroying or repelling enemy reconnaissance and security forces in the enemy security area, it must report this to the commander and await further instructions. Offensive covering force tasks include—

- Performing zone reconnaissance along the main body's axis of advance or within the AO.
- Clearing or bypassing enemy forces within the AO in accordance with bypass criteria.
- Denying the enemy information about the strength, composition, and objective of the main body.

12-99. Covering tasks against a defending enemy include—

- Penetrating the enemy's security area to locate enemy main defensive positions.

- Determining enemy strengths and dispositions.
- Locating gaps or weaknesses in the enemy's defensive scheme.
- Defeating or repelling enemy forces as directed by the higher commander.
- Deceiving the enemy into thinking the main body has been committed and causing him to launch counterattacks prematurely.
- Fixing enemy forces to allow the main body to maneuver around enemy strengths or through weaknesses.

12-100. In a meeting engagement, covering tasks include–

- Destroying enemy reconnaissance, the advance guard, and the lead elements of the main body.
- Determining the location of enemy assailable flanks.
- Fixing enemy forces to allow the main body to maneuver around enemy strengths or through weaknesses.

12-101. Planning for offensive covering force operations is similar to planning for zone reconnaissance or movement to contact. Mission analysis using the products of the IPB process helps determine the width of the area to cover and areas (NAIs and TAIs) or routes of special importance. The commander determines specific missions for subordinate elements and assigns boundaries. The covering force commander retains a reserve, which is ready to deploy anywhere in the covering force area. This reserve may be centrally located; it typically locates itself on the most dangerous or critical avenue of approach in the security area.

12-102. The covering force advances on a broad front, normally with its subordinate ground maneuver elements abreast (except for the reserve). This force should clear the enemy's security area of small combat elements while penetrating into the enemy's main defenses. Air cavalry normally reconnoiters forward of advancing ground covering force elements. On enemy contact, the air cavalry reports the enemy's location to the appropriate ground unit and maintains contact. Once the air cavalry makes contact, the covering force rapidly develops the situation. It reports enemy dispositions immediately to the main body commander so he can exploit enemy weaknesses. The covering force fixes encountered enemy forces and destroys them using fire and movement. The covering force does not bypass enemy forces without the permission of the main body commander.

12-103. If the covering force discovers a gap in the enemy's defenses, it prepares to exploit the weakness and disrupt the integrity of that defense. The covering force commander immediately reports this to the main body commander so he can divert main body follow-on forces to support the penetration. The main body commander synchronizes the penetration by the covering force with the arrival of other maneuver units, CS, and CSS to prevent counterattacking enemy forces from isolating and destroying the penetrating elements of the covering force.

12-104. When the covering force can advance no farther, it defends and prepares to assist the forward passage of lines of main body units. It continues to perform reconnaissance of enemy positions to locate gaps or assailable flanks. The covering force may guide main body units as they attack through or around the covering force. If the covering force has accomplished its mission, the main body commander will attack the enemy's weak point with previously uncommitted main body forces at the appropriate time.

FLANK COVER

12-105. When the main body commander perceives a significant threat to one of his flanks, he normally establishes a flank covering force. That force conducts its mission in much the same way as a flank guard performs its mission. The main differences between the two missions are the scope of operations and the distance the covering force operates away from the main body.

12-106. Just as in a flank guard, the flank covering force must clear the area between its route of advance and the main body. It must also maintain contact with an element of the main body specified by the main body commander. This element is normally part of the advance guard for the flank unit of the main body.

DEFENSIVE COVER

12-107. A defensive covering force prevents the enemy from attacking at the time, place, and combat strength of his choosing. (See Figure 12-16.) Defensive cover gains time for the main body, enabling it to deploy, move, or prepare defenses in the MBA. It accomplishes this by disrupting the enemy's attack, destroying his initiative, and establishing the conditions for decisive operations. The covering force makes the enemy deploy repeatedly to fight through the covering force and commit his reserve or follow-on forces to sustain momentum.

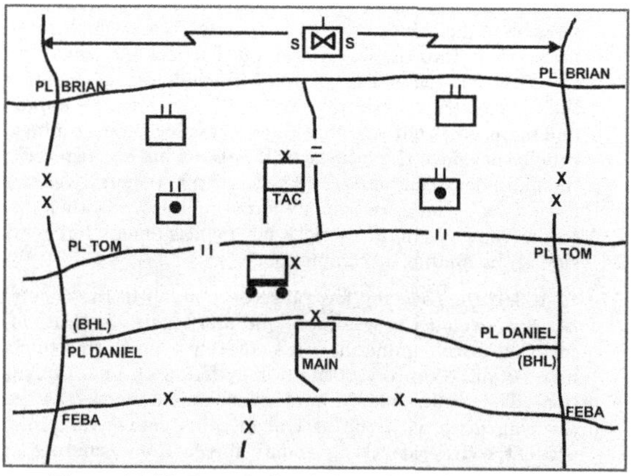

Figure 12-16. Generic Depiction of a Defensive Cover

12-108. Unless the commander orders otherwise, a defensive covering force performs certain tasks within its capabilities. If a unit does not have the time or other resources to complete all of these tasks, it must inform the commander assigning the mission of the shortfall and request guidance on which tasks to complete or the priority of tasks. After starting the mission, if the unit determines that it cannot complete an assigned task, such as defeat enemy advance guard formations, it must report this to the commander and await further instructions. A defensive covering force emphasizes the following tasks—

- Prevent the main body from being surprised and becoming engaged by direct-fire weapons.
- Defeat enemy advance guard formations.
- Maintain continuous surveillance of high-speed avenues of approach into the security area.
- Defeat all enemy reconnaissance formations before they can observe the main body.
- Cause the deployment of the enemy main body.
- Determine the size, strength, composition, and direction of the enemy's main effort.
- Destroy, defeat, or attrit enemy forces within its capacity.
- Deprive the enemy of his fire support and air defense umbrellas, or require him to displace them before he attacks the MBA.
- Deceive the enemy regarding the location of main body and main defensive positions.
- Avoid being bypassed.

12-109. The defensive covering force may be required to defend, delay, or counterattack. If the covering force area is not occupied, the force may have to reconnoiter and clear the area before establishing the cover. As in offensive operations, aerial reconnaissance is necessary to extend the area covered. Aviation units can screen less threatened areas and rapidly reinforce with their fires when other elements of the covering force are heavily engaged.

12-110. Whatever the command relationships may be at the outset, as the defensive covering force battle progresses, the covering force will be forced back toward the MBA. At this time, some or all of the covering force units fall under the control of the brigades charged with defending the MBA. Once the defensive covering force completes its mission, ground maneuver task forces reinforcing the covering force can do one of three things, separately or in combination. They can take up positions in the MBA, undergo reconstitution, or become part of the echelon reserve. The commander may use cavalry and other reconnaissance elements from the covering force as flank or rear security forces. Alternatively, he may use them to locate and follow the movement of the enemy's follow-on forces. They only establish BPs in the MBA as a last resort.

12-111. The conduct of a rearward passage of lines is an inherent part of the conduct of a defensive cover with its associated requirement to transfer responsibility for the battle between units. The commander must thoroughly plan this complex task as an integral part of the covering force mission. Passage of lines may not occur simultaneously for all covering force units. As some units begin passage, others may still be taking advantage of offensive opportunities in other parts of the security area. The covering force commander prepares to continue fighting in those portions of the security area where his forces are successful to set up offensive opportunities for the main body.

12-112. The covering force commander must exercise caution when issuing orders within the covering force. Commanders at each echelon will have a different perspective of the battle. This is never truer than in a covering force action. For example, while the covering force commander may be told to delay forward of a river line for 72 hours, he may tell his task force commanders to defend in certain BPs, perhaps for a specified period of time. Once the period expires, the covering force should not automatically retire from the covering force area. It must create enough resistance to force the enemy to deploy his main forces. Commanders at each echelon must precisely state the mission to their subordinate commanders without telling them how to do it. (This is mission command. See FM 6-0.) All too often, a small-unit commander, when told to delay, yields to an urge to *shoot too little, pull back too early*, and *move back too far*. Thus, it is imperative that each commander conveys to his subordinates precisely what their mission is in the context of the overall mission. Within a covering force, company teams and troops are mainly involved in a series of defensive operations.

AREA SECURITY

12-113. Area security operations may be offensive or defensive in nature. They focus on the protected force, installation, route, or area. Forces to protect range from echelon headquarters through artillery and echelon reserves to the sustaining base. Protected installations can also be part of the sustaining base or they can constitute part of the area's infrastructure. Areas to secure range from specific points (bridges and defiles) and terrain features (ridge lines and hills) to large population centers and their adjacent areas.

12-114. Operations in noncontiguous AOs require commanders to emphasize area security. During offensive and retrograde operations, the speed at which the main body moves provides some measure of security. Rapidly moving units in open terrain can rely on technical assets to provide advance warning of enemy forces. In restrictive terrain, security forces focus on key terrain such as potential choke points.

12-115. A commander executes rear area and base security as part of an echelon's sustaining operations responsibilities or as part of stability operations and support operations. (Rear area and base security and route and convoy security operations are the subject of Appendix E.) During conventional operations, area security operations are normally economy-of-force measures designed to ensure the continued conduct of sustaining operations designed to support the echelon's decisive and shaping operations. All area security operations take advantage of the local security measures performed by all units regardless of their location within the AO.

12-116. Since civilians are normally present within the AO, a unit restrains its use of force when conducting area security operations. However, the commander always remains responsible for protecting his force and considers this responsibility when establishing his rules of engagement. Restrictions on conducting operations and using force must be clearly explained and understood by everyone. Soldiers must understand that their actions, no matter how minor, may have far-reaching positive or negative effects. They must realize that either friendly or hostile media and psychological operations organizations can quickly exploit their actions, especially the manner in which they treat the civilian population.

12-117. Sometimes area security forces must retain readiness over long periods without contact with the enemy. This occurs most often during area security operations when the enemy knows that he is seriously overmatched in terms of available combat power. In this case, he normally tries to avoid engaging friendly forces unless it is on his terms. Forces conducting area security should not develop a false sense of security even if the enemy appears to have ceased operations within the secured area. The commander must assume that the enemy is observing his operations and is seeking routines, weak points, and lax security for the opportunity to strike with minimum risk.

LOCAL SECURITY

12-118. Local security includes any local measure taken by units against enemy actions. It involves avoiding detection by the enemy or deceiving the enemy about friendly positions and intentions. It also includes finding any enemy forces in the immediate vicinity and knowing as much about their positions and intentions as possible. Local security prevents a unit from being surprised and is an important part of maintaining the initiative. The requirement for maintaining local security is an inherent part of all operations. Units perform local security when conducting full spectrum operations, including tactical enabling operations.

12-119. Units use both active and passive measures to provide local security. Active measures include—

- Using OPs and patrols.
- Establishing specific levels of alert within the unit. The commander adjusts those levels based on the factors of METT-TC.
- Establishing stand-to times. The unit SOP should detail the unit's activities during the conduct of stand-to.

12-120. Passive local security measures include using camouflage, movement control, noise and light discipline, and proper communications procedures. It also includes employing available ground sensors, night-vision devices, and daylight sights to maintain surveillance over the area immediately around the unit.

COMBAT OUTPOSTS

12-121. **A *combat outpost* is a reinforced OP capable of conducting limited combat operations.** (See Figure 12-17.) Using combat outposts is a technique for employing security forces in restrictive terrain that precludes mounted security forces from covering the area. They are also used when smaller OPs are in danger of being overrun by enemy forces infiltrating into and through the security area. The commander uses a combat outpost when he wants to extend the depth of his security area, when he wants his forward

OPs to remain in place until they can observe the enemy's main body, or when he anticipates that his forward OPs will be encircled by enemy forces. Both mounted and dismounted forces can employ combat outposts.

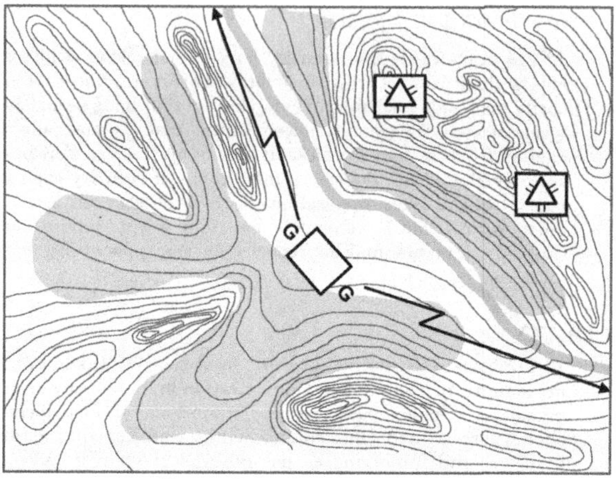

Figure 12-17. Combat Outposts

12-122. While the factors of METT-TC determine the size, location, and number of combat outposts established by a unit, a reinforced platoon typically occupies a combat outpost. A combat outpost must have sufficient resources to accomplish its designated missions, but not so much as to seriously deplete the strength of the main body. It is usually located far enough in front of the protected force to preclude enemy ground reconnaissance elements from observing the actions of the protected force.

12-123. The commander organizes a combat outpost to provide an all-around defense to withstand a superior enemy force. When the enemy has significant armored capability, a combat outpost may be given more than a standard allocation of antitank weapons. Forces manning combat outposts can conduct aggressive patrolling, engage and destroy enemy reconnaissance elements, and engage the enemy main body prior to their extraction. The commander should plan to extract his forces from the outpost before the enemy has the opportunity to overrun them.

Chapter 13

Reconnaissance Operations

You can never have too much reconnaissance.

General George S. Patton Jr., *War As I Knew It*, 1947

Reconnaissance operations **are those operations undertaken to obtain, by visual observation or other detection methods, information about the activities and resources of an enemy or potential enemy, or to secure data concerning the meteorological, hydrographical or geographical characteristics and the indigenous population of a particular area.** Reconnaissance primarily relies on the human dynamic rather than technical means. Reconnaissance is a focused collection effort. It is performed before, during, and after other operations to provide information used in the intelligence preparation of the battlefield (IPB) process, as well as by the commander in order to formulate, confirm, or modify his course of action (COA). The four forms of reconnaissance are route, zone, area, and reconnaissance in force.

13-1. Reconnaissance identifies terrain characteristics, enemy and friendly obstacles to movement, and the disposition of enemy forces and civilian population so the commander can maneuver his forces freely and rapidly. Reconnaissance prior to unit movements and occupation of assembly areas is critical to protecting the force and preserving combat power. It also keeps the force free from contact as long as possible so that it can concentrate on its decisive operation.

CONTENTS

RECONNAISSANCE OBJECTIVE

13-2. The commander orients his reconnaissance assets by identifying a reconnaissance objective within the area of operation (AO). **The *reconnaissance objective* is a terrain feature, geographic area, or an enemy force about which the commander wants to obtain additional information.** The reconnaissance objective clarifies the intent of the reconnaissance effort by specifying the most important result to obtain from the reconnaissance effort. Every reconnaissance mission must specify a reconnaissance objective. The commander assigns a reconnaissance objective based on his priority intelligence requirements (PIR) resulting from the IPB process and the reconnaissance asset's capabilities and limitations. The reconnaissance objective can be information about a specific geographical location, such as the cross-country trafficability of a specific area, a specific enemy activity to be confirmed or denied, or a specific enemy unit to be located and tracked. When the reconnaissance unit does not have enough time to complete all the tasks associated with a specific form of reconnaissance, it uses the reconnaissance objective to guide it in setting priorities.

13-3. A commander may need to provide additional detailed instructions beyond the reconnaissance objective, such as the specific tasks he wants accomplished or the priority of tasks. He does this by issuing additional guidance to his reconnaissance unit or by specifying these instructions in his tasks to his subordinate units in the operation order. For example, if, based on all technical and human intelligence (HUMINT) sources, a division G2 concludes that the enemy is not in an area and the terrain appears to be trafficable without obstacles, the division commander may decide he does not need a detailed reconnaissance effort forward of his unit. He may direct his cavalry squadron to conduct a zone reconnaissance mission with guidance to move rapidly and report by exception terrain obstacles that will significantly slow the movement of his subordinate maneuver brigades. Alternatively, when the objective is to locate an enemy force, the reconnaissance objective would be that force, and additional guidance would be to conduct only that terrain reconnaissance necessary to find the enemy and develop the situation.

RECONNAISSANCE FUNDAMENTALS

13-4. The seven fundamentals of successful reconnaissance operations are as follows:

- Ensure continuous reconnaissance.
- Do not keep reconnaissance assets in reserve.
- Orient on the reconnaissance objective.
- Report information rapidly and accurately.
- Retain freedom of maneuver.
- Gain and maintain enemy contact.
- Develop the situation rapidly.

ENSURE CONTINUOUS RECONNAISSANCE

13-5. Effective reconnaissance is continuous. The commander conducts reconnaissance before, during, and after all operations. Before an operation, reconnaissance focuses on filling gaps in information about the enemy and the terrain. During an operation, reconnaissance focuses on providing the commander with updated information that verifies the enemy's composition, dispositions, and intentions as the battle progresses. This allows the commander to verify which COA is actually being adopted by the enemy and determine if his plan is still valid based on actual events in the AO. After an operation, reconnaissance focuses on maintaining contact with the enemy to determine his next move and collecting information necessary for planning subsequent operations. When information regarding the current operation is adequate, reconnaissance focuses on gathering information for branches and sequels to current plans. As a minimum, reconnaissance is conducted continuously as an integral part of all security missions, including the conduct of local security for forces not in contact. (See Chapter 12.)

13-6. Reconnaissance operations over extended distances and time may require pacing reconnaissance assets to maintain the effort, or rotating units to maintain continuous coverage. The human and technical assets used in the reconnaissance effort must be allowed time for rest, resupply, troop leading procedures, additional and refresher training, and preventative maintenance checks and services. The commander must determine not only where, but also when he will need his maximum reconnaissance effort and pace his reconnaissance assets to ensure that adequate assets are available at critical times and places.

DO NOT KEEP RECONNAISSANCE ASSETS IN RESERVE

13-7. Reconnaissance assets, like artillery assets, are never kept in reserve. When committed, reconnaissance assets use all of their resources to accomplish the mission. This does not mean that all assets are committed all the time. The commander uses his reconnaissance assets based on their capabilities and METT-TC to achieve the maximum coverage needed to answer the commander's critical information requirements (CCIR). At times, this requires the commander to withhold or position reconnaissance assets to ensure that they are available at critical times and places. The rest required by reconnaissance assets to sustain the reconnaissance effort is not to be obtained by placing them in reserve. However, all reconnaissance assets should be treated as committed assets with specific missions assigned at all times. Units with multiple roles, specifically armored and air cavalry, that can conduct reconnaissance, security, and other combat missions in an economy-of-force role may be kept as a reserve for security or combat missions.

ORIENT ON THE RECONNAISSANCE OBJECTIVE

13-8. The commander uses the reconnaissance objective to focus his unit's reconnaissance efforts. Commanders of subordinate reconnaissance elements remain focused on achieving this objective, regardless of what their forces encounter during the mission. When time, limitations of unit capabilities, or enemy action prevents a unit from accomplishing all the tasks normally associated with a particular form of reconnaissance, the unit uses the reconnaissance objective to focus the reconnaissance effort.

REPORT INFORMATION RAPIDLY AND ACCURATELY

13-9. Reconnaissance assets must acquire and report accurate and timely information on the enemy, civil considerations, and the terrain over which operations are to be conducted. Information may quickly lose its value. Reconnaissance units report exactly what they see and, if appropriate, what they do not see. Seemingly unimportant information may be extremely important when combined with other information. Negative reports are as important as reports of enemy activity. Failure to report tells the commander nothing. The unit information management plan ensures that unit reconnaissance assets have the proper communication equipment to support the integrated intelligence, surveillance, and reconnaissance (ISR) plan.

RETAIN FREEDOM OF MANEUVER

13-10. Reconnaissance assets must retain battlefield mobility to successfully complete their missions. If these assets are decisively engaged, reconnaissance stops and a battle for survival begins. Reconnaissance assets must have clear engagement criteria that support the maneuver commander's intent. They must employ proper movement and reconnaissance techniques, use overwatching fires, and standing operating procedures (SOP). Initiative and knowledge of both the terrain and the enemy reduce the likelihood of decisive engagement and help maintain freedom of movement. Prior to initial contact, the reconnaissance unit adopts a combat formation designed to gain contact with the smallest possible friendly element. This provides the unit with the maximum opportunity for maneuver and enables it to avoid having the entire unit become decisively engaged. The IPB process can identify anticipated areas of likely contact to the commander. Using indirect fires to provide suppression and obscuration as well as destroy point targets is a method reconnaissance assets use to retain their freedom of maneuver.

GAIN AND MAINTAIN ENEMY CONTACT

13-11. Once a unit conducting reconnaissance gains contact with the enemy, it maintains that contact unless the commander directing the reconnaissance orders otherwise or the survival of the unit is at risk. This does not mean that individual scout and reconnaissance teams cannot break contact with the enemy. The commander of the unit conducting reconnaissance is responsible for maintaining contact using all available resources. That contact can range from surveillance to close combat. Surveillance, combined with stealth, is often sufficient to maintain contact and is the preferred method. Units conducting reconnaissance avoid combat unless it is necessary to gain essential information, in which case the units use maneuver (fire and movement) to maintain contact while avoiding decisive engagement.

DEVELOP THE SITUATION RAPIDLY

13-12. When a reconnaissance asset encounters an enemy force or an obstacle, it must quickly determine the threat it faces. For an enemy force, it must determine the enemy's composition, dispositions, activities, and movements and assess the implications of that information. For an obstacle, it must determine the type and extent of the obstacle and whether it is covered by fire. Obstacles can provide the attacker with information concerning the location of enemy forces, weapon capabilities, and organization of fires. In most cases, the reconnaissance unit developing the situation uses actions on contact. (See Chapter 4 for a discussion of actions on contact.)

HISTORICAL EXAMPLE

13-13. Military history contains numerous examples of the importance of reconnaissance operations. The following historical example illustrates the major role of reconnaissance operations in ensuring the success of an operation. This non-US, medieval example illustrates that the study of other armies and other times has a great deal to contribute in helping the tactician understand the art and science of tactics.

The Battle of the Sajo River

Reconnaissance was critical in determining enemy dispositions and taking advantage of the terrain in this and many other Mongol battles. The Mongol army conducted continuous reconnaissance with a definite reconnaissance objective, and a significant part of their success resulted from their reconnaissance operations. During operations, light cavalry preceded each of their army's main columns performing reconnaissance. They reported on terrain and weather conditions as well as the enemy's size, location, and movements. If a Mongol column met an enemy force that it could defeat, it did so. If it could not, its light cavalry maintained contact with the enemy, developed the situation to its advantage, and maintained freedom of movement. The Mongol light cavalry inflicted casualties and disrupted the enemy's movements while the main Mongol army deployed for action.

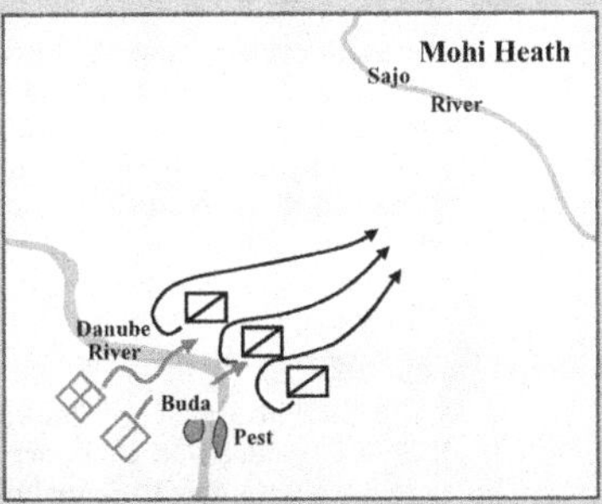

Figure 13-1. Mongol Army Route

In March 1241, a Mongol army of some 70,000 crossed the Carpathian Mountains from Russia into the Hungarian Plain. By mid-April, its light cavalry located the 100,000-man Hungarian army near the cities of Buda and Pest on the Danube River. In response, the Mongol army concentrated its previously dispersed columns as it approached the Danube. Once that the Mongols knew that they had been detected by the Hungarians, they deliberately withdrew about 100 miles northeast and led the Hungarians to a previously selected spot, Mohi Heath, on the Sajo River. The Mongols crossed the Sajo using an existing stone bridge and camped east of the river. The Hungarians followed and halted on the west bank, built a camp, took the stone bridge, and left a bridgehead on the east bank. Mongol reconnaissance discovered the location and dispositions within the Hungarian camp as well as a river-crossing site north of the camp. After dark, the main body of the Mongol army moved to cross the river at the crossing site. In addition to using the ford, the Mongols constructed a bridge to aid their crossing.

The next morning, the remainder of the Mongol army conducted a supporting attack on the Hungarian force at the stone bridge, drawing the Hungarian army out of its camp to fight. While the supporting Mongol forces succeeded in recrossing the Sajo via the stone bridge, the fighting was hard and they nearly lost their battle while waiting for the main body to come to their support. After 2 hours, the Mongol main body fell on the Hungarian rear and flank, driving the Hungarians back into their camp. As was Mongol practice, they deliberately left an escape route from the enemy camp open. The ensuing Mongol pursuit destroyed the Hungarian army when they tried to withdraw from their camp.

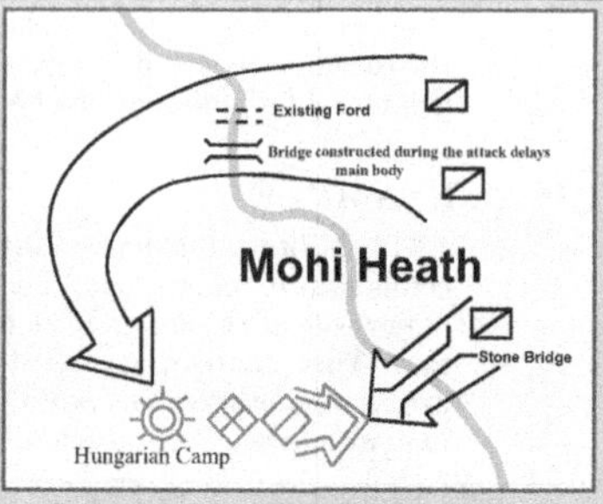

Figure 13-2. Mongol Army Pursuit

CHARACTERISTICS OF RECONNAISSANCE ASSETS

13-14. The responsibility for conducting reconnaissance does not reside solely with specifically organized units. Every unit has an implied mission to report information about the terrain, civilian activities, and friendly and enemy dispositions, regardless of its battlefield location and primary function. Frontline troops and reconnaissance patrols of maneuver units at all echelons collect information on enemy units with which they are in contact. In rear areas, reserve maneuver forces, fire support assets, air defense, military police, host nation agencies, combat support, and combat service support elements observe and report civilian and enemy activity. Although all units conduct reconnaissance, those specifically trained in reconnaissance tasks are ground and air cavalry, scouts, long-range reconnaissance units, and Special Forces. Some branches, such as the Corps of Engineers and the Chemical Corps, have specific reconnaissance tasks to perform that complement the force's overall reconnaissance effort. However, the corps and division commanders will primarily use their organic cavalry and intelligence elements to conduct reconnaissance operations.

13-15. At battalion level and above, the commander assigns missions to his ISR assets based on their organization, equipment, and training. The commander must know the capabilities and limitations of his available reconnaissance assets to ensure the employment of these assets within their capabilities and on missions for which they have been trained and equipped. Table 13-1 on page 13-6 shows the typical nesting of ISR assets available at different tactical echelons.

Table 13-1. Typical ISR Assets Available

	Platoon	Co/Tm	BN/TF	Brigade	Division	Corps	EAC
Observation Post	XXX	XXX	XXX	XXX	XXX	XXX	XXX
Reconnaissance Patrol	XXX	XXX	XXX	XXX	XXX	XXX	XXX
Combat Outpost	AAA	AAA	XXX	XXX	XXX	XXX	XXX
Scout Platoon	AAA	AAA	XXX	XXX			
Brigade Recon Troop		AAA	AAA	XXX	XXX		
Cavalry Troop (Sep Bde)		AAA	AAA	XXX	XXX	XXX	XXX
Chemical Reconnaissance		AAA	XXX	XXX	XXX	XXX	XXX
FA COLT Team	AAA	AAA	XXX	XXX			
FA Target Acq Systems			AAA	AAA	XXX	XXX	
ADA Target Acq Systems			AAA	AAA	XXX	XXX	XXX
Grd Surveillance Radars		AAA	XXX	XXX			
Other MI Collection Sys			AAA	XXX	XXX	XXX	XXX
Division Cavalry Squadron				AAA	XXX	XXX	
Air Cavalry				AAA	XXX	XXX	XXX
Unmanned Aerial Vehicles			AAA*	XXX*	XXX	XXX	XXX
Cavalry Regiment					AAA	XXX	XXX
Long-Range Surveillance Unit					AAA*	XXX	XXX
SOF (SF/RGR)					AAA	AAA	XXX
Technical Surveillance Platforms			AAA	AAA	AAA	AAA	XXX
XXX = Echelon controls or routinely tasks the asset.							
AAA = Echelon can routinely expect the information from that source to be made available to it.							
* Can be found in some divisions.							

13-16. A commander primarily conducts reconnaissance with a combination of manned ground and air assets supported by technical systems. Acting in concert, these assets create a synergy, using the strengths of one system to overcome the weaknesses of another. To produce this synergy, the commander must delineate reporting procedures for all units to pass on information gathered during reconnaissance operations. This facilitates rapid mission execution.

13-17. Dedicated reconnaissance assets are easily overtasked and overextended. The commander uses all available resources, not just reconnaissance units, to satisfy his information requirements. Ground reconnaissance can involve assets not specifically tailored for the mission. Engineer reconnaissance units collect information on how the terrain affects the movement of enemy and friendly forces. Nuclear, biological, and chemical (NBC) reconnaissance teams can determine the presence or absence of NBC contamination and the extent of that contamination. Artillery forward observers, fire support teams, and combat observation and lasing teams (COLTs) report combat information as they observe the battlefield. Air defense units observe and report enemy aircraft and air corridors in use.

13-18. Ground reconnaissance elements are generally limited in the depth to which they can conduct reconnaissance. However, they can operate under weather conditions that prohibit air reconnaissance operations.

13-19. Reconnaissance conducted by manned Army aviation platforms complements ground reconnaissance by greatly increasing the speed and depth with which reconnaissance operations can be conducted over a given area. Air reconnaissance can operate easily over terrain that hinders ground operations, such as swamps, extremely rugged terrain, or deep snow. Aviation assets can operate at a considerable depth, far in advance of the normal capability of dedicated ground reconnaissance elements normally focused on the close fight. Thus, they provide the commander with additional time to attack or otherwise react to the enemy's presence. Scout and attack helicopters use their optics, video, thermal imaging, and communications capabilities to detect and report the enemy. All types of aviation units generate pilot reports in the course of conducting their primary missions. These reports are often a source of valuable combat information.

13-20. While several technical systems can perform reconnaissance, the majority of these types of systems can be more accurately described as surveillance platforms. Surveillance complements reconnaissance by cueing the commitment of reconnaissance assets against specific locations or specially targeted enemy units. Surveillance provides information while reconnaissance answers the commander's specific questions.

13-21. Military intelligence (MI) assets conduct both surveillance and reconnaissance missions. They provide intelligence and electronic warfare (IEW) support, such as electronic intercept, ground surveillance radars, unmanned aerial vehicles (UAVs), and remotely emplaced sensors. Theater and national reconnaissance and surveillance systems provide broadcast dissemination of information and intelligence to the commander and can provide near real-time imagery as a part of an integrated ISR effort. Artillery and air defense target acquisition radars can complement MI surveillance systems as a part of the ISR effort. HUMINT collection occurs through face-to-face interrogation of captured enemy soldiers, screening of the civilian population, and debriefing of friendly soldiers, such as scouts and SOF.

FORMS OF RECONNAISSANCE

13-22. The four forms of reconnaissance operations are—

- Route reconnaissance.
- Zone reconnaissance.
- Area reconnaissance.
- Reconnaissance in force (RIF).

Table 13-2 shows what types of dedicated reconnaissance units are typically assigned the missions of conducting the four forms of reconnaissance operations.

Table 13-2. Dedicated Reconnaissance Units and Forms of Reconnaissance Operations

	SCOUT PLATOON	TROOP/CO TEAM	AIR CAV TROOP	AR CAV SQD/ BN	AR CAV REGT/ BDE	DIV
Route	X	X	X			
Zone	X	X	X	X	X	
Area	X	X	X	X	X	
Recon in Force				X	X	X

ROUTE RECONNAISSANCE

13-23. ***Route reconnaissance*** **is a form of reconnaissance that focuses along a specific line of communication, such as a road, railway, or cross-country mobility corridor.** It provides new or updated information on route conditions, such as obstacles and bridge classifications, and enemy and civilian activity along the route. A route reconnaissance includes not only the route itself, but also all terrain along the route from which the enemy could influence the friendly force's movement. The commander normally assigns this mission when he wants to use a specific route for friendly movement.

Organization of Forces

13-24. The commander may assign a route reconnaissance as a separate mission or as a specified task for a unit conducting a zone or area reconnaissance. A scout platoon can conduct a route reconnaissance over only one route at a time. For larger organizations, the number of scout platoons available directly influences the number of routes that can be covered at one time. Integrating ground, air, and technical assets assures a faster and more complete route reconnaissance.

13-25. A ground reconnaissance effort is essential if the mission is to conduct detailed reconnaissance of the route or the mission requires clearing the enemy from an AO that includes the route and the terrain around the route. The forces assigned to conduct this ground reconnaissance must be robust enough to handle expected enemy forces in the AO. If the commander expects them to make contact with enemy forces possessing more combat power than that typically found in enemy reconnaissance elements, he ensures that his forces conducting ground reconnaissance have access to readily available fire support. If the commander requires detailed information on the route, engineer reconnaissance assets can determine the classification of critical points along the route more quickly and accurately than scouts can. If the commander anticipates significant obstacles, combat engineers should be included as part of the force. If NBC contamination is expected, NBC reconnaissance assets should accompany the force conducting ground reconnaissance because they can detect and determine the extent of contamination more accurately and quickly than scouts can. Air reconnaissance can be used if the reconnaissance mission must be completed quickly. However, aerial reconnaissance can rarely clear an enemy force from a location where it can affect movement on the route and aircraft cannot breach obstacles. When time is limited, air reconnaissance is essential to determine which areas are clear of enemy forces and obstacles, and to cue ground reconnaissance regarding where to focus its efforts.

Control Measures

13-26. Control measures for a route reconnaissance create an AO for the unit conducting the reconnaissance. (See Figure 13-3.) The commander places lateral boundaries on both sides of the route, far enough out to allow reconnaissance of all terrain from which the enemy could dominate the route. He places a line of departure (LD) perpendicular to the route short of the start point (SP), allowing adequate space for the unit conducting the reconnaissance to deploy into formation. The LD creates the rear boundary of the AO. A limit of advance (LOA) is placed far enough beyond the route's release point (RP) to include any terrain from which the enemy could dominate the route. A SP and a RP define that section of the route where the unit collects detailed information. He may add phase lines (PLs) and checkpoints to maintain coordinated reconnaissance, control movement, or designate critical points. He places additional control measures to coordinate indirect and direct fire as necessary. He

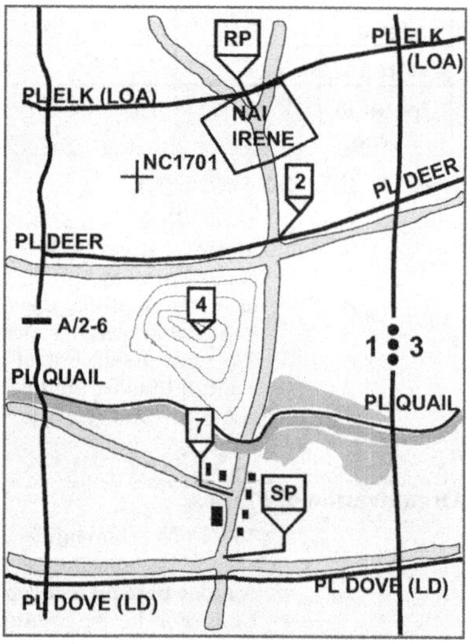

Figure 13-3. Route Reconnaissance Control Measures

places these control measures on terrain features that are identifiable from both the ground and the air to assist in air-to-ground coordination.

Tasks

13-27. Unless the commander orders otherwise, the unit conducting a route reconnaissance performs specific tasks within the limits of its capabilities. If a unit does not have the time or resources to complete all of these tasks, it must inform the commander assigning the mission. He must then issue further guidance on which tasks the unit must complete or the priority of each task, which is usually clear from the reconnaissance objective. If, after starting the reconnaissance, the unit determines that it cannot complete an assigned task, such as clearing the enemy or reducing obstacles to create lanes as required to support the maneuver of the main body along the route, it must report and await further instructions.

13-28. Route reconnaissance tasks are as follows:

- Find, report, and clear within capabilities all enemy forces that can influence movement along the route.
- Determine the trafficability of the route; can it support the friendly force?
- Reconnoiter all terrain that the enemy can use to dominate movement along the route, such as choke points, ambush sites, and pickup zones, landing zones, and drop zones.
- Reconnoiter all built-up areas, contaminated areas, and lateral routes along the route.
- Evaluate and classify all bridges, defiles, overpasses and underpasses, and culverts along the route.
- Locate any fords, crossing sites, or bypasses for existing and reinforcing obstacles (including built-up areas) along the route.
- Locate all obstacles and create lanes as specified in execution orders.
- Report the above route information to the headquarters initiating the route reconnaissance mission, to include providing a sketch map or a route overlay.

(See FM 3-34.212 and FM 3-20.95 for additional information concerning route reconnaissance.)

ZONE RECONNAISSANCE

13-29. ***Zone reconnaissance* is a form of reconnaissance that involves a directed effort to obtain detailed information on all routes, obstacles, terrain, and enemy forces within a zone defined by boundaries.** Obstacles include both existing and reinforcing, as well as areas with NBC contamination. The commander assigns a zone reconnaissance mission when he needs additional information on a zone before committing other forces in the zone. It is appropriate when the enemy situation is vague, existing knowledge of the terrain is limited, or combat operations have altered the terrain. A zone reconnaissance may include several route or area reconnaissance missions assigned to subordinate units.

13-30. A zone reconnaissance is normally a deliberate, time-consuming process. It takes more time than any other reconnaissance mission, so the commander must allow adequate time to conduct it. A zone reconnaissance is normally conducted over an extended distance. It requires all ground elements executing the zone reconnaissance to be employed abreast of each other. However, when the reconnaissance objective is the enemy force, a commander may forgo a detailed reconnaissance of the zone and focus his assets on those named areas of interest (NAI) that would reveal enemy dispositions and intentions. A reconnaissance unit can never disregard terrain when focusing on the enemy. However, it minimizes its terrain reconnaissance to that which may influence an NAI.

Organization of Forces

13-31. Considerations for organizing a zone reconnaissance are the same as for organizing a route reconnaissance except that several subordinate units, rather than just one unit, operate abreast during the zone reconnaissance. If the commander expects significant enemy forces to be found within the zone, he should provide the force conducting the zone reconnaissance with a reserve. This reserve should have adequate combat power to extract elements of the reconnaissance force from decisive engagement. In an armored cavalry squadron of an armored cavalry regiment, the tank company normally performs this task. If a unit conducts a zone reconnaissance out of supporting range of the main body, the commander ordering the zone reconnaissance provides the reconnaissance unit with adequate fire support assets that can move with the reconnaissance unit.

Control Measures

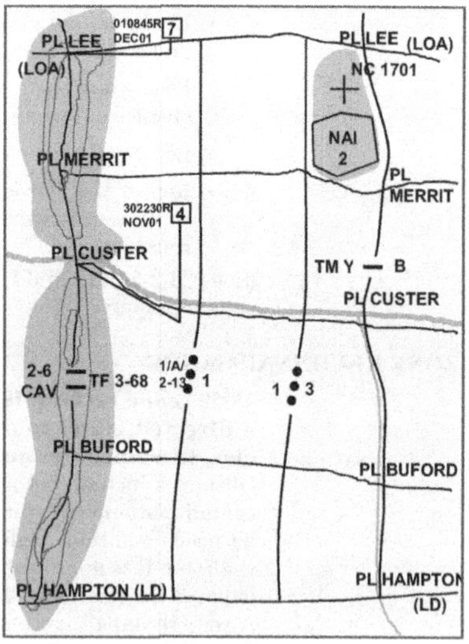

Figure 13-4. Zone Reconnaissance Control Measures

13-32. The commander controls a zone reconnaissance by assigning an AO to the unit conducting the reconnaissance. (See Figure 13-4.) The lateral boundaries, a LD, and a LOA define this AO. Within the AO, the force conducting the zone reconnaissance further divides the AO with additional lateral boundaries to define subordinate unit AOs. Subordinate AOs are not necessarily the same size. Phase lines and contact points, located where the commander determines that it is necessary for adjacent units to make physical contact, are used to coordinate the movement of elements operating abreast. He may further designate the time that this physical contact takes place. He uses checkpoints to indicate critical terrain features and to coordinate air and ground teamwork. He may use fire support coordinating measures to control direct and indirect fires. He uses additional control measures as necessary. In addition, the commander assigning the zone reconnaissance mission must specify the route the reconnaissance unit must use to enter the AO. All control measures should be on recognizable terrain when possible.

Tasks

13-33. Unless the commander orders otherwise, a unit conducting a zone reconnaissance performs the following tasks within the limits of its capabilities. If a unit does not have the time or resources to complete all of these tasks, it must inform the commander assigning the mission. He must then issue further guidance on which tasks the unit must complete or the priority of tasks, which is usually clear from the reconnaissance objective. After starting the reconnaissance, if the unit determines that it cannot complete an assigned task, such as clear enemy or reduce obstacles in zone to create lanes as required to support the main body's maneuver, it must report and await further instructions.

13-34. Zone reconnaissance tasks are as follows:

- Find and report all enemy forces within the zone.
- Clear all enemy forces in the designated AO within the capability of the unit conducting reconnaissance.
- Determine the trafficability of all terrain within the zone, including built-up areas.
- Locate and determine the extent of all contaminated areas in the zone.
- Evaluate and classify all bridges, defiles, overpasses, underpasses, and culverts in the zone.
- Locate any fords, crossing sites, or bypasses for existing and reinforcing obstacles (including built-up areas) in the zone.
- Locate all obstacles and create lanes as specified in execution orders.
- Report the above information to the commander directing the zone reconnaissance, to include providing a sketch map or overlay.

AREA RECONNAISSANCE

13-35. ***Area reconnaissance* is a form of reconnaissance that focuses on obtaining detailed information about the terrain or enemy activity within a prescribed area.** This area may include a town, a ridgeline, woods, an airhead, or any other feature critical to operations. The area may consist of a single point, such as a bridge or an installation. Areas are normally smaller than zones and are not usually contiguous to other friendly areas targeted for reconnaissance. Because the area is smaller, an area reconnaissance moves faster than a zone reconnaissance.

Organization of Forces

13-36. Considerations for the organization of forces for an area reconnaissance are the same as for organizing a zone reconnaissance. (See paragraphs 13-31 to 13-33.)

Control Measures

13-37. The commander assigning an area reconnaissance specifies the area for reconnaissance with a single continuous line to enclose the area to reconnoiter. Alternatively, he may designate the area by marking lateral boundaries, a LD, and a LOA. An area reconnaissance mission always specifies the route to take in moving to the area. The commander of the unit conducting the area reconnaissance mission can use control measures for a zone reconnaissance within the AO to control the operation of his subordinate elements. (See Figure 13-5.)

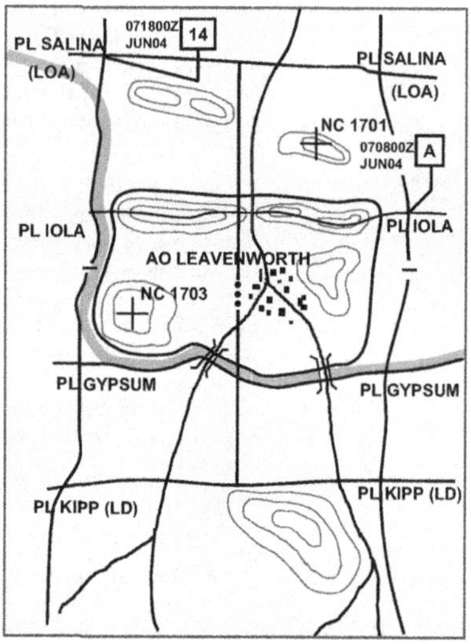

Figure 13-5. Area Reconnaissance Control Measures

Tasks

13-38. The tasks for an area reconnaissance are also the same as for a zone reconnaissance. (See paragraph 13-34.)

RECONNAISSANCE IN FORCE

13-39. **A *reconnaissance in force* is a deliberate combat operation designed to discover or test the enemy's strength, dispositions, and reactions or to obtain other information.** Battalion-size task forces or larger organizations usually conduct a reconnaissance in force (RIF) mission. A commander assigns a RIF mission when the enemy is known to be operating within an area and the commander cannot obtain adequate intelligence by any other means. A unit may also conduct a RIF in restrictive-type terrain where the enemy is likely to ambush smaller reconnaissance forces. A RIF is an aggressive reconnaissance, conducted as an offensive operation with clearly stated reconnaissance objectives. The overall goal of a RIF is to determine enemy weaknesses that can be exploited. It differs from other reconnaissance operations because it is normally conducted only to gain information about the enemy and not the terrain.

Organization of Forces

13-40. While specifically trained and equipped units usually conduct the other forms of reconnaissance operations, any maneuver force can conduct a RIF. The force conducting a RIF is organized as if it is conducting offensive operations. However, the lack of enemy information dictates that the force be large and strong enough to develop the situation, protect the force, cause the enemy to react, and put the enemy at some risk. The less known about the enemy, the stronger the force conducting the RIF must be. Because of the lack of information about the enemy, a commander normally conducts a RIF as a movement to contact or a series of frontal attacks across a broad frontage.

Control Measures

13-41. The control measures for a RIF are the same as for offensive operations. The operation is conducted as an movement to contact with limited objectives. (Chapter 4 discusses the conduct of a movement to contact.)

Tasks

13-42. A unit conducting a RIF performs the following tasks within the limits of its capabilities. If a unit does not have the time or resources to complete all of these tasks, it must inform the commander assigning the mission. He must then issue further guidance on which tasks the unit must complete or the priority of tasks, which is usually clear from the reconnaissance objective. After starting the RIF, if the unit determines that it cannot complete an assigned task, it must report and await further instructions. Reconnaissance in force tasks are—

- Penetrating the enemy's security area and determining its size and depth.
- Determining the location and disposition of enemy main positions.
- Attacking enemy main positions and attempting to cause the enemy to react by using local reserves or major counterattack forces, employing fire support assets, adjusting positions, and employing specific weapon systems.
- Determining weaknesses in the enemy's dispositions to exploit.

PLANNING A RECONNAISSANCE

13-43. Reconnaissance contributes significantly to a commander's battlefield visualization. It supports the overall integrated ISR plan, which in turn supports the commander's decision making process.

13-44. The commander must make judicious yet aggressive use of his reconnaissance assets. Reconnaissance planning ensures that available reconnaissance assets produce the greatest results. Because there are never enough assets to accomplish all tasks, the commander must set priorities. Generating many unfocused missions rapidly wears down assets, making them ineffective. Improperly using assets can also leave an enemy vulnerability undiscovered.

13-45. The commander ensures the coordination and synchronization of his reconnaissance effort at all echelons. Since the need for reconnaissance cuts across all parts of the operational framework and core functions, reconnaissance operations demand an integrated approach to planning, preparation, and execution. The two habitual participants in the reconnaissance planning process are the echelon operations and intelligence staff officers. The echelon operations staff officer (G3 or S3) has primary staff responsibility for reconnaissance planning, allocating, and tasking resources. Normally, he has staff responsibility for ground and air reconnaissance assets, which includes engineers, NBC, and artillery. The echelon intelligence staff officer (G2 or S2) has primary responsibility for ground surveillance systems and special electronics mission aircraft. The commander ensures these two staff elements adopt an integrated combined arms approach to planning, preparing, executing, and assessing reconnaissance.

INTELLIGENCE, SURVEILLANCE, AND RECONNAISSANCE PLAN

13-46. The commander closely integrates reconnaissance missions with other intelligence-collection efforts to ensure that each ISR asset is used to its best advantage. The echelon staff, primarily the intelligence staff officer, identifies gaps in the intelligence available, based on the initial IPB and the situationally dependent CCIR. The IPB process helps determine factors that impact on the reconnaissance effort, such as—

- Avenues of approach that support friendly movement and exploit enemy weaknesses.
- Key terrain, choke points, obstacles, and danger areas.
- Enemy positions, especially flanks that can be exploited.
- Observation points.

The reconnaissance effort and the IPB process are interactive and iterative, each feeding the other. (See FM 2-0 for more information on the intelligence cycle. FM 2-01.3 addresses the IPB process.)

13-47. The intelligence staff officer develops an initial integrated ISR plan to acquire information to help answer those PIR based on available reconnaissance and surveillance assets. The ISR plan assigns specific intelligence acquisition tasks to specific units for action. It integrates surveillance and reconnaissance into the overall intelligence-collection plan.

13-48. The echelon operations staff officer uses the initial ISR plan as the base in preparing the ISR annex to the operation order. The ISR annex provides for the flexible execution of reconnaissance tasks, including providing for adequate command and control, indirect fires, and logistics when completed. (FM 5-0 discusses reconnaissance and the military decision making process.)

RECONNAISSANCE-PULL VERSUS RECONNAISSANCE-PUSH

13-49. In reconnaissance-pull, the commander uses the products of the IPB process in an interactive and iterative way. He obtains combat information from his reconnaissance assets to determine a preferred COA for the tactical situation presented by the factors of METT-TC. In reconnaissance-push, the commander uses the products of the IPB process in an interactive, but not iterative, way with combat information obtained from his reconnaissance assets in support of a previously determined COA. The time available to a commander is normally the chief reason for preferring one method over the other.

13-50. The time required to develop a preferred COA can give the enemy enough time to recover and prepare so that an objective which could be obtained with few casualties one day will cost far more to seize the next day.

There is no available model that a commander can use to determine how much is enough; that determination is part of the tactical art.

RECONNAISSANCE MANAGEMENT

13-51. No single reconnaissance asset can answer every intelligence requirement, and there are rarely enough reconnaissance assets to cover every requirement. The echelon staff uses as mix of reconnaissance management methods, such as cueing, mixing, redundancy, and task organizing, in an attempt to use limited assets most effectively and collect the most critical information with the fewest assets as quickly as possible,.

13-52. ***Cueing* is the integration of one or more types of reconnaissance or surveillance systems to provide information that directs follow-on collecting of more detailed information by another system.** Cueing helps to focus limited reconnaissance assets, especially limited ground reconnaissance assets, which can rarely examine every part of a large area closely. Electronic, thermal, visual, audio, and other technical assets with wide-area surveillance capabilities, often working from aerial platforms, can quickly determine areas of enemy concentration or areas where there is no enemy presence. These assets may cue ground and air reconnaissance assets to investigate specific areas to confirm and amplify information developed by technical assets. For example, joint surveillance target attack radar system (JSTARS) and Guardrail-equipped aircraft can cover large areas and cue ground reconnaissance or UAVs once an enemy force is identified. The commander may dispatch ground reconnaissance or UAVs to verify the information and track the enemy for targeting purposes. Similarly, a ground reconnaissance asset could cue surveillance assets. The key point is to use reconnaissance assets based on their capabilities and use the complementary capabilities of other assets to verify and expand information available.

13-53. ***Mixing* is using two or more different assets to collect against the same intelligence requirement.** Employing a mix of systems not only increases the probability of collection, but also tends to provide more complete information. For example, a JSTARS aircraft may detect and locate a moving enemy tactical force, while the G-2 analysis and control element uses organic and supporting assets to determine its identity, organizational structure, and indications of future plans. Employing a mix of systems is always desirable if the situation and available resources permit. Mixing systems can also help uncover deception attempts by revealing discrepancies in information reported by different collectors.

13-54. ***Redundancy* is using two or more like assets to collect against the same intelligence requirement.** Based on the priority of the information requirement, the commander must decide which NAI justifies having more than one asset covering it. When more than one asset covers the same NAI, a backup is available in the event that one asset cannot reach the NAI in time, the first asset suffers mechanical failure, or the enemy detects and engages the first asset. Redundancy also improves the chances that the required information will be collected.

13-55. To increase the effectiveness and survivability of a reconnaissance asset, the commander may task organize it by placing additional assets under the control of the unit. For example, to conduct an area reconnaissance of possible river crossing sites at extended distances from a division's current location, a ground reconnaissance troop of the division cavalry squadron could be task-organized with a COLT, a signal retransmission element, an engineer reconnaissance element, and a mechanized infantry platoon. The engineers would provide additional technical information on proposed crossing sites; the signal retransmission elements would allow the reconnaissance troop's combat net radios to reach the division tactical command post. The COLT provides additional observation, lazing, and fire coordination capabilities. Last, the infantry platoon would provide additional protection for the reconnaissance troop.

SUSTAINMENT

13-56. Sustaining reconnaissance assets before, during, and after their commitment is a vital part of maintaining the commander's capability to conduct reconnaissance. Because the way that a commander deploys his reconnaissance assets in a given situation depends on the factors of METT-TC, the methods he employs to sustain those assets are equally situationally dependent. He must address them as part of the planning process for each reconnaissance operation.

13-57. Reconnaissance elements frequently operate in locations distant from their organic sustaining base. In this event, reconnaissance elements must either carry a large enough basic load or be task organized with those assets necessary to ensure their sustainment until they can be relieved. With either COA, casualty evacuation remains a problem. An alternative solution would be to plan and coordinate their sustainment from units near their operating locations.

EXECUTING A RECONNAISSANCE

13-58. Reconnaissance can be characterized as either stealthy or aggressive. Depending on how they are employed, scout helicopters and other aerial platforms, as well as mounted and dismounted ground reconnaissance, can be characterized as either stealthy or aggressive.

13-59. A key factor in reconnaissance execution is the time available to conduct the reconnaissance mission. The commander must recognize that he accepts increased risk to both the reconnaissance element and the main body when he accelerates the pace of reconnaissance. This risk can be somewhat offset by employing air reconnaissance and technical means to cover open terrain or areas of lower threat.

13-60. Aggressive reconnaissance is characterized by the speed and manner in which the reconnaissance force develops the situation once it makes contact with an enemy force. A unit conducting aggressive reconnaissance uses both direct- and indirect-fire systems and movement to rapidly develop the situation. Firepower, aggressive exploitation of actions on contact, operations security, and training are required for the unit to survive and accomplish its mission when conducting aggressive reconnaissance. Mounted reconnaissance is normally characterized as aggressive.

13-61. Stealthy reconnaissance emphasizes avoiding detection and engagement by the enemy. It is more time consuming than aggressive reconnaissance. Stealthy reconnaissance takes maximum advantage of covered and concealed terrain and the reduced battlefield signatures associated with systems that typically conduct stealthy reconnaissance, such as dismounted scouts. However, stealth cannot be guaranteed. As a result, units attempting to conduct stealthy reconnaissance must also be drilled to react correctly once the enemy makes contact, and they must have immediate access to supporting fires.

13-62. The commander considers the factors of METT-TC to determine whether to conduct mounted or dismounted reconnaissance. Conditions that may result in a decision to conduct mounted or aerial reconnaissance include—

- Time is limited.
- Detailed reconnaissance is not required.
- Air units are available to perform coordinated reconnaissance with the ground assets.
- The IPB process has provided detailed information on the enemy.
- Terrain is relatively open.
- Environmental conditions permit this type of reconnaissance. Deep snow and muddy terrain greatly hinder mounted reconnaissance.
- Dismounted reconnaissance cannot complete the mission within existing time constraints, while mounted reconnaissance can.

13-63. The following conditions may result in the commander directing a dismounted reconnaissance effort:

- Time is available.
- Detailed reconnaissance is required.
- Stealth is required.
- The IPB process indicates close proximity to enemy positions.
- The reconnaissance force encounters danger areas.
- Restrictive terrain limits the effectiveness of mounted reconnaissance.

FM 3-21.92 describes dismounted patrolling in detail.

13-64. Typically, air reconnaissance operates in concert with ground reconnaissance units. (Friendly ground forces in an area offer additional security to aircrews.) Aviation units can insert surveillance teams at observation posts. Aircraft can observe and provide security on station for extended times using rotation techniques if they have detailed requirements in advance. Dismounting an aircrew member to evaluate bridges, fords, or crossing sights is a last alternative because of the danger to the aircrew and the aircraft. Before resorting to this, the aircrew uses the sophisticated systems on the aircraft to avoid risk and to avoid drawing attention to the area of interest.

13-65. Reconnaissance by fire is a technique in which a unit fires on a suspected enemy position to cause the enemy to disclose his presence by movement or return fire. This technique is appropriate when time is critical and stealthy maneuver to further develop the situation is not possible. The fires may be either direct, indirect, or a combination. The advantage of indirect fire is that it does not give away friendly locations and usually causes the enemy to displace from the impact area. However, reconnaissance by fire may not cause a seasoned or prepared enemy force to react. Reconnaissance by fire is always characterized as aggressive.

13-66. Smoke and battlefield obscuration, fog, rain, and snow all result in reduced visibility. Generally, reconnaissance during limited-visibility conditions takes more time. However, these conditions provide for better stealth and enhance the survivability of reconnaissance assets. A commander frequently employs dismounted reconnaissance patrols at night. These patrols use light amplification and thermal observation devices, electronic surveillance devices, and surveillance radars to compensate for reduced visibility conditions.

13-67. In limited visibility, mounted reconnaissance tends to focus on road networks. The enemy can detect engine and track movement noises of friendly mounted reconnaissance elements at considerable distances at night, which makes them susceptible to ambush. Strict sound and light discipline, along with masking sounds, such as artillery fires, helps a mounted reconnaissance force from being compromised or ambushed.

13-68. High winds, extreme temperature, and loose topsoil or sand may adversely affect aerial reconnaissance. Air reconnaissance units plan their missions in much the same way as ground units. They use the same type of operations graphics and consider the same critical tasks. The air reconnaissance commander organizes his assets to accomplish his mission by considering the same IPB aspects as those associated with ground forces. He focuses on air hazards to navigation and anticipated enemy air defense capabilities. (The effects or weather and atmosphere conditions are discussed in FM 2-01.3.)

RECUPERATION AND RECONSTITUTION OF RECONNAISSANCE ASSETS

13-69. When any small unit is employed continuously for an extensive period of time, it can become ineffective. When this occurs, restoring the unit to an acceptable level of effectiveness may require either recuperation or reconstitution. Recuperation—a short break for rest, resupply, and maintenance—is often sufficient to return the unit to the desired degree of combat effectiveness. Leaders in reconnaissance units probably need more rest than their subordinates. If the recuperation period is extended, it can also be used to conduct refresher training, new equipment training, or any required specialized training for the next mission.

13-70. Units and systems performing reconnaissance are vulnerable to detection, engagement, and destruction by the enemy. When this occurs and the unit can no longer perform its primary mission, the commander must determine whether to reconstitute, by either regenerating or reorganizing the unit. (See FM 4-100.9 for additional information concerning reconstitution.)

13-71. Regenerating a unit requires significant resources. The organization two echelons above the unit being regenerated conducts the procedure. For example, a battalion task force can regenerate its scout platoon. In the regeneration process, the battalion could use a combination of weapon system replacement operations, battle damage assessment and repair, normal replacement operations, and medical returnees to provide the needed resources. These resources, combined with training, could be used to regenerate the scout platoon. Alternatively, the commander could designate one of his line platoons as the task force's new scouts. This approach has significant training implications and requires adjustments to the line platoon's table of organization and equipment.

13-72. A unit commander can reorganize his unit with the approval of the next higher commander. For example, an armored cavalry troop commander could reorganize his two scout and two tank platoons into three platoons containing a mix of scouts and tanks. This approach to reconstitution also requires training time and other equipment resources to ensure the combat effectiveness of the resulting composite organization.

Chapter 14

Troop Movement

Aptitude for war is aptitude for movement.

Napoleon I, *Maxims of War,* 1831

***Troop movement* is the movement of troops from one place to another by any available means.** The ability of a commander to posture his force for a decisive or shaping operation depends on his ability to move that force. The essence of battlefield agility is the capability to conduct rapid and orderly movement to concentrate the effects of combat power at decisive points and times. Successful movement places troops and equipment at their destination at the proper time, ready for combat. The three types of troop movement are administrative movement, tactical road march, and approach march.

METHODS OF TROOP MOVEMENT

14-1. Troop movements are made by foot marches, motor transport, rail, water, air, and various combinations of these methods. The method employed depends on the situation, the size and composition of the moving unit, the distance the unit must cover, the urgency of execution, and the condition of the troops. It also depends on the availability, suitability, and capacity of the different means of transportation. Troop movements over extended distances have extensive logistics considerations.

DISMOUNTED MARCHES

14-2. ***Dismounted marches*, also called foot marches, are movements of troops and equipment, mainly by foot, with limited support by vehicles.** They increase the number of maneuver options available to a commander. Their positive characteristics include combat readiness—all soldiers

CONTENTS

can immediately respond to enemy attack without the need to dismount, ease of control, adaptability to terrain, and their independence from the existing road network. Their limitations include slow rate of movement and increased personnel fatigue—soldiers carrying heavy loads over long distances or large changes in elevation get tired. A unit conducts a dismounted march when the situation requires stealth, the distance to travel is short, transport or fuel is limited, or the situation or terrain precludes using a large number of vehicles. (FM 3-25.18 has more information on the techniques and procedures for conducting dismounted marches.)

MOUNTED MARCHES

14-3. A unit conducts mounted marches when it employs combat and tactical vehicles to move all of its personnel and equipment. Armored and mechanized units routinely conduct mounted marches. The speed of the march and the increased amounts of supplies that can accompany the unit characterize this march method. Heavy maneuver units are normally self-sufficient to conduct mounted marches over short distances. Light maneuver units and most combat support (CS) and combat service support (CSS) units are not completely motorized and need assistance from transportation elements to conduct mounted marches. Considerations for mounted marches over extended distances include:

- The ability of the route network to support the numbers, sizes, and weights of the tactical and combat vehicles assigned to or supporting the unit making the move.
- Available refueling and maintenance sites and crew-rest areas.
- The need for recovery and evacuation assets.

(FM 4-01.30 discusses considerations for mounted marches.)

ARMY AIR MOVEMENTS

14-4. ***Army air movements*** **are operations involving the use of utility and cargo rotary-wing assets for other than air assaults.** The commander conducts air movements to move troops and equipment, to emplace systems, and to transport ammunition, fuel, and other high-value supplies. He may employ air movements as a substitute for a ground tactical movement. Army air movements are generally faster than ground tactical moves. The same general considerations that apply to air assault operations also apply to Army air movements. (See Appendix C for air assault operations. See FM 3-04.113 for additional information concerning air movement.)

14-5. Tactical forces can use rail and water modes to conduct troop movement if they are available within an area of operations (AO). Their use can provide flexibility by freeing other modes of transport for other missions. Their use normally involves a mixture of military and commercial assets, such as defense freight railway interchange railcars pulled by privately owned diesel-electric engines to transport tanks along railroad right of ways from one rail terminus to another. Responsibility for coordinating the use of railroads and waterways resides within the ARFOR headquarters within the theater of operations. (FMs 4-01.41 and 4-01.50 provide additional information concerning these two transportation modes.)

14-6. In cases of tactical necessity, a unit can accelerate its rate of movement by conducting a forced march so that it arrives at its destination quickly. Both heavy and light units can conduct a forced march. Forced marches require speed, exertion, and an increase in the number of hours marched or traveled by vehicles each day beyond normal standards. Soldiers cannot sustain forced marches for more than a short period. In a forced march, a unit may not halt as often nor for as long as recommended for maintenance, rest, feeding, and fuel. The commander must understand that immediately following a long and fast march, his soldiers and combat vehicles experience a temporary deterioration in their physical condition. The combat effectiveness and cohesion of his unit also temporarily decreases. His plan must accommodate the presence of stragglers and address the increased number of maintenance failures.

ADMINISTRATIVE MOVEMENT

14-7. *Administrative movement* is a movement in which troops and vehicles are arranged to expedite their movement and conserve time and energy when no enemy interference, except by air, is anticipated (JP 1-02). The commander conducts administrative movements only in secure areas. Examples of administrative movements include rail and highway movement within the continental United States. Once a unit deploys into a theater of war, administrative movements are the exception, not the norm. Since these types of moves are nontactical, the echelon logistics officer (the G4 or S4) usually supervises the moves. (FM 4-01.40 discusses administrative movement and convoy planning.)

TACTICAL ROAD MARCH

14-8. A *tactical road march* is a rapid movement used to relocate units within an area of operations to prepare for combat operations (FM 3-0). Security against enemy air attack is maintained and the unit is prepared to take immediate action against an enemy ambush, although contact with enemy ground forces is not expected.

14-9. The primary consideration of the tactical road march is rapid movement. However, the moving force employs security measures, even when contact with enemy ground forces is not expected. Units conducting road marches may or may not be organized into a combined arms formation. During a tactical road march, the commander is always prepared to take immediate action if the enemy attacks. (See Figure 14-1, page 14-4.)

ORGANIZATION FOR A TACTICAL ROAD MARCH

14-10. The organization for a tactical road march is the march column. **A *march column* consists of all elements using the same route for a single movement under control of a single commander.** The commander organizes a march column into four elements: reconnaissance, quartering party, main body, and trail party. A brigade conducting a tactical road march is an example of a march column. The subordinate elements of a march column are a march serial and a march unit. **A *march serial* is a major subdivision of a march column that is organized under one commander**

who plans, regulates, and controls the serial. An example is a battalion serial formed from a brigade-size march column. **A *march unit* is a subdivision of a march serial. It moves and halts under the control of a single commander who uses voice and visual signals.** An example of a march unit is a company from a battalion-size march serial.

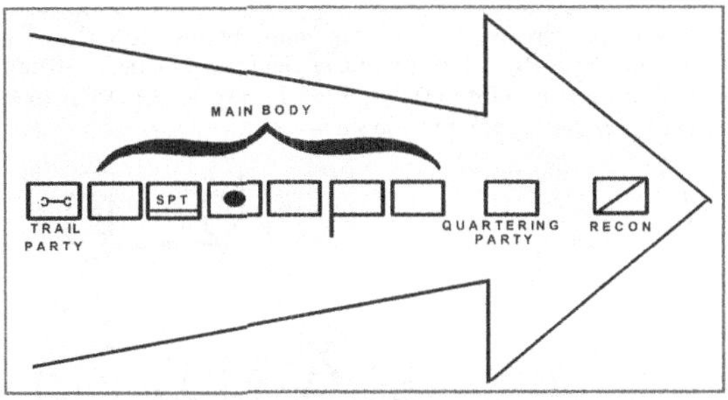

Figure 14-1. Tactical Road March

14-11. A march column provides excellent speed, control, and flexibility, but sacrifices flank security. It provides the ability to deploy forces to the front of the column. The commander uses a march column when speed is essential and enemy contact is unlikely. However, the commander spaces combat support (CS) elements, such as air defense and engineers, throughout the column to protect and support the movement. Reconnaissance elements augmented by engineer, nuclear, biological, and chemical reconnaissance, and other CS assets, as appropriate, conduct a route reconnaissance of the march routes. This reconnaissance confirms and supplements the data obtained from map studies and other headquarters.

14-12. A unit quartering party usually accompanies the reconnaissance effort to the designated assembly area (AA). Unit standing operating procedures (SOP) establish the exact composition of the quartering party and its transportation, security, and communications equipment needs, and its specific duties. The quartering party secures, reconnoiters, and organizes an area for the main body's arrival and occupation. It typically reconnoiters and confirms the tentative locations selected by the commander of its parent element, based on a map reconnaissance. When necessary, the quartering party changes previously assigned unit locations within the AA. The quartering party guides the main body into position from the release point (RP) to precise locations within the AA.

14-13. The main body of the march column consists of the remainder of the unit, including attachments minus the trail party. The trail party is the last march unit in a march column and normally consists of primarily maintenance elements in a mounted march. It maintains communications with the main body. The function of the trail party is to recover disabled vehicles or control stragglers in a dismounted march. If the trail party cannot repair a disabled vehicle immediately, it tows the disabled vehicle and moves its crew and passengers to a unit maintenance collection point (UMCP) located at a secure area near the movement route.

GRAPHIC CONTROL MEASURES

14-14. The commander directing a tactical road march often uses a strip map or overlay to graphically depict critical information about the route to his subordinates. The overlay or strip map should show the route of march, start points (SPs), RPs, checkpoints, critical points (such as bridges), light line, and traffic control points (TCPs). (See Figure 14-2.) Other graphic control measures include AA sand phase lines.

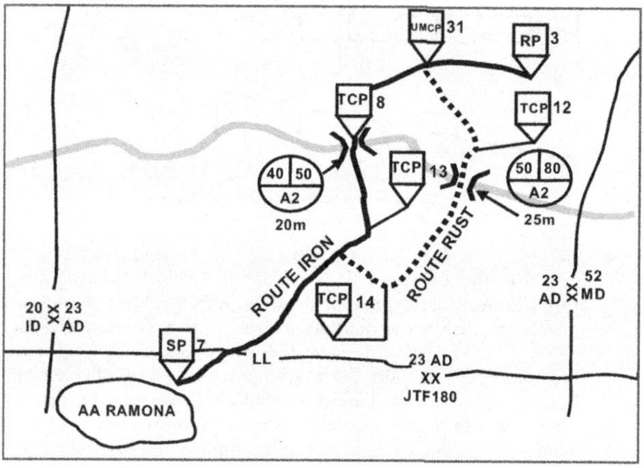

Figure 14-2. Route Control Measures

The *start point* is a location on a route where the marching elements fall under the control of a designated march commander. Figure 14-3 shows SP 7. All routes must have a designated SP and it must be easily recognizable on the map and on the ground, such as a road junction. It must be far enough from the AA to allow units to organize and move at the prescribed speed and interval when they reach the SP.

Figure 14-3. Start Point

14-16. **A *release point* is a location on a route where marching elements are released from centralized control.** Figure 14-4, page 14-6, shows RP 11. Each SP must have a corresponding RP, which must also be easy to recognize on the ground. Marching units do not stop at the RP; instead, as they move through the RP, unit guides meet each march unit and lead it to AAs.

14-17. The commander designates checkpoints along the route to assist marching units in gauging their compliance with the timetable. Also, the movement overlay identifies critical points along the route where interference with movement might occur. The commander positions manned TCPs along the route to prevent congestion and confusion. They may be manned by MPs or unit personnel. These soldiers report to the appropriate area movement control organization when each convoy, march column, and march serial arrives at and completes passage of its location.

Figure 14-4. Release Point

14-18. **A *light line* is a designated phase line, forward of which vehicles are required to use blackout lights during periods of limited visibility.** Commanders at either corps or division echelon establish it based on the risk that the enemy will be able to detect moving vehicles using white light. Figure 14-5 depicts the light line for the 2nd Armored Division as the division rear boundary. (FM 4-01.40 details other march control measures, such as the bridge classification symbols depicted in Figure 14-2.)

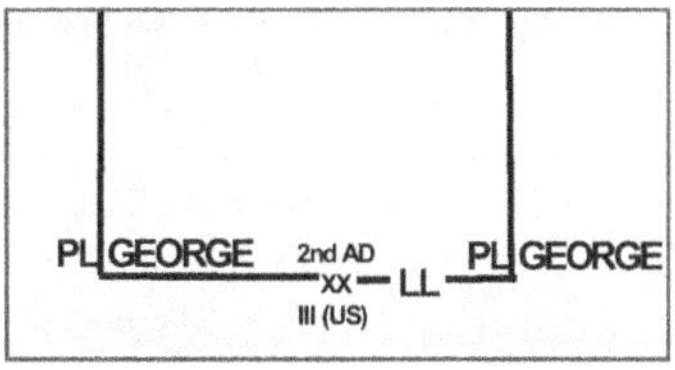

Figure 14-5. Light Line

TACTICAL MARCH TECHNIQUES

14-19. Units conducting tactical road marches employ three tactical march techniques: open column, close column, and infiltration. Each of these techniques uses scheduled halts to control and sustain the road march. The factors of METT-TC require adjustments in the standard distances between vehicles and soldiers. During movement, elements within a column of any length may encounter many different types of routes and obstacles simultaneously. Consequently, parts of the column may be moving at different speeds, which can produce an undesirable accordion-like effect. The movement order establishes the order of march, rate of march, interval or time gaps between units, column gap, and maximum catch-up speed. Unless the commander directs them not to do so for security reasons, march units report when they have crossed each control point. During the move, the commander maintains air and ground security.

Open Column

14-20. In an open column, the commander increases the distance between vehicles and soldiers to provide greater dispersion. The vehicle distance varies from 50 to 100 meters, and may be greater if required. The distance between dismounted soldiers varies from two to five meters to allow for dispersion and space for marching comfort. Any distance that exceeds five meters between soldiers increases the length of the column and hinders control. The open column technique is normally used during daylight. It may also be used at night with infrared lights, blackout lights, or passive night-vision equipment. Using an open column roughly doubles the column's length and thereby doubles the time it takes to clear any given point when compared to a close column. The open column is the most common movement technique because it offers the most security while still providing the commander with a reasonable degree of control. In an open column, vehicle density varies from 15 to 20 vehicles per kilometer. A single light infantry company, with intervals between its platoons, occupies roughly a kilometer of road or trail.

Close Column

14-21. In a close column, the commander spaces his vehicles about 20 to 25 meters apart. At night, he spaces vehicles so each driver can see the two lights in the blackout marker of the vehicle ahead. The commander normally employs a close column for marches during darkness under blackout driving conditions or for marches in restricted terrain. This method of marching takes maximum advantage of the traffic capacity of a route but provides little dispersion. Normally, vehicle density is from 40 to 50 vehicles per kilometer along the route in a close column.

14-22. The dismounted equivalent to the close column is a limited-visibility march. The distance between individual soldiers is reduced to one to three meters to help maintain contact and facilitate control. Limited-visibility marches are characterized by close formations, difficult command and control (C2) and reconnaissance, a slow rate of march, and good concealment from enemy visual observation and air attack.

Infiltration

14-23. The commander dispatches vehicles in small groups, or at irregular intervals, at a rate that keeps the traffic density down and prevents undue massing of vehicles during a move by infiltration. Infiltration provides the best possible passive defense against enemy observation and attack. It is suited for tactical road marches when there is enough time and road space and when the commander desires the maximum security, deception, and dispersion. The disadvantages of an infiltration are that more time is required to complete the move, column control is nearly impossible, and recovery of broken-down vehicles by the trail party is more protracted when compared to vehicle recovery in close and open columns. Additionally, unit integrity is not restored until the last vehicle arrives at the destination, complicating the unit's onward deployment. Infiltration during troop movement should not be confused with infiltration as a form of maneuver as discussed in Chapter 3.

14-24. During extended road marches, halts are necessary to rest personnel, service vehicles, and adjust movement schedules as necessary. The march order or unit standing operating procedures (SOP) regulates when to take halts. In motor movements, the commander schedules short halts for every two to three hours of movement and may last up to an hour. Long halts occur on marches that exceed 24 hours and last no more than two hours. Long halts are not scheduled at night, which allows maximum time for night movement. During halts, each unit normally clears the march route and moves to a previously selected AA to prevent route congestion and avoid being a lucrative target. Units establish security and take other measures to protect the force. Unit leaders receive prompt notification of the time and approximate length of unscheduled halts.

14-25. The commander emphasizes the need to maintain security during halts. Once a unit stops moving, there is a natural tendency for soldiers to let their guard down and relax their vigilance. The commander addresses this problem by explicitly defining unit actions in his SOP for various types of halts, such as maintenance halts, security halts, and unexpected halts.

APPROACH MARCH

14-26. An *approach march* is the advance of a combat unit when direct contact with the enemy is intended (FM 3-0). However, it emphasizes speed over tactical deployment. Both heavy and light forces conduct tactical road marches and approach marches.

14-27. The commander employs an approach march when the enemy's approximate location is known, since it allows the force to move with greater speed and less physical security or dispersion. (See Figure 14-6.) Units conducting an approach march are task-organized before the march begins to allow them to transition to an on-order or a be-prepared mission without making major adjustments in organization. For example, artillery units march within their supported unit's columns, while engineer units are well forward to facilitate mobility. Air defense units may leapfrog short-range and medium-range assets to ensure continuous coverage. The approach march terminates in a march objective, such as an attack position, AA, or assault position, or can be used to transition to an attack. Follow-and-assume and reserve forces may also conduct an approach march forward of a LD.

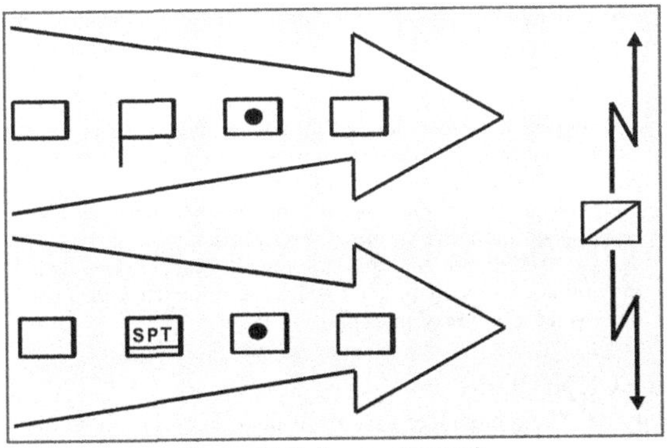

Figure 14-6. Approach March

14-28. Based on the products of his intelligence preparation of the battlefield (IPB) process, the overall commander should assign an AO or an axis of advance in combination with routes to the unit conducting the approach march. These routes, AOs, or axes should facilitate the force's movement and maximize its use of available concealment. Within the approach march, the commander should assign the force conducting the decisive operation and forces conducting each shaping operation separate routes, AOs, or axes of advance unless an individual unit has the task of either follow-and-assume or follow-and-support.

14-29. As the approach march nears areas of likely enemy interference, the commander divides his main body into smaller, less vulnerable columns that move on additional multiple routes or cross-country while continuing to employ security elements. The commander takes advantage of successful reconnaissance and security operations to increase the distance traveled before the main body must transition to a tactical formation. As discussed in Chapter 12, the advance and any flank guards remain within supporting distance of the main body, which stays in these smaller columns to facilitate rapid movement.

14-30. Tactical road marches and approach marches occur within a theater of war when contact with the enemy is possible or anticipated. This style of movement emphasizes tactical considerations such as security and de-emphasizes efficiency and ease of movement. The commander organizes his unit to conduct combat operations in a tactical movement. A unit generally maintains unit integrity throughout its movement. It plans for enemy interference either en route to or shortly after arrival at its destination. Units conducting either a tactical road march or an approach march use formations and techniques consistent with the factors of METT-TC. The unit may conduct them over unsecured routes if there are no friendly forces between the foremost elements of the moving force and the enemy. The echelon operations officer (the G3 or S3) is the primary staff officer responsible for planning these tactical movements, with input from other staff members.

14-31. There are several differences between an approach march and a tactical road march. A force conducting an approach march employs larger security forces because of its greater exposure to enemy attack. Units conducting an approach march arrange their systems into combined arms organizations. An approach march allows the commander to disperse his task-organized force into a tactical formation in unrestricted terrain without being constrained to existing roads and trails. On the other hand, road marches can organize their columns for administrative convenience; for example, vehicles of similar type, speed, and cross-country capabilities move together. Units conducting an approach march establish appropriate tactical intervals between vehicles; they do not normally employ a close column. They also use more routes than units conducting road marches.

MOVEMENT TECHNIQUES

14-32. The commander uses the combat formations described in Chapter 3 in conjunction with three movement techniques: traveling, traveling overwatch, and bounding overwatch. Figure 14-7 on page 14-10 shows when a unit is most likely to use each technique.

14-33. Movement techniques limit the unit's exposure to enemy fire and position it in a good formation to react to enemy contact. The commander selects the appropriate movement technique based on the chance of enemy contact. While moving, individual soldiers and vehicles use the terrain to protect themselves anytime enemy contact is possible or expected. They use natural cover and concealment to avoid enemy fires. The following rules apply to soldiers and vehicle crews using terrain for protection:

- Do not silhouette yourself against the skyline.
- Cross open areas quickly.
- Do not move directly forward from a concealed firing position.
- Avoid possible kill zones because it is easier to cross difficult terrain than fight the enemy on unfavorable terms.
- Avoid large, open areas, especially when they are dominated by high ground or by terrain that can cover and conceal the enemy.
- Take active countermeasures, such as using smoke and direct and indirect fire, to suppress or obscure suspected enemy positions.

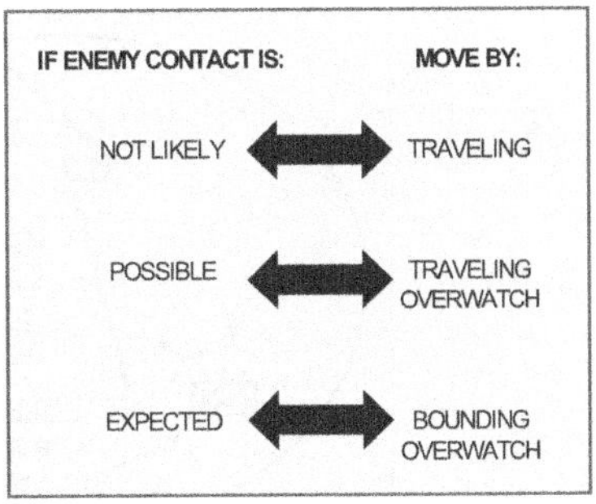

Figure 14-7. Movement Techniques

TRAVELING

14-34. The commander uses the traveling movement technique when speed is necessary and contact with enemy forces is not likely. All elements of the unit move simultaneously. The commander or small-unit leader locates where he can best control the situation. Trailing elements may move in parallel columns to shorten the column and reaction time. (See Figure 14-8.)

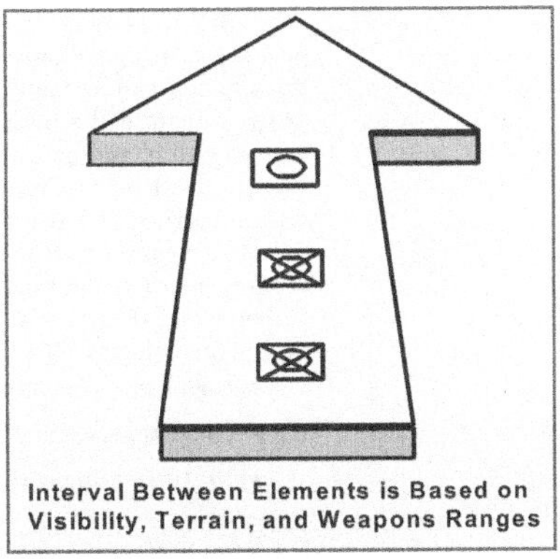

Figure 14-8. Traveling

TRAVELING OVERWATCH

14-35. The commander uses the traveling overwatch movement technique when contact with enemy forces is possible, but speed is important. The lead element is continuously moving, while the trailing elements move at variable speeds, sometimes pausing to overwatch movement of the lead element. (See Figure 14-9.) The trailing elements key their movement to the terrain, overwatching from a position where they can support the lead element if it engages the enemy. The trailing elements overwatch from positions and at distances that will not prevent them from firing or moving to support the lead element.

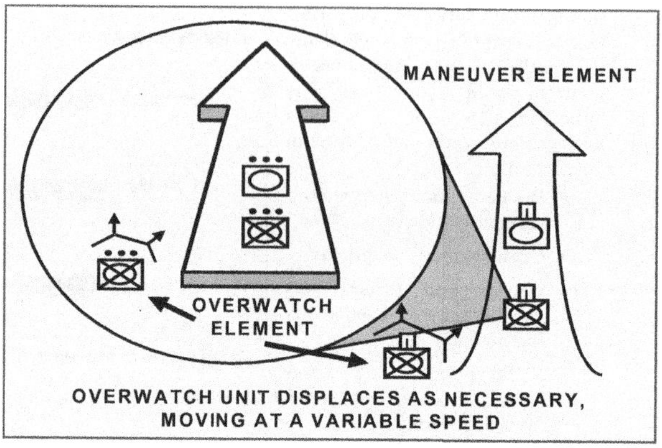

Figure 14-9. Traveling Overwatch

BOUNDING OVERWATCH

14-36. The commander uses the bounding overwatch movement technique when he expects to make contact with enemy forces. There are two variations of this technique: alternate bounds and successive bounds. In both cases, the overwatching elements cover the bounding elements from covered, concealed positions with good observation and fields of fire against possible enemy positions. They can immediately support the bounding elements with maneuver or fires alone if the bounding elements make contact. Unless they make contact en route, the bounding elements move via covered and concealed routes into the next set of support-by-fire positions. The length of the bound is based on the terrain and the range of overwatching weapons. The commander can use the uncommitted part of the force whenever he feels it is needed as part of an immediate and controlled reaction to any threat to the bounding force. In bounding overwatch, all movement keys on the next support-by-fire position, which should offer at least some of the following advantages:

- Cover and concealment.
- Good observation and fields of fire.
- Protection for stationary weapon platforms.

14-37. If the unit uses alternate bounds, the lead element moves forward, halts, and occupies a support-by-fire position that is covered at all times by the rear overwatching element. That former rear overwatching element advances past the former lead element and takes an overwatch position. The initial lead element then advances past the initial trail element and occupies a new support-by-fire position. One element moves at a time. This method is usually more rapid than successive bounds. (See Figure 14-10.)

14-38. If the unit uses successive bounds, the lead element, covered by the trail element, advances and occupies a support-by-fire position. The trail element advances to a support-by-fire position abreast of the lead element and halts. The lead element moves to the next position and the move continues.

Only one element moves at a time, and the trail element avoids advancing beyond the lead element. (See Figure 14-11.)

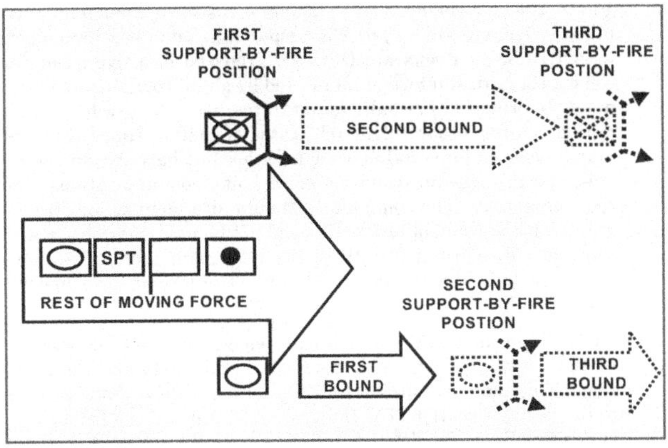

Figure 14-10. Bounding Overwatch—Alternate Bounds

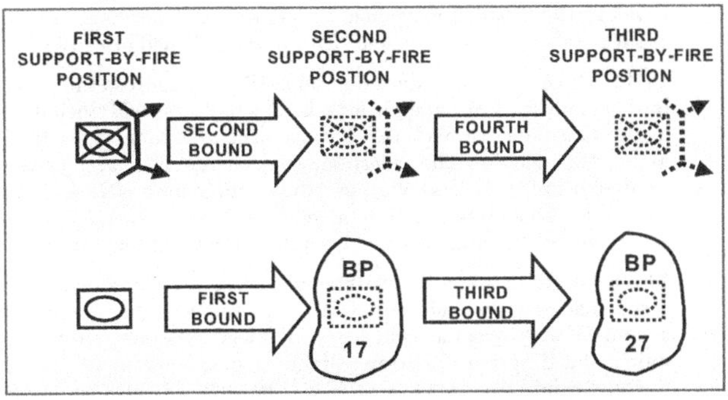

Figure 14-11. Bounding Overwatch—Successive Bounds

PLANNING A TROOP MOVEMENT

14-39. The commander plans, prepares, executes, and assesses troop movements to ensure the organized and uninterrupted flow of tactical units throughout the AO. The objective of a successful move is for the unit to arrive at its destination in a condition suitable to its probable employment. The goal of all movement planning is to retain flexibility to execute a variety of plans to meet ever-changing conditions. The commander ensures that his unit's movement SOP contain specifics, and he conducts rehearsals to ensure that his soldiers and subordinate leaders understand them. The SOP should use a standard task organization to simplify planning, provide flexibility, and allow greater responsiveness. Such SOP allow for smoother cooperation by establishing habitual relationships between the unit's subordinate elements and outside supporting elements.

14-40. The movement order is the end result of the commander's planning process. The movement order is prepared as an annex to an operations order (OPORD) or as a separate OPORD. Prepared in a five-paragraph format, it consists of critical information needed by a unit to plan and execute the movement. Information normally found in the movement order includes the destination, routes, orders of march, rates of march, times that each serial (or march element for serial movement orders) will arrive and clear its SP, intervals, speeds, scheduled maintenance halts, communications, and location of the commander. The commander should also identify logistics sites and services in his movement order. He need not include information and procedures contained in movement SOP in the movement order. The movement order should include a strip map or overlay. (The format for a movement order is covered in FM 5-0.)

14-41. The commander bases his movement order on the best available information on the enemy, terrain, weather, unit capabilities, and civil considerations. This plan establishes how the unit will move from its current location to the desired location. The integration of and support from combat and CS—such as artillery, air defense, intelligence, military police, and engineers—are critical for a successful tactical movement. The commander's operations staff develops the detailed movement order, with the assistance of the commander's logistics staff, in accordance with his established priorities.

14-42. The movement order and unit SOP must address the possibility of ambushes, indirect fires, and air attacks. A small-unit SOP includes drills for reacting to these circumstances. Passive measures to mitigate the effects of an air attack include route selection, vehicle intervals, and movement during limited visibility. In case of attack, the commander has an evacuation plan for casualties. This plan takes into account SOP items, such as using combat lifesavers and dispersing medical evacuation assets throughout the convoy.

14-43. For units that are not 100-percent mobile in organic vehicles, such as a corps headquarters and many CSS units, the commander can either conduct a shuttle with organic vehicles or request assistance from transportation units. Shuttling requires transporting troops, equipment, and supplies by a series of round trips with the same vehicles. It may also be performed by carrying successive parts of a load for short distances while the remaining soldiers continue on foot.

14-44. The higher headquarters logistics staff normally coordinates the provision of logistics support to moving units, although units carry sufficient fuel and lubricants in their unit trains to conduct local movements. In coordination with the engineers, the logistics staff ensures that routes are adequate to support the movement of the types and numbers of vehicles and supplies projected for movement. The commander must be aware of the load-carrying capability of each route and the distances over which forces can be supported. His logistics operators determine if any logistics assets should displace to support the mission. The commander also establishes halts for refueling as part of his movement plan. Halt times should be long enough and locations large enough to allow the entire march unit to refuel.

14-45. The simplest troop movement scenario to plan and conduct is one where the commander directing the movement controls the entire AO. In this situation, he can use his normal C2 system. The headquarters ordering the tactical road march schedules the movement times and approves the routes, while its movements control organization allocates the required space and time on the approved routes. If the movement results in a unit going outside its parent headquarters' AO, coordination through various movement control centers is required. Otherwise, a higher headquarters must plan and control the movement.

14-46. Whenever possible, the commander should use multiple routes to move his unit. This reduces the length of columns, the vulnerability to enemy air attack, and the amount of time the routes are not available to other units. Multiple routes provide the commander with the flexibility to react to unexpected situations and permit more rapid concentration of combat power. The two primary disadvantages of using multiple routes are difficulty in exercising C2, and the unit may not have enough resources to provide logistic and maintenance support on all routes.

14-47. The echelon transportation officer uses route classification components, such as route widths, route types, military load classifications, overhead clearance, route obstructions and special conditions, as he determines his traffic circulation plan. A supporting engineer terrain detachment provides his the majority of this information. Engineer reconnaissance obtains necessary information not contained in existing databases. FM 3-34.212 and FM 3-34.310 defines these components and describes how to use them.

14-48. The staff depicts the echelon traffic circulation plan on overlays using transportation control measures. The traffic circulation plan takes into account—

- The most restrictive route features and route designations.
- Direction of movement over each route.
- Location of boundaries, units, highway regulation points, TCPs, and principal supply points.
- Major geographic features and light lines, if applicable.
- Routes designated for one-way traffic.
- Separate routes for CSS and tactical units.
- Current data on traffic regulation and control restrictions, obstructions, detours, defiles, capacities, surface conditions, and enemy activities that affect the highway net.

From information contained in the traffic circulation plan, a traffic control plan is prepared—usually by the provost marshal—from information contained in the traffic circulation plan. The traffic control plan normally is prepared in the form of an overlay. The commander primarily uses available aviation, movement regulating teams, and MP units to assist in traffic control, but can assign this mission to other units, such as battalion scout platoons.

PREPARING A TROOP MOVEMENT

14-49. Reconnaissance precedes unit movement. Before a unit starts any march, a reconnaissance element from that unit should reconnoiter the route from its current location to the SP and determine how long it will take the unit to reach the SP. This reconnaissance effort continues beyond the start point and carefully examines the route's trafficability, including the impact of weather, such as ice, snow, and rain. This reconnaissance should also include alternative routes and choke points, such as defiles, bridges, and fords, which could slow the march. This reconnaissance effort complements map and technical reconnaissance and provides the commander with important information about the terrain, obstacles, and potential enemy forces within his AO. He can then take steps to establish TCPs at critical locations along the route or mark the route where it becomes confusing.

14-50. A quartering party often accompanies reconnaissance elements to mark routes and battle positions. The party may also secure new positions with observation posts or limited forces until the unit conducting the movement arrives.

14-51. The unit begins a tactical movement, such as a road march, fully supplied. The unit should refuel at every opportunity, such as at halts and on arrival at the final destination. The transportation of fuel and the security of existing stockpiles are major factors in any mounted road march. The commander may choose to conduct a refuel on the move (ROM) to extend the range of his vehicles. Refuel on the move is a technique in which the commander positions tankers just off the route of the march to refuel combat and tactical vehicles rapidly, but only in the previously established quantities necessary to extend their range to the desired length.

14-52. Based on the form of movement selected and the march and movement techniques adopted, the commander may have to pre-position CSS assets to conduct rapid and efficient refueling and resupply. Generally, a column formation is the easiest movement technique to support. Any other formation requires increased logistics planning. Night movements require special preparation because not all soldiers have night-vision devices. These special preparations include marking vehicles and equipment for easy identification by friendly forces and repositioning vehicles and soldiers closer together so they can detect each others' movement.

EXECUTING A TROOP MOVEMENT

14-53. A unit's ability to execute movement depends on its march discipline and ability to maintain required movement standards and procedures as prescribed by its movement SOP and movement order. This includes staying on the given route and maintaining start, passage, and clear times. March discipline is absolutely essential throughout the movement. Any deviation from the movement order may interfere with the movements of other units and may have serious consequences. However, march discipline can only be maintained when the plan matches conditions and the unit's ability to move.

14-54. The strength and composition of the moving unit's security elements vary, depending on the factors of METT-TC. The commander employs his organic assets and any supporting security assets to protect his forces from enemy activities. He positions them to the front, rear, and flanks of his formations while moving and at the halt to provide all-around security for the main body. He can also enhance security by adopting a march formation and movement technique that facilitate applying combat power in the direction he expects to make contact with the enemy.

14-55. Higher-echelon CSS organizations may support some tactical movements. When the situation permits, CSS organizations establish maintenance, ambulance exchange, and supply points along the route. While procedures, amounts, and types of external support vary among major commands, each logistics organization ensures that these sites are operational at the designated times and locations. External CSS along the route may include aeromedical evacuation, maintenance, water, and POL. Maintenance sites generally consist of UMCPs where disabled vehicles can be moved for limited maintenance and Class IX supplies. Vehicles unable to continue the movement remain at a UMCP and join their parent organization when repaired. The troop movement is complete when the last march unit clears the RP.

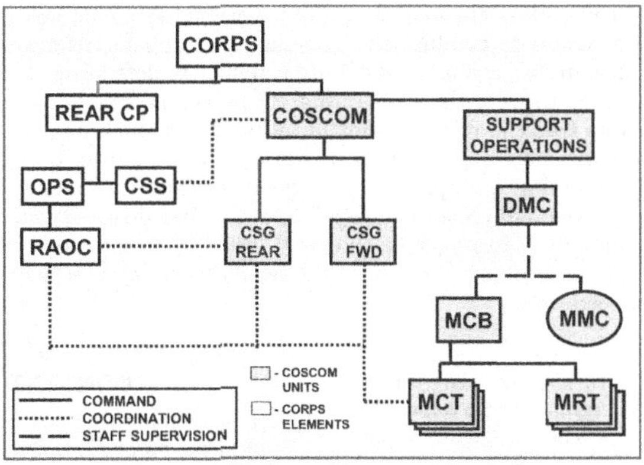

Figure 14-12. Corps Movement Control

MOVEMENT CONTROL

14-56. *Movement control* is the planning, routing, scheduling, and control of personnel and cargo movements over lines of communications (JP 1-02). It is a continuum that involves coordinating and integrating logistics, movement information, and programs that span the strategic, operational, and tactical levels of war. The balancing of requirements against capabilities and assigning resources based on the commander's priorities guides the conduct of movement control. Movement control gives the commander the ability to deconflict the movement of units—troop movement—and the distribution of supplies and services inherent in the provision of CSS. It is not a simple system as is shown in Figure 14-12 by the number of different agencies involved in corps movement control. (FM 4-01.30 discusses movement control.)

Chapter 15

Relief in Place

... the necessity for conservation of the fighting power of the troops requires provision for the periodic relief of units in line.

FM 100-5, *Field Service Regulations: Operations*, 22 May 1941

A *relief in place* is a tactical enabling operation in which, by the direction of higher authority, all or part of a unit is replaced in an area by the incoming unit. The directing authority transfers the responsibilities for the mission and the assigned area of operations (AO) from the replaced elements to the incoming unit. A commander conducts a relief in place as part of a larger operation, primarily to maintain the combat effectiveness of committed units. The higher headquarters directs when and where to conduct the relief and establishes the appropriate control measures. Normally, the unit relieved is defending. However, a relief may set the stage for resuming offensive operations. A relief may also serve to free the relieved unit for other tasks, such as decontamination, reconstitution, routine rest, resupply, maintenance, or specialized training. Sometimes, as part of a larger operation, a commander wants the enemy force to discover the relief, because that discovery might cause it to do something in response that is prejudicial to its interest, such as move reserves from an area where the friendly commander wants to conduct a penetration.

CONTENTS

15-1. There are three techniques for conducting a relief: sequentially, simultaneously, or staggered. A sequential relief occurs when each element within the relieved unit is relieved in succession, from right to left or left to right, depending on how it is deployed. A simultaneous relief occurs when all elements are relieved at the same time. A staggered relief occurs when the commander relieves each element in a sequence determined by the tactical situation, not its geographical orientation. Simultaneous relief takes the least time to execute, but is more easily detected by the enemy. Sequential or staggered reliefs can take place over a significant amount of time.

15-2. A relief is either deliberate or hasty, depending on the amount of planning and preparations. The major differences are the depth and detail of planning and, potentially, the execution time. Detailed planning generally facilitates shorter execution time by determining exactly what the commander believes he needs to do and the resources needed to accomplish the mission. Deliberate planning allows him and his staff to identify, develop, and coordinate solutions to most potential problems before they occur and to ensure the availability of resources when and where they are needed.

ORGANIZATION OF FORCES

15-3. Both units involved in a relief in place should be of similar type—such as mounted or dismounted—and task organized to help maintain operations security (OPSEC). The relieving unit usually assumes as closely as possible the same task organization as the unit being relieved. It assigns responsibilities and deploys in a configuration similar to the relieved unit.

15-4. The relieving unit establishes advance parties to conduct detailed coordination and preparations for the operation, down to the company level and possibly to the platoon level. These advance parties infiltrate forward to avoid detection. They normally include the echelon's tactical command post, which co-locates with the main headquarters of the unit being relieved. The commander may also attach additional liaison personnel to subordinate units to ensure a smooth changeover between subordinate units.

CONTROL MEASURES

15-5. Control measures associated with a relief in place are generally restrictive to prevent fratricide. As a minimum, these control measures include the AO with its associated boundaries, battle positions, contact points, start points, routes, release points, assembly areas (AAs), fire support coordinating measures, and defensive fire coordination measures, such as target reference points and engagement areas. (See Figure 15-1.) Expanded discussions of all these control measures appear elsewhere in this manual. A commander may use any control measure he feels is necessary to conduct a relief in place.

PLANNING A RELIEF IN PLACE

15-6. Once ordered to conduct a relief in place, the commander of the relieving unit contacts the commander of the unit to be relieved. The co-location of unit command posts also helps achieve the level of coordination required. If the relieved unit's forward elements can defend the AO, the relieving unit executes the relief in place from the rear to the front. This facilitates movement and terrain management.

15-7. In a deliberate relief, units exchange plans and liaison personnel, conduct briefings, perform detailed reconnaissance, and publish orders with detailed instructions. In a hasty relief, the commander abbreviates the planning process and controls the execution using oral and fragmentary orders. In both cases, the relieved unit designates liaison personnel from its combat, combat support (CS), and combat service support (CSS) elements to remain with the relieving unit until completing the necessary plans. The relieving unit receives current intelligence, operations, and logistics information from the unit being relieved, as well as from common higher headquarters, adjacent units, and subordinate elements. The complexity of a relief in place requires extensive liaison and reconnaissance. Exchanging information about the enemy and civilian situations, friendly dispositions, terrain analysis, and fire support and obstacle plans, coupled with reconnaissance, helps the relieving commander plan and execute his mission.

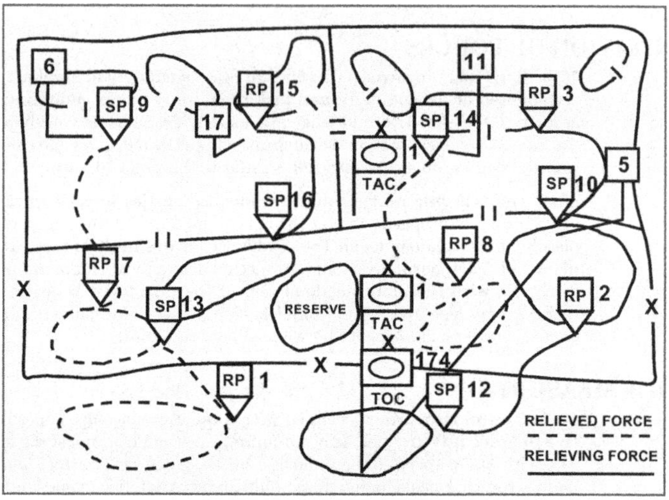

Figure 15-1. Overlay of a Brigade Relief in Place

15-8. The relief is a tactically vulnerable operation. The units involved must concentrate on security while preparing for and executing the operation. The intent of the operation is to complete the relief without discovery by the enemy. Consequently, commanders typically plan reliefs for execution during periods of reduced visibility, such as night or fog. Concealment of the relief from the enemy is a primary concern when the unit is conducting the relief as part of an economy of force measure to free forces for other operations. The enemy should perceive only one unit's command structure in operation—that of the unit being relieved—until completing the operation. This requires a detailed knowledge of friendly vulnerabilities. A counterintelligence assessment of enemy collective capabilities directed against the friendly forces involved in the relief can provide that detailed knowledge.

15-9. Generally, as soon as the mutual higher headquarters issues the warning order, the commander of the relieving unit co-locates one of his command posts with the command post of the unit being relieved. As a minimum, he establishes communications and liaison with that unit. The warning order designates the time of relief, relieving and relieved units, and sequence of events. It specifies the future missions of the relieved force, route priorities, any restrictions on advance parties, any extraordinary security measures, and the time and place for issuing the complete order.

15-10. During a relief, commanders and leaders from the relieving unit should conduct reconnaissance of the area for which they will assume responsibility. This leaders' reconnaissance should include the lowest-echelon leader allowed by the tactical situation. The reconnaissance should focus on the route into the position the unit is to occupy, the positions themselves, the current disposition of the unit being relieved, and any obstacles that could affect troop movement.

15-11. The two commanders must decide on a time or an event that initiates the passage of command. This allows the smooth transition of command and control from one commander to another. Normally, this occurs when the frontline subordinate commanders have assumed responsibility for their respective AOs and the incoming commander has sufficient communication facilities in operation to control the operation. Regardless of their parent organization, all units in the AO come under the operational control of the AO commander if the AO comes under attack or when a specified event occurs during the relief.

15-12. The fire support coordinators coordinate fire support coordinating measures and identify those artillery and other fire support units that are available to support the relief. The relieving unit adopts the fire plan of the unit being relieved. The fire support assets of both units support the relief. This is critical if the enemy detects the relief and tries to exploit the situation. Units plan their fires to deceive the enemy and expedite the relief. Units should maintain normal activity patterns. For example, a unit should continue to expend the same average number of artillery rounds per day during the relief that it expended prior to the initiation of the relief. The commander should not relieve fire support and other CS and CSS units at the same time as the maneuver units they support. The commander relieves these organizations at other times.

15-13. The relief plan must specify the method to use in relieving artillery units. If terrain allows, relieving artillery units should not occupy previously used firing positions. Instead, relieving firing units should establish firing positions nearby those firing positions of the relieved unit and carefully integrate their fire with that of the relieved unit. Occupying firing positions at night or during periods of poor visibility enhances OPSEC.

15-14. Priority of the air defense effort is to protect identified choke points, battle positions, routes to conduct the operation, and AAs. The air defense assets of both units support the relief. The air defense unit supporting the relieving force coordinates with the replaced force's supporting air defense unit. This coordination covers, but is not limited to, air intelligence preparation of the battlefield, rules of engagement, current air activity, present fire unit positions, Army airspace command and control information, the operation plan, logistics, and communications. Higher-echelon and joint air defense organizations may also support the relief. Provisions to obtain local air superiority reduce the vulnerability of the forces during the relief in place when the units involved cannot avoid congestion on the ground.

15-15. The relieving unit verifies the obstacle records of the unit being relieved. Handover of obstacles is a complex procedure. Initially, the engineer priority is on mobility to get the relieving unit into the AO. It focuses on those routes and lanes leading into the AO. Once the relief occurs, priority of the mobility and survivability effort transitions to support the relieving unit's continuing mission. The commander may require his engineers to assist with survivability tasks to support the relieving force.

15-16. Force-intermingling inherent in a relief, places an increased burden on command and control systems. The consequences of mutual interference between the units and the complexity associated with such areas as traffic control, fire support coordination, obstacle plans, and communications require close coordination between all headquarters involved. Establishing early liaison between the stationary and the relieving forces is critical.

15-17. The relieving unit is responsible for all sustaining operations. As the support elements of the unit being relieved displace, they leave the relieving unit supply stocks according to previously coordinated arrangements. If the units conducting the relief have different modified tables of organization and equipment (MTOEs), mission analysis must be conducted to determine how the relieving unit will meet all of its responsibilities and what weapon systems will be used. The unit logistics staff must determine any special support requirements the relieving unit will have and address supporting those requirements with the available supporting organizations. The unit logistics staff ensures that both commanders know of any CSS constraints that might affect the relieving unit. The two units' rear command posts also co-locate and a single headquarters coordinates traffic movement into and out of the AO.

PREPARING A RELIEF IN PLACE

15-18. The commander conceals the relief from the enemy for as long as possible. At the first indication that a relief is necessary, which is usually the warning order for the relieving unit, both the relieved unit and the relieving unit review their OPSEC plans and procedures. Commanders may use deception measures when conducting a relief in place to maintain secrecy. To maintain security during the relief in place, the relieving unit makes maximum use of the relieved unit's radio nets and operators. Both units involved in the relief operate on the command frequencies and encryption variables of the relieved unit at all levels. The relieved unit's signal officer is in charge of communications throughout the relief operation.

15-19. To enhance security, commanders impose light and noise discipline and electromagnetic emission control measures, such as radio silence or radio-listening silence. In joint and multinational operations, the senior commander specifies the frequency bands and equipment types affected. Radio silence is a condition when the commander turns off all or specific radio equipment. Radio-listening silence is a situation in which combat net radios remain turned on and monitored, with strict criteria governing when a station on the radio network is allowed to break silence. An example of radio-listening silence would be, "Maintain radio listening silence until physical contact with the enemy is made."

15-20. The units conduct rehearsals to discover any weaknesses in the plan and familiarize all elements of both forces with the plan. Finding time for rehearsals requires commanders and staffs to focus on time management.

15-21. Reconnaissance elements of the relieving unit precede its movement with a route reconnaissance to the AA. They conduct reconnaissance of the routes leading from the AAs to the positions of the unit being relieved. The commander of the relieving unit normally conducts a leader's reconnaissance before starting the operation.

15-22. The commander must allocate time to construct individual vehicle fighting positions if a heavy unit is relieving a light unit. In a similar fashion, preparations for an armor heavy unit to relieve a mechanized infantry heavy unit must include expanding individual vehicle fighting positions to accommodate the larger tanks.

15-23. While the units involved plan, prepare, and execute the relief in place, their common higher headquarters and other units continue actions to mask the relief. These include using demonstrations, feints, smoke, and harassing and interdiction fires. The common higher headquarters executes operations to attack and disrupt the enemy's uncommitted and reserve forces during the relief. Its intent is to fix or distract the enemy so that he does not detect or interfere with the relief.

EXECUTING A RELIEF IN PLACE

15-24. In situations where the commander desires to conceal the relief from the enemy, such as during a sequential or staggered relief, the relieving unit may occupy the same positions as the unit it relieves. Alternatively, it may establish more favorable positions within the vicinity of the relieved unit's location. Occupying different positions makes early discovery by the enemy more likely. Any increase in activity in forward positions can reveal the relief to the enemy. Friendly intelligence, surveillance, and reconnaissance systems attempt to detect if the enemy can discover the relief before its completion.

15-25. The enemy can usually detect a relief effort because of the increased activity resulting from the movement of soldiers and equipment out of position by the relieved unit and into position by the relieving unit. Additionally, after any period of combat, there are differences in the types and amount of equipment between the relieving unit and the relieved unit, even if they have the same MTOEs. These differences can also reveal the relief to the enemy. The two units establish guidelines for exchanging compatible equipment and supplies to limit these differences. In addition, it may be necessary to exchange certain weapons, supplies, equipment, and occasionally, vehicles between units. When major differences in the number of combat systems between the units exist—for example, a tank-heavy task force relieves a mechanized infantry-heavy task force—inoperable equipment or visual simulators may assist in hiding the change of units.

15-26. In a simultaneous relief, the relieving unit begins moving from its current location to AAs in the AO of the unit being relieved. Once the relief begins, all elements involved execute the relief as quickly as possible. Both units are vulnerable to enemy attack because of the concentration, movement, and intermingling of forces in a simultaneous relief. Any unnecessary delay during execution provides the enemy additional time to acquire and engage the forces involved. All units in the AO come under the operational control of the relieving unit commander at the time or triggering event previously established by the plan for the operation.

15-27. As the first relieving element arrives from the AA to assume the position, it establishes a screen of the relieved unit's positions as the tactical situation permits. The remainder of the relieving unit moves forward to positions behind the unit being relieved. The relieving unit may use the relieved unit's alternate and supplementary defensive positions to take advantage of any previous defensive preparations. At the previously established time or event, passage of command takes place. At that point, if possible, the commander of the relieving unit informs all units involved in the relief of the passage of command.

15-28. The relieved unit continues to defend. The relieving unit's advance parties coordinate procedures for the rearward passage of the relieved unit. On order, the relieved unit begins withdrawing through the relieving unit and moves to AAs. Crew-served weapons are usually the last elements relieved after exchanging range cards. The relieving unit replaces them on a one-for-one basis to the maximum extent possible to maintain the illusion of routine activity. The relieved unit's CS and logistics assets assist both the relieved unit and the relieving unit during this period.

15-29. A relief does not normally require artillery units to relieve weapon system for weapon system unless the terrain limits the number of firing positions available. Generally, the relieved unit's artillery and other fire support assets remain in place until all other relieved elements displace and are available to reinforce the fires of the relieving unit in case the enemy tries to interfere. If the purpose of the relief is to continue the attack, the artillery of both forces generally remains in place to support the subsequent operation.

15-30. Multiple main supply routes that allow only one-way traffic can simplify the forward and rearward movement of both units. The relieving unit's rear command post controls both units' military police and any other traffic management assets. (The main command post performs these functions if the echelon does not have a rear command post.) The commander uses these assets to help control unit and convoy movement on lines of communications, main supply routes, and movement routes throughout his AO.

15-31. In the future, it is likely that conflicts will involve the relief of an allied or coalition force. The commander should consider the following additional points when such reliefs occur:

- Dissimilar unit organizations may require special adjustments in assigned areas.
- Control of fire support may require special liaison.
- Language difficulties may require an increased use of guides and translators.
- Using relieved unit communications requires special signal arrangements and additional operators.
- Ammunition and equipment incompatibility may make exchanging assets more difficult.
- Impact of civilians on the operations.

Chapter 16

Passage of Lines

The principal task involved in a passage of lines is the preparation for continuing the attack.

FM 1005, *Field Service Regulations: Operations*, 22 May 1941

***Passage of lines* is a tactical enabling operation in which one unit moves through another unit's positions with the intent of moving into or out of enemy contact.** A commander conducts a passage of lines to continue an attack or conduct a counterattack, retrograde security or main battle forces, and anytime one unit cannot bypass another unit's position. The conduct of a passage of lines potentially involves close combat. It involves transferring the responsibility for an area of operations (AO) between two commanders. That transfer of responsibility usually occurs when roughly twothirds of the passing force has moved through the passage point. If not directed by higher authority, the unit commanders determine—by mutual agreement—the time to pass command. They disseminate this information to the lowest levels of both organizations.

16-1. The commander's reasons for conducting a passage of lines are to—

- Sustain the tempo of an offensive operation.
- Maintain the viability of the defense by transferring responsibility from one unit to another.
- Transition from a delay or security operation by one force to a defense.
- Free a unit for another mission or task.

The headquarters directing the passage of lines is responsible for determining when the passage starts and finishes.

16-2. A passage of lines occurs under two basic conditions. **A *forward passage of lines* occurs when a unit passes through another unit's positions while moving toward the enemy. A *rearward passage of lines* occurs when a unit passes through another unit's positions while**

CONTENTS

moving away from the enemy. Ideally, a passage of lines does not interfere with conducting the stationary unit's operations.

ORGANIZATION OF FORCES

16-3. A unit may participate in a passage of lines as either the passing or stationary force. Except for co-locating command posts and providing for guides by the stationary force, conducting a passage of lines does not require a special task organization. Both the passing force and the stationary force maintain their previous combat organization during the passage. Usually, if the stationary unit has the capability, it is responsible for conducting operations against uncommitted enemy forces. However, operations directed against uncommitted enemy forces may be the responsibility of a higher echelon, depending on the echelon at which the passage takes place.

16-4. A forward passing unit's order of march is generally reconnaissance and security elements first. The ground combat force move next, followed by combat support (CS) and combat service support (CSS) units. The commander integrates his artillery, air defense, and engineers into the order of march in accordance with the factors of METTTC. The passing unit reverses this order of march in a rearward passage of lines. The stationary unit normally provides the moving unit with guides to expedite the passage. Attack helicopters and air cavalry are useful in providing security.

CONTROL MEASURES

16-5. Control measures associated with a passage of lines are generally restrictive to prevent fratricide. As a minimum, they include the AO, assembly areas (AAs), attack positions, battle handover line (BHL), contact points, passage points, passage lanes, routes, gaps, phase lines, and recognition signals. The headquarters directing the passage designates or recommends contact points, passage lanes, AAs, routes, and start and end times for the passage. The commander may also use start points, release points, fire support coordinating measures, such as coordinated fire lines (CFLs), and other control measures as necessary to conduct this task. (See Figure 16-1, page 16-2.) Unless the higher headquarters of the two units establishes the necessary graphic control measures, the stationary unit establishes them for the passage. However, the stationary unit commander must coordinate them with the passing unit commander. The stationary unit establishes these measures because it owns the terrain, it knows where the obstacles are, and it knows the tactical plan. If the control measures dictated by the higher headquarters are not sufficient—because they do not contain enough passage points, lanes, and so forth—the two units can agree to add the necessary measures.

16-6. **A *passage point* is a specifically designated place where the passing units pass through the stationary unit.** The location of this point is where the commander wants subordinate units to physically execute a passage of lines. In a forward passage of lines, the passage point marks the location where the passing unit is no longer bound by the restrictions placed on it by the stationary force. On the other hand, in a rearward passage of lines, the passage point marks the location where the stationary unit can restrict the movement and maneuver of the passing force. Between the contact

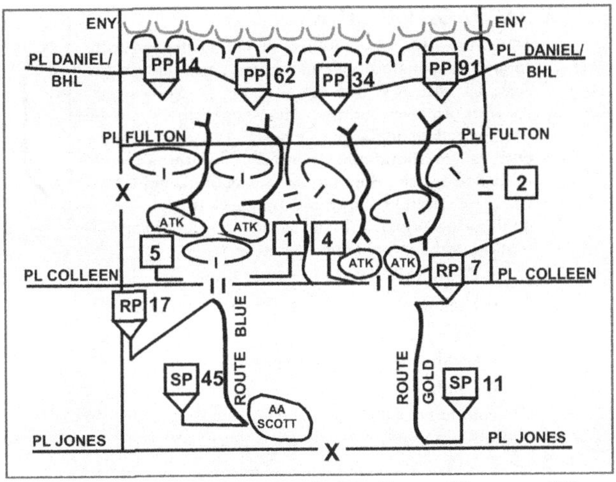

Figure 16-1. Control Measures Associated with a Forward Passage of Lines

point and the passage point, the stationary unit controls the passing force's movement. Figure 16-2 depicts the graphic control measure for passage point 8.

Figure 16-2. Passage Point 8

16-7. **A *passage lane* is a lane through an enemy or friendly obstacle that provides safe passage for a passing force.** The lane may be cleared, including being reduced and proofed, as part of a breach operation, or it may be included as part of the design of a friendly obstacle. It is a clear route all the way through an obstacle. Passage lanes normally end where a route begins. That route should allow the passing unit to move rapidly through the stationary unit's area. Figure 16-3 depicts the graphic control measure for a lane.

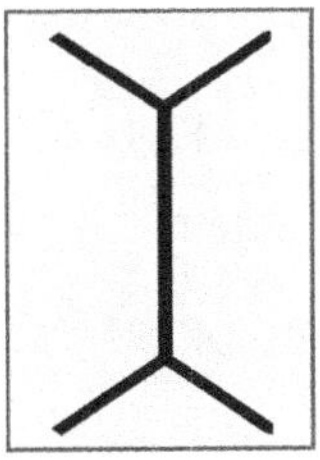

Figure 16-3. Lane

16-8. **A *gap* is an area free of armed mines or obstacles whose width and direction allow a friendly force to pass through the area containing obstacles while dispersed in a tactical formation**. The presence of gaps prevents inadvertent concentrations of soldiers and equipment around the entry points of lanes. Figure 16-4 depicts the graphic control measure for a gap.

PLANNING A PASSAGE OF LINES

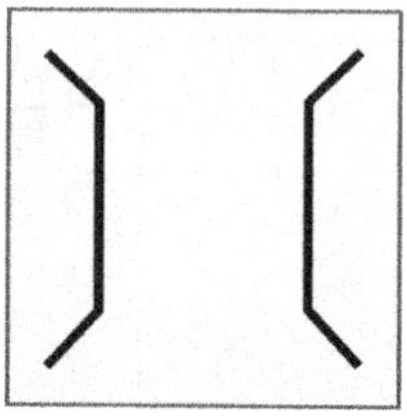

Figure 16-4. Gap

16-9. As with any activity involving transferring combat responsibility from one unit to another, the complex nature of a passage of lines involves risk. As with other operations, a passage of lines may be categorized as deliberate or hasty. During a passage of lines, the commander normally maintains the established tempo. Sustaining that established tempo requires detailed planning and preparations for a deliberate passage of lines. In this case, both the stationary and moving force have time to—

- Publish written orders.
- Exchange plans, intelligence information, databases, and liaison personnel.
- Conduct briefings and detailed reconnaissance.
- Conduct rehearsals.

The commander uses oral and fragmentary orders to conduct a hasty passage of lines.

16-10. In a passage of lines, the headquarters directing the passage is responsible for designating—

- Subsequent missions for both forces.
- When and under what conditions passage of command takes place.
- Start and finish times for the passage.
- Contact points between the units involved.
- Common maneuver control measures and graphics.

The directing headquarters normally establishes this information in either the warning order or the order directing the passage. In the absence of higher-echelon guidance, close coordination and understanding between the commanders and staffs of the two units are essential to a smooth passage.

16-11. The unit commanders plan the passage of lines to maintain enemy contact and provide constant fires on the enemy. Commanders reduce risk and ensure synchronization through detailed planning and decentralized execution. With forces intermingling during the passage, the need for positive control increases. The passage requires close coordination, clearly understood control measures, liaison between all headquarters and echelons involved in the passage, and clear identification of the moment or event that causes one force to assume responsibility for the AO from another.

16-12. After receiving the warning order that directs a passage of lines, the passing unit's commander and key staff representatives generally co-locate with the command post of the stationary unit to facilitate in planning the passage and establishing common situational understanding. If the passing unit cannot co-locate one of its command posts to help plan the passage, it conducts extensive liaison with the stationary unit. The planning focus for both the passing unit and the stationary unit is on operations following the passage. While this occurs, the two units involved coordinate the following:

- The exchange of intelligence and combat information.
- Current friendly dispositions and tactical plans, especially deception and obstacle plans.
- Direct and indirect fires and close air support plans.
- Any necessary maneuver control measures and graphics not directed by the higher headquarters, such as boundary changes, the BHL, emergency CSS points, and AA and firing positions for artillery, air defense, and other units.
- Long-range and short-range recognition symbols and vehicle markings to reduce the probability of fratricide.
- When and under what conditions control of the AO transfers from one headquarters to the other, if not previously established.
- Provisions for movement control, including contact points, start and release points, primary and alternate routes, route selection, priorities for using routes and facilities, passage points, and provision for guides.
- Reconnaissance by elements of the passing unit.
- Signal operating instruction details, such as call signs, frequencies, and recognition signals.
- Security measures during the passage, including nuclear, biological, and chemical reconnaissance or biological detection systems.
- Fires, obscurants, and any other combat, CS, and CSS provided by the stationary unit.
- Measures to reduce both units' vulnerability to attack by enemy weapons of mass destruction.
- Operations security measures required before or during the passage.
- Allocation of terrain for use by the passing force.
- Air defense cover—up to and forward of the BHL.
- Logistics support for the passing unit provided by the stationary unit, especially fuel, maintenance, and medical treatment.

16-13. The fire support elements of both the stationary and the passing unit must agree on allocating firing positions. The AO commander controls the allocation of firing positions in case of disagreement. These positions must be far enough forward to support the operation without having to redeploy during critical stages of the battle. The fire support elements normally position in areas not identified by the enemy.

16-14. Detailed air defense planning is essential for a passage of lines. Moving units tend to move slowly and often in some type of column formation during the passage. Vehicle congestion presents lucrative targets to enemy aircraft. In most cases, the stationary air defense elements can protect the passing force, allowing the air defense units supporting the passing force to move with the passing force. Dissemination of early warning and Army airspace command and control information reduces the risk of fratricide to friendly aviation assets while increasing the probability of the timely detection of enemy air. Strict adherence to identification, friend-or-foe (IFF) procedures among pilots and air defense fire units is critical, especially during periods of limited visibility. Local air superiority also reduces the vulnerability of the two forces when congestion cannot be avoided on the ground.

16-15. Once a passage of lines begins, it occurs quickly. Where possible, the operation takes place when the enemy has the least capability to detect it, such as at night or during periods of reduced visibility. In any passage of lines, the commander considers using smoke to screen friendly movement, even at night.

16-16. The passing unit prefers to conduct the passage through a gap in the stationary unit's positions rather than through a lane or a route that traverses those positions. This reduces the vulnerability that results from concentrating forces when one unit passes directly through the occupied positions of another unit. It also avoids the danger of concentrating the passing unit into passage lanes.

16-17. In a forward passage of lines, when there are no gaps through the stationary unit's positions, each battalion task force normally needs at least two passage lanes. In a rearward passage of lines, each battalion needs at least one passage lane. In both cases, a brigade needs at least one additional lane for its tactical vehicles. The routes and lanes provide cover, concealment, and rapid movement of the passing force. The commander may designate alternative routes and lanes for elements of the moving force that are contaminated. They should not disrupt the combat capability of the stationary unit. The commander seeks additional lanes to speed the process if the terrain and enemy situation allow.

16-18. The passing unit normally has priority of route use to and within the stationary unit's AO. Clearing and maintaining passage routes up to the BHL are the responsibility of the stationary force. The stationary force must provide an obstacle overlay of its obstacles. The passing unit must be prepared to help maintain these routes, and it positions its engineer equipment accordingly. The stationary unit is responsible for traffic control within its AO until the passing unit assumes control. During the passage, the passing unit augments the traffic-control capability of the stationary unit as required.

16-19. Based on the commander's concept and intent, the passing force focuses its planning effort on two general areas: coordination with the stationary force and guidance to subordinate units conducting the passage. These planning efforts occur simultaneously. If the enemy attacks during the passage, the plan probably requires modification to prevent hampering friendly maneuver.

16-20. Executing a passage of lines successfully requires effective communication between the two units. The commanders build redundancy of communication signals and means into their passage plans, such as using mobile subscriber equipment and combat net radios. The commanders also designate contact points to ensure effective communication between the two forces at the lowest tactical level.

FORWARD PASSAGE OF LINES

16-21. The purpose of a forward passage of lines is to move forces forward to conduct operations. It ensures the maintenance of enemy contact while allowing the relief of previously committed forces. The stationary force must control and secure the AO far enough to its front that the moving force can pass through the stationary force and reform into a combat formation prior to contact with an enemy force. Generally, the stationary unit supports the passing unit until the passing unit masks the stationary unit's direct fires. The stationary unit continues to support the passing force with its fire support systems until the passing unit moves beyond the supporting range of the stationary force. The stationary unit is also responsible for the security of the line of departure of the forward passing unit until it is able to assume that responsibility. The boundaries of the forward passing force after it completes its passage do not have to coincide with the boundaries of the stationary force. (See Figure 16-5.)

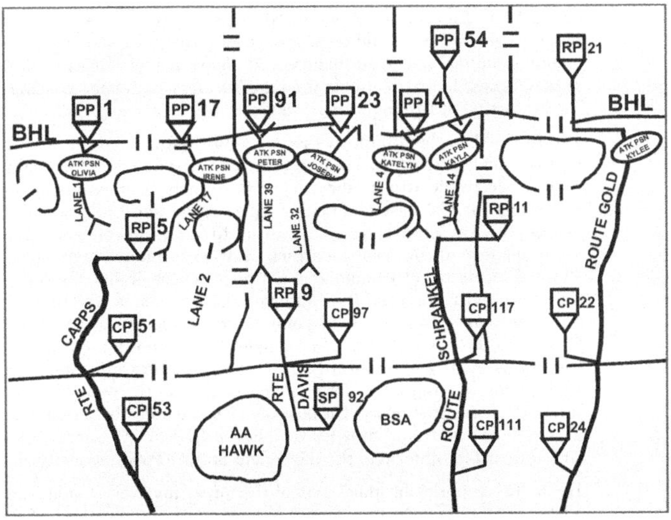

Figure 16-5. Forward Passage of Lines

PREPARING A FORWARD PASSAGE

16-22. The passing unit conducts reconnaissance from its current location to its designated AAs, which are generally located to the rear of the stationary unit. After completing its reconnaissance, the passing unit occupies these AAs.

16-23. The commander should organize the passing force for its subsequent mission before initiating the forward passage of lines. The passing force avoids regrouping in forward AAs or attack positions.

EXECUTING A FORWARD PASSAGE

16-24. When the passing force moves forward, it should move without a halt through the stationary unit while deployed in a combat formation. That minimizes the time the two forces are concentrated in the forward area, making them less vulnerable to enemy attack.

16-25. Support by the stationary force ends when the combat elements of the moving force, including the reserve, have moved beyond direct-fire range. However, artillery, air defense, and other long-range systems may remain to support the passing unit until a previously designated event occurs or a higher headquarters directs another mission.

16-26. When executing the forward passage, the passing unit's reconnaissance elements operate forward of the release points and establish a screen in front of the passing unit. The stationary unit continues to conduct aggressive security operations throughout the passage of lines. The movement of main body forces begins from their AAs to attack positions, where the passing unit conducts its final preparations for the passage of lines and the attack. The passing unit moves to and occupies attack positions when observation by the enemy is unlikely. The stationary unit clears any obstacles from designated passage gaps, lanes, or routes, and guides elements of the passing unit from the contact point through the passage points.

16-27. The direct and indirect-fire assets of the stationary unit normally support the movement of the passing unit. Offensive information operations—especially electronic attack—directed against enemy command and control (C2) nodes disrupt his dissemination of information and his reaction to friendly operations. Any preparatory or covering fires should coincide with the passing unit's movement from the attack position to the passage lanes. After the forward moving unit commander assumes responsibility for the AO, he coordinates all fire support. Depending on the situation at the time, the passing commander may continue to use only the fire support assets of the stationary force until the passage of lines is complete. This allows the passing unit's fire support assets to move forward, in the case of artillery, or remain available to support the passing unit's forward movement, in the case of attack helicopters and close air support. On passage of command, the passing commander also assumes control of fires forward of the BHL. For example, he moves the CFL forward to conform to the movement of his forward security elements.

16-28. The superior headquarters of the forces involved should exercise overall C2 of the passage. In a forward passage, the commander of the passing force normally assumes responsibility for conducting operations beyond the BHL once the attack begins. In practice, however, it is useful to complete the transfer of responsibility, including fire support, just before starting the operation. During the passage, two parallel chains of command are operating in one area simultaneously, and the possibility of confusion exists. A successful passage of lines requires clear C2 responsibilities. The passing unit's command post passes through the lines as soon as possible after the lead elements complete their passage and locates where it can best control operations.

16-29. The stationary unit furnishes the passing unit with any previously coordinated or emergency logistics assistance within its capabilities. These typically include—

- Evacuating casualties and enemy prisoners of war.
- Controlling dislocated civilians.
- Using areas and facilities such as water points and medical facilities.
- Controlling routes and traffic management.
- Recovering disabled vehicles and equipment.

The passing force normally assumes full responsibility for its CSS support forward of the BHL.

16-30. When dissimilar units, such as light infantry and mounted forces, are involved in a passage of lines, the principles involved are the same; however, the execution is different. For example, the type and amount of support provided by the stationary unit will change. In some cases, the higher headquarters ordering the passage needs to provide assets to support the passage.

REARWARD PASSAGE OF LINES

16-31. A rearward passage of lines is similar in concept to a forward passage of lines. It continues the defense or retrograde operation, maintaining enemy contact while allowing for recovery of security or other forward forces. This operation may or may not be conducted under enemy pressure. Counterintelligence analysis provides an assessment of enemy collection against friendly forces, specified by gaps and vulnerabilities, and countermeasures to enemy collection. Additionally, that analysis provides the commander with a view into the enemy's decision making and intelligence cycles and the time period in which the enemy may discover the movement.

PLANNING A REARWARD PASSAGE

16-32. Planning procedures for a rearward passage of lines closely resemble the planning procedures for a forward passage of lines. However, rearward movement is likely to be more difficult because of the following:

- The enemy probably has the initiative, which tends to reduce the time available to conduct liaison and reconnaissance and make detailed plans.
- If the rearward moving force has been in action, its soldiers are tired and possibly disorganized to some degree.
- The enemy may be applying pressure on the passing force.
- Friendly forces may be more difficult to recognize because enemy forces may be intermixed with them.

16-33. Close coordination between the two commanders is crucial to successfully executing the rearward passage and subsequent transfer of responsibility. This requirement for close coordination is even more critical when the tactical situation results in a staggered or incremental rearward passage across an AO. The passing commander relinquishes control of his elements remaining in contact at the time of the transfer of responsibility to the stationary commander. Generally, the stationary unit assumes control of the AO forward of the BHL after twothirds of the passing force's combat elements move through the passage points.

16-34. After receiving the warning order, the passing unit begins coordination and establishes communication with the stationary unit. The commanders of these units coordinate the same details as those outlined for a forward passage of lines. For example, the stationary commander coordinates for fires to support the rearward passing force. The two staffs coordinate those control measures necessary to support retrograde operations and their associated rearward passage of lines. (See paragraphs 16-5 to 16-8.) The commanders establish a probable time to initiate passage. The stationary commander assigns responsibility for closing and executing obstacles.

16-35. The stationary unit identifies multiple routes through its AO and across its rear boundary to AAs. The passing unit begins reconnaissance of these routes as soon as possible. The stationary unit must physically show all obstacles and routes and gaps through them to the passing unit. It provides guides for the passing unit—especially through obstacles—and mans contact points and passage points. The passing unit begins to reconnoiter its routes to the established contact points with the stationary unit's troops. The stationary unit establishes a security area in which responsibility transitions from the moving force to the stationary force. Normally, a BHL designates the forward edge of this area. The BHL is within direct-fire range and observed indirect-fire range of the stationary force.

PREPARING A REARWARD PASSAGE

16-36. The command posts of both units involved should move to a position where they can co-locate as part of the preparations for the rearward passage. This co-location reduces the risk associated with a passage because it makes it easier to coordinate between the two units. If circumstances prevent the units' command posts from co-locating, they must exchange liaison teams to ensure thorough coordination. If necessary, fire support assets from the stationary force occupy positions forward of their primary positions to give maximum coverage of forces of rearward moving unit.

EXECUTING A REARWARD PASSAGE

16-37. The passing unit maintains command of its subordinate elements throughout the retrograde and rearward passage. The normal order of march in a rearward passage of lines is CSS elements, main command post, CS elements, tactical command post, and combat units. The passage point marks the location where the passing unit comes under the control of restrictions placed by the stationary unit. (See Figure 16-6.) Note that the unit on the far right does not have a passage point because of the gap existing at that location. If the enemy continues to press his attack during the passage, the passing unit controls the battle from co-located command posts while the stationary unit monitors and controls the passage of lines until battle handover occurs. The passing unit's command post passes through the lines as soon as possible after the lead elements complete their passage. On passage of command, the stationary unit assumes the defense of the AO.

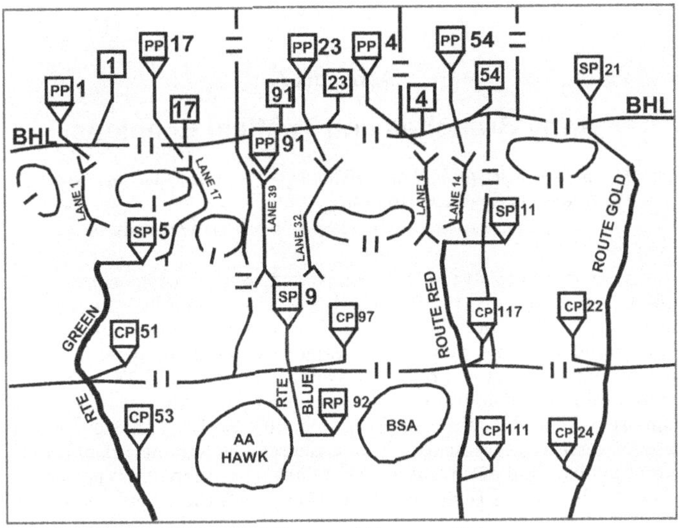

Figure 16-6. Rearward Passage of Lines

16-38. The stationary unit provides the passing unit with as much assistance as possible. Pivotal to the success of the rearward passage of lines is providing indirect and direct fire support by the stationary unit to the passing unit. This is especially important in covering the withdrawal of elements left in contact during a delay. The stationary unit's fire support assets answer calls for fire from the passing unit until battle handover occurs. The passing unit's fire support assets echelon rearward to provide continuous fire support for the passing unit until it successfully disengages. Once the passing unit hands over control of the battle to the stationary unit, the stationary unit initiates and clears calls for all fires forward of its location. The same procedure applies to the dedicated air defense assets of the passing and stationary units.

16-39. The stationary unit's engineer assets provide support to prepare the defense and execute the passage. Priority of effort initially ensures that the passing unit is able to move through passage lanes around the stationary unit's defensive positions. It shifts to close these passage lanes once the passing unit and any security elements disengage and withdraw through the security area and obstacles.

16-40. The stationary unit provides the passing unit with the previously coordinated CSS as far forward as possible. The stationary unit concentrates on providing the passing unit with emergency medical, recovery, and fuel supplies to enable the passing unit to rapidly move through the stationary unit's positions.

Appendix A

Army Branches and Tactical Echelons

It is not so much the mode of formation as the proper combined use of the different arms which will insure victory.

Antoine Henri, Baron de Jomini: *Precis de l'Art de la Guerre*, 1838

The Army consists of the active component, reserve components, and civilians acting in concert with other US services and allies. Its flexibility, versatility, and adaptability are based on a broad range of branch capabilities and echelons that can be rapidly tailored for deployment and task-organized for the prevailing conditions of METT-TC. The Army groups its force structure into three general categories: combat, combat support (CS), and combat service support (CSS). Each category incorporates diverse capabilities of varying degrees of lethality, deployability, sustainability, and survivability. There are more than 400 types of Army units. Each category complements and reinforces the others and the joint force. Appropriate combinations provide a balanced and versatile force mix, maximizing the commander's freedom of action in virtually any METT-TC condition.

A-1. There is no primary or dominant branch or arm. One branch reinforces or complements the effects of another, such as heavy forces in a support-by-fire position reinforcing an assault by light forces with their large-caliber, direct-fire weapons. Some branches have capabilities that can be described as combat arms, CS, or CSS, depending on the specific situation. Much of a

CONTENTS

commander's effectiveness during an operation relies on his ability to rapidly reinforce and complement weapon systems' effects in symmetrical and asymmetrical ways.

COMBAT ARMS

A-2. ***Combat arms* are units and soldiers who close with and destroy enemy forces or provide firepower and destructive capabilities on the battlefield.** Combat branches of the US Army include Air Defense Artillery, Armor, Aviation, Engineers, Field Artillery, Infantry, and Special Forces (SOF). Figure A-1 lists some types of units that deploy to support operations. The Army classifies combat arms units as heavy, light, or special operations forces. However, the Army is currently developing a medium weight force capable of increased strategic responsiveness in full spectrum operations.

- **AIR DEFENSE ARTILLERY**
  - **Short Range**
  - **Theater Air Defense**
  - **Theater Ballistic Missile Defense**
- **ARMOR**
  - **Armor**
  - **Armored Cavalry**
  - **Light Cavalry**
- **AVIATION**
  - **Attack**
  - **Assault**
  - **Air Cavalry**
- **ENGINEERS**
  - **Mobility**
  - **Countermobility**
  - **Survivability**
- **INFANTRY**
  - **Light**
  - **Mechanized**
  - **Motorized**
  - **Air Assault**
  - **Airborne**
- **FIELD ARTILLERY**
  - **Target Acquisition**
  - **Cannon Artillery**
  - **MLRS/ATACMS**
- **SPECIAL OPERATIONS FORCES**
  - **Ranger**
  - **Special Forces**
  - **Special Operations Aviation**

Figure A-1. Combat Arms Capabilities

A-3. Heavy forces employ a combination of armored and mechanized forces that use their tactical mobility, protection, and firepower to close with and destroy the enemy, seize and hold terrain, and conduct reconnaissance. They consist of armor, mechanized infantry, aviation, and armored cavalry. Heavy forces employ tanks, armored fighting vehicles, attack and utility helicopters, and dismounted infantry to form the nucleus of a combined arms team that delivers mobile, protected firepower to create tremendous shock effect. The combination of mobile, protected firepower with dismounted infantry achieves complementary and reinforcing effects that neither can attain separately. However, restricted and urban terrain place limits on where the commander can employ them.

A-4. Light forces close with and destroy the enemy, seize and hold terrain, and gain information. Light forces traditionally include light infantry, airborne, and air assault forces. Light forces are particularly suited to operations in restricted and urban terrain, where they have a mobility advantage over heavy forces. Light forces are limited by a relative lack of protection against direct and indirect fires and limited firepower compared with heavy forces. They also have limited organic tactical mobility once deployed into an area of operations (AO) compared to heavy forces. While light forces may be lifted into engagement areas (EAs) by truck, helicopter, or airplane, they fight mainly on foot. Depending on the factors of METT-TC, light forces either complement or are complemented by heavy forces.

A-5. Special operations are actions conducted by specially organized, trained, and equipped military forces to achieve military, diplomatic, economic, or psychological objectives by unconventional means. Special operations forces can reinforce, augment, and complement conventional forces, heightening the effectiveness of the total effort. These forces can also conduct independent operations in situations that demand a small, discrete, highly trained force. The principal SOF missions are—

- Unconventional warfare.
- Foreign internal defense.
- Psychological operations. (See Paragraph A-37.)
- Civil affairs. (See Paragraph A-36.)
- Information operations.
- Direct action.
- Special reconnaissance.
- Combating terrorism.
- Counterproliferation.

Five types of units compose Army SOF. Army special operations aviation (SOA), rangers, and special forces (SF) units are combat arms forces. Civil affairs (CA) and psychological operations (PSYOP) units are CS organizations. FM 3-05 is the capstone manual for Army SOF.

A-6. The Army is currently fielding two initial brigade combat teams (IBCTs) as the prototypes of a medium weight force designed for increased strategic responsiveness across all types of military operations. After experimentation, these two IBCTs will transform into an interim, strategically responsive force, which may expand to encompass almost a third of the regular Army's ground maneuver brigades. The desired end state of the objective force is an operational force strategically responsive through increased deployability and greatly reduced sustainability requirements when compared with current heavy forces. The goal for the objective force is to deploy a brigade anywhere in the world and close within 96 hours, given sufficient ports and strategic lift. The objective force will also retain or increase the tactical agility, lethality, and survivability of current heavy forces and the versatility of current light forces. Developing the objective force's combat and tactical vehicles requires scientific and technological breakthroughs.

AIR DEFENSE ARTILLERY

A-7. Air defense artillery (ADA) units provide force protection against air and missile threats. Tactical air defense supports the overall objectives of divisions and corps. Air defense artillery air and missile defense units protect maneuver forces and vital assets. Divisional ADA units provide short-range air defense (SHORAD) protection for units conducting tactical combat operations. Corps ADA brigades have both SHORAD and high-to-medium altitude air defense (HIMAD) missile defense units to protect corps assets and reinforce divisional ADA units.

A-8. Air defense artillery units contribute to intelligence and information operations by gathering and disseminating information about the enemy air order of battle. They also contribute by denying the enemy the ability to operate his own reconnaissance and command and control (C2) aircraft. (FM 3-01 is the capstone ADA manual. FM 3-01.12 discusses Army theater missile defense.)

ARMOR

A-9. The tank is the primary offensive ground weapon in mounted warfare. Its firepower, protection from enemy fire, and speed create the shock effect necessary to disrupt or defeat the enemy. Tanks can destroy enemy armored vehicles, infantry, and antitank guided missile carriers. Tanks can break through suppressed defenses, exploit the success of an attack by striking deep into the enemy's rear area, and pursue defeated enemy forces. Armored units can also blunt enemy attacks and launch counterattacks as part of a defense.

A-10. The primary missions of cavalry units are reconnaissance and security. A cavalry unit's ability to find the enemy, develop the situation, and provide the commander with reaction time and security also makes it ideal for operating in an economy-of-force role. Cavalry forces can delay an attacking enemy and assist in a withdrawal. (There is no capstone manual for armor operations. Doctrine on heavy combined arms echelons is found in FMs 3-91.1, 3-91.2, 3-91.3, and 3-93. FM 3-20.95 addresses cavalry operations.)

AVIATION

A-11. The firepower, agility, and speed of Army aviation permit ground commanders to close with and defeat a wide range of enemy forces. Attack helicopters are ideally suited for rapid reaction during decisive, shaping, and sustaining operations where the terrain restricts or prohibits ground force occupation. Attack helicopters can influence the battle when ground forces are decisively engaged. The tactical mobility provided to airborne, air assault, and light infantry by assault aviation is limited by weather and lift asset availability.

A-12. Air cavalry platoons and troops can reconnoiter and maintain surveillance coverage over a much larger area in a shorter period of time than ground platoons and troops, but with less detail on ground features. During security operations, air cavalry reconnoiters, screens forward and to the flanks of ground forces, and acts as a rapid-reaction force. Scout helicopters provide a wide range of reconnaissance and security capabilities. Air cavalry scouts are essential in detecting and identifying enemy forces throughout the battlefield—an important source of real-time battlefield information. Onboard radars and digital communications are key to the rapid dissemination of combat information that these systems obtain. (FM 3-04.100 is the capstone manual for aviation doctrine.)

ENGINEERS

A-13. The commander task organizes combat engineer units with maneuver units and integrates them into a combined arms formation. The engineer unit provides demolition and breaching capabilities to the combined arms team. The engineer unit also can employ direct-fire weapon systems to aid in employing demolition and breaching assets. These breaching assets include assault bridges, rafts, and mechanized breaching systems. Regardless of the mission, armored engineer vehicles are combat vehicles and provide a significant contribution to the combat power of the entire formation. To accomplish the mission, engineers fire and move under the direction of the formation commander, as necessary, using engineer skills where appropriate. (FM 3-34 is the capstone engineer manual.)

A-14. When involved in an assault, engineers fight dismounted on the objective, focusing on breaching enemy close-in protective obstacles, as well as demolishing fighting positions and dug-in vehicles. Demolition charges produce significant shock and concussion effects on defenders, as well as destroying critical positions, munitions, and combat vehicles.

A-15. Combat engineers employed on reserve demolition targets in the defense mainly execute the technical procedures necessary to ensure target destruction. However, the engineer demolition party responds to enemy contact. They assist the demolition guard in securing the target by holding it open or gaining time to ensure that it is destroyed. The engineer force may assist in target defense by installing command-detonated mines to support the defensive scheme.

A-16. Combat-engineer units have a secondary mission to fight as infantry. While engineers fight continually as engineers, employing them as infantry requires serious considerations. Engineers employed as infantry require augmentation, including crew-served weapons and medical support. Any commander who owns engineers in a command relationship has the authority to employ them as infantry unless otherwise prohibited. A commander must carefully weigh the gain in infantry strength against the loss of engineer support. Engineers provide far more combat power in their primary mission than when configured as infantry. Stopping the engineer work may reduce the combat power of the commander's entire force. Because of the long-term impact, a commander notifies the next higher headquarters when he employs his engineers as infantry.

FIELD ARTILLERY

A-17. Field artillery (FA) is the commander's principal means for providing indirect-fire support to his maneuver forces. Self-propelled or towed FA units contain cannon or multiple rocket launchers. Field artillery can neutralize, suppress, or destroy enemy direct-fire forces, attack enemy artillery and mortars, and deliver scatterable mines to isolate and interdict enemy forces or protect friendly operations. The commander may use artillery fires to cover key terrain, flanks, obstacles, and dead space to reduce his risk when maneuver forces are not available. Field artillery elements within maneuver organizations serve as the integrating center for all fire support elements. Field artillery units contribute both to attacking the enemy throughout the depth of his formations and suppressing enemy air defense systems to facilitate ground and air operations. Artillery fires can provide simultaneous precision strikes of targets at long ranges that other means cannot attack without significant risk. As mobile as the maneuver force it supports, FA systems provide continuous fires in support of the commander's scheme of maneuver. (FM 3-09 is the capstone FA manual.)

INFANTRY

A-18. There are many different types of infantry units, such as airborne, air assault, light, long-range reconnaissance, mechanized, and ranger. Each different type of infantry unit has its own unique skills and organizational design, but all share the common mission: "To close with and destroy the enemy by means of close combat, fire, and movement." Regardless of their mode of conveyance to the battlefield—aircraft, tracked or wheeled armored fighting vehicle, truck, or foot—they all serve as a key source of combat power in close combat. (There is not a capstone infantry manual. Echelon and unit specific manuals, such as FMs 3-21.7, 3-21.8, 3-21.10, 3-21.20, and 3-21.30, describe infantry operations at different tactical echelons.)

A-19. Airborne infantry units have the greatest capability for large-scale force-projection operations. They rapidly deploy over great distances and conduct combined arms parachute or air-landing assaults to seize and secure vital objectives. The commander can insert these units on virtually any objective area under almost any weather conditions. Once on the ground, their capabilities and lethality are similar to other nonmechanized infantry units.

A-20. Air assault infantry units have great tactical mobility and train to fight across the range of military operations. Their significant antiarmor capability—coupled with strategic deployability—makes them well suited as an early deploying force in contingency operations against heavy forces. They train and fight as a team in combination with artillery, attack, and lift aviation. They can penetrate deep into enemy territory to cut lines of communications, seize airfields, destroy C2 nodes, block reinforcing units, or seize key terrain. Because of their agility and mobility, air assault infantry units are well suited for covering force operations in appropriate terrain.

A-21. Light infantry units can operate effectively in most terrain and weather conditions. They may be the dominant arm in fast-breaking operations because of their rapid strategic deployability. In such cases, they can wrest the initiative early, seize and hold ground, and mass fires to stop the enemy. They are particularly effective in urban terrain, where they can infiltrate and move rapidly to the rear of enemy positions. The commander can enhance their tactical mobility by using helicopters and tactical airlift.

A-22. Mechanized and motorized infantry forces integrate mobile, survivable, and lethal, vehicle-mounted direct-fire and indirect-fire weapon systems and dismounted infantry skills into an effective fighting system that enhances the striking power of the combined arms force. Mechanized infantry has the same mobility as armor forces, but less firepower and protection. Armor and mechanized infantry train and fight as a team to defeat enemy heavy forces. When equipped with infantry fighting vehicles, mechanized infantry can accompany tanks in mounted assaults. The commander must carefully determine if, when, and where his infantry must dismount to accomplish its mission. Mechanized infantrymen can act as fixing forces in an attack and serve as pivot points for maneuvering tank-heavy forces in the defense.

SPECIAL OPERATIONS FORCES

A-23. Special operations are actions conducted by specially organized, trained, and equipped personnel to achieve military, political, economic, or psychological objectives by nonconventional means in hostile, denied, or politically sensitive areas. They are conducted in peace, conflict, and war, independently or in coordination with operations of conventional forces. Politico-military considerations frequently shape special operations, requiring clandestine, covert, or low-visibility techniques and oversight at the national level. Special operations usually differ from conventional operations in their degree of risk, operational techniques, mode of employment, independence from friendly support, and dependence on operational intelligence and indigenous assets. (The Army's capstone manual on special operations is FM 3-05.)

A-24. Army SOA units are specialized aviation assets dedicated to conducting special operations missions. They provide a mix of short-, medium-, and long-range lift and light attack capabilities to support all principal and collateral mission areas and conduct autonomous special operations.

A-25. Ranger units are rapidly deployable, airborne-capable, and trained to conduct joint strike operations with or in support of special operations units of all services in any environment. They plan and conduct special military operations to support national policies and objectives. They also conduct direct-action missions to support conventional operations and operate as conventional light infantry units when integrated with other combined arms elements. (FM 3-21.85 is the capstone manual for ranger operations.)

A-26. Special Forces units plan, conduct, and support special operations activities in all operational environments, throughout the range of military operations. The US Army organizes, trains, and equips SF units to perform seven primary missions:

- Unconventional warfare.
- Foreign internal defense.
- Information operations.
- Counterproliferation.
- Direct action.
- Special reconnaissance.
- Combating terrorism.

Mission priorities vary from theater to theater. Special Forces missions are dynamic because politico-military considerations affect them directly. A change in national security policy or national military strategy may radically alter the nature of a SF mission. (FM 3-05.20 is the capstone SF manual.)

COMBAT SUPPORT

A-27. ***Combat support*** **encompasses critical combat functions provided by units and soldiers, in conjunction with combat arms units and soldiers, to secure victory.** Those functions include Army Aviation, Chemical Corps, Engineers, Military Intelligence, Military Police Corps, Signal Corps, and Special Operations Forces (CA and PSYOP units). Figure A-2 lists some CS units and their capabilities that support tactical operations.

- **AVIATION**
  - Air Traffic Services
  - C2 Aircraft
- **CHEMICAL CORPS**
  - Staff Support
  - Decontamination
  - NBC Reconnaissance & Surveillance
  - Smoke & Obscuration
- **ENGINEERS**
  - Mobility
  - Countermobility
  - Survivability
  - Topographic Support
- **SIGNAL CORPS**
  - Signal Support
  - Combat Camera
- **MILITARY INTELLIGENCE**
  - Counterintelligence
  - Analysis
  - HUMINT
  - IMINT
  - MASINT
  - SIGINT
  - TECHINT
  - Electronic Warfare
- **MILITARY POLICE CORPS**
  - Criminal Investigation
  - EPW Support
  - Military Police CS
- **SPECIAL OPERATIONS FORCES**
  - Civil Affairs
  - Psychological Operations

Figure A-2. Combat Support Capabilities

AVIATION

A-28. Army aviation units provide CS to the commander in the following areas:

- C2, communications, and intelligence.
- Air movement of combat power.
- Aerial delivery of mines.
- Search and rescue.
- Air traffic services.

CHEMICAL CORPS

A-29. The chemical corps provides CS to the Army through two major activities. The first is chemical staff support organic from the battalion through the corps echelon. Integrating organizational equipment and individual technical expertise permits units to continue operations with minimal degradation of capability. The second activity is the support provided by chemical units at corps and higher levels, including nuclear, biological, and chemical (NBC) reconnaissance, decontamination, smoke and obscuration, and further staff augmentation. Both activities are critical; the former optimizes unit effectiveness under NBC conditions, the latter augments and reinforces unit combat power and sustainability. (FM 3-11 is the capstone manual for chemical operations.)

ENGINEERS

A-30. Engineers multiply the effectiveness of friendly forces on a lethal battlefield. Four of the five primary engineer functions are CS. Combat engineers operate as an integral member of the combined arms team to provide a full range of mobility, countermobility, survivability, and topographic capabilities. Engineers advise the maneuver commander on the effective use of terrain; construct, improve, and maintain routes, bridges, and airfields; and reorganize to fight as infantry when required.

A-31. In offensive operations, combat engineers concentrate their efforts to support maneuver by breaching and crossing obstacles, assisting in the assault of fortified positions, and emplacing obstacles to protect the flanks of friendly attacking forces. In defensive operations, engineers reinforce the terrain to anchor the defense in critical areas, maximize the effects of defenders' fires, provide maximum protection to friendly fighting positions, and facilitate the movement of counterattack forces. Topographic engineer units furnish detailed terrain analysis products, maps, and digital terrain data. Topographic engineering products help the commander identify avenues and routes, obstacle locations, EAs, unit positions, and possible target areas.

MILITARY INTELLIGENCE

A-32. Military intelligence (MI) units provide the commander with early warnings of enemy intentions, intelligence preparation of the battlefield products, aid in the development of his situational understanding, and assist in target development, force protection, and battle damage assessment. They participate in offensive information operations, as well as provide critical counterintelligence support to friendly command force protection programs. (FM 2-0 is the capstone intelligence manual.)

A-33. Military intelligence involves four intelligence disciplines—human intelligence, imagery intelligence, measurement and signature intelligence, and signals intelligence. There are two multidiscipline intelligence functions—counterintelligence and technical intelligence. However, rarely will a single discipline or function produce a comprehensive picture of the enemy. Each of these disciplines and functions complement and cue each other. Each discipline or function produces bits and pieces of information that analysts use to create the enemy and environment portion of the common operational picture and other products that facilitate the commander's situational understanding. The commander should compare information gained by one discipline or function with information provided by the other disciplines and functions whenever the situation permits to avoid deception by enemy information operations. United States Code prohibits collecting intelligence information on US personnel or institutions by MI assets except under specific situations.

MILITARY POLICE CORPS

A-34. Military police units provide the commander with a versatile, responsive force capable of performing a wide range of combat, CS, and CSS missions that include—

- Maneuver and mobility operations—MP units enhance maneuver and mobility by expediting and monitoring the flow of personnel and materials throughout the depth and breadth of the battlefield.
- Area security operations—MP units provide security to critical personnel and facilities within their assigned AO because of their tactical mobility, firepower, and communications capabilities.
- Internment and resettlement operations—MP units conduct internment and resettlement operations for US military prisoners, enemy prisoners of war, and dislocated civilians to relieve the tactical commander of the burden they impose.
- Police intelligence operations—MP and MI units collect, analyze, and disseminate police intelligence, aiding commanders in identifying and defeating threats from criminals, saboteurs, and terrorists.
- Law and order operations—MP units assist the commander in maintaining law and order in both garrison and the field.

Military police units perform these operations independently or in combination with other units. United States Code prohibits MP and other Army units from performing routine law enforcement operations within US territory. State and federal law allow Army National Guard units to conduct law enforcement operations within their sponsoring state when performing their mission as state militia. (FM 3-19.1 is the capstone MP manual.)

SIGNAL CORPS

A-35. The signal corps provides worldwide information systems and networks for real-time C2 of Army, joint, and multinational forces. Signal corps units enable effective control systems to operate. In force-projection operations, signal units make split-based operations possible by employing satellite downlink equipment. (FM 6-02 is the capstone signal manual.)

SPECIAL OPERATIONS FORCES

A-36. Civil affairs units are organized, trained, and equipped specifically to conduct CA activities and to support civil-military operations. These operations focus on the civil center of gravity and establish, maintain, influence, and support the commander's moral obligations and legal responsibilities with government organizations, nongovernmental organizations, and international organizations. Civil Affairs activities include populace and resources control, foreign nation support, humanitarian assistance, military civil action, civil defense, civil assistance, and civil administration in recently liberated friendly territory and in occupied territory. Civil Affairs efforts enhance the relationships between military forces, civilian authorities, and populations in friendly, neutral, or hostile AOs through these activities. They provide opportunities to use local human and material resources to support the assigned mission. Civil Affairs forces facilitate military operations by reducing civilian interference with military operations and gaining popular understanding, support, and compliance with measures required to accomplish the mission. In the aftermath of combat, CA units conduct activities that stabilize disrupted areas. They are often tasked to create a military government to temporary control institutions, populations, and resources. Civil Affairs functional specialist's capabilities reside in the reserve component and are additional areas of expertise, normally acquired through civilian education, training, and experience. These special functions teams include—

- Emergency services
- Public administration
- Public education
- Public health
- Public safety
- Public welfare
- Civilian supply
- Economic development
- Food and agriculture
- Public communications
- Public transportation
- Public works and utilities
- Civil information
- Cultural relations
- Dislocated civilians
- International law
- Environmental management

(FM 3-57 is the capstone manual for civil affairs operations.)

A-37. Psychological operations are planned operations to convey selected information and indicators to foreign audiences to influence their emotions, motives, objective reasoning, and ultimately the behavior of foreign governments, organizations, groups, and individuals. The purpose of psychological operations is to induce or reinforce foreign attitudes and behavior favorable to the originator's objectives (JP 3-53). PSYOP is a functional area and not a branch. PSYOP units provide the commander with the ability to communicate information to non-US audiences via radio, television, leaflets, and loudspeakers. The PSYOP soldier's language skills, regional orientation, and knowledge of communication media provide a means of delivering critical information to host nation, neutral, and enemy audiences. Joint direction from the operational-level commander characterizes PSYOP. Federal law prohibits using PSYOP against US citizens or organizations. (FM 3-05.30 is the capstone PSYOP manual.)

COMBAT SERVICE SUPPORT

A-38. The primary role of Army tactical CSS units is to sustain Army forces. The CSS effort is successful only when it concentrates and supports forces by focusing on the sustainment and reconstitution of tactical units. Warfare consumes massive amounts of resources. The CSS system must provide resources in a way that minimizes constraints on the commander. Tactical-level CSS support must be responsive. (Figure A-3 on page A-12 illustrates CSS branch capabilities.)

A-39. While not traditional Army branches, forward elements of the US Army Materiel Command (USAMC) and the Military Traffic Management Command within a theater of operations assist in providing that combat service support necessary to sustain tactical forces. These forward elements provide logistic assistance, conduct port operations, sustainment at Army field support centers, control Army prepositioned stocks, and manage USAMC contract support. Those forward commands or subordinate logistics support elements deploy to support operations. Logistics assistance personnel, who are part of USAMC forward elements, deploy with major combat units. They assist and advise the unit; they also can reach back to the continental United State for required additional augmentation of sustainment capabilities.

- **ADJUTANT GENERAL**
  - **Personnel Services**
  - **Postal**
  - **Replacement**
- **AVIATION (GS AVN)**
- **CHAPLAIN**
- **ENGINEERS (GEN ENG)**
- **FINANCE CORPS**
  - **Support To Local Procurement**
  - **Pay Services**
  - **Resource Management**
- **JUDGE ADVOCATE GENERAL CORPS**
  - **Court Martial Trial & Defense**
  - **Legal Assistance**
  - **Contract Advisory Assistance**
- **MEDICAL**
  - **Combat Stress**
  - **Dental**
  - **Hospital**
  - **Medical Evacuation/Support/ Surveillance**
  - **Veterinary**
- **ORDNANCE CORPS**
  - **Ammunition Maintenance**
  - **Ammunition Supply**
  - **Electronic Maintenance**
  - **Explosive Ordinance Disposal**
  - **Mechanical Maintenance**
  - **Missile Maintenance**
- **QUARTERMASTER CORPS**
  - **Supply & Field Services**
- **TRANSPORTATION CORPS**
  - **Highway/Rail/Water Transportation**
  - **Movement Control**
  - **Intermodel (Terminal) Operations**

Figure A-3. Combat Service Support Capabilities

ADJUTANT GENERAL CORPS

A-40. The Adjutant General Corps helps to build and sustain combat readiness through planning, operating, and managing all human resources support activities. Human resources support functions include manning the force; personnel services; personnel readiness management; personnel accounting; replacement management; personnel information management; casualty operations management; postal operations management; morale, welfare, and recreation; and band support. (FM 1-0 is the capstone manual for military personnel activities.)

AVIATION

A-41. Aviation units provide CSS to sustain combat forces, primarily through air movement of personnel, supplies, and equipment and performing aeromedical evacuation and aviation maintenance. Air movement CSS missions include the aerial movement of personnel and supplies, to include—

- Over-the-shore logistics operations.
- Aerial preplanned and immediate resupply.
- Air movement of critically short or sensitive supplies.

- Pre-positioning and movement of fuel and ammunition.
- Air movement of security forces.
- Medical evacuation.

CHAPLAIN CORPS

A-42. The chaplain corps advises the commander on matters of religion, morals, and morale as affected by religion; to include the impact of indigenous religions on the mission. The role of the chaplain is the personal delivery of religious support—worship care and counseling—to the soldier. Chaplains assist in reducing the rate and severity of psychiatric casualties by working with combat stress teams. (FM 1-05 is the capstone doctrinal manual for chaplains and chaplain assistants.)

FINANCE CORPS

A-43. Financial management operation provides a commander with—

- Financial support for soliders.
- Financial support for non-US military personnel and all Department of Defense civilians.
- Pay for local hire civilians.
- Contract support.
- Payments to commercial vendors that provide services and supplies in support of military forces on an immediate or recurring basis.
- Disbursements of public funds to support an Army presence on an area-support basis.
- Accounting and disclosing expended funds.

(FM 1-06 is the capstone doctrinal manual for financial management support.)

ENGINEERS

A-44. Sustainment engineering is the primary CSS engineer function. It involves a variety of roles, such as civil engineering, fire fighting, maintaining lines of communications, managing inland waterways, prime power, and environmental protection.

MEDICAL CORPS

A-45. The Army's health care team provides all types of medical support to the commander and his soldiers. The Army's health care team comprises six medical corps: Medical Corps, Dental Corps, Medical Service Corps, Nurse Corps, Veterinary Corps, and Medical Specialist Corps. (FM 4-02, is the capstone doctrinal manual for health service support.)

A-46. Ordnance units sustain the commander's weapon systems, ammunition, missiles, and ground-mobility materiel. Ordnance soldiers are trained to repair and manage tank-automotive/ground-mobility materiel, missile materiel, and ammunition materiel, including explosive ordnance disposal. (There is no capstone manual addressing all the activities of the ordnance corps, but FM 4-0 provides references and information on ordnance support.)

A-47. Future maintenance concepts call for the consolidation of the current four levels of maintenance into two levels—field and sustainment. Field maintenance combines the current organizational and direct support levels of maintenance. Field maintenance includes those tasks performed by operators and maintainers at the point of breakdown or the point of repair. This maintenance level focuses on the repair of vehicles through the replacement of major system components. Sustainment maintenance consists of tasks normally focused on the repair of component items, and it combines the current general support and depot maintenance levels. Additionally, it includes some current direct support-level tasks. At this level of maintenance, maintainers focus on the repair of component items, such as major assemblies, line-replaceable units, and reparable items and their return to the distribution system.

ORDNANCE CORPS

A-48. Ordnance units sustain the commander's weapon systems, ammunition, missiles, and ground-mobility materiel. Ordnance soldiers are trained to repair and manage tank-automotive/ground-mobility materiel, missile materiel, and ammunition materiel, including explosive ordnance disposal. There is no capstone manual addressing all the activities of the ordnance corps, but FM 4-0 provides references and information on ordnance support.

A-49. Future maintenance concepts call for the consolidation of the current four levels of maintenance into tow levels-field and sustainment. Field maintenance combines the current organizational and direct support levels of maintenance. Field maintenance includes those tasks performed by operators and maintainers at the point of breakdown or the point of repair. This maintenance level focuses on the repair of vehicles through the replacement of major system components. Sustainment maintenance consists of tasks normally focused on the repair of component items and combines the current general support and depot maintenance levels. Additionally, it includes some current direct-support-level tasks. At this level of maintenance, maintainers focus on the repair of component items, such as major assemblies, line-replaceable units, and reparable-type items and their return to the distribution system.

QUARTERMASTER CORPS

A-50. The quartermaster corps arranges for or provides supplies, materiel management, distribution, procurement, and field services to support and sustain soldiers, units and their equipment. These support functions are broken down into two areas: supply and field services. Supplies provided by quartermaster units include—

- Class I—substance and commercial bottled water.
- Class II—clothing, individual equipment, tentage, tool sets, and administrative and housekeeping supplies equipment including nonclassified maps.
- Class III—petroleum, oils, and lubricants (POL), including petroleum fuels, lubricants, hydraulic and insulating oils, preservatives, liquid and compressed gases, checmical products, and coolants.

- Class IV—construction materials, including installed equipment and all fortification and barrier materials.
- Class VI—health and comfort packs/support to tactical field exchanges.
- Class VII—major end items
- Class IX—repair parts
- Class X—miscellaneous supplies

A-51. Field services provided by the quartermaster corps includes—

- Airdrop—parachute packing, air-item maintenance, slingload operations, and rigging of supplies and equipment for airdrop.
- Field feeding—the standard is three quality meals each day, with the capability to distribute, prepare, and serve at least one heated unitized group rations, with semi-perishable components (UGR-A), meal per day to include supplements in accordance with the factors of METT-TC.
- Laundry—soldiers will each be provided with up to 15 pounds of laundry support a week.
- Shower—the army standard is one shower per soldier per week, with a goal of two showers per week.
- Water purification. (Note that storage and distribution of purified water are supply functions.)
- Mortuary affairs—recovery, tentative identification, and processing of human remains.
- Clothing and light textile repair—minor repairs made at the laundry.

(FM 4-20 is the capstone manual for quartermaster units.)

TRANSPORTATION CORPS

A-52. The transportation corps provides transportation services to the commander. These services include—

- Performing transportation unit operations, to include truck, boat, rail and trailer/cargo transfer operations. Planning, scheduling, and supervising the use of each mode of transportation for effectively moving personnel and cargo.
- Providing terminal services for all modes of transportation and stevedoring services at fixed ports and unimproved beach sites.
- Providing transportation engineering services.
- Providing direct maintenance and supply for marine and rail equipment.

(FM 4-01 is the capstone doctrinal manual for the transportation corps.)

TACTICAL ECHELONS

A-53. The Army echelons its broad array of capabilities to perform diverse functions. These functions vary with the type of unit and, particularly at operational echelons, with the organization of the theater, the nature of the conflict, and the number of friendly forces committed to the effort. For a discussion of operational echelons, including ARFOR, joint force land component, and joint task force (JTF) see FMs 3-0 and 3-93.

A-54. At each echelon, the commander task organizes his available capabilities to accomplish the mission. The commander's purpose in task organization is to maximize subordinate commanders' abilities to generate a combined arms effect consistent with the concept of operations. Commanders and staffs work to ensure the distribution of capabilities to the appropriate components of the force to weight the decisive operation. The relationships between units within and supporting an echelon are described in terms of command and support relationships. (See FM 5-0 for a discussion of these relationships.)

COMPANIES, BATTERIES, AND TROOPS

A-55. **A *company* is a unit consisting of two or more platoons, usually of the same type, with a headquarters and a limited capacity for self-support. A *troop* is a company-size unit in a cavalry organization. A *battery* is a company-size unit in a field artillery or air defense artillery organization.** A company normally consists of more than 75 soldiers. Exceptions to this rule are some aviation and tank companies. Companies and air defense and artillery batteries are the basic elements of battalions. Armored, light, and air cavalry troops are the basic elements of squadrons. Companies, batteries, and troops may also be assigned as separate units of brigades and larger organizations. Some companies, such as SF companies, have subordinate detachments, instead of platoons, which are organized and trained to operate independently for extended periods.

A-56. Company-size combat units can fight in mass or by subordinate platoons. In attack helicopter battalions, companies fight as integral units. Cavalry troops fight more frequently with their platoons in separate areas. In infantry and armor battalions, companies fight either as integral units or as task-organized teams reinforced with close-combat platoons of the same or different types. **A *company team* is a combined arms organization formed by attaching one or more nonorganic tank, mechanized infantry, or light infantry platoons to a tank, mechanized infantry, or light infantry company, either in exchange for, or in addition to, its organic platoons.** These company teams can include other supporting squads or platoons, such as short-range air defense (SHORAD) and ground surveillance radar teams. Company teams are task-organized for a specific mission. Such teams can match capabilities to missions with greater precision. However, the attachment of different units at the company level demands thorough training to achieve the maximum complementary effects. Whenever possible, platoons and detachments should train together before they are committed.

A-57. FA batteries are the basic firing units of FA battalions. They are organized with firing platoons, a headquarters, and limited support sections. They may fire and displace together or by platoons. Normally, batteries fight as part of their parent battalions, but the commander can attach them to other batteries or FA battalions. In rare cases they respond directly to a maneuver battalion or company. Multiple Launch Rocket System (MLRS) batteries usually operate independently. Regimental armored cavalry squadrons have organic howitzer batteries.

A-58. ADA batteries operate as the fighting elements of ADA battalions or, if they are SHORAD batteries, in direct support of maneuver brigades or battalions. Separate SHORAD batteries exist in separate brigade-size organizations.

A-59. Combat engineer companies control three or four engineer platoons. Their parent battalion may employ them in a variety of tasks, or they may support maneuver brigades or battalions.

A-60. Most CS and CSS units organize as separate companies with greater self-sustainment capabilities than normally found in comparable size combat arms organizations. However, they may receive unit-level sustainment support on an area basis. Such CS and CSS companies vary widely in size, employment, and assignment.

BATTALIONS AND SQUADRONS

A-61. **A *battalion (or a cavalry squadron)* is a unit consisting of two or more company-, battery-, or troop-size units and a headquarters.** Most combat arms battalions are organized by branch, arm, or service and, in addition to their line companies, contain a headquarters company that gives them the ability to perform some administrative and logistic services. Typically, battalions have three to five companies in addition to their headquarters.

A-62. The commander can reinforce his maneuver battalions with other combat and CS companies to form task forces for special missions. **A *battalion task force* is a combat arms battalion-size unit consisting of a battalion headquarters, at least one assigned company-size element, and at least one attached company-size element from another combat arms or combat support unit.** Task organization increases the capability of maneuver battalions. For example, based on a brigade commander's understanding of the factors of METT-TC, he may task-organize tank, mechanized infantry, and light infantry battalions by cross-attaching companies between these units. (See Figure A-4.) FA battalions can be reinforced with batteries of any kind to form artillery task forces. The commander can reinforce engineer battalions with the same or different types of engineer companies and platoons to form engineer task forces.

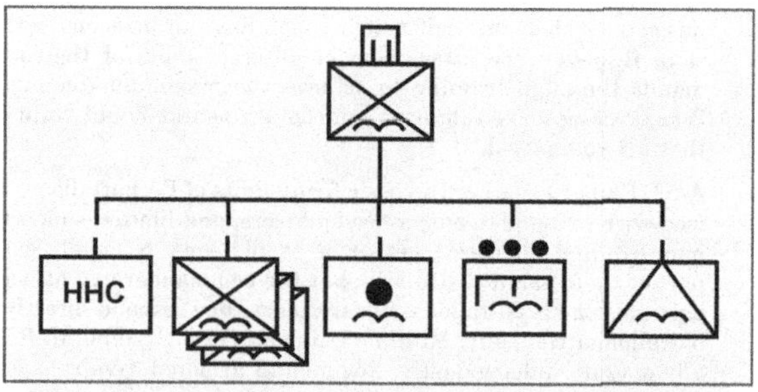

Figure A-4. Airborne Battalion-Size Task Force

A-63. CS and CSS battalions vary widely in type and organization. They may be separate divisional or nondivisional battalions, performing functional services for a larger supported unit within that unit's AO. All battalions are capable of short-term limited, self-defense. ADA and signal battalions assigned to or supporting divisions routinely operate throughout the division AO. Their commanders also perform the additional duties of division special staff officers.

BRIGADES, REGIMENTS, AND GROUPS

A-64. **A *brigade* is a unit consisting of two or more battalions and a headquarters.** Its capacity for independent action varies by type. A commander can use separate infantry, armor, FA, ADA, engineer, and aviation brigades and armored cavalry regiments to reinforce corps or divisions and shift these units from one division or corps to another to tailor forces for combat. (See Figure A-5.)

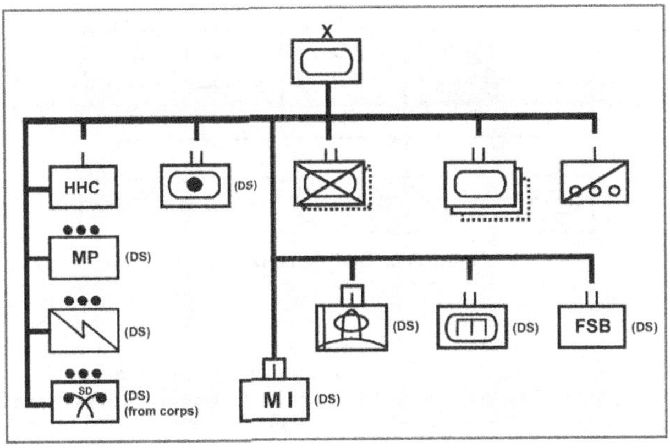

Figure A-5. Heavy Brigade Combat Team

A-65. Maneuver brigades are the major combat units of divisions; they can employ any combination of maneuver battalions. Division commanders adjust the organization of their brigades and change their task organization into brigade combat teams as required by the factors of METT-TC. **A *brigade combat team* is a semipermanent combined arms organization consisting of a brigade headquarters, at least two combat arms battalions, and all necessary supporting combat support and combat service support, formed for the purpose of carrying out a continuing specific task.** They normally include a direct support FA battalion, a combat engineer battalion (in heavy divisions and heavy separate brigades), forward support battalions, and smaller combat, CS, and CSS units. Brigades combine the efforts of their battalions and companies to fight engagements and battles and perform tactical tasks within division-level battles and engagements. Their chief tactical responsibility is synchronizing the plans and actions of their subordinate units to accomplish tasks for the division or corps.

A-66. Separate maneuver brigades and armored cavalry regiments have a fixed organization with organic cavalry, engineer, air defense, FA, MI, chemical, and CSS units. The commander can use separate brigades and armored cavalry regiments to reinforce corps or divisions, but they are capable of operating as independent units. The two initial brigade combat teams (IBCTs) at Fort Lewis, Washington, are organized as separate brigades even though they are designated as divisional brigades. (See Figure A-6.)

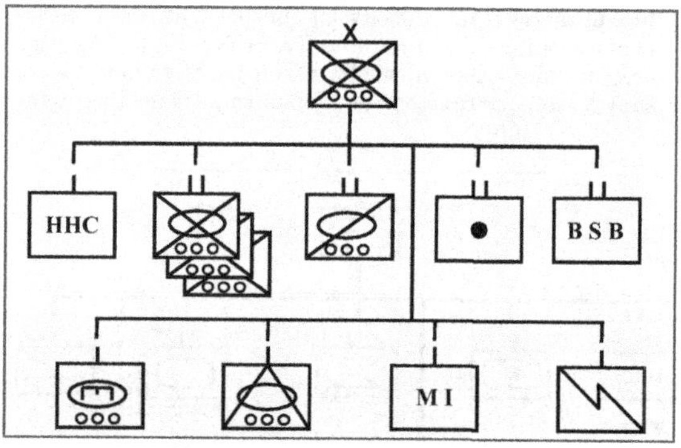

Figure A-6. IBCT Organization

A-67. The commander organizes other combat, CS, and CSS brigades and groups to control capabilities for divisions, corps, and larger units. Engineer, ADA, signal, aviation, MP, and transportation brigades are typical of such units. They may also be the building blocks of large-unit support structures, such as corps and theater army support commands, and CS commands, such as engineer commands. When the span of control for a brigade exceeds seven battalion-size subordinate units, the commander can establish a group headquarters under the brigade as an intermediate headquarters for two or more CS and CSS battalions. Organic, brigade-size division artillery (DIVARTY), an aviation brigade, and a division support command (DISCOM) of CSS battalions and separate companies provide support for divisions. Heavy divisions have an organic engineer brigade.

DIVISION

A-68. The division possesses great flexibility. ***Divisions*** **are the largest fixed organizations in the Army that train and fight as tactical teams, and are organized with varying numbers and types of combat, CS, and CSS units.** Their commanders task organize their organic brigades and attached forces for specific combat missions. The division commander may attach or place any or all divisional CS and CSS battalions and separate companies in support of one or more brigades for particular missions. (Figure A-7 on page A-20 depicts a heavy division's organic elements.) They are capable of performing any tactical mission over a wide range of environments and are largely self-sustaining for up to 72 hours. The Army's division organizational concept embraces six types of divisions—armored, mechanized infantry, medium, light infantry, airborne and air assault—with each having specific capabilities and resources for conducting military operations. The Army continues to examine its divisional design to ensure it possesses forces that are decisive across the full spectrum of operations. The Army is currently experimenting with Force XXI and Interim Division variants in a search for designs that have greater strategic responsiveness and effectiveness.

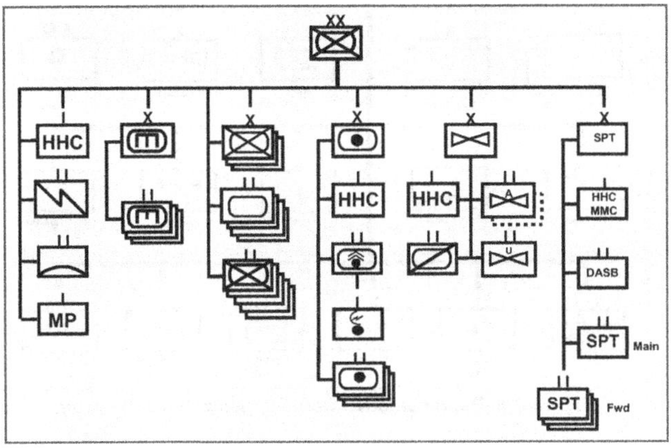

Figure A-7. Heavy Division

A-69. Divisions perform major tactical operations for the corps and conduct sustained engagements. With significant augmentation, a corps or a numbered army may direct a division to perform tasks of operational importance. These may include exploiting tactical advantages to seize objectives in depth, moving to gain contact with enemy forces, or moving by air to seize positions behind an enemy force.

A-70. Critical division roles include—

- Deployment.
- Conduct of full spectrum operations as part of a corps.
- Airborne and air assault divisions able to conduct force entry operations.

With augmentation a division headquarters can perform as the ARFOR or joint force land component headquarters. (FM 3-91 describes how divisions conduct full spectrum operations.)

CORPS

A-71. **The *corps* is the Army's largest tactical unit and the instrument by which higher echelons of command conduct maneuver at the operational level.** There is no standard organizational structure for a corps, although every corps typically has the components shown in Figure A-8. Operational headquarters tailor corps to the theater and mission for which they are deployed. Once tailored a corps contains all the combat, CS, and CSS capabilities required to sustain its operations for an extended period.

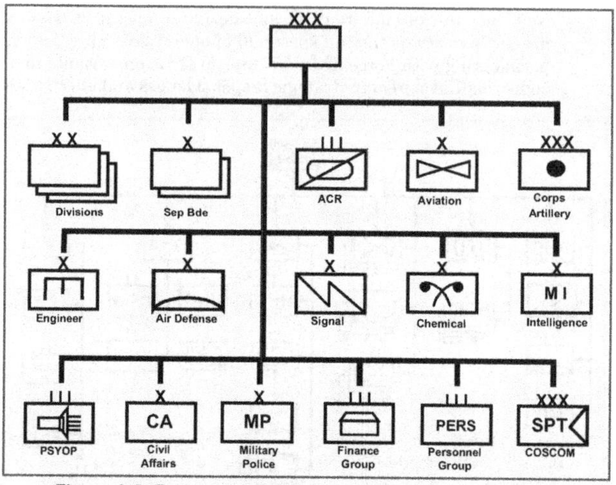

Figure A-8. Representative Corps Organizational Diagram

A-72. A corps is normally tailored to comprise two to five divisions of any type and combination required by the theater and the mission. It possesses organic support commands and is assigned combat and CS organizations based on its needs for a specific operation. Armored cavalry regiments and FA, engineer, ADA, and aviation brigades are the nondivisional units commonly available to the corps to weight its decisive operation and to perform special combat functions. The commander may also assign separate infantry or armor brigades to corps. Signal brigades, MI groups, and MP groups are the CS organizations normally present in a corps.

A-73. A corps plans and conducts battles and major operations. Corps operations are combined arms operations that synchronize tactical activities, including the maneuver of divisions, the fires of artillery units and supporting aerial forces, and the actions of CS and CSS units. A corps may operate under the control of a higher echelon, such as a numbered army. Its headquarters may provide the headquarters for a land component command of a unified or subunified command or JTF. The corps headquarters could also be employed as the ARFOR headquarters of a JTF. A corps commander may serve as a JTF commander. A corps normally exercises a mixture of operational and tactical responsibilities. It may have a key role in translating strategic and operational objectives of higher echelons into the specific and detailed tactics used to achieve those objectives.

A-74. Critical corps roles in the planning, preparing, executing, and assessing full spectrum operations include—

- Planning, preparing, and executing operations with other elements of the joint force to achieve campaign objectives, including force projection of corps units to the AO.
- Integrating available joint, interagency, and multinational assets in support of sustained land operations, including intelligence, target acquisition, target attack, electronic warfare, suppression of enemy air defenses, and CSS.
- Collecting intelligence, and anticipating enemy activities and intentions.
- Providing the information pipelines over which the corps distributes the common operational picture to all corps units.
- Planning, preparing, executing, and assessing shaping and sustaining operations by corps troops that support the corps decisive operations.

With augmentation a corps headquarters can perform as an ARFOR, joint force land component, or JTF headquarters. (FM 3-92 describes how corps conduct full spectrum operations.)

Appendix B

Tactical Mission Tasks

Doctrine provides a military organization with a common philosophy, a common language, a common purpose, and a unity of effort.

General George H. Decker, address at Fort Leavenworth, 16 Dec 1960

The tactical mission tasks in this appendix describe the results or effects the commander wants to achieve—the *what* and *why* of a mission statement not previously addressed in this manual. A mission statement contains the *who, what, when, where,* and *why* associated with a specific operation. The *what* and *why* of a mission statement are not the same thing and both are needed. The *what* is an effect that is normally measurable. The *why* of a mission statement provides the mission's purpose or reason. These tasks have specific military definitions that are different from those found in a dictionary. In some cases, the discussion includes more than just a definition; it includes results or effects in relation to the enemy, terrain, or friendly forces not associated with any specific type or form of an operation. Tasks that identify a friendly action rarely provide sufficient clarity for a mission statement.

B-1. As shown in Figure B-1, there is no definitive list of words or terms to describe the *what* and the *why* of a mission state-ment. The commander is not limited to the tactical mission tasks listed in this appendix in specify-ing what actions he wants from his subordinates in an OPORD or OPLAN. Many of the words and terms used to describe the *what* and *why* of a mission statement do not have special connotations beyond their common English

CONTENTS

language meanings. However, both the commander and the subordinate must have a common understanding of the *what* and *why* of the operation. It should be noted that tasks involving only actions by friendly forces rarely provide sufficient clarity for a mission statement thus the addition of a solid purpose coupled with the task adds understanding and clarity. The commander ensures that the missions he assigns his subordinate units are consistent with his scheme of maneuver and the resources allocated to his subordinates. For example, a defending unit requires far greater effort (resources) to destroy an enemy force than to defeat it. Likewise, an attacking unit requires more combat power to clear the enemy from a given area than to contain him in that same area.

B-2. Many of the tactical mission tasks in this appendix have a tactical mission graphic associated with them. Tacticians use these graphics in conjunction with course of action development.

ACTIONS BY FRIENDLY FORCE		EFFECTS ON ENEMY FORCE
Assault*	Follow and Assume	Block
Attack-by-Fire	Follow and Support	Canalize
Breach	Linkup*	Contain
Bypass	Occupy	Defeat
Clear	Reconstitutioin**	Destroy
Combat Search and Rescue	Reduce	Disrupt
Consolidation & Reorganization*	Retain	Fix
Control	Secure	Interdict
Counterreconnaissaince	Seize	Isolate
Disengagement	Support-by-Fire	Neutralize
Exfiltrate	Suppress	Penetrate
		Turn

TYPES AND FORMS OF OPERATIONS	
MOVEMENT TO CONTACT*	RETROGRADE OPERATIONS*
Search and Attack*	Delay*
ATTACK*	Withdrawal*
Ambush*	Retirement
Demonstration*	RECONNAISSANCE OPERATIONS**
Feint*	SECURITY OPERATIONS*
Raid*	INFORMATION OPERATIONS**
Spoiling Attack*	COMBINED ARMS BREACH OPERATIONS**
EXPLOITATION*	PASSAGE OF LINES*
PURSUIT*	RELIEF IN PLACE*
FORMS OF OFFENSIVE MANEUVER*	RIVER CROSSING OPERATIONS**
Envelopment*	TROOP MOVEMENT*
Frontal Attack*	Administrative Movement*
Infiltration*	Approach March*
Penetration*	Road March*
Turning Movement*	
AREA DEFENSE*	*Defined elsewhere in this manual
MOBILE DEFENSE**	**Defined in other manuals

Figure B-1. Tactical Mission Tasks

ACTIONS BY FRIENDLY FORCE

ATTACK-BY-FIRE

B-3. ***Attack-by-fire* is a tactical mission task in which a commander uses direct fires, supported by indirect fires, to engage an enemy without closing with him to destroy, suppress, fix, or deceive him.** A commander assigning this task to a subordinate must also state the desired effect on the enemy, such as neutralize, fix, or disrupt. A commander normally employs this task when the mission does not dictate or support close combat and occupation of a geographical objective by another friendly force. The commander may assign the force conducting an attack by fire a battle position with either a sector of fire or an engagement area (EA), or he may assign it an axis of advance and a force-oriented objective. The enemy may be stationary or moving. Figure B-2 shows the tactical mission graphic for attack by fire. The arrow points at the targeted force or objective, and the commander places the base of the arrow in the general area from which he wants to deliver the attack.

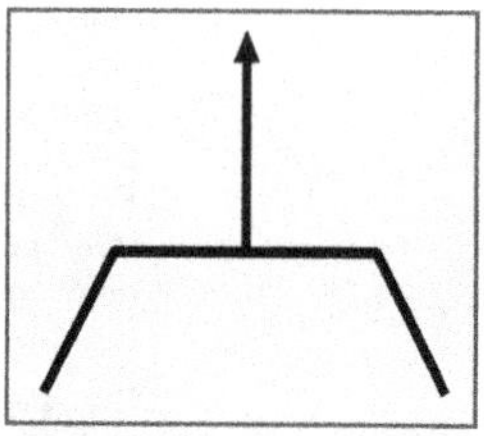

Figure B-2. Attack-by-Fire Tactical Mission Graphic

B-4. An attack by fire closely resembles the task of support by fire. The chief difference is that one unit conducts the support-by-fire task to support another unit so it can maneuver against the enemy. The attack by fire task includes—

- Assigning sectors of fire or EAs to each subordinate weapon system to include the enemy's defensive positions or avenues of approach.
- Designating control measures to allow massing, distributing, and shifting of direct and indirect fires.
- Designating battle positions, area of operations (AO), or axis of advance to allow the friendly force to engage the enemy.
- Providing for security and all-around defense, including control measures to ensure tie-in of subordinate elements and maximum use of hide positions.
- OPSEC to deceive the enemy about movement, occupation, and intent of the operation.
- Reconnaissance, preparing and securing movement routes and firing positions before the movement of the main body, and stocking Class V items.
- Movement instructions to the initial battle positions.

BREACH

B-5. ***Breach* is a tactical mission task in which the unit employs all available means to break through or secure a passage through an enemy defense, obstacle, minefield, or fortification.** A commander attempts to bypass and avoid obstacles and enemy defensive positions to the maximum extent possible to maintain tempo and momentum. Breaching enemy defenses and obstacle systems is normally his last choice. When they occur, they are a synchronized combined arms operation under the control of the maneuver commander. (Figure B-3 shows the control graphic for a breach.) The area located between the arms of the graphic shows the general location for the breach. The length of the arms extend to include the entire depth of the area that must be breached. Breaching operations may be required to support an attack anywhere along the continuum from deliberate to hasty attack. Regardless of where the attack falls along the continuum, the breaching tenets—intelligence, breaching fundamentals, breaching organization, mass, and synchronization—apply when conducting breaching operations in support of an attack. (FM 3-34.2 gives detailed information concerning breaching operations.)

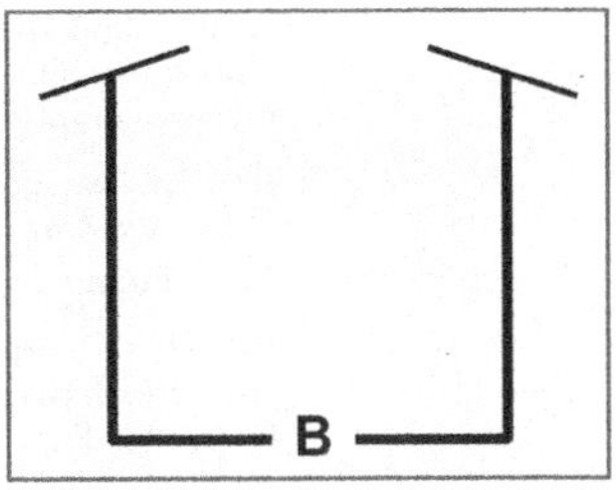

Figure B-3. Breach Tactical Mission Graphic

BYPASS

B-6. **Bypass is a tactical mission task in which the commander directs his unit to maneuver around an obstacle, position, or enemy force to maintain the momentum of the operation while deliberately avoiding combat with an enemy force.** A commander orders a bypass and directs combat power toward mission accomplishment. A bypass can take place in offensive or defensive actions. (Figure B-4 shows the tactical mission graphic for a bypass.) The arms of the graphic go on both sides of the location or unit that will be bypassed.

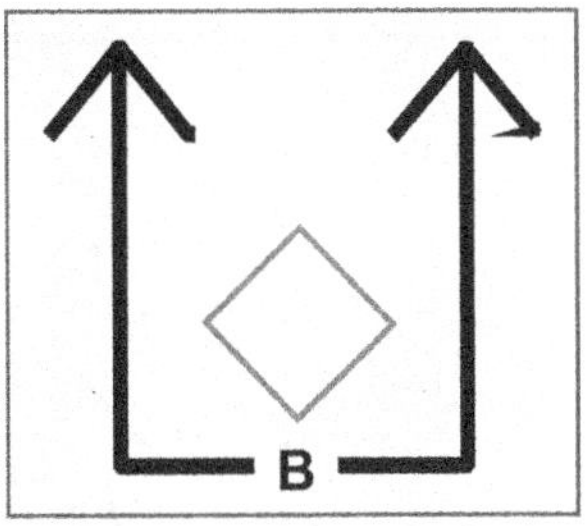

Figure B-4. Bypass Tactical Mission Graphic

B-7. The commander bases his bypass decision on—

- The requirement to maintain momentum and aggressive action.
- Knowledge of enemy strength, intent, or mission.
- The degree to which the bypassed enemy can interfere with the advance.
- The general state of the enemy force; for example, if enemy resistance is crumbling, the friendly force can take greater risks.
- Any bypass criteria established by a higher headquarters.

B-8. The force conducting the bypass immediately reports any bypassed obstacles and enemy forces to its higher headquarters. The force normally keeps the bypassed enemy under observation until relieved by another force unless as part of a raid. A senior commander does not normally delegate authority to bypass below the battalion task force level. He establishes bypass criteria to limit the size of the enemy force that can be bypassed without the authority of the next higher commander. Before approving the bypass, the commander ensures that the bypassing force checks the bypass route for enemy presence and trafficability. At no time can the bypassing force allow the bypassed enemy force to interfere with the moving friendly force.

B-9. The two bypass techniques that the force can employ are—

- Avoiding the enemy totally.
- Fixing the enemy in place with fires and then conducting the bypass.

B-10. If the force cannot avoid the enemy, the bypassing force must fix the enemy with part of its maneuver elements and bypass with the balance of the force. (See Figure B-5.) Generally, a commander will not attempt to bypass an enemy force if more than a third of his combat power is required to fix the enemy. The commander assigns one subordinate unit the mission of fixing the enemy in this situation, reinforcing the fixing force as required by the factors of METT-TC. The fixing force coordinates with the unit assigned to relieve him as soon as possible and provides the new commander with all available information about the enemy and terrain. The relieving unit is normally another unit assigned a follow-and-support task. Once relieved, the force fixing the enemy either rejoins its parent organization or becomes part of the following element and comes under its control.

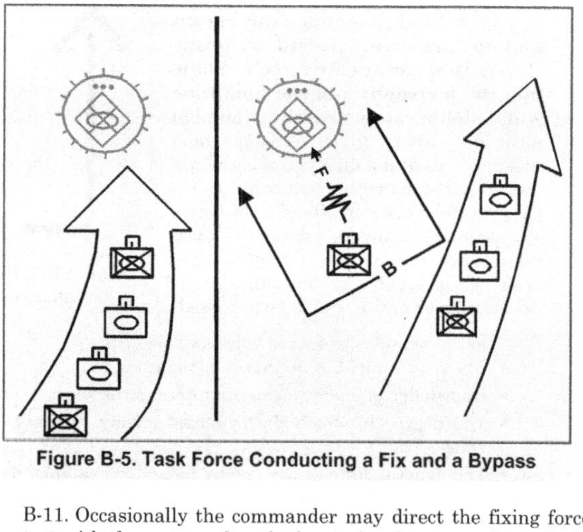

Figure B-5. Task Force Conducting a Fix and a Bypass

B-11. Occasionally the commander may direct the fixing force to break contact with the enemy after the bypassing force completes the bypass. This occurs when the bypassing force has no requirement to maintain an uninterrupted logistics flow, such as in a raid. In this case, the fixing force fixes the enemy by employing defensive and limited offensive actions in synchronization with all available fire support until ordered to rejoin the bypassing force.

CLEAR

B-12. ***Clear*** **is a tactical mission task that requires the commander to remove all enemy forces and eliminate organized resistance within an assigned area.** The force does this by destroying, capturing, or forcing the withdrawal of enemy forces so they cannot interfere with the friendly unit's mission. In all cases, this task requires a thorough reconnaissance to discover the enemy's locations. After discovering the location, the clearing force maneuvers against the enemy force. (Figure B-6 shows the tactical mission graphic for clear.) The bar connecting the arrows designates the desired limit of advance for the clearing force. The bar also establishes the width of the area to clear.

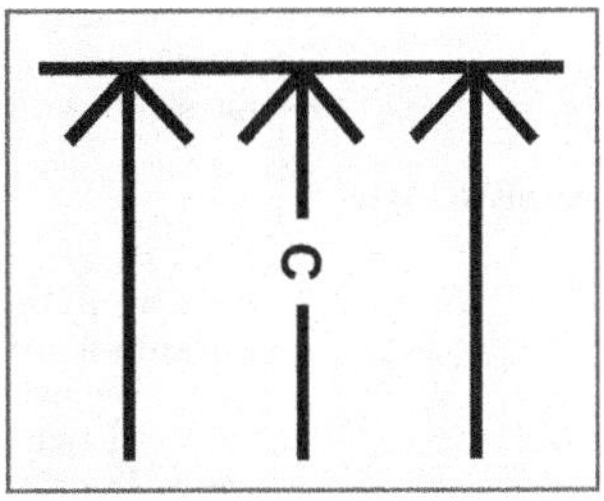

Figure B-6. Clear Tactical Mission Graphic

B-13. This task requires significant time and other resources. In his mission statement, a commander can modify the objective associated with this task to destroying, capturing, or forcing the withdrawal of only enemy forces larger than a stated size. In this case, the clearing force keeps smaller enemy forces under observation while the rest of the friendly force bypasses them.

CONTROL

B-14. ***Control*** **is a tactical mission task that requires the commander to maintain physical influence over a specified area to prevent its use by an enemy or to create conditions necessary for successful friendly operations.** That influence can result from friendly forces occupying the specified area or dominating that area by their weapon systems. Control of an area does not require the complete clearance of all enemy soldiers from the specified area. The tactical mission task of control differs from that of secure because secure does not allow enemy fires to impact on the secured area. The enemy can engage targets within the controlled area but cannot move his ground forces through that area.

B-15. Control may also mean a command relationship or a function commanders exercise through their C2 system. (See FMs 3-0, 3-52, and 6-0.)

COUNTERRECONNAISSANCE

B-16. ***Counterreconnaissance*** **is a tactical mission task that encompasses all measures taken by a commander to counter enemy reconnaissance and surveillance efforts. Counterreconnaissance is not a distinct mission, but a component of all forms of security operations.** It prevents hostile observation of a force or area. Counterreconnaissance is an element of all security operations and most local security measures. It involves both active and passive elements and includes combat action to destroy or repel enemy reconnaissance units and surveillance assets.

B-17. Destroying enemy ground reconnaissance assets while denying the enemy information through other collection systems allows friendly force commanders to operate against an enemy who is operating blindly. The enemy commander's inability to see the battlefield eventually desynchronizes his actions and renders his command vulnerable to aggressive action by friendly forces. (See Chapter 12 for additional information on counterreconnaissance.)

DISENGAGE

B-18. ***Disengage* is a tactical mission task where a commander has his unit break contact with the enemy to allow the conduct of another mission or to avoid decisive engagement.** It involves moving to a location where the enemy cannot engage the friendly force with either direct fires or observed indirect fires. Disengaging from the enemy while displacing from one position to the next is a difficult procedure. A disengagement plan includes—

- The maneuver concept of operations for tactical elements after disengagement, along with the movement routes for each subordinate unit.
- Fires to suppress the enemy and cover the unit's movement.
- Screening smoke to conceal the unit's movement, as part of a deception operation, or to cover passage points.
- Contact and passage points if moving through friendly lines. (See Chapter 15.)
- The time disengagement starts.
- The earliest time that CS and CSS elements move.

B-19. The senior headquarters conducts operations to support the disengaging forces and relieve pressure on units in contact with the enemy. For example, if a division is conducting a delay, the division commander uses his aviation assets to help a ground maneuver brigade disengage from the close fight. Simultaneously, the division uses its long-range artillery, rocket, and EW systems to destroy or disrupt enemy follow-on echelons to prevent them from interfering with the disengagement. The intent is to create conditions that allow the unit to disengage while avoiding decisive combat.

B-20. To facilitate disengagement, the commander suppresses the enemy in contact by bombarding him with large volumes of both direct and indirect fires provided by forces other than the disengaging unit. In open terrain, the unit generally moves its short-range systems first. In close terrain, it generally moves its long-range systems first to support by fire positions. The time involved to move a system to its next position also affects when that system moves. Small unit leaders usually direct this movement because of the limited range of combat net radios and the fact that the tactical situation varies across a unit's front. The process repeats as necessary. Once disengagement starts, units must complete it rapidly. The commander can employ supporting units or reserves to protect the disengaging unit's flanks and assist in freeing any closely engaged elements. The unit then moves to its next position using the appropriate movement techniques. (See Chapter 14.)

B-21. If enemy combat systems have not closed within direct-fire range of the friendly disengaging unit, all its elements may be able to move simultaneously under the cover of intense fires and smoke. Speed of execution and continued coordination are essential to the success of this task.

EXFILTRATE

B-22. ***Exfiltrate* is a tactical mission task where a commander removes soldiers or units from areas under enemy control by stealth, deception, surprise, or clandestine means.** Friendly forces exfiltrate when they have been encircled by enemy forces and cannot conduct a breakout or be relieved by other friendly forces. Forces returning from a raid, an infiltration, or a patrol behind enemy lines can also conduct an exfiltration. The commander exfiltrates an encircled force to preserve a portion of the force; it is preferable to the capture of the entire force. A force exfiltrates only after destroying or incapacitating all equipment (less medical) it must leave behind. Only as a last resort, when the alternative is the capture of the entire force, does a force conducting an exfiltration leaves its casualties in place with supplies, chaplain support, and medical attendants.

B-23. Exfiltration is most feasible through rough or difficult terrain in areas lightly covered by enemy observation and fire. These conditions often allow undetected movement of small elements, when movement of the entire force would present more risk. Exfiltration requires resourcefulness, a high degree of discipline, expert land navigational skills, and motivation. It is unlikely that the entire force will be able to exfiltrate, since part of it may have to create a diversion. Good, small-unit leadership is essential in this type of operation.

B-24. The exfiltrating force first establishes its rally points and exfiltration lanes. It coordinates its linkup plans with other friendly units. The commander designates exfiltration lanes as restricted fire areas (RFAs) or no-fire areas (NFAs). The exfiltrating force uses preparatory fires to cover its movement and to expend stockpiled ammunition. Based on reconnaissance and available intelligence, the exfiltrating force subdivides into small groups and exfiltrates during periods of limited visibility, passing through or around enemy defensive positions. If detected, it tries to bypass the enemy. Exfiltration may be more difficult with combat and tactical vehicles because of the noise they make and the limitations they impose on exfiltration routes, make detection more likely.

FOLLOW AND ASSUME

B-25. ***Follow and assume* is a tactical mission task in which a second committed force follows a force conducting an offensive operation and is prepared to continue the mission if the lead force is fixed, attritted, or unable to continue.** The follow-and-assume force is not a reserve but is committed to accomplish specific tasks. Figure B-7 shows the tactical mission graphic for follow and assume. The commander places the box part of the graphic around the symbol of the unit being assigned this task.

B-26. Tasks for a follow-and-assume force include—

- Preparing to execute all missions of the followed unit.
- Maintaining contact with the trail elements of the leading force.
- Preparing to conduct a forward passage of lines through the force it is following.
- Monitoring all combat information and intelligence being provided to and from the force it is following.

- Avoiding engaging enemy forces bypassed by the force it is following.

B-27. A commander assigns a follow-and-assume mission to ensure that he can maintain the momentum of his offensive operation. The follow-and-assume force ensures that it can immediately execute a forward passage of lines and assume the mission of the lead force.

B-28. The commander assigning a unit the task of follow and assume has two options in establishing the relationship between the lead and trail units. He normally retains command of both units and requires that all requests for support from the supported unit to the supporting unit pass through his headquarters. Alternatively, in situations where the commander will not be able to maintain control over both units, he places the supporting unit in a standard command relationship with the supported unit, such as attached or operational control. An example of this occurs when both units are trying to encircle a retrograding enemy force and the commander remains with the direct-pressure force.

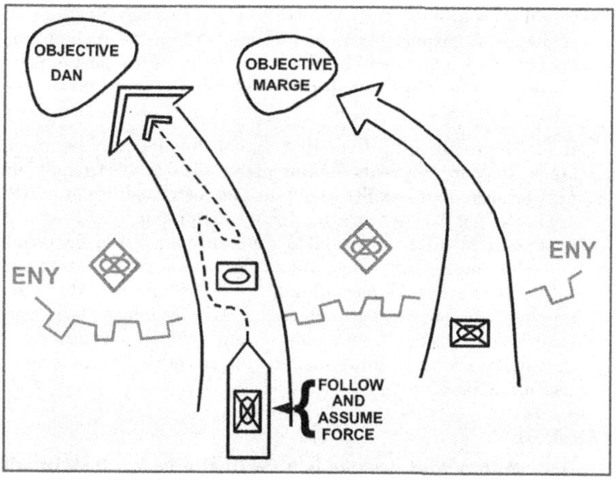

Figure B-7. Follow and Assume Tactical Mission Graphic

FOLLOW AND SUPPORT

B-29. ***Follow and support* is a tactical mission task in which a committed force follows and supports a lead force conducting an offensive operation.** The follow-and-support force is not a reserve but is a force committed to specific tasks. (Figure B-8 on page B-10 shows the tactical mission graphic for follow and support.) The commander places the box part of the graphic around the symbol of the unit being assigned this task.

B-30. Tasks for a follow-and-support force include—

- Destroying bypassed enemy units when the lead unit does not clear the AO as it advances.
- Blocking movement of enemy reinforcements.

- Relieving in place any direct-pressure or encircling force halted to contain the enemy.
- Securing lines of communication.
- Clearing obstacles.
- Guarding prisoners, key areas, and installations.
- Recovering friendly battle losses.
- Securing key terrain.
- Controlling dislocated civilians.

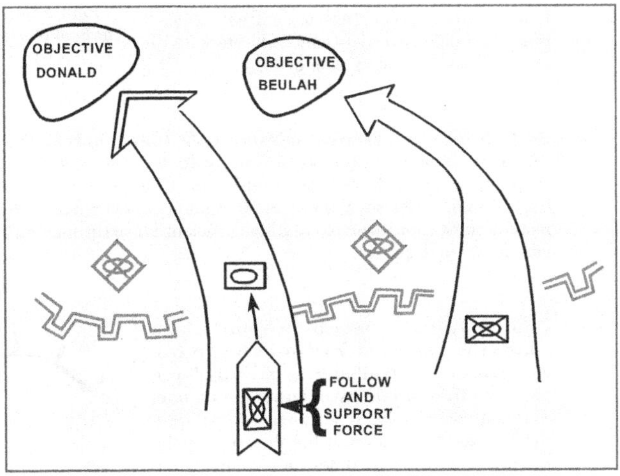

Figure B-8. Follow and Support Tactical Mission Graphic

B-31. A commander assigns a unit the task of follow and support to keep the supported force from having to commit its combat power to tasks other than the decisive operation, which would slow the offensive operation's momentum and tempo. The follow-and-support force must accomplish its tasks to prevent the enemy, obstacles, and other factors from interfering with offensive operations, especially along the lines of communications.

B-32. The commander assigning the follow-and-support task has two options in establishing the relationship between the supported and the supporting units. He can place the follow-and-support unit in a standard command relationship with the supported unit, such as attached or operational control. Alternatively, he can retain command of the follow-and-support force and require that all tasking request from the supported unit go through his headquarters.

OCCUPY

B-33. ***Occupy*** **is a tactical mission task that involves moving a friendly force into an area so that it can control that area. Both the**

force's movement to and occupation of the area occur without enemy opposition. A unit can control an area without occupying it, but not vice versa. That is the difference between the tactical mission tasks of occupy and control. (Figure B-9 shows the occupy tactical mission graphic.) The *X* on the tactical mission graphic has no significance, but the graphic should encompass the entire area that the commander desires to occupy. Units typically occupy assembly areas, objectives, and defensive positions. (Chapters 2, 3, and 8 discuss this process in detail.)

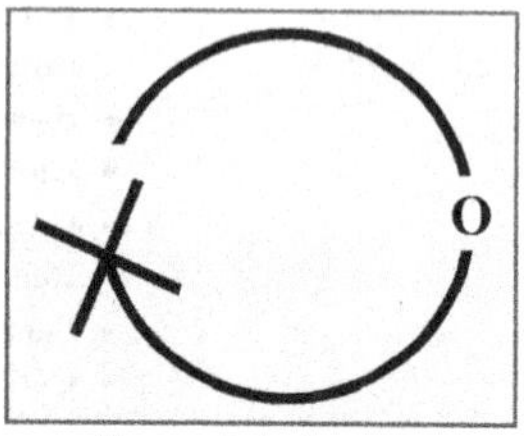

Figure B-9. Occupy Tactical Mission Graphic

REDUCE

B-34. ***Reduce* is a tactical mission task that involves the destruction of an encircled or bypassed enemy force.** There is no tactical mission graphic for this task. This task can occur at any location on the battlefield. (Appendix D discusses the reduction of an encircled enemy.) Reduce is also a mobility task that involves creating sufficient lanes through an obstacle to negate its intended effect.

RETAIN

B-35. ***Retain* is a tactical mission task in which the commander ensures that a terrain feature controlled by a friendly force remains free of enemy occupation or use.** The commander assigning this task must specify the area to retain and the duration of the retention, which is time- or event-driven. While a unit is conducting this task, it expects the enemy to attack and prepares to become decisively engaged. A unit tasked to retain a specific piece of terrain does not necessary have to occupy it. (Figure B-10 shows the tactical mission graphic for retain.) The direction of the arrow has no significance, but the graphic should include the entire area the commander wants to retain.

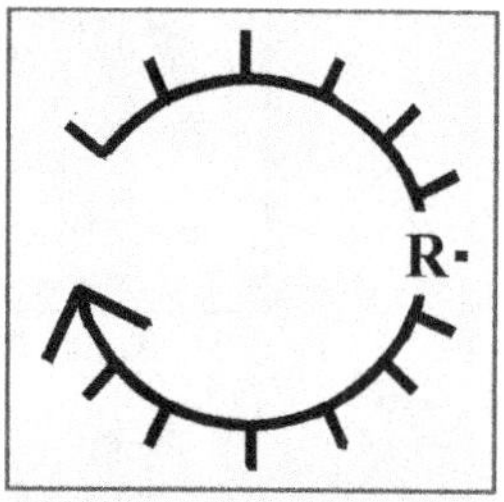

Figure B-10. Retain Tactical Mission Graphic

SECURE

B-36. ***Secure* is a tactical mission task that involves preventing a unit, facility, or geographical location from being damaged or destroyed as a result of enemy action.** This task normally involves conducting area security operations. (See Chapter 12.) A force given the mission of securing a unit, facility, or geographical location not only prevents enemy forces from over-running or occupying the secured location, but also prevents enemy direct fires and observed indirect fires from impacting the secured location. This is the primary difference between control and secure. The control tactical mission task allows enemy direct and indirect fires to affect the location being controlled. A unit does not have to physically occupy the area immediately around the unit, facility, or geographical location it is securing if it can prevent the enemy from occupying or firing at that location by other means. The commander states the mission duration in terms of time or event when assigning a mission to secure a given unit, facility, or geographical location. (Figure B-11 shows the tactical mission graphic for secure.) The direction of the arrow has no significance, but the graphic should include the entire area the commander wants to secure.

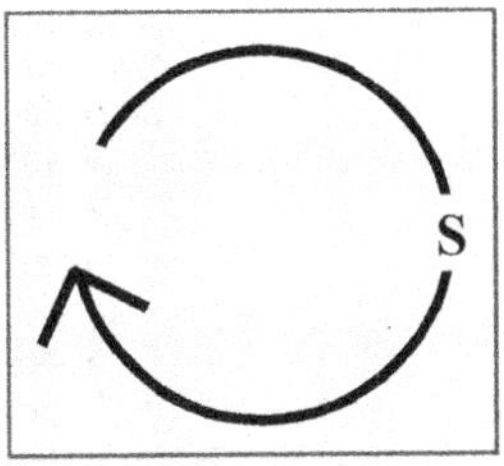

Figure B-11. Secure Tactical Mission Graphic

SEIZE

B-37. ***Seize*** **is a tactical mission task that involves taking possession of a designated area by using overwhelming force.** (Figure B-12 shows the tactical mission graphic for seize.) An enemy force can no longer place direct fire on an objective that has been seized. The arrow points to the location or objective to seize. This task differs from secure because it requires offensive action to obtain control of the designated area or objective. It differs from the task of occupy because it involves overcoming anticipated enemy opposition. Once a force seizes a physical objective, it clears the terrain within that objective by killing, capturing, or forcing the withdrawal of all enemy forces.

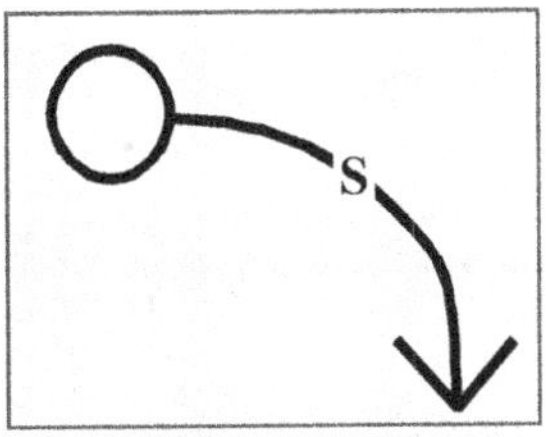

Figure B-12. Seize Tactical Mission Graphic

SUPPORT-BY- FIRE

B-38. ***Support-by-fire*** **is a tactical mission task in which a maneuver force moves to a position where it can engage the enemy by direct fire in support of another maneuvering force.** The primary objective of the support force is normally to fix and suppress the enemy so he cannot effectively fire on the maneuvering force. The secondary objective is to destroy the enemy if he tries to reposition. The commander must specify the desired effect on the enemy when assigning this task to a subordinate.

B-39. A unit conducting the task of support by fire does not maneuver to capture enemy forces or terrain. The commander gives this task to another unit as part of a larger maneuver. When assigning a support-by-fire mission, the commander designates the enemy, when to attack, the general location from which to operate, the friendly force to support, and the purpose of the task, such as fix or suppress. (Figure B-13 shows the tactical mission graphic for support by fire.) The ends of the arrows should point in the general direction of the targeted unit or location. The base of the area indicates the general area from which to deliver fires.

B-40. Once the commander gives an element the task of support by fire, it should occupy support by fire positions that have cover and concealment, good observation, and clear fields of fire. Elements occupying support-by-fire positions should—

- Check the security of the position.
- Search for targets.
- Orient weapons on likely or suspected enemy positions.
- Assume fighting positions that provide some degree of protection. Heavy forces occupy hull-down firing positions, while light forces use trees, natural berms, buildings, and similar existing terrain features.

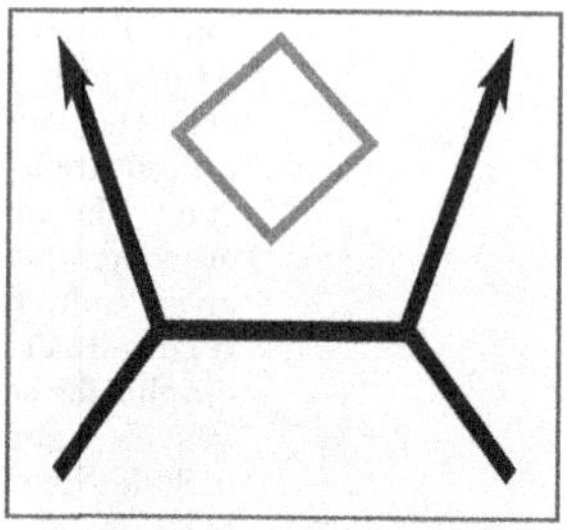

Figure B-13. Support-by-Fire Position Tactical Mission Graphic

- Assign observation sectors to each soldier or weapon system in the support-by-fire element.
- Use its available thermal sights to locate heat sources not visible to the naked eye, such as vehicles concealed in tree lines or other wooded areas or personnel serving at OPs.

B-41. Support by fire closely resembles the task of attack by fire. The difference is that support by fire supports another force so it can maneuver against the enemy, while an attack by fire does not support the maneuver of another friendly force.

EFFECTS ON ENEMY FORCE

BLOCK

B-42. ***Block* is a tactical mission task that denies the enemy access to an area or prevents his advance in a direction or along an avenue of approach.** A blocking task normally requires the friendly force to block the enemy force for a certain time or until a specific event has occurred. The line perpendicular to the enemy's line of advance indicates the limit of enemy advance. A blocking unit may have to hold terrain and become decisively engaged. (Figure B-14 illustrates the tactical mission graphic for a blocking task.)

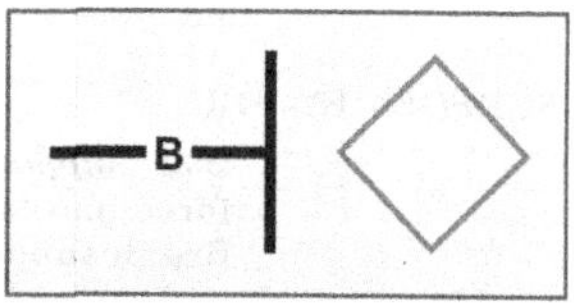

Figure B-14. Block Tactical Mission Graphic

B-43. ***Block* is also an engineer obstacle effect that integrates fire planning and obstacle effort to stop an attacker along a specific avenue of approach or prevent him from passing through an engagement area.** The vertical line in the obstacle effect graphic indicates the limit of enemy advance. They also indicate where the obstacle ties in to restricted terrain. (Figure B-15 on page B-14 illustrates the block obstacle effect graphic.) A blocking force may employ blocking obstacles to assist in the task. Blocking obstacles are complex, employed in depth, and integrated with fires to prevent the enemy from proceeding along an avenue of approach, or to proceed only at unacceptable cost. When employed, blocking obstacles should serve as a limit, not allowing the enemy beyond that point. Obstacles alone cannot accomplish a blocking task. (FM 3-34.1 describes the block engineer obstacle effect.)

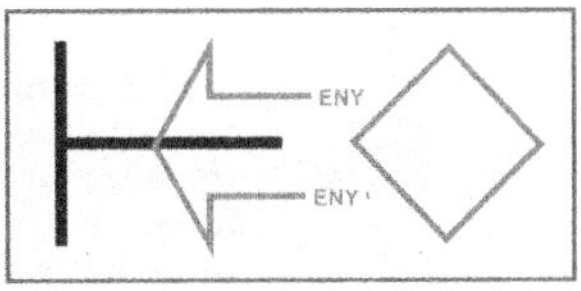

Figure B-15. Block Obstacle Effect Graphic

B-44. Block as a tactical mission task differs from the tactical mission task of fix because a blocked enemy force can move in any direction other than the obstructed one, while a fixed enemy force cannot move in any direction.

CANALIZE

B-45. ***Canalize*** **is a tactical mission task in which the commander restricts enemy movement to a narrow zone by exploiting terrain coupled with the use of obstacles, fires, or friendly maneuver.** (See Figure B-16.) Figure B-17 shows how successful canalization results in moving the enemy formation or individual soldiers and weapon systems into a predetermined position where they are vulnerable to piecemeal destruction by direct and indirect fires.

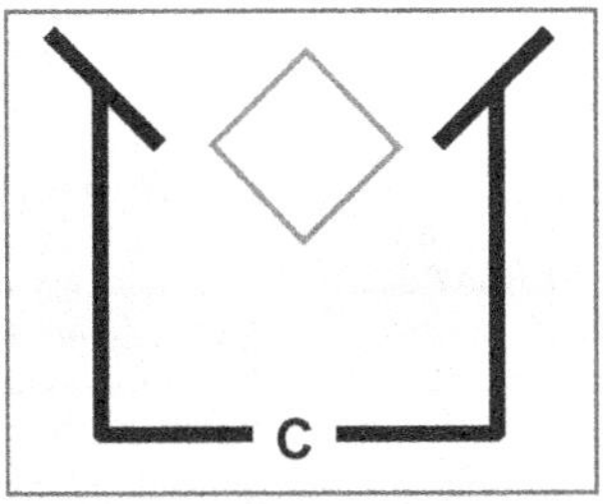

Figure B-16. Canalize Tactical Mission Graphic

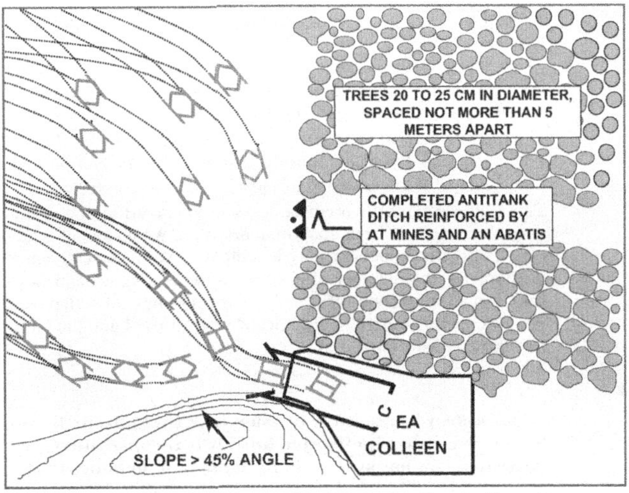

Figure B-17. Canalizing Terrain Enhanced by Obstacles Used with an Engagement Area

CONTAIN

B-46. ***Contain*** **is a tactical mission task that requires the commander to stop, hold, or surround enemy forces or to cause them to center their activity on a given front and prevent them from withdrawing any part of their forces for use elsewhere.** Containment allows an enemy to reposition himself within the designated geographical area, whereas fixing an enemy does not. Geo-graphic terms or time may express the limits of the containment. The contain graphic encompasses the entire geographical area in which the commander desires to contain the enemy during the development of alternative courses of action. (Figure B-18 shows the tactical control graphic for contain.)

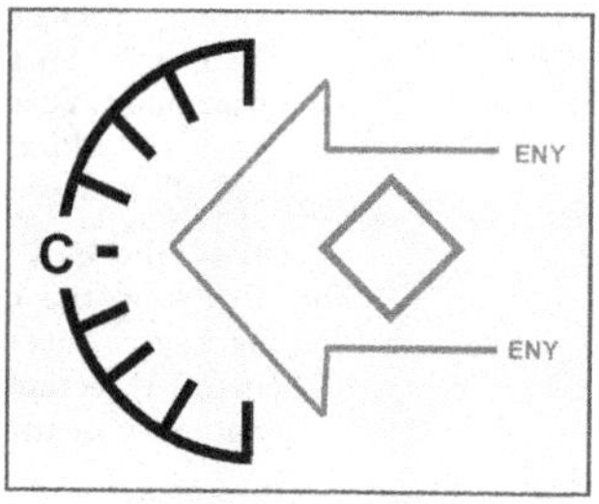

Figure B-18. Contain Tactical Mission Graphic

DEFEAT

B-47. ***Defeat*** **is a tactical mission task that occurs when an enemy force has temporarily or permanently lost the physical means or the will to fight. The defeated force's commander is unwilling or unable to pursue his adopted course of action, thereby yielding to the friendly commander's will and can no longer interfere to a significant degree with the actions of friendly forces. Defeat can result from the use of force or the threat of its use.**

B-48. A commander can generate different effects against an enemy to defeat him:

- Physical. The enemy loses the physical means to continue fighting. He no longer has the personnel, weapon systems, equipment, or supplies to carry out his assigned mission.
- Psychological. The enemy loses the will to fight. He becomes mentally exhausted, and his morale is so low that he can no longer continue to carry out his assigned mission.

These effects typically occur as a result of catastrophic losses inflicted over a very short time or from sustained attrition. An opponent who is not ideologically motivated opponent may be defeated psychologically on observing preparations for the delivery of clearly overwhelming combat power on his position. Defeat manifests itself in some sort of physical action, such as mass surrenders, abandonment of significant quantities of equipment and supplies, or retrograde operations.

DESTROY

B-49. ***Destroy*** **is a tactical mission task that physically renders an enemy force combat-ineffective until it is reconstituted. Alternatively, to destroy a combat system is to damage it so badly that it cannot perform any function or be restored to a usable condition without being entirely rebuilt.** The amount of damage needed to render a unit combat-ineffective depends on the unit's type, discipline, and morale. Destroying armored or dug-in targets with area fire weapons requires considerable ammunition and time, so forces do not normally attempt it unless they have terminally guided munitions. (Figure B-19 shows the tactical mission graphic for destroy.)

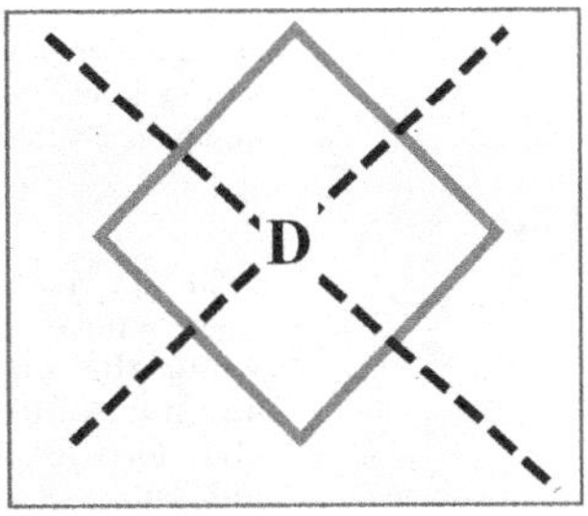

Figure B-19. Destroy Tactical Mission Graphic

DISRUPT

B-50. ***Disrupt* is a tactical mission task in which a commander integrates direct and indirect fires, terrain, and obstacles to upset an enemy's formation or tempo, interrupt his timetable, or cause his forces to commit prematurely or attack in a piecemeal fashion.** This increases the enemy's vulnerability to friendly fires. It may temporarily knock a unit out of the battle. Disruption is never an end; it is the means to an end. (Figure B-20 shows the tactical mission graphic for disrupt. The center arrow points toward the targeted enemy unit.)

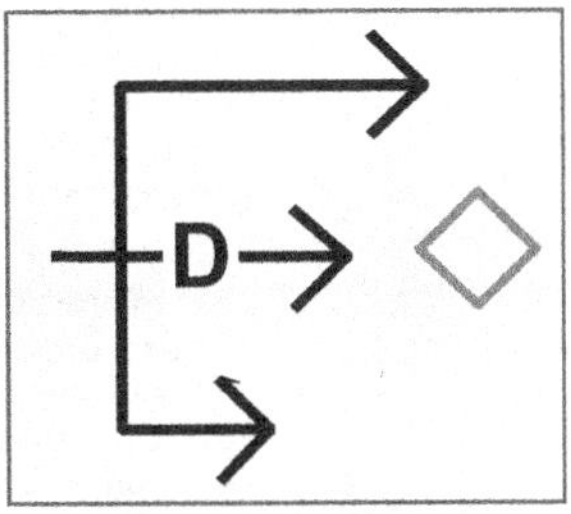

Figure B-20. Disrupt Tactical Mission Graphic

B-51. The maneuver force attempting to disrupt an enemy must attack him with enough combat power to achieve desired results with one mass attack or sustain the attack until it achieves the desired results. It may involve attacking the enemy while he is still in his assembly areas or in an approach march before he can deploy into a combat formation. The commander deter-mines the amount of risk he is willing to accept based on anticipated friendly losses, the location of the attack, and the number of attacks.

B-52. ***Disrupt* is also an engineer obstacle effect that focuses fire planning and obstacle effort to cause the enemy to break up his formation and tempo, interrupt his timetable, commit breaching assets prematurely, and attack in a piecemeal effort.** It also helps to deceive the enemy concerning the location of friendly defensive positions, to separate combat echelons, or to separate combat forces from their logistic support. As shown in Figure B-21, the short arrow(s) in the obstacle-effect graphic indicates where obstacles impact the enemy's ability to maneuver. The longer arrow(s) indicate where the commander allows the enemy to bypass the obstacle effect so he can attack him with fires. The arrows indicate the direction of enemy attack. A defending

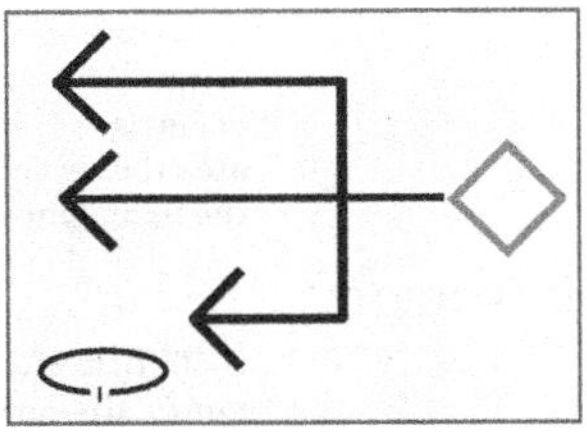

Figure B-21. Disrupt Obstacle Effect Graphic

commander normally uses the disrupt obstacle effect forward of his EAs. Obstacles alone cannot disrupt an enemy unit. (FM 3-34.1 describes the disrupt engineer obstacle effect.)

FIX

B-53. ***Fix* is a tactical mission task where a commander prevents the enemy from moving any part of his force from a specific location for a specific period.** This may occur by engaging him to prevent his withdrawal for use elsewhere, or by using deception, such as transmitting false orders. The commander uses fix in offensive and defensive actions; it is always a shaping operation. (Figure B-22 shows the tactical mission graphic for fix.) The commander points the arrow toward the enemy unit that he desires to fix. The broken part of the arrow indicates the desired location for that event to occur.

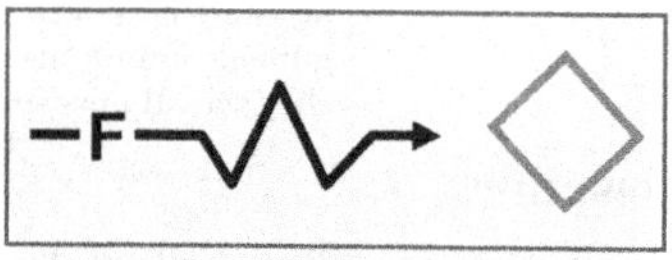

Figure B-22. Fix Tactical Mission Graphic

B-54. Fixing an enemy force does not mean destroying it. The friendly force has to prevent the enemy from moving in any direction. This task usually has a time constraint, such as fix the enemy reserve force until OBJECTIVE FALON, the decisive operation, is secured. The tactical mission task of fix differs from that of block in that a fixed enemy force cannot move from a given location, but a blocked enemy force can move in any direction other than the one obstructed.

B-55. ***Fix* is also an engineer obstacle effect that focuses fire planning and obstacle effort to slow an attacker's movement within a specified area, normally an engagement area.** Primary use of this effect is to give the friendly unit time to acquire, target, and destroy the attacking enemy with direct and indirect fires throughout the depth of an EA or avenue of approach. The irregular part of the arrow in the obstacle-intent graphic indicates the location where the enemy's rate of advance will be slowed by complex obstacles. The arrow indicates the direction of enemy advance. (See Figure B-23.) (FM 3-34.1 describes the fix engineer obstacle effect.)

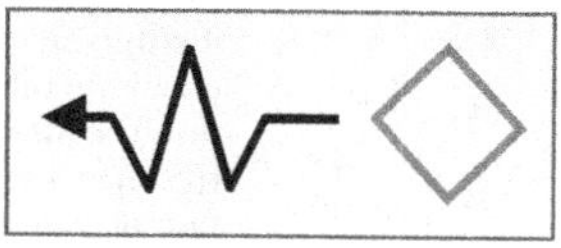

Figure B-23. Fix Obstacle Intent Graphic

INTERDICT

B-56. ***Interdict* is a tactical mission task where the commander prevents, disrupts, or delays the enemy's use of an area or route.** Interdiction is a shaping operation conducted to complement and reinforce other ongoing offensive or defensive operations. (Figure B-24 on page B-18 shows the tactical mission graphic for interdict.) The two arrows should cross on the unit or location targeted for interdiction. An interdiction tasking must specify how long to interdict, defined as a length of time or some event that must occur before the interdiction is lifted, or the exact effect desired from the interdiction.

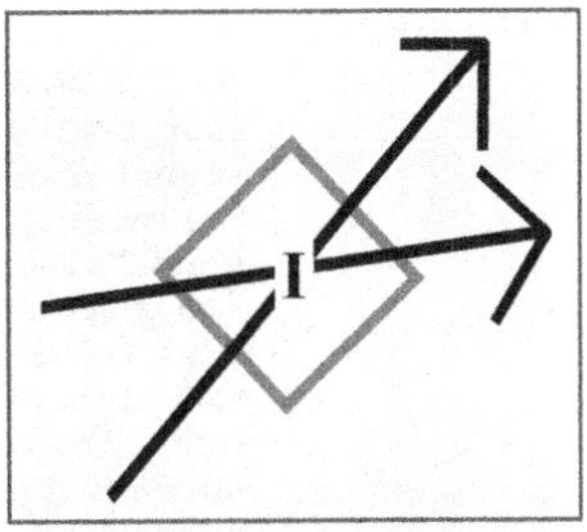

Figure B-24. Interdict Tactical Mission Graphic

B-57. The depth at which the attacking force conducts the interdiction generally determines the friendly force's freedom of action. Increasing the depth of operations reduces the danger of fratricide to air and surface forces, reduces the coordination required, and allows increasingly flexible operations. With more freedom of action, aerial forces leave the enemy with no location immune from attack.

B-58. The depth at which interdiction takes place also determines the speed with which its effects are observed. Normally, ground maneuver units first focus on targets close to the forward of line own troops (FLOT). Interdiction efforts there have immediate impact on enemy forces near the interdiction target but do not affect the enemy's ability to mass force effects. Attacks at greater distances from the FLOT have a delayed impact on close combat but eventually degrade the enemy's ability to mass effects.

B-59. The friendly force's capability to interdict may have a devastating impact on the enemy's plans and ability to respond to friendly actions. For example, interdiction efforts that result in the enemy's maneuver being delayed or disrupted enhances the friendly force's ability to achieve tactical advantages. Delaying or disrupting enemy resupply efforts limits his ability to sustain intense, high-tempo offensive or defensive operations and restricts the mobility of his forces.

B-60. Interdicting the movement of enemy units can be extremely effective in assisting their encirclement and eventual destruction. Fixed enemy ground forces—or those trapped by the loss of their mobility—provide lucrative targets. The commander should plan to interdict withdrawing enemy forces to enhance his pursuit. While interdiction can contribute to success by hampering reinforcement and resupply, it can also contribute by trapping enemy forces or canalizing their maneuvers, leading to their destruction in detail.

ISOLATE

B-61. ***Isolate* is a tactical mission task that requires a unit to seal off—both physically and psychologically—an enemy from his sources of support, deny him freedom of movement, and prevent him from having contact with other enemy forces.** A commander does not allow an isolated enemy sanctuary within his present position but continues to conduct offensive actions against him. (Figure B-25 shows the tactical mission graphic for isolate.) The position or direction of the arrow has no significance, but the graphic should surround the targeted enemy unit.

Figure B-25. Isolate Tactical Mission Graphic

NEUTRALIZE

B-62. ***Neutralize*** **is a tactical mission task that results in rendering enemy personnel or materiel incapable of interfering with a particular operation.** (Figure B-26 shows the neutralize tactical mission graphic.) The two lines cross over the symbol of the unit or facility targeted for neutralization. When assigning a task to neutralize, the commander must specify the enemy force or materiel to neutralize and the duration, which is time- or event-driven. The neutralized target may become effective again when casualties are replaced, damage is repaired, or effort resulting in the neutralization is lifted. The commander normally uses a combination of lethal and nonlethal fires to neutralize enemy personnel or materiel. The assets required to neutralize a target vary according to the type and size of the target and the weapon and munitions combination used.

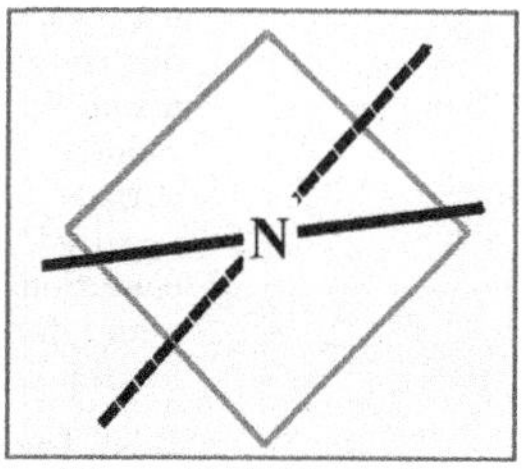

Figure B-26. Neutralize Tactical Mission Graphic

SUPPRESS

B-63. ***Suppress*** **is a tactical mission task that results in the temporary degradation of the performance of a force or weapon system below the level needed to accomplish its mission.** It occurs when a commander employs direct or indirect lethal fires, offensive information operations, or smoke on enemy personnel, weapons, and equipment to prevent or degrade enemy fires, sensors, and visual observation of friendly forces. As opposed to the neutralization task, the original target regains its effectiveness without needing to reconstitute once the effects of the systems involved in the suppression effort lift or shift to another target.

TURN

B-64. ***Turn*** **is a tactical mission task that involves forcing an enemy element from one avenue of approach or movement corridor to another.** The commander relates obstacles, fires, and terrain to improve his tactical situation while degrading the enemy's situation. For example, in the offense, a commander might want to turn an enemy force he is pursuing to place it in a position where he can destroy it. In the defense, a commander might want to turn an attacking enemy force to allow him to conduct a counterattack into its flank. (Figure B-27 shows the turn tactical mission graphic.) The place where the arrow breaks indicates the general location of the obstacle complex that will force the enemy to move from one avenue of approach to another.

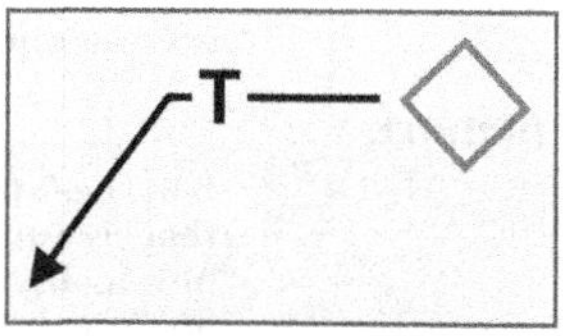

Figure B-27. Turn Tactical Mission Graphic

B-65. ***Turn*** **is also a tactical obstacle effect that integrates fire planning and obstacle effort to divert an enemy formation from one**

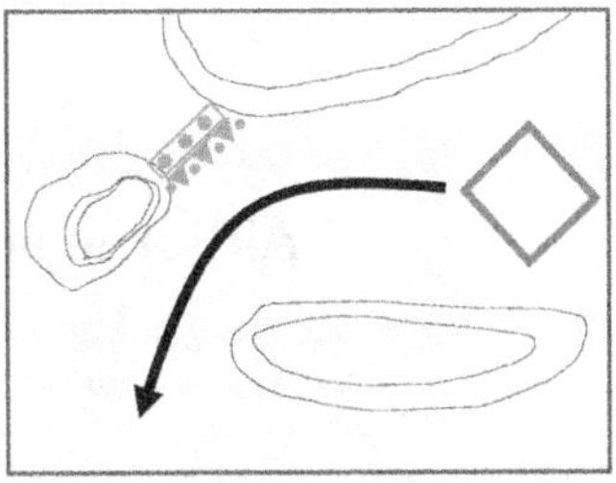

Figure B-28. Turn Obstacle Effect

avenue of approach to an adjacent avenue of approach or into an engagement area. Its development requires well-defined mobility corridors and avenues of approach. To achieve this effect, the obstacles have a subtle orientation relative to the enemy's approach as shown in Figure B-28. The obstacles and their associated fires allow bypasses in the direction desired by the friendly scheme of maneuver. Finally, the obstacles tie into restrictive terrain at the initial point of the turn. A commander normally uses the turn effect on the flanks of an EA. The direction of the arrow indicates the desired direction of turn. (See FM 3-34.1 for more information on tactical obstacle effects.)

Appendix C

Airborne and Air Assault Operations

A landing against organized and highly trained opposition is probably the most difficult undertaking which military forces are called upon to face.

General of the Army George C. Marshall, 1943

Joint force commanders conduct entry operations at the operational level of war. Commanders operating at the tactical level conduct airborne or air assault operations to gain a positional advantage or to envelop or turn the enemy. Airborne and air assault operations are types of entry operations that use a vertical envelopment to insert a force into an area of operations (AO). An enemy may or may not be in a position to oppose the operation. While the commander should attempt to achieve an unopposed landing, he must prepare for the presence of opposition.

C-1. The capability to conduct airborne and air assault operations allows the commander to—

- Threaten enemy rear areas, causing the enemy to divert combat elements to protect vital installations and hold key terrain.
- Overcome distances quickly, overfly barriers, and bypass enemy defenses.
- Extend the area over which he can exert his influence.
- Disperse his reserve forces widely for force protection reasons while maintaining their capability for effective and rapid response.
- Exploit his combat power by increasing tactical mobility.

COMMON FACTORS

C-2. Factors common to airborne and air assault operations are the use of the reverse planning process, condition setting, and the impact of meteorological conditions (weather and light data).

CONTENTS

REVERSE PLANNING PROCESS

C-3. An inverse sequence of detailed planning and joint coordination characterizes both operations. As a minimum, airborne and air assault plans include a–

- Ground tactical plan.
- Landing plan.
- Movement plan.
- Marshaling plan. (The air assault terminology for this last plan is the staging plan.)

Intelligence regarding the enemy and terrain characteristics of the objective area is vital to this process.

C-4. Airborne and air assault operations result in establishing positions that support completing the force's assigned mission. The ground tactical plan is the first plan completed. It must address the early destruction of any enemy forces that pose an immediate threat to the lodgment area. Commanders and their staffs normally develop the landing plan from the ground tactical plan. From the landing plan, they develop the movement plan. This continues until the staff completes the marshaling plan.

C-5. The ground tactical plan is the basis for planning throughout the planning process. However, each plan affects the others, and changes in one plan can require adjustments in the other plans. The commander must determine if such adjustments entail acceptable risk. If the risk is unacceptable, the concept of operations must change. For example, the amount of lift available determines the feasibility of the ground tactical plan. If there are not enough lift systems to put all the required forces in place at the required time, the commander should adjust the ground tactical plan as well as the other plans. Therefore, planning for airborne and air assault operations requires the staff to obtain vital planning data, such as the availability of lift systems and the technical and tactical capabilities of those systems as early as possible.

C-6. Commanders ensure continuous coordination between the parallel echelons of the assaulting combat force and the unit or service providing the transportation from the beginning of an operation until its completion or abandonment. Units jointly coordinate and staff each detail before initiating operations. The commander makes maximum use of combined arms capabilities to ensure the assault force has sufficient power to accomplish its mission and protect itself. Short planning times often require staffs to modify existing contingency plans and standing operating procedures to meet the exact situation while still ensuring adequate coordination.

CONDITION SETTING

C-7. Setting conditions is also a common factor necessary for the success of air assault and airborne operations. Condition setting postures the air assault or parachute force for success with minimal or acceptable losses. The commander determines the exact conditions required in accordance with the factors of METT-TC, to include the degree of risk he is willing to accept with regard to each condition. Setting conditions is not limited to conducting suppression of enemy air defense and preparatory fires. It requires the participation of numerous staffs, units, cells, and boards in different echelons and services.

C-8. Condition setting is an iterative process. The commander through his situational understanding, decides what part of the situation must change to ensure the success of the vertical envelopment. He tasks his intelligence, surveillance, and reconnaissance (ISR) assets to detect the location of those enemy systems that unacceptably endanger the success of the operation. This allows his fire support systems to target and deliver fire effects against those enemy systems. He tasks his other battlefield operating systems to continue planning and preparing for the operation. The commander requests assistance from his higher headquarters if he does not have sufficient organic assets and information to accomplish the mission. He then assesses the progress of his battlefield operating systems. This process repeats until he is satisfied with the result or operational necessity forces him to either cancel or conduct the vertical envelopment.

METEOROLOGICAL CONDITIONS

C-9. Meteorological conditions affect airborne and air assault operations more than they affect any other type of operation. Long-range forecasts are vital to planning. As part of the planning process, commanders determine what adverse weather conditions would delay or cancel operations.

C-10. Commanders consider current and forecasted weather conditions in terms of their impact on tactical operations and aircraft performance. To issue the execution order that initiates the operation, the commander must know the current weather information at departure sites and pickup zones (PZs), along approach routes, and in the objective area. Operations conducted during marginal weather conditions may enhance the element of surprise, but they also increase the risk of accidents. The commander may have to postpone a planned operation or reduce the tempo of an ongoing operation when the risk becomes unacceptable because of deteriorating weather conditions.

C-11. Weather conditions affect aircraft performance and influence the conduct of operations. These conditions include: wind shears, crosswinds, and the ambient temperatures throughout the course of the operation. High temperature and altitude degrade aircraft lift performance. The combination of these factors results in trade-offs in the operating parameters of all missions. For example, a commander may insert dismounted reconnaissance teams on mountainsides in the cool of the morning, but be unable to execute the same mission in the noonday heat.

AIRBORNE OPERATIONS

C-12. An *airborne operation* is an operation involving the air movement into an objective area of combat forces and their logistic support for execution of a tactical, operational, or strategic mission. The means employed may be any combination of airborne units, air transportable units, and types of transport aircraft, depending on the mission and the overall situation (JP 3-17). The objective area is known as the airhead. The airhead contains enough drop zones (DZs) and landing zones (LZs) to allow airborne forces to mass effects on their objectives. The airhead should also contain extraction zones (EZs), interior lines of communications (LOCs), and terrain that allows for conducting a defense in depth. An administrative air movement of personnel, supplies, or equipment is not an airborne operation, although some procedures used in an airborne operation may apply. (JP 3-18 and FM 3-18.11 provide the doctrinal basis for airborne operations.)

C-13. Airborne operations are joint operations because of the interservice links of modern command and control (C2) systems, the multiservice structure of the defense transportation system, and the broad range of forces and support involved. Airborne operations require secure staging and departure areas coupled with the need to maintain operations security (OPSEC). OPSEC measures may include establishing intermediate support bases within tactical airlift range. The operation begins and ends on the order of the commander who establishes the joint airborne force.

ORGANIZATION OF FORCES

C-14. Once the commander determines the principal components of the ground tactical plan and the maneuver and fire support schemes, the airborne force organizes to execute its assigned mission. The commander balances the immediate need for combat power with the need to ensure force sustainability over time. To ensure unity of effort, part or all of the assigned forces' subordinate units can form into one or more temporary tactical groupings, such as teams or task forces. Each tactical group has a designated commander. Doctrine cannot prescribe in advance a standard organization to meet all conditions. However, airborne forces generally divide into one of three echelons: the assault echelon, the follow-on echelon, and the rear echelon.

C-15. The *assault echelon* is the element of a force that is scheduled for initial assault on the objective area (JP 4-01.2). In an airborne assault it normally comprises those forces capable of insertion by parachute in a single drop by the available lift systems. The assault echelon is a combined arms organization with only limited sustainment capabilities. The commander cross-loads vital assets, such as commanders, principal staff, communication systems, reconnaissance and security forces, and crew-served weapons among the various transportation systems so the loss of a single air frame will not compromise the operation. (Cross-loading also applies to air assault operations.)

C-16. **The *follow-on echelon* contains those additional forces moved into the objective area after the assault echelon.** They provide the combat power necessary to expand the initial airhead, secure the lodgment area, and establish one or more air or seaports of debarkation. The composition of the follow-on echelon depends on the factors of METT-TC. It can consist of heavy and light combined arms formations, field and air defense artillery assets, and combat engineers, as well as significant combat support (CS) and combat service support (CSS) elements. Introducing this echelon can extend over several days and involve multiple sorties by individual lift systems. Usually, this echelon does not require cross-loading of its allocated lift systems. This increases the carrying capacity of the lift systems delivering this echelon. This echelon contains increased sustainment capabilities.

C-17. The rear echelon contains those elements of the force that are not required in the objective area. It may remain at home station or at an intermediate staging base or intermediate support base throughout short-duration operations. This echelon generally contains the airborne unit's long-term sustainment capabilities.

CONTROL MEASURES

C-18. The commander has the full range of graphic control measures to help control his operation. As a minimum, the commander must assign each subordinate unit an AO. The airborne operation commander also use DZs, LZs, EZs, assault objectives, and the airhead line to focus the efforts of his subordinates.

C-19. Selecting DZs and LZs is a joint responsibility. The mission commander is responsible for delivering personnel and cargo to the DZ or LZ and for selecting approaches to the landing area. Both the joint and component commanders must base their decisions on their knowledge of respective problems and on the needs of the overall operation. The nature and location of landing areas are important when preparing the scheme of maneuver. The mission governs the general area where they should be established. At higher echelons, commanders can assign landing areas in broad terms. At lower echelons, they must describe these locations specifically. The commander selects his DZs only after conducting a detailed analysis. He uses the information provided by his intelligence system and Army pathfinders as he considers the following factors when selecting DZs and LZs:

- Ease of identification.
- Straight-line approach.
- Suitable for the weather and terrain.
- Out of range of enemy air defenses, strong ground defenses, and suppressive indirect fires.
- Close to or on top of an assault objective.

Since the last two entries conflict, he must decide which consideration has priority. (FM 3-17 provides detailed information regarding the desired characteristics of DZs and LZs.)

C-20. When assigning objectives and boundaries in airborne operations, the commander must consider other factors in addition to those inherent in conventional operations. He selects specific assault objectives based on an analysis of the situation. (See Figure C-1, page C-6.) The assault objectives dictate the size and shape of the airhead, although the commander develops the airhead line and determines the assault objectives concurrently. He selects assault objectives for his subordinate elements. Concurrently, the commanders of these subordinate elements decide the size, type, or disposition of the force that they commit to gain and maintain control of their objectives.

C-21. Selecting assault objectives should allow forces to accomplish mission-essential tasks while meeting the commander's intent. However, they may not include those objectives that must be seized to secure the airhead line. An appropriate assault objective is one that the force must control early in the assault to accomplish the mission or enhance the security of the airborne force. This can include key terrain within the airhead or terrain required for linkup. The airborne force is vulnerable from the time it lands until follow-on forces arrive at the airhead. A mounted enemy unit that attacks the airhead immediately following the airborne assault can completely disrupt the operation or even cause it to fail. Therefore, the assault objectives selected by the commander are terrain locations that dominate high-speed enemy avenues of approach into the airhead. He can also select enemy positions that threaten the mission and are within the airhead. The unit must seize its assault objectives immediately to establish the airhead and provide security for follow-on forces.

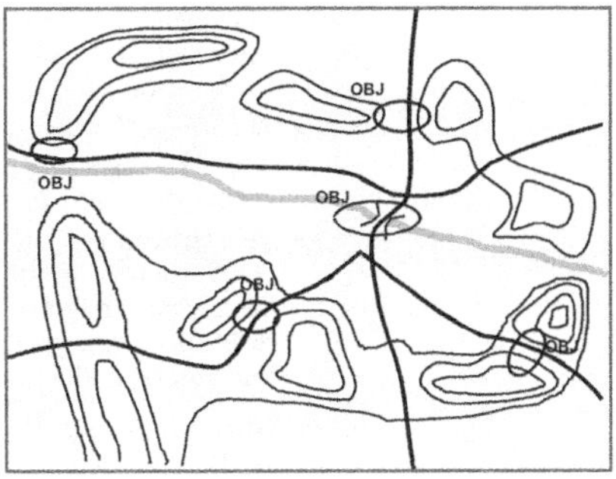

Figure C-1. Assault Objectives

C-22. The commander ranks the assault objectives based on the most likely threat or mission requirements. The airborne force secures its assault objectives before it establishes a perimeter defensive line along the trace of the airhead. It clears the terrain within the airhead of organized enemy resistance and positions forces to secure the airhead line.

C-23. The commander selects assault objectives at the same time as he considers the extent of the airhead. He draws the airhead line to delineate the specific area to seize and designate the airhead. An airhead line resembles a forward edge of the battle area in that security and other forces operate outside of the airhead line. The airhead acts as a base for further operations and as the lodgment to allow the airborne force to build up combat power. Once the assault force secures the airhead, it clears all enemy forces within it, not just organized enemy units. The following factors determine the location, extent, and form of the airhead line:

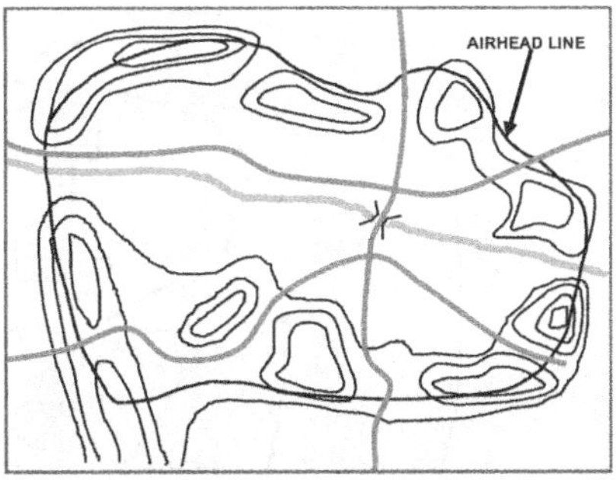

Figure C-2. Airhead Line

- The actual trace of the airhead line reflects the control of key or critical terrain essential to the mission. (See Figure C-2.) The air-head line should place the arrival airfield and any LZs or DZs out of the range of enemy direct fires and observed indirect fires.
- The airhead line is anchored on obstacles, and the airhead takes advantage of existing natural and man-made obstacles.

- The airhead contains enough DZs, LZs, and EZs to ensure the force has interior LOCs and to permit one massed parachute assault by the entire assault echelon rather than piecemeal insertion.
- The airhead allows enough space to disperse units and supplies to reduce the airhead's vulnerability to nuclear, biological, and chemical weapons if they are a threat.
- The airhead must be large enough to provide for defense in depth, yet small enough for the airborne unit to defend. Although this depends largely on the factors of METT-TC, a battalion can defend an airhead 3 to 5 kilometers in diameter. A brigade can occupy an airhead 5 to 8 kilometers in diameter.

C-24. When assigning boundaries and subordinate AOs in airborne operations, the commander considers several factors beyond those affecting more routine operations. Ideally, each unit's AO should include at least one DZ and one LZ to enable the unit and its attachments to land within its assigned AO during the assault. Each unit's presence also facilitates resupply and evacuation of enemy prisoners of war and casualties. Establishing a LZ and a DZ reduces coordination requirements with adjacent units. The commander assigns boundaries that should not require a unit to defend in more than one direction at the same time. Boundaries should extend as far as necessary beyond forward security forces to coordinate fires. This enables subordinate security units to operate forward of the airhead with minimal coordination. (See Figure C-3.)

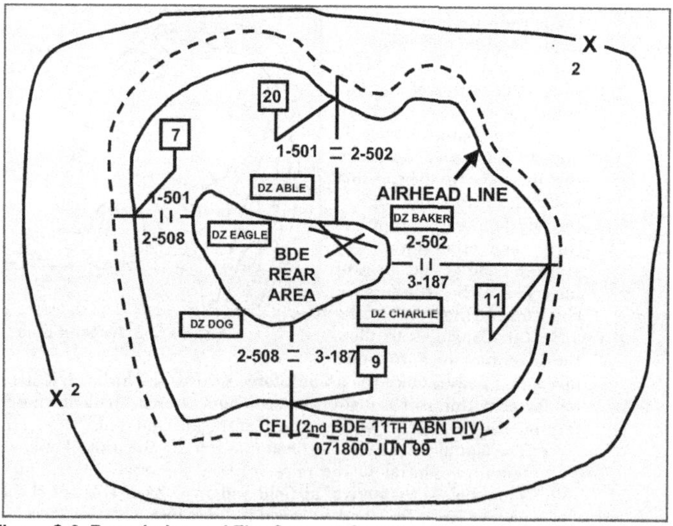

Figure C-3. Boundaries and Fire Support Coordinating Measures for an Airhead

PLANNING CONSIDERATIONS

C-25. The airborne force commander begins planning when he receives an initiating directive or a warning order. JP 3-18 describes the contents of an initiating directive or warning order as including the—

- Mission for subordinate units.
- Higher commander's concept of the operation.
- Command structure for the operation.
- Time and duration of the operation, including D-day and H-hour (execution time).
- Intelligence and security requirements.
- Allocation and distribution of airlift assets.
- Unit deployment list and sequence.
- Departure airfields, remote marshaling bases, and intermediate staging bases.
- Signal requirements and instructions.
- Linkup and withdrawal concept.

C-26. In an airborne operation that envisions an early linkup with conventional ground maneuver forces, the airborne unit defends the airhead until completing the linkup. After linkup, the airborne force either resumes the offensive within the commander's concept of the operation or prepares for subsequent operations. Tactical airborne operations begin with an initial assault followed by independent operations. They then transition to the defense of the established airhead until enough forces can be delivered to the objective area to break out of the established lodgment or linkup with ground forces.

C-27. The flexibility of airborne forces gives the commander wide latitude to select approach routes and objective areas. Airborne forces bypass ground obstacles and enemy positions to strike objectives in otherwise inaccessible areas. The ability of airborne forces to move rapidly and land on or near their objectives increases the element of surprise. It also facilitates the massing of relative combat power because airborne forces can attack the objective from any direction, which leads to the dissipation of the enemy's defenses. The presence of airborne forces also constitutes a threat that affects the enemy's capability to mass. These forces compel him to disperse combat power to protect vital sustainment installations and other key locations.

C-28. The primary prerequisites to conducting successful airborne operations are moving forces to an objective area without incurring unacceptable losses and supplying these forces with the required combat power, CS, and CSS. Steps taken to attain these objectives include—

- Obtain and maintain air superiority.
- Suppress enemy air defense capabilities and ground fires.
- Provide adequate air defense in the marshaling area and en route to and within the objective area.

C-29. In an airborne operation, the commander's primary initial sources of long-range fire support are air support and rocket or missile fires that can range the airhead. Other sources may include naval surface fires. The commander may insert airborne artillery cannon units and attack helicopters to provide close supporting fires during the initial assault. He may also introduce additional cannon, MLRS, and helicopter assets into the airhead during subsequent phases of the operation.

C-30. The capacity and availability of aircraft limit the size and amount of equipment and supplies available for movement to the objective area. The commander can insert his heavy equipment into the objective area by heavy parachute airdrops or an air landing. The limited number of vehicles in airborne units reduces the unit's tactical mobility in open terrain compared with that of armored and mechanized formations. However, airborne units may gain considerable mobility by using helicopters. Units also make concerted efforts to capture and exploit enemy supplies, equipment, weapons, vehicles, and petroleum, oils, and lubricants (POL).

C-31. Because of the displacement range of forces and the need for air LOCs, airborne operations magnify the problems normally inherent in sustaining a combat force. Therefore, commanders must emphasize planning for resupply, equipment maintenance, casualty evacuation, graves registration, and prisoner of war handling. Prepackaging company- and battalion-size resupply sets can ease these problems when support units must push supplies to the combat units.

C-32. Commanders consider all supplies and equipment required for mission accomplishment as part of their tactical planning. The initial combat requirements dictate the quantities and types of supplies and equipment carried by assault forces in the operation. Commanders ensure that only supplies required to meet the immediate needs of the assault force initially deploy into the objective area. Excess supplies and equipment can constitute a burden on the assault force. Staffs establish and maintain required levels of supply by phasing supplies into the objective area on an accompanying, follow-on (automatic and on-call), and routine basis. Ammunition, water, and POL products normally constitute the major tonnage items in airborne operations.

C-33. As part of the preparation for the airborne operation, soldiers receive briefings on the plan of their unit, adjacent units, and higher echelons including contingencies. It is particularly important that all personnel understand the commander's intent of the next two higher echelons. This helps units or soldiers landing in unplanned areas direct their efforts toward accomplishing the mission.

EXECUTING AIRBORNE OPERATIONS

C-34. Airborne operations may precede, accompany, or follow other types of operations. An airborne unit conducts day or night operations; each has its advantages and disadvantages, such as ease of target acquisition and identification of DZs. Initially, as part of preparatory fires, available fires destroy or suppress enemy systems and units that pose an immediate danger to the airborne assault. Using precision munitions increases the probability of achieving the desired effect. At the same time, it reduces the number of friendly fire support systems required to achieve this effect.

C-35. Executing the ground tactical plan involves initially seizing DZs and LZs in and around an airfield, or actually seizing an airfield. The assault echelon lands as close as possible to its objective by parachute and immediately assembles. Its initial assault emphasizes the coordinated action of small units to seize initial objectives before the advantage of surprise has worn off. Aggressive small-unit actions characterize this critical phase. Small-unit leader initiative is a key factor in a unit's ability to accomplish the mission. As assault forces seize assault objectives, the airborne force directs its efforts toward consolidating the airhead.

C-36. Tactical surprise and detailed planning should enable units to seize their assault objectives and to establish the airhead before the enemy has time to react in force. This ensures the uninterrupted landing of air-transported troops, equipment, and supplies. The commander changes the missions of his units as necessary in response to the enemy's actions. Units can expect the enemy to launch uncoordinated attacks quickly along major avenues of approach, using his locally available forces. Since the degree of coordination and strength of these attacks increases progressively, the airborne force must develop correspondingly greater strength in its defensive positions and prepare to defend against a mounted counterattack.

C-37. Units assigned to perform reconnaissance and security missions land in early serials so they can establish roadblocks, locate enemy forces, and disrupt enemy communication facilities. Since ground reconnaissance by unit commanders is seldom possible before the airborne operation, it must begin as soon as the unit lands. The flow of information must be continuous. The airborne commander's information requirements do not vary greatly from those of other light force commanders. However, his unit's method of arriving into the combat area makes immediate and thorough reconnaissance and transmission of combat information to higher headquarters necessary.

C-38. If the initial assault objectives are heavily defended, the bulk of the force has the task of seizing them. When initial objectives are lightly defended, the bulk of the force can clear assigned AOs and prepare defensive positions in depth. The commander initiates extensive patrolling as soon as possible between adjacent defensive positions within the airhead line and between the airhead and the forward trace of his security area. He uses his scout helicopters to support this patrolling effort. In most cases, the commander establishes contact with any special operations forces or friendly irregular forces in the area through a special operations command and control element that accompanies the assault force. Advanced ISR and digital C2 systems can assist this process by detecting the location of enemy forces within the airhead line and rapidly disseminating an accurate and timely common operational picture to all command posts involved in the operation. This precludes the necessity of conducting a zone reconnaissance of the entire area within the airhead line by foot mobile soldiers.

C-39. Sufficient communications personnel and equipment must move into the airhead before, (or simultaneously with) the assault command post to ensure the timely installation of vital communications. As soon as communications and the tactical situation permit, the commander establishes—

- Command fire control channels within the airborne forces.
- Communications with supporting air and naval forces.
- Communications with airlift forces concerned with buildup, air supply, and air evacuation.

- Communications with bases in friendly territory.
- Communications between widely separated airborne or ground forces with a common or coordinated mission, such as link-up forces.

C-40. The commander influences the action by shifting or reallocating available fire support means. He may also—

- Move forces.
- Modify missions.
- Change objectives and boundaries.
- Employ reserves.
- Move to a place from which he can best exercise personal influence, especially during the initial assault.

C-41. With initial objectives secured, subordinate units seize additional objectives to expedite establishing a coordinated defense or conducting future operations. The commander then organizes defensive positions, supplements combat net radio communications as required, and establishes a reserve. These, as well as other measures, prepare the force to repel enemy counterattacks, minimize the effects of weapons of mass destruction, or resume the offensive.

C-42. The reserve prepares and occupies defensive positions pending its commitment. The commander commits his reserve to exploit success, take over the mission of a unit delivered to the wrong locations, deal with unexpected opposition in seizing assault objectives, and secure the initial airhead. He reconstitutes a reserve on the commitment of his initial reserve, in accordance with the factors of METT-TC.

C-43. After the force makes the initial assault landing and accomplishes its first missions, the commander must organize his airhead line. The situation dictates how units occupy and organize the airhead line. The commander adjusts the disposition of his units and installations to fit the terrain and the situation. Units take reconnaissance and security measures, which usually include reinforcing the security area. The mission, enemy capabilities, and defensive characteristics of the terrain determine the degree to which the airhead line is actually occupied and organized for defense.

C-44. Introducing follow-on echelon forces in the buildup of the airhead proceeds concurrently with the seizure and organization of the airhead line. The intent of the buildup is to provide a secure operating logistics base for forces working to move the airhead away from the original point of attack. As additional combat troops arrive, they reinforce the airhead defensive positions, secure additional requisite terrain features and maneuver space as required by the mission, constitute reserves, and prepare for offensive operations. Follow-on ground operations exploit the advantages provided by the airhead. After firmly establishing the airhead or lodgment area, or after executing a linkup with ground forces, a higher commander will usually relieve airborne units to allow them to prepare for subsequent airborne assaults. If they cannot be relieved immediately, he provides them with additional combat power and sustainment capabilities.

AIR ASSAULT OPERATIONS

C-45. *Air assault* operations are those in which assault forces (combat, combat support, and combat service support) using the firepower, mobility, and total integration of helicopter assets, maneuver on the battlefield under the control of the ground or air maneuver commander to engage and destroy enemy forces or to seize and hold key terrain (FM 3-97.4). They are often high-risk, high-payoff operations.

C-46. An air assault task force (AATF) can dramatically extend the commander's ability to influence operations within his AO and to execute operations in locations beyond the capability of more conventional forces. The air assault force retains the flexibility to rapidly redeploy to conduct subsequent offensive or defensive operations. Air assault operations closely resemble airborne operations. Air assault forces are most vulnerable during the takeoff from PZs and the landing at LZs in unsecured areas.

C-47. The assault force uses the firepower, mobility, and total integration of helicopter assets to maneuver throughout the AO. Its purpose is to engage and destroy enemy forces or to seize and hold key terrain. Joint doctrine regards air assault operations as a subset of airborne operations. Air assault operations are not administrative movements of soldiers, weapons, and materiel by Army aviation units. An air assault is a deliberate, precisely planned, and vigorously executed combat operation designed to allow friendly forces to strike over extended distances and terrain barriers to attack the enemy when and where he is most vulnerable. The commander plans these operations using the previously described reverse planning process. (The primary references for air assault operations are FMs 3-04.113 and 3-97.4.)

C-48. The substantial mobility of an air assault force enables its commander to achieve surprise and deception and to conduct operations throughout his AO. However, air assault operations conducted in locations geographically remote from supporting forces may place the air assault force at increased risk if its ISR systems do not accurately detect enemy forces positioned to disrupt the air assault. Air assault units are well suited for use as reaction forces and in search and attack operations when information about the enemy's location, strength, and disposition is vague.

C-49. The large-scale use of helicopters in air assault operations greatly multiplies the mobility of ground units and contributes directly to an increase in combat effectiveness. Their use allows the ground commander to take advantage of the speed and flexibility of Army aircraft to accomplish a variety of tasks. For example, during a river-crossing operation, an air assault can help secure the crossing site or bridgehead line.

ORGANIZATION OF FORCES

C-50. Air assault operations employ AATFs. An AATF is a combined arms force specifically designed to hit fast and hard. It is under the command of a single headquarters. Tactical commanders use an AATF in situations that provide a calculated advantage because of surprise, terrain, threat, or mobility. An AATF consists of infantry, attack helicopters, fire support, electronic warfare, and logistic assets. The ground or air maneuver commander is designated as the AATF commander.

C-51. The lowest-echelon headquarters capable of controlling and coordinating the entire air assault operation exercises control of the aircraft in accordance with the overall plan. As a minimum, this is normally a battalion headquarters. This headquarters must coordinate airspace with other users, including artillery, air defense, air support, and other Army aviation units. It must also coordinate the air assault force's plans for maneuver and logistics with those of higher, subordinate, and adjacent units.

C-52. The airlift unit is either in direct support of the ground combat unit or under the operational control (OPCON) of the AATF. The AATF commander determines—with the air mission commander's input—when the OPCON relationship begins and ends. The commander does not attach the airlift unit to the AATF, because it is unlikely that a ground unit can control the aviation unit and supply the aviation-specific munitions and large amounts of fuel required by aviation units. Direct support (DS) and OPCON command relationships place no logistics responsibility for the supporting unit on the supported unit. Consequently, DS or OPCON is usually the desired relationship between air and ground units in air assault operations.

CONTROL MEASURES

C-53. The control measures that apply to an airborne operation also apply to an air assault operation. As a minimum, the commander assigns each subordinate unit an AO. The AATF and aviation staffs select LZs that support the ground tactical plan and offer the best survivability for the AATF. As in airborne operations, designating LZs within the unit's AO simplifies the provision of additional support to the unit. The AATF commander also uses assault objectives and the airhead line to focus the efforts of his subordinates. As necessary, the commander uses those attack control measures introduced in Chapter 5 to help control the force's maneuver once it enters the AO.

C-54. In air assault operations, the commander makes extensive use of Army airspace command and control (A2C2) measures to control the movement of the assault, attack, special electronic mission, and cargo aircraft. For example, Figure C-4 on page C-14 shows flight routes as depicted on an overlay. (FM 3-52 details A2C2 measures.)

PLANNING CONSIDERATIONS

C-55. Integrating aviation and infantry does not fundamentally change the nature of combat operations. The air assault force continues to fight as a combined arms team. However, the tempo and distance involved in such operations change dramatically. Missions normally assigned to an AATF should take advantage of its superior mobility. However, the commander should not employ an AATF without a detailed resupply plan in operations that require sustained ground combat. Once the air assault is complete, the aviation unit can continue to support the infantry by conducting aerial movement of systems and critical supplies.

C-56. There are several basic air assault planning operational guidelines—

- Assign a mission that takes advantage of the AATF's mobility.
- Task organize the AATF as a combined arms team.
- Ensure the air assault plan supports the AATF commander's intent.

- Allow extra time for planning and preparing for limited-visibility and adverse weather air assaults.
- Maintain small-unit integrity throughout the air assault to ensure the ability to fight as a cohesive unit immediately upon landing.
- Plan and posture fire support to provide suppressive fires along flight routes, on LZs, and on enemy air defense systems.

C-57. The foundation of a successful air assault operation is the commander's ground tactical plan. The AATF staff prepares this plan based on input from all task force elements. All aircrews must be familiar with the ground tactical plan and the ground commander's intent.

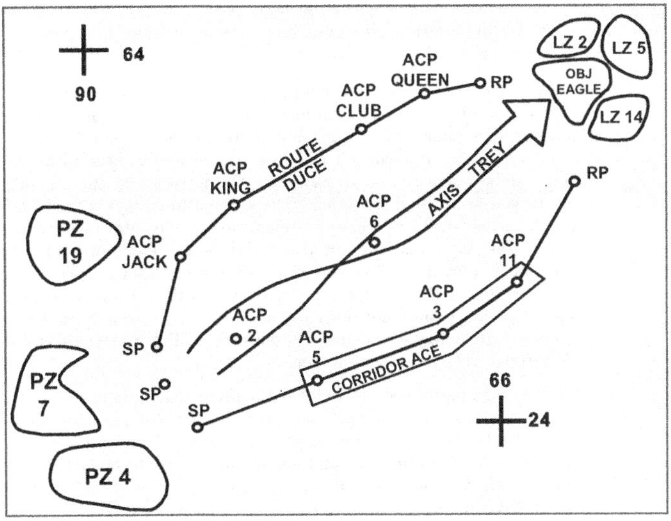

Figure C-4. Flight Routes Depicted on an Overlay

C-58. The ground tactical plan for an air assault operation comprises essentially the same elements as any other infantry attack but differs in the requirements for speed and mobility. The plan places task-organized assault units on or near the objective so they are capable of seizing objectives immediately and consolidating quickly. If the commander cannot introduce adequate combat power quickly into the objective area, the air assault force must land away from the objective and build up combat power. This force then assaults like any other infantry unit; however, this diminishes the effectiveness of the air assault operation. The scheme of maneuver may take many forms depending on the situation and the factors of METT-TC.

C-59. The ground tactical plan addresses—

- Assault objectives for subordinate elements.

- Designating the LZs available for each subordinate element, considering the distance from each unit's LZ to the assault objective.
- D-day and H-hour.
- Special tasks.
- Task-organization and command relationship of all organic and supporting units.
- Fire support during the assault, such as close air support, field artillery, mortars, and jammers.
- Flight corridors.
- Air defense suppression.
- Subsequent operations, such as defense, linkup, and withdrawal.
- Enemy locations, including air defense positions and type.
- Combat service support.

C-60. To achieve tactical surprise, the commander may decide to make the initial assault without preparatory fires. However, he always plans fires to support helicopter assault and combat operations on each LZ so that they are rapidly available if needed. These fires are normally short with a high volume to maximize surprise and shock effect. All indirect fires should end just before the first assault element lands. The commander uses attack helicopters to suppress and destroy enemy systems during the interim between when indirect fires stop impacting and the initial assault element lands and prepares to conduct operations. Fire support planning provides suppressive fires along flight routes and near LZs to help ensure the air assault unit lands as planned. The commander normally assigns a high priority to the suppression of enemy air defense systems. The location of those systems is critical information needed by the commander.

C-61. A unit maintains its tactical integrity throughout the air assault. All members of a squad load onto the same aircraft, and platoons are in the same serial. Both ensure unit integrity upon landing. The commander cross-loads key weapons, ammunition, and command groups to ensure that the loss of one aircraft does not result in losing a given weapon system or disrupting the chain of command.

C-62. The ground commander uses aviation resources to the maximum degree of effectiveness. He should not retain aircraft under his direct control without viable aircraft mission requirements. The air mission commander must have the flexibility to shift idle aircraft to support other combat units, conduct required maintenance, or allow for crew rest. Plans to commit preplanned reaction forces should include provisions for their airlift to be on standby or alert status. The AATF commander makes the decision to release supporting aviation resources. The air mission commander ensures that the AATF commander is aware of subsequent or competing missions for his aviation resources. At times, the AATF commander may need to retain aviation support beyond the original time planned. In this case, he must inform higher headquarters immediately. The air mission commander continues to provide aircraft support until the AATF commander releases his unit.

C-63. The commander plans and organizes his CSS operations to support a rapid tempo of highly mobile and widely dispersed operations. Traditional doctrinal supporting distances and support responsibilities do not always apply to air assault operations. The air assault logistics planner recognizes this from the outset and adapts the plan using available resources. Just as the commander tailors the AATF for combat operations by air, the logistics system must tailor itself to support by air. Medical evacuation, resupply, and reinforcement airlifts may be necessary to sustain the force's combat operations. Lift restrictions limit what can enter the airhead by helicopter. However, careful planning by the aviation staff provides methods for inserting reinforcements and most equipment lines and supplies.

EXECUTING AIR ASSAULT OPERATIONS

C-64. At the prescribed time, units move from the assembly area to the holding area via a route designated by the AATF commander. Each unit commander notifies the PZ control party when his unit arrives in the holding area. The PZ control officer (PZCO) coordinates the arrival of aircraft and troops so that they arrive at their respective loading points just before the aircraft land. This prevents congestion, facilitates security, and reduces vulnerability to enemy actions during PZ operations.

C-65. When the aircraft are loaded and ready, the PZCO signals the flight leader. Lift-off should be at the time prescribed in the air-movement table. However, aircraft will not loiter in the PZ. If they are early, they lift off and later speed to cross the start point (SP) or first ACP on time.

C-66. The air movement commander predetermines the enroute flight speed, and the flight leader paces the flight to ensure it crosses the SP on time. Commanders remain oriented throughout the flight by following and verifying the flight route using terrain observation, maps, global positioning systems, and other aids.

C-67. Attack helicopters and air cavalry assets assist in providing security for the air assault force. Under the control of the air mission commander, these helicopters provide reconnaissance of the routes and LZs, provide security for the lifting helicopters en route to the LZ, and protect the lifted ground maneuver force as it assembles on the LZs and moves toward its objective. At the conclusion of the air assault phase of the mission, attack helicopters may remain OPCON to the ground maneuver force and provide reconnaissance and security operations in the objective area.

C-68. After passing the release point (RP), serials proceed to assigned LZs. The commander uses the RP crossing to time the lifting and shifting of fire support assets. The RP is also where aircraft shift to LZ formation (if required) and the commander initiates preparatory fires.

C-69. Incendiary ordnance is not normally used on an LZ and its immediate vicinity just prior to landing because foliage fire and smoke could endanger aircraft or hamper the mission. However, helicopters equipped with smoke generators can provide a smoke screen.

C-70. The AATF lands as planned unless last-minute changes in the tactical situation force the commander to abort or alter the landing. Aviation crews keep soldiers in their aircraft informed of the situation, especially of any changes to the original plan. The commander wants his unit to land simultaneously to place the maximum number of soldiers on the ground in a given area in the shortest possible time. Individual soldiers are most vulnerable during landing; they disembark rapidly and deploy to carry out assigned missions.

C-71. At the LZ, leaders at each command echelon account for all personnel and equipment and submit appropriate reports to higher headquarters. After the unit completes its consolidation of the LZ, the commander reorganizes it as necessary. The ground combat operations of an air assault unit are no different from those conducted by other infantry units.

Appendix D

Encirclement Operations

When the enemy is driven back, we have failed, and when he is cut off, encircled and dispersed, we have succeeded.

Field Marshal Prince Aleksander V. Suvorov

***Encirclement operations* are operations where one force loses its freedom of maneuver because an opposing force is able to isolate it by controlling all ground lines of communication and reinforcement.** A unit can conduct offensive encirclement operations designed to isolate an enemy force or conduct defensive encirclement operations as a result of the unit's isolation by the actions of an opposing force. Encirclement operations occur because combat operations involving modernized forces are likely to be chaotic, intense, and highly destructive, extending across large areas containing relatively few units as each side maneuvers against the other to obtain positional advantage.

OFFENSIVE ENCIRCLEMENT OPERATIONS

D-1. The commander intends offensive encirclements to isolate an enemy force. Typically, encirclements result from penetrations and envelopments, or are an extension of exploitation and pursuit operations. As such, they are not a separate form of offensive operations but an extension of an ongoing operation. They may be planned sequels or result from exploiting an unforeseen opportunity. They usually result from the linkup of two encircling arms conducting a double envelopment. However, they can occur in situations where the attacking commander uses a major obstacle, such as a shoreline, as a second encircling force. Although a commander may designate terrain objectives in an encirclement, isolating and defeating enemy forces are the primary goals. Ideally, an encirclement results in the surrender of the encircled force. This minimizes friendly force losses and resource expenditures.

CONTENTS

ORGANIZATION OF FORCES

D-2. An encirclement operation usually has at least two phases—the actual encirclement and actions taken against the isolated enemy. The commander should consider adjusting his task organization between phases to maximize his unit's effectiveness in each phase. The first phase is the actual encirclement that results in the enemy force's isolation. The organization of forces for an encirclement is similar to that of a movement to contact or an envelopment. The commander executing an encirclement operation organizes his forces into a direct pressure force and one or more encircling arms. Armor, mechanized and motorized infantry, aviation, air assault, and airborne units are especially well suited for use as an encircling arm since they have the tactical mobility to reach positions that cut enemy lines of communications (LOCs). The presence of bypassed and encircled enemy forces on the flanks and rear of advancing friendly forces requires all-around security, which includes local security measures and security forces.

D-3. One commander should direct the encirclement effort. However, there must also be unity of command for each encircling arm. The encircling force headquarters may name one of its subordinate units as the headquarters for an encircling arm. Alternatively, that force headquarters may create a temporary command post from organic assets, such as its tactical command post, to control one or more arms of the encirclement. If that encircling arm has subordinate inner and outer arms, each of them also requires separate subordinate commanders. The missions and spatial orientation between the inner and outer encircling arms are sufficiently different; therefore, one force cannot act in both directions at once. (See Figure D-1.)

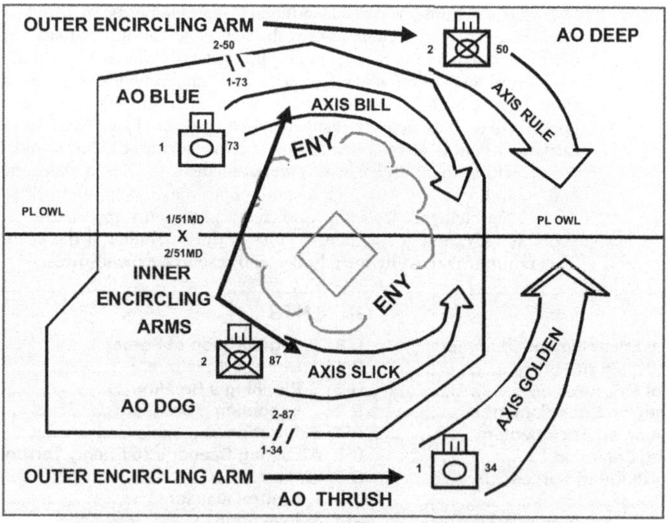

Figure D-1. Inner and Outer Arms of an Encirclement

D-4. The commander organizes only an inner encircling arm if there is no possibility of the encircled forces receiving relief from enemy forces outside the encirclement. If there is danger of an enemy relief force reaching the encircled enemy force, the commander organizes both inner and outer encircling arms. He assigns the outer encircling arm a security mission, an offensive mission to drive away any enemy relief force, or a defensive mission to prevent the enemy relief force from making contact with the encircled enemy force. Once the encirclement is complete, these inner or outer encircling arms form a perimeter.

D-5. The second phase of an encirclement operation involves actions taken against an isolated enemy. The commander's decision on whether to fix, contain, or destroy isolated enemy forces affects his task organization, as will enemy attempts to break out from the encirclement or linkup with the encircled force. All these possible outcomes require resources in terms of units and supplies, but some require more resources than others do. If the commander's mission is to contain or fix an isolated enemy, he organizes his forces for defensive action and arranges them around the enemy's perimeter. If the commander's mission is to reduce or destroy that same enemy, he organizes his forces for offensive action. A higher commander often assigns either mission to the commander of a follow-and-support force.

D-6. Regardless of whether the commander decides to fix, contain, or destroy the enemy, he conducts reconnaissance to maintain contact and monitor enemy actions in response to the encirclement. This allows him to respond effectively to any enemy movement. The most effective reconnaissance combines ground, aerial, and surveillance systems to provide constant coverage and multiple assessments of enemy activities throughout the encircled area.

CONTROL MEASURES

D-7. Control measures for an encirclement are similar to those of other offensive operations, especially an envelopment, but with a few additional considerations. (See Figure D-2.) If the commander uses both an inner and an outer encircling arm, he must establish a boundary between them. He should place the boundary so that each element has enough space to accomplish the mission. The inner force must have enough space to fight a defensive battle to prevent the encircled force from breaking out. The outer force must have adequate terrain and enough depth to its area of operations (AO) to defeat any attempt to relieve the encircled force.

D-8. The commander who controls both converging forces establishes a restricted fire line (RFL) between them. The commander may also establish a free fire area (FFA), which encloses the area occupied by a bypassed or encircled enemy forces. (Chapter 2 discusses the use RFLs, FFAs, and other fire support coordinating measures.)

PLANNING AN ENCIRCLEMENT

D-9. Encirclement operations may require allocating large forces and significant resources. They take a great deal of time and usually slow an advance. If the mission of the encircling force is to maintain contact with a bypassed enemy force, the following general planning considerations apply:

- Determine the best available assets that gain and maintain contact with the enemy.
- Keep the enemy isolated and incapable of receiving intelligence, logistics, and fire support from enemy formations outside of the encirclement.
- Use intelligence, surveillance, and reconnaissance (ISR) assets so that the commander knows the capabilities of the encircled force and, as much as possible, its commander's intentions.
- Retain freedom of maneuver.

D-10. The commander applies the general defensive planning considerations outlined in Chapter 8 if the mission is to contain or fix the encircled enemy force in a given location. If the mission is to attack and destroy the encircled enemy force, he applies the planning considerations outlined in Chapters 3 and 5. Commanders should plan to rotate the forces involved in reducing the encircled pocket to maintain constant pressure on the enemy.

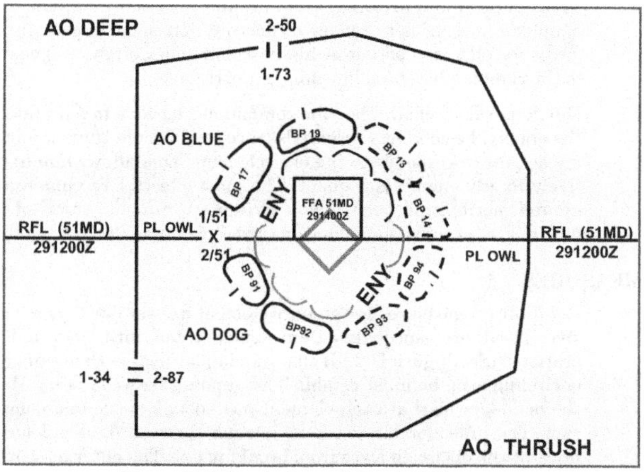

Figure D-2. Encirclement Control Measures

D-11. Every encircled enemy unit reacts differently. Initially, some become demoralized and cannot offer any serious resistance. However, if left undisturbed, most enemy units recover and attempt to break out and regain contact with their main force or attack the flank and rear of advancing friendly units. The encircling force must plan for the enemy's most probable reactions.

D-12. If the enemy force is not reduced and it can be resupplied, or it has access to considerable supply stocks, it continues to be a serious threat to the commander in future operations. The encircling force must be approximately equal in size to this type of encircled force to fix or contain it. This situation occurred when German forces occupied various fortified French ports after Allied armies liberated the rest of France in 1944. Each encircled German division took approximately one Allied division to maintain its isolation. Conversely, an enemy force isolated without adequate supplies either surrenders or faces containment by considerably smaller forces. This situation occurred in Egypt during the 1973 Arab-Israeli War when an Israeli division isolated the Egyptian 3rd Army. Planning considerations for the linkup of encircling forces, such as command and control relationships, are outlined later in this appendix.

EXECUTING AN ENCIRCLEMENT

D-13. When feasible, the encircling force advances parallel to the enemy's direction of movement. It attempts to reach defiles, bridges, and other critical points before the main enemy force reaches them. When the encircling force cannot outdistance the enemy, it engages his flanks to force him to fight under the most unfavorable conditions possible, ultimately in two or more directions simultaneously. Engineer units rapidly breach obstacles in the path of the encircling force. Friendly forces emplace obstacle complexes, supported by fires, to block probable avenues of escape as they counter attempted enemy breakouts from encirclement. The commander may use air assault and airborne forces to seize defiles or other critical terrain objectives to cut enemy LOCs. He completes the encirclement when all enemy ground LOCs are cut. This generally occurs when the two arms of a double envelopment complete their linkup.

D-14. A commander usually creates intervals between the advancing units of an enveloping force to provide protection from enemy weapons of mass destruction (WMD). They can also occur during combat operations as the result of different rates of advance by combat formations that face dissimilar degrees of enemy resistance and different terrain. The encircled enemy attempts to discover intervals and take advantage of them as he tries to escape from or breakout of the encirclement. Once the enveloping force completes the linkup that actually creates the encirclement, it must close these intervals as quickly as possible to prevent the enemy from exploiting them.

D-15. The enemy may attempt to cut off the encircling force and extend his flank beyond the area of the friendly attack. If the commander attempts to outflank such a hostile extension, it may lead to his own overextension or to a dangerous separation of the enveloping force from support. It is usually better to take advantage of the enemy's extension and subsequent weakness by penetrating his thinly held front rather than overextending in an effort to completely outflank his position. Alternatively—in response to the unfolding encirclement—the enemy may attempt a frontal, spoiling attack. In this case, the friendly force in contact defends itself or engages in a delaying operation while the enveloping force continues the envelopment or moves directly toward the enemy force in a counterattack.

D-16. The commander of a highly mobile force forming the inner encircling arm may choose not to establish a continuous series of positions around an encircled enemy. He may order his forces to occupy only key terrain from which they can strike at the encircled enemy to prevent him from concentrating forces and to further isolate him. To effectively isolate the enemy, a commander who adopts this technique must be able to detect enemy attempts to breakout and concentrate sufficient combat power against these attempts to thwart them. The commander of the outer encircling arm prevents additional enemy forces from reinforcing the isolated enemy force or interfering with the activities of the inner encircling arm.

D-17. Other operations may result in the encirclement of enemy forces. These include offensive operations that bypass large enemy forces to maintain the momentum of the force. Reconnaissance and security missions conducted by the main body must focus on detecting and reporting bypassed units. The main body should conduct these missions not only to its flanks, but also to its rear to discover if enemy forces move in behind them. Unit ISR assets should watch for measures taken by the enemy's main body to relieve or assist its bypassed or encircled forces.

D-18. Once the commander decides to destroy an encircled enemy force, he reduces the enemy as rapidly as possible to free resources for use elsewhere. The reduction of an encircled enemy force should continue without interruption, using the maximum concentration of forces and fires, until the encircled enemy force's complete destruction or surrender. A commander may destroy encircled enemy forces by fires alone or by a combination of fire and movement. The five main methods for reducing an encircled enemy are fire strike, squeeze, hammer and anvil, wedge, and escape route.

D-19. **A *fire strike* is the massed, synchronized, and nearly simultaneous delivery of precision-guided munitions.** It is the preferred method for destroying an encircled enemy force. The initial targets for these munitions are systems that present the greatest danger to the encircling force, such as the enemy's WMD, command posts, fire support and air defense systems, and field fortifications. However, the commander's ability to use precision-guided munitions in mass may be limited by the ability of the combat service support (CSS) system to supply them. Therefore, fixed-wing and rotary-wing aircraft and conventional artillery continue to play an important role in the destruction of encircled forces. The commander also conducts offensive information operations—such as psychological operations (PSYOP) and electronic warfare (EW)—against the encircled enemy force. In some situations, fire strikes result in the rapid destruction of the encircled enemy. However, destruction is not guaranteed. In most cases, reducing the enemy pocket requires using ground maneuver forces.

D-20. The squeeze technique uses simultaneous, coordinated blows on the enemy from various directions. (See Figure D-3.) Following the initial encirclement, the capture or destruction of the enemy force is methodical and thorough. The commander uses fire and movement together in a controlled contraction of the encirclement. As the enemy's perimeter contracts, the commander removes units from the inner perimeter and adds them to his reserve depending on the terrain and other factors of METT-TC. This technique is effective against battalion or smaller groups of encircled enemy forces.

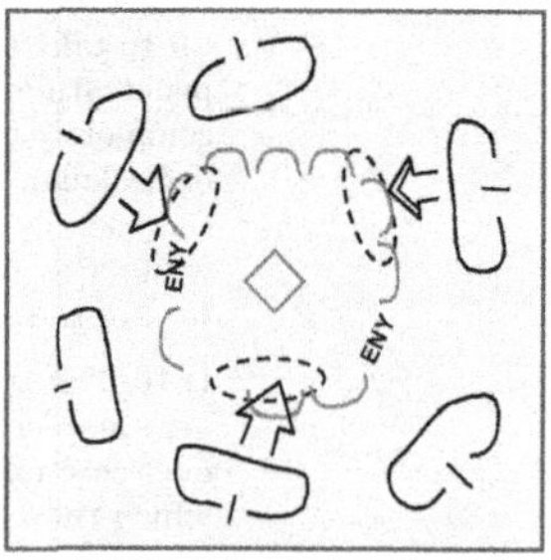

Figure D-3. Squeeze Technique

D-21. The squeeze technique promotes the enemy's confusion and rapid dispersion of combat power and prevents him from using his reserves in a decisive manner. The commander should shape the operation by initially concentrating on destroying enemy command nodes, air defense systems, artillery systems, and CSS capabilities. These CSS capabilities include any drop zones, landing zones, or airstrips available to the enemy that would allow him to receive support from outside the encirclement.

D-22. The hammer and anvil technique employs a stationary blocking force as an anvil on one or more sides of the inner perimeter while other elements of the encircling force use offensive action as a hammer to force the encircled enemy force against the blocking force. (See Figure D-4.) Either the anvil or the hammer can destroy the enemy. Usually the hammer, as the attacking element, accomplishes this task. This technique is most effective when the blocking force is located on or to the rear of a natural terrain obstacle. On favorable terrain, an airborne or air assault force can be used as an anvil or a blocking element.

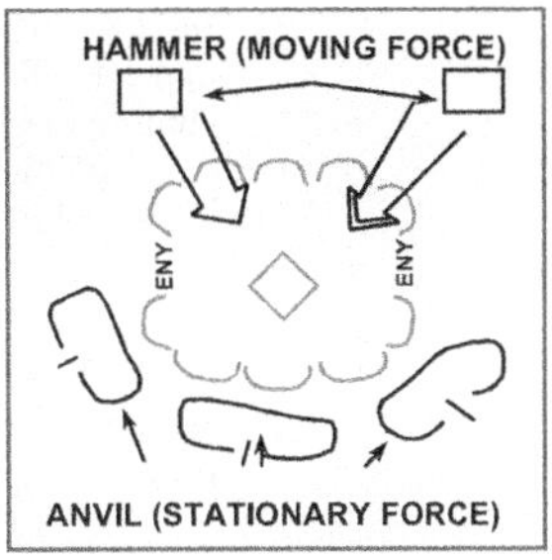

Figure D-4. Hammer and Anvil Technique

D-23. The wedge technique uses a unit to divide enemy forces within the pocket while the rest of the encircling force remains in place. (See Figure D-5.) This technique allows the commander to concentrate against a small portion of the encircled enemy. However, the encircling force must maintain pressure on other encircled enemy forces to prevent them from reinforcing or supporting the threatened area. It is important that the unit dividing the pocket conduct sudden and swift attacks immediately after the end of supporting preparatory fires.

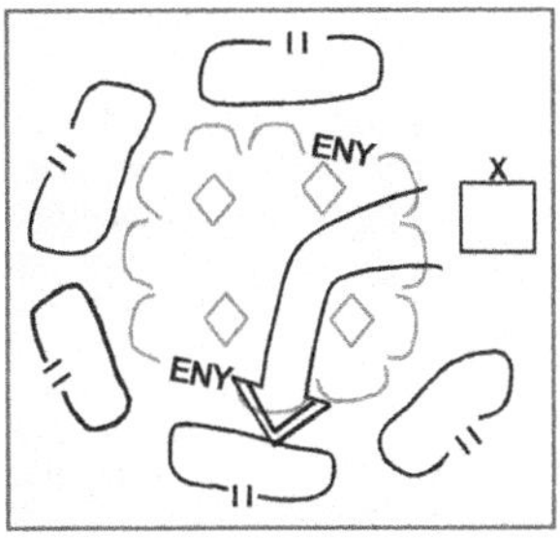

Figure D-5. Wedge Technique

D-24. The escape route technique involves leaving one or more gaps in the inner encircling arm to entice the enemy to attempt a breakout. Once the enemy starts moving, and is no longer sheltered in defensive positions, he is more vulnerable to acquisition, attack, and destruction. A commander using this technique should use PSYOP and constant offensive action to demoralize the escaping enemy force.

D-25. The negative aspect of these techniques is that they require considerable forces and supplies, which are not always available. Therefore, at times the encircling force has to limit itself to less decisive measures. These include temporarily containing or fixing bypassed enemy forces until resources become available to enable the encircling force to destroy the enemy. Continued isolation of the encircled force can only be guaranteed when the enemy cannot strengthen his forces by inserting additional units and supplies by air.

Even total, long-term isolation does not necessarily lead to decisive defeat of the encircled enemy. It is a temporary measure designed to provide the attacking force additional time.

DEFENDING ENCIRCLED

D-26. An encircled force can continue to defend encircled, conduct a breakout, exfiltrate toward other friendly forces, or attack deeper into enemy-controlled territory. The commander's form of maneuver once his unit becomes encircled depends on his senior commander's intent and the factors of METT-TC, including the—

- Availability of defensible terrain.
- Relative combat power of friendly and enemy forces.
- Logistic status of the encircled force and its ability to be resupplied, including the ability to treat and evacuate wounded soldiers.
- Morale and fighting capacity of the soldiers.

D-27. Encirclement of a friendly force is likely to occur during highly mobile fluid operations, or when operating in restricted terrain. A unit may find itself encircled as a result of its offensive actions, as a detachment left in contact, when defending a strong point, when occupying a combat outpost, or when defending an isolated defensive position. The commander must anticipate becoming encircled when he has a mission as a stay-behind force, or when he occupies either a strong point or a combat outpost. He must then make the necessary preparations.

D-28. The senior commander within an encirclement assumes command over all encircled forces and takes immediate action to protect them. In the confusion leading to an encirclement, it may be difficult to even determine what units are being encircled, let alone identify the senior commander. However, the senior commander must be determined as quickly as possible. When that commander determines he is about to be encircled, he must decide quickly what assets stay and what assets leave. He immediately informs his superior of the situation. Simultaneously, he begins to accomplish the following tasks:

- Establish security.
- Reestablish a chain of command.
- Establish a viable defense.
- Maintain morale.

D-29. The commander positions his security elements as far forward as possible to reestablish contact with the enemy and provide early warning. Vigorous patrolling begins immediately. Each unit clears its position to ensure that there are no enemy forces within the perimeter. Technical assets, such as JSTARS and EW systems, augment local security and locate those areas along the perimeter where the enemy is deploying additional forces.

D-30. The commander reestablishes unity of command. He reorganizes any fragmented units and places soldiers separated from their parent units under the control of other units. He establishes a clear chain of command throughout the encircled force, reestablishes communications with units outside the encirclement, and adjusts support relationships to reflect the new organization.

ORGANIZATION OF FORCES

D-31. The commander of the encircled force establishes a perimeter defense. (Chapter 8 discusses conducting a perimeter defense.) He must be aware of the unique capabilities and limitations of the different units within the encirclement. Therefore, he designs his defense to maximize the capabilities of his available forces. Forward units establish mutually supporting positions around the perimeter and in depth along principal avenues of approach. Units occupy the best available defensible terrain. It may be necessary to attack to seize key or decisive terrain so that it is incorporated within the perimeter defense. Once the commander assigns defensive AOs and battle positions, preparations are the same as in the defense. (See Figure D-6.) Encircled units make their defensive positions as strong as possible given time and resource constraints. The defensive scheme must anticipate that the enemy will attempt to split the defenses of the encircled force and defeat it in detail.

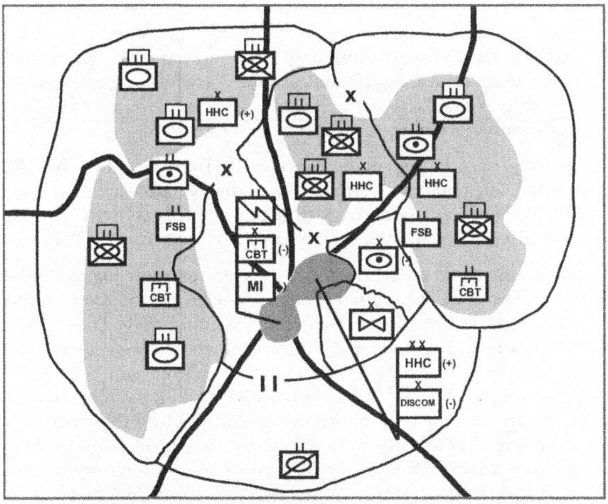

Figure D-6. Encircled Armor Division's Perimeter Defense

D-32. The encircled force commander establishes a reserve, which must have sufficient mobility to react in a timely manner. Therefore, given the availability of sufficient fuel, the commander uses armored and mechanized infantry units as his reserve. He centrally positions them to take advantage of interior lines, which exist if he can shift his forces' locations or reinforce faster than the enemy can shift location or reinforce. He can achieve interior lines through central position (with operations diverging from a central point), from superior lateral LOCs, or greater tactical mobility. If only dismounted infantry forces are available, the commander should establish small local reserves to react to potential threats. He organizes a mobile antiarmor element from the best available antiarmor systems. If possible, subordinate echelons should also retain a reserve.

D-33. While defending encircled, the commander may use his reserve to limit penetrations along the perimeter. It may conduct spoiling attacks or vigorous counterattacks. He initiates a counterattack at the decisive moment and location as the enemy force attempts to penetrate the defensive positions.

BOS CONSIDERATIONS

D-34. Divisions and corps may consider relocating their aviation systems to locations that are not encircled. Aviation can rapidly bring additional firepower to bear on the encircling enemy force or rapidly move reaction forces to threatened locations along the defensive perimeter. Generally, aviation assets fly out of the encirclement when it becomes small enough to allow the enemy's artillery to range throughout the area.

D-35. The commander centrally controls his fire support systems, such as artillery, to provide support at numerous points along the perimeter and mass the effects of his fires. Designating a fire support coordinator for all fire support systems is a technique for centrally controlling his fires. At lower levels, mortars from various units may be co-located under centralized control, especially if there are insufficient artillery assets. The encircled commander also centrally controls his air defense assets, ensuring that the forward units have sufficient short-range air defense coverage.

D-36. Generally, engineers concentrate first on countermobility, then survivability. An encircled force is particularly vulnerable to the enemy's use of WMD. Dispersal is difficult in a perimeter-type defense; therefore, the next best alternative is position hardening by constructing field fortifications.

D-37. Encircled units must closely monitor their logistic assets, especially if they cannot be resupplied for an extended period. Conservation and centralized control of available resources are imperative. The commander may force his forward units to virtually cease all vehicle movement to allocate remaining fuel assets to the reserve. He retains essential CSS capabilities to sustain his operations. They fall under the control of a senior logistician. When possible, the commander positions these units and their assets out of the reach of potential penetrations in protected and concealed locations. He may incorporate other CSS units into defensive positions in depth or around key facilities. He may choose to use soldiers from CSS units as fillers for combat units, although this action may affect his sustainment capabilities.

D-38. Casualty evacuation and mortuary affairs pose particular challenges for the encircled force. The commander evacuates his wounded from the encirclement whenever possible for humanitarian reasons. This also reduces the logistic burden of providing long-term medical care to wounded soldiers.

D-39. Soldiers have an inherent fear of being encircled by the enemy. Unchecked, this fear can lead to a degradation in morale and discipline. When encircled, soldiers under the firm control of their leaders can withstand the mental strain. Discipline can disintegrate rapidly in an encirclement. Officers and NCOs must uphold the highest standards of discipline. Their personal conduct sets the example. The commander must be seen frequently by his troops and display a calm and confident manner.

D-40. Soldiers in the encirclement must not regard their situation as desperate or hopeless. Commanders and leaders at all levels maintain the confidence of soldiers by resolute action and a positive attitude. They must keep soldiers informed to suppress rumors. The commander counters enemy PSYOP by conducting defensive information operations.

BREAKOUT FROM AN ENCIRCLEMENT

D-41. A breakout is an offensive and a defensive operation. An encircled force normally attempts to conduct breakout operations when one of the following four conditions exist:

- The commander directs the breakout or the breakout falls within the intent of a higher commander.
- The encircled force does not have sufficient relative combat power to defend itself against enemy forces attempting to reduce the encirclement.
- The encircled force does not have adequate terrain available to conduct its defense.
- The encircled force cannot sustain itself long enough to be relieved by forces outside the encirclement.

ORGANIZATION OF FORCES

D-42. Units typically task organize into a rupture force, follow-and-assume force, main body, and rear guard to conduct a breakout attack. (Figure D-7). If sufficient forces exist within the encirclement, the commander can organize a reserve and a separate diversionary force from his available resources. Some encircled units will be weakened, and if sufficient combat power does not exist to resource each of these forces, the commander must prioritize

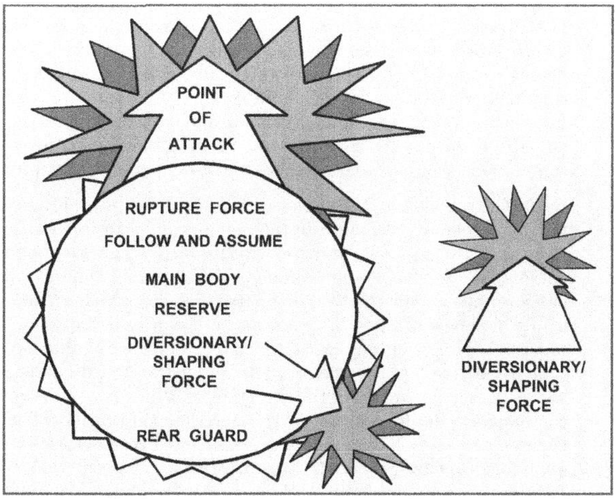

Figure D-7. Organization of Forces for a Breakout Operation

which ones to resource. Normally, his first priority is to resource the rupture force. He assigns the responsibilities of those unresourced forces to the forces he can resource. For example, the follow-and-assume force could receive a be-prepared mission to help extract the rear guard, a mission generally given to the reserve. Forces located outside the encirclement can assist the breakout by conducting shaping operations. Above all, the encircled force must maintain the momentum of the attack; otherwise, it is more vulnerable to destruction than it was before the breakout attempt.

D-43. The force must reorganize based on available resources to conduct the breakout. Without resupply, armored and mechanized infantry units may not be able to move all of their vehicles during the breakout attack. Priority of support may be limited to the rupture force and the rear guard, with the remaining force keeping only sufficient transportation assets to move the wounded and critical assets and supplies. The breakout plan should outline the commander's destruction criteria for equipment or supplies left behind. All vehicles, critical munitions and other supplies, and equipment—less medical—that cannot be moved should be destroyed as soon as possible.

D-44. An encircled force attacks by using the rupture force to penetrate the enemy defensive positions in at least one location. The commander must produce overwhelming combat power at each breakout point. The commander assigns the rupture force, which varies in size from one-third to two-thirds of the total encircled force, the mission to penetrate the enemy's encircling position, widen the gap, and hold the shoulders of the gap until all other encircled forces can move through. The rupture force must have sufficient strength to penetrate the enemy line. This force must use surprise, mobility, and firepower to achieve a favorable combat power ratio over the enemy at the point of attack. (Chapter 3 discusses the penetration as a form of maneuver.)

D-45. Initially, the rupture force is the decisive operation. The attack occurs at a location where the commander anticipates a successful rupture of the enemy's inner ring, which facilitates subsequent operations by enabling the commander to attack enemy units from their flanks and rear. The rupture force commander most likely has additional assets attached to his unit, such as air defense artillery assets or additional engineer soldiers. The commander should integrate these assets to achieve the rupture.

D-46. The follow-and-assume force follows the rupture attack and is committed, as necessary, to maintain the momentum of the attack and secure objectives past the rupture. After the rupture force secures a gap in the enemy encirclement, the follow-and-assume force normally conducts the decisive operation until completing linkup operations with another friendly force. When a unit receives a follow and assume mission in a breakout, its commander must coordinate closely with the rupture force commander regarding the location of the gap, the enemy situation at the rupture point, and the enemy situation, if known, along the direction of attack past the rupture point. The commander should not assign this force supporting shaping tasks, such as clear routes and fix bypassed enemy forces, if those tasks would dissipate its available combat power. If executing these support tasks is vital to the success of the breakout and resources permit, the commander should designate a separate follow and support force to perform these tasks.

D-47. The main body consists of the main command post, the bulk of the CSS, the unit's casualties, and some CS assets. It contains combat forces not required for other missions and has sufficient combat power to protect itself. The commander should place one individual in charge of the various elements of the main body to ensure orderly movement. Typically, the main body establishes a flank security force that deploys once the main body passes through the point of penetration and performs flank screen or a guard mission.

D-48. The rear guard consists of soldiers and equipment left on the perimeter to provide protection for the rupture attack and any shaping operations, such as diversionary forces. Forces left in contact must conduct a vigorous delaying operation on the perimeter so that no portion of the rear guard gets cut off. Under a single commander, the rear guard protects the main body from attack while it moves from the area. In addition to providing security, the rear guard deceives the enemy about the intentions of the encircled force, simulating its activities until the main body clears the gap.

D-49. The primary purpose of a reserve is to retain flexibility through offensive action. The commander makes every attempt to keep a small portion of the encircled force uncommitted so he can employ it at the decisive moment to ensure the success of the breakout. The situation may preclude establishing a separate reserve force because of the need to resource either the rupture force, the follow-and-assume force, or the rear guard. In this event, the commander assigns and prioritizes various be-prepared missions to the follow-and-assume force.

D-50. A successful diversion is important to the success of any breakout operation. If the diversion fails to deceive the enemy regarding the intentions of the encircled force, he could direct his full combat power at the rupture point. On the other hand, the diversionary force may rupture the enemy's lines. If a rupture occurs, the diversion force commander must know the intent of the commander of the encircled force. The encircled force commander may choose to exploit the success of forces conducting a diversion, or he may have to disengage them for use elsewhere in the breakout attempt.

CONTROL MEASURES

D-51. As a minimum, a commander uses boundaries; a line of departure (LD) or line of contact; time of the attack; phase lines; axis of advance or direction of attack; objectives; and a limit of advance (LOA) to control and synchronize the breakout. (Chapter 2 describes using boundaries and phase lines. Chapter 3 discusses using axis of advance, direction of attack, objectives, LD or line of contact, LOA, and time of attack.) The commander imposes only those control measures necessary to synchronize his operations.

PLANNING A BREAKOUT

D-52. The commander should initiate a breakout attack as quickly as possible after the enemy encircles his force. While detailed combat information about the enemy's disposition is probably not available, the enemy is normally disorganized at that point in time and is least likely to respond in a coordinated manner. The enemy has not yet brought in sufficient combat power to encircle the friendly force in strength, and weak points exist in his perimeter.

However, sometimes the commander will not attempt a breakout until all other options fail.

D-53. Early in an encirclement, there are gaps between or weaknesses in the enemy's encircling forces. The commander uses his available ISR assets to provide information that increases the accuracy of his situational understanding and determines enemy weak points. The commander plans for the breakout attack to capitalize on identified weak points. Although the resulting attack may be along a less-direct route or over less-favorable terrain, it is the best course of action (COA) because it avoids enemy strength and increases the chance for surprise.

D-54. An encircled force may be operating under adverse conditions and may not have all of its ISR systems operating. This forces the commander to operate with low levels of intelligence regarding enemy strengths, weaknesses, and intentions. Within this environment, he should conduct aggressive reconnaissance to gather information on the enemy. The commander should also obtain information from long-range surveillance units, stay-behind units, and special operations forces in the area. If the enemy is in close contact, the commander may be forced to conduct a reconnaissance in force to ascertain information on enemy strengths. In either case, he must select a COA quickly and develop a plan accordingly.

D-55. A shaping operation, such as a diversionary attack, can assist a breakout by diverting enemy attention and resources away from the rupture effort. The force conducting shaping operations may be located either inside or outside the encirclement area. The enemy must regard the efforts of this force as credible and a threat to the continuity of his maneuver plan. The commander should direct the force's efforts to a point where the enemy might expect a breakout or relief effort. The diversionary force is as mobile as available vehicles, fuel stocks, and trafficability allow so it can reposition to take part in the breakout or maneuver elsewhere to support the breakout. Mobile, self-propelled weapon systems suit the needs of forces conducting shaping operations. Additionally, the probability of a successful breakout increases measurably if another friendly force attacks toward the encircled force as it attempts to breakout.

D-56. The commander conducts offensive information operations to assist the breakout attempt. Deception operations mislead the enemy about the intentions of the encircled force, especially the location of the breakout attempt. If it is not possible to breakout immediately, the commander attempts to deceive the enemy regarding the time and place of the breakout by concealing his preparations and changing positions. He can also give the appearance that the force will make a resolute stand and await relief.

D-57. The commander can use dummy radio traffic for the enemy to monitor or landlines that he might tap to convey false information. The breakout should not be along the obvious route toward friendly lines unless there is no other alternative. In this respect, the preparations for a breakout mirror the preparations for any other type or form of offensive operations. As in other offensive actions, secrecy, deception, and surprise allow for success. The other planning considerations for the breakout are the same as for any other attack.

EXECUTING A BREAKOUT

D-58. The commander exploits darkness and limited visibility during a breakout if his encircled forces have superior night-operations capabilities. The cover of darkness, fog, smoke, or severe weather conditions favor the breakout because the encircling enemy weapons are normally less effective then. The enemy has difficulty following the movements of the breakout force during conditions of limited visibility. However, if the encircled force commander waits for darkness or limited visibility, the enemy may have time to consolidate his containment positions. If friendly forces enjoy air superiority, they may initiate a breakout attack during daylight to fully exploit the capabilities of close air support.

D-59. The unit takes all possible precautions to deceive the enemy about the location of the decisive operation. The rupture force minimizes occupation of attack positions before starting the breakout. A commander may require one or more shaping operations to assist the rupture force in penetrating enemy positions and expanding the shoulders. He may use feints and demonstrations to deceive the enemy concerning the location and time of the decisive operation. However, diversionary attacks need not always occur first.

D-60. The commander organizes and controls his rupture force as he would an attack or movement to contact. (See Figure D-8.) The rupture force generates overwhelming combat power at the point of penetration and attempts to rapidly overwhelm enemy positions and expand the penetration. A commander hard pressed to generate sufficient combat power for both the rupture force and the perimeter defense can thin his defensive perimeter in certain areas by using a detachment left in contact in conjunction with a withdrawal prior to executing the attack. He may also shorten the perimeter's length, which reduces the size of the area occupied by the encircled force.

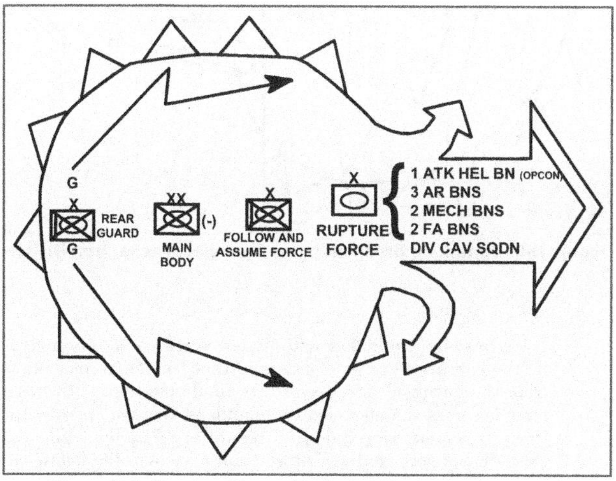

Figure D-8. Breakout by an Encircled Mechanized Division

D-61. The rupture forces applies the breaching fundamentals of suppress, obscure, secure, reduce, and assault to ensure its success at the point of penetration. These fundamentals always apply, but their application will vary based on the prevailing factors of METT-TC. FM 3-34.2 defines these breaching fundamentals and provides guidance regarding the organization of forces, control measures, and planning, preparation, execution, and assessment considerations of combined arms breaching operations.

D-62. If enemy forces at the point of penetration have roughly the same combat power as the rupture force, the commander orders the rupture force to hold the shoulders of the penetration while the follow-and-assume force moves forward. It then becomes the decisive operation. (See Figure D-9.) If the enemy is not in strength, the commander may have the rupture force continue its attack. If there are no identified enemy formations beyond the penetration, the rupture force may transition to a movement to contact. After the encircled friendly force breaks out, it moves toward other friendly forces and links up with them. The next section addresses the control measures and considerations associated with conducting a linkup.

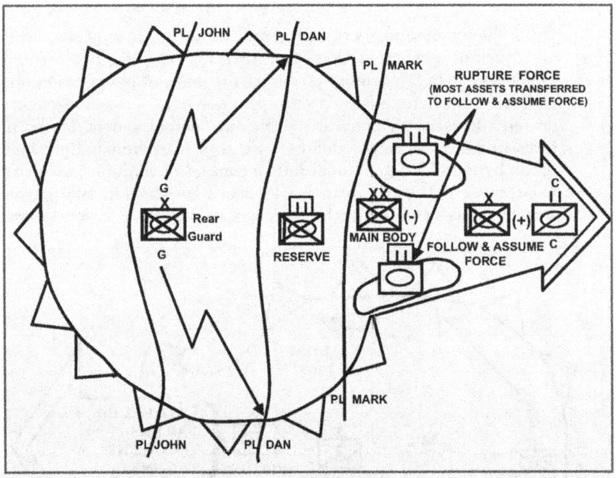

Figure D-9. Continued Breakout by an Encircled Mechanized Division

D-63. Initially, the follow-and-assume force passes through the gap created by the rupture force. It is essential that this force continue to move rapidly from the encircled area toward its final objective. If the follow-and-assume force becomes the encircled commander's decisive operation, it cannot allow itself to become bogged down. Preparatory fires by artillery, Army aviation, close air support, and air interdiction may help the follow-and-assume force in maintaining momentum out of the encircled area.

D-64. Once the breakout attack starts, the rear guard and any diversion forces disengage or delay toward the area of the rupture. Perimeter forces integrate smoothly into the rear of the breakout column. The commander shifts his priority of fires as required by METT-TC once the breakout occurs.

D-65. As other encircled units support or move through the area of penetration, the rear guard commander must spread his forces over an extended area. This requires flexibility and mobility by the rear guard. The perimeter must withstand enemy pressure. If the enemy succeeds in destroying or encircling the original rear guard in the breakout process, the commander must reconstitute a new rear guard.

D-66. The main body follows the follow-and-assume force. It moves rapidly as a single unit on multiple routes in an approach march or road march formation immediately behind the follow-and-assume force, protected on its flanks by security elements. It contains sufficient combat power to protect itself and reinforce the flank or rear security forces if they come under attack.

D-67. Normally, the rear guard initially conducts a withdrawal to break contact with the enemy forces around the perimeter. It contracts the perimeter as it delays back behind the main body. If the enemy closely pursues the breakout force, the efforts of the rear guard may become the decisive operation for the encircled force. The commander should position the reserve where it can also support the rear guard.

D-68. Initially, the priority for fire support is with the rupture force and should focus on suppressing and obscuring the point of penetration. Fire support assets move as part of the main body and rear guard so security forces have adequate fire support. Target identification difficulties resulting from close proximity and intermixing of forces, as well as the rapidly changing ground situation during the execution of a breakout, make the provision of close air support difficult.

D-69. Engineers with the rupture force focus on mobility operations. Engineers with the follow-and-assume force or the reserve improve routes as necessary. Engineers supporting flank security elements focus on conducting countermobility operations. The rear guard must also have enough engineers to conduct countermobility operations.

D-70. The commander prioritizes his air defense assets to protect the rupture force, the rear guard, and the main body. The rear guard is second in priority of protection to help prevent it from being overrun by an enemy pursuit targeted at the main body. The commander must dedicate air defense systems to cover critical points through which the encircled force will pass.

D-71. The commander can relieve his logistics shortfalls by using aerial resupply, ordering external forces to establish support areas, and by using captured supplies. All units and vehicles carry the maximum supplies possible, with emphasis on carrying POL and ammunition. The encircled force only takes vehicles it can support. It may be possible for the higher headquarters of the encircled force to establish an intermediate support base as the breakout attack moves toward a linkup.

EXFILTRATION

D-72. If the success of a breakout attack appears questionable, or if it fails and a relief operation is not planned, one way to preserve a portion of the force might be through organized exfiltration. (Appendix B describes exfiltration as a tactical mission task.)

ATTACKING DEEPER INTO ENEMY TERRITORY

D-73. A COA that the enemy is not likely to expect from an encircled force is to attack deeper to seize key terrain. It involves great risk but may offer the only feasible COA under some circumstances. Attacking may allow the encircled unit to move to a location where it can be extracted by other ground, naval, or air forces. It is only feasible if a unit can sustain itself while isolated, although that sustainment can come from aerial resupply and enemy supply stocks.

D-74. When the enemy is attacking, an encircled friendly force that attacks deeper into the enemy rear may disrupt his offense and provide an opportunity for linkup from another direction. If the enemy is defending and the attacking force finds itself isolated through its own offensive action, it may continue the attack toward its assigned objective or a new objective located on more favorable defensive terrain.

LINKUP

D-75. **A *linkup* is a meeting of friendly ground forces, which occurs in a variety of circumstances.** It happens when an advancing force reaches an objective area previously seized by an airborne or air assault; when an encircled element breaks out to rejoin friendly forces or a force comes to the relief of an encircled force; and when converging maneuver forces meet. Both forces may be moving toward each other, or one may be stationary. Whenever possible, joining forces exchange as much information as possible before starting an operation.

D-76. The headquarters ordering the linkup establishes—

- A common operational picture.
- Command relationship and responsibilities of each force before, during, and after linkup.
- Coordination of fire support before, during, and after linkup, including control measures.
- Linkup method.
- Recognition signals and communication procedures to use, include-ing pyrotechnics, armbands, vehicle markings, gun-tube orientation, panels, colored smoke, lights, and challenge and passwords.
- Operations to conduct following linkup.

CONTROL MEASURES

D-77. The commander establishes minimum control measures for units conducting a linkup. He assigns each unit an AO defined by lateral boundaries and a RFL that also acts as a LOA. The commander establishes a no-fire area around one or both forces and establishes a coordinated fire line beyond the area where the forces linkup. The linkup forces use the linkup points established by the commander to initiate physical contact. The commander designates alternate linkup points since enemy action may interfere with the primary linkup points. He adjusts such control measures during the operation to provide for freedom of action as well as positive control.

EXECUTION

D-78. There are two methods of conducting a linkup. The preferred method is when the moving force has an assigned LOA near the other force and conducts the linkup at predetermined contact points. Units then coordinate further operations. The commander uses the other method during highly fluid mobile operations when the enemy force escapes from a potential encirclement, or when one of the linkup forces is at risk and requires immediate reinforcement. In this method, the moving force continues to move and conduct long-range recognition via radio or other measures, stopping only when it makes physical contact with the other force.

D-79. When one of the units involved is stationary, the commander usually locates the linkup points near the RFL/LOA. (See Figure D-10.) The linkup points are also located near the stationary force's security elements. Stationary forces assist in the linkup by opening lanes in minefields, breaching or removing selected obstacles, furnishing guides, and designating assembly areas. When a moving force is coming to relieve an encircled force, it brings additional logistics assets to restore the encircled unit's combat effectiveness to the desired level.

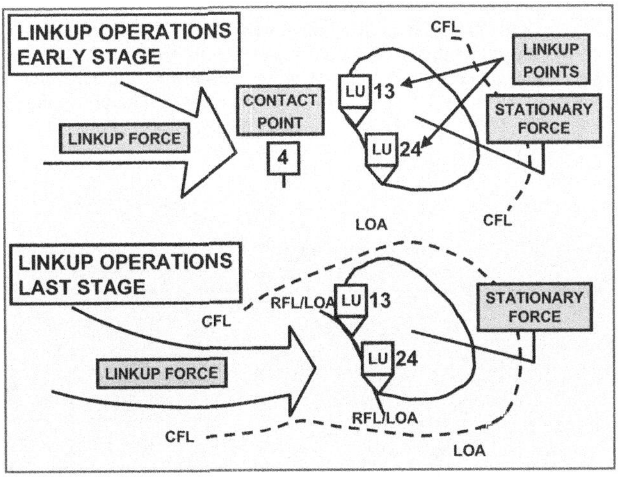

Figure D-10. Linkup of a Moving Force and a Stationary Force

D-80. Linkup between moving units is one of the most difficult operations. The commander establishes a LOA to prevent fratricide. He establishes primary and alternate linkup points for the moving forces near the LOA. Fire support considerations are similar to when a stationary and moving force linkup. Leading elements of each force should exchange liaison teams and be on a common radio net. (See Figure D-11.)

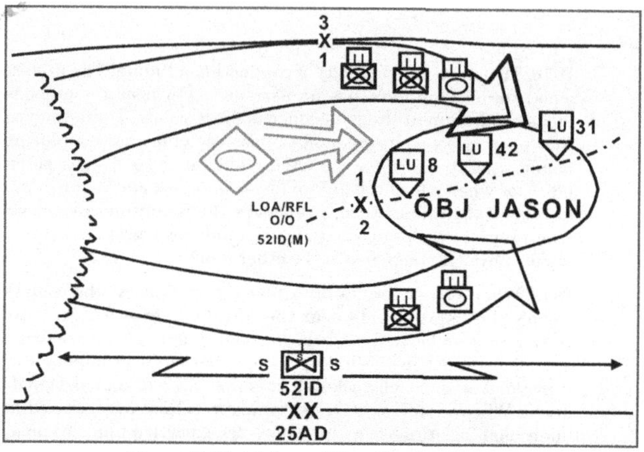

Figure D-11. Linkup of Two Moving Forces

D-81. The commander must carefully coordinate linkup operations with forces of other nations. This is especially true if the two armies are not both members of an alliance with established internationally standardized procedures, or if the units involved have not previously established the necessary procedures.

Appendix E

Rear Area and Base Security

The base commander's plan of action must achieve adequate protection to ensure accomplishment of missions by base elements with as small a force as necessary, since any drain of time and personnel from operational activities will adversely affect the accomplishment of their missions.

Preface, FM 31-81 (Test), *Base Defense*, March 1970

This appendix provides basic guidance for commanders and staff officers on the organization of forces, control measures, and planning, preparing, and executing considerations pertaining to rear area and base security and route and convoy security operations. The tactics and techniques expressed in this appendix are applicable across the full spectrum of operations (offense, defense, stability, and support). However, commanders must interpret them in accordance with the enemy's capabilities. For example, the physical layout of a base will be vastly different if the enemy has a significant air capability than if the enemy has no air capability. The purpose of rear area and base security operations is to prevent interruption of current and future operations. Rear area and base security operations constitute one of the five elements that constitute sustaining operations. (See FM 3-0 for a discussion of sustaining operations as part of the battlefield framework.)

CONTENTS

COMMAND RESPONSIBILITIES

E-1. One of the fundamental considerations in rear area and base security is the resolution of command authority. The echelon commander must clearly define command relationships before initiating any operations. A conflict of interest can occur if command relationships and responsibilities for rear area and base security are not clearly identified. All tenant units in a specific geographical area are responsible to the commander of that area for rear area and base security. Therefore, a rear area commander has tactical control (TACON) over tenant units for conducting rear area and base security. The echelon commander resolves conflicts between the tactical and functional command elements. (See Figure E-1.)

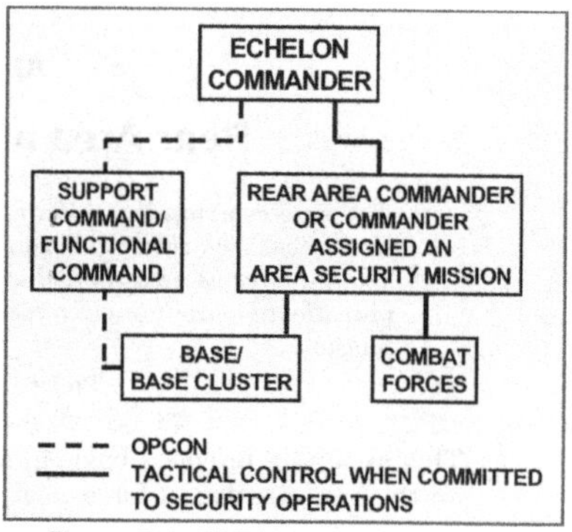

Figure E-1. OPCON versus TACON in Rear Area and Base Security

E-2. There are units subordinate to commanders other than the echelon commander located throughout an echelon's rear area. Many units operating in the rear areas of a corps, division, or brigade are not organic to those echelons. For example, units from the theater support command and other theater organizations, such as the theater signal command or theater engineer command, operate within the corps rear area. The division has elements of corps support groups (forward) and other corps organizations, such as a corps signal battalion, operating within the division rear area. The brigade many have elements of theater, corps, and division organizations operating within its rear area.

E-3. However, an echelon rear area, base cluster, or base commander is responsible for the rear area and base security of all units in his assigned area. The echelon rear area commander has the authority for terrain management within the echelon rear area and assigns every unit to a base or base cluster. All units assigned to a base or base cluster are TACON to the commander of that base or base cluster when an emergency or tactical situation requires the implementation of rear area and base security measures unless specifically exempted by order of the echelon commander.

E-4. The echelon commander establishes the relationships between the subordinate forces of different countries occupying the same geographical area. This structure unifies the different forces to the maximum degree feasible. Resulting language, doctrinal, and philosophical differences may be resolved by—

- Recognizing national affinities and permitting homogenous elements to operate together.

- Combining units of different nationalities only where a clear requirement exists.
- Cross-assigning liaison personnel to improve understanding.
- Organizing area responsibilities along national lines.
- Establishing a mutually supporting warning system.

E-5. Using host nation forces presents a distinct advantage to a commander responsible for rear area and base security because of their knowledge of the area, its language, and customs. Commanders involved in sustaining operations take all possible efforts to use these assets effectively.

DOD CIVILIANS AND CONTRACTORS

E-6. Civilian employees supporting US military forces are subject to hostile action. Protecting DOD civilians and contractors providing battlefield support to Army units is the responsibility of the commander of the host base. The supported base or military unit assures the protection of its operations and personnel. DOD civilians and contractor personnel cannot take an active role in conducting security operations, but retain the inherent right to self-defense.

E-7. Commanders and staff planners assess the need for protecting these individuals and designate forces to provide security when appropriate. The mission, threat, and location where DOD civilians or contractors operate determine the degree of security required.

E-8. Protecting DOD civilians and contractors involves not only active protection through the use of armed military forces to provide escort or perimeter security, but also training and equipping of these personnel for self-protection. Base cluster commanders may train and equip DOD civilian and contractor personnel, if they agree, to work in a nuclear, biological, and chemical (NBC) environment and to carry individual, military specification weapons for personal protection when the threat exists and the geographic combatant commander grants approval.

E-9. The base commander identifies specific security and protection requirements for DOD civilians and contractors located on or transiting the base and designates resources and forces to address those requirements. He includes these decisions in the base's OPLANs/OPORDs and their supporting plans and standing operating procedures (SOP). This allows the base's tenant units and civilians to properly integrate these protection requirements into their operational procedures.

E-10. DOD civilians and contractors are responsible for following all protection requirements and policies—as stated in the contractual agreement. This includes ensuring they receive training in NBC protection and weapons familiarization, when authorized, and informing them of local threats, facility and travel security requirements, and off-limits restrictions directed by the base commander. Failure to adhere to protection requirements and policies could result in disciplinary actions for DOD civilians and possible contract termination for contractors.

COMMAND RELATIONSHIPS

E-11. The success of rear area and base security hinges on binding together diverse resources. The commander must meet the following requirements to eliminate a potential confusion and to achieve a cohesive effort—

- There must be a single responsible commander.
- Geographical areas of operations (AOs) and thus responsibility must be clearly defined.

E-12. Overall responsibility for conducting sustaining operations in a commander's AO rests with the commander. Because the commander cannot afford to focus strictly on sustaining operations he assigns that responsibility to another officer and gives that individual a staff to help him accomplish that operation. The individual responsible for a tactical echelon's sustaining operations is called the rear area commander.

E-13. A corps commander normally assigns the mission of sustaining operation to the deputy corps commander. In a similar manner, the assistant division commander (support) is responsible to the division commander for sustaining operations in the division rear area. The forward support battalion commander supporting the brigade is normally responsible to the maneuver brigade commander for sustaining operations in the brigade rear area. While the echelon commander has the authority to designate someone else as the echelon's rear area commander, doing so complicates staff relationships.

E-14. The responsible rear area commander at each of these three echelons uses the echelon rear command post to monitor the echelon's sustaining operations. This assures the synchronization of the five elements of sustaining operations and an orderly, effective, and timely escalation when conducting rear area and base security. This encompasses the use of host nation and organic combat support (CS) and combat service support (CSS) units through employing a tactical combat force (TCF). The echelon rear command post develops plans integrating the rear area and base security activities of all units located within the echelon rear area. Each echelon rear area commander has the authority to further divide the echelon rear area into subordinate AOs, base clusters, and bases.

E-15. A rear operations center (ROC) provides the corps or division rear area commander with the nucleus of an operations cell within the corps or division rear command post. It is a reserve component organization since the functions it performs are not required in garrison. It contains no organic life support capabilities, such as mess, supply, and maintenance, and relies on the echelon headquarters' commandant for those functions. The soldiers assigned in peacetime to these organization lose their separate identity after being activated and deployed. They are assigned to the corps or division headquarters company and become the rear command post representatives of the echelon G2 and G3.

E-16. A corps rear area commander generally subdivides the corps rear area into subordinate AOs that correspond to the AOs of his subordinate corps support groups. A rear area operations center (RAOC) provides the corps support group commander with a greatly increased plans and operations capability for his operations center to assist in conducting terrain management, rear area and base defense, movement control and area damage control (ADC) within his AO. It also provides him a limited capability to conduct liaison with subordinate base cluster defense operations centers (BCOCs) and base defense operations centers (BDOCs). (See Figure E-2.)

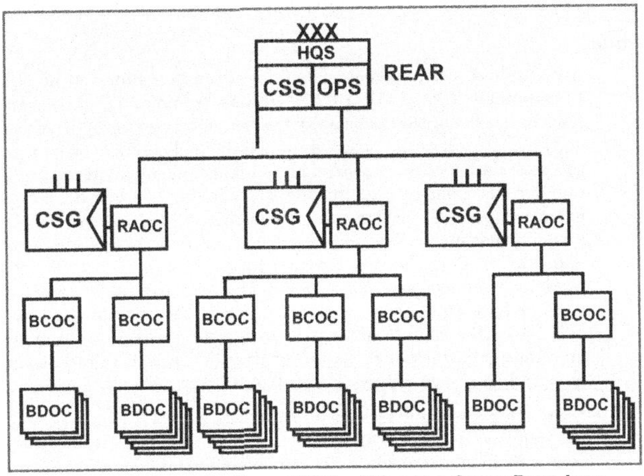

Figure E-2. Organization of C2 Facilities in a Corps Rear Area

E-17. Like a ROC, a RAOC is a reserve component organization that contains no organic life support capabilities, such as mess, supply, and maintenance. It must rely on its assigned CS/CSS brigade/group headquarters to which they are assigned for those functions. The soldiers assigned to a RAOC maintain their separate identity after being activated and deployed, unlike those assigned to a ROC. Therefore, the corps rear area commander can move them as necessary to provide additional command and control (C2) capabilities within the corps rear area.

E-18. Responsibility for overall rear area and base security, as well as defense against large enemy attacks, belongs to the echelon rear area commander. He charges each base and base cluster commander with responsibility for his local and internal security to safeguard the base/base cluster and its facilities from infiltration and small-scale attacks. Each base and base cluster defense plan is coordinated with, and becomes a part of, the echelon rear area defense plan.

BASE

E-19. A *base* is a locality from which operations are projected or supported (JP 1-02). It can contain one unit or multiple units. It has a defined perimeter, established access controls, and is situated and configured to take advantage of natural and man-made terrain features. Missions and security considerations determine the selection of a base site. To nullify or reduce the effectiveness of enemy attack, the rear area commander assigns all units located in the rear area that are not actively involved in operations to a base.

Bases may be further organized into base clusters or organized as a separate bases that report directly to the ROC or RAOCs controlling rear area and base security for the AO where the base is located. A commander may name his base or use a mixture of alphanumeric characters to designate it.

Base Commander

E-20. In base defense operations, the *base commander* is the officer assigned to command a base (JP 1-02). The base commander is responsible for defending his base. He is normally the senior Army competitive category officer on the base unless the base contains only medical units. He exercises C2 over the base and necessary control of resident and transient units. The base commander's responsibilities include establishing the overall base defense organization as well as planning, preparing, executing, and assessing all local security measures. This includes the tactical control of US Navy, Air Force, and Marine Corps forces assigned to the base primarily for the purpose of local base defense, such as a US Air Force security flight. Forces of other services or functional components assigned or attached to the base for primary purposes other than local base defense will support local base defense during an imminent attack or threat of an attack. For more information on joint rear area and base security see JPs 3-10 and 3-10.1.

E-21. The base commander uses all assets within the base perimeter to create the required level of local security. He has the authority to position units assigned to his base in mutually supporting positions. He directs employing of obstacles, such as wire- and command-detonated antipersonnel mines, to defend each base within the guidance of his rules of engagement (ROE). He identifies shortages in materiel or weaknesses in required defensive capabilities early and requests the necessary reinforcements and supplies to correct those weaknesses.

E-22. A base commander is also usually a battalion commander, and the BDOC forms around the nucleus provided by his battalion staff. The echelon rear area commander normally does not assign the function of base command to MP commanders because of their need to exercise command over MP elements responding to Level II threats within the echelon rear area.

Base Defense Operations Center

E-23. The base defense force (BDF) commander needs a BDOC to ensure the integration of defense plans and maximum effectiveness of the total base defense effort. A *base defense operations center* is a command and control facility established by the base commander to serve as the focal point for base security and defense. It plans, directs, integrates, coordinates, and controls all base defense efforts, and coordinates and integrates into area security operations with the rear area operations center/rear tactical operations center (JP 3-10.1). The nature of the BDOC depends on the combination of forces involved and may be joint, interagency, or multinational depending on the combination of forces involved.

E-24. The BDOC is similar in many respects to the tactical operations center found in maneuver brigades and battalions. Representatives of the intelligence, maneuver, and fire support battlefield operating systems (BOS) staff the BDOC. The base commander provides other BOS representatives to his BDF commander on an "as required" basis. Multiservice, other agency, and multinational representation, with decision making authority, is present in the BDOC when elements of their armed forces, police, or paramilitary forces are directly involved in the overall base defense effort.

E-25. The center consists of two primary sections—the command section and the plans and operations section—with additional sections as deemed necessary. These additional sections could include a CSS section to plan the provision of services and support to the base, and an ADC section that provides inspection, planning, and control of the base's ADC resources. This center is fully manned and operates continuously. It primary functions are—

- Keeping the base commander informed of the current situation within his AO, including enemy, friendly, and civilian activities.
- Preparing comprehensive plans to implement the commander's overall base defense and security guidance.
- Ensuring the participation of all units within the base perimeter in conducting active and passive defense and security measures.
- Reducing any conflicting interests inherent in multiservice or multinational BDF environment.
- Providing centralized collection and processing of information from various intelligence and operational sources and disseminating resultant intelligence products as appropriate.
- Providing the essential C2 organization necessary to conduct coordinated defensive operations.

E-26. The base commander establishes the BDF from available base CS and CSS assets if MP and combat arms units are not available to provide the necessary local security. He designates the BDF commander to avoid confusion and provides that individual necessary guidance.

BASE CLUSTER

E-27. A *base cluster* in base defense operations, is a collection of bases, geographically grouped for mutual protection and ease of command and control (JP 3-10). A base cluster normally does not have a defined perimeter or established access points. The base cluster is normally the next higher tactical C2 headquarters of the base. A base cluster typically contains five bases but can range from two to seven bases. The rear area commander designates each base cluster and assigns it either a name or an alphanumeric designation.

Base Cluster Commander

E-28. The senior Army competitive category officer in each cluster becomes the commander of that cluster. Like a base commander, he establishes a continuously operating BCOC to support the tactical chain of command. This officer is normally the commander of a CS or CSS brigade or group and his staff forms the BCOC nucleus. The BCOC integrates each base defense plan into the base cluster defense plan.

Base Cluster Operations Center

E-29. The BCOC keeps the base cluster commander informed of the situation in his area and of the resources available to cope with emergencies. It represents the commander's planning capability and provides him a C2 system to control the execution of rear area and base security missions. The BCOC can provide functional elements with current information they may require to conduct their normal CS and CSS functions. This includes—

- Highway road conditions, new unit or activity locations, security and support activities, refueling points, emergency maintenance and recovery facilities, medical facilities, rest stops, and reroutes.
- Intelligence data, including weather, terrain, and civil considerations.
- Location of area communications sites that units, convoys, or soldiers on the move can link into to maintain a current common operational picture.
- Location and level of contamination from enemy use of weapons of mass destruction or chemical contamination resulting from the use of conventional munitions and industrial accidents.
- Location of hazards resulting from battlefield activities of both sides.
- Available explosive ordnance disposal (EOD) support.

CONTROL EXERCISED BY US AREA COMMANDERS

E-30. There are three basic considerations in determining command policies governing the authority and limitations of the echelon rear area commander. These three factors are—

- The threat levels and situation within the echelon rear area.
- The utility of different locations within the echelon rear area.
- Political considerations.

THREAT LEVELS

E-31. Threats to bases in the rear area are categorized by the three levels of defense required to counter them. Any or all levels may exist simultaneously in the rear area. Emphasis on specific base defense and security measures may depend on the anticipated threat level. **A *Level I threat* is a small enemy force that can be defeated by CS and CSS units operating in the rear area or by the perimeter defenses established by friendly bases and base clusters.** A Level I threat for a typical base consists of a squad-size unit or smaller groups of enemy soldiers, agents, or terrorists. Typical objectives for a Level I threat include supplying themselves from friendly supply stocks; disrupting friendly C2, logistics, and facilities; and interdicting friendly lines of communication (LOC).

E-32. **A *Level II threat* is enemy activities that can be defeated by a base or base cluster augmented by a response force.** A typical response force is a MP platoon; however, it can be a combat arms maneuver element. Level II threats consist of enemy special operations teams, long-range reconnaissance units, mounted or dismounted combat reconnaissance teams, and partially attritted small combat units. Typical objectives for a Level II threat include the destruction, as well as the disruption, of friendly C2 and logistics and commercial facilities, and the interdiction of friendly LOCs.

E-33. A ***Level III threat* is beyond the defensive capability of both the base and base cluster and any local reserve or response force.** It normally consists of a mobile enemy force. The friendly response to a Level III threat is a TCF. Possible objectives for a Level III threat include seizing key terrain, interfering with the movement and commitment of reserves and artillery, and destroying friendly combat forces. Its objectives could also include destroying friendly CSS facilities, supply points, command post facilities, airfields, aviation assembly areas, arming and refueling points, and interdicting LOCs and major supply routes.

UTILITY OF DIFFERENT LOCATIONS

E-34. The utility of a given location is a vital factor to the rear area commander. Some areas are critical to conducting the echelon's sustaining operations. Others are priority areas important to conducting the echelon's sustaining operations. Still others are areas where he must exercise some general control but which are neither critical nor priority areas. Finally, there are areas of no importance in the echelon's sustaining operations.

E-35. A critical area is an area or geographical point essential to accomplishing the echelon's sustaining operations. The loss, neutralization, or interdiction of this area allows no alternate operations and adversely affects sustaining operations. An example of a critical area would be the only available port within an AO.

E-36. A priority area is also an area or geographical point essential to accomplishing the echelon's sustaining operation except that within an acceptable amount of time sustaining operations functions can be transferred to an alternate location without an immediate significant impact on the total CSS effort. An example of a priority area would be a supply point containing the ammunition designed to support an offensive operation or an installation performing an important function, such as repairing thermal sights. Another example of a priority area would be a bridge over a major waterway.

E-37. Areas retained under general control are normally the best areas for their designated purpose, but sustaining operations conducted within them can move to other locations without major problems. Examples of areas over which the rear area commander must have some general control are LOCs and main supply routes.

E-38. The rear area commander coordinates with the host nation, using available civil affairs (CA) and operational law personnel to obtain the necessary authority to conduct security measures required by the sensitivity of the area and the situation. These security measures include conducting patrols outside of base perimeters, establishing procedures for clearing indirect fires, and conducting counterintelligence operations. Negotiations also address such measures as establishing a curfew, imposing circulation control over civilian traffic, and similar policies for the host nation civilian populace as dictated by the factors of METT-TC.

E-39. At the division and corps echelons, there could be areas within the echelon rear area commander's AO that do not affect the echelon's sustaining operations. The rear area commander prefers not to commit his scarce intelligence, surveillance, and reconnaissance resources in those areas but leaves the responsibility for detecting and countering enemy activity in those areas to the host nation.

POLITICAL CONSIDERATIONS IMPACTING THE REAR AREA

E-40. The degree of control that the rear area commander has over host nation authorities and the civilian population depends primarily on the political conditions under which US forces entered the AO. Generally, US military presence occurs in one of two ways:

- Entry by invitation of the host nation.
- Forcible entry.

The extent of and limitations on military activities are prescribed by customary international law or by international agreements to which the US is a party, or by both in either environment.

E-41. The rear area commander may encounter a range of situations extending from a host-guest relationship to total US authority under conditions of military government. He may have no power over the host nation authorities and civilian population except what he can exert through influence and coordination in some cases. Populace controls, mutual support between civilian and military resources, and the coordination of civilian and military security and damage control measures are of utmost importance to rear area and base security. The commander must constantly be aware of and understand his relationship with host nation authorities and the civilian population. A clear understanding of his authority is essential to the rear area commander in exercising that degree of control necessary to ensure security and safety to all friendly military forces and the civilian population located within the AO.

Assisted Unopposed Entry

E-42. Through legitimate agreement by the interested governments, a host nation may request US forces to assist the host nation for a mutually agreed reason. Liaison and coordination are the primary means of establishing acceptable military-civilian communications. The commander's initial requirement is the provision of maximum civilian support for and the prevention of civil interference with military operations. Host nation forces may be effective in rear area and base security because of their knowledge of the area, its language, and customs. US military advice and assistance rendered in accordance with applicable laws and regulations may enhance their effectiveness.

Forcible Entry

E-43. Forcible entry entails the seizure of terrain and facilities from enemy forces. Depending on circumstances, the operation may require the removal of existing civil or martial governmental apparatus or the restoration of a suppressed government. In the absence of civil authority, Army forces establish measures to control the civilian population and provide basic services. When US combat forces enter enemy territory to occupy it, initial disclosure of the nature and extent of US authority should include basic occupation directives in accordance with the laws of land warfare and CA doctrine. (See FMs 1-27.10 and 3-57.) These actions include issuing necessary proclamations concerning US authority as it relates to civil government, curfew, movement control, public order, and similar matters. Civil affairs units initiate, coordinate, and supervise local compliance with these directives.

US-Host Nation Relationships

E-44. The rear area commander must have the authority to protect and secure his installations and protect resources traveling along LOCs within his AO. Since this control affects the host nation governmental agencies and the civilian populace, he must not use that authority in a manner that alienates those audiences to the US mission and objectives.

E-45. Within the limits of the authority granted to them under local law, host nation civil officials may delegate political authority over local government to the military commanders of their own nation. The rear area commander must coordinate his operations with the host nation military area commander in conformance with US policy directives and requirements when this condition exists. Otherwise, the rear area commander coordinates his operations with the appropriate host nation agency.

E-46. The host nation military commander normally assumes responsibility for reconnaissance and security outside the perimeter of US military bases. The base commander provides for interior guard and control within the confines of his base and the security of his perimeter. The rear area commander may not have exclusive authority over local national civilian employees within his installation. Authority to hire and fire, investigative jurisdiction in matters of sabotage or espionage, and other vital matters of mutual concern to both the US command and the host nation must be delineated and agreed on for effective rear area and base security. The US rear area commander must maintain authority to control access to his bases when the situation warrants. Counterintelligence units are responsible for investigating sabotage, espionage, and subversion. Military police criminal investigation teams assist them as appropriate. In response to events having significant high-level interest, other governmental agencies, such as the FBI, may participate in these investigations.

BASE AND BASE CLUSTER DEFENSE

E-47. Base and base cluster defense is the cornerstone of an echelon's successful rear area and base security effort. The following paragraphs outline the organization of forces, control measures, and considerations pertaining to planning, preparing, and executing base and base cluster defense.

ORGANIZATION OF FORCES

E-48. The rear area commander uses a combination of base and base cluster defense forces, designated response forces, and TCF, to provide the required security within the echelon rear area. Special operations forces (SOF) may conduct counterterrorism operations within the rear area on an exception basis.

Base Defense Force

E-49. A ***base defense force*** **is a security element established to provide local security to a base.** It normally consists of the combined security assets provided by each unit on the base toward the ongoing security requirements of the base. These assets include crews, weapon systems, and radios of combat vehicles temporarily located on the base for maintenance or other reasons. It may consist of MP, infantry, armor, or cavalry units when the echelon commander's mission focuses on stability or support operations. The mission of the BDF is to conduct reaction operations to deter, resist, or destroy an enemy Level I force attacking the base. The BDF must be sufficiently flexible to defend the base perimeter, provide for the base's interior guard requirements, and conduct local offensive operations beyond the base perimeter if allowed by the host nation.

E-50. The base commander normally appoints a BDF commander to assist in executing base defense functions. He tasks units located within the base to provide soldiers, equipment, and materiel to form the BDF. The principal BDF tactical increments are normally provisionally organized security platoons. (See Figure E-3.) The base commander organizes his BDF as required by the intensity of enemy activity within the rear area.

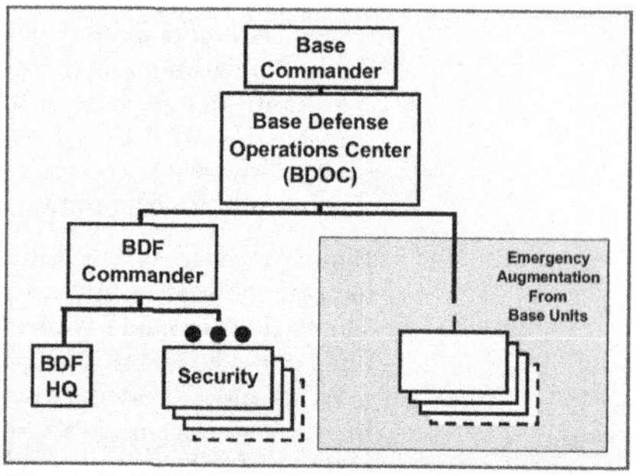

Figure E-3. Typical Base Defense Organization

E-51. A BDF should have a high degree of direct-fire lethality provided by a mixture of small arms, automatic weapons, and antitank systems. It should also have access to supporting indirect fires, a high degree of tactical mobility, and a reasonable span of control. It should be capable of—

- Conducting reconnaissance patrols for detecting and reporting the location, strength, and capabilities of enemy forces located near the base.
- Developing positions within and outside the base from which enemy advances can be stopped or destroyed.
- Using reserve elements to attack relatively small enemy units that threaten to penetrate the base perimeter.
- Providing internal security for critical capabilities located on the base.

The base commander must reconcile any issues about personnel and equipment resources shortfalls affecting the BDF and the desired level of security.

E-52. Most CS and CSS companies can provide a platoon-size security element toward the defense of their assigned base. (Figure E-4 on page E-12 shows a possible organization of such a security platoon.) A base having more than one security platoon also establishes a provisional security company headquarters for the BDF. The BDF commander will be the commander of this provisional security company. Most CS and CSS battalion headquarters can provide the resources for a provisional security company headquarters. The BDF commander designates a portion of the available security force as his reserve. In an emergency, the BDF is augmented by the defensive efforts of all units assigned to or transiting the base.

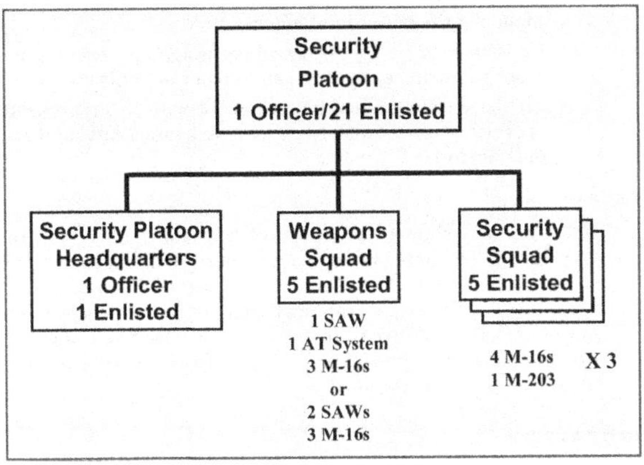

Figure E-4. Security Platoon

E-53. Operational requirements frequently modify a unit's ability to contribute resources to the defense effort of its base. Responsibility for reporting such modifications rests with the individual unit. The base commander brings these unrelated and unfamiliar groups together when creating his BDF. Every effort must be made to retain the unit integrity of these security platoons while assigning security tasks to the BDF.

E-54. Providing a BDF from CS and CSS organizations results in two undesirable conditions—

- Reduced operational effectiveness of the units from which the resources are drawn.
- Soldiers performing duties other than those for which they are specifically trained.

The unit commanders providing these security platoons assign personnel to them. Soldiers assigned to the BDF should remain there if operations permit, in order to facilitate training, provide for continuity of defense operations, and ensure force efficiency.

E-55. The base commander considers all available Army, joint, and multinational resources and forces in determining the exact organization of the BDF. Host nation police, paramilitary, and military elements will frequently be available to assist in providing rear area and base security. Early clarification of the positions, responsibilities, and authority of these elements reduces possible confusion and duplication of effort. The degree of cooperation the base commander attains from local host nation or third nation officials largely determines the effectiveness of those countries resources as they apply to base defense, regardless of the existence or absence of agreements made between forces at the national level. Special considerations for US employment of host nation elements include—

- Designated host nation security force must be responsive to the demands of the base commander.
- Caution may be necessary when assigning host nation security forces tasks requiring specialized equipment or skills.
- The base commander should consult with the appropriate civil-military operations office to clarify any restrictions that might apply to the use of these forces.

Base Cluster Defense Force

E-56. The base cluster defense force consists of all of the BDFs within the base cluster. The base cluster commander can direct the employment of BDFs within the base cluster or segments of those BDFs to counter identified Level II threats within the geographical area of the base cluster. The more normal course of action (COA) is to have a response force engage these Level II threats. Base commanders must reconstitute those portions of their BDFs committed to these missions.

Rear Area Security Forces

E-57. The purpose of rear area security forces is to facilitate a base's ability to return to mission accomplishment as quickly as possible rather than devoting CS and CSS resources to limited tactical operations. Rear area security forces can be further broken down into response forces and TCFs.

E-58. **Response Forces.** A *response force* is a mobile force with appropriate fire support designated, usually by the area commander, to deal with Level II threats in the rear area (JP 1-02). It usually consists of MP forces supported by available fire support and Army aviation assets. They are usually either platoon- or company-size forces. Military police elements may be temporarily or permanently unavailable in sufficient strength for response force operations because of their commitment to other missions. The commander should designate alternative response forces before this occurs. Other possible response force options include engineer units, chemical units, transiting combat units, elements of the reserve, or host nation assets. The rear command post coordinates with the echelon G3/S3 before committing forces other than MP units to response force missions.

E-59. The MP chain of command directs the size and composition of the response force when the echelon is primarily focused on conducting offensive and defensive operations. The nature and size of the threatening enemy force influences the size and number of MP elements that make up the response force. So, too, does the current rear intelligence preparation of the battlefield (IPB) and the echelon's rear area commander's risk assessment. The echelon provost marshal, in coordination with his rear area commander, considers—

- The priority of ongoing operations.
- The criticality of the base under attack.
- The amount of time needed for given elements to consolidate.

The rear area commander continuously assesses the situation and, if appropriate, commits more response force assets to deal with the threat.

E-60. **Tactical Combat Forces.** A *tactical combat force* (TCF) is a combat unit, with appropriate combat support and combat service support assets, that is assigned the mission of defeating Level III threats (JP 1-02). These threats can result from enemy forces infiltrating friendly positions or penetrating friendly defensive positions and moving into the friendly rear area, or conducting airborne/air assault operations. A successful defense against such a Level III threat requires planning and preparing for the TCF's employment. The echelon commander can dedicate a TCF to deal with such a threat, designate another force as the on-order TCF, or accept the risk of not having a designated TCF. A dedicated TCF has the necessary assets, such as artillery and combat engineers, allocated to it to accomplish the mission. The echelon commander decides the exact composition of the TCF after weighing the risk of allocating forces to the TCF and thus decreasing the combat power available for his decisive operation.

E-61. The echelon G3 designates the number of TCFs in accordance with the situation and the echelon commander's guidance. Designating more than one TCF provides flexibility of response to competing needs. The primary advantage of having a dedicated, rather than an on-order, TCF is the assurance that it will be available when needed. Other advantages include the ability of the TCF staff to focus the force's planning and preparing activities on just one mission. This includes establishing liaison and communications with supported bases or base clusters and echelon Level II response forces. It also allows the dedicated TCF to rehearse its plans. When the commander assigns a subordinate unit an on-order TCF mission, he must establish criteria on when to commit that unit as the TCF.

E-62. A TCF can expect support from the following types of assets:

- MPs with working dogs.
- Military intelligence (includes counterintelligence).
- Field artillery.
- Engineers.
- Army aviation (attack, air cavalry, air assault, C2, and possibly special electronic mission aircraft).
- Public affairs.
- Air defense artillery.
- Area signal nodes.
- Civil affairs assisted by psychological operations (PSYOP) elements.
- CSS elements required to sustain the force conducting area security.

Any or all of these types of assets may not be immediately available to the TCF on its commitment. They are limited assets and may be engaged in other missions. The force conducting the rear area security mission may also receive support from other US and allied services, including the host nation.

E-63. The factors of METT-TC determine the size of the TCF. For a corps, this normally results in at least one battalion task force being assigned the mission. A division typically designates a reinforced company team as its TCF. The echelon commander configures his TCF to conduct offensive operations. The TCF must be extremely mobile and capable of moving by air and ground modes because of limited reaction time and extended distances between units common to the rear area. The TCF must be as mobile as the enemy. It should be capable of destroying armor-protected vehicles and dismounted infantry and of suppressing enemy shortrange air defense (SHORAD) systems. Consequently, a TCF typically consists of infantry, Army aviation (attack and utility helicopters), and air cavalry elements with engineer and field artillery support. The commander may also organize a TCF with armored cavalry, armor, and mechanized infantry units if the situation so dictates.

E-64. The commander strives to obtain tactical self-sufficiency within both the response force and the TCF. He does this by attaching or placing in direct support adequate fire support, CS, and CSS elements so that they can perform semiindependent operations. In addition, he must obtain resources, such as CA, PSYOPS, and interpreters, for interacting with the civilian population of the area being secured.

Special Operations Forces

E-65. Special operations forces other than CA and PSYOP may combat terrorism within the AO of a conventional unit. A unit conducting rear area and base security may be tasked by a higher headquarters to provide conventional support to SOF elements conducting counterterrorism operations within its AO. If SOF operate in the area, a special operations command and control element (SOCCE) will normally be deployed at the Army forces or corps echelon, with liaisons attached lower for a specific mission to the unit responsible for the appropriate AO for the duration of the SOF mission.

CONTROL MEASURES

E-66. Control measures in base defense operations are the same as those used in other defensive operations. (See Figure E-5, page E-16.) The headquarters establishing the base or base cluster designates the base's AO using rear, lateral, and forward boundaries. The AO for the base may or may not be contiguous to the AO of other units. The echelon rear area commander may further subdivide his assigned area into subordinate AOs, bases, and base clusters. He can assign maneuver forces to assembly areas and battle positions. He establishes phase lines, contact points, objectives, and checkpoints as necessary to control his maneuver. He establishes fire support coordinating measures (FSCM) to permit or restrict fires in and around the base. (See Chapters 2, 3, and 8 and FM 3-09 for a discussion of FSCM.) No-fire areas may be required to protect civilians, prevent disruption of sustaining operations, or protect combat outposts, observation posts, and patrols from friendly fire. All established control graphics are coordinated with host nation agencies to minimize interference, misunderstandings, and unnecessary collateral damage. The BDF commander, in coordination with the base commander, designates the base perimeter, target reference points, and sectors of fire to organizations located within the base.

PLANNING CONSIDERATIONS

E-67. The echelon commander designs his rear area and base security system to ensure—

- Maximum effectiveness of BDFs provided by CS and CSS units.

- Minimum degradation to unit mission performance while providing the unit's contribution to the BDF.
- Combat arms units are not diverted to provide dedicated rear area and base security forces.

E-68. Commanders in the rear area can improve their capability to effectively use available manpower and materiel by establishing logical priorities. The commander must allocate resources available to protect personnel, supplies, and facilities in consonance with the degree of risk involved. Sufficient personnel and materiel will not usually be available in the rear area to provide simultaneous protection to all of the echelon's sustaining operations.

E-69. Base defense plans are flexible and allow for differing degrees of security based on the probability of threat activity to maximize unit mission accomplishment. The common defensive planning considerations outlined in Chapter 8 apply to planning for base security. The following paragraphs explain fundamental principles of rear area and base security. They also address BOS specific planning considerations that apply to base security.

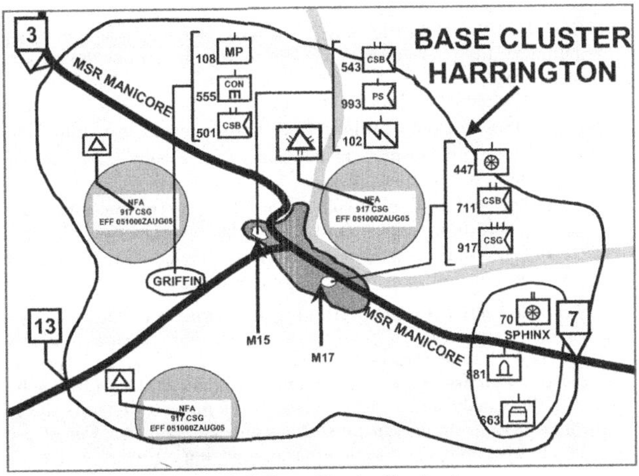

Figure E-5. Base Cluster Harrington

Fundamentals of Base Defense

E-70. The fundamentals of base defense are—

- Understand the enemy.
- See the AO.
- Use the defender's advantages.
- Concentrate at critical times and places.
- Conduct counterreconnaissance and counterattacks.
- Coordinate critical defense assets.
- Balance base security with political and legal constraints.

- Know the ROE.
- Responsiveness.

E-71. **Understand the Enemy.** Defenders must be familiar with the capabilities and limitation of enemy forces, weapons, equipment, and tactics. The base commander also must have access to the latest intelligence concerning probable enemy intent.

E-72. **See the AO.** Intelligence operations are key to assembling an accurate picture of the AO. The IPB process provides the commander a continuous, integrated, and comprehensive analysis of the effects of enemy and neutral capabilities, terrain, and weather on operations. It helps the commander anticipate events and develop the priority intelligence requirements and information requirements tied to those events. Intelligence and multidiscipline counterintelligence estimates are prepared, continuously updated, and integrated into the base commander's staff decision making process. Space-based, airborne, and ground sensors, patrols, and counterintelligence activities are all means that rear area commanders can employ to detect possible indicators of planned enemy action.

E-73. **Use the Defender's Advantages.** Defender's advantages may permit a numerically inferior force to defeat a much larger one. Some of these advantages are:

- The ability to fight from cover.
- More detailed knowledge of local terrain and environment.
- The ability to prepare positions, routes between them, obstacles, and fields of fire in advance.
- The ability to plan communications, control measures, indirect fires, and logistic support to fit any predictable situation.
- The ability to deceive enemy forces about friendly defensive capabilities, dispositions, and execution of operations.

E-74. **Concentrate at Critical Times and Places.** A base commander must measure and continually evaluate the various risks to his base. An enemy usually concentrates his resources on the most sensitive or vulnerable targets. Defense of a base is normally conducted along interior lines, permitting the timely and secure movement of forces to engage the most critical threats. The commander must mass enough combat power at points of decision by economizing in some areas, employing a reserve, and maneuvering to gain local superiority at critical points.

E-75. **Conduct Counterreconnaissance and Counterattacks.** A major contributing factor to vulnerability in the rear area is the relatively fixed nature of operations. An enemy can study the operational patterns of semifixed units and installations to determine their existing defensive measures. Further, fixed bases having well-established perimeters usually have limited depth. Counterreconnaissance and counterattack add depth to the battle outside the perimeter, allowing the base to continue its primary mission with minimal interference.

E-76. **Coordinate Critical Defense Assets.** Synchronization of indirect fires, air and missile defense resources, tactical aircraft, engineers, dismounted troops, armored vehicles, naval surface fire support, and helicopters can produce a combined arms effect. Synchronizing forces and fires produces a synergy capable of defeating a larger enemy force. This synergy results from making enemy movement difficult or impossible and by causing a reaction that may make enemy forces more vulnerable to other friendly capabilities. Periodic rehearsals of security plans ensure effectiveness of the alert system, proficiency of key individuals, and identification of potential problem areas.

E-77. **Balance Base Security with Political and Legal Constraints.** Base security may have to be designed around numerous political constraints.

E-78. **Know the ROE.** The base commander and his subordinates must comply with established ROE. He ensures the reconciliation of any inconsistencies between service component and multinational ROE.

E-79. **Responsiveness.** The echelon rear area commander and his subordinate base cluster and base commanders have the authority and the capability to communicate operational requirements directly to units located within their AOs. The chain of command frequently changes from the functional to the tactical chain of command after implementing a rear area and base security plan. Therefore, the rear area commander must critically analyze and test his communication systems that filter through various levels of command to assure that they are responsive under given emergency conditions. The echelon rear area commander establishes direct communication links between the echelon rear command post and its subordinate base cluster and base operations centers if existing information systems are not adequate to transmit rear area and base security-related information to those centers in a timely manner.

E-80. Responsiveness may be attained by many different procedures and capabilities, some of which are listed below:

- Clearly established authority granted to base cluster and base commanders.
- A wide-area information system that provides a flexible warning system for a large variety of situations.
- Continuous identification and tracking of resources that can assist in providing rear area and base security, such as maintenance float weapon systems and replacement crews.
- Response forces have communication systems capable of using the existing wide-area information systems to communicate directly with the threatened BDOC. They also can communicate with the BCOC located in the threatened base cluster and the echelon rear command post.
- MP forces responding to Level II rear area threats have combat vehicles with a mixture of firepower, armor protection, and a high degree of tactical mobility.
- Identifying dual-capable units. For example, a rear area security force identified to respond to Level I or Level II threats also has capabilities to perform rescue or provide labor in an ADC mission. However, a rear area security element is not concurrently identified as an ADC element.

Other Planning Considerations

E-81. When the rear area commander commits his forces to semifixed installations or sites, the rear area security mission may become defensive in nature. Therefore, he must carefully balance his approach with the need for offensive action. Detecting enemy activity early is critical when conducting rear area security; it provides the commander with time to react to any threat. Proper integration of the BOS is key to successful operations.

E-82. **Intelligence.** A review is made of all available information about the general area before occupying a site. The IPB process addressing the rear area includes not only enemy information, but terrain, weather, and information about the civilian population and infrastructure. Intelligence, cavalry, scout, and CA organizations obtain relevant information about the area from host nation governmental organizations and the civilian population.

E-83. In the commander's visualization of the factors of METT-TC for the rear area security mission, he focuses on the enemy situation to determine how to organize and conduct the mission. Tactical intelligence is the key to defeating enemy operations directed against friendly sustaining operations. It provides the commander with information on current and possible future enemy locations, activities, strengths, weaknesses, and plans, which helps him seize the initiative. The commander should exploit all relevant sources of information within his AO, including information provided by civilian agencies and individuals. The IPB process is a vital part of answering the commander's intelligence requirements which include—

- Size and capabilities of enemy forces within the commander's AO.
- Natural defensive characteristics of the terrain.
- Movement corridors throughout the AO for mounted and dismounted operations.
- Conditions of existing roads, railways, pipelines, and waterways for use as both military LOCs and civilian commerce.
- Proximity of enemy forces to critical facilities and installations, such as airfields, power generation plants, petroleum tank farms, and civic buildings.

E-84. **Maneuver.** The commander should not employ any organization in a role or configuration for which it is not principally designed unless such action is unavoidable. Rear area commanders employ the defensive potential of CS and CSS units to counter enemy actions. The defensive potential of each unit is that portion of a CS or CSS unit that may be used to provide rear area and base security with minimum degradation to overall mission capability. The echelon commander does not divert his combat forces to provide rear area and base security unless required by the factors of METT-TC. The rear area commander takes maximum advantage of host nation capabilities to provide response forces and TCFs.

E-85. Site selection is also an important factor in planning rear area and base security. Combat support and CSS operational requirements do not correspond to security requirements. Supporting units need a good road net, ready source of labor, storage areas, and other facilities that are available in established commercial areas. However, such placement decreases the security capabilities. The best site, from a security standpoint—given no air threat—is in a cleared area, occupying high ground with sufficient vegetation to conceal base facilities from observation. The worst site contains complex terrain and large numbers of noncombatants. The former extremely limits observation and fields of fire from the base and the latter limits the firepower that can be employed to defend the base. The most common tendency is to occupy the low ground through which most roads run, a balance between operational requirements and security requirements. A commander can justify the cost, time, and manpower needed to relocate a site to more defensible terrain rather than secure a poor position. The base commander locates his base in a place that enhances its physical security. The base cluster and echelon rear area commanders must approve the location.

E-86. Forces engaged in rear area security must be flexible and able to adapt to rapidly changing tactical situations. Rear area and base security are normally the domain of small units. The organization of forces for base defense also differs, depending on how the commander plans to accomplish his mission. The base commander task-organizes his forces into combined arms elements to counter the level of enemy capabilities as determined by his intelligence assets. This enables him to find the enemy with the smallest possible element, maintain contact without risking the destruction of the element in contact before it can receive outside support, and mass his combat power rapidly to destroy the detected enemy.

E-87. All security measures must have an inherent offensive capability. The echelon TCF, response forces provided by echelon MP and other assets, and local security measures undertaken by bases and base clusters contain the rear area commander's offensive capabilities. The base commander uses the potential combat power available within CS and CSS units located within his AO to provide those local security measures. He organizes and trains that identified potential combat power to conduct limited offensive operations against confirmed enemy Level I threats that threaten his bases and base clusters.

E-88. Absent a direct enemy threat, there is a tendency to relax security measures after completing the base's fighting and survivability positions. However, well-established security patterns allow enemy elements to reconnoiter the base and determine its points of vulnerability with little risk.

E-89. The base commander designates reconnaissance and patrol areas outside the base's perimeter to provide additional security if allowed to do so by the host nation. The objective is to provide the base additional early warning.

E-90. The factors of METT-TC and unit capabilities determine the specific security element tasks of subordinate maneuver units. For example, mounted forces are more suited to secure key points, such as major road junctions, bridges, tunnels, canal locks, dams, and power plants; while dismounted forces are well-suited for securing areas containing restricted terrain or large numbers of civilians.

E-91. The base commander should plan to use only the minimum firepower required to accomplish the base security mission. While his right to self-defense is never denied, it may be limited in some manner. For example, unlimited use of firepower that negatively affects civilians in the secured area may cause them to embrace the enemy's cause or withhold support to the friendly force. Soldiers must understand this and follow strict ROE when conducting sustaining operations.

E-92. Because of the decentralized nature of sustaining operations and the intermingling of military forces and civilians, the commander must address fratricide prevention when planning for rear area and base security. Fires that miss their intended targets can affect other forces, routes, and installations in the secured area. When considering a COA, the commander weighs the risk of fratricide against losses to enemy fire.

E-93. **Fire Support.** Artillery and mortars, if available, are the fastest means of responding to intruders beyond the base perimeter in the absence of maneuver forces positioned there. The commander desires the designation of a free-fire area beyond the base perimeter to allow for the optimum employment of available fire support systems. He positions no-fire or restricted fire areas over locations occupied by civilians and friendly forces operating outside the base, such as patrols and observation posts. These areas are large enough to provide the necessary degree of standoff protection.

E-94. **Air Defense.** Air and missile attacks are perhaps the greatest threats to units in the rear area. Rear area bases normally rely on passive air defense for protection against enemy air threats. Most rear area bases will not be supported by SHORAD systems. They will normally be located within the coverage provide by high- to medium-altitude air defense (HIMAD) systems. They will not normally control the positioning of HIMAD systems. (Chapter 8 discusses passive and active air defense considerations. Ballistic missile defense is an operational-level task addressed in FM 3-01.12.)

E-95. **Mobility/Countermobility/Survivability.** The commander directs the construction and improvement of perimeter barriers. This includes establishing clear fields of fire and implementing physical security measures, such as checkpoints on vehicle and pedestrian routes leading into the base, to improve overall base security. Barriers, natural obstacles, and aggressive offensive actions deny enemy access to the area immediately surrounding the base. If the force keeps the enemy at a distance, he cannot launch damaging attacks toward the base. The commander continues to direct improvement in his base defenses, as time and other resource permit, to ensure a stable security system.

E-96. Force protection considerations against terrorist attack are part of base defense planning. The base commander assesses the vulnerability of the base to terrorist attack and other nonbattlefield threats. He must assess the base's vulnerability to each identified threat. Risk and vulnerability assessments are the foundation for priorities of work and resource distribution.

E-97. The base defense plan includes those ADC measures necessary to minimize damage to the base's critical facilities resulting from fire, natural disaster, or enemy action. The commander dedicates resources to protect areas that, if destroyed, could cause catastrophic loss or mission failure since force protection resources are limited. The plan delineates responsibilities for ADC activities, such as fire fighting, light rescue and labor, and decontamination, and assigns appropriate priorities for assistance furnished by each unit on the base.

E-98. **Combat Service Support.** Since base security forces generally operate in close proximity to the base, the commander tailors the load carried by his dismounted elements to the immediate requirements of METT-TC. Resupply of dismounted elements takes place using ground or air modes.

E-99. **Command and Control.** Planning for rear area and base security operations, like all planning, requires the rear area commander to have an accurate common operational picture of his AO. This includes the strength and security potential of available units, terrain and environmental data, and intelligence information. The rear area commander uses existing echelon rear command posts and information systems to provide him and his subordinate commanders that common operational picture.

E-100. In planning for base security, a simple, flexible defensive plan disseminated to the lowest level of command is normally the best COA. The plan should include the maximum use of standing operating procedures (SOPs) and battle drills at the user level. It should incorporate adequate control measures to ensure the safety of friendly troops and civilians. Such a plan minimizes the impact of the inevitable miscues that accompany any combat operation.

E-101. The base commander integrates the base security plans with those of its base cluster and supporting rear area security forces. He periodically reviews the base defense plan. Alterations in the units assigned to the base and changes to their defense capabilities because of mission requirements also require updating the base defense plan. Frequent alterations of base security procedures reduce the base's vulnerability. The continuous monitoring of plans also allows adjustments based on experience to be incorporated into those plans.

E-102. The base commander establishes communications with the BCOC (or the echelon rear command post if the base in not part of a base cluster). He also establishes communications with the headquarters of the designated Level II response force.

E-103. Communication problems can occur in information exchanges during the planning for base security involving multinational forces. The commander must address any language barriers because units operating in the rear area normally need to communicate with various host nation governmental agencies and the civilian population. The ability to communicate with these different agencies and the local population helps to protect the force.

E-104. The echelon rear area commander maintains simplicity by making the minimum number of plans necessary. Rear area and base security is a secondary mission for CS and CSS units. The echelon rear command post, BCOC, and the BDOC must ensure that rear area and base security is not neglected in favor of the parent organization's primary mission. A BDOC sends out liaison teams to each base in its base cluster, with RAOC assistance, to ensure that each base's defensive plans are workable, coordinated with the BCOC, and routinely practiced.

PREPARING A BASE DEFENSE

E-105. Preparing for base defense begins before base units arrive, if possible. Normally, combat units provide the initial defense in the base area, as they are the first forces into an area. An exception may occur when conducting force projection operations supported by a host nation. In this case, the host nation's forces are the first forces into the area. These forces can be conventional military forces or police and internal security forces.

E-106. The rear area commander uses his available resources to initiate an area reconnaissance to determine the physical characteristics of the area and the presence or absence of enemy forces. Since rear area and base security is normally an economy of force measure, it is unlikely that sufficient forces will be available to secure the echelon's entire rear area at one time. Therefore, they concentrate on those areas immediately adjacent to the protected element, such as friendly forces. It is essential that base, base cluster, and other rear area security forces conduct continuous reconnaissance operations. They may focus their efforts on people, materiel, buildings, or terrain.

E-107. Reconnaissance operations that involve only US forces may not be effective if language problems prevent communication with the local population. In these instances, interpreters—US Army or host nation—should accompany forces conducting reconnaissance. When possible, host nation authorities, such as the civil police, should also accompany the reconnaissance force. Distributing leaflets and handbills before or during an area reconnaissance and using loudspeaker teams during the reconnaissance can aid the reconnaissance effort by informing civilians of what information the US reconnaissance force is seeking and how they can report that information to US forces.

E-108. The rear area commander directs units within his AO to establish bases and base clusters even while this initial area reconnaissance is ongoing. These bases and base clusters provide security to units located within them and can provide perimeter security around critical locations. Perimeters vary in shape depending on the factors of METT-TC. The perimeter shape conforms to the terrain features that offer the best use of observation and field of fire.

E-109. When the units constituting the base arrive, they immediately start organizing the defense of the base. Many tasks occur simultaneously, but some may require priority. The base commander specifies the sequence for preparing the defense system. (Chapter 9 provides a recommended sequence in its discussion of the preparation for an area defense.)

E-110. Construction of personnel shelters throughout the billeting, administrative, and maintenance areas provides individual protection against standoff attacks. These shelters may vary in construction. Shipping containers, dugouts, and double-walled plywood shelters with sand or gravel fill, all with sandbag reinforcement and overhead cover, provide acceptable protection. These shelters are close to the billets and work areas to permit rapid access. (To the degree possible, construction will be in consonance with the guidance provided in FM 3-34.34.)

E-111. Units may construct fighting positions or prefabricate and move them into position for assembly. These bunkers are constructed to withstand a direct hit by a medium antitank missile on the front and sides and a direct hit by a mortar on the top. Installing chain link fences around these positions can prematurely detonate shaped-charge warheads.

E-112. Constructing revetments for critical resources provides protection against mortar or rocket fragmentation. These revetments may be sand-filled, double-walled construction with either plywood or steel plate sides. Units within the base provide overhead cover for critical facilities and functions when possible. The base commander's preventive measures to avoid or minimize effects of enemy attacks also contribute to his ADC efforts by providing units, soldiers, facilities, and supplies on the base protection from major accidents and natural disasters as well.

E-113. The commander uses tactical wire barriers within the perimeter to limit and canalize penetrations by enemy groups or individuals. These interior barriers can be as simple as a single roll of concertina wire. Generally, the commander should place them to prevent a direct approach to vital facilities within the base. He should make provisions to cover these barriers by automatic weapons. The base should construct inconspicuous barriers and periodically relocate them to prevent counterplanning by the enemy. Further, the barriers should not be so intensive as to preclude freedom of movement by the reaction or reserve force. These forces, as well as all soldiers, should be familiarized with the location of all barriers through daylight and night drills.

E-114. Base defense training is an essential element of preparing to conduct a base defense. Base defense requires integrating the BDF and its emergency augmentation by all units on the base. Because they rarely function together, the commander can form an efficient fighting entity only through proper training. The BDF commander carefully coordinates the training of base units for base defense operations with the support operations of the base. All individuals receive training to allow them to participate, at least to a limited degree, in the base's defense. All units receive training to provide at least limited local security for the facilities they operate. Conducting rehearsals that test the base defense plan is an essential part of base defense training. Some equipment deemed necessary for base defense or for training purposes may not be available. The BDF commander must use his initiative to overcome these shortfalls.

E-115. Most of the training required to support base defense operations is currently a part of individual and unit training programs. Soldiers designated to take any part in base defense operations will probably require additional training in the following areas as applicable to their roles in the base defense effort:

- Ambush techniques and defensive measures against enemy ambushes.
- Use of hearing, sight, and smell as detection means.
- Police-type patrolling and operating roadblocks and checkpoints.
- Limited-visibility operations to include using night vision devises, sensors, and techniques for using sign and countersign procedures.
- Marksmanship, especially night firing.

- Observation post operations with emphasis on security, sound and light discipline, and reporting procedures.
- Operation and operator maintenance of base defense equipment, such as sensors, night observation devices, radios, and computer systems.
- Cross-training on all communications equipment available within the base and communication techniques.
- Barrier construction and, if authorized, using mines and boobytraps.
- Patrolling.
- Small-unit attacks, such as the employing movement formations, rushes, and a base of fire.
- Fire control and fire distribution of available weapons.

E-116. The existence of a common operational picture and the commander's situational understanding are the keys to knowing how and when a response force will be needed. Separate bases and base clusters establish continuous communications with the echelon rear command post to allow timely response and information dissemination.

E-117. The commander of the response force has a copy of the base defense plan so he can effect necessary coordination between the base and the response force once the latter is activated for the response mission. This coordination occurs through the BDOC. The response force commander coordinates with the base to ensure that he understands the base defense plan, to include—

- Method of contacting the BDOC or BCOC, including call signs and frequencies.
- Base defense plans and layouts to include—
  - Positions of critical internal assets, external coordination points, and no-fire areas.
  - Locations of any obstacles or mines near the base.
  - Locations and direction of fire of crew-served weapons.
  - Locations of target reference points and preplanned fires.
  - Locations of OPs and friendly patrols if employed by the base.
  - Signal for final protective fires.
- How to obtain available fire support.

E-118. The response force commander designated for each base must have the capability to mass the effect of supporting fires and support TCF operations. He must know which fire support targets are approved for engagement and the locations of the nearest medical treatment facility, NBC decontamination site, and ammunition supply point. He must be able to communicate with—

- Any supporting artillery and Army aviation units tasked to respond.
- The TCF, BDOC, BCOC, and the echelon rear command post.

EXECUTING A BASE DEFENSE

E-119. The preferred way to conduct a base defense is as an offensive operation designed to clear the area of enemy forces. Detecting and destroying significant enemy forces are responsibilities of the echelon rear area commander. However, the individual conducting rear area security may be forced to adopt a primarily defensive posture because of—

- The size of the rear area.
- A lack of resources.
- Inability to detect and track the enemy.
- Enemy capabilities in relation to the friendly force capabilities.

E-120. Forces engaged solely in base defense operations conduct aggressive patrols, develop and occupy defensive positions within their assigned AOs, and provide security forces to counter enemy attacks. These actions may be multiservice or multinational depending on the composition of the base.

E-121. Rear area commanders at all echelons request ground maneuver forces to use as BDFs when the net effect of enemy activities in the echelon's rear area is beyond the limited defensive capabilities of CS or CSS resources. If a threat exceeds a base's ability to defend itself, the base commander requests response force support through his BCOC or the echelon rear command post if he commands a separate base.

E-122. The conduct of base defensive operations against a ground attack is the same as for conducting a perimeter defense. (Chapter 8 discusses the perimeter defense.) The paragraphs below discuss those execution considerations unique to base defense.

Patrols

E-123. Base defense operations to counter small enemy forces include aggressive, frequent patrolling by squad- and platoon-size forces to detect and capture or destroy small enemy bands. Host nation personnel are well suited to assist BDF patrols. Their knowledge of the terrain, inherent ability to operate effectively in the environment, language skills, and familiarity with local customs increases the effectiveness of these patrols.

E-124. Small, highly mobile units conduct patrols by moving on foot or by land, water, or air vehicles during daylight and darkness. They search populated areas contiguous to the base and establish surprise checkpoints along known or suspected routes used by enemy forces.

E-125. The BDF mans dug-in or concealed night ambush sites located outside of the base's external barrier system on a random basis. Host nation security personnel should accompany patrols traveling through populated areas. The base develops fire support targets to support the ambush in accordance with their availability and the ROE. If the area adjacent to the base perimeter is a free fire area, it is seldom necessary to occupy ambush sites there. When local restrictions preclude establishing a free-fire area, ambush sites are manned forward of the perimeter and the BDF is prepared to assist those elements conducting the ambush as necessary.

E-126. Patrols can install sensors in locations enemy forces threatening the base would cross to provide early warning of enemy movements. Such patrols also observe and report enemy activities, such as assembling personnel; moving weapons, amunition, or other supplies; and preparing firing positions for indirect-fire weapons. In addition to the acquisition of specific targets, patrols may confirm or deny the presence of enemy activity in named areas of interest located near to base.

E-127. The base uses any available small boats to patrol nearby rivers and coasts. An infiltrating enemy takes advantage of waterways to transport heavy weapons because of their weight and bulk. Detecting these weapon systems is an important river and coastal surveillance mission for the BDF commander. The Army has type-classified small watercraft suitable for use in this type of patrol, as does the Navy and Coast Guard.

E-128. Military working dogs are a valuable asset in base defense operations. Their employment should habitually be considered in planning all patrol and security operations. Successfully using these dogs depends on the skill of the handler and the knowledge and full understanding of the capabilities and limitations of the dogs by employing commanders and staffs. (Army Regulation 190-12 provides this information.)

E-129. During base security operations, the base commander monitors the location of friendly troops and their relationship to friendly fires. The base commander must be constantly aware of the relative location of enemy, friendly, and neutral personnel. Subordinate elements, such as patrols, must understand the importance of accurately reporting their positions. Automation and navigation aids, such as the global positioning system (GPS) tied to combat net radios, assist in maintaining a common operational picture. When employing artillery, mortars, and air support, the commander must exercise positive control, whereby an observer must be able to see the target area. Unit SOP at all levels must address specific procedures for clearing indirect fires directed against identified targets located within the rear area.

Static Positions

E-130. The static positions of the base defense consist primarily of the bunkers and towers forming the base's defensive positions. Positioning bunkers and towers correctly affords maximum observation and mutually supporting fires over the area forward of the perimeter to include the perimeter barrier and sensor system.

E-131. The base commander establishes communications between the BDOC, the command posts of the units occupying the base, and the base's bunkers, towers, and reserve. In addition, each node be able to communicate laterally with adjacent units and defensive positions.

E-132. **Bunkers.** Although full-time observation and all-around defense of the base are essential, the BDF commander may reduce the number of soldiers conducting static defense mission by designating key bunkers around the perimeter to be manned at all times, with the remainder to be fully manned during periods of reduced visibility and increased enemy threat. The base prepares individual fighting positions near the bunkers to provide supporting fires. Night vision devices, binoculars, automatic weapons, grenade launchers, and hand grenades are commonly available within the bunker positions, and antitank systems cover possible ground vehicle approaches.

E-133. **Towers.** Towers provide increased visibility, which may be desired in specific situations. Elevated platforms enhance the capability of detecting perimeter infiltration when coupled with night vision devices, binoculars, and spotter scopes. They may enhance the base's ability to determine the firing positions of enemy indirect weapon systems. Towers can be constructed according to standard military design using standard construction materiels or constructed from local materiels. Multistory buildings, such as grain silos, can also serve as towers. Installating sandbags or steel plating around the observation platform provides protection against automatic weapons and small-arms fire. Constructing a ground-level bunker near the base of the tower provides additional protection when the enemy directs fires against the tower. Access to the bunker from the tower may be by means of a fireman's pole or a ladder. Towers should incorporate the following safety measures—

- Lightning arresters.
- Built to withstand strong winds and to support two observers and their equipment.
- Enclosed mounting ladder.
- Provision of safety nets around tall towers.
- Painted an appropriate color to reduce reflection from moonlight.
- Installation of a suitable roof to shield observers from the elements without interference to observation. (A double-roof design can cause some types of mortar rounds to detonate at a height that affords some protection to the observers positioned in the tower.)

The base commander's decision to use towers is a result of his analysis of the factors of METT-TC. For example, he must decide if the additional security they provide is worth the effort to construct them in a fluid situation or the risk that they will attract additional enemy attention to the base.

Reaction Operations

E-134. Reaction operations are operations conducted by the BDF, operating from the base or from positions outside the base, for the purpose of countering local enemy activities. They are offensive in nature and conducted as either spoiling attacks or counterattacks.

E-135. The commander uses his BDF to destroy any small enemy units detected in the vicinity of the base. The base requests assistance if its forces cannot destroy or contain the enemy force in its current location. Elements of the BDF do not pursue a retrograding enemy outside of supporting distance from the base, because it could be part of a diversion designed to weaken the base's defensive capabilities. Pursuit of such fleeing enemy forces is turned over to rear area response forces instead.

E-136. Reaction operations are simple, planned, and rehearsed because the majority of actions are required at night. Designating primary and alternate checkpoints within the base facilitates response to multiple contingencies. Within security limitations, the force uses actual checkpoints during rehearsals to promote familiarity with the area and the reaction plan.

Response Force Operations

E-137. Response force operations are offensive operations conducted by the response force to countering Level II attacks. The response force moves quickly to counter the enemy before he can extensively damage the base. The base commander lifts or shifts base defense fires to support the maneuver of the response force. The response force attempts to delay and disrupt the threat until a TCF arrives if it cannot destroy or deter the enemy. The base commander notifies his BCOC if the base defense and the response forces engage an enemy they cannot defeat. They maintain contact with the enemy force until the TCF arrives. The TCF commander has operational control (OPCON) of all base and response forces within his designated AO on commitment of the TCF.

Host Nation and Third Country Forces

E-138. The BDF commander integrates available host nation and third country forces in the overall base defense effort. Particular emphasis is placed on integrating host country forces in patrol and population control activities. Both host nation and third country forces provide local security for their own units. However, to ensure maximum benefit, all such local plans should be coordinated with and integrated in the base defense plan. The actual degree of host nation and third country force participation in base defense depends on the orders and guidance of their respective governments and the personal relationship their commanders have with the base commander.

Defense Against an Enemy Conducting an Attack-by-Fire

E-139. An enemy may conduct an attack by fire against a base when he knows that he is unable to penetrate the base perimeter. He wants to place a large volume of fire on the base to inflict casualties and destroy resources. Such an attack is normally of short duration, from 10 to 20 minutes. He may also direct harassing fires toward the base to accomplish the same purpose. The weapon systems delivering these attacks normally displace after completing their mission to prevent their destruction by friendly maneuver, and direct fires, and counterfires.

E-140. The base develops an aggressive reconnaissance and surveillance plan to detect the infiltration of enemy mortars, artillery, rockets, and antitank systems and to locate firing positions. The BDF commander considers likely firing positions and routes to them, intelligence reports, and available resources when formulating this plan. He plans offensive opertions to locate and destroy the enemy force prior to, during, and after such attacks and passive defensive actions to reduce friendly casualties and damage associated with them.

E-141. Primary active measures employed against these types of attack are airborne observers, establishing a critical friendly zone (CFZ) over the base, and using reaction forces. An unmanned aerial vehicle or attack helicopter may provide the airborne observer. The airborne observer orbits the base area, checking likely or suspected positions and noticeable changes in the terrain. The base acts against an enemy weapon system on its discovery by employing friendly fire support systems or maneuver forces against it. A CFZ over the base allows friendly counterfires against the enemy weapon system as soon as it attacks the base. In the absence of radar coverage, the base may use visual observation of the muzzle blast from two or more points to provide an intersection. This requires these observation posts to have properly oriented azimuth boards so the direction of sightings can be rapidly determined and reported to the BDOC. Using a reaction force deployed by helicopter or tactical vehicle to engage and destroy the enemy weapon system is a primary consideration. The reaction force must be careful to prevent establishing a pattern, such as always using the same route or landing zone. If the enemy is aware of the pattern, he can set up an ambush to destroy the reaction force.

Sniper Teams

E-142. Sniper teams are useful when forces come under harassing small-arms fire by a small enemy element firing from, among, or near areas where civilians or cultural monuments protect it from the return fires of more non-discriminating weapons. These teams allow the BDF commander to fix the enemy and move other BDF elements into positions where they can engage the enemy without endangering civilian lives or causing excessive collateral damage. The commander can employ smoke and, if authorized, riot-control agents to aid in this maneuver. If the force cannot engage the enemy without endangering civilian life or property, the BDF commander can try to disengage his forces and move them to positions that block the enemy's escape routes. However, the base commander must use all available means to protect his forces, even when it may place civilians at risk.

E-143. These teams may be provided from infantry units. Alternatively, this capability may be provided by giving special training to selected BDF soldiers to maximized the capabilities of their assigned individual weapons or specialized equipment provided to them.

Passive Defensive Measures

E-144. Passive defense measures are always practiced within the confines of the base. In addition, the commander employs CA and PSYOP programs to ensure the cooperation of the local civilian population. The BDOC evaluates all patrol, guard, and intelligence reports indicating an attack. Preparation of firing or assault positions, movement of weapons or ammunition, unusual actions or movement of the civilian population, and reports by host nation personnel may be indications of an attack.

Area Damage Control

E-145. When an attack occurs, the objective is to resume operations, including the maintaining or restoring of control, evacuating casualties, isolating danger areas, and reducing personnel and materiel losses. The base commander conducts an immediate survey of the damage and reports his assessment to the BCOC. Simultaneously, he initiates actions to isolate the danger areas and prevent extension or continuation of the damage, such as fighting fires and minimizing flooding.

E-146. Casualties receive self-, buddy-, and first aid care. If possible, medical personnel and vehicles evacuate casualties. However, timely transportation of casualties is important. The situation may require using nonmedical vehicles for mass casualties. If possible, medical personnel accompany those patients being transported in nonmedical vehicles to provide en route patient care.

E-147. The base establishes traffic control using MP or other elements to ensure fire-fighting equipment gains access to the area and ambulances and evacuation vehicles clear the area. The BDOC notifies the BCOC of blocked routes and necessary traffic diversions. The element conducting traffic control can also temporarily provide for controlling displaced civilians and stragglers and provide a degree of local security for the damaged area.

E-148. The BDOC coordinates with the BCOC and echelon rear command post for engineer support to restore critical facilities and any specialized fire fighting capabilities required. Engineer support is normally needed to clear debris and rubble to support the base damage control mission and extinguish extensive fires. The BDOC also coordinates for any necessary explosive ordnance disposal (EOD) and decontamination support.

E-149. The base commander takes precautions by maintaining local security and interior guard posts when receiving ADC support from host nation resources to ensure the security of the base. He cannot afford confusion surrounding the base's damage control efforts to provide an opportunity for saboteurs and other hostile personnel to penetrate the base perimeter.

ROUTE SECURITY

E-150. ***Route security operations*** **are a specialized kind of area security operations conducted to protect lines of communication and friendly forces moving along them.** Enemy attempts to interdict LOCs may have little immediate impact on ongoing decisive and shaping operations because of unit basic loads and previously positioned caches. However, the security of those routes over which the echelon's sustaining operations flow is critical to sustained land operations, regardless of whether the commander is attempting to exploit success or recoup from failure.

E-151. The security of routes and LOCs, whether rail, pipeline, highway, or waterway, presents one of the greatest security problems in an echelon's rear area. Route security operations are defensive in nature and are terrain-oriented. A route security force prevents an enemy force from impeding, harassing, or destroying traffic along the route or portions of the route itself. Having a base permits the concentration of security resources. LOC security requires dedicated resources at almost infinite points or reaction forces to counter possible enemy action. Units performing missions that require the habitual use of LOCs, such as MP, transportation, supply, and service units, can perform route security operations in conjunction with their primary activities. The echelon rear area commander can employ the following techniques to provide route security:

- Passive security.
- Route reconnaissance.
- Cordon security.
- Combat security.

While the scope of these operations depends on the factors of METT-TC, route security operations tend to require significant resources. A combination of passive and route reconnaissance is commonly used over secured routes such as a LOC.

PASSIVE SECURITY

E-152. The passive security technique for provide route security includes measures initiated to achieve security without a significant expenditure of manpower or resources. It includes—

- Camouflage
- Formation and march control of convoys so they present the least lucrative target possible under prevailing conditions.
- Proper selection of routes.
- Capitalizing on security offered by other activities not related to the route's security requirements. These activities include aircraft traversing the route, maintenance activities taking place along the route, training exercises or troop movements adjacent to or along the route, military and host nation police traffic control activities, and the activities of the civilian population.
- Defensive information operations.

E-153. The commander uses passive security for all conditions or situations and as an adjunct to any other technique of route security employed. They are the products of a long-term and continually improved program for an integrated security system.

ROUTE RECONNAISSANCE

E-154. Route reconnaissance addresses not only the route itself, but also all terrain along the route from which the enemy could influence a friendly force's movement. Route reconnaissance takes place at irregular intervals to avoid developing a regular pattern that an enemy could exploit. (Chapter 13 discusses route reconnaissance.)

CORDON SECURITY

E-155. It is normally not feasible to secure all points on a route by physically allocating resources to each point. ***Cordon security*** **is the security provided between two combat outposts positioned to provide mutual support.** A commander can assign air assault, mechanized, or motorized units to combat outposts established at critical locations along the route. He locates these combat outposts within supporting distance of each other whenever possible. Units assigned to these combat outposts provide response forces in the event of enemy activity along the route within their assigned AO. Forces based at these combat outposts conduct reconnaissance patrols and offensive operations to counter enemy activities between these two points on a frequent but irregular schedule. Cordon security requires considerable manpower and other resources. The commander uses it only when a clear and evident requirement exists.

E-156. Normally, each combat outpost contains at least a platoon-size element equipped with automatic weapons, communications, and sensors and supported by those fire support assets available to the rear area commander. (Chapter 12 contains additional information on the combat outpost.) The outpost commander rotates his soldiers between duty at the combat outpost and duty on patrol. The higher commander contacts each of his deployed combat outposts frequently to check its status.

E-157. The establishment of cordon security will do little to eliminate the rear area threats that require its adoption. A commander directing the establishment of cordon security should also direct a series of parallel corrective actions, to include—

- Searching for a new route.
- Vigorous search and attack operations to destroy the enemy within the echelon rear area.
- Rigorously enforcing circulation control measures in coordination with the host nation over the civilian population.
- Removing cover from areas offering concealment to an enemy.
- Constructing new routes to bypass dangerous areas.
- Repositioning supply activities to permit better route selection.
- Mass evacuation of the civilian population from towns and villages along the route. (This is an extreme measure and requires consultation with host nation authorities and adherence to the laws of land warfare.)

E-158. The commander must recognize the loss of complete control over any given route early. The continued investment of resources into a "lost route" merely serves the enemy. Those resources and energy should be diverted to security measures elsewhere whenever feasible.

COMBAT SECURITY

E-159. The combat security technique for providing route security includes all measures taken by combat arms units to seize and secure the terrain necessary to permit use of the route in question. Units conduct combat security along a route to prevent enemy ground forces from moving into direct fire range of the protected route. The route and its adjoining terrain compose the AO. A route security force operates on and to the flanks of a designated route.

E-160. Combat security is the most costly technique for providing route security and draws on the strength of the maneuver force. For example, because of the distances involved, a commander usually assigns at least one battalion-size unit, if not more, to secure one route through a division-size AO. The maneuver force establishes a perimeter around the route or cordons sections of it to search suspected enemy locations while establishing roadblocks and checkpoints along the main route and lateral routes to stop and search vehicles and people as they enter or leave the route. Special considerations include coordinating with segments manned by rear area security forces and the echelon rear command posts integrating the maneuver force's operations with other rear area and base defense activities, and clarifying of C2 arrangements.

CONVOY SECURITY

E-161. ***Convoy security operations*** **are specialized area security operations conducted to protect convoys.** Units conduct convoy security operations anytime there are not enough friendly forces to continuously secure LOCs in an AO and there is a significant danger of enemy ground action directed against the convoy. The commander may also conduct them in conjunction with route security operations. A convoy security force operates to the front, flanks, and rear of a convoy element moving along a designated route. Convoy security operations are offensive in nature and orient on the force being protected. (Chapter 14 discusses convoy movement operations; this section only addresses convoy security concerns.)

E-162. To protect a convoy, the convoy escort element must accomplish the following critical tasks:

- Reconnoiter the convoy route.
- Clear the route of obstacles or positions from which the enemy could affect movement along the route.
- Provide early warning of enemy presence along the route.
- Prevent the enemy from attacking the convoy.

ORGANIZATION OF A CONVOY ESCORT

E-163. The size of the escort needed to conduct convoy security operations is METT-TC dependent. However, as a general rule, an average-size convoy of 50 vehicles normally requires at least a maneuver company, an MP company, or air defense artillery battery equipped with self-propelled cannons to conduct the mission. Without such an escort, the transportation unit will have to provide security using its organic assets. This results in the reduction of cargo-hauling capabilities of each transportation unit to the degree that it must convert cargo vehicles into ad hoc gun trucks and leave other trucks in a motor pool because of their drivers being diverted to provide the necessary degree of security.

E-164. Cavalry, tank, mechanized infantry units, and MP elements equipped with tracked or wheeled armored vehicles are well suited to protect a convoy because of their organic reconnaissance capabilities and combat power. The commander may reinforce the convoy security force with engineers, dedicated air defense, and other assets as required. Usually the convoy's trail party also provides CSS support to the convoy escort, such as vehicle recovery and medical evacuation. If the convoy's trail element cannot provide that support, then the convoy escort provides its own CSS support. Other METT-TC considerations, such as restrictive terrain and limited time, dictate how to use aviation assets as part of the coordinated effort.

E-165. The convoy security force organizes into several elements to accomplish its tasks. (See Figure E-6.) The advance guard performs a route reconnaissance forward of the convoy. The security element provides early warning and security to the convoy's flanks and rear. This element can also perform the duties of the escort element. The escort element provides close-in protection to the convoy. It may also provide an immediate response force to assist in repelling or destroying enemy forces if they engage the convoy. The rear guard prevents an enemy from overrunning the convoy from the rear. It can also act as an immediate response force to enemy contact made on either flank of the convoy. Finally, a reaction force is available to respond to emergencies. Since the reaction force is a committed force, and not a reserve, it is task organized to include available fire support assets and other combat multipliers. It provides firepower and support to all convoy elements to help develop the situation, conduct a hasty attack, or prevent the convoy's destruction. The headquarters directing the convoy, rather than the convoy commander, controls the reaction force. It monitors the progress of the convoy and responds if the convoy encounters an enemy unit the escort element cannot defeat or repel. It must be able to respond within 15 minutes. The reaction force always anticipates an ambush when moving to the relief of a convoy under enemy attack.

E-166. When a lack of resources prevents establishing all five elements, the commander normally first resources his forward reconnaissance or advance guard. He then resources his escort, flank screening, and rear guard elements in that order. It is the responsibility of the headquarters directing the convoy to designate the response element.

E-167. If available, air cavalry assets can participate in the convoy security operation by screening the convoy's movement as it moves along the route of march. Alternatively, they can assist in clearing the route ahead of the convoy as it moves along the route of march, or clearing the route in conjunction with the route reconnaissance element. Air cavalry can also assist by controlling indirect-fire support and coordinating with forward air controllers from other services for close air support. The supporting air cavalry unit must know the maneuver intentions of the ground element if contact with the enemy occurs. If heavy forces are not available for use as a convoy escort, the commander can use light infantry to augment the self-defense capabilities of transportation units.

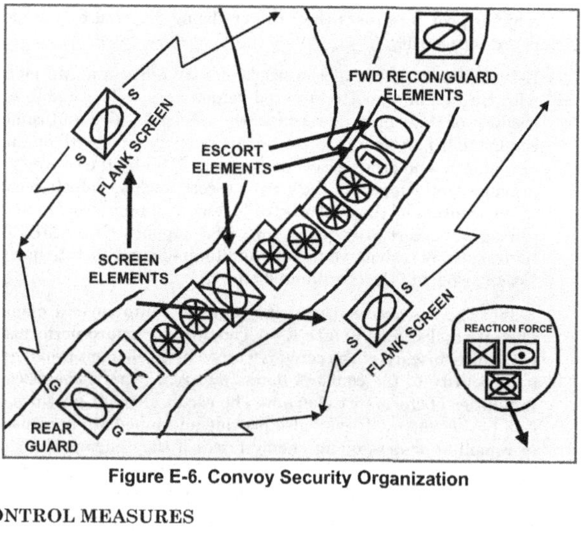

Figure E-6. Convoy Security Organization

CONVOY CONTROL MEASURES

E-168. As shown in Figure E-7 on page E-36, convoy security control measures typically include the following:

- Routes (primary and alternate).
- Start point.
- Release point.

- Checkpoints.
- Phase lines.
- Observation posts.
- Target reference points.
- Areas of operation for various units that affect the convoy.
- Staging or marshaling areas.
- Known obstacles and lanes.

E-169. Although convoy size, passage time, and march speed are not graphic control measures, they determine how long it takes to execute the convoy security mission. Rules of engagement are also not a graphic control measure, but they greatly influence how a commander conducts convoy security operations.

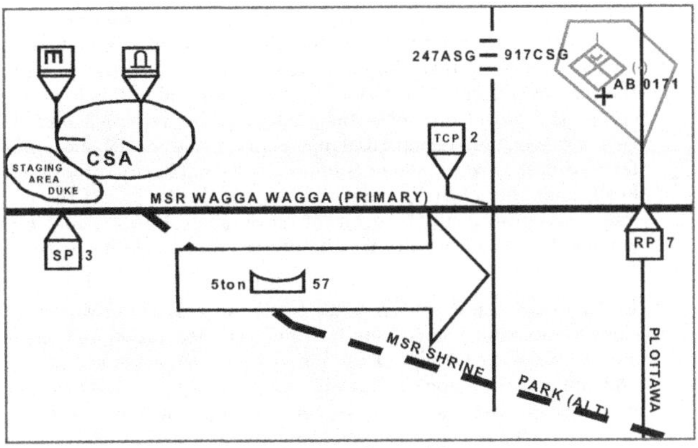

Figure E-7. Convoy Control Measures

PLANNING CONVOY OPERATIONS

E-170. Because of the inherent dangers of convoy operations, the commander emphasizes security measures during the planning process. The best defense is to avoid being ambushed. These security measures include—

- Secrecy when planning and disseminating orders.
- Strict noise and light discipline during movement.
- Varying routes and schedules.
- Avoiding routes with known danger areas.
- Conducting route reconnaissance using aerial and ground systems.
- Using current intelligence information concerning the condition of the route and any enemy or other forces that may impact on the convoy's use of that route.
- Coordinating with supporting air liaison officers to ensure the scheduling of close air support to assist the movement if the convoy warrants committing these assets.

- Fire support elements to provide close and continuous fire support for the movement.
- Training in immediate actions drills, including actions at danger areas and in case of a near ambush, a far ambush, booby traps, encounters with enemy tactical combat vehicles, sniper contact, aerial attack, and indirect fire.
- Communications and coordination with supporting units and other units along the route, adjacent host nation forces, and higher headquarters, including airborne radio relay.
- Various locations for leaders, communications, medical support, and weapon systems within the movement formation.
- Questions asked of local civilians along the movement route for intelligence information, including possible enemy ambush sites.

E-171. The convoy commander must consider the unique requirements of convoy security when formulating his plan. He briefs his subordinates on the latest information regarding the enemy situation and the area through which the convoy will pass. He formulates his plan and issues his order, which includes the movement formation, intervals between echelons and vehicles, rate of travel, and a detailed plan of action if the convoy encounters an enemy force. Since there is seldom time to issue complicated orders during an ambush, subordinate commanders must plan the actions of the escort element and reaction force in the event of an ambush. Units should rehearse these actions before movement and execute them as drills if enemy contact takes place.

E-172. Communications are vital to the success of convoy movements. The convoy commander must plan for radio communication and ensure the availability and compatibility of communication means between convoy elements and indirect-fire support and air cavalry assets, as well as with units and host nation agencies in areas along the movement route. The convoy commander prearranges visual and sound signals, such as colored smoke, identification panels, whistles, or horns. Many of these signals can be SOP items. He must ensure that all convoy members understand these signals and rehearse the actions required by each. This is necessary because of the limited number of radios and other communications equipment in tactical vehicles.

E-173. The commander's fire support plan covers the entire convoy route, paying special attention to known danger areas and potential choke points. Convoy security elements may have mortars organic to the escort element or indirect fires provided by fire support elements positioned to range all or portions of the route. Coordinating with fire direction centers before the convoy's departure enables fire support teams in the convoy security element to enter the appropriate fire control nets, calling for and adjusting fires as necessary.

E-174. The convoy security commander must carefully plan for the appropriate CSS to support the operation. He should include fuel, maintenance, and recovery elements in the convoy or pre-position them in secure areas along the route. He also plans for casualty evacuation support that covers the entire route.

E-175. Convoy security operations in an urban environment or built-up area require different emphasis and techniques than those in rural areas. The population density and characteristics of the area may require using nonlethal weapons and carefully employing lethal weapons. When applying minimum-essential force to minimize loss of life and destruction of property, subordinate commanders must conduct detailed planning, coordination, and control. Whenever possible, convoys should move through populated areas when they are the least congested and, therefore, less dangerous to the security of the convoy. Convoy operations may require assistance from military or local police and other governmental agencies to secure the route before the convoy enters a built-up area.

PREPARING CONVOY SECURITY OPERATIONS

E-176. The convoy commander must rehearse maintaining march intervals and actions on contact for a variety of scenarios, including air attack, artillery or indirect fire, snipers, and ambush. Since the purpose of a convoy is to reach its destination—not to destroy the enemy in a movement to contact—the goal of these responses should be to free the convoy from enemy contact so it can continue its mission. All soldiers and vehicles traveling in the convoy should undergo precombat inspections to ensure they know and adhere to unit SOP, and that vehicles have a reasonable expectation of completing the convoy without mechanical failure. During this stage, drivers act to harden their vehicles, such as adding sandbags and Kevlar blankets and cover loads to prevent an enemy from identifying the cargo. The convoy commander spaces prime targets throughout the convoy and cross-loads his trucks as much as feasible to avoid losing all of one type of supply, such as all artillery fuzes, if a single vehicle is destroyed.

E-177. If possible, the convoy commander and his key subordinates perform a route reconnaissance. Shortly before the convoy's departure, MPs or scouts should perform another route reconnaissance to determine current conditions. This last route reconnaissance should also remove or neutralize individual mines emplaced by the enemy that endanger the convoy.

EXECUTING CONVOY SECURITY OPERATIONS

E-178. Unless the security force encounters the enemy during the convoy, the convoy conducts operations as discussed in Chapter 14. If the forward reconnaissance element, advance guard, or flank screen first encounters the enemy, it performs those actions on contact discussed in Chapter 4. (Chapter 8 discusses the active and passive air defense measures a convoy can take.)

Artillery or Indirect Fire

E-179. An enemy may use his fire support systems in an attempt to destroy convoys or to harass and interdict the movement of supplies and personnel. Artillery fires are either preplanned fires or fires called in and adjusted on a target of opportunity by a forward observer. The adjusted fires present the most complex problem, as the enemy can adjust the effects of his artillery fires to follow the actions of the convoy. The two active measures that a convoy can take against enemy artillery fires are using direct or indirect fires against the enemy forward observer, if his location can be identified and calling for counterfire if the direction and approximate distance to the enemy artillery system can be estimated.

E-180. The convoy commander has three passive options when the convoy comes under enemy artillery fire: halt in place, continue to march, or disperse quickly to concealed positions. Regardless of the option selected, the unit SOP should cover the action taken and the signal directing the action. The primary consideration is the immediate departure from the impact area. The convoy only halts when the artillery concentration is ahead of the convoy. The convoy commander looks for an alternate route around the impact area and the convoy should remain prepared to move out quickly. The mission or the terrain may require the convoy to continue through the beaten area. In this case, the convoy increases its speed of movement and spreads out to the maximum extent allowed by the terrain.

Sniper Fire

E-181. The convoy commander takes caution when receiving sniper fire to ensure that any return fire does not harm friendly troops or civilians in the area. The best reactive measures against sniper attack by a convoy are passive. Ensure all soldiers wear helmets and available body armor. Vehicles move through the area without stopping. Convoy escort personnel act as necessary to locate, suppress, and kill snipers in accordance with the ROE.

Ambush

E-182. The very nature of an ambush—a surprise attack from a concealed position—places an ambushed convoy at a disadvantage. No single defensive measure, or combination of measures, prevents or effectively counters all ambushes. The training of the convoy elements directly relates to the effectiveness of counterambush measures. Immediate reaction and aggressive leadership are essential to limit casualties and damage to vehicles, cargo, and personnel. Small enemy forces are seldom able to contain an entire convoy in a single kill zone. This is due to the extensive road space occupied by even platoon-size convoys and because of the presence of security elements escorting the convoy. More often, an enemy is able to ambush only a part of a convoy—either the head, trail, or a section of the main body. Fire in the kill zone may be from only one side of the road, with a small holding force on the opposite side. An enemy frequently places mines and booby traps on the holding force side to contain the convoy in the kill zone. The part of the convoy that is in the kill zone and receiving fire must exit as quickly as possible if the road ahead is open. The security escort takes care in assaulting the main ambush forces because an enemy will commonly use mines and booby traps to protect his flanks.

E-183. The convoy leaves disable vehicles behind or, if they are blocking the road, pushes them out of the way with following vehicles. Combat vehicles providing escort to the convoy must not block convoy vehicles by halting in the traveled portion of the road to return fire. If the main body splits into two or more elements because of the ambush, those elements forward of the ambush site move to a secure location. Vehicles that have not entered the kill zone must not attempt to do so. They should stop, their personnel should dismount, take up a good defensive position, and await instructions. Since escort vehicles may have left the road to attempt to overrun a hostile position, elements of the convoy should not fire on suspected enemy positions without coordinating with the escort forces.

E-184. Other actions that the convoy commander can take to neutralize the ambushing enemy force include—

- Calling for artillery fire on enemy positions.
- Calling for attack helicopters to engage the enemy positions.
- Directing the crew-served weapons mounted on convoy vehicles and the individual weapons of nondriving personnel to suppress the ambush force while the rest of the convoy's vehicles move out of the kill zone.
- Calling for the reaction force.

E-185. A convoy with a limited escort is seldom able to defeat an ambushing enemy force and should not attempt to do so. When part of the convoy is isolated in the kill zone, vehicles that have not entered the ambush area must not attempt to do so. They should stop in a herring bone formation; personnel should dismount, take good defensive positions, and await instruction until supporting forces clear the ambush area. Normally, the transportation unit conducting the convoy will not maneuver to attack the enemy ambush force unless it is necessary for self-preservation. It relies on its escort force, fire support, and reaction forces.

E-186. Convoy personnel will dismount, take cover, and return fire on enemy positions when an element of a convoy halts in the kill zone and is unable to proceed because of disabled vehicles, damaged bridges, or other obstacles, Personnel dismount their vehicles away from the direction of enemy fire. Escorting security forces that have passed through the kill zone or have not yet entered the kill zone maneuver to destroy the ambushing element. The convoy commander calls for the reaction force as soon as the ambush occurs.

E-187. The convoy clears the route and movement resumes as soon as possible after the ambushing enemy force is defeated or withdraws. Wounded soldiers are evacuated using the fastest possible mode. When the convoy cannot tow disabled vehicles, it should distribute their cargo among other vehicles if time permits. The convoy destroys vehicles and cargo on order of the convoy commander if it is not feasible to evacuate them. The convoy recovers combat net radios and other critical items before it destroys the vehicles. These items should not fall into enemy hands.

Source Notes

These are the sources quoted or used as sources for historical examples in this publication. They are listed by page number. Boldface indicates the titles of historical vignettes.

1-1 *"War is, above all things, an art…"*: Francis V. Greene, "The Important Improvements in the Art of War During the Past Twenty Years and Their Probable Effect on Future Military Operations," *Journal of the Military Service Institution of the United States* 4, no. 13 (1883): 41.

1-5 **The Human Dimension—Combat at Han-sur-Nied**: Hugh M. Cole, *The Lorraine Campaign*, volume 5, United States Army in World War II—The European Theater of Operations (Washington, D.C.: Historical Division, United States Army, 1950), 358–361. The action took place on 11 November 1944. The American units were from Combat Command A, 6th Armored Division. They were opposed by elements of the German 48th and 17th SS Panzer Genadier Divisions.

1-6 **Cowpens, 17 January 1781**: John Moncure, *The Cowpens Staff Ride and Battlefield Tour* (Fort Leavenworth, Kans.: Combat Studies Institute, US Army Command and General Staff College, 1996.

2-1 *"In war, obscurity…"*: The Infantry Journal, *Infantry in Battle*, 2d edition (Washington, D.C.: Infantry Journal, 1939; reprint, Fort Leavenworth, Kans.: US Army Command and General Staff College, 1981), 16.

3-0 *"To move swiftly,…"*: Thomas J. "Stonewall" Jackson, cited in *Dictionary of Military and Naval Quotations*, compiled by Robert Debs Heinl (Annapolis: US Naval Institute Press, 1966 [hereafter cited as Heinl]), 131.

3-1 **The 101st Airborne Division (Air Assault), 24–28 February 1991**: Command Report, 101st Airborne Division (Air Assault) for Operations Desert Shield and Desert Storm, 2 August 1990 through 1 May 1991 (Fort Leavenworth, Kans.: Combined Arms Research Library Archives, 1991), 17–26, 46–60; *see also* Robert H. Scales, *Certain Victory: The US Army in the Gulf War* (Washington, D.C.: Office of the Chief of Staff, US Army, 1993; reprint, Fort Leavenworth, Kans.: US Army Command and General Staff College Press, 1994); Richard M. Swain, *"Lucky War": Third Army in Desert Storm* (Fort Leavenworth, Kans.: US Army Command and General Staff College Press, 1994).

4-1 *"When armies approach each other…"*: Winston Churchill, (1949), cited in Heinl, 186.

4-2 **The Soviet Manchuria Campaign, August 1945**: David M. Glantz, *August Storm: The Soviet 1945 Strategic Offensive in Manchuria*, Leavenworth Papers No. 7; *August Storm: Soviet Tactical and Operational Combat in Manchuria, 1945*, Leavenworth Papers No. 8 (Fort Leavenworth Kans.: Combat Studies Institute, US Army Command and General Staff College, 1983).

5-0 *"I approve of all methods of attacking…"*: Frederick the Great (1747), cited in Heinl, 20.

5-39 **The Raid in Mogadishu, 3–4 October 1993**: Kent DeLong, *Mogadishu!: Heroism and Tragedy* (Westport, Conn.: Praeger, 1994); *see also* Kenneth C. Allard, *Somalia Operations: Lessons Learned* (Washington, D.C.: National Defense University Press, 1995); John L. Hirsch, *Somalia and Operation Restore Hope: Reflections on Peacemaking and Peacekeeping* (Washington, D.C.: United States Institute of Peace Press, 1995).

6-1 *"The most important goal…"*: Field Marshal Prince Mikhail I. Kutuzov (1745-1813), cited in *Warriors' Words: A Quotation Book*, compiled by Peter G. Tsouras (London: Arms and Armour Press, 1992 [hereafter cited as Tsouras]), 134.

7-1 *"In pursuit you must…"*: Field Marshal Viscount Allenby of Meggido (1917), cited in *A Dictionary of Military Quotations*, compiled by Trevor Royle (New York: Simon & Schuster, 1989), 4.

8-1 *"So the defensive form…"*: Carl von Clausewitz, *On War*, edited and translated by Michael Howard and Peter Paret (Princeton, N.J.: Princeton University Press, 1984), 357.

8-2 **The Battle of Kursk, July 1943**. David M. Glantz, "Soviet Defensive Tactics at Kursk," CSI Report No. 11 (Fort Leavenworth Kans.: Combat Studies Institute, US Army Command and General Staff College, 1986).

9-1 *"Regardless of the considerations..."*: FM 100-5, *Field Service Regulations: Operations* (Washington, D.C.: War Department, 22 May 1941; reprint, Fort Leavenworth, Kans.: US Army Command and General Staff College Press, 1992 [hereafter cited as FM 100-5 (1941)]), 152 (paragraph 652).

10-1 *"A swift and vigorous transition..."*: Carl von Clausewitz, cited in Heinl, 83.

10-2 **Manstein's Donbas Operation, February 1943**: David M. Glantz, *From the Don to the Dnepr: Soviet Offensive Operations, December 1942–August 1943,* Cass Series on Soviet Military Experience (London: Frank Cass, 1991), 82-150.

11-1 *"The withdrawal should be..."*: B.H. Liddell Hart (1944), cited in Heinl, 362.

11-2 **The Atlanta Campaign, 1864**: Albert Castel, *Decision in the West: The Atlanta Campaign of 1864* (Lawrence: University Press of Kansas, 1992), 151–181, 188.

12-0 *"The officers and men who..."*: D.H. Hill (1863), cited in Heinl, 317.

12-3 **Operation Bagration, 1944**. William M. Connor, "Analysis of Deep Attack Operations: Operation Bagration, Belorussia, 22 June–29 August, 1944" (unpublished paper, Combat Studies Institute, US Army Command and General Staff College, Fort Leavenworth, Kans., March, 1987).

13-0 *"You can never have too much reconnaissance."*: George S. Patton Jr., in *War As I Knew It* (Boston: Houghton-Mifflin Co., 1947), 400.

13-4 **The Battle of the Sajo River**: Simon Goodenough, *Tactical Genius in Battle* (Oxford, England: Phaidon Press, Limited, 1979), 123-126.

14-1 *"Aptitude for war..."*: Napoleon I (1831), cited in Heinl, 198.

15-1 *"...the necessity for conservation..."*: FM 100-5 (1941), 158 (paragraph 669).

16-0 *"The principal task involved..."*: FM 100-5 (1941), 130 (paragraph 572).

A-1 *"It is not so much..."*: Antoine Henri, Baron de Jomini, *Precis de l'Art de la Guerre* (1838), cited in Heinl, 58.

B-1 *"Doctrine provides a military organization..."*: George H. Decker (1960), cited in Heinl, 95.

C-1 *"A landing against organized..."*: George C. Marshall (1943), cited in Heinl, 11.

D-0 *"When the enemy is driven..."*: Field Marshal Prince Aleksandr V. Suvorov (1729-1800), cited in Tsouras, 134.

E-0 *"The base commander's plan..."*: FM 31-81 (Test), *Base Defense*, 17 March 1970, i.

Glossary

The glossary lists acronyms and terms with Army or joint definitions, and other selected terms. Where Army and joint definitions are different, *(Army)* follows the term. Terms for which FM 3-90 is the proponent manual (the authority) are marked with an asterisk (*). The proponent manual for other terms is listed in parentheses after the definition.

A2C2 Army airspace command and control

AA assembly area

AATF air assault task force

AATFC air assault task force commander

ACA airspace coordination area

ACM airspace control measure

ACP air control point

acq acquisition

ACR armored cavalry regiment

***actions on contact** a series of combat actions, often conducted simultaneously, taken on contact with the enemy to develop the situation

active air defense direct defensive action taken to nullify or reduce the effectiveness of hostile air action. It includes such measures as the use of aircraft, air defense weapons, weapons not used primarily in an air defense role, and electronic warfare. (FM 3-01.8)

AD armored division

ADA air defense artillery

ADC area damage control

administrative movement a movement in which troops and vehicles are arranged to expedite their movement and conserve time and energy when no enemy interference, except by air, is anticipated (JP 1-02)

AG adjutant general

AGL above ground level

AH attack helicopter

AI air interdiction

air assault operations those operations in which assault forces (combat, combat support, and combat service support)—using the firepower, mobility, and total integration of helicopter assets—maneuver on the battlefield under the control of the ground or air maneuver commander to engage and destroy enemy forces or to seize and hold key terrain (FM 3-97.4)

airborne operation an operation involving the air movement into an objective area of combat forces and their logistic support for execution of a tactical, operational, or strategic mission. The means employed may be any combination of airborne units, air transportable units, and various types of transport aircraft, depending on the mission and the overall situation. (JP 3-18)

air control point an easily identifiable point on the terrain or an electronic navigational aid used to provide necessary control during air movement. ACPs are generally designated at each point where the flight route or air corridor makes a definite change in direction and at any other point deemed necessary for timing or control of the operation. (FM 3-52)

air corridor a restricted air route of travel specified for use by friendly aircraft and established for the purpose of preventing friendly aircraft from being fired on by friendly forces (JP 1-02)

air defense battlefield operating system the employment of all active measures designed to nullify or reduce the effectiveness of attack by hostile aircraft and missiles after they are airborne (FM 7-15)

airspace coordination area a three-dimensional block of airspace in a target area, established by the appropriate ground commander, in which friendly aircraft are reasonably safe from friendly surface fires. The airspace coordination area may be formal or informal. (JP 3-09.3)

alt altitude

***alternate position** a defensive position that the commander assigns to a unit or weapon for occupation when the primary position becomes untenable or unsuitable for carrying out the assigned task

ambush a form of attack by fire or other destructive means from concealed positions on a moving or temporarily halted enemy (FM 3-0)

AO area of operations

APOD aerial port of debarkation

approach march (Army) the advance of a combat unit when direct contact with the enemy is intended (FM 3-0)

AR armor

area defense a type of defensive operation that concentrates on denying enemy forces access to designated terrain for a specific time rather than destroying the enemy outright (FM 3-0)

area of influence a geographical area wherein a commander is directly capable of influencing operations by maneuver or fire support systems normally under the commander's command or control (JP 1-02)

area of interest that area of concern to the commander, including the area of influence, areas adjacent thereto, and extending into enemy territory to the objectives of current or planned operations. This area also includes areas occupied by enemy forces who could jeopardize the accomplishment of the mission. (JP 2-03)

area of operations an operational area defined by the joint force commander for land and naval forces. Areas of operations do not typically encompass the entire operational area of the joint force commander, but should be large enough for component commanders to accomplish their missions and protect their forces. (JP 1-02)

***area reconnaissance** a form of reconnaissance that focuses on obtaining detailed information about the terrain or enemy activity within a prescribed area

***area security** a form of security operations conducted to protect friendly forces, installations, routes, and actions within a specific area

ARFOR the senior Army headquarters and all Army forces assigned or attached to a combatant command, subordinate joint force command, joint functional command, or multinational command (FM 3-0)

***Army air movements** operations involving the use of utility and cargo rotary-wing assets for other than air assaults

Army airspace command and control those actions that ensure the synchronized use of airspace and enhance the command and control of forces using airspace (FM 3-52)

***art of tactics** The art of tactics consists of three interrelated aspects: the creative and flexible array of means to accomplish assigned missions, decision making under conditions of uncertainty when faced with an intelligent enemy, and understanding the human dimension—the effects of combat on soldiers

ASCC Army service component command

ASG area support group

aslt assault

***assailable flank** a flank exposed to attack or envelopment

assault echelon the element of a force that is scheduled for initial assault on the objective area (JP 4-01.2) [The joint definition includes a sentence relating to amphibious operations that is omitted.]

***assault position** a covered and concealed position short of the objective, from which final preparations are made to assault the objective

***assault time** the moment to attack the initial objectives throughout the geographical scope of the operation

***assembly area** an area a unit occupies to prepare for an operation

AT antitank

ATACMS Army Tactical Missile System

atk attack

ATS air traffic service

attack an offensive operation that destroys or defeats enemy forces, seizes and secures terrain, or both (FM 3-0)

***attack by fire** a tactical mission task in which a commander uses direct fires, supported by indirect fires, to engage an enemy without closing with him to destroy, suppress, fix, or deceive him

***attack-by-fire position** the general position from which a unit conducts the tactical mission task of attack by fire

***attack position** the last position an attacking force occupies or passes through before crossing the line of departure

***avenue of approach** (Army) the air or ground route leading to an objective (or key terrain in its path) that an attacking force can use

avn aviation

***axis of advance** the general area through which the bulk of a unit's combat power must move

AXP ambulance exchange point

base a locality from which operations are projected or supported (JP 1-02)

base cluster in base defense operations, a collection of bases geographically grouped for mutual protection and ease of command and control (JP 3-10)

base commander In base defense operations, the officer assigned to command a base (JP 1-02)

***base defense force** A base defense force is a security element established to provide local security to a base

base defense operations center a command and control facility established by the base commander to serve as the focal point for base security and defense. It plans, directs, integrates, coordinates, and controls all base defense efforts, and coordinates and integrates into area security operations with the rear area operations center/rear tactical operations center. (JP 3-10.1)

***battalion** A battalion (or a squadron) is a unit consisting of two or more company-, battery-, or troop-size units and a headquarters.

***battalion task force** a combat arms battalion-size unit consisting of a battalion headquarters, at least one assigned company-size element, and at least one attached company-size element from another combat arms or combat support unit

***battery** a company-size unit in a field artillery or air defense artillery organization

battle a set of related tactical engagements that last longer and involve larger forces than an engagement (FM 3-0)

battlefield organization the allocation of forces in the area of operations by purpose. It consists of three all-encompassing categories of operations: decisive, shaping, and sustaining. (FM 3-0)

***battle handover line** a designated phase line on the ground where responsibility transitions from the stationary force to the moving force and vice versa

***battle position** a defensive location oriented on a likely enemy avenue of approach

BCD battlefield coordination detachment

BCOC base cluster operations center

BCT brigade combat team

BDAR battle damage assessment and repair

BDOC base defense operations center

bde brigade

BDF base defense force

BFV Bradley fighting vehicle

BG brigadier general

BHL battle handover line

***block** a tactical mission task that denies the enemy access to an area or prevents his advance in a direction or along an avenue of approach. Block is also an engineer obstacle effect that integrates fire planning and obstacle effort to stop an attacker along a specific avenue of approach or prevent him from passing through an engagement area

bn battalion

Bn TF battalion task force

BOS battlefield operating systems

boundary a line that delineates surface areas for the purpose of facilitating coordination and deconfliction of operations between adjacent units, formations, or areas (JP 0-2)

BP battle position

***breach** a tactical mission task in which the unit employs all available means to break through or secure a passage through an enemy defense, obstacle, minefield, or fortification

***brigade combat team** a semipermanent combined arms organization, consisting of a brigade headquarters, at least two combat arms battalions, and all necessary supporting combat support and combat service support, formed for the purpose of carrying out a continuing specific task

***brigade**	a unit consisting of two or more battalions and a headquarters
BSA	brigade support area
***bypass**	a tactical mission task in which the commander directs his unit to maneuver around an obstacle, position, or enemy force to maintain the momentum of the operation while deliberately avoiding combat with an enemy force
C	cover (used on graphics to indicate a security mission. See paragraph 12-13)
C2	command and control
CA	civil affairs
calculated risk	an exposure to chance of injury or loss when the commander can visualize the outcome in terms of mission accomplishment or damage to the force, and judges the outcome as worth the cost
***canalize**	(Army) a tactical mission task in which the commander restricts enemy movement to a narrow zone by exploiting terrain coupled with the use of obstacles, fires, or friendly maneuver
CAS	close air support
cbt	combat
CCIR	commander's critical information requirements
CD	cavalry division counterdrug
CFL	coordinated fire line
CFZ	critical friendly zone
***checkpoint**	a predetermined point on the ground used to control movement, tactical maneuver, and orientation
CI	counterintelligence
***clear**	a tactical mission task that requires the commander to remove all enemy forces and eliminate organized resistance within an assigned area
close area	where forces are in immediate contact with the enemy and the fighting between the committed forces and readily available tactical reserves of both combatants is occurring, or where commanders envision close combat taking place. Typically, the close area assigned to a maneuver force extends from its subordinates' rear boundaries to its own forward boundary. (FM 3-0)
close combat	combat carried out with direct-fire weapons, supported by indirect fires, air-delivered fires, and nonlethal engagement means. Close combat defeats or destroys enemy forces or seizes and retains ground. (FM 3-0)
CNR	combat net radio

co company

COA course of action

COL colonel

COLT combat observation and lasing team

***combat arms** units and soldiers who close with and destroy enemy forces or provide firepower and destructive capabilities on the battlefield

***combat formation** an ordered arrangement of forces for a specific purpose and the general configuration of a unit on the ground

combat information (Army) unevaluated data gather by or provided to a commander that, due to its highly perishable nature or the critical nature of the tactical situation, cannot be processed into tactical intelligence or other staff products in time to meet the commander's information requirements (FM 6-0)

***combat outpost** a reinforced observation post capable of conducting limited combat operations

combat power the total means of destructive and/or disruptive force which a military unit/formation can apply against the opponent at a given time (JP 1-02)

combat service support battlefield operating system the support and services provided to sustain forces during war and military operations other than war (FM 7-15)

***combat support** (Army) critical combat functions provided by units and soldiers in conjunction with combat arms units and soldiers to secure victory

combined arms the synchronized or simultaneous application of several arms—such as infantry, armor, artillery, engineers, air defense, and aviation—to achieve an effect on the enemy that is greater than if each arm was used against the enemy separately or in sequence (FM 3-0)

command and control battlefield operating system all tasks associated with the exercise of authority and direction by a properly designated commander over assigned and available forces in the accomplishment of the mission (FM 7-15)

***company** a unit consisting of two or more platoons, usually of the same type, with a headquarters and a limited capacity for self-support

***company team** a combined arms organization formed by attaching one or more nonorganic tank, mechanized infantry, or light infantry platoons to a tank, mechanized infantry, or light infantry company, either in exchange for, or in addition to, its organic platoons

***consolidation** the process of organizing and strengthening a newly captured position so that it can be defended

***contact point** an easily identifiable point on the terrain where two or more ground units are required to make physical contact

***contain** (Army) a tactical mission task that requires the commander to stop, hold, or surround enemy forces or to cause them to center their activity on a given front and prevent them from withdrawing any part of their forces for use elsewhere

***contiguous area of operations** when all of a commander's subordinate forces' areas of operations share one or more common boundaries

COP common operational picture

***control** a tactical mission task that requires the commander to maintain physical influence over a specified area to prevent its use by an enemy or to create conditions necessary for successful friendly operations

CONUS continental United States

***convoy security operations** a specialized kind of area security operations conducted to protect convoys

coordinated fire line a line beyond which conventional, direct, and indirect surface fire support means may fire at any time within the boundaries of the establishing headquarters without additional coordination. The purpose of the CFL is to expedite the surface-to-surface attack of targets beyond the CFL without coordination with the ground commander in whose area the targets are located. (JP 3-09)

***cordon security** the security provided between two combat outposts positioned to provide mutual support

***corps** the Army's largest tactical unit and the instrument by which higher echelons of command conduct maneuver at the operational level

COSCOM corps support command

co/tm company/team

counterattack (Army) a form of attack by part or all of a defending force against an enemy attacking force, with the general objective of denying the enemy his goal in attacking (FM 3-0)

***counterreconnaissance** (Army) a tactical mission task that encompasses all measures taken by a commander to counter enemy reconnaissance and surveillance efforts. Counterreconnaissance is not a distinct mission, but a component of all forms of security operations

***cover** a form of security operation whose primary task is to protect the main body by fighting to gain time while also observing and reporting information and preventing enemy ground observation of and direct fire against the main body

***covering force** a self-contained force capable of operating independently of the main body, unlike a screening or guard force

CP checkpoint; command post

***critical friendly zone** an area, usually a friendly unit or location, that the maneuver commander designates as critical to the protection of an asset whose loss would seriously jeopardize the mission

CS combat support

CSG corps support group

CSS combat service support

***cueing** the integration of one or more types of reconnaissance or surveillance systems to provide information that directs follow-on collecting of more detailed information by another system

DAG division artillery group

DASB division aviation support battalion

***decisive terrain** key terrain whose seizure and retention is mandatory for successful mission accomplishment

decisive engagement in land and naval warfare, an engagement in which a unit is considered fully committed and cannot maneuver or extricate itself. In the absence of outside assistance, the action must be fought to a conclusion and either won or lost with the forces at hand. (JP 1-02)

deep area an area forward of the close area that commanders use to shape enemy forces before they are encountered or engaged in the close area (FM 3-0)

***defeat** a tactical mission task that occurs when an enemy force has temporarily or permanently lost the physical means or the will to fight. The defeated force's commander is unwilling or unable to pursue his adopted course of action, thereby yielding to the friendly commander's will and can no longer interfere to a significant degree with the actions of friendly forces. Defeat can result from the use of force or the threat of its use.

***defeat in detail** Defeat in detail is achieved by concentrating overwhelming combat power against separate parts of a force rather than defeating the entire force at once.

defensive operations operations to defeat an enemy attack, buy time, economize forces, or develop conditions favorable for offensive operations. Defensive operations alone normally cannot achieve a decision. Their purpose is to create conditions for a counteroffensive that allows Army forces to regain the initiative. (FM 3-0)

delay a form of retrograde [JP 1-02 uses *an operation*] in which a force under pressure trades space for time by slowing down the enemy's momentum and inflicting maximum damage on the enemy without, in principle, becoming decisively engaged (JP 1-02, see delaying operation)

***delay line** a phase line where the date and time before which the enemy is not allowed to cross the phase line is depicted as part of the graphic control measure

***deliberate operation** an operation in which a commander's detailed intelligence concerning the situation allows him to develop and coordinate detailed plans, including multiple branches and sequels. He task-organizes his forces specifically for the operation to provide a fully synchronized combined arms team. He conducts extensive rehearsals while conducting shaping operations to set the conditions for his decisive operation.

demonstration (Army) a form of the attack designed to deceive the enemy as to the location or time of the decisive operation by a display of force. Forces conducting a demonstration do not seek contact with the enemy. (FM 3-0)

***denial operations** actions to hinder or deny the enemy the use of space, personnel, supplies, or facilities

***destroy** a tactical mission task that physically renders an enemy force combat-ineffective until it is reconstituted. Alternatively, to destroy a combat system is to damage it so badly that it cannot perform any function or be restored to a usable condition without being entirely rebuilt.

***detachment left in contact** an element left in contact as part of the previously designated (usually rear) security force while the main body conducts its withdrawal

***direction of attack** a specific direction or assigned route a force uses and does not deviate from when attacking

DISCOM division support command

***disengage** a tactical mission task where a commander has his unit break contact with the enemy to allow the conduct of another mission or to avoid decisive engagement

***disengagement line** a phase line located on identifiable terrain that, when crossed by the enemy, signals to defending elements that it is time to displace to their next positions

***dismounted marches** Dismounted marches, also called foot marches, are movements of troops and equipment, mainly by foot, with limited support by vehicles.

***disrupt** a tactical mission task in which a commander integrates direct and indirect fires, terrain, and obstacles to upset an enemy's formation or tempo, interrupt his timetable, or cause his forces to commit prematurely or attack in a piecemeal fashion. Disrupt is also an engineer obstacle effect that focuses fire planning and obstacle effort to cause the enemy to break up his formation and tempo, interrupt his timetable, commit breaching assets prematurely, and attack in a piecemeal effort.

div division

DIVARTY division artillery

DIVCAV division cavalry

***divisions** (Army) the largest fixed organizations in the Army that train and fight as tactical teams, and are organized with varying numbers and types of combat, CS, and CSS units

DLIC detachment left in contact

DOD Department of Defense

DS direct support

DSA division support area

DZ drop zone

EA engagement area

EAC echelon above corps

eff effective

***encirclement operations** operations where one force loses its freedom of maneuver because an opposing force is able to isolate it by controlling all ground lines of communication and reinforcement

engagement a small, tactical conflict between opposing maneuver forces, usually conducted at brigade level and below (FM 3-0)

***engagement area** where the commander intends to contain and destroy an enemy force with the massed effects of all available weapons and supporting systems

***engagement criteria** protocols that specify those circumstances for initiating engagement with an enemy force

***engagement priority** the order in which the unit engages enemy systems or functions

envelopment (Army) a form of maneuver in which an attacking force seeks to avoid the principal enemy defenses by seizing objectives to the enemy rear to destroy the enemy in his current positions. At the tactical level, envelopments focus on seizing terrain, destroying specific enemy forces, and interdicting enemy withdrawal routes. (FM 3-0)

ENY enemy (graphics)

EOD explosive ordnance disposal

EPW enemy prisoner of war

ESB enhanced separate brigade

evac evacuation

EW electronic warfare

***exfiltrate** a tactical mission task where a commander removes personnel or units from areas under enemy control by stealth, deception, surprise, or clandestine means

exploitation a type of offensive operation that usually follows a successful attack and is designed to disorganize the enemy in depth (JP 1-02)

EZ extraction zone

FA field artillery

FARP forward arming and refueling point

FCL final coordination line

FEBA forward edge of the battle area

feint (Army) a form of attack used to deceive the enemy as to the location or time of the actual decisive operation. Forces conducting a feint seek direct fire contact with the enemy but avoid decisive engagement. (FM 3-0)

FFA free-fire area

***field of fire** the area that a weapon or group of weapons may cover effectively from a given position

***final coordination line** a phase line close to the enemy position used to coordinate the lifting or shifting of supporting fires with the final deployment of maneuver elements

final protective fires an immediately available preplanned barrier of fires designed to impede enemy movement across defensive lines or areas (JP 3-09)

***fire strike** the massed, synchronized, and nearly simultaneous delivery of precision-guided munitions

***fire superiority** that degree of dominance in the fires of one force over another that permits that force to conduct maneuver at a given time and place without prohibitive interference by the enemy

fire support battlefield operating system the collective and coordinated use of target-acquisition data, indirect-fire weapons, fixed-wing aircraft, offensive information operations, and other lethal and nonlethal means against targets located throughout an area of operations (FM 7-15)

fire support coordination line a FSCM that is established and adjusted by appropriate land or amphibious force commanders within their boundaries in consultation with superior, subordinate, supporting, and affected commanders. The FSCL facilitates the expeditious attack of surface targets of opportunity beyond the coordinating measure. A FSCL does not divide an area of operations by defining a boundary between close and deep operations or a distinct area [JP 3-09 uses *zone*] for close air support. The FSCL applies to all fires of air-, land-, and sea-based weapon systems using any type of ammunition. Forces attacking targets beyond an FSCL must inform all affected commanders in sufficient time to allow necessary reaction to avoid fratricide. Supporting elements attacking targets beyond the FSCL must ensure that the attack will not produce adverse effects on, or to the rear of, the line. Short of an FSCL, all air-to-ground and surface-to-surface attack operations are controlled by the appropriate land or amphibious force commander. The FSCL should follow well-defined terrain features. Coordination of attacks beyond the FSCL is especially critical to commanders of air, land, and special operations forces. In exceptional circumstances, the inability to conduct this coordination will not preclude the attack of targets beyond the FSCL. However, failure to do so may increase the risk of fratricide and could waste limited resources. (JP 3-09)

fire support target see target

***fix** a tactical mission task where a commander prevents the enemy from moving any part of his force from a specific location for a specific period to of time. Fix is also an engineer obstacle effect that focuses fire planning and obstacle effort to slow an attacker's movement within a specified area, normally an engagement

***flanking position** a geographical location on the flank of a force from which effective fires can be placed on that flank

***flanks** the right or left limits of a unit

FLE forward logistics element

FLOT forward line of own troops

FM field manual

FOB forward operating base

***follow and assume** a tactical mission task in which a second committed force follows a force conducting an offensive operation and is prepared to continue the mission if the lead force is fixed, attrited, or unable to continue

***follow and support** a tactical mission task in which a committed force follows and supports a lead force conducting an offensive operation

***follow-on echelon** those additional forces moved into the objective area after the assault echelon

foot march see dismounted marches

forward edge of the battle area the foremost limits of a series of areas in which ground combat units are deployed, excluding the areas in which the covering or screening forces are operating, designated to coordinate fire support, the positioning of forces, or the maneuver of units (JP 1-02)

forward line of own troops a line which indicates the most forward positions of friendly forces in any kind of military operation at a specific time. The forward line of own troops normally identifies the forward location of covering and screening forces. (JP 1-02)

***forward passage of lines** when a unit passes through another unit's positions while moving toward the enemy

FPF final protective fires

FRAGO fragmentary order

free-fire area a specific area into which any weapon system may fire into without additional coordination with the establishing headquarters (JP 3-09)

front an echelon of the Soviet Army two levels above corps, the equivilent of a US army group

frontal attack (Army) a form of maneuver in which an attacking force seeks to destroy a weaker enemy force or fix a larger enemy force in place over a broad front (FM 3-0)

FSB forward support battalion

FSCL fire support coordination line

FSCM fire support coordinating measure

FSCOORD fire support coordinator

G guard (used on graphics to indicate a security mission. See paragraph 12-13)

G1 assistant chief of staff, personnel

G2 assistant chief of staff, intelligence

G3 assistant chief of staff, operations

G4 assistant chief of staff, logistics

G5 assistant chief of staff, civil affairs

G6 assistant chief of staff, communications

***gap** (Army) an area free of armed mines or obstacles whose width and direction allow a friendly force to pass through the area containing obstacles while dispersed in a tactical formation

GEN general

GPS global positioning system

GRC guards rifle corps

grd ground

GS general support

GSR ground surveillance radar

***guard** (Army) a form of security operations whose primary task is to protect the main body by fighting to gain time while also observing and reporting information and preventing enemy ground observation of and direct fire against the main body. Units conducting a guard mission cannot operate independently because they rely upon fires and combat support assets of the main body.

***hasty operation** an operation in which a commander directs his immediately available forces, using fragmentary orders, to perform activities with minimal preparation, trading planning and preparation time for speed of execution

hel helicopter

HHC headquarters and headquarters company

HIMAD high- to medium-altitude air defense

HQ headquarters

HUMINT human intelligence

IBCT initial brigade combat team

ID infantry division

IFF identification, friend or foe

IMINT imagery intelligence

IN infantry

infiltration (Army) a form of maneuver in which an attacking force conducts undetected movement through or into an area occupied by enemy forces to occupy a position of advantage in the enemy rear while exposing only small elements to enemy defensive fires (FM 3-0)

***infiltration lane** a control measure that coordinates forward and lateral movement of infiltrating units and fixes fire planning responsibilities

intelligence battlefield operating system the activity to generate knowledge of and products portraying the enemy and environmental features required by a commander in planning, preparing, executing, and assessing operations (FM 7-15)

***interdict** a tactical mission task where the commander prevents, disrupts, or delays the enemy's use of an area or route

IO information operations

IPB intelligence preparation of the battlefield

ISB intermediate support base

***isolate** a tactical mission task that requires a unit to seal off—both physically and psychologically—an enemy from his sources of support, deny an enemy freedom of movement, and prevent an enemy unit from having contact with other enemy forces

ISR intelligence, surveillance, and reconnaissance

JAAT joint air attack team

JAG judge adcocate general

JAOC joint air operations center

JFACC joint force air component commander

JFC joint force commander

JFLCC joint force land component commander

JP joint publication

JSOA joint special operations area

JSTARS	joint surveillance target attack radar system
JTF	joint task force
***key terrain**	any locality or area, the seizure or retention of which affords a marked advantage to either combatant in a given course of action
***kill zone**	that part of an ambush site where fires are concentrated to isolate, fix, and destroy the enemy
LC	line of contact
LD	line of departure
***Level I threat**	a small enemy force that can be defeated by CS and CSS units operating in the rear area or by the perimeter defenses established by friendly bases and base clusters
***Level II threat**	enemy activities that can be defeated by a base or base cluster augmented by a response force
***Level III threat**	a threat beyond the defensive capability of both the base and base cluster and any local reserve or response force
lg	long
***light line**	a designated phase line, forward of which vehicles are required to use blackout lights during periods of limited visibility
***limit of advance**	a phase line used to control forward progress of the attack. The attacking unit does not advance any of its elements or assets beyond the LOA, but the attacking unit can push its security forces to that limit.
***line of contact**	a general trace delineating the location where friendly and enemy forces are engaged
***line of departure**	a phase line crossed at a prescribed time by troops initiating an offensive operation
line of communication	a route, either land, water, and/or air, which connects an operating military force with a base of operations and along which supplies and military forces move (JP 1-02)
***linkup point**	a point where two infiltrating elements in the same or different infiltration lanes are scheduled to meet to consolidate before proceeding on with their missions
***linkup**	a meeting of friendly ground forces, which occurs in a variety of circumstances
LL	light line
LLTR	low-level transit route
LOA	limit of advance
LOC	line of communication
***local security**	low-level security operations conducted near a unit to prevent surprise by the enemy

***logistic package** a grouping of multiple classes of supply and supply vehicles under the control of a single convoy commander

LOGPAC logistic package

LRP logistics release point

LRSU long-range surveillance unit

LTC lieutenant colonel

LU linkup

LZ landing zone

M mechanized

meters

***main battle area** the area where the commander intends to deploy the bulk of his combat power and conduct his decisive operation to defeat an attacking enemy

maneuver the employment of forces on the battlefield through movement in combination with fires, or fires potential, to achieve a position of advantage with respect to the enemy in order to accomplish the mission (JP 1-02)

maneuver battlefield operating system the movement of forces to achieve a position of advantage with respect to enemy forces. This system includes the employment of forces on the battlefield in combination with direct fire or fire potential. This system also includes the conduct of tactical tasks associated with force projection. (FM 7-15)

***march column** all elements using the same route for a single movement under control of a single commander

***march serial** a major subdivision of a march column that is organized under one commander who plans, regulates, and controls the serial

***march unit** a subdivision of a march serial. It moves and halts under the control of a single commander who uses voice and visual signals.

MASINT measurement and signature intelligence

max maximum

MBA main battle area

MCC movement control center

MD mechanized division

MDMP military decision making process

MECH mechanized infantry

MED medical

meeting engagement (Army) a combat action that occurs when a moving force engages an enemy at an unexpected time and place (FM 3-0)

METL mission-essential task list

METT-TC mission, enemy, terrain and weather, troops and support available, time available, and civil considerations (FM 6-0)

MG major general

MI military intelligence

military gamble a situation in which a commander risks the force without a reasonable level of information about the outcome (FM 6-0)

min minimum

***mixing** using two or more different assets to collect against the same intelligence requirement

MLRS Multiple Launch Rocket System

MMC materiel management center

mobile defense (Army) a type of defensive operation that concentrates on the destruction or defeat of the enemy through a decisive attack by a striking force (FM 3-0)

mobility/countermobility/survivability battlefield operating system Mobility operations preserve the freedom of maneuver of friendly forces. Countermobility operations deny mobility to enemy forces. Survivability operations protect friendly forces from the effects of enemy weapon systems. (FM 7-15)

MOPP mission-oriented protective posture

mort mortar

movement control the planning, routing, scheduling, and control of personnel and cargo movement over lines of communication (JP 4-01.3)

movement to contact a type of offensive operation designed to develop the situation and establish or regain contact (FM 3-0)

MP military police

MRB motorized rifle battalion

MRR minimum risk route

MRT movement regulating team

MSR main supply route

MTC movement to contact

mutual support support that units render to each other against an enemy because of their assigned tasks, their position relative to each other and to the enemy, and their inherent capabilities (JP 1-02)

NAI named area of interest

***named area of interest** the geographical area where information that will satisfy a specific information requirement can be collected

NBC nuclear, biological, and chemical

***neutralize** (Army) a tactical mission task that results in rendering enemy personnel or materiel incapable of interfering with a particular operation

NFA no-fire area

NGO nongovernmental organization

no-fire area a land area designated by the appropriate commander into which fires or their effects are prohibited (JP 3-09)

***noncontiguous area of operations** when one or more of a commander's subordinate forces' areas of operation do not share a common boundary

OAKOC observation and fields of fire, avenues of approach, key terrain, obstacles, and cover and concealment

O/O on-order

OBJ objective

***objective** (Army) a location on the ground used to orient operations, phase operations, facilitate changes of direction, and provide for unity of effort

***objective rally point** a rally point established on an easily identifiable point on the ground where all elements of the infiltrating unit assemble and prepare to attack the objective

obstacle any obstruction designed or employed to disrupt, fix, turn, or block the movement of an opposing force, and to impose additional losses in personnel, time, and equipment on the opposing force. Obstacles can exist naturally or can be manmade, or can be a combination of both. (JP 1-02)

obstacle belt a brigade-level command and control measure, normally given graphically, to show where within an obstacle zone the ground tactical commander plans to limit friendly obstacle employment and focus the defense (JP 1-02)

***obstacle control measures** specific measures that simplify the granting of obstacle-emplacing authority while providing obstacle control

obstacle restricted area a command and control measure used to limit the type or number of obstacles within an area (JP 1-02)

***obstacle groups** one or more individual obstacles grouped to provide a specific obstacle effect

obstacle zone a division-level command and control measure, normally done graphically, to designate specific land areas where lower echelons are allowed to employ tactical obstacles (JP 1-02)

***occupy** a tactical mission task that involves moving into an area so that it can control that entire area. Both the force's movement to and occupation of the area occur without enemy opposition.

offensive operations	Offensive operations aim at destroying or defeating an enemy. Their purpose is to impose US will on the enemy and achieve decisive victory. (FM 3-0)
OH	observation helicopter
OP	observation post
OPCON	operational control
operation	a military action or the carrying out of a strategic, operational, tactical, service, training, or administrative military mission (JP 1-02)
operational framework	the arrangement of friendly forces and resources in time, space, and purpose with respect to each other and the enemy or situation. It consists of the area of operations, battlespace, and the battlefield organization. (FM 3-0)
OPLAN	operation plan
OPNS	operations
OPORD	operation order
OPS	operations
OPSEC	operations security
ORA	obstacle restricted area
ord	ordnance
ORP	objective rally point
PAA	position area for artillery
***passage lane**	a lane through an enemy or friendly obstacle that provides a safe passage for a passing force
***passage of lines**	(Army) a tactical enabling operation in which one unit moves through another unit's positions with the intent of moving into or out of enemy contact
***passage point**	a specifically designated place where the passing units pass through the stationary unit
passive air defense	all measures, other than active air defense, taken to minimize the effects of hostile air action. They include cover, concealment, camouflage, dummy positions, dispersion, and protective construction. (FM 3-01.8)
PD	point of departure
***penetration**	(Army) a form of maneuver in which an attacking force seeks to rupture enemy defenses on a narrow front to disrupt the defensive system (FM 3-0)
phase line	a line utilized for control and coordination of military operations, usually a terrain feature extending across the operational area [JP 1-02 uses *zone of action*] (JP 1-02)

***piecemeal commitment** the immediate employment of units in combat as they become available instead of waiting for larger aggregations of units to ensure mass, or the unsynchronized employment of available forces so that their combat power is not employed effectively

PL phase line

PLD probable line of deployment

***point of departure** the point where the unit crosses the line of departure and begins moving along a direction of attack

POL petroleum, oils, and lubricants

***position area for artillery** an area assigned to an artillery unit where individual artillery systems can maneuver to increase their survivability. A PAA is not an area of operations for the artillery unit occupying it.

PP passage point

***primary position** the position that covers the enemy's most likely avenue of approach into the AO

***probable line of deployment** a phase line that a commander designates as the location where he intends to completely deploy his unit into assault formation before beginning the assault

***procedures** standard and detailed courses of action that describe how to perform tasks

psn position

PSYOP psychological operations

pursuit an offensive operation designed to catch or cut off a hostile force attempting to escape, with the aim of destroying it (JP 1-02)

PZ pickup zone

PZCO pickup zone control officer

PzDiv panzer division

raid a form of attack [JP 1-02 uses *operation*], usually small scale, involving a swift entry into hostile territory to secure information, confuse the enemy, or destroy installations. It ends with a planned withdrawal from the objective area upon mission completion. (FM 3.0)

***rally point** an easily identifiable point on the ground at which units can reassemble and reorganize if they become dispersed. Alternatively, it is an easily identifiable point on the ground at which aircrews and passengers can assemble and reorganize following an incident requiring a forced landing.

RAOC rear area operations center

RC rifle corps

RD rifle division

rear area (Army) The rear area for any command extends from its rear boundary forward to the rear boundary of the next lower level of command. This area is provided primarily for the performance of support functions and is where the majority of the echelon's sustaining operations occur. (FM 3-0)

***rearward passage of lines** when a unit passes through another unit's positions while moving away from the enemy

recon reconnaissance

***reconnaissance in force** (Army) a deliberate combat operation designed to discover or test the enemy's strength, dispositions, and reactions or to obtain other information [The Army definition adds *dispositions, and reactions* to the joint definition.]

***reconnaissance objective** a terrain feature, geographical area, or an enemy force about which the commander wants to obtain additional information

***reconnaissance operations** (Army) those operations undertaken to obtain, by visual observation or other detection methods, information about the activities and resources of an enemy or potential enemy, or to secure data concerning the meteorological, hydrographical or geographical characteristics and the indigenous population of a particular area. [The Army definition adds *and the indigenous population* to the joint definition.]

reconstitution those actions that commanders plan and implement to restore units to a desired level of combat effectiveness commensurate with mission requirements and available resources. Reconstitution operations include regeneration and reorganization. (FM 4-100.9)

***reduce** a tactical mission task that involves the destruction of an encircled or bypassed enemy force

***redundancy** using two or more like assets to collect against the same intelligence requirement

***release point** a location on a route where marching elements are released from centralized control

***relief in place** (Army) a tactical enabling operation in which, by the direction of higher authority, all or part of a unit is replaced in an area by the incoming unit

reorganization action taken to shift internal resources within a degraded unit to increase its level of combat effectiveness (FM 4-100.9)

response force a mobile force with appropriate fire support designated, usually by the area commander, to deal with Level II threats in the rear area (JP 1-02)

restrictive fire area an area in which specific restrictions are imposed and into which fires that exceed those restrictions will not be delivered without coordination with the establishing headquarters (JP 3-09)

restrictive fire line a line established between converging friendly surface forces that prohibits fires or their effects across that line (JP 3-09)

***retain** a tactical mission task in which the commander ensures that a terrain feature already controlled by a friendly force remains free of enemy occupation or use

retirement a form of retrograde [JP 1-02 uses *operation*] in which a force out of contact with the enemy moves away from the enemy (JP 1-02)

retrograde (Army) a type of defensive operation that involves organized movement away from the enemy (FM 3-0)

RFA restricted fire area

RFL restricted fire line

RGR ranger

RIF reconnaissance in force

RISTA reconnaissance, intelligence, surveillance, and target acquisition

rly rally

ROC rear operations center

ROE rules of engagement

ROM refuel on the move

route the prescribed course to be traveled from a specific point of origin to a specific destination (JP 1-02)

***route reconnaissance** a form of reconnaissance that focuses along a specific line of communication, such as a road, railway, or cross-country mobility corridor

***route security operations** a specialized kind of area security operations conducted to protect lines of communication and friendly forces moving along them

RP release point

rte route

rules of engagement directives issued by competent military authority which delineate the circumstances and limitations under which United States forces will initiate and/or continue combat engagement with other forces encountered (JP 1-02)

S screen (used on graphics to indicate a security mission. See paragraph 12-13)

S1 battalion/brigade personnel staff officer

S2 battalion/brigade intelligence staff officer

S3 battalion/brigade operations staff officer

S4 battalion/brigade logistics staff officer

S5 battalion/brigade civil-military operations staff officer

S6 battalion/brigade communications staff officer

SAAFR standard use Army aircraft flight route

SAW squad automatic weapon

***science of tactics** the understanding of those military aspects of tactics—capabilities, techniques, and procedures—that can be measured and codified

***screen** (Army) a form of security operations that primarily provides early warning to the protected force

SEAD suppression of enemy air defenses

search and attack a technique for conducting a movement to contact that shares many of the characteristics of an area security mission (FM 3-0)

***sector of fire** (Army) that area assigned to a unit, a crew-served weapon, or an individual weapon within which it will engage targets as they appear in accordance with established engagement priorities

***secure** a tactical mission task that involves preventing a unit, facility, or geographical location from being damaged or destroyed as a result of enemy action

***security operations** those operations undertaken by a commander to provide early and accurate warning of enemy operations, to provide the force being protected with time and maneuver space within which to react to the enemy, and to develop the situation to allow the commander to effectively use the protected force

***seize** a tactical mission task that involves taking possession of a designated area using overwhelming force

sep separate

SF special forces

SHORAD short-range air defense

SIGINT signals intelligence

SOA special operations aviation

SOCCE special operations command and control element

SOF special operations forces

SOI signal operating instruction

SOP standing operating procedures

SP start point

SPF special-purpose forces

spoiling attack (Army) a form of attack that preempts or seriously impairs an enemy attack while the enemy is in the process of planning or preparing to attack (FM 3-0)

spt support

sqd	squadron
***start point**	a location on the route where the marching elements fall under the control of a designated march commander
STAVKA	acronym for the Soviet high command during World War II
***stay-behind operation**	an operation in which the commander leaves a unit in position to conduct a specified mission while the remainder of his forces withdraw or retire from an area
striking force	a committed force organized to conduct the decisive attack in a mobile defense. It normally comprises the maximum combat power available to the commander at the time of the attack. (FM 3-0)
***strong point**	(Army) a heavily fortified battle position tied to a natural or reinforcing obstacle to create an anchor for the defense or to deny the enemy decisive or key terrain
***subsequent position**	a position that a unit expects to move to during the course of battle
***supplementary position**	a defensive position located within a unit's assigned AO that provides the best sectors of fire and defensive terrain along an avenue of approach that is not the primary avenue along which the enemy is expected to attack
***support-by-fire position**	the general position from which a unit conducts the tactical mission task of support by fire
***support by fire**	a tactical mission task in which a maneuver force moves to a position where it can engage the enemy by direct fire in support of another maneuvering force
***supporting distance**	the distance between two units that can be traveled in time for one to come to the aid of the other. For small units, it is the distance between two units that can be covered effectively by their fires. (FM 3-0)
***supporting range**	the distance one unit may be geographically separated from a second unit, yet remain within the maximum range of the second unit's indirect-fire weapon systems (FM 3-0)
***suppress**	(Army) a tactical mission task that results in the temporary degradation of the performance of a force or weapon system below the level needed to accomplish its mission
surveil	surveillance
sys	system
TAA	tactical assembly area
tactical combat force	a combat unit, with appropriate combat support and combat service support assets, that is assigned the mission of defeating Level III threats (JP 1-02)

tactical level of war the level of war at which battles and engagements are planned and executed to accomplish military objectives assigned to tactical units or task forces. Activities at this level focus on the ordered arrangement and maneuver of combat elements in relation to each other and to the enemy to achieve combat objectives. (JP 1-02)

***tactical mobility** the ability to move rapidly from one part of the battlefield to another, relative to the enemy

tactical road march a rapid movement used to relocate units within an area of operations to prepare for combat operations (FM 3-0)

***tactician** an individual devoted to mastering the science and art of tactics

tactics (Army) the employment of units in combat. It includes the ordered arrangement and maneuver of units in relation to each other, the terrain, and the enemy to translate potential combat power into victorious battles and engagements. (FM 3-0)

TAI targeted area of interest

target in the fire support context, an area designated and numbered for future firing (JP 1-02)

***target reference point** an easily recognizable point on the ground, such as a building or a road junction, used in conjunction with engagement areas and sectors of fire to initiate, distribute, and control fires

***targeted area of interest** the geographical area or point along a mobility corridor where successful interdiction will cause the enemy to abandon a particular course of action or require him to use specialized engineer support to continue. It is where he can be acquired and engaged by friendly forces.

TCF tactical combat force

TCP traffic control point

TECHINT technical intelligence

***techniques** the general and detailed methods used by troops and commanders to perform assigned missions and functions, specifically, the methods of using equipment and personnel

TF task force

***time of attack** the moment the leading elements of the main body cross the line of departure, or in a night attack, the point of departure

tm team

TOC tactical operations center

TOW tube-launched, optically-tracked, wire-guided (refers to the Army's heavy, antitank missile system)

trans transportation

***trigger line** a phase line used to initiate and mass fires into an engagement area or an objective at a predetermined range for all or like weapon systems

***troop** a company-size organization in a cavalry organization

***troop movement** the movement of troops from one place to another by any available means

TRP target reference point

TTP tactics, techniques, and procedures

***turn** a tactical mission task that involves forcing an enemy force from one avenue of approach or movement corridor to another. Turn is also a tactical obstacle effect that integrates fire planning and obstacle effort to divert an enemy formation from one avenue of approach to an adjacent avenue of approach or into an engagement area

turning movement (Army) a form of maneuver in which the attacking force seeks to avoid the enemy's principal defensive positions by seizing objectives to the enemy rear and causing the enemy to move out of his current positions or divert major forces to meet the threat (FM 3-0)

UAV unmanned aerial vehicle

UMCP unit maintenance collection point

US United States

USAF United States Air Force

USMC United States Marine Corps

USN United States Navy

withdrawal a planned operation in which a force in contact disengages from an enemy force (JP 1-02) [The Army considers it a form of retrograde.]

WMD weapons of mass destruction

***zone reconnaissance** a form of reconnaissance that involves a directed effort to obtain detailed information on all routes, obstacles, terrain, and enemy forces with a zone defined by boundaries

Bibliography

JOINT PUBLICATIONS

Most joint publications are online at http://www.dtic.mil/doctrine/jel/

(S) CJCSI 3121.01A. *Standing Rules of Engagement for US Forces* (U). 15 January 2000.

JP 0-2. *Unified Action Armed Forces (UNAAF).* 24 February 1995.

JP 1-02. *DOD Dictionary of Military and Associated Terms.* 12 April 2001. Available online at http://www.dtic.mil./doctrine/jel/doddict/

JP 2-0. *Doctrine for Intelligence Support to Joint Operations.* 09 March 2000.

JP 2-03. *Joint Tactics, Techniques, and Procedures for Geospatial Information and Services Support to Joint Operations.* 31 March 1999.

JP 3-09. *Doctrine for Joint Fire Support.* 12 May 1998.

JP 3-09.3. *Joint Tactics, Techniques, and Procedures for Close Air Support (CAS).* 01 December 1995.

JP 3-10. *Joint Doctrine for Rear Area Operations.* 28 May 1996.

JP 3-10.1. *Joint Tactics, Techniques, and Procedures for Base Defense.* 23 July 1996.

JP 3-17. *Joint Tactics, Techniques, and Procedures for Theater Airlift Operations.* 18 July 1995.

JP 3-18. *Joint Doctrine for Forcible Entry Operations.* TBP.

JP 3-52. *Doctrine for Joint Airspace Control in the Combat Zone.* 22 July 1995.

JP 3-53. *Doctrine for Joint Psychological Operations.* 10 July 1996.

JP 4-01.2. *Joint Tactics, Techniques, and Procedures for Sealift Support to Joint Operations.* 09 October 1996.

JP 4-01.3. *Joint Tactics, Techniques, and Procedures for Movement Control.* 21 June 1996.

ARMY PUBLICATIONS

Most Army doctrinal publications are online at http://155.217.58.58/atdls.htm

AR 190-12. *Military Working Dogs.* 30 September 1993. Available online at http://www.usapa.army.mil/gils/epubs1.html

FM 1-0 (12-6). *Personnel Doctrine.* 09 September 1994.

FM 1-02 (101-5-1). *Operational Terms and Graphics.* 30 September 1997.

FM 1-05 (16-1). *Religious Support.* 26 May 1995.

FM 1-06 (14-100). *Financial Management Operations.* 07 May 1997.

FM 1-27.10 (27-10). *The Law of Land Warfare.* 18 July 1956.

FM 2-0 (34-1). *Intelligence and Electronic Warfare Operations.* 27 September 1994.

FM 2-01.3 (34-130). *Intelligence Preparation of the Battlefield.* 08 July 1994.

FM 3-0 (100-5). *Operations.* 14 July 2001.

FM 3-01 (44-100). *US Army Air and Missile Defense Operations.* 15 June 2000.

FM 3-01.8 (44-8) *Combined Arms for Air Defense.* 01 June 1999.

FM 3-01.12 (100-12). *Army Theater Missile Defense Operations.* 31 March 2000.

FM 3-04.100 (1-100). *Army Aviation Operations.* 21 February 1997.

FM 3-04.113 (1-113). *Utility and Cargo Helicopter Operations.* 12 September 1997.

FM 3-05 (100-25). *Doctrine for Army Special Operations Forces.* 01 August 1999.

FM 3-05.20 (31-20). *Doctrine for Special Forces Operations.* 20 April 1990.

FM 3-05.30 (33-1). *Psychological Operations.* 19 June 2000.

FM 3-07 (100-20). *Military Operations in Low Intensity Conflict.* 05 December 1990. When revised, FM 3-07 will be retitled *Stability Operations and Support Operations.*

FM 3-09 (6-20). *Fire Support in the Airland Battle.* 17 May 1988.

FM 3-09.12 (6-121). *Tactics, Techniques, and Procedures for Field Artillery Target Acquisition.* 25 September 1990.

FM 3-11 (3-100). *Chemical Operations Principles and Fundamentals.* 08 May 1996.

FM 3-11.9 (3-9). *Potential Military Chemical/Biological Agents and Compounds.* 12 December 1990.

FM 3-11.50 (3-50). *Smoke Operations.* 04 December 1990.

FM 3-12 (100-30). *Nuclear Operations.* 29 October 1996.

FM 3-13 (100-6). *Information Operations.* 27 August 1996.

FM 3-17 (100-27). *USA/USAF Doctrine for Joint Airborne and Tactical Airlift Operations.* 31 January 1985. FM 3-17 will be incorporated into FM 3-52.

FM 3-18.11 (90-26). *Airborne Operations.* 18 December 1990.

FM 3-19.1 (19-1). *Military Police Operations.* 22 March 2001.

FM 3-20.95 (17-95). *Cavalry Operations.* 24 December 1996.

FM 3-21.7 (7-7). *Mechanized Infantry Platoon and Squad (APC).* 15 March 1985.

FM 3-21.8 (7-8). *Infantry Rifle Platoon and Squad.* 22 April 1992.

FM 3-21.10 (7-10). *The Infantry Rifle Company.* 14 December 1990.

FM 3-21.20 (7-20). *The Infantry Battalion.* 06 April 1992.

FM 3-21.30 (7-30). *The Infantry Brigade.* 03 October 1995.

FM 3-21.85 (7-85). *Ranger Unit Operations.* 09 June 1987.

FM 3-21.92. (7-92). *The Infantry Reconnaissance Platoon and Squad (Airborne, Air Assault, Light Infantry)* 23 December 1992.

FM 3-24.3 (20-3) *Camouflage, Concealment, and Decoys.* 30 August 1999.

FM 3-25.18 (21-18). *Foot Marches.* 01 June 1990.

FM 3-34 (5-100). *Engineer Operations.* 27 February 1996.

FM 3-34.1 (90-7). *Combined Arms Obstacle Integration.* 29 September 1994.

FM 3-34.2 (90-13-1). *Combined Arms Breaching Operations.* 31 August 2000.

FM 3-34.112 (5-103). *Survivability.* 10 June 1985.

FM 3-34.212 (5-170). *Engineer Reconnaissance.* 05 May 1998.

FM 3-34.310 (5-34). *Engineer Field Data.* 30 August 1999.

FM 3-52 (100-103). *Army Airspace Command and Control in a Combat Zone.* 07 October 1987. When revised, FM 3-52 will be retitled *Army Airspace Command and Control.*

FM 3-55. *Intelligence, Surveillance, and Reconnaissance Operations.* TBP.

FM 3-57 (41-10). *Civil Affairs Operations.* 14 February 2000.

FM 3-91 (71-100). *Division Operations.* 28 August 1998.

FM 3-91.1 (71-1). *Tank and Mechanized Infantry Company Team.* 26 January 1998.

FM 3-91.2 (71-2). *The Tank and Mechanized Infantry Battalion Task Force.* 27 September 1998.

FM 3-91.3 (71-3). *The Armored and Mechanized Infantry Brigade.* 08 January 1996.

FM 3-92 (100-15). *Corps Operations.* 29 October 1996.

FM 3-93 (100-7). *Decisive Force: The Army in Theater Operations.* 31 May 1995.

FM 3-97.4 (90-4). *Air Assault Operations.* 16 March 1987.

FM 3-97.13 (90-13). *River-Crossing Operations.* 26 January 1998.

FM 3-100.14 (100-14). *Risk Management.* 23 April 1998.

FM 4-0 (100-10). *Combat Service Support.* 03 October 1995.

FM 4-01 (55-1). *Transportation Operations.* 03 October 1995.

FM 4-01.30 (55-10). *Movement Control.* 09 February 1999.

FM 4-01.40 (55-30). *Army Motor Transport Units and Operations.* 27 June 1997.

FM 4-01.41 (55-20). *Rail Transport in a Theater of Operations.* 01 June 2000.

FM 4-01.50 (55-60). *Army Terminal Operations.* 15 April 1996.

FM 4-02 (8-10). *Health Service Support in a Theater of Operations.* 01 March 1991.

FM 4-20 (10-1). *Quartermaster Principles.* 11 August 1994.

FM 4-100.9 (100-9). *Reconstitution.* 13 January 1992.

FM 5-0 (101-5). *Staff Organizations and Operations* 31 May 1997. When revised, FM 5-0 will be retitled *Army Planning and Orders Production.*

FM 6-0. *Command and Control.* TBP.

FM 6-02 (24-1). *Signal Support to the Army.* TBP.

FM 6-22 (22-100). *Army Leadership.* 31 August 1999.

FM 6-22.5. *Combat Stress.* 23 June 2000.

FM 6-99.2 (101-5-2). *US Army Report and Message Formats.* 29 June 1999.

FM 7-0 (25-100). *Training the Force.* 15 Nov. 1988.

FM 7-15. *The Army Universal Task List.* TBP.

Index

Entries are by paragraph number unless page (p.), figure (fig.), or table (tbl.) is specified.

C

D

G

R

S

U–V

W–Z

TABLE OF CONTENTS

PART I. SCOUTING

PART I. SCOUTING

CHAPTER 1. FUNDAMENTALS OF SCOUTING

Scouting involves observing terrain and/or the enemy, and accurately reporting those observations. Scouting requires proficiency in the use of weapons, cover and concealment, route selection, and skill in unobserved day or night movement.

1001. PURPOSE

When an infantry unit is not actively fighting the enemy, it should be actively searching for the enemy, which is the primary purpose of scouting. The unit attempts to keep the enemy off balance while making preparations for further attacks. Physically locating and keeping the enemy off balance are normally accomplished by small units ranging from a two-person scouting party to a squad-size patrol.

Infantrymen are sent out as scouts or as members of a patrol because the commander needs information about the enemy, terrain, and the location of friendly troops. The lives of the entire unit may depend upon the success or failure of a scout or patrol and the accuracy and timeliness of the report. The success of the scout or patrol will depend upon their training, preparation by the commander, and understanding their mission and the commander's requirements.

To wage combat successfully, a commander must have accurate, detailed, and timely information about the enemy, the terrain, and adjacent friendly units. Well-trained scouts and capably led patrols are among the most effective means the commander has for acquiring the information necessary to plan tactical actions and make decisions in execution.

1002. REQUIRED SCOUTING SKILLS

To be effective, a scout must be able to—

- Recognize terrain features.
- Read a map and determine direction.
- Practice and implement the principles of cover and concealment.
- Fully utilize movement and route selection.
- Know the enemy (estimate enemy unit composition and strength).
- Observe and report information accurately.
- Select routes and move through numerous types of terrain.

CHAPTER 2. TERRAIN, MAPS, AND DIRECTION

A scout must understand map symbols, identify elevations from contour interval lines, scale distance on a map, relate natural and man-made features shown on the map to the actual features on the ground, plot a course from one point to another, and locate his current position. To relate a map to the actual terrain and its features, a scout must be able to orient it to the ground using a compass, two points, a watch, and the sun or the stars.

2001. TERRAIN FEATURES

Since the infantry works and fights on the ground, terrain ashore information that scouts gather and report on is of great importance. Hills, valleys, woods, and streams are the forms and growths commonly referred to as natural land features. Artificial or man-made features include houses, bridges, and railroads. Figure 2-1 shows some important terrain features.

2002. THE LENSATIC COMPASS

The best method of finding direction, during both day and night, is with a compass. The lensatic and M2 are the two types of compasses issued to infantry units. Both work on the same principle.

The standard compass for general use in the Marine Corps is the pivot-mounted lensatic compass, so called because azimuths are read through a magnifying lens in the eyepiece. Figure 2-2, on page 2-2, shows the lensatic compass and its nomenclature. The meter graphic scale on the side of the compass is 1:50,000, which is the most used scale in military mapping. The graphic scale is useful in the field as a straightedge, an aid in orienting the map, and a means of reading map grid coordinates. The plastic dial is graduated in both degrees and mils. Numbers on the dial are printed in black. There are luminous markings on the bezel, floating dial, and on both ends of the sighting wire, plus a 3-degree bezel serration and clicking device that permit reading azimuths at night. The compass is carried in a nylon case that may be attached to the cartridge belt.

Magnetic compasses are affected by the presence of iron or magnetic fields. Consequently, the scout should not be within the influence of local magnetic

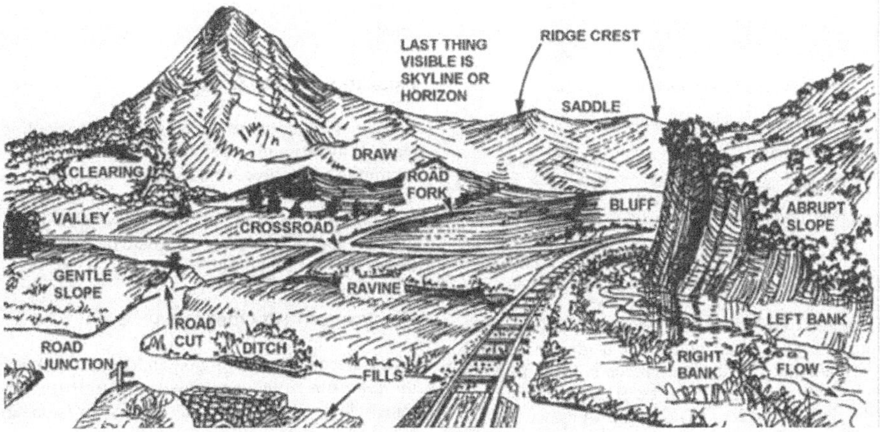

Figure 2-1. Natural and Artificial Terrain Features.

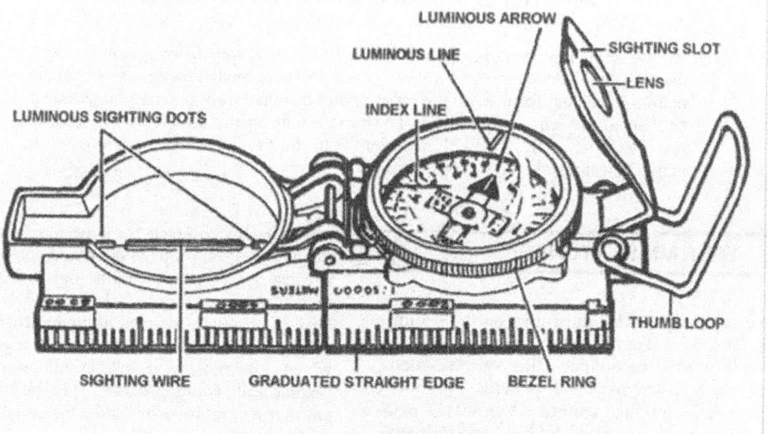

Figure 2-2. Lensatic Compass.

attraction while using a compass to determine direction. The rifle, pistol, and other metal objects must be laid aside when reading the compass. The minimum distances the scout should be from visible masses of iron and electrical fields of magnetism for the compass to provide accurate readings are provided below.

Magnetic Fields	Minimum Distances (meters)
High tension powerlines	60
Field guns	20
Vehicles (wheeled or tracked)	20
Telephone and telegraph wires	10
Barbed wire	10
Machine gun	3
Rifle, pistol	1

Reading an Azimuth

To read an azimuth to any point, the cover of the compass is raised to an angle of 90 degrees in relation to the index face, and the eyepiece is lifted to a 45-degree angle in relation to the bezel, or so the numbers on the dial can be seen. The thumb of either hand is placed in the thumb loop, the index finger extended along the side of the compass case, and the remainder of the hand closed. The closed hand and wrist are grasped with the other hand. The elbows are drawn in close to the body, forming a firm foundation for the compass. The eye is placed to the lens on the eyepiece. The compass is pointed at the object or point to which the azimuth is to be read. A sighting is taken through the sighting slot in the eyepiece, and the point is lined up with the sighting wire in the cover. The compass is held until the dial steadies; then the reading is taken through the lens of the eyepiece. This reading is the magnetic azimuth of the line from the observer to the point.

Reading a Back Azimuth

A back azimuth is the direction opposite the line of sight. If the azimuth is less than 180 degrees, the back azimuth is obtained by adding 180 degrees. If the azimuth is greater than 180 degrees, the back azimuth is obtained by subtracting 180 degrees. Back azimuths are used to determine a return route or to resection to determine a current position.

Circumventing Obstacles

When a scout is traveling on an azimuth and comes upon an obstacle—such as a contaminated area, minefield or swamp—the following steps (sometimes referred to as the *90-degree offset* method) is employed to go around or circumvent the obstacle and resume movement along the original azimuth (see fig. 2-3). The steps are as follows:

1. Move up to the obstacle and make a full 90-degree turn to the right (or left).
2. Walk beyond the obstacle, keeping track of the distance in paces or meters.
3. Stand at the end of the obstacle, face in the original direction of march, and follow that azimuth until the obstacle has been passed.
4. Make a 90-degree turn to the left (or right) and move the distance previously measured to return to the original line of march.

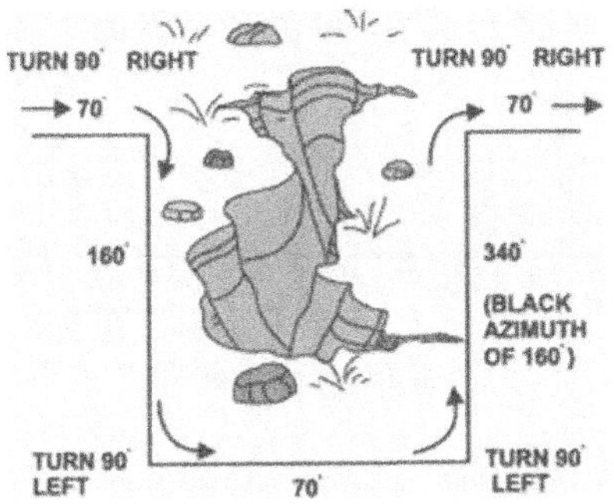

Figure 2-3. Circumventing an Obstacle.

Following an Azimuth During the Day

The eye is placed to the lens in the eyepiece and the compass moved until the desired azimuth reading is visible beneath the fixed index. Without moving the compass, the vision is shifted from the lens through the sighting slot in the eyepiece, and a sighting is taken out beyond the sighting wire in the cover. A prominent terrain feature on this line of sight is selected, the compass closed, and the landmark approached. When the scout reaches the landmark, the procedure is repeated.

Following an Azimuth at Night

It is necessary to prepare and set the compass before departing on a night movement because at night only the luminous parts of the compass can be seen. To prepare the compass for night use, the luminous parts must be fully charged by sunlight or artificial light, such as a flashlight. To set a compass—

1. Move the compass so that the desired azimuth on the dial is directly under the index line on the lower glass.
2. Rotate the upper movable glass so that the luminous line is directly above the north arrow of the dial.
3. Set the compass for marching at night on the specified azimuth.

—OR—

1. Face the general direction of movement.
2. Line up the north arrow and the luminous line on the bezel with the luminous sighting dots.
3. Hold the compass still with one hand and grip the knurled bezel ring with the other hand.
4. Turn the bezel ring the prescribed number of clicks in the proper direction, remembering that each click equals 3 degrees. For example, to set an azimuth of 21 degrees, the bezel ring would be turned seven clicks to the left.
5. Turn the whole compass until the north needle lines up with the luminous line. The compass is then set on the desired azimuth. The azimuth is the line formed by the two luminous sighting dots on the inside of the cover.

To march on a preset azimuth during night movement, open the compass and move it so the north arrow is directly below the luminous line (see fig. 2-4). Move

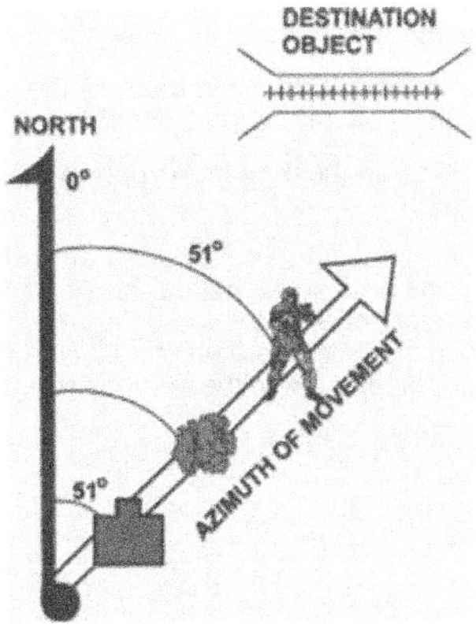

Figure 2-4. Following a Night Azimuth.

in the direction of the line formed by the two luminous sighting dots. It is necessary to refer to the compass more frequently at night than during the day. If stars are visible, find a prominent star along the azimuth of movement to use as a reference point. When the view of the sky is restricted by overcast conditions or vegetation, send a scout forward along the azimuth of movement to the limit of visibility. This scout is guided along the azimuth of movement by a stationary navigator. When the scout reaches the limit of visibility, the navigator moves to the scout's location. This process is repeated until the destination is reached.

A more rapid method for reaching the scout's destination is to equip the navigator with a compass. The navigator can set the compass as explained earlier and the scout proceeds providing security 180 degrees to the front on the specified azimuth, receiving right and left corrections from the navigator while both are on the move. The point scout must stay within visual range of the navigator. If available, a strip of white or luminous tape on the back of the point scout's helmet will assist.

Intersection

Intersection is the location of an unknown point by successively occupying at least two, preferably three known positions and sightings on the unknown point. It is used to locate features not depicted on the map or not readily identifiable. To determine an intersection, perform the following steps (see fig. 2-5):

1. Orient the map using the compass.
2. Locate and mark your position on the map.
3. Measure the magnetic azimuth to the unknown position; then convert to grid azimuth.
4. Draw a line on the map from your position on this grid azimuth.
5. Move to a second known position from which unknown point is visible. Locate this position on the map and again orient the map using the compass. The second unknown position should be a minimum of 30 degrees offset from the first position.
6. Repeat steps 4 and 5.

Figure 2-5. Intersection.

To check accuracy, move to a third position and repeat steps 1 thorough 4. Where the lines cross is the location of the unknown position. Using three lines, a triangle is sometimes formed—called the *triangle of error*—instead of an intersection. If the triangle is large, recheck your work to find the error. Do not assume that the position is at the center of the triangle.

Resection

Resection is the location of the user's unknown position by sighting on two or three known features that are identifiable on the map. To determine a resection, perform the following steps (see fig. 2-6):

1. Orient the map using the compass.
2. Locate two or three known positions on the ground and mark them on the map.
3. Measure the magnetic azimuth to a known position then convert to grid azimuth.

4. Change the grid azimuth to a back azimuth and draw a line on the map from the known position back toward the unknown position.

5. Repeat step 3 and step 4 to determine a second known position.

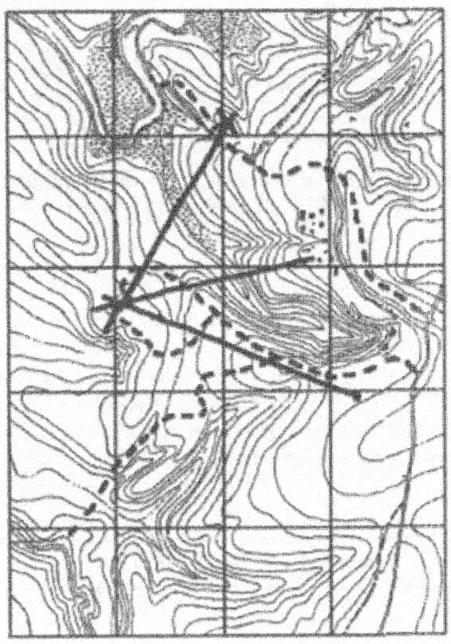

Figure 2-6. Resection.

To check accuracy, repeat the steps above for a third known position. The intersection of the lines is your location. Using three lines, a triangle of error may be formed. If the triangle is large, recheck.

2003. STARS, SUN, AND OTHER FEATURES

In rare cases when a scout is without a compass, the following examples are alternate means to determine direction. When using constellations to determine direction, identify your location's Temperate Zone. A Temperate Zone is the area between the tropics and the polar circles.

At night, the stars provide an excellent means of maintaining a line of march. In the North Temperate Zone (north of the equator), the Big Dipper constellation is one key to determining direction of true north. It is made up of seven fairly bright stars in the shape of a dipper with a long curved handle (see fig. 2-7). The two stars that form the side of the cup farthest from the handle, used as pointers, are situated in the direction of a bright star that is about five times the distance between the two stars of the dipper cup. This bright star is the North Star and is directly over the North Pole. The pointers always designate the North Star, which is the direction of true north.

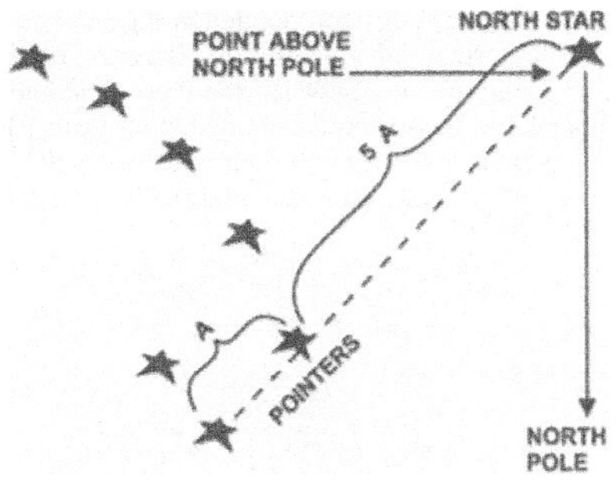

Figure 2-7. Locating the North Pole.

In the Southern Hemisphere, true south is determined in relation to the Southern Cross, a constellation composed of five stars. Two bright pointer stars in the vicinity of the Southern Cross serve as locators to help locate true south (see fig. 2-8). The outer four stars are

Figure 2-8. Locating the South Pole.

fairly bright and form a cross. This cross is imagined as the frame of a kite. A straight tail, four and one half times as long as the length of the kite itself, is put on the kite using finger widths for a measuring stick. The end of this tail will be close to a position directly over the South Pole. Usually, it will not be possible to see a star in the immediate vicinity, because there is no bright star visible directly above the South Pole.

During daylight hours, a watch and the sun can be used to determine direction within 8 degrees. In the North Temperate Zone, the watch is held horizontally, face up, and the hour hand pointed at the sun (see fig. 2-9). The north-south line and the direction of south can be found midway between the hour hand and the number 12, if the watch is set on standard time. If in daylight savings time, the direction of south is found midway between the hour hand and the number 1.

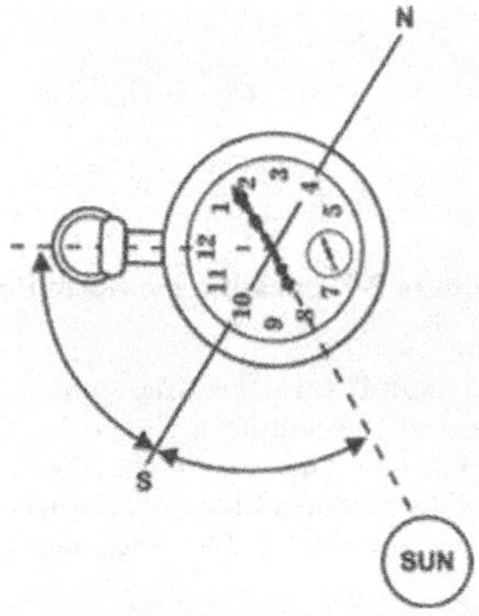

Figure 2-9. Determining Direction by Watch and Sun (North Temperate Zone).

In the South Temperate Zone, if the watch is set on standard time, the number 12 on the watch is pointed at the sun; if the watch is set on daylight savings time, the number 1 is pointed at the sun. North is midway between 12 (or 1) and the hour hand (see fig. 2-10).

When laying in a north-south line, if any doubt exists as to which end of the line is north, remember that the sun is in the east before noon and in the west in the afternoon.

In addition to the sun and stars, other methods a scout without a compass can use to determine direction include determining prevailing wind direction and

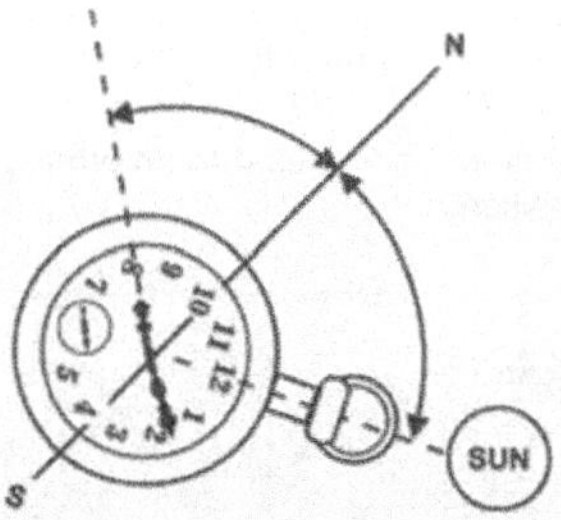

Figure 2-10. Determining Direction by Watch and Sun (South Temperate Zone).

using a mountain for orientation. By previous study of maps and photographs, a scout can keep informed of location and direction by using a distinctive edge of woods, a deep ravine or the direction of a stream's flow. A scout should constantly evaluate and memorize both the immediate terrain and general area for prominent features and landmarks.

2004. RANGE DETERMINATION

Range determination is the method of finding the distance between an observer and an enemy target or an object. By accurate range determination, the members of a given unit can set their sights correctly and place effective fire on enemy targets. The degree of accuracy is dependent on several factors, such as terrain relief, time available, and experience of the observer.

Mental Estimation

A mental distance estimate is made using a known unit of measure. Distance is estimated to the nearest 100 meters by determining the number of known units of measure between the observer's position and a target. For example, a football field, which is 100 yards, can be used as a known unit of measure for determining the distance between an observer's position and a target. For longer distances, progressive estimation may be necessary. To do this, the observer determines the number of units of measure to an intermediate point and doubles the value. The observer should consider the effects in table 2-1 in estimating distances.

Table 2-1. Effects to Consider in Mental Estimation of Distances.

Objects Appear Nearer	Objects Appear More Distant
In bright light.	In poor light or in fog.
In clear air at high altitude.	Only a small part of the object can be seen.
The background is in contrast with the color of the object.	The background is similar in color to that of the object.
The observer is looking down from a height.	The observer is looking over a depression, most of which is visible.
The observer is looking over a depression, most of which is hidden.	The observer is kneeling or sitting, especially on a hot day, when the ground is moist.
The observer is looking down a straight feature such as a road.	
The observer is looking over water, snow, or a uniform surface such as a cultivated field or desert.	

Estimating in Good Visibility

When visibility is good, distances can be estimated by using the appearance of tree trunks, branches, and foliage (as seen by the naked eye) in comparison with map data. Table 2-2 is a guide for wooded terrain. Table 2-3 is a guide for urban environments.

Estimating From a Terrain Study

The Marine should always use terrain/map analysis to assist in estimating distances. When the Marine is looking in a specific direction, the estimation of distance can be enhanced by studying the terrain and comparing it with the map. Particular emphasis should be given to color contrasts of terrain features seen along the observer-target line (OTL). For example, the distance across successive ridge lines or depressions in the distance may be identifiable by only slight changes of color to the eye. Different colors of grass might reveal a hidden terrain feature such as a stream.

Table 2-2. Estimating Distance in Wooded Terrain.

Distance in Meters	Tree Description
1,000	Trunk and main branches are visible. Foliage appears in cluster-like shape. Daylight may be seen through the foliage.
2,000	Trunk visible, main branches distinguishable, foliage appears as smooth surface. Outline of foliage of separate trees distinguishable.
3,000	Lower half of trunk visible. Branches blend with foliage. Foliage blends with adjoining trees.
4,000	Trunk and branches blend with foliage and appears as a continuous cluster, smooth in appearance. Movement of foliage due to wind cannot be deleted.
5,000 and beyond	Whole area covered by trees and appears smooth and dark.

Table 2-3. Estimating Distance in Urban Terrain.

Distance in Meters	Object Identified by the Unaided Eye
1,000	Lone tree trunk
1,500	Individuals and horsemen
3,000	Chimneys on rooftops
4,000	Windows in houses
4,000–5,000	Individual houses in populated area
8,000–9,000	Villages and individual houses
15,000–18,000	Large houses, towers, and steeples

CHAPTER 3. ENEMY ACTIVITY

A commander often acts on information furnished by scouts. Therefore, scouts must aim at absolute accuracy in reporting enemy activity. This chapter discusses estimating enemy strengths, interpreting signs and tracks, and knowing the enemy.

3001. ESTIMATING ENEMY STRENGTHS

If troops cannot be counted, their strength may be estimated by: noting the length of time it takes various types of moving columns to pass given point, the area required of a unit in camp or bivouac, or the front on which they are deployed. When the ground is dry, infantry on the march raise a low, thick cloud of dust, and motor vehicles or mechanized units raise a thick, rapidly moving cloud. Additionally, through practice, a scout may gain information as to the strength and composition of enemy forces by listening to noises and observing lights, fires, and smoke. A scout gains valuable experience in estimating enemy strengths by observing friendly forces in camp, on the march, and deployed. The knowledge scouts gain during field exercises of the appearance and tactical dispositions of squads, platoons, companies, and larger units will be of great assistance in estimating the strength and composition of enemy units observed under various conditions.

3002. INTERPRETING SIGNS AND TRACKS

In addition to estimates made through direct observation, a scout may often be able to estimate size, composition, direction, rate of movement, condition, discipline, state of training, and morale of enemy forces through signs and tracks left behind.

Signs

The examination of vacated enemy positions provides valuable information. The size of a bivouac or defense area ordinarily indicates the number of enemy occupants. Clothing, ration containers, dumps, etc., further indicate the quantity of the departed enemy force. The condition of the bivouac area and amount of material abandoned give an indication of the enemy morale, training, and discipline. A well-policed area indicates good discipline. Rubbish, ration and smoking residue, and nonessential personal items of equipment adrift indicate a lower state of morale, training, and discipline. Stores and material left behind in good condition may indicate a hasty movement or withdrawal. Burned or destroyed materials indicate a deliberate, orderly withdrawal or movement. Letters, insignia, and other articles may reveal the identity of the enemy unit.

In the case of a moving enemy, the distance between periodic halts indicates the rate of march if enemy habits relative to marches and halts are known. Condition of the halt areas indicates the state of morale, training, and discipline.

The physical condition of enemy dead and wounded and their personal equipment and weapons are reported. The general condition and state of maintenance of destroyed or abandoned vehicles should also be reported.

Tracks

A track is a mark left on the ground by the passage of a person or object. Examination of tracks reveals information about the enemy.

Troops

A few tracks overlapping each other on both sides of a road or trail may indicate a patrol in staggered formation. A large number of tracks indicates troops in column formation. A large column will wear a dry road smooth and flat. In damp terrain, a freshly made track will have sharp edges; ordinarily, signs of moisture will disappear in about 15 minutes. A runner's toes are dug into the ground; a walker's footprint is fairly even.

Vehicles

The type of track indicates whether the vehicle is wheeled or tracked. A scout acquires the necessary experience to make the proper determination by observing vehicle tracks during training.

The direction of travel can be determined by the way tracks pass across ruts, by impressions on the edges of holes in the ground, how water is splashed from puddles, or by the way grass, twigs, and branches are bent; for example—

- A vehicle (wheeled or tracked) entering a rut pushes dirt into the rut and leaves an indentation on the exit side of the rut.
- A wheel going over holes in the ground leaves a deeper impression on the edge toward the direction of travel.
- The side of a puddle with the greater splash indicates the direction of travel of the vehicle.
- When traveling cross-country, the direction in which grass is bent and/or twigs, branches, and bushes are broken indicates direction of travel.

A general rate of speed can be estimated by the amount of water or mud splattered. A fast-moving vehicle will throw larger amounts of water or mud a greater distance to the front and sides than a slow-moving vehicle, and it leaves a deeper impression on the exit edges of holes. The faster the travel, the deeper the impression.

3003. KNOWING THE ENEMY

A scout should learn as much as possible about enemy psychology, habits, organization, and tactics. The more knowledge gained about the enemy, particularly the enemy's normal security measures, the better the scout's chances are to observe and obtain accurate information with minimum risk to the mission's success. Scouts gain much of this knowledge through experience, but they also gain a great deal of their preliminary information, particularly that pertaining to enemy organization and tactics, during training and may be updated by unit commanders and intelligence officers.

CHAPTER 4. DAYLIGHT SCOUTING

A scout must be able to operate in all types of terrain and under all conditions of visibility. He must be thoroughly familiar with the principles for using cover and concealment, camouflage, individual movement, and route selection, both to and from the objective.

4001. COVER AND CONCEALMENT

Cover is protection from the fire of hostile weapons. Concealment is protection from observation or surveillance from hostile air and ground observation, but not from hostile fire. Both cover and concealment are divided into two main categories: natural and artificial. Natural cover includes small hills, ditches, rocks or vegetation. Fighting holes, bunkers, and brick walls are examples of artificial cover. Some features, such as buildings, provide both cover and concealment. In deciding whether to seek cover or concealment, a scout must make the best choice to complete the mission (see fig. 4-1).

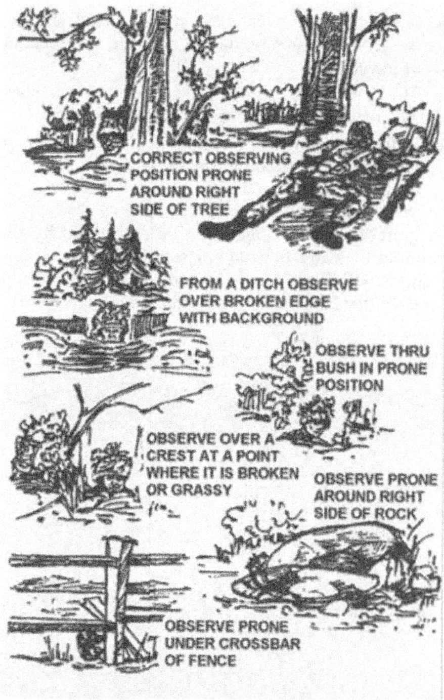

Figure 4-1. Correct Use of Cover.

Concealment Principles

Concealment principles are as follows:

- Remain motionless while observing. Anything in motion attracts the eye.
- Use all available concealment.
- Observe from the prone position (it offers a low silhouette and makes detection by the enemy difficult.
- Expose nothing that reflects light.
- Blend with the background because contrasting colors are noticeable.
- Remain in the shade because moving shadows attract attention.
- Distort or change the regular outline of objects. Most military objects have distinctive shapes that make obvious shadows and silhouettes.
- Avoid the skyline. Figures on the skyline can be seen from great distances and are easily identified by their outlines.

Concealment Techniques

Concealment techniques are as follows:

- When observing, the scout looks around an object's side (unless it is transparent) and prepares to fire, if necessary, around the side of or, if possible, through an object.
- Looking or firing over an object can make the scout an easily visible target for the enemy. If the scout must fire over the top of concealment or cover, the outline of the head or helmet should be broken or distorted.
- Upon the approach of an airplane, the scout takes a prone position, turns face-down, and remains motionless. If surprised by an airplane, the scout remains in place and does not look up.

- The scout covers exposed body parts such as the face, back of the neck, and hands with grease paint, mud or other materials to reduce sun reflection.
- Camouflage for equipment can be improvised from garnishing or sandbags to prevent sun reflection.
- In snowy terrain, white overgarments are worn.
- The helmet cover outline should be distorted.

4002. CAMOUFLAGE

Camouflage is the use of concealment and disguise to minimize the possibility of detection and/or identification of troops, material, equipment, and installations. The purpose of camouflage is to provide concealment of military objects from enemy observation. Camouflage is also used to conceal an object by making it look like something else. A scout's mission usually requires individual and equipment camouflage. If natural camouflage is not adequate, the position is camouflaged. In using camouflage, remember that objects are identified by their form (outline), shadow, texture, and color. The principal purpose of camouflage in the field is to prevent direct observation and recognition.

Individual Camouflage

Successful individual camouflage involves the ability to recognize and take advantage of all forms of natural and artificial concealment available (vegetation, soil, debris, etc.) and knowledge of the proper use of artificial camouflage materials.

Aids to Individual Camouflage

A scout must recognize the terrain's dominant color and pattern and must change the appearance of clothing and equipment accordingly in order to blend and not contrast with the terrain (see fig. 4-2).

The helmet is camouflaged by breaking up its shape, smooth surface, and shadow. Use of a helmet cover works best. In the absence of a helmet cover, mud can be irregularly blotched on the helmet to disguise its form and dull the surface. A helmet cover may be improvised from irregularly colored cloth or burlap to blend with the background. Foliage can be draped to prevent the visor of the helmet from casting a dark shadow across the face. Foliage should not stick up like plumes because any head movement will give away the position.

A small, thin bush in the shadow of a large bush makes a good observation point. Lone trees, rocks, fence corners, and outstanding landmarks are easily picked up by the enemy as obvious observation posts.

If camouflage clothing is not available, other available clothing can be attached in irregular splotches of appropriate colors.

Exposed skin reflects light and attracts the enemy's attention. Even very dark skin will reflect light because of its natural oil. The buddy system is recom-mended when applying camouflage. Standard Marine Corps issue camouflage face paint sticks are two toned:

- Loam and light green for light-skinned troops, in all but snow regions.
- Sand and light green for dark-skinned troops.
- Loam and white for troops in snow-covered terrain.

Shiny areas (forehead, cheekbones, nose, and chin) are painted with a dark color. Shadow areas (around the eyes, under the nose, and under the chin) are painted

Figure 4-2. Avoid Contrasting Backgrounds.

with a light color. Skin that is exposed on the back of the neck and hands is painted with a two-color combination in an irregular pattern (see fig. 4-3).

When standard issue face paint sticks are not available, burnt cork, charcoal or lamp black can be used to tone down exposed areas of skin.

Mud is used only in an emergency because it changes color as it dries and may peel off, leaving the skin exposed. Since mud may contain harmful bacteria, mud should be washed off as soon as possible.

Any equipment that reflects light should be covered with a nonreflective material that aids in the concealment of the weapon (for example, black electrical tape or mud). The straight line of the rifle or other infantry weapons may be very conspicuous to an enemy observer. The barrel and hand guard should be wrapped with strips of contrasting colored cloth or

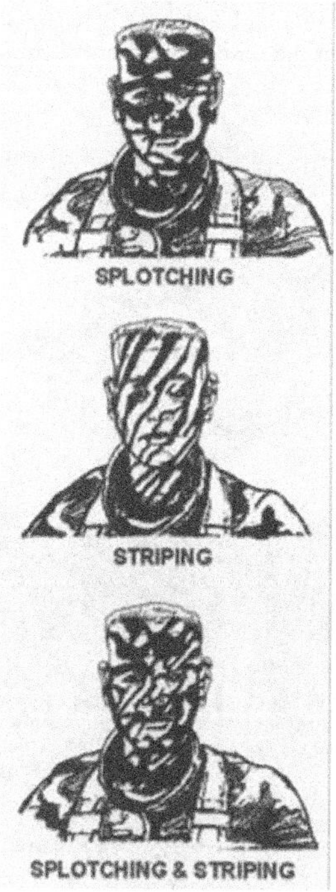

Figure 4-3. Face Camouflage.

tape to break the regular outline. Mud or dirt dulls the reflecting surface of the stock, barrel, and bayonet where coloring has been worn. Lamp black may also be used on metal parts. The function of the weapon must not impaired.

If time, material, and surroundings permit, a ghillie suit should be constructed. (Refer to MCWP 3-15.3, *Scout Sniping*.)

Aids to Camouflage a Position

To successfully camouflage a position, the scout must remember to—

- Camouflage the position as soon as it is occupied.
- Avoid using too much material for camouflage. Even though natural materials are used, too much may make the object and its shadow stand out from its surroundings, thus attracting the attention of a hostile observer.
- Inspect completed camouflage work from the enemy's point of view to check effectiveness.

Continuous Camouflage

Camouflage around and on the scout's position must be maintained in a fresh condition as wilted and dead foliage can give the position away. If the mission dictates that the position should be occupied for longer periods, wilted foliage should be replaced during periods of reduced visibility.

4003. INDIVIDUAL MOVEMENT

Principles

The principles of individual movement are as follows:

- Scouts move from one concealed position to another. When not changing positions, they remain motionless.
- The scout's head is lifted slowly but steadily, without abrupt movements, to search for a new position.
- Scouts select the next stopping place before moving and ensures it is not contained by the enemy.
- Scouts change position on the run: springs up, runs with the body bent low, zigzags, quickly drops to the ground slightly to the right or left of the objective, then rolls or crawls to the desired position.

(Remember the phrase, "I'm up—he sees me—I'm down.")

Rushing

When starting from the prone position—

- Raise the head slowly and steadily and select a new position.
- Lower the head slowly, draw arms inward, cock right leg forward, and prepare to rush.
- Use one movement to raise the body by straightening both arms.
- Spring to your feet, stepping off with the left foot.
- Bend forward as low as possible when running. Never advance directly to the next position; always zigzag.

When hitting the deck—

- Stop.
- Plant both feet in place.
- Drop quickly to the knees and slide the hand to the heel of the rifle.
- Fall forward, breaking your fall with the butt of the rifle. (To confuse the enemy, roll over after hitting the deck and roll into firing position with feet, knees, and stomach flat on the ground.)
- Keep head down if you do not intend to fire.

When rolling over—

- Hit the deck and assume the prone position.
- Bring the rifle in close to the body, placing the rifle butt in the crotch.
- Roll over swiftly to confuse any enemy observers as to final intended location. Never reappear at the same place you went down.

Low Crawl

The low crawl is used when—

- Cover and concealment are scarce.
- The enemy has good observation over the area in which the scout is moving.
- Speed is not essential.

To perform the low crawl, keep the body as flat as possible against the ground. Grasp the rifle sling at the upper sling swivel. Let the balance of the rifle rest on the forearm and let the butt of the rifle drag on the ground. Keep the muzzle off the ground.

To start forward, push arms forward and pull right leg forward. To move forward, pull with arms and push with right leg. Change the pushing leg frequently to avoid fatigue.

High Crawl

The high crawl is used when—

- Cover and/or concealment are available.
- Poor visibility reduces enemy observation
- Greater speed of movement is required.

To perform the high crawl, keep body off the ground. Rest weight on forearms and lower legs. Cradle rifle in arms, keeping the muzzle off the ground. Keep knees well behind the buttocks to stay low.

Move forward, alternately advancing right forearm and left knee; then left forearm and right knee.

Movement Aids

Aids to movement include—

- Carrying only necessities. Additional weight causes premature fatigue and impedes free movement.
- Not disturbing birds or animals whose flight would betray your presence. If birds or animals are alerted, remain motionless under cover for a few minutes, as the enemy's attention may also be attracted.
- Moving during an incident that diverts attention, such as an airplane flight, a distant disturbance or sudden bursts of fire.
- Fog, smoke, or even light haze offer concealment for movement; however, the enemy may have thermoimagery and night vision devices. Therefore, darkness and smoke cannot be used as easily.
- Following a stream or road by staying as far away from them as possible while still keeping them in sight. Keep close to the dune line when moving along a beach.
- When moving through tall grass or similar growth, move when the wind blows, changing direction frequently. A straight route will be more readily noticed.
- Whenever possible, avoid areas of soft ground so as not to leave tracks.

- When crossing a road or water obstacle, choose crossing sites where the enemy's observation is restricted (an area in shadows or near a bend) and cross rapidly.

4004. ROUTE SELECTION

Prior to Movement

A scout and the immediate commander conduct a map reconnaissance before starting on a mission. This assists them in selecting the route according to available cover and concealment and any indicated enemy activity.

Prior to and during the course of the mission, move to an observation point to visually reconnoiter the terrain for movement and select the tentative route. It may be necessary to make wide detours around open spaces or those containing enemy patrols or other enemy activity.

Carefully study the country to be traversed and pay close attention to the general features, streams, ridges to be crossed, and their relation to the general direction to be taken (see fig. 4-4).

Make notes of terrain features and landmarks along the proposed route and rely on notes for guidance (see fig. 4-5 on page 4-6). Additionally, determine the compass direction and readings for each change of direction at the start. Finally, learn the location of unit boundaries and observation/listening posts as well as general location of other friendly or scouting parties. Be sure to avoid man-made and natural obstacles as they will slow progress and overall success of the mission. If possible, use the local populous as a source of intelligence. When returning to friendly lines, avoid using the same route.

En Route

En route, the actual advance will be a series of movements from one observation point to the next. The distance and route will depend on cover and terrain. Assess the cover, terrain, and any enemy or civilian activity to determine whether or not to modify the approach or return routes. Unless the mission requires it, avoid danger areas (for example, houses, villages, potential assembly or bivouac areas, roads, and streams) that may give away your position by being

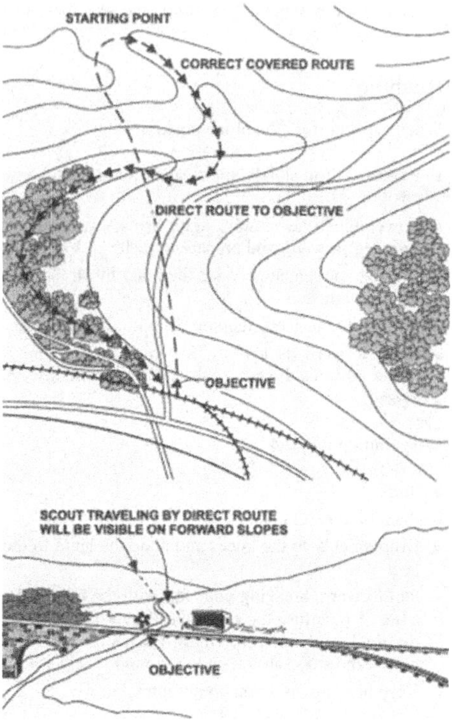

Figure 4-4. Choosing a Concealed Route of Advance from a Map.

observed by the enemy. When required to reconnoiter danger areas, choose a covered approach and return, and make entry or passage as quietly and quickly as possible. If part of a larger effort, the approach and return should be covered by observation and fires of the other members of the scouting party or patrol.

Stream Crossings

When the crossing does not appear to be held by the enemy, advance upon it rapidly. If there are two or more scouts, one crosses while the other(s) provide protection. Note the length, width, depth, and approaches to a crossing. Observe the condition of the road or trail that crosses the stream, and report on the suitability of the crossing for use by tracked and wheeled vehicles. If the crossing is under observation by enemy, seek another crossing site or dash across to avoid detection.

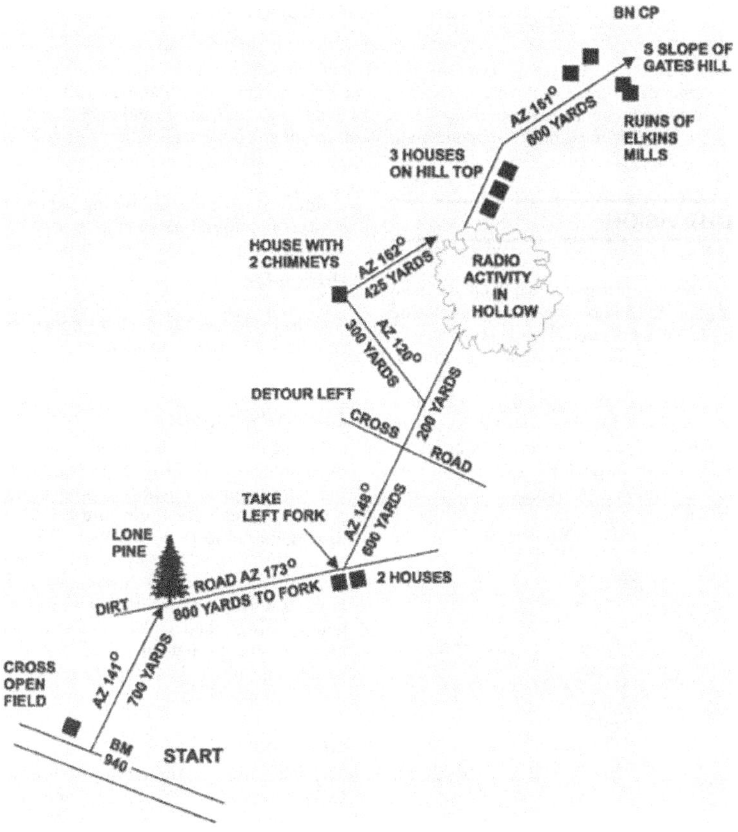

Figure 4-5. Proposed Route Sketch.

CHAPTER 5. NIGHT SCOUTING

Night scouting presents many of the same problems encountered in day operations-such as cover, concealment, movement, and camouflage-as well as additional considerations. Knowledge of human eye construction and operation will enable maximum advantage under night conditions or poor visibility.

5001. NIGHT VISION

Certain parts of the eye correspond to parts of a simple camera (see fig. 5-1). The lens focuses light entering the eye similar to a camera lens. The iris (colored part of eye) corresponds to the diaphragm of a camera, opening and closing to regulate the amount of light entering the eye through the pupil. The retina corresponds to camera film. Light rays strike the retina, form an image, and cause an impression to be transmitted to the brain through the optic nerve. In a camera, the image is formed and fixed on film.

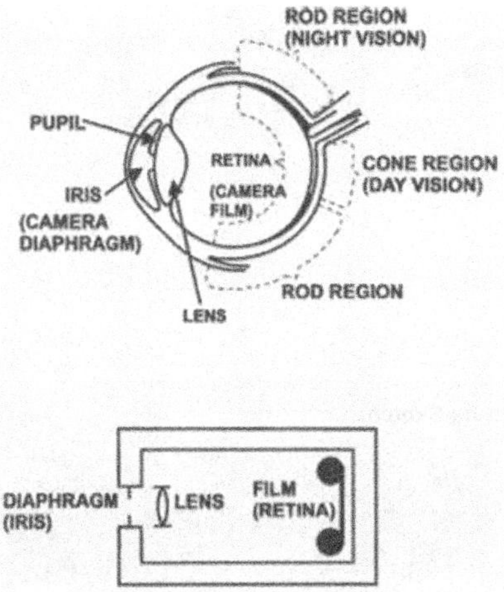

Figure 5-1. The Eye is Like a Camera.

The retina is composed of cone cells and rod cells, so-called because of their shapes. Cone cells distinguish color, shape, and sharp contrast. Because they are activated by light conditions, they are blind during periods of low illumination. Rod cells produce a chemical substance called visual purple that makes them active in darkness, low illumination or night conditions. Rod vision distinguishes black, white, shades of gray, and general outlines.

Principles

To effectively "see" at night, the principles of night vision dark adaptation, off-center vision, and scanning are applied.

Dark Adaptation

Allowing the eyes to become accustomed to low levels of illumination is called dark adaptation. It takes the rod cells about 30 minutes to produce enough visual purple to activate them and enable the eye to distinguish objects in dim light. This may also be accomplished by staying in a red-lighted area, or by wearing red goggles for 20 minutes, followed by 10 minutes in darkness (which allows the pupils to open wide). This method saves valuable time by allowing Marines to be in a lighted area to receive orders, check equipment, or perform some other function before moving into darkness.

Off-Center Vision

The technique of focusing on an object without looking directly at it is called off-center vision. When looking directly at an object, the image is formed on the cone region, which is not sensitive at night (see fig. 5-2 on page 5-2). When looking slightly to the left, right, above or below an object, the image is formed on the area of the retina containing rod cells, which are sensitive in darkness. The most sensitive area varies in individuals, but is usually found by looking 6 to 10 degrees away from an object; in effect, out of the corner of the eye (see fig. 5-3 on page 5-2).

Scanning

Off-center vision used to observe an area or an object is called scanning. When using rod vision, the visual purple in the rod cells bleaches or blacks out in 4 to 10 seconds and the object observed disappears. As the visual purple in the rod cells in one area bleaches out, the eyes must slightly shift to use fresh rod cells. Eyes

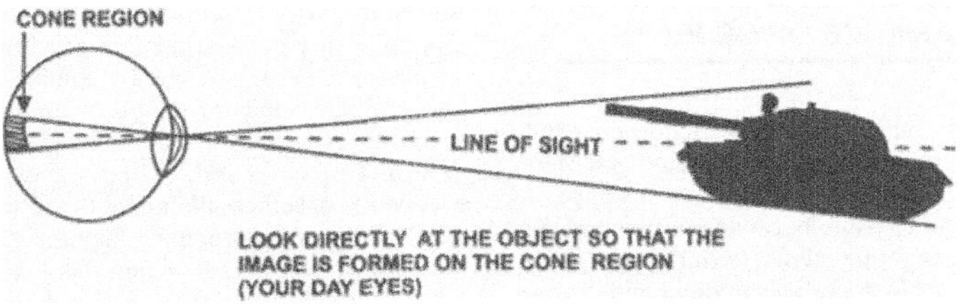

Figure 5-2. Day Vision.

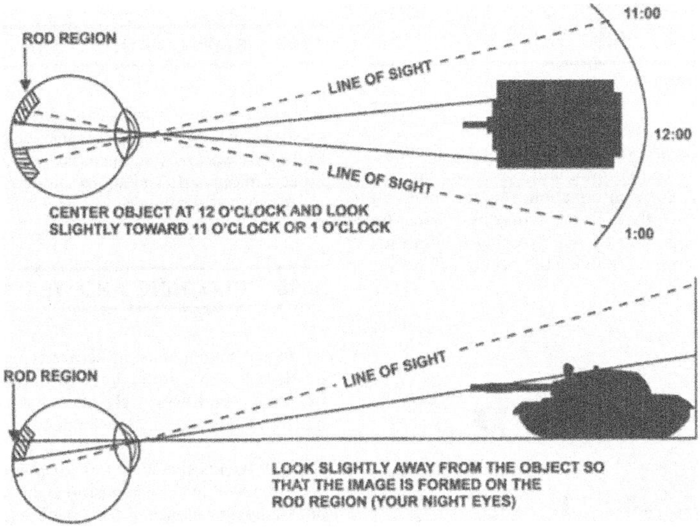

Figure 5-3. Night Vision.

should be moved in short, abrupt, irregular movements over and around the target (see fig. 5-4).

Preserving Night Vision

Night vision is quickly destroyed if bright light is allowed to enter the eye. When entering a lighted area or when observing in a temporarily lighted area (illumination, flares), one eye should be closed and covered to preserve its night vision. When the light goes off, fades or the lighted area is exited, the night vision retained by the protected eye enables it to see until the other eye adapts to the darkness. Red light helps preserve night vision, but like white light, it can be observed at long distances.

Factors that decrease night visual acuity include fatigue, lack of oxygen, long exposure to sunlight, alcohol, nicotine (within the past 48 hours), and age. When night vision has been attained, straining will not improve effectiveness; however, practice identifying objects at night will improve perception.

5002. APPEARANCE OF OBJECTS

Darkness not only makes objects difficult to see but also changes their appearance, distorts size, and blots out details. A tree visible against the night sky appears smaller than in the daytime because the twigs at the end of branches cannot be seen. A scout must train to identify objects by block outlines at night and cannot rely on details visible in daylight. Binoculars enlarge objects or parts of objects otherwise too small to be seen and help identify objects already spotted. Night observation devices increase night visibility and should be used whenever possible.

5003. SOUNDS

At night, sounds become very important. By listening, a scout gains information about the enemy and by exercising care, keeps information from the enemy. A scout stops frequently to listen. Scouts must listen for long periods in perfect silence. Hearing is amplified with the mouth open; removing the helmet will reduce sound distortion. Sounds are transmitted a greater distance in wet weather and at night than in dry weather and in the daytime. By holding the ear close to the ground sounds of people walking and vehicles moving can be heard. Sound travels approximately 370 meters a second. When a flash from a fired weapon is observed, the range to the weapons can be easily estimated by counting the time interval between the flash and hearing the report. For example, counting to three (one thousand one, one thousand two, one thousand three), indicates the distance is 1,110 meters. The cadence is determined by actual practice at known ranges.

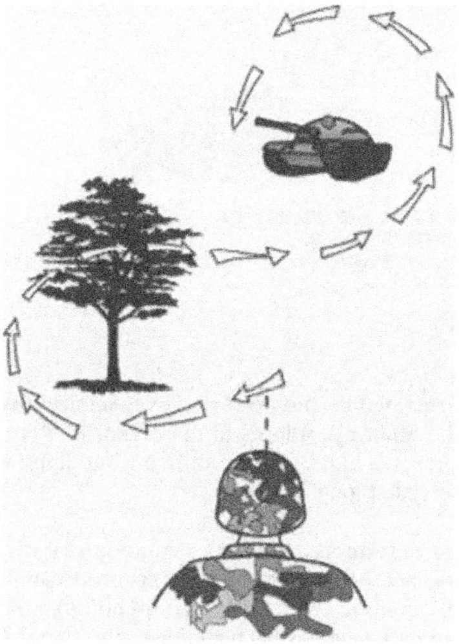

Figure 5-4. Scanning

5004. SMELLS AND TOUCH

A scout's sense of smell can warn of enemy fires, cooking, motor parks, gasoline and diesel engines, and bodies of water. A scout must feel and recognize objects in the dark, and adjust and operate equipment quietly by sense of touch.

5005. CLOTHING AND WEAPONS

All loose clothing must be secured (string or tape can be used) to prevent snagging on barbed wire, brambles, and brush. Helmet covers are worn to muffle sounds made by low branches.

The belt buckle should be turned around to the side in order to move in a prone position without scraping the buckle against stones or hard surfaces. Identification tags can be taped together to prevent rattling. Hands, face, and neck can be blackened so skin does not reflect light or appear as white spots in the darkness. (Refer to para. 4002.)

Scouts are normally armed with rifles. Rifle slings should be taped to prevent rattling. All weapons parts should be checked for glare elimination measures.

5006. CONCEALMENT

Although total darkness provides concealment, scouts must observe the same principles of concealment during moonlight conditions as in the daytime. Scouts should assume enemy employment of night observation devices and observe the principles of night movement such that presence will not be disclosed by noise when close to the enemy.

5007. AIDS TO NIGHT SCOUTING

Aids to night scouting include the following:

- Carry out scouting missions close to or within hostile positions on dark or rainy nights.
- Stifle a threatening sneeze by pressing fingers upward against the nostrils.
- Stifle a threatening cough by applying slight pressure with the finger on the Adam's apple.
- Stop a ringing sound that interferes with hearing by yawning.
- Speak softly rather than whispering if voice communication is necessary.
- Move boldly and rapidly when taking advantage of any sound-such as shelling, rustling wind or distant firing-to push forward if firing is taking place.
- Avoid shell craters and depressions in damp weather conditions if the enemy has employed chemical munitions.
- Move the eyes constantly; concentrating on one object too long will strain them.
- Take notice of the enemy's use of flares. When the enemy employs flares, few enemy patrols are apt to be out; when flares are not employed, the enemy's patrols are likely to be numerous.
- Drop to a prone position upon hearing a flare being fired and before it illuminates. Remain motionless while it is burning. If open or moving when a flare bursts in the air, freeze or drop quickly in the split second after the flare illuminates while the enemy is blinded. You are an easy target for the enemy if the flare bursts in the air or on the ground behind you. Never look at a flare. If you activate a trip flare, drop to the ground and crawl away from the illuminated area.
- Consider all patrols or individuals encountered as hostile until proven friendly. If encountering someone, crouch low, silhouetting the approaching person against the sky. At the same time, make yourself an indistinct target in case the person encountered is an enemy.
- Return fire only to avoid capture if fired on when close to enemy positions.

5008. AIDS TO NIGHT MOVEMENT

- Aids to night movement include the following:
- Move silently.
- Advance in stealthy legs. Each leg should follow some terrain feature that serves as a guide. When there are no terrain features to serve as guides, move in a straight or nearly straight line from one defined point to another, or maintain direction by using a compass.
- Avoid running, except in an emergency.
- Take advantage of sounds that may distract the enemy.
- Fall silently without making an outcry.

Walking

When walking at night—

- Place the heel down first. Balance the weight of the body on the rear foot until a secure spot is found.
- Lift the forward foot high to clear any stiff grass, brush, or other obstruction.
- Continue to balance body weight on the rear foot, lower the forward foot gently, toe first, to explore the ground for objects that might make noise. Step over fallen logs and branches, not on them.
- Lower the heel of the forward foot slowly; gradually transferring body weight to that foot.

Creeping

The low crawl and high crawl are not suitable at night when very near the enemy because an easily heard shuffling noise results. Creeping is the recommended method of movement:

- Creep at night on the hands and knees.
- Use your hands to feel for twigs, leaves or other substances that might make a noise. Clear a spot to place your knee. Keeping your hand at that spot,

bring your knee forward until it meets your hand. Then place your knee on the ground and repeat the action with the other hand and knee.

- Lay the rifle on the ground at your side and clear an area for it. Lift the rifle up and move it forward. Movement is slow and tedious, since it must be done silently.

Hitting the Deck at Night (Right-Handed Shooter)

To safely hit the deck at night from the standing position—

- Advance your left leg, place the butt of the rifle in your right armpit with the hand remaining on the pistol grip, and grasp it with the right hand at the balance.
- Quietly drop down on the right knee and left hand.

- Move the left leg carefully to the rear, and then move the right leg to the rear.

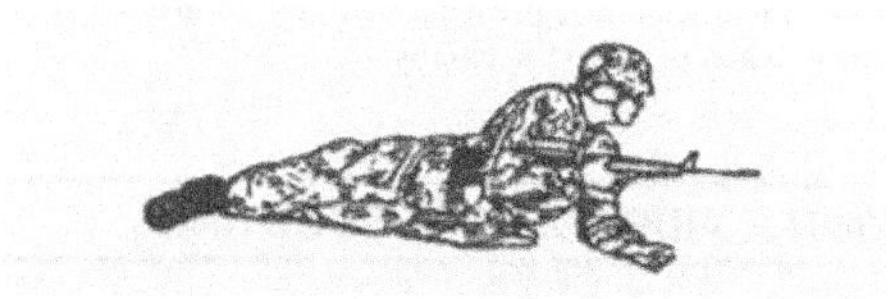

- Lie flat on the ground, or take up a firing position if necessary.

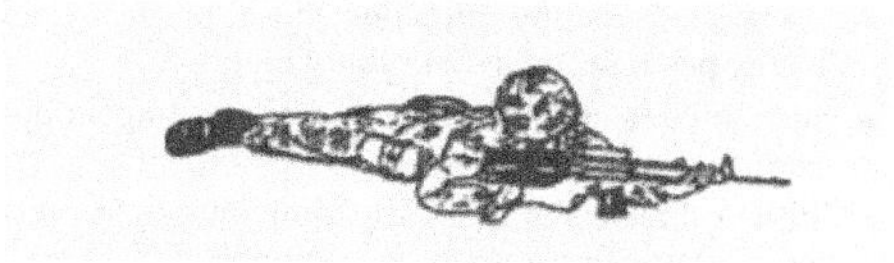

Wire Obstacles

A mission often requires a scout to pass through and work behind enemy positions. To accomplish this, the scout must be able to quietly pass through enemy wire obstacles and cross trenches. Cutting a gap in wire is time-consuming. If possible, walk over the low bands of enemy wire and crawl under the high bands (see fig. 5-5). Avoid movement along wire barriers, as enemy

Figure 5-5. Crossing Wire Silently at Night.

covering fires are generally planned parallel to them to take advantage of canalization and enfilade fire.

To step over low wire at night, crouch low to view the strands against the sky. Grasp the top strand with one hand; with the other hand, reach forward and feel for a clear spot for foot placement without stepping on other strands or any object apt to make a noise. Raise the body up, still grasping the top strand of wire. To avoid catching the foot in another strand, lift the foot up and over, passing it close to the hand grasping the wire.

If a high wire obstacle is encountered at night and wire cutters are not available, pass under the wire with your back on the ground. Grasp the lowest strands in your hands and hold them clear of the body while you slide under them.

When cutting wire and working solo, cut a wire near a post (see fig. 5-6), then dispose of all but one loose end. Grasp the wire close to a post and cut between your hand and the post, muffling the sound and keeping the loose wire in your grasp. When cutting wire in tandem, one firmly holds the wire with the hands positioned close to the cutters, in order to muffle the sound and prevent the loose ends from flying back, while the other one cuts. In both instances, the loose ends of the wire are bent back to form a passage.

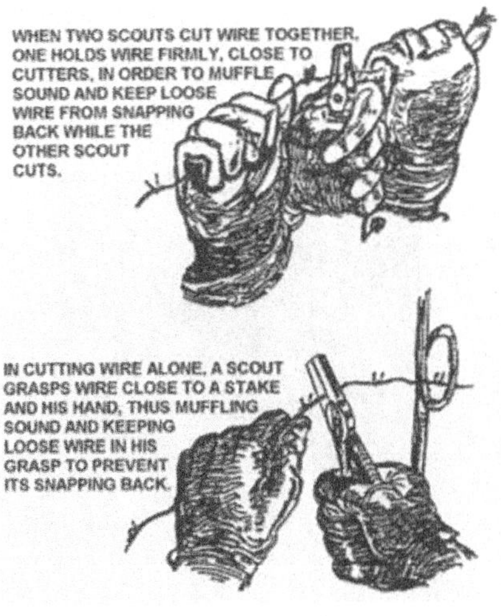

Figure 5-6. Cutting Wire Silently at Night.

Wrap a sandbag around the wire cutters and wire to deaden the sound.

Do not cut a complete gap in the wire; cut only the bottom wire(s). Leave the top wire(s) intact to lessen the chance of discovery by the enemy.

Crossing Trenches

Before approaching a trench, wait outside the trench for awhile and listen. Do not enter or cross a trench near its junction with a communication trench. Crawl silently up to the edge of the trench and look into it. Remove all loose dirt and rocks from the edge. If it is a narrow trench, spring up and jump across, sinking quietly to the ground on the other side and remaining there a moment to listen before proceeding. If the trench is wide, climb silently and slowly down into it and out the other side, using the revetment for support (see fig. 5-7). Do not enter enemy trenches unless it is absolutely necessary in order to accomplish the mission. Ordinarily, work is better accomplished from outside the trench. Sentries usually pay more attention to sounds in front of them; therefore, if it is necessary to enter a trench, cross it first at the place where enemy observation is restricted, then approach from the rear.

5009. LOCATING AND PLOTTING THE ENEMY AT NIGHT

For night work, a scout must understand the use of a lensatic compass. (Refer to para. 2002.) Using the lensatic compass, the scout can guide the platoon into position, locate adjoining elements of the command, keep direction when on patrol, determine the location of gaps in the enemy wire and the position of enemy out guards.

Locating Gaps in Enemy Wire

When searching for gaps in enemy wire, at least two lensatic compasses are needed: one to register the gap in the wire and the other for navigation. When a gap in enemy wire is located, lie outside the gap, keeping a distance of 10 meters from the barbed wire. Sight with the lensatic compass on a prominent point on the skyline in line with the gap. In selecting the prominent point in the skyline, pick one that appears on the map (i.e., hill mass, house, road junction). If the only prominent point available is one not identifiable on a map (i.e., a tree, destroyed vehicle, enemy position),

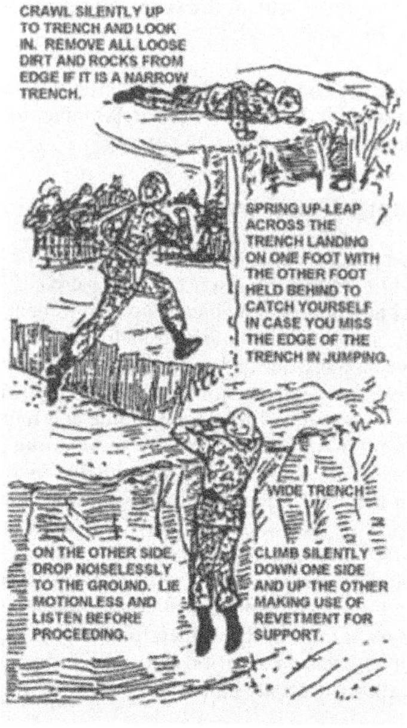

Figure 5-7. Crossing Trenches Silently at Night.

use it. The exact location of this point can be fixed the next day by visual reconnaissance of the area from an observation point. When the needle rests—

- Clamp it in place by lowering the eyepiece to the closed position.
- Rotate the movable bezel ring until the luminous line is directly over the north end of the needle. The azimuth of the gap from the prominent point is now registered.

The compass is carried back without further adjustment. The azimuth setting can be recorded later on a map (see fig. 5-8).

Locating Enemy Out Guards

At night, locate enemy guards by their sounds and failure to maintain light discipline. As sounds of the enemy are heard, and/or observations of the enemy made, shoot an azimuth with one compass. When the needle comes to rest, clamp the compass by lowering the eyepiece to the closed position.

Figure 5-8. Locating a Gap in Enemy Wire at Night.

Note the time and nature of each sound, the estimated distance, and which compass was used to fix the location. Plot this data on a map then wait until the debrief to turn in the notes and compasses. See figure 5-9.

Figure 5-9. Locating Enemy Outguards at Night.

50010. ROUTES OF MOVEMENT

Prior to beginning a night missions, a scout studies the ground in detail from an observation point, air photos, and a map during daylight. The route of advance should be below the skyline. Avoid becoming a silhouette (see fig. 5-10).

Unless the moon is bright, avoid, if possible, passing through woods, ditches, ravines, and brush, because noises of movement may lead to discovery. If the enemy is known to have night observation device capability, avoiding these kinds of terrain may not be possible. To avoid enemy ambushes, return by a different route from the advance and change routes on successive nights.

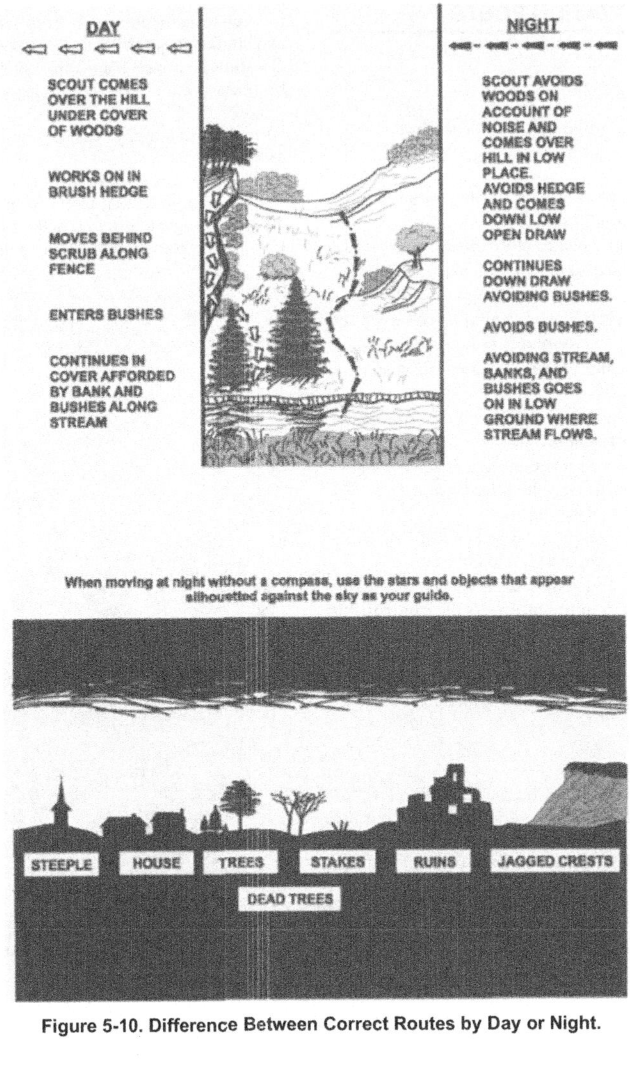

Figure 5-10. Difference Between Correct Routes by Day or Night.

CHAPTER 6. OBSERVING AND REPORTING

When sent out on a mission, a scout's duties are to observe and report. the first five chapters covered scout protection measures and movement in enemy territory. This chapter discusses performing the assigned mission and reporting the mission after its completion.

6001. OBSERVATION POSTS

Positions

The following guidelines apply to observation posts. See figure 6-1.

- When selecting observation posts, scouts should chose the least prominent position.
- Scouts may occupy one or more observation posts.
- An observation post should not be manned for more than 24 hours.
- A selected observation post should be observed for 10 to 15 minutes to ensure it is not occupied.
- Scouts move to the chosen observation post by a concealed route.
- If the post is located on a hill, crawl to a position where the skyline is broken.
- If a tree is used, the position should have a background so as not to be silhouetted against the sky while climbing or observing.
- When leaving the observation post, a different route from that of the approach should be used.
- If a radio is used, its antenna should be located to provide clear communication to the controlling commander but masked from enemy observation and direction-finding equipment. Upon departure, scouts should remove the antennae from the observation post so as not to give away the position.

Observing

Using all senses available, be particularly alert for movement, objects, sounds, and smells inappropriate to the surroundings. While observing, avoid all unnecessary movement. If observing from a building, keep back from doors and windows.

In daylight, look first at the ground nearest you. Begin observing close to your post and search a narrow strip 50 meters or less deep, going from right to left parallel to your front; then search from left to right a second and similar strip farther away but overlapping the first. Continue to observe until the entire field of view has been searched (see fig. 6-2 on page 6-2).

At night, use a night observation device. If one is not available, search the horizon with short, jerky movements, and short pauses. Look a little to one side of an object and then to the other. Lower the head close to the ground to view the object more clearly. Use low-powered field glasses to increase sight range.

Figure 6-1. Method of Approaching an Observation Post.

Figure 6-2. Method of Searching Ground.

6002. REPORTING

It is imperative that the scout accurately and completely report who, where, when, and what was observed to the proper authority upon mission completion. In most cases, a scout will not be equipped with a radio.

Verbal Reports

Verbal reports should be made when writing is impractical, when the information is not complicated or when the enemy is likely to intercept a messenger carrying a written message.

Written Messages

Written messages, preferred to verbal reports, are recorded in message book blanks issued for that purpose (see fig. 6-3) and delivered to a higher authority as soon as possible. The NATO spot report (SPOTREP) should be reviewed as the written message is a NATO format.

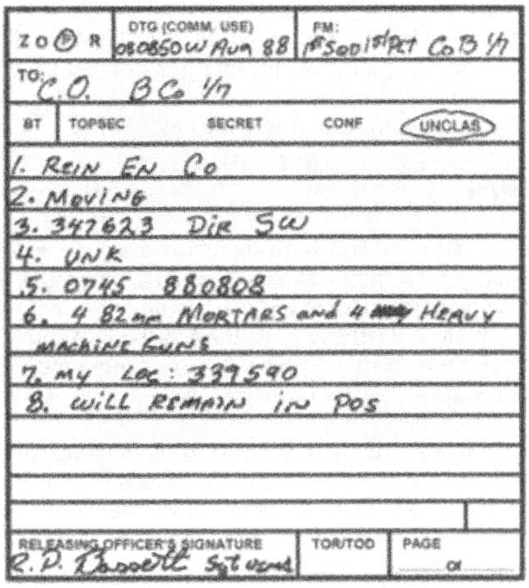
Z O (P) R | DTG (COMM. USE) 080850W Aug 88 | FM: 1st Sqd 1st Plt Co B 1/7

TO: C.O. B Co 1/7

BT | TOPSEC SECRET CONF (UNCLAS)

1. REIN EN CO
2. MOVING
3. 347623 DIR SW
4. UNK
5. 0745 880808
6. 4 82mm MORTARS and 4 HEAVY MACHINE GUNS
7. MY LOC: 339590
8. WILL REMAIN IN POS

RELEASING OFFICER'S SIGNATURE R.P. Bassett Sgt USMC | TOR/TOD | PAGE ___ OF ___

Figure 6-3. Sample Field Message.

The message body is brief, accurate, and clear; facts and opinions are distinguished. If secondhand information is reported, its source is included. Reports include all information of value, first about the enemy, and then about the originator. Information about the enemy should cover—

- Size and/or strength.
- Actions or activity.
- Location and direction of movement.
- Unit identification. (The designation of the enemy unit may be derived from unit markings, uniforms worn or through prisoner interrogation.)
- Time of observation.
- Equipment and weapons.

Messages are printed in block letters. Individual items of information are numbered and separated into paragraphs. If doubt exists as to message receipt by the commander, a summary of its contents is included in the next message. Information about the originator or writer should cover—

- Location at the time of enemy observation (reference to an important terrain feature, by map coordinates, by the back azimuth from each of two definitely located points, or the back azimuth and distance from one known point).
- Intentions. (Remain in position? Continue on the mission? Take other action?)

The message is carefully reread and if possible read by another person to ensure understandability. If a messenger is used, the messenger must read and understand the message in order to answer any questions the commander might have.

Sketch

Information difficult to describe may be given accurately on a simple sketch. The sketch may give all the necessary information or it may be used to supplement a written message. A military sketch is generally one of two types: simple or panoramic. The simple sketch is easily made and read.

Figure 6-4 shows a simple sketch that has been included in the message itself. The sketch may be on a separate sheet of paper, but all of the necessary information must be contained in either the sketch, the message or both.

The panoramic sketch is a picture of the terrain's elevation in perspective, as seen from one point of observation. Although a panoramic sketch is not difficult to create, skill and training are necessary to enhance usefulness. Figure 6-5 on page 6-4 illustrates how to make a panoramic sketch.

Overlay

The same information sent back to higher headquarters on the sketch may be sent on an overlay, if the sender and the person to whom the message is to be sent have copies of the same map. Figure 6-6 on page 6-6 illustrates a simple overlay. The overlay is drawn on transparent paper as follows:

- Orient the map and place it on a hard, flat surface.
- Place the transparent paper over the part of the map of the object or information to be transmitted and hold the paper in this position.
- Orient the overlay to the map by tracing in the intersecting grid lines at two opposite corners of the overlay. Write the correct number designation on the overlay. The cross made by the intersection is called a *tick mark* and enables the receiver to locate the exact area on the map covered by the overlay.
- Sketch the object seen or the information to be transmitted on the tracing paper (the sheet on top of the map) in the exact location it would appear on the map (the sheet underneath the tracing paper).

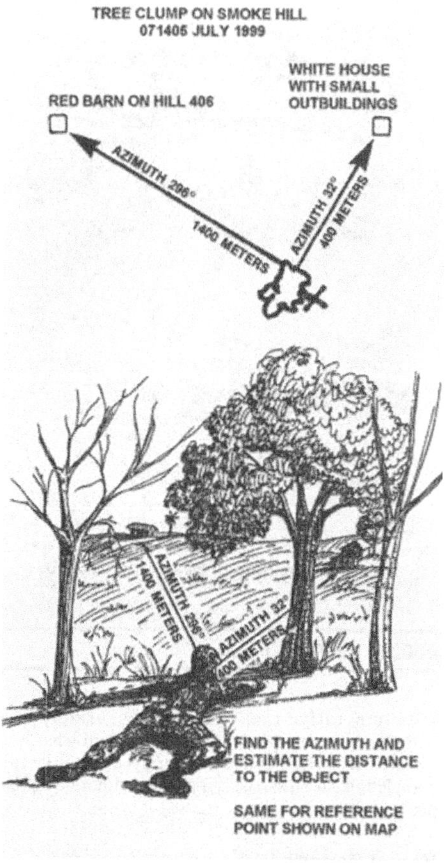

Figure 6-4. Making a Simple Sketch.

Explanatory notes are annotated in the overlay's margin, arrows point to the objectives.

- Indicate with an "X" and an appropriate explanation the position from which the observer saw the object or obtained information.
- Include title and scale of the map from which the overlay was made, date and hour the information was obtained, and signature of the observer on the overlay in the lower right-hand corner.

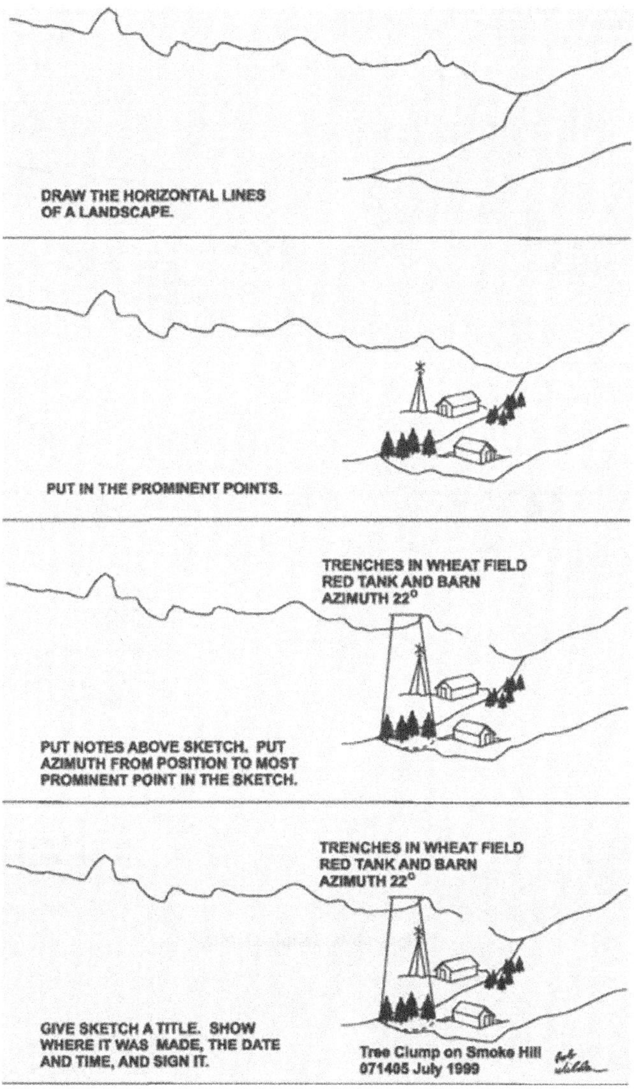

Figure 6-5. Panoramic Sketch.

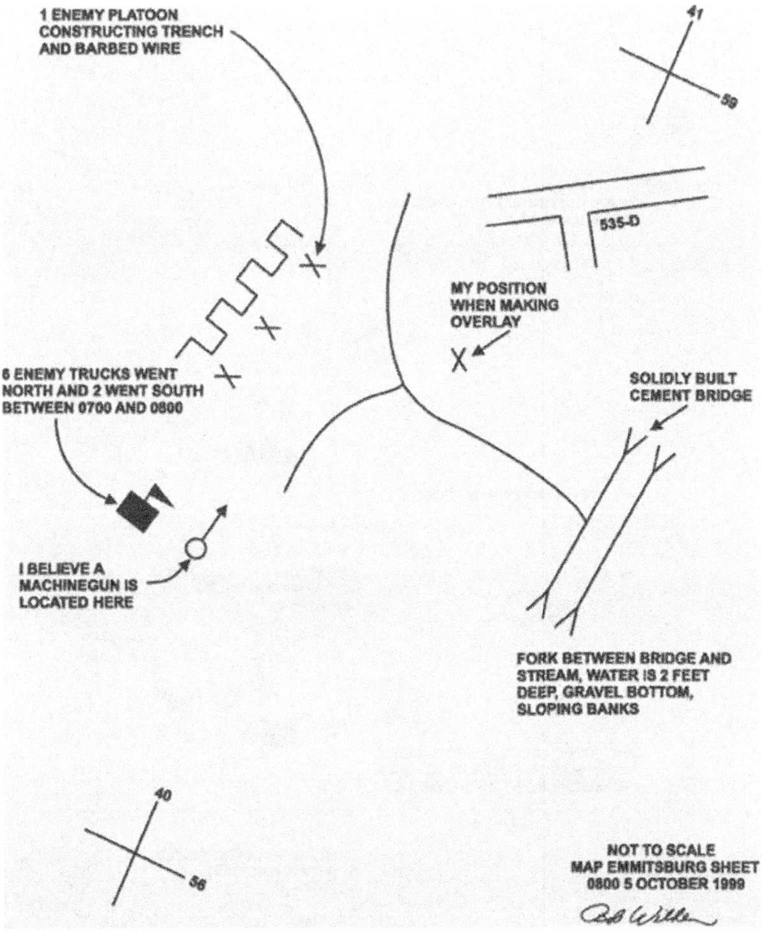

Figure 6-6. Simple Overlay.

CHAPTER 7. SCOUTING FIRE TEAMS

Upon completion of individual training, a Marine adapts this training to the functioning of the unit. The basic tactical unit for scouting is the fire team. Within the fire team, scouts normally work in pairs to furnish security and gather information. Information gathered by the individual scouts is passed to the fire team leader.

7001. POSITIONING

When a rifle platoon in the approach march is not preceded by adjacent units (left, right, front, and rear), it employs its own scouting elements. The scouting element for a platoon is usually one fire team; however, an entire squad may be used.

A fire team used as a scouting element is called a scouting fire team and is controlled by the platoon commander, assisted by the squad leader. A squad leader whose squad is providing the scouting fire teams normally marches near the platoon commander to assist in the control of the scouting fire teams.

A scouting fire team moves aggressively to cover the front of the advancing platoon and to locate the enemy's position(s). It generally moves in a wedge or skirmisher's formation. Normally, a scouting fire team is deployed on a frontage of 50 to 75 meters (10 to 17 meters between each individual scout). The entire squad may be employed to cover a wider frontage. The platoon commander coordinates the movement of the scouting fire team(s) so as to protect the main body of the platoon from enemy fire from points within 400 to 600 meters away, or in close terrain from points within the limits of enemy observation.

Scouting fire team(s) should have enough firepower to overcome resistance from small enemy advanced posts and patrols; the intent is to make enemy riflemen and machine gunners open fire and disclose their positions. Without scouts in advance, the platoon may move into areas where enemy fire may prevent further advance or maneuver and inflict heavy casualties (see fig. 7-1).

Scouting fire teams are covered by the platoon or, when the platoon is masked, the fire team leader ensures individual scouts are maneuvered and coordinated so that the fire team covers its own advance. The fire team leader constantly watches for signals from the platoon commander and remains in visual contact at all times.

The distance between the scouting fire team and the platoon is terrain-dependent. The scouting fire team should not be beyond visibility of the platoon. In open terrain, the platoon commander usually directs the scouting fire team to move by bounds along a succession of locations designated by the platoon commander as intermediate objectives.

Individual scouts should advance as stealthily as possible, while remaining consistent with their mission of reconnaissance to the front, taking advantage of cover without delaying the advance. An occasional glimpse of scouts constantly advancing over a wide front can make the enemy uneasy. It is this activity, and not the target the scouts offer, that may cause the enemy to open fire and disclose its location.

When fired upon, scouts must drop to cover and return fire only when necessary to complete the mission. If

Figure 7-1. Position of Scouts Preceding an Attacking Platoon.

necessary, one scout reports back to the fire team leader who informs the platoon commander.

7002. LOCATING ENEMY POSITIONS

The key terrain of defensive positions are those points that afford extended observation over the ground where the attack must advance. The enemy will place machine guns and infantry to defend critical points.

Members of a scouting fire team preceding an attacking platoon identify the probable enemy infantry and machine gun positions (see fig. 7-2). They use concealment and cover to conduct their advance in order to discover the exact location of enemy positions.

7003. ACTION WITH AN ATTACKING PLATOON

The scouting fire team reconnoiters to the front of the advancing platoon. As soon as the scouting fire team leader indicates the area is secure, the platoon advances and the scouting fire team moves forward. Squads within the platoon advance by bounds; at least one squad is positioned to support the other(s) by fire. Successive positions along the line of advance are selected and designated by the platoon commander as intermediate objectives, and reconnoitered by the scouting fire team before occupation. By conducting proper reconnaissance, surprise by the enemy or movement in the wrong direction may be prevented.

Movement

The distance between the scouting fire team and the front of the main body of the platoon is dependent on the mission, enemy, terrain and weather, troops and support available, and time available (METT-T). In close terrain, such as dense woods, the scouting fire team's movements closely resemble those used for night operations. In approaching houses, woods, and villages, one scout of each pair covers the other while the latter reconnoiters (see fig. 7-3).

A scouting fire team moves forward aggressively to cover the front of the advancing platoon, usually adopting either skirmisher's or wedge formation in

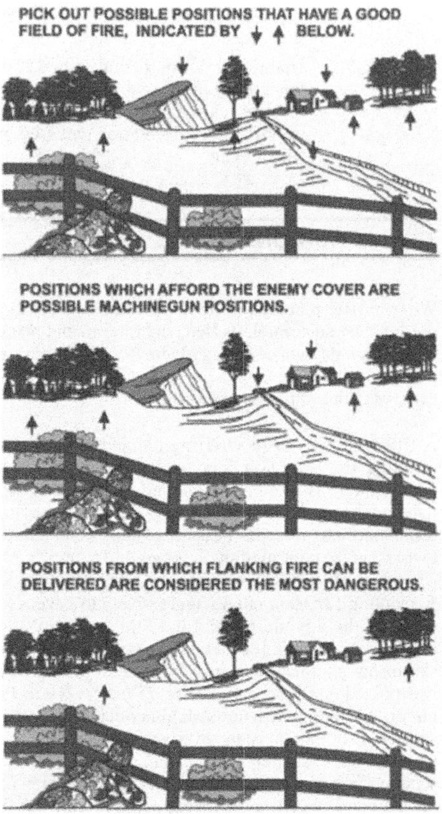

Figure 7-2. Assessing Probable Enemy Machine Gun Positions.

order to be prepared to go into action immediately and to cover a wide frontage of 50 to 75 meters.

As a scouting fire team advances in open terrain, it is supported when possible by elements of the platoon; in close terrain, by mutual support within each fire team. Mutual support within the fire team is accomplished by the fire team leader and the automatic rifleman forming a team that supports by fire the advance of the rifleman and assistant automatic rifleman until they reach a location designated by the fire team leader. The rifleman and assistant automatic rifleman then support by fire the movement of the team leader and automatic rifleman. These successive points to which the sub elements of the fire team move are normally designated as fire team intermediate objectives by the fire team leader. The fire team leader sets as many fire team intermediate objectives as necessary to maintain mutual support within the team. This process is repeated until the team can be covered by other elements of the rifle platoon.

A scouting fire team takes advantage of available cover and concealment without delaying its advance. The orders of the platoon commander govern the distance at which it precedes the platoon. The terrain and the probable position of the enemy affect the scouting fire team's distance in front of the platoon. It may be as much as 400 to 600 meters in advance of the platoon. In open terrain, the platoon commander usually directs that the scouting fire team move by bounds to a succession of intermediate objectives. In close terrain or conditions of limited visibility, the scouting fire team is normally ordered to precede the platoon at the limit of visibility, maintaining visual contact with the platoon commander.

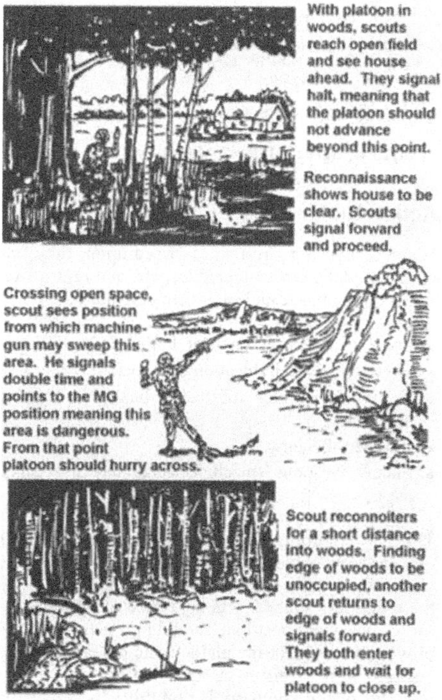

Figure 7-3. Conduct of Scouts Preceding a Platoon.

Action in Woods

When a scouting fire team is directed to advance over open ground to the edge of a woodline, two members of the team, preferably the rifleman and assistant automatic rifleman, reconnoiter inside the woodline while the remainder of the fire team covers them. It is not recommended they separate until finished with their reconnaissance of the far side of the danger area. Both members staying together can cover the same area using a zigzag reconnaissance and they are better equipped to overcome any opposition.

In heavy underbrush and/or poor visibility, the rifleman and assistant automatic rifleman proceed into the woodline together for 50 to 60 meters. The two then separate, searching out either flank to the first high ground or limits of observation, probably 50 to 100 meters (see fig. 7-4 on page 7-4). After the initial search and out posting the limit of advance, the remaining scout signals the fire team forward.

In light underbrush and/or good visibility, the assistant automatic rifleman remains at the edge of the woodline while the rifleman searches the woodline. (see fig. 7-5 on page 7-5). The rifleman searches the woodline in a zigzag pattern, reports back to the assistant automatic rifleman, then moves to an outpost position at the limit of advance. In turn, the fire team leader signals the platoon commander that it is safe for the platoon to move forward.

The fire team leader then moves the remainder of the fire team into the woods, joining up with the forward scout manning the outpost. The scouting fire team occupies and holds a line 50 to 75 meters within the woods and observes toward the direction of movement until the platoon closes up. The scouting fire team leader awaits further word from the platoon commander before moving the team further into the woods.

When directed, the scouting fire team leader moves the team forward until they reach the far edge of the woods. The team is held at the edge of the woods and the fire team leader notifies the platoon commander of the situation. The platoon commander moves the platoon to a position where it can cover the scouting

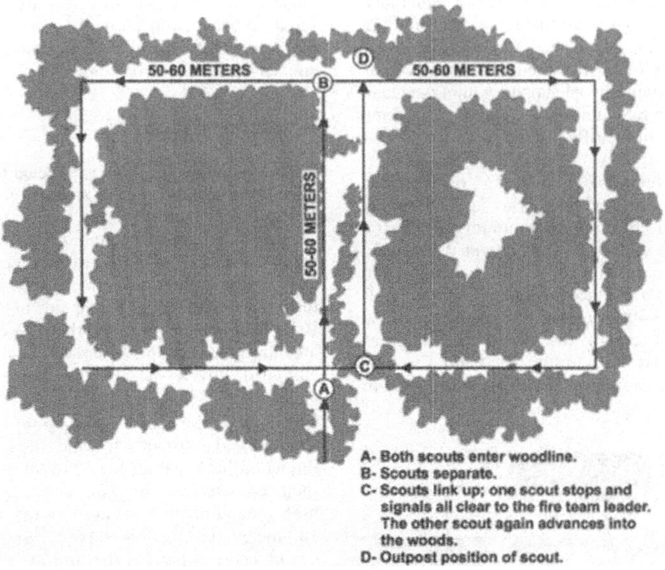

Figure 7-4. Searching Edge of a Woodline (Dense Underbrush/Poor Visibility).

fire team as it exits the woods and directs the team leader to move out and continue the scouting mission.

A scouting fire team passing through woods ahead of its unit maintains a distance allowing visual and oral communications. If an obstacle is encountered, reconnaissance to its front and flanks must be carried out. When advancing along a road or path, scouts precede the platoon to provide necessary protection and to prevent surprise fire on the platoon. When crossing a road or path, they reconnoiter well to the flanks before signaling "all clear" to the platoon.

The scouting fire team will not exit the woods until the arrival of the platoon commander, who will then be given an opportunity to alter the disposition or direction of march. The point where the platoon exits the woods is considered a danger area where the platoon is vulnerable to enemy fire. The scouting fire team is sent ahead to reconnoiter the danger area, as well as the next area to be occupied by the platoon. They signal back whether conditions require a halt, an advance or a quick rush across the open area. The scouting fire team leader must be continually on the lookout for signals from the rear.

Action Under Fire

When a scouting fire team is fired upon, they immediately take cover, locate targets, and return fire. The scouting fire team leader then determines—

- Location of enemy (range and reference points).
- Extent of position (location of flanks).
- Types of positions (obstacles, bunkders, fighting holes, etc.).
- Number of enemy.
- Enemy weapons (machine guns, mortars, tanks, etc.).

The platoon commander assesses the situation as quickly as possible based on the limited information obtained. Usually the platoon commander brings up the remaining squads, sets up a base of fire, and assaults the enemy position. Should the enemy position prove too strong for the platoon, the platoon remains engaged with the enemy as a base of fire until the remainder of the company is committed to clear the enemy resistance.

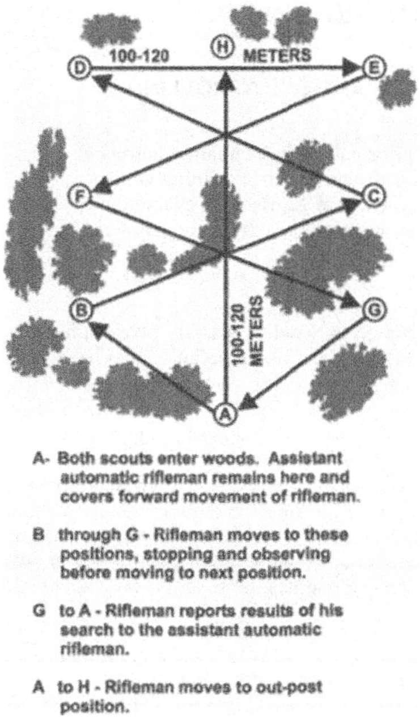

Figure 7-5. Searching Edge of a Woodline (Light Underbrush/Good Visibility).

7004. ACTION WITH AN ENVELOPING UNIT

When a platoon is given the mission to envelop an enemy position, a scouting fire team is employed for protection and reconnaissance in the same way as when the platoon is advancing in the approach march.

PART 2. INFANTRY PATROLLING

CHAPTER 8. FUNDAMENTALS OF INFANTRY PATROLLING

This chapter begins Part 2 and provides basic information about infantry patrols; specifically, their purpose, types, and missions. Infantry patrol training is also addressed and keys to successful patrolling are presented. Subsequent chapters of this part cover patrol organization, preparation, movement, and reconnaissance actions.

8001. DEFINITIONS

A patrol is a detachment of ground, sea or air forces sent out for the purpose of gathering information or carrying out a destructive, harassing, mopping-up or security mission (Joint Publication [JP] 1-02). The mission to conduct a patrol may be given to a fire team, squad, platoon or company.

8002. RELATION OF PATROLLING TO SCOUTING

Each patrol member must be knowledgeable in the principles of scouting and maintain membership of a larger team. To develop the teamwork skills required among the members of a patrol, additional training beyond the basic principles is necessary to become a well-trained scout. A patrol member must respond quickly to the decisions and orders of the patrol leader. There must be complete confidence among all members of the patrol and the confidence that they, as a team, will be successful in their mission.

8003. PURPOSE

A commander must have current information about the enemy and the terrain in order to employ the unit effectively. Patrols are an important means of gaining this information and are used to destroy enemy installations, capture enemy personnel, perform security missions or prevent the enemy from gaining information. Modern warfare places a high premium on effective patrolling because units have larger areas of operations and can be threatened from all directions. As distances between units increase, more patrolling becomes necessary to prevent infiltration by guerrillas or small enemy units, as well as to maintain contact with friendly adjacent units. Active patrolling by numerous small groups is needed to locate the enemy and gather information on the enemy's disposition, strength, morale, and weapons, as well as gather and confirm information about the terrain.

8004. TYPES OF PATROLS

Classification as to Mission

Reconnaissance Patrol

Reconnaissance patrols gather information about the enemy, terrain or resources. Relying on stealth rather than combat strength, they gather this information and fight only when necessary to complete the mission or to defend themselves. The distance covered by reconnaissance patrols varies based on the terrain and mission. The squad is ideally suited for reconnaissance patrol missions because of its relative small size and its experience of working together.

Combat Patrol

A combat patrol is a fighting patrol assigned missions that require engagement with the enemy in combat. Larger and more heavily armed than reconnaissance patrols, combat patrols have a mission to capture enemy documents, provide security, and capture or destroy enemy equipment and installations. Such action is ordinarily followed by a return to friendly positions. Regardless of the mission, the patrol reports any information concerning the enemy and terrain acquired during the accomplishment of the assigned mission. There are four types of combat patrols: raid, contact, ambush, and security (normally conducted by a Marine rifle platoon). A rifle platoon reinforced with crew-served weapons is normally considered the minimum size for contact, economy of force or ambush patrols. In some situations, such as the capture of a small enemy outpost, a rifle platoon could conduct a raid. However, a raid is a complex mission and, due to the organization of a raid force (command, reconnaissance, assault, support, security, and reserve elements), a rifle company is normally the smallest force assigned to a raid.

Classification as to Means of Movement

Foot Patrols

Movement by foot is the most common means; however, there are inherent disadvantages. Foot patrols travel slowly and carry limited quantity and types of equipment and supplies. Range and area coverage is relatively restricted. Foot patrols also have apparent advantages in that they have fewer restrictions as to terrain that can be covered; are more difficult for the enemy to detect; provide thorough coverage within limits of range; and are generally not inhibited by weather.

Motorized Patrols

Where terrain and road networks permit, a motorized patrol overcomes the inherent disadvantages of the foot patrol. Mechanized forces require patrolling units that can keep pace with them. However, motorized patrols are restricted to certain types of terrain, and tend to bypass areas that may be advantageous to and occupied by enemy infantry.

Waterborne Patrols

Waterborne patrols move over seas, lakes, rivers and streams, canals, and other inland waterways. The water is either used as a medium of entry to an objective area or is the actual patrol route. Waterborne patrols are limited by the location of water routes in the terrain and tend to bypass areas that may be advantageous to and occupied by the enemy.

Helicopterborne Patrols

Where terrain is extremely difficult or the enemy situation precludes the use of vehicle or motorized patrols, helicopterborne patrols are a method or means to conduct a patrol.

8005. TRAINING

Training is essential to successful patrolling. Premature and unordered actions by members of the patrol destroy coordination and control. Leaders are trained to issue their orders calmly to inspire confidence and discipline, and to avoid misunderstanding. Patrol members must work together and fight as a team. Training should develop the following skills:

- Expertise in handling individual and special weapons, and familiarity with enemy weapons that may be captured.
- Recognize camouflaged personnel, equipment, and defensive positions; ability to pick up fleeting targets fire the rifle from any firing position.
- Understand fire discipline and, after weapon firing, immediately change position (see fig. 8-1).
- Quick and accurate observation skills, and the ability to recall and transmit clearly and briefly, both orally and in writing.

Figure 8-1. Changing Position After Firing.

- Recognize and quickly respond to improvised signals, visually or by sound.
- Ability to swim with weapon and equipment.
- Use issued or improvised camouflage suits and garnish helmet in order to blend with the surroundings. Smudge face, hands, and any bright surfaces of weapons and equipment with some substance, such as mud or charcoal, to prevent the reflection of light.
- Silence self, equipment, and weapon.
- Use antimalarial and water purification tablets.
- Acclimation to temperature extremes.
- Develop a sense of direction and learn how to follow a course by compass, stars, sun, flow of streams, prominent terrain features, and by observing other natural phenomenon. Learn to determine the distance traveled from a known point and to keep a record of azimuths and the distance traveled on each azimuth (dead reckoning).
- Call for and adjust indirect fire assets.
- Familiarization with all communications assets and the use of field expedient antennae.

8006. KEYS TO SUCCESSFUL PATROLLING

Regardless of the category or means of conducting a patrol, the keys to successful patrolling are—

- Detailed planning. Every portion of the patrol must be planned, all possible contingencies considered.
- Productive, realistic rehearsals. Each phase of the patrol is rehearsed, beginning with actions in the objective area. Similar terrain and environmental conditions are used when conducting rehearsals.
- Thorough reconnaissance. Ideally, the patrol leader will physically conduct a reconnaissance of the route and objective. Photographs and/or maps will be used to supplement the reconnaissance.
- Positive control. The patrol leader must maintain positive control, this includes supervision during patrol preparations.
- All-around security. Security must be maintained at all times, particularly near the end of the patrol where there is a natural tendency to relax.

Chapter 9. Patrol Organization

Organizing a patrol is a two-step process: the general organization of the entire patrol and the task organization of various patrol elements. Normally, the nature of patrolling does not permit long preparation periods and rehearsals to specifically build a unit for each mission. Accordingly, the patrol leader must combine unit integrity considerations with proven concepts of patrol organization.

9001. General Organization

The patrol leader establishes a patrol headquarters and elements to accomplish the mission.

The headquarters is composed of the patrol leader and the personnel who provide support for the entire patrol, such as a forward observer, corpsman, and radio operator.

The major subdivisions of reconnaissance and combat patrols are elements. The existing infantry structure (squads and fire teams) is reinforced as required. For example, a reinforced platoon tasked to conduct a combat patrol that will raid an enemy outpost could be organized as follows:

- Platoon headquarters (command element):
  - Patrol leader (platoon leader).
  - Assistant patrol leader (platoon sergeant).
  - Navigator.
  - Radio operator (company tactical net).
  - Radio operator (patrol tactical net).
  - Corpsman.
- The first squad (security element) provides security en route to the objective area (the point, flank security, and rear security) and at the objective area (the flanks and the objective rally point).
- The second squad (support element) supports by fire for the attack, covering fire for the withdrawal, and supporting fires to cover the crossing of danger areas.
- The third squad (assault element) provides the assault force to attack and seize the objective; searchers to clear the objective; pacers, compass man, navigator, and the assistant patrol leader en route and back from the objective area.

Any attachments a patrol may have (i.e., demolition team, scout snipers, and machine gun squad) will be added to the element that supports its function. For example, the demolition personnel should go with the unit conducting the attack, and scout snipers and machine gunners should stay with the support squad.

9002. Task Organization

The preceding paragraphs described the elements necessary for a patrol to accomplish its mission. These elements reflect the internal functions or tasks required for the patrol to succeed. Depending on the METT-T, there are various methods of grouping these elements together. Task-organization is the further subdivision of patrol elements into teams that are required to perform essential tasks. In creating teams, unit integrity of infantry units should be maintained.

The patrol is organized so each individual, team, and element is assigned a specific task, but capable and prepared to perform other tasks. This may not be possible for certain specialist tasks requiring a trained technician.

CHAPTER 10. PATROL PREPARATION

For a patrol to succeed, all members must be well trained, briefed, and rehearsed. The patrol leader must have a complete understanding of the mission and a thorough understanding of the enemy and friendly situations. The patrol leader should make a complete reconnaissance of the terrain to be covered (either visual or map), and must issue an order to the patrol, supervise preparations, and conduct rehearsals.

10001. MISSION

The mission assigned to a patrol must be clear and oriented toward one objective with a specific task and purpose. More than one primary objective or indefinite missions invites confusion, casualties, and failure.

10002. FACTORS INFLUENCING PATROL SIZE

The circumstances under which patrolling by infantry units is conducted make it necessary for combat patrols to be able to fight offensively, security patrols to defend themselves, and reconnaissance patrols to move quickly and only fight if necessary to break contact or defend themselves. The size of a patrol depends on METT-T.

Generally, a patrol should contain the least number of members needed to accomplish the mission. Combat missions ordinarily require larger patrols than reconnaissance missions.

10003. COMMANDER'S DUTIES

Determining Patrol Requirements

The need for conducting patrols derives from the commander's stated mission (issued by higher level commander) and other specified and implied tasks (secondary or supporting efforts necessary to accomplish the stated mission). This requires analyzing the unit's mission and determining the necessary reconnaissance and/or combat tasks that must be performed. By considering the mission, along with time available, the commander develops the overall concept of operations to include the patrol plan and the specific tasks for each patrol.

Assigning Units

When assigning patrol missions, maintaining the integrity of the existing unit while considering the skills and experience of the unit and its leader are critical factors to the infantry commander. To provide operational depth and equitable apportionment of hazardous assignments, the commander ensures that each of the subordinate leaders and units develops the skills and experience necessary to conduct successful patrols. Assignment of patrol units must consider the commander's concept of operations as a whole and the plans for subsequent employment of assigned forces after completion of the assigned patrolling mission.

Providing Adequate Time

The commander must allow the patrol sufficient preparation time by completing the mission analysis, estimate, and preparation of orders. The commander should use warning orders to alert subordinates to possible requirements and afford them the opportunity for concurrent planning.

Providing the Patrol Leader Information

The following information should be provided:

- A simple, straightforward explanation of the mission, particularly for night patrols.
- General routes (defined by checkpoints) or exact routes (defined by avenues of approach or other terrain features) to follow.
- Enemy composition, disposition, and strength.
- Location and activities of friendly troops.
- Outposts or other security elements through which the patrol is to pass.
- Terrain conditions.
- Missions and routes of other patrols.

- Time patrol is to depart and return.
- Method of reporting information while on patrol (radio, messenger), place where messengers are to be sent, and place where the patrol leader is to report upon completion of the patrol.
- The challenge and password to be used for exiting and reentering friendly lines.
- Special instructions such as locations to be avoided and essential elements of information higher headquarters is seeking.
- Report barrier/obstacle plan of friendly units if known; support available from friendly unit (e.g., medical evacuation [MEDEVAC], react force).
- Fire support available.

Providing Special Skills and Equipment

There will be situations when the unit assigned to conduct a patrol does not have the necessary technical skills or equipment organic to it to successfully accomplish the patrol mission. In these cases, additions (attachments) are made to the unit, such as—

- Machine gun and/or shoulder-launched, multipurpose assault weapon (SMAW) team and/or squad.
- Forward observers (mortar/artillery).
- Radio operator(s).
- Combat engineers.
- Tracked vehicle crewman to assess the ability to traverse the terrain.
- Corpsman.
- Other personnel (snipers, translators) as required.

Providing Miscellaneous Support

The commander must ensure that the patrol leader is provided with the food, water, ammunition, radios and batteries, maps, special clothing, and any other items required by the unit (includin attachments) needed for the mission. Post-patrol support such as debriefings by intelligence personnel must also be planned.

Reviewing the Patrol Leader's Plan and Preparations

Once the patrol leader receives the mission, conducts visual and/or map reconnaissance, and develops the plan, the commander confirms the patrol leader's understanding of the mission and plan for accomplishing it. This discussion between the patrol leader and commander ensures that the patrol leader understands the commander's intent and is properly prepared to provide the patrol leader an opportunity to ask for clarification or additional support, if required.

Debriefing the Patrol

Upon return of the patrol, the commander receives the patrol report at a debriefing attended by the patrol leader and all patrol members. The debriefing should be conducted as soon as possible following the patrol's return, while information is still fresh in the minds of the patrol members. A patrol report, based on the information collected during the debrief, is generated and forwarded to the next higher com-mander.

10004. PATROL LEADER DUTIES

The patrol leader organizes and prepares the patrol by using the six troop-leading steps to make the best use of resources available. These steps are to—

1. Begin planning.
2. Arrange for reconnaissance and coordination.
3. Make reconnaissance.
4. Complete the plan.
5. Issue the order.
6. Supervise.

Begin Planning

The patrol leader begins by evaluating all factors affecting the mission. He looks for possible courses of action that lead to a decision, and then transforms this decision into an order. The first step includes making an initial assessment and decision on using available time, issuing a warning order and initial preparatory tasks, and initiating his estimate.

The initial planning effort assesses the time, assistance, and information available, and plans the proper use of each. Time allowances include reconnaissance; completion of the estimate and order; troop preparation; and such briefings, rehearsals, and inspections as required before beginning the patrol.

The patrol leader reviews the mission and the attachments and/or support available and decides what preparatory efforts must begin immediately. Proper use of subordinates to manage these initial tasks during this period reduces preparation time and frees the patrol leader for proper planning and reconnaissance. A warning order is issued using a modified five-paragraph order format (situation, mission, execution, administration and logistics, and command and signal [SMEAC]) as a checklist. (Refer to app. A for specific information that a patrol warning order should contain within each checklist item.

Once the initial preparations are set in motion, the patrol leader begins estimating by analyzing the mission; considering the friendly, enemy, and operating environments; considering each course of action available against what the enemy might do; comparing the courses of action in terms of mission accomplishment, capabilities, and probable casualties; and choosing one that becomes the basis for concept and order. While the patrol leader should use and organize notes, estimates must be done quickly and accurately, particularly for immediate situations. To organize thinking, the patrol leader uses METT-T, which consists of:

- Mission—the mission assigned to the patrol and how it relates to the mission of the commander who is sending the patrol.
- Enemy—what is known or suspected of enemy presence and capabilities, habits and characteristics, and fighting techniques.
- Terrain and weather—including ground, vegetation, drainage, weather, and visibility.
- Troops and support available—friendly situation and support available.
- Time available—the constraints and impact of time on preparation and mission accomplishment.

The estimate begins with mission analysis, which is the most important part of the entire planning process. Here, the patrol leader considers the specified tasks of the mission assigned and identifies other significant actions (specified and implied tasks) that must be undertaken to accomplish the stated mission. These tasks are arranged in sequence of accomplishment. The tasks and sequence create the framework for developing courses of action for the patrol concept of operations.

Arrange for Reconnaissance and Coordination

The patrol leader arranges a personal reconnaissance to observe as far forward as possible and also coordinates with the appropriate commanders for the patrol's "passage of lines" (see paragraph 11001 for complete definition) and supporting fires. The patrol leader also coordinates with other patrol leaders who may be operating in the same or adjacent areas and requests that the commander assigning him the patrol mission coordinate the patrol action with adjacent commanders, local security, and night defensive fires, as appropriate. The patrol leader may delegate any or all of these arrangements to the assistant patrol leader if the patrol leader requires the time for planning.

Make Reconnaissance and Complete the Estimate

The patrol leader uses personal reconnaissance to answer questions that arise from the map reconnaissance and METT-T evaluation. Specific points include passage points, lanes through obstacles, locations of friendly listening posts and observation posts, possible approach and return routes, enemy positions (if any), and intermediate observation points on the way to the objective.

In selecting approach and return routes, the patrol leader chooses routes that best use concealment and avoid opposition and obstacles. To lessen the chances of ambush by the enemy, the return trip is planned along a different route. In addition to personal reconnaissance and review of the map and aerial photographs, the advice of other patrol leaders who already are familiar with the terrain and the objective area should be considered.

After compiling information about the situation and possible time constraint, the patrol leader completes an estimate. The first step is developing courses of action, each of which will provide for movement to the objective area, mission accomplishment, and the return, based on the tasks and their sequencing identified in mission analysis. While the eventual concept of operations is presented in order of occurrence, the patrol leader must develop the courses of action by either backward or forward planning. In situations where the objective is well defined and there is sufficient information to plan the action for mission accomplishment (reconnaissance or combat), the patrol leader begins the scheme for accomplishing the mission at the objective and then, planning backwards, considers the options for getting there and back.

The following sections on movement to and return from the objective area, reconnaissance missions, and combat missions discuss methods and options available to the patrol leader in developing the courses of action. The principal variables between courses of action will be who, where, and how in the following:

- Patrol task organization.
- Routes to the objective area.
- Observation point(s) (reconnaissance patrol), ambush site, form of maneuver, type of ambush (combat patrol), fire support plan.
- Return routes.

The patrol leader then mentally considers the progress of each course of action (a map or simple sketch is a useful aid) against expected and unexpected enemy action. By comparing the options against each other and prospective enemy opposition, the patrol leader chooses the course of action that has the best chance of success. Includes in this mental preview process is the time to determine the patrol plan for unexpected contingencies (enemy attack/counterattack, casualty handling). These contingency actions, together with the selected course of action, become the patrol concept of operations.

Once the patrol leader determines the scheme of maneuver, the fire support required to accomplish the mission is addressed including the fire power organic to the patrol and what additional indirect fire support will have to be provided by other units. When planning for indirect fire support, the patrol leader considers the following questions:

- Will artillery, mortar, or close air support be required at the objective area (combat patrols)?
- What artillery and mortar targets exist along the routes to and from the objective area that can be employed by the patrol if it encounters the enemy during movement (reconnaissance patrols and combat patrols)?
- What additional fire support will be required to cover the patrol's movement from the objective area back to the friendly area once the enemy is aware of the patrol's actions at the objective area (combat patrols)?

The effect that casualties have upon the patrol depends upon many factors. Generally, more casualties can be expected in a combat patrol than in a reconnaissance patrol. A patrol may continue on to the objective carrying its casualties, send them back with a detail of Marines, abort the mission and return the entire patrol with the casualties, or call their parent unit for assistance.

Some factors that determine what action the patrol leader takes are: patrol's mission; unit's standing operating procedure for handling wounded; number of casualties and nature of their injuries; availability of aid, helicopters or other means of casualty evacuation. Helicopter evacuation should only be used for the most serious casualties. For infantry units conducting patrols in proximity to the enemy, helicopter evacuation of casualties may compromise the patrol's mission and force the patrol to return to friendly positions before the mission is completed.

The patrol leader determines the requirement for nuclear, biological, and chemical (NBC) defense equipment. Gas masks should always be carried due to the availability of riot control agents (RCAs) to the enemy. If chemical or biological agents have been employed in the area that the patrol must pass through, protective garments will have to be worn by patrol members for part of or the entire patrol. Wearing extra clothing and carrying extra equipment affects the speed of the patrol's movement. A contingency plan for post patrol decontamination must be developed.

Complete the Plan

At this point, the patrol leader has completed the basic thinking necessary for accomplishing the assigned mission. The patrol leader prepares the patrol order to spell out the details, assign tasks to subordinates, and explain the entire endeavor for ease of understanding by the other members of the patrol.

Prepare the Order

The patrol leader's order contains more detailed information than discussed in the warning order. Orders follow the prescribed five-paragraph order format but contain greater detail. The patrol order is a modified 5-paragraph order; the major modifications are to paragraphs 3a and 3c. The format for the order is contained in appendix B.

Issue the Order

The patrol leader asks for a status report on the initial preparatory tasks assigned to subordinate leaders and specialists when the warning order was issued. When the patrol leader has completed planning and initial preparations have progressed to the point where the patrol order may be issued, the members of the patrol are assembled. Roll call is taken to ensure all patrol members are present, then the prepared order is issued.

This will be the only opportunity for the patrol leader to issue detailed instructions. The mission, in particular, must be unmistakably clear so that once the patrol is committed, all subordinate leaders can act with unity of purpose.

Whenever possible, the patrol leader should have a Marine, such as the navigator, build a terrain model using dirt, sand, twigs, etc., explaining the concept of operations for movement to the objective area, actions at the objective area, and the return. Terrain models provide patrol members with a clear and simple layout of the area of operations and key terrain.

Supervise

Inspections and rehearsals, vital to proper preparation, are conducted even though the patrol leader and patrol members are well experienced in patrolling. Inspections determine the patrol's state of physical and mental readiness.

The patrol leader inspects before rehearsals to ensure completeness and correctness of uniform and equipment. The following areas are checked:

- Camouflage.
- Identification tags, Geneva Convention cards.
- Prescribed equipment, weapons, and ammunition are available and serviceable.
- Tape and other items are used to "silence" equipment (prevent noise produced during movement).
- Items that could provide information to the enemy (e.g., letters and papers) remain behind.
- Unnecessary equipment and excess weight remain behind.

The patrol leader questions each patrol member to ensure the following is known:

- The mission, planned routes (primary and alternate), and the fire support plan of the patrol.
- The individual's role: what to do and when to do it.
- What others are to do and how their actions impact.
- Challenges and passwords, codes, reporting times, radio call signs, frequencies, and any other pertinent details.

There is usually a period of time between final rehearsal and departure. The patrol leader reinspects just before departure to ensure all equipment is still in working order and the unit is ready to embark on the mission.

Rehearsals ensure the operational proficiency of the patrol. Plans are checked and needed changes are made. The patrol leader verifies the suitability of equipment. It is through rehearsals that patrol members become thoroughly familiar with the actions to take during the patrol.

If the patrol will operate at night, both day and night rehearsals are conducted. Terrain similar to that over which the patrol will operate is used. All actions are rehearsed. If time is limited, the most critical phases are rehearsed. Action at the objective is the most critical phase of the patrol and is always rehearsed.

An effective method is to talk the patrol through each phase, describing the actions and having each member perform individual duties. When satisfied, the patrol leader walks the patrol through all phases of the patrol using only the signals and commands to be used during the actual conduct of the patrol. Rehearsals continue until the patrol is thoroughly familiar with the plan. The rehearsal is also used to test the soundness of the patrol order and patrol organization.

After the rehearsal, the patrol leader makes final adjustments to the plan and patrol organization based on what was learned during the rehearsal and from other sources, such as the S-2 and adjacent patrols. When this is completed, the patrol leader issues final instructions to subordinate leaders noting any changes made in the patrol organization or plan. While the subordinate leaders are briefing the remainder of the patrol members, the patrol leader reports to the commander stating that the patrol is ready to begin the mission. The patrol leader also coordinates the location and time that the patrol can test fire all weapons prior to departure.

CHAPTER 11. MOVEMENT TO AND RETURN FROM THE OBJECTIVE AREA

This chapter provides guidance to patrol leaders for movement to and return from the objective area. Action in the objective area depends on whether the patrol is assigned a reconnaissance mission, combat mission or security mission. Chapters 12 and 13 provide detailed guidance on these types of missions.

11001. PASSAGE OF LINES

A passage of lines is an operation in which a force moves forward or rearward through another force's combat positions with the intention of moving into or out of contact with the enemy. (JP 1-02)

During the initial preparation for the patrol, the patrol leader selects a patrol assembly area and reconnoiters the area of passage designated by the commander. In coordination with the unit commander responsible for the area of passage, the patrol leader identifies gaps or lanes in minefields and wire obstacles and locates local security elements through which the patrol will pass. The patrol leader also checks the route from the patrol assembly area to the passage point or contact point where the patrol will depart friendly lines. If possible, both the route to the passage point and the route through the frontlines should be concealed from the enemy's view.

The patrol leader also reconnoiters the area for return passage of lines and coordinates with the unit commander responsible for the area of passage for passage points and lanes as necessary. The patrol leader observes these points from the direction that the patrol will use upon return to friendly lines, if possible, to aid in recognition upon return. The patrol leader provides the forward unit with information about the size of the patrol, general route, and expected time of return. The manner of challenge and recognition of the returning patrol should be coordinated in detail.

Upon return to friendly forward local security squads and/or frontlines, the patrol leader leaves the patrol in a covered position and moves forward with a radio operator and at least one Marine for security to make contact with the friendly unit in the manner agreed. After contact is made and recognized, the patrol leader rejoins the patrol and takes them to the passage point, personally checking in each member.

11002. ORGANIZATION FOR MOVEMENT

The patrol's task organization establishes the elements and teams needed to accomplish the mission in the objective area and to and from the objective. The patrol leader determines the formation(s) in which the patrol moves to the objective area.

Formations

The proper use of patrol formations is critical to the patrol's success. The squad and fire team formations described in FMFM 6-5 (proposed MCWP 3-11.2), *Marine Rifle Squad*, also apply to infantry patrolling formations. Because the movement of the patrol must be concealed from the enemy, the patrol normally moves through terrain that provides concealment. Control of the patrol in this type of terrain is difficult; thus the column formation, which is easily controlled, is normally used. However, as various types of terrain are encountered, the patrol leader uses the same considerations in determining the appropriate formation used in other infantry operations.

The standard squad and fire team formations are adaptable to a patrol. The patrol may change formations en route to match the situation and terrain. The patrol leader may have to sacrifice some control for better dispersion or give up some speed for greater stealth and more security. Other considerations include—

- Visibility, weather, terrain, and vegetation will influence dispersion and control of individuals and units. These factors may also affect the enemy; if visibility is good for the patrol, it is also good for the enemy. Two pieces of luminous tape worn on the back of the collar will aid in control and movement on dark nights. The collar is turned down when near the enemy. The tape can also be worn on the back of the cap, but should be covered or removed when near the enemy.

- Preserving the integrity of fire units (fire teams and, if attached, machine-gun teams) is of primary importance. If team members are detached from a machine-gun team, the unit integrity is lost and effectiveness is reduced. The patrol leader must also position fire units so as not to mask their fires.
- All-around defense of the patrol must not be sacrificed under any circumstances. The conventional squad and platoon formations provide adequate firepower in any direction required. When attachments are made to the patrol, the attachments are positioned within the formation to enhance the firepower of the patrol. If a fire unit, such as a machine-gun team or squad, is attached, it is incorporated into the all-around defense of the patrol by modifying the conventional formation (see fig. 11-1). It is permissible to employ machine-gun teams individually during movement. However, assault rocket squads and teams should be employed primarily as rifle fire units for movement. Firing the SMAW or AT-4 from within a patrol formation can present a serious back-blast danger to patrol members and should only be fired on the patrol leader's direction.
- Time allotted for mission accomplishment is also a major consideration. In selecting the formations, the patrol leader must consider the speed of the movement required to meet the time constraints (if any) imposed on the patrol. If required to meet a time schedule, a formation that permits rapid movement should be used. Speed, however, must never be permitted to force the patrol leader to make rash tactical decisions.

Exercise of Control

The patrol leader is positioned for best control over the patrol. The assistant patrol leader moves at or near the rear of the patrol. Other subordinate leaders move with their elements. All patrol members assist by staying alert and passing on signals and orders. A signal to halt may be given by any patrol member, but the signal to resume movement is given only by the patrol leader.

Arm-and-hand signals are the primary means of communication within a patrol and should be used exclusively when near the enemy. All members must know the standard infantry signals (refer to FMFM 6-5), as well as any special signals required, and be alert to receive and pass them to other members.

The patrol leader should speak just loudly enough to be heard. At night, or when close to the enemy, the patrol leader halts the patrol and has subordinate leaders come forward. They speak in a low voice and then pass the information to their subordinates by moving from member to member.

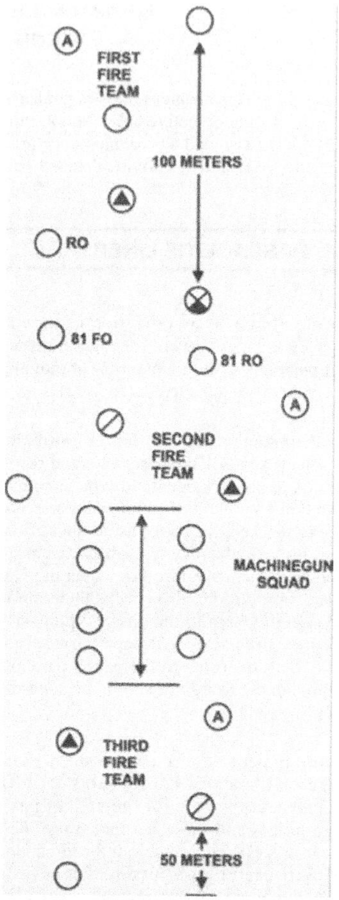

Figure 11-1. Squad Patrol with Attached Mortar Forward Observer Team and Machine Gun Squad.

Radios provide a means of positive control within a large patrol; however, radios should be used only when arm-and-hand signals or face-to-face contact between the patrol leader and subordinate leaders is impractical. When close to the enemy, words are spoken into the microphone with hands cupped over it in a low voice.

Other sound signals may be used if the patrol leader is sure they serve the purpose intended. Planned sound signals are rehearsed before starting on the patrol. Sound signals used must be simple, natural sounds that are few in number and easily understood. Bird and animal calls are seldom satisfactory.

Infrared equipment, such as the sniper scope and infrared filters for the flashlight, may be used as means of sending and receiving signals and maintaining control at night.

Luminous tape may be used to assist in control at night. Small strips on the back of the cap or collar of patrol members aid in keeping visual contact with the front member. However, the luminous tape must be covered when near the enemy.

An important aspect of control is accounting of personnel, especially after crossing danger areas, enemy contacts, halts, and exiting and re-entering friendly lines. The patrol leader may arrange for the last member to send up the count automatically after crossing danger areas, enemy contact, and halts. In large patrols or when moving in a formation other than a column, subordinate leaders check members and report the count to the patrol leader by the quickest method appropriate to the circumstances.

Navigation

One or more members are assigned as navigators for the patrol. Their function is to assist the patrol leader in maintaining direction by the use of the compass.

The patrol leader assigns at least two members as pacers to keep track of the distance from point to point. The average of their count is used to approximate the distance traveled. Pacers are separated so they do not influence each other's count. Pacers must know how to convert their own paces to meters.

The route is divided into increments, with each increment starting at a recognizable point on the ground. The pacers begin their counts from zero at the beginning of each leg. This makes the pace count easier to keep and checks for accuracy.

The pace count is sent forward when the patrol leader turns to the member behind and in a low voice says, "send up the pace" or uses the arm-and-hand signal of tapping his boot. This signal is passed to both pacers, who in turn send up the pace count in meters; for example, "two-hundred" or "one-seven-five."

Patrol members must understand that the counts of both pacers are sent forward. The patrol leader must know the counts of both men in order to check them.

Security

The patrol is dispersed consistent with control, visibility, cover, and concealment. Scouts are employed to the front, flanks, and rear of the patrol to provide security. For the patrol members traveling in the main body of the patrol, areas of responsibility are assigned to the front, flanks, and rear. Scouts are the eyes and ears of the patrol leader. They move when and where directed by the patrol leader and maintain contact with the patrol leader at all times, except when momentarily obscured by vegetation or other terrain features.

Front

Small patrols (squad size) may employ from one scout up to a fire team as the point, depending on the enemy situation, terrain, and route being followed by the patrol. Normally, squad-size patrols will use two scouts as the point. The point is responsible for investigating the route of advance immediately to the front of the patrol.

The point moves as far ahead of the patrol as visibility and terrain permit. When visibility is good, the point may precede the main body by as much as 100 meters. The point travels right and left ahead of the patrol, searching the area over which the patrol will pass.

The point maintains direction by knowledge of the general route to be followed and visual contact with the patrol leader. The patrol leader or the navigator ensures that the point is proceeding correctly.

The point, which stays far enough ahead of the patrol to provide security, is not a trail breaker for the patrol. If the point loses contact with the patrol, the point waits for the main body to catch up or moves rearward if contact is not quickly regained.

One of the navigators may be positioned with the point. One or more members works as the point while the other is the navigator.

Flank

Flank security for a patrol of squad size or less may be provided by using one or two members on either flank. If two scouts are assigned to a flank, one is positioned to observe the patrol leader and the other works farther out from the patrol. The scout who must observe the patrol leader remains within a maximum distance of 100 meters. The scout farther out remains in sight of the inside scout but normally does not move more than 20 or 25 meters away and remains prepared to relieve flank security regularly. Moving through dense woods or jungle may render the use of flank security impractical because of reduced visibility. In such cases, it moves with the patrol itself, but maintains observation to its assigned flanks.

Rear

A small patrol normally has only one rifleman assigned as rear security. An interval between the member assigned as rear security and the last Marine of the patrol is maintained at the limit of visibility, up to 50 meters. This member maintains rear security for the patrol by constantly observing to the rear.

Halting

Speed of movement is slower at night than in day patrols and reduces the danger of a Marine becoming separated from the patrol. The patrol occasionally halts to observe and listen for enemy activity; this is called a security halt. Upon signal, when reaching a danger area and periodically throughout movement en route, every member freezes in place, remains quiet, observes, and listens. It may be necessary to call a security halt just after departing friendly areas and just before entering friendly areas.

The patrol may halt briefly to send a message, eat, rest, check direction, or make a reconnaissance. The area selected should provide adequate concealment and cover, as well as favor the defense. All-round security is established and the patrol leader ensures all members move out when the patrol resumes movement. (For extended halts, see paragraph 11005.)

Infiltration

The disposition of enemy forces may sometimes prevent a patrol from entering the enemy occupied area as a unit; however, pairs of scouts or fire teams may slip through without being discovered. (Refer to FMFM 6-5 for infiltration techniques and procedures).

11003. CONTROL MEASURES FOR MOVEMENT

Checkpoints

A checkpoint is a predetermined point on the surface of the Earth used as a means of controlling movement, a registration target for fire adjustment or reference for location (JP 1-02). Checkpoints are a means of control between the parent unit and the patrol. These locations are decided upon and coordinated before the patrol leaves, so that both the patrol members and parent unit will know the patrol's location when it reports in. The parent unit can follow the progress of the patrol without transmitting coordinates over the radio that the enemy could monitor.

Rally Points

A rally point is an easily identifiable point on the ground where units can reassemble and reorganize if they become dispersed (MCRP 5-12C). It should provide cover and concealment, be defensible for at least a short time, and be easily recognized and known to all patrol members. All rally points are considered and identified as tentative rally points until they are occupied, found to be suitable, and designated as rally points by the patrol leader.

Initial Rally Point

The initial rally point is within friendly lines where the patrol can rally if it becomes separated before departing the friendly area or before reaching the first en route rally point. It may be the assembly area where the patrol waits while the patrol leader contacts the last friendly position through which the patrol will pass. The location of the initial rally point must be coordinated with the forward unit commander.

En route Rally Points

En route rally points are between the initial rally point and the objective rally point; and from the objective rally point back to the point where the patrol reenters friendly lines. They are determined as the patrol passes through a likely area that is suitable for a rally point.

Objective Rally Point

The objective rally point is located nearest the objective where the patrol makes final preparations prior to approaching the objective. It also serves as a location where the patrol reassembles after completing actions on the objective. The objective rally point must be suitable to accommodate those activities accomplished prior to actions on the objective. This position must provide the patrol concealment from enemy observation and, if possible, cover from enemy fires. It may be located short of, to a flank or beyond the objective. It should be out of sight, sound, and small-arms range of the objective area. The patrol leader's reconnaissance of the objective is made from this position; it is the release point from which patrol elements and teams move to the objective to accomplish the mission.

Rally Point Selection

The patrol leader selects likely locations for tentative rally points during reconnaissance or map study. A tentative rally point must be confirmed and announced after examination proves suitability. A tentative initial rally point and a tentative objective rally point are always selected and identified in the patrol order. If necessary, the patrol leader selects additional rally points en route as suitable locations are reached. When the patrol reaches a danger area that cannot be bypassed, such as an open meadow or stream, the patrol leader selects a rally point on both the near and far side. If good locations are not available, the patrol leader designates the rally points in relation to the danger area. The patrol leader will say, for example, "50 meters this side of the trail" or "50 meters beyond the stream."

Rally Point Use

The initial rally point and en route rally points are designated to enable the patrol to reassemble if it is unavoidably separated or dispersed. Identifying features are pointed out. The patrol leader ensures that the information is passed to all patrol members. When crossing a danger area, a near side rally point and a far side rendezvous point are designated. These rally points should only be used when all other methods of retaining control of the patrol have failed. The success of the patrol may be jeopardized if it is dispersed and cannot rally expeditiously.

If the patrol has left the friendly area and becomes dispersed, patrol members return to the last designated rally point (the initial or an en route rally point) unless the patrol leader gives other instructions.

As previously noted, the patrol leader selects two rally points at the near and far sides of danger areas that cannot be bypassed. If the patrol becomes separated or dispersed at a danger area, and there has been no enemy contact, the patrol should reassemble at the rally point on the far side of it. If, however, the patrol is separated or dispersed at a danger area as a result of enemy contact, members who have already crossed the danger area assemble and reorganize at the rally point on the far side; members who have not crossed assemble and reorganize at the rally point on the near side. In the absence of the patrol leader and unless directed otherwise, the senior member at the rally point on the near side takes charge, attempts to move the rallied patrol members to the rally point on the far side, and rejoins the remainder of the patrol.

Rally Point Actions

The patrol leader plans the actions to be taken at rally points and instructs the patrol accordingly in the patrol order. Planned actions at the initial rally point and en route rally points must provide for the continuation of the patrol as long as there is a reasonable chance to accomplish the mission. Plans for actions at rally points should provide for—

- Recognition signals for assembly at rally points.
- Minimum number of members and maximum amount of waiting time required before the senior member at the rally point moves the rallied patrol members onward toward the objective or returns to friendly lines.
- Instructions for patrol members who find themselves alone at a rally point.

11004. PRECAUTIONS AT DANGER AREAS

A danger area is any place where the patrol is vulnerable to enemy observation or fire (open areas, roads, trails, and obstacles such as barbed wire, minefields, rivers and streams, and lakes). Any known or suspected enemy position the patrol must pass is also a danger area. The patrol leader plans for crossing each danger area and includes these plans in the order.

The patrol reconnoiters the near side of a danger area first, then the patrol leader sends scouts to reconnoiter the far side. Once the scouts report that the far side is clear of the enemy, the remainder of the patrol crosses the danger area. As each individual or group crosses the danger area, they are covered by those remaining and by those who have successfully crossed. Enemy obstacles are avoided since they are usually covered by fire.

In crossing a river, the near bank is reconnoitered first; then the patrol is positioned to cover the far bank. Scouts are sent across to the far bank. After the far bank has been reconnoitered and the scouts report that it is clear of the enemy, the patrol crosses as rapidly as possible. This may be done individually or in pairs. If crossing the river requires swimming, the patrol uses improvised rafts to float equipment, weapons, and ammunition across. (Refer to MCRP 3-02C, *Water Survival Handbook.)*

A road or trail is crossed at or near a bend or where the road is narrow. Observation is restricted and, if the enemy is present, the patrol is exposed as short a time as possible. The near side is reconnoitered first, then scouts are sent across to reconnoiter the far side. This includes reconnoitering the tentative rally point on the far side. Once the scouts report "all clear," the remainder of the patrol crosses rapidly and quietly.

If the patrol must pass close to an enemy position, it takes advantage of battlefield noises to cover the sounds of movement. If supporting fires are available, the patrol leader can call for them to divert the enemy's attention as the patrol passes.

11005. HIDE

When a patrol is required to halt for an extended period in an area not protected by friendly troops, the patrol moves into a location which, by the nature of the surrounding terrain, provides passive security from enemy detection. Such an assembly area is termed a *hide*. To establishment a hide—

- Cease all movement during daylight hours to avoid detection.
- Hide the patrol for an extended period while the patrol leader conducts a detailed reconnaissance of the objective area.
- Rest and reorganize after extended movement.
- Reorganize after a patrol has infiltrated the enemy area in small groups.

The patrol leader's plan must include tentative hide locations when the patrol's mission dictates an extended halt within enemy areas. These tentative locations must be confirmed by actual ground reconnaissance prior to occupation by the patrol. The plan for a hide includes both passive and active security measures.

Passive security measures are—

- Avoid built-up areas.
- Select an area remote from all human habitation.
- Avoid known or suspected enemy positions.
- Avoid ridge lines, topographic crests, valleys, lakes, and streams.
- Avoid roads and trails.
- Avoid open woods and clearings.
- Select areas offering dense vegetation, preferably bushes and trees that spread out close to the ground.

Active security measures—

- Establish security covering all likely avenues of approach into the site.
- Establish communications (wire, radio, signal, runner) with posted security to provide early warning of enemy approach.
- Select an alternate area for occupation if the original hide is compromised or found unsuitable.
- Plan for withdrawal in the event of discovery.
- Establish an alert plan with a certain percent of the personnel awake at all times.
- Organize the elements of the patrol so necessary activities can take place with a minimum amount of movement.

The size of the area physically occupied by a patrol in a hide and the number of security posts required are governed by the terrain, quantity and quality of cover and concealment, and size of the patrol.

If the situation permits, a hide can also be used as the final preparation position and/or objective rally point.

11006. IMMEDIATE ACTIONS UPON ENEMY CONTACT

A patrol may make contact with the enemy at any time. Contact may be through observation, a meeting engagement or ambush. Contact may be visual, in which the patrol sights the enemy but is not itself detected. When this is the case, the patrol leader can decide whether to make or avoid physical contact, basing his decision on the patrol's assigned mission and capability to successfully engage the enemy unit.

When a patrol's assigned mission prohibits physical contact (except that necessary to accomplish the mission), its actions are defensive in nature. Physical contact, if unavoidable, is broken as quickly as possible and the patrol, if still capable, continues its mission.

When a patrol's assigned mission permits or requires it to seek or exploit opportunities for contact (as in the case of a combat patrol), its actions are offensive in nature, immediate, and positive.

In patrolling, contacts (visual or physical) are often unexpected at very close ranges, and short in duration. Effective enemy fire often provides leaders little or no time to fully evaluate situations and issue orders. In these situations, immediate action provides a means for swiftly initiating positive offensive or defensive action, as appropriate.

Two types of physical contact with the enemy are meeting engagement and ambush. Meeting engagement is a combat action that occurs when a moving force, incompletely deployed for battle, engages an enemy at an unexpected time and place. It is an accidental meeting where neither the enemy nor the patrol expect contact and are not specifically prepared to deal with it. An ambush is a surprise attack from a concealed position.

Immediate Actions

Immediate actions are designed to provide swift and positive small unit reaction to visual or physical contact with the enemy. They are simple courses of action in which all Marines are well trained. Minimal signals or commands are required and they are developed as needed for the combat situation. The signals can, in many cases, be initiated by any member of the unit. It is not feasible to attempt to design an immediate action drill to cover every possible situation. It is better to know the immediate action drill for each of a limited number of situations that may occur during a patrol. Arm-and-hand signals associated with immediate actions—such as FREEZE, ENEMY IN SIGHT, and HASTY AMBUSH RIGHT or LEFT—are contained in FMFM 6-5 (proposed MCWP 3-11.2).

Immediate Halt

When the patrol detects the enemy but is not itself detected, the situation requires the immediate, in-place halt of the patrol. The first member visually detecting the enemy gives the silent signal for FREEZE. Every member halts in place, weapon at the ready, and remains absolutely motionless and quiet until further signals or orders are given.

Air Observation and/or Attack

These actions are designed to reduce the danger of detection from aircraft and casualties from air attack.

When an enemy or unidentified aircraft that may detect the patrol is heard or observed, the appropriate immediate action drill is FREEZE. The first member hearing or sighting an aircraft that may be a threat signals FREEZE. Every member halts in place until the patrol leader identifies the aircraft and gives further signals or orders. Members of the patrol must not look up at the aircraft as sunlight can reflect off their faces even when camouflaged.

When an aircraft detects a patrol and makes a low level attack, the immediate action drill air attack is used. The first member sighting an attacking aircraft shouts, "AIRCRAFT," followed by the direction of the incoming attack: FRONT, LEFT, REAR or RIGHT. The patrol moves quickly into line formation, well spread out, at right angles to the aircraft's direction of travel. As each member comes on line, the member hits the ground, using available cover, then positions the body perpendicular to the aircraft's direction of travel, to present the shallowest target possible (see fig. 11-2 on page 11-8). Between attacks (if the aircraft returns or if more than one aircraft attacks), patrol members seek better cover. Attacking aircraft are fired upon only on command of the patrol leader.

Meeting Engagement

Hasty Ambush. This immediate action is used to avoid contact and to prepare to initiate an unplanned ambush on the enemy. It may often be a subsequent action after the command freeze. When the signal HASTY AMBUSH is given (by the point member, patrol leader or another authorized patrol member), the entire patrol moves quickly to the right or left of the line of movement, as indicated by the signal, and takes up the best available concealed firing positions (see fig. 11-3 on page 11-9). The patrol leader initiates the ambush by opening fire and shouting, "FIRE"; thus ensuring initiation of the ambush if the weapon misfires. If the patrol is detected before this, the first member aware of detection initiates the ambush by firing and shouting. The patrol leader may decide not to initiate the ambush in order to avoid contact unless the patrol is detected. When used as an offensive

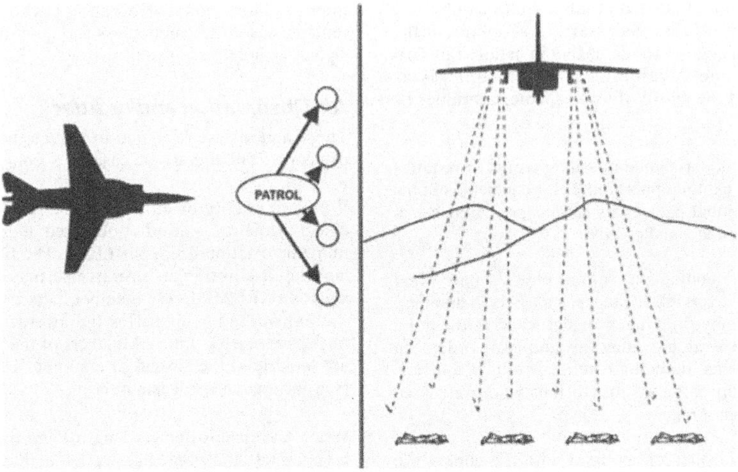

Figure 11-2. Immediate Action, Air Attack.

measure, the enemy is allowed to advance until he is in the most vulnerable position before the ambush is initiated. An alternate means for initiating the ambush is to designate an individual (for example, point or last member) to open fire when a certain portion of the enemy unit reaches or passes that member.

Immediate Assault. This immediate action drill is used defensively to make and quickly break undesired but unavoidable contact (including ambush) and offensively to decisively engage the enemy (including ambush). When used in a meeting engagement, members nearest the enemy open fire and shout, "CONTACT," followed by the direction of the incoming attack: FRONT, LEFT, REAR or RIGHT. The patrol moves swiftly into line formation and assaults (see fig. 11-4).

Defensive Measures

When used defensively, the assault is stopped if the enemy withdraws and contact is broken quickly. If the enemy stands fast, the assault is carried through the enemy positions and movement is continued until contact is broken.

Offensive Measures

When used offensively, the enemy is decisively engaged. Escapees are pursued and destroyed until orders to break contact are given by the patrol leader. If the patrol is fired upon from beyond 50 meters, the patrol must break contact as quickly as possible and continue the mission. If it engages the enemy any longer than necessary to break contact, it may put the mission in jeopardy.

Fire and Maneuver

Fire and maneuver is one means to break contact. One portion of the patrol returns the enemy fire while another portion moves by bounds away from the enemy. Each portion of the patrol covers the other by fire until contact is broken by all.

Clock System

The clock system is another means to break contact. Twelve o'clock is the direction of movement of the patrol. The patrol leader shouts a direction and a distance. For example: "TEN O'CLOCK-TWO HUNDRED," means the patrol should move in the direction of ten o'clock for 200 meters. Patrol members keep their same relative positions as they move so the original formation is not disrupted. Subordinate leaders must be alert to ensure that the members of their elements and teams receive the correct order and move as directed.

Counter Ambush

When a patrol is ambushed, the immediate action drill used depends on whether the ambush is a near ambush

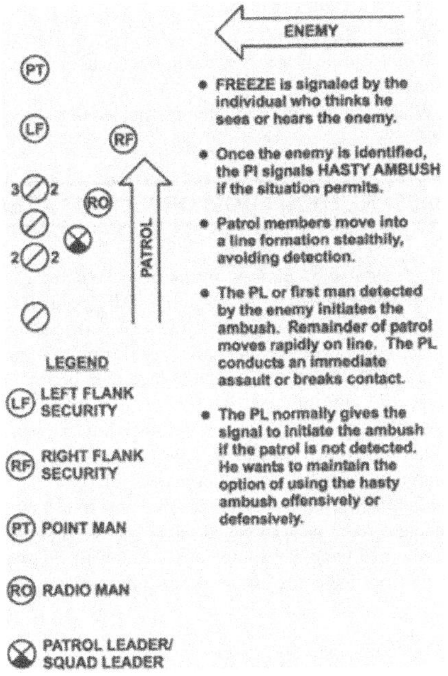

Figure 11-3. Immediate Action, Hasty Ambush.

(the enemy is within 50 meters of the patrol) or a far ambush (the enemy is beyond 50 meters of the patrol). Fifty meters is considered the limit from which the ambushed patrol can effectively launch an assault against the enemy.

Near Ambush. In a near ambush, the killing zone is under very heavy, highly concentrated, close range fires. There is little time or space for members to maneu-ver or seek cover. The longer they remain in the killing zone, the greater the chance they will become casualties. Therefore, if members in the kil-ling zone are attacked by a near ambush, they immediately assault without order or signal directly into the ambush position, occupy it, and continue the assault or break contact, as directed. This action moves them out of the killing zone, prevents other elements of the ambush from firing on them without firing on their own members, and provides positions from which other actions may be taken (see fig. 11-5 on page 11-10). Members not in the killing zone

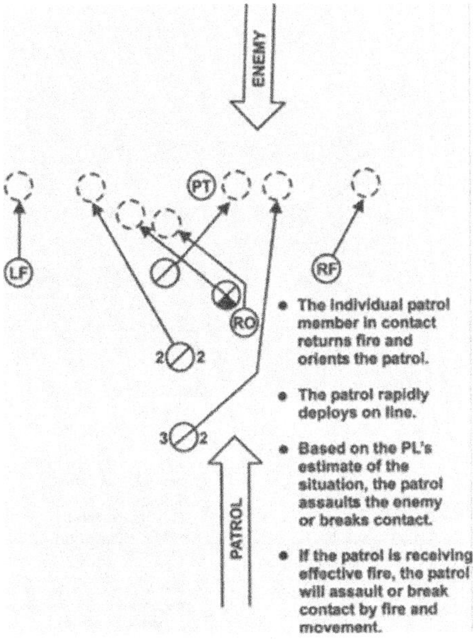

Figure 11-4. Immediate Action, Immediate Assault.

maneuver against the enemy as directed. The assault continues until all patrol members are outside of the killing zone.

Far Ambush. In a far ambush, the killing zone is also under very heavy, highly concentrated fires, but from a greater range. This greater range provides members in the killing zone maneuver space and some opportunity to seek cover at a lesser risk of becoming a casualty. If attacked by a far ambush, members in the killing zone, without order or signal, immediately return fire, take the best available positions, and continue firing until directed otherwise. Members not in the killing zone maneuver against the ambush force, as directed (see fig. 11-6 on page 11-10). The assault is continued against the enemy or until the order to break contact is given.

In each situation, the success of the counter ambush employed depends on the members being well trained in quickly recognizing the distance from which an ambush is initiated and well rehearsed in the proper reaction.

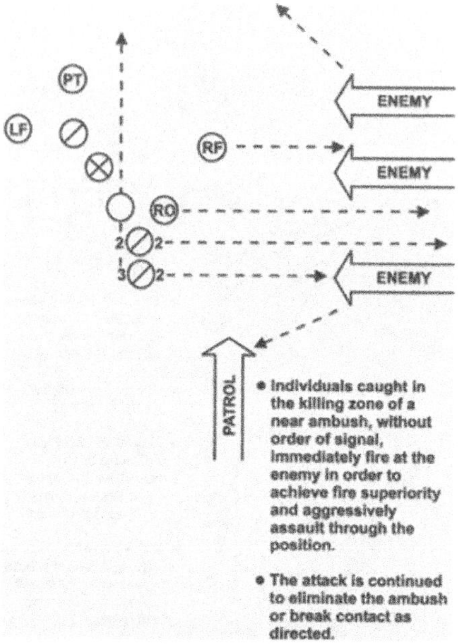

Figure 11-5. Near Ambush.

11007. PATROL LEADER'S ACTION IN A DEVELOPING SITUATION

While good patrolling depends on good planning, the patrol leader's plan must be flexible. Every combat situation develops differently than expected, and the patrol leader must be ready to quickly adapt to the situation as it develops. One of the most difficult tasks in battle is to recognize the correct moment for making a decision. Generally, it is more difficult to determine the moment for making a decision than it is to formulate the decision itself. When the situation demands, decisions must be made promptly without waiting for more information. In a developing situation, the patrol leader should use the following questions as a guide for battlefield decisionmaking:

- How has the situation changed?
- How does the change affect mission accomplishment and the immediate superior's mission?
- Must a decision be made now?
- What are the options?
- Which option best serves the mission and the unit as a whole?
- Which option offers the greatest chance of success?

11008. RETURN FROM OBJECTIVE AREA

After performing actions in the objective area, the patrol reassembles at the objective rally point. This phase of the patrol is perhaps the most difficult and dangerous. Patrol members are experiencing fatigue, emotional letdown, and wounds; they may be low on water and ammunition. Above all, the enemy was likely alerted if not in pursuit. At this point, the patrol leader must move the patrol rapidly but carefully and maintain patrol security at a high level. In returning to friendly lines, the patrol neither uses nor travels near the same route used to get to the objective area. The enemy may have that route covered. For information on reentry of friendly lines, refer to paragraph 11001.

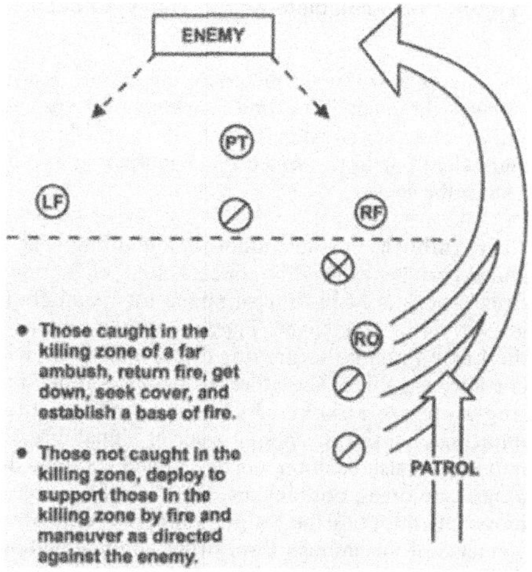

Figure 11-6. Far Ambush.

Chapter 12. Reconnaissance Patrols

The commander needs accurate, timely information about the enemy and the terrain the enemy controls to assist in making tactical decisions. Reconnaissance patrols are one of the most reliable means for obtaining this information. These patrols engage the enemy only when necessary to accomplish their mission or for protection. They generally avoid combat, accomplish the mission by stealth, and do not maintain contact with the unit that sent them out. A reconnaissance patrol is capable of carrying the search for information into the area occupied by enemy forces—usually beyond the range of vision of friendly ground observation posts—and examining objects and events at close range.

12001. GENERAL MISSIONS

General missions for reconnaissance patrols include gathering information about location and characteristics of friendly or hostile positions and installations, terrain (routes, stream crossings), and obstacles.

12002. SPECIFIC MISSIONS

Locate the Enemy

Reconnaissance patrols try to determine the—

- Location of enemy forces, installations, and equipment.
- Identification of enemy units and equipment.
- Strength of enemy forces.
- Disposition of enemy forces.
- Movement of enemy personnel and equipment.
- New or special types of weapons.
- Presence of mechanized units.
- Unusual enemy activity.
- Presence of NBC equipment.

Reconnaissance of Enemy Wire Obstacles

A patrol with the mission of investigating enemy-emplaced wire obstacles employs a formation providing for all-around security and takes precautions against being observed by the enemy. The patrol leader and one Marine inspect each gap and establish its location by means of compass bearings to prominent objects in the rear of enemy or friendly positions or through the use of global positioning satellite receivers.

Reconnaissance of Contaminated Areas

A patrol with the mission of investigating a contaminated area reconnoiters and marks the area's boundaries. Patrol members wear protective clothing and field protective masks. The patrol leader reports the extent of the area, the ttype of agent used, the terrain and vegetation and the method of marking the contaminated area. A sketch of the contaminated area, prepared by the patrol leader, should be included in the patrol report.

Reconnaissance of Enemy Minefields

Patrols assigned to reconnoiter enemy minefields are often composed of specially trained personnel. The type of mine is identified and the possibility of bypassing the area determined.

Reconnaissance of Terrain

The reconnaissance patrol is particularly suited for gathering information about the terrain within their area of operations. (Refer to MCRP 2-15.3B, *Reconnaissance Reports Guide*.) Commanders must know the location and condition of bridges, streams, and roads in order to make proper plans for the movement of troops. The following information should be obtained by a reconnaissance patrol assigned such missions:

- Bridges:
  - Maximum load capacity.
  - Material used in construction (wood, stone, concrete or steel).

- Material used for abutments (dirt, wood, masonry, concrete).
- Type and number of supporting members of the bridge.
- Condition and dimensions of bridge flooring.
- Presence or absence of mines and demolition charges on the bridge and at entrances/exits.
- Terrain crossed (ravine, stream, road).
- Primary use (rail, auto, foot).
- Location and type of an alternate crossing if bridge is unusable.
- Streams and fords:
  - Width and depth of stream.
  - Composition of the bottom (mud, sand, gravel, rock).
  - Speed of current in miles per hour.
  - Composition and gradient of banks.
  - Surrounding terrain and vegetation cover.
- Roads:
  - Width.
  - Composition of surface.
  - Condition.
  - Road blocks.
  - Gradient/degrees of slopes.
  - Curves (location and length).
  - Pitch and culvert locations.
  - Primary use and frequency of use.

12003. TYPES OF RECONNAISSANCE

Route

Route reconnaissance is a directed effort to obtain detailed information of a specified route and all terrain from which the enemy could influence movement along that route. (Refer to MCRP 5-12C.) It is also described as a form of reconnaissance focused along a specific line of communications—such as a road, railway, or waterway—to provide new or updated information on route conditions and activities along the route. (Refer to MCRP 5-12A, *Operational Terms and Graphics*.)

A route reconnaissance normally precedes the movement of forces. It provides detailed information about a specific route and the surrounding terrain that could be used to influence movement along that route. Considerations include trafficablility, danger areas, critical points, vehicle weight and size limitations, and locations for friendly obstacle emplacement.

The objective area is normally defined by a line of departure, a route, and a limit of advance.

Area

An area reconnaissance is a directed effort to obtain detailed information concerning the terrain or enemy activity within a prescribed area such as a town, ridge line woods or other features critical to operations. (Refer to MCRP 5-12C.) An area reconnaissance could also be made of a single point, such as a bridge or installation. (Refer to MCRP 5-12A.)

Emphasis is placed on reaching the area without being detected. Enemy situations encountered en route are developed only enough to allow the reconnoitering unit to report and bypass.

Zone

A zone reconnaissance is a directed effort to obtain detailed information concerning all routes, obstacles (to include chemical or radiological contamination), terrain, and enemy forces within a zone defined by boundaries. A zone reconnaissance is normally assigned when the enemy situation is vague or when information concerning cross-country trafficability is desired. (Refer to MCRP 5-12C.)

The commander specifies specific routes or areas of interest within the zone. The zone to be reconnoitered usually is defined by a line of departure, lateral boundaries, and a limit of advance.

12004. TASK ORGANIZATION

A reconnaissance patrol is organized around the current structure of the Marine rifle squad with one or more of the squad's fire teams assigned as the reconnaissance element to reconnoiter or maintain surveillance over the objective. At least one fire team acts as a security element whose functions are to—

- Secure the objective rally point (see para. 11003).
- Give early warning of enemy approach.
- Protect the reconnaissance unit.

Reinforcing the squad is based on METT-T. If two squads are required, then the mission should be assigned to a platoon.

A company may send a reconnaissance patrol to specifically reconnoiter a given area (a riverbed or a bridge), or to maintain surveillance on a particular streambed for the next four nights. Reconnaissance patrols perform three basic functions en route to and from the objective:

- Provide control.
- Provide security while moving and in the objective area.
- Conduct reconnaissance or surveillance.

Depending on the size of the reconnaissance patrol, an element must be tasked to reconnoiter the area or zone. In the case of a surveillance mission, two teams are recommended to share the same task (so one could lay up and rest, while the other lays low and hides to maintain strict silence and no movement while surveillance is being maintained on the objective). Whatever the case, each unit that has a surveillance mission or reconnaissance mission must be thoroughly briefed as to what essential elements of information are to be collected as well as their location in the objective area.

The patrol should be organized with one or two fire teams to actually conduct the reconnaissance mission and the remaining fire team to provide security. Each fire team should be prepared to assume either mission. A small area reconnaissance patrol needs only one fire team for the assigned mission. A patrol with a wider area reconnaissance mission should use two fire teams to physically conduct the mission and one fire team for cover and/or security.

The security element for a reconnaissance patrol should be organized to cover the likely avenue of approach into the objective area, to protect the units conducting the reconnaissance, and to cover the objective rally point.

The variations of the special organization for reconnaissance patrols illustrate the patrol leader's flexibility in organizing the patrol to meet mission requirements. The patrol leader is not limited to the variations reflected herein but can choose any special organization as long as that task organization accommodates the requirement for command and control, reconnaissance, and security.

12005. SIZE OF RECONNAISSANCE PATROLS

A reconnaissance patrol should be kept to the minimum number of personnel required to accomplish the mission. A mission requiring a patrol to remain away from its unit for a considerable period of time, or one requiring a patrol to send back information by messenger, increases the size of the patrol. Reconnaissance patrols seldom exceed a squad in strength. Unit integrity should be preserved whenever possible. Intelligence personnel, interpreters, and other specialists, such as radio operators or engineers, are assigned to a patrol if the particular mission requires.

12006. RECONNAISSANCE EQUIPMENT

Patrol members are armed and equipped as necessary for accomplishing the mission. The automatic rifle in each fire team provides a degree of sustained firepower in case of enemy contact. The patrol should have at least two: pairs of binoculars, pairs of wire cutters, maps, compasses, and watches. Night observation equipment may be used. Pencils and small notebooks are carried so notes and sketches can be made. A message book with message blanks and overlay paper is mandatory.

12007. RECONNAISSANCE PATROL ACTIONS AT THE OBJECTIVE AREA

Route

The patrol leader halts and conceals the patrol near the objective area in the final preparation position. The patrol leader then conducts leader's reconnaissance to confirm the plan for positioning the security teams and employing units assigned to the reconnaissance mission. The patrol leader returns to the patrol and positions the security to provide early warning of enemy approach and secure the objective rally point. The reconnaissance unit(s) then reconnoiters the objective area (route). The reconnaissance unit may move to several positions, along or adjacent to the specific route, in order to conduct a thorough reconnaissance. After completing the reconnaissance, each reconnaissance team moves to the objective rally point and reports to the patrol leader. The patrol then returns to friendly lines and the leader makes a full report.

Area

The patrol leader halts and conceals the patrol near the objective area in the final preparation position. The patrol leader then conducts leader's reconnaissance to pinpoint the objective and confirm the plan for positioning the security teams and employing units assigned the reconnaissance mission. The patrol leader returns to the patrol and positions security to provide early warning of enemy approach and secure the objective rally point. The reconnaissance unit(s) then reconnoiters the objective area. The reconnaissance unit may move to several positions, perhaps making a circle around the objective area, in order to conduct a thorough reconnaissance. When the reconnaissance is completed, the patrol leader assembles the patrol and tells members what has been observed and heard. Other patrol members contribute observations. The patrol then returns to friendly lines and the patrol leader makes a full report.

Zone

The patrol leader halts the patrol at the final preparation position, confirms the plan, and conducts leader's reconnaissance. The patrol leader positions the security team and sends out the reconnaissance team. When the entire patrol is used to reconnoiter the zone, it provides its own security. After completing the reconnaissance, each reconnaissance team moves to the objective rally point and reports to the patrol leader. The patrol then returns to friendly lines.

CHAPTER 13. COMBAT PATROLS

Combat patrols are assigned missions that usually include engaging the enemy. They are fighting patrols. Every combat patrol has a secondary mission: gaining information about the enemy and terrain. Combat patrols are employed in both offensive and defensive combat operations and they assist the parent unit in accomplishing its mission by inflicting damage on the enemy; establishing and/or maintaining contact with friendly and enemy forces; denying the enemy access to key terrain; and probing enemy positions to determine the nature and extent of enemy presence.

13001. TASK ORGANIZATION

A combat patrol is organized around the current structure of the Marine rifle platoon. A combat patrol leader should use the unit's normal organization (fire team, squad, and platoon) in assigning functions, patrol missions, and chain of command. Combat patrols must be able to perform the following four basic functions en route to and from the objective:

- Provide control.
- Provide security.
- Provide support by fire.
- Attack or assault the objective.

A rifle platoon could task-organize as a combat patrol as follows:

- Platoon headquarters (patrol headquarters).
- First squad (security).
- Second squad (support).
- Third squad (assault).

Every combat patrol must—

- Provide a control mechanism in the form of a headquarters.
- Designate a unit (a fire team or squad) to provide security while moving en route to the objective or while at the objective. At the objective area, this unit isolates the objective area, secures the objective rally point, and covers the withdrawal.
- Designate a unit to act as support. This unit provides the base of fire in the attack or covers withdrawals or advances.
- Designate a unit(s) to conduct the attack or assault. This unit(s) engages the enemy at the objective area by fire and maneuver or movement. It also operates immediately in the objective area (searching, demolition, prisoners of war, etc.). Paragraph 9001 outlines the general organization of combat patrols. As in the case with reconnaissance patrols, the task organization of a combat patrol depends on the specific mission assigned. If any special requirements are generated because of the specific mission, the patrol is task-organized to fit the needs of the mission.

13002. EQUIPMENT

Combat patrols are armed and equipped as necessary for accomplishing the mission. In addition to binoculars, wire cutters, compasses, and other equipment generally common to all patrols, it usually carries a high proportion of automatic weapons and grenades. Communications with higher headquarters is important as success of the mission may depend on being able to call for supporting fires. Also, internal radio communications with the units and teams may be useful. However, the patrol must not be so overburdened with equipment as to impede movement or mission accomplishment.

13003. RAID PATROLS

A raid is a surprise attack on an enemy force or installation with the attacking force withdrawing after accomplishing its mission. Raids destroy or capture enemy personnel or equipment, destroy installations, or free friendly personnel who have been captured by the enemy. Patrolling techniques are used in planning and when moveing to and from the objective. (Refer to MCWP 3-41.2, *Raids*.) Surprise, firepower, and violence of action are the keys to a successful raid. Patrols achieve surprise by attacking—

- When the enemy is least prepared (e.g., during periods of poor visibility such as darkness, rain, fog, or snow).
- From an unexpected direction. (This might be accomplished by approaching through a swamp or other seemingly impassable terrain.)
- With concentration of firepower at critical points within the objective.

Planning

A successful raid requires detailed planning. The leader of a combat patrol engaged in raiding must anticipate probable situations and decide upon definite courses of action to meet them. Rehearsals are imperative.

A raid patrol conducts such missions as destroying an enemy outpost or seizing prisoners from an observation post or lightly defended position.

While preparing for the mission, the patrol leader requests fire support required for the accomplishment of the mission. If practical, artillery and mortars should be employed to isolate the objective to prevent movement of enemy reinforcements into the area.

Execution

The leader's plan must be detailed and complete. All of the considerations outlined in chapter 11 must be covered. Patrol formations must provide for ease of control and all-around security while moving to and from the objective area and provide for rapid and coordinated deployment of the various units once the objective area is reached. The leader's plan usually includes the encirclement of the hostile position—either physically or by fire—in order to isolate it during the assault.

The final simultaneous assault against the objective develops when enemy defensive fires at the objective are suppressed by either friendly fire superiority or surprise. The assault is covered by the fire of the unit assigned the function of support by fire.

Grenades, SMAWs, and demolitions are most effective for clearing bunkers.

Security units are posted to isolate the objective. The patrol leader signals them when the withdrawal begins. As a minimum, security is on each flank and to the rear (at the objective rally point).

Actions at the Objective

The patrol leader halts the patrol near the objective at the final preparation position. Security is established and the leader's reconnaissance is made with appropriate subordinate leaders. When the leaders return to the patrol, they confirm previous plans or announce any changes. Movements are arranged so all units reach their positions simultaneously. This improves the patrol's capability for decisive action, if prematurely detected by the enemy.

The teams of the security element move to positions to secure the objective rally point, give early warning of enemy approach, block avenues of approach into—and prevent enemy escape from—the objective area. As the assault element moves into position, the security element informs the patrol leader of all enemy activity, firing only if detected or on the patrol leader's order. Once the assault element commences action, the security element prevents enemy entry into or escape from the objective area. The security element covers the withdrawal of the assault element (and support element, if employed) to the objective rally point, withdrawing only upon order or prearranged signal.

As the assault element approaches the objective, it deploys early enough to permit immediate assault if detected by the enemy. Each team uses stealth while moving into proper position. On command, or if one or more of the assault element is detected and fired upon by the enemy, the support element opens fire to neutralize the objective, then ceases or shifts fire according to prearranged plans and signals. As supporting fires cease or shift, the assault element assaults the objective. Demolition, search, and other teams are protected by the assault element while they work. On order, the assault element withdraws to the objective rally point.

If a support element is employed, its leader deploys teams to provide fire support for the assault element. Each member of the support element must know the scheme of maneuver to be used by the assault element, specific targets or areas to be neutralized by fire, and the signals that will be employed to commence, shift, and cease fires. The support element withdraws on order of the patrol leader. At the objective rally point, the patrol leader quickly reorganizes the patrol and begins the return movement to friendly lines.

13004. CONTACT PATROLS

General

Contact patrols establish and/or maintain contact to the front, flanks or rear by—

- Establishing contact with an enemy force when the definite location of the force is unknown.
- Maintaining contact with enemy forces through direct and/or indirect fires, or observation.
- Avoiding decisive engagement with the enemy.

Task Organization and Equipment

Task organization and equipment depend on the known enemy situation and anticipated enemy contact. A patrol sent out to establish contact with an enemy force is organized, armed, and equipped to overcome resistance of light screening forces in order to gain contact with the main enemy force. It is not organized and equipped to engage the main enemy forces in combat. Communication is paramount; radios must be reliable over the entire distance covered.

Actions at the Objective

The patrol leader selects a series of objectives. Once an objective is reached, the patrol leader initiates a planned set of actions in order to establish and maintain contact with the enemy. These plans and actions are guided by the missions to establish or maintain contact—not to engage in decisive combat. Contact with the enemy is maintained for the purposes of surveillance, applying pressure, and preventing seizure of the initiative. If the contact patrol becomes decisively engaged with the enemy, many of the tasks originally assigned to the patrol cannot be accomplished, since the enemy has seized the initiative and friendly forces are not forced to react.

13005. AMBUSH PATROLS

General

An ambush is a surprise attack from a concealed position upon a moving or temporarily halted target. It is one of the oldest and most effective types of military actions. Ambush patrols conduct ambushes of enemy patrols, resupply columns, and convoys. The intent of an ambush is to place the enemy in a dilemma where staying in the kill zone or attempting to move out of it prove equally lethal. The ambush may include an assault to close with and decisively engage the enemy, or the attack may be by fire only.

Purpose of Ambushes

Ambushes are executed for the general purpose of reducing the enemy's over-all combat effectiveness and for the specific purpose of destroying its units. The cumulative effect of many small ambushes on enemy units lowers enemy troop morale and harasses the enemy force as a whole.

Destruction is the primary purpose of an ambush because loss of men killed or captured, and loss of equipment and supplies destroyed or captured, reduces the overall combat effectiveness of the enemy.

Harassment, though less apparent than physical damage, is a secondary purpose of ambushes. Frequent ambushes force the enemy to divert men from other missions to guard convoys, troop movements, and carrying parties. When enemy patrols fail to accomplish their mission because they are ambushed, the enemy is deprived of valuable information. A series of successful ambushes causes the enemy to be less aggressive and more defensive minded. His men become apprehensive, overly cautious, reluctant to go on patrols, seek to avoid night operations, are more subject to confusion and panic if ambushed, and in general, decline in effectiveness.

Classification of Ambushes

A *deliberate ambush* is one in which prior information about the enemy permits detailed planning before the patrol departs for the ambush site. Information needed to plan a deliberate ambush includes the size, composition, and organization of the force to be ambushed; how the force operates; and the time it will pass certain points or areas. A deliberate ambush may be planned for such targets as—

- Any force if sufficient prior information is known.
- Enemy patrols that establish patterns by frequent use of the same routes or habitually depart and re-enter their own areas at the same point.
- Logistic columns.
- Troop movements.

An *ambush of opportunity* is conducted when available information does not permit detailed planning before the patrol departs. This is the type of ambush that an infantry unit normally conducts. An ambush of opportunity should not be confused with a hasty ambush. An ambush of opportunity is a planned ambush; a hasty ambush is an immediate action. In planning for an ambush of opportunity, the patrol must be prepared to execute any of several courses of action based on the types of targets that may be ambushed and must rehearse prior to departure. The course of action taken is determined when the opportunity for ambush arises.

The patrol leader may be directed to reconnoiter an area for a suitable ambush site, set up at the site selected, and execute an ambush against the first profitable target that appears.

The patrol may depart just after dark, move to a specific point, observe until a designated time, ambush the first profitable target after that time, and return before daylight.

A *hasty ambush* is an immediate action where the patrol makes visual contact with an enemy force and has time to establish an ambush without being detected. The actions for a hasty ambush must be well rehearsed and accomplished through the use of hand and arm signals given from the patrol leader.

Types of Ambushes

There are two types of ambushes: point and area. The *point ambush* is one where forces are deployed to attack along a single killing zone. The *area ambush* is one where forces are deployed as multiple related point ambushes.

A point ambush, whether independent or part of an area ambush, is positioned along the enemy's expected route of approach. Formation of the forces conducting the ambush is important because, to a great extent, the formation determines whether a point ambush is able to deliver the heavy volume of highly concentrated fire necessary to isolate, trap, and destroy the enemy.

The ambush formation to be used is determined by careful consideration of possible formations and the advantages and disadvantages of each in relation to—

- Terrain.
- Visibility.
- Forces.
- Weapons and equipment.
- Ease or difficulty of control.
- Target to be attacked.
- Combat situation.

For a detailed discussion of ambush formation, see appendix D.

Ambush Operation Terms

The *ambush site* is the location where an ambush is established.

The *killing zone* is that portion of an ambush site where fires are concentrated to trap, isolate, and destroy the target. On little-traveled roads, an obstacle placed in a defile, in the woods, on a bridge or on a steep upgrade can be used effectively to force vehicles to halt, and thus render the occupants vulnerable to attack. Antitank mines may be emplaced and the occupants of the wrecked vehicle killed or captured while still dazed by the explosion.

A *near ambush* is a point ambush where the attacking force is located within reasonable assaulting distance of the killing zone (50 meters is a guide). A near ambush is most often conducted in close terrain, such as a jungle or heavy woods.

A *far ambush* is a point ambush where the attack force is located beyond reasonable assaulting distance of the killing zone (beyond 50 meters is a guide). A far ambush may be more appropriate in open terrain offering good fields of fire or when the target will be attacked by fire only.

Factors for a Successful Ambush Patrol

There are many factors that give the ambush its best chance of success. The ideal situation would be to position the ambush on favorable terrain and have detailed planning completed beforehand.

The patrol leader of an ambush looks for the most favorable terrain in which the enemy is canalized between two obstacles with limited opportunity to attack or escape. Suitable areas include defiles, small clearings, bends in trails, and steep grades. Dense undergrowth adjacent to the ambush site permits observation from concealed positions. The ambush patrol should have maximum cover and concealment, not only for the firing positions, but for the routes of withdrawal. The enemy should be in an area offering as little protection from fire as possible. Favorable fields of fire include stretches of road, trail or open ground of at least 100 meters for machine guns and 15 meters for rifle fire and grenades. The ambush site can be improved by constructing obstacles—such as felled trees, wire, land mines, or booby traps—to impede the enemy.

Planning

A deliberate ambush or an ambush of opportunity requires thorough planning.

A deliberate ambush plan is based on extensive knowledge of the enemy and terrain, and is planned and rehearsed in great detail. A physical reconnaissance of the ambush site is made during the preparation phase and information gained is incorporated into the plan. All likely immediate actions of the enemy when ambushed are examined. Planned counteractions are developed and rehearsed.

In planning an ambush of opportunity, any available information on the enemy and terrain is used. A tentative plan for the ambush that incorporates all anticipated actions is developed and rehearsed. However, the bulk of planning is done concurrently during the patrol leader's reconnaissance of the prospective ambush site. In a rapidly developing situation, hasty ambush immediate action is employed.

The route and ambush site considerations apply to both deliberate ambushes and ambushes of opportunity. A primary route that allows the patrol to enter the ambush site from the rear is planned. Entering the prospective killing zone is avoided. If the killing zone must be entered to place mines or explosives, care is taken to remove any tracks and signs that might alert the enemy and compromise the ambush. If mines or explosives are to be placed on the far side of the ambush site, or if the appearance of the site from the enemy's viewpoint is to be checked, a wide detour is made around the killing zone. Care is taken to remove any tracks that might reveal the ambush. An alternate route from the ambush site to the objective rally point, as in other patrols, is planned.

Maps and aerial photographs are used to carefully analyze the terrain. When possible, an on-the-ground reconnaissance of the ambush site is made prior to occupation. Obvious ambush sites are avoided as the element of surprise is even more difficult to achieve in these areas. An ambush site must provide for—

- Favorable fields of fire.
- Occupation and preparation of concealed positions.
- Canalization of the target into the killing zone. (An ideal killing zone restricts the enemy on all sides, confining him to an area where he can be quickly and completely destroyed. Natural obstacles, such as cliffs, streams, embankments, or steep grades, are used whenever possible to force vehicles to slow down. Man-made obstacles, such as barbed wire, mines, and craters in the roads, are used to supplement natural obstacles.)
- Covered routes of withdrawal that enable the ambush force to break contact.
- Avoidance of enemy pursuit by fire.

Occupation of Ambush Site

The surrounding area is searched for enemy patrols prior to occupation of the ambush site. Ambush formations are used to physically deploy the patrol in a manner to inflict maximum destruction upon the enemy and to provide maximum security to the patrol. Ambush formations are contained in appendix D.

Positions

The patrol is moved into the ambush site from the objective rally point. Security is positioned first to prevent surprise while the ambush is being established. Automatic weapons are then positioned so each can fire along the entire killing zone. If this is not possible, overlapping sectors of fire are provided to cover the entire killing zone. The patrol leader then selects his position, located so he can tell when to initiate the ambush. Riflemen and grenadiers are positioned and sectors of fire are assigned to cover any dead space left by the automatic weapons. The patrol leader sets a time by which positions are to be prepared. Patrol members clear fields of fire and prepare positions in that order, with attention to camouflage for both.

Suitable Objective Rally Point

An easily located objective rally point is selected and made known to all patrol members. The objective rally point is located far enough from the ambush site so that it will not be overrun if the enemy assaults the ambush. Routes of withdrawal to the objective rally point are reconnoitered. Situation permitting, each man walks the route he is to use and picks out checkpoints. When the ambush is to be executed at night, each man must be able to follow his route in the dark. After the ambush has been executed, and the search of the killing zone completed, the patrol is withdrawn quickly but quietly, on signal, to the objective rally point where it reorganizes for the return march. If the ambush was not successful and the patrol is pursued, withdrawal may be by bounds. The last group may arm mines, previously placed along the withdrawal route, to further delay pursuit.

Local Security

Security must be maintained. Security elements do not usually participate in the initial attack, but protect the rear and flanks, and cover the withdrawal.

Patience

The Marines of the ambush force must control themselves so that the ambush is not compromised. Patience and self-discipline are exercised by remaining still and quiet while waiting for the target to appear, particularly if the patrol occupies the ambush site well ahead of the arrival of the enemy. Patience is necessary so as not to alert the enemy to the presence of the ambush.

Surprise

Surprise must be achieved, or the attack is not an ambush. If complete surprise cannot be achieved, it must be so nearly complete that the target is not aware of the ambush until too late for effective reaction. Surprise is achieved by careful planning, preparation, and execution so that targets are attacked when, where, and in a way for which they are least prepared.

Coordinated Fires

Properly timed and delivered fires contribute heavily to the achievement of surprise, as well as to destruction of the enemy. The lifting or shifting of fires must be equally precise; otherwise, the assault is delayed and the enemy has an opportunity to recover and react. All weapons, mines, and demolitions are positioned and all fires, including those of available artillery and mortars, are coordinated to achieve the following results:

- Isolation of the killing zone to prevent the enemy's escape or reinforcement.
- Surprise delivery of a large volume of highly concentrated fires into the killing zone.

Control

Close control of the patrol is maintained during movement to, occupation of, and withdrawal from the ambush site. This is best achieved through rehearsals and establishment and maintenance of good communications. When the enemy approaches, the temptation to open fire before the signal is given is resisted. The patrol leader must effectively control all elements of the ambush force. Control is most critical at the time the enemy approaches the killing zone. Control measures must provide for—

- Early warning of enemy approach.
- Fire control. Withhold fire until the enemy has moved into the killing zone, then open fire at the proper time.
- Initiation of appropriate action, if the ambush is prematurely detected. Individual patrol members must be prepared to react if detected by the enemy prior to the initiation of the ambush.
- Timely and orderly withdrawal of the ambush force from the ambush site and movement to the objective rally point.

It is important to remember that an ambush patrol should have four distinct signals: one to open fire (with an alternate signal to open fire to be used at the same time as the primary); a signal to cease fire or shift fire; a signal to assault or search the killing zone; and a signal to withdraw. The signal to open fire should meet two criteria: first, it should be the firing of a weapon that will kill the enemy; secondly, it should be a weapon reliable in any weather condition. A good primary signal is a Claymore mine, and an alternate signal would be a closed bolt weapon (M16A2). Open bolt weapons (M240G, M249) should not be relied upon to initiate an ambush.

Execution of an Ambush

The manner in which the patrol executes an ambush depends primarily on whether the ambush's purpose is harassment or destruction. To a lesser degree, the execution of the ambush is determined by whether the ambush is deliberate or an ambush of opportunity.

When the primary purpose is harassment, the patrol seals off the area with security teams to prevent enemy reinforcement and escape. Maximum damage is inflicted with demolitions and automatic weapons fire. The patrol delivers a very heavy volume of fire for a short time and withdraws quickly and quietly. The patrol avoids being seen by the enemy.

When the primary purpose of the patrol is destruction, the area is sealed off with security units. Maximum damage is inflicted with demolitions, antitank weapons, and automatic weapons fire from the support team or element. When these fires cease or shift, an assault is launched into the killing zone with heavy fire and violence to complete destruction. The assault unit provides security, while designated teams search and/or capture personnel and destroy vehicles and equipment. On the patrol leader's command, or by prearranged signal, all units withdraw to the objective rally point and move out quickly.

When the patrol's primary purpose is to obtain supplies or capture equipment, security units seal off the area. Demolitions and weapons are used to disable vehicles. The assault unit must use care to ensure its fire does not damage the desired supplies or equipment. Designated teams secure the desired items; other teams then destroy enemy vehicles and equipment.

The most successful ambush is one where the attacker is deployed and concealed in such a way that the enemy will unknowingly be surrounded by fire. The usual method is for the attackers to deploy themselves along a trail or route the enemy will travel. The enemy is permitted to pass by the center of the attacker's force so that the attack can be made from the front, flank, and/or rear. One or two men are posted well forward and to the rear along the route to prevent any enemy from escaping. All fires should be delivered simultaneously on a prearranged signal.

An effective method of luring the enemy is for an ambushing patrol to cut communication or electrical wire. The patrol then deploys and ambushes the line repair crew when it arrives. Since the line crew may be protected by riflemen, the attackers must be careful to engage the entire party.

Vehicles and foot personnel moving on well-established transportation routes can sometimes be captured by altering or moving directional signs so as to divert the enemy into an area where he can be more readily attacked. The attack can best be accomplished at an obstacle, such as a stream or gully, that forces the enemy to stop or slow down.

After the enemy has been ambushed and destroyed, the unit quickly withdraws over a prearranged route to the objective rally point. Speed is very important, since the noise of the ambush could alert other nearby enemy units.

13006. SECURITY PATROLS

General

Security patrols are assigned missions that may or may not require them to engage the enemy. They are used in proximity to defensive positions, on the flanks of advancing units or in rear areas. Purposes of security patrols are to detect infiltration by the enemy, destroy infiltrators, and protect against surprise and ambush.

In any situation where there is a threat of attack, such as a rear area threatened by guerrillas or a facility that is under threat of a terrorist attack, all Marines, not just the infantry, must know how to conduct a security patrol.

In just the offensive operations, infantry units provide security patrols to screen their flanks, areas, and routes. Whereas, in defensive operations, security patrols are used to prevent the enemy from infiltrating an area, detect and destroy infiltrators, and prevent surprise attacks. In rear areas, particularly when there is guerrilla or terrorist threat, the requirement to conduct security patrols increases for all Marine air-ground task force (MAGTF) units ashore, particularly aviation and combat service support units.

Task Organization and Equipment

Generally, a Marine rifle squad or similar sized organization is considered ideal for security patrols. Communications are important to higher headquarters so that they receive information from the patrol; and communications are important to the patrol to request fire support, etc. The radio the patrol carries must have the range necessary for higher headquarters to be able to receive transmissions from anywhere along the patrol route, and the patrol must have a secondary means of reporting (i.e., flare signals upon contact).

Patrol Procedures

All of the procedures presented in previous sections are to be used in security patrols.

Patrol Planning

Security patrol planning includes—

- Rehearsing prior to departing friendly lines.
- Maintaining communications.
- Support by organic weapons.
- Reinforced if necessary.
- Using varied routes and never establishing a routine pattern.
- Staying within proximity of friendly units.

Patrol Techniques

Within rear areas, an irregular pattern of patrol is established and changed daily. Outside of friendly lines it would be prudent to establish a definite preplanned route for the patrol, of which all adjacent units know the route. The parent unit commander establishes frequent checkpoints for control. If checkpoints are designated, the patrol leader treats them as individual objectives to be searched and cleared.

The patrol has a definite plan as to what to do if contact with enemy is made, how to break contact, how to defend itself, and how to call for supporting fires. It is imperative that patrol members know what to do if they become split or separated; i.e., location of rally points and how to be recovered.

13007. URBAN PATROLS

General

As national strategy continues to focus on regional vice global conflicts, the Marine Corps will continue to conduct urban operations in various operational environments. Cities and towns are often the center of economic and political power and are therefore extremely vulnerable to urban insurgent activities and violence. The lessons learned from recent operations in Somalia, as well as experiences gained by British forces in Northern Ireland, provide the foundations of urban patrolling.

Enemy Forces

Enemy forces in urbanized areas range from organized military forces to low intensity engagements with insurgents, such as terrorists or local gangs. For Marine Corps doctrine and tactics, techniques, and procedures (TTP) for dealing with the higher intensity threat in urbanized areas, refer to MCWP 3-35.3, *Military Operations on Urbanized Terrain.*

Common Threat Tactics

Cities provide cover and concealment for both friendly forces and enemy forces. However, enemy forces generally find active support only in certain areas of the town or city. The urban-based insurgent or terrorist usually lives in a friendly community or in one where the people are too frightened to withhold support or inform anyone about the situation. The insurgent or terrorist normally maintains close contact with leaders and others friendly to the cause. The enemy will often have efficient communication and intelligence systems, sometimes involving women and children to provide cover for its activities.

The urban insurgent normally cannot, like his rural counterpart, establish bases and recruit large military units. He is generally an individual or a member of a relatively small group. He relies on the cover afforded by the city's populace and terrorizing them to coerce loyalty or support.

Urbanized areas tend to give the insurgent and/or terrorist many opportunities to initiate action and gain advantage. The normal presence of large numbers of people in cities provides the foe an opportunity to mass crowds quickly and manipulate demonstrations easily. The presence of women and children during mass demonstrations may restrict the courses of action available to friendly forces. Major incidents stemming from overreaction or excessive use of force by friendly forces may provide the insurgent with propaganda material. Publicity is easily gained in an urbanized area because major incidents can't be completely concealed from the local population. Insurgent successes can be exploited to discredit the ability of host nation police, friendly forces, and the civil government, and gain recognition for the insurgents' cause.

The urban insurgent or terrorist can usually be expected to operate more boldly than his rural counterpart. This is reflected in the enemy's tactics. A single sniper or bomber may be the norm in the urbanized area, whereas the rural threat is generally the more conventional ambush. In urbanized areas, explosive devices can be easily emplaced and used effectively against large groups or select individuals. Enemy forces may be expected to employ the following tactics in urbanized areas:

- Using local communications, such as radio and newspapers, for propaganda purposes.
- Disrupting industry and public services through strikes and sabotage.
- Generating widespread disturbances designed to stretch the resources of the security force.
- Creating incidents or massing crowds in order to lure the patrol or reaction force into a trap.
- Provoking security forces in the hope that they may react improperly, therefore discrediting the security force by means of propaganda.
- Sniping at roadblocks, outposts, sentries, and patrols.
- Attacking friendly bases with rockets and mortars.
- Planting explosive devices, either against specific targets or indiscriminately, to cause confusion and destruction, and lower public morale and confidence.
- Using ambush patrols.
- Firing on friendly helicopters.

Principles of Urban Patrolling

Patrolling in an urban environment often presents conditions considerably different and often more complex than those encountered in rural and less inhabited areas. While the principles of patrolling are still relevant in an urban situation, the nature of urban patrolling has led to the development of six specific urban patrolling principles. They are—

1. **Depth.** The restrictive, canalizing nature of urbanized terrain usually limits a patrol's ability to disperse laterally. To prevent the patrol from bunching up, patrols normally maintain dispersion along the length of a patrol formation.

2. **Mutual Support.** The positioning of units in-depth within the patrol enables one unit to cover another unit's movement and facilitates immediate action during various situations. Aircraft, vehicles, and snipers also provide good mutual support.

3. **Deception and Pattern Avoidance.** Deception and pattern avoidance are normally a planning consideration of the headquarters directing the entire patrolling plan. By varying patrol routes, durations, and departure times, hostile actions commonly used against urban patrols—such as enemy ambushes, and roadblocks—are more difficult to plan and may be preempted.

4. **Intra-patrol Communication.** Elements of an urban patrol must have the means to communicate with each other. Ideally, each element will possess a radio enabling it to remain in continuous communication that facilitates rapid response and reporting to higher headquarters, reaction force coordination, and coordination of actions with other patrols or fire support agencies.

5. **Establishment of a Reaction Force.** The nature of urbanized terrain (its compartmentalization) makes urban patrols more vulnerable to a wider range of hostile actions. The requirement for immediate, coordinated reinforcement of a patrol is best satisfied by employment of an established reaction force. The reaction force requires superior mobility (relative to the enemy's) and fire support to be effective for this task.

6. **Three-Dimensional Threat**. Patrolling in an urban environment requires constant attention to its three-dimensional aspect; hostile actions can originate from rooftops, streets, subsurface levels or combinations of all levels at once.

Classification of Urban Patrols

Mission

The vast majority of urban patrols are overt in nature, with their presence readily apparent to the local populace. Most urban patrols are combat vice reconnaissance patrols. The vast majority of urban patrols are security patrols. However, units may be assigned secondary tasks of reconnoitering specific or general areas along the patrol routes.

Raids normally involve a swift penetration of an objective to secure information, confuse the enemy or destroy installations. Raids include a planned withdrawal upon completion of the assigned mission.

Movement

Means of movement are as follows:

- Dismounted: movement on foot.
- Vehicular: movement by motorized, mechanized or armored vehicle.
- Helicopterborne: movement by helicopter, however, helicopterborne patrols will usually involve dismounted or vehicular movement after patrol insertion.
- Combination: movement using a combination of methods.

Task-Organization

Units task-organize to meet the specific requirements of the mission and situation. The Marine rifle squad is ideally suited for urban patrolling and can easily integrate attached specialists required by specific missions. Specialists who may accompany urban patrols include—

- Interrogator-translator team (ITT) and counterintelligence team (CIT) Marines.
- Explosive ordnance disposal (EOD) personnel.
- Members of host nation or allied military forces.
- Interpreters.
- Local community leaders.
- Local law enforcement officers.
- Public affairs personnel escorting media representatives.

Dismounted Patrol Organization

Patrols should maintain unit integrity (fire team and/or squad) in organizing elements for an urban patrol. Similar to reconnaissance patrols, the vulnerability of urban patrols necessitates that all elements must provide for their own security in addition to the combat aspect of the mission. Combined assault and security teams are an effective method to organize for all-around security.

Urban Patrol Base Operations

Urban patrols may operate from an established patrol base that may be located within the unit's assigned area of operations or an area designated for a patrol. The patrol base should be located in a building used exclusively for this purpose. Patrol bases may also be located within a larger site that houses other agencies, such as a higher headquarters' command echelon.

If located within a larger site, the patrol base will be included in the overall facility security plan. If isolated, the patrol base must consider the following security factors:

- External security:
  - Barrier plan.
  - Sentry posts.
  - Local security patrols.
- Internal security:
  - Covered positions for all Marines.
  - Contingency plan for hostile actions against patrol base.

The Reaction Force

The high probability of an urban patrol becoming involved in a hostile or volatile incident requires the establishment of a dedicated reaction force for rapid reinforcement, support or extraction of the patrol. Ideally, the reaction force is—

- Large enough and task-organized in a manner that it can meet and quickly defeat the expected threat. Reaction forces are normally tiered with a lead element (normally one-third of the unit's size) and a main body force (the other remaining two-thirds of the force). Reaction force response times routinely are determined in advance by higher headquarters.
- Ready to respond immediately.
- Motorized or mechanized and supported by close air support and other fire support.
- Familiar with the area of operations.
- Briefed on the patrol's plans and monitors the status of patrols in progress.
- Task-organized to be multimission capable.
- Able to communicate with the higher headquarters, fire support assets, patrol base, and the patrol.
- Controlled by higher headquarters, once employed.

Patrol Preparation

Planning

Higher headquarters will—

- Designate the area for patrol.
- Provide intelligence briefs and updates.
- Ensure liaison with allied forces and the civilian populace.
- Provide special equipment and personnel required for the mission (scout snipers, public affairs officer, interpreters, etc.)
- Provide urban maps, photos, terrain models as required.
- Consider deception and pattern avoidance when issuing mission.
- Prescribe rules of engagement (ROE).

Intelligence Brief

An intelligence brief is conducted by the S-2 officer or representative prior to a patrol conducting its mission. The brief adresses the situation relevant to the specific patrol (e.g., routes, areas, updated enemy situations).

Coordination

Higher headquarters will effect liaison with adjacent and allied forces, as well as civil authorities and other agencies, having a possible effect on the patrol. The patrolling unit generally follows the same procedures as those used during patrol planning and execution in a jungle or forest environment.

The Urban Patrol Order or Warning Order

The Urban Patrol or Warning Orders use the same format and considerations as noted in this publication for patrolling. They rely heavily on a detailed terrain model, photographs, and subterranean construction to ensure complete understanding of the plan.

Rehearsals

The limited size of the patrol base usually precludes the need for full-scale rehearsals. Immediate action drills, such as crossing danger areas, are rehearsed in as much detail as possible, despite the limited available space.

Inspections

Initial and final inspections are conducted in the same manner as other patrols. Attached personnel must be fully integrated into the patrol and familiar with the plan and unit standing operating procedures (SOPs).

Conducting an Urban Patrol

Movement

Individual and unit movement considerations are generally the same as those for other patrols. However, urban environments require consideration of additional factors. Because of these factors, an urban patrol leader should—

- Ensure that each movement within a patrol takes place under the observation or cover of another individual or element of the patrol.
- Know where cover can be taken in the event of a hostile incident or action.
- Be prepared for contact with civilians, especially children, during the patrol and be aware that they may intentionally attempt to distract patrol members.
- Expect the presence of vehicles (both moving and stationary) along the patrol route.
- Expect members of the patrol to be approached by dogs and what action to take if threatened.

Patrol Formations

Squad-sized Patrols. The need for immediate fire power outweighs the dangers of becoming canalized. In contrast to other types of patrols, the headquarters element of an urban patrol will normally locate at the lead of the patrol column. This allows the patrol leader greater flexibility through control of two combined assault and security (A&S) teams. The leaders of these elements tactically stagger their members on each side of the road (see fig. 13-1 on page 13-12).

A&S teams follow in trace of the headquarters element and maintain unit integrity on separate sides of the street. One unit will remain slightly to the rear to create a staggered interval between Marines on either side of the street. This allows A&S teams to take lateral routes in support of headquarters element without having to cross a street to do so.

Platoon-sized Patrols. Squads will generally travel abreast of each other, moving along parallel routes. The interval between squad-sized units and/or teams is situation-dependent, but is usually between 100 and 150 meters (roughly two city blocks; this often prevents visual contact between the units). The intent is to create less of a target to an aggressor, yet still allow the patrol to quickly react to an incident. Individuals within units or teams will move in a staggered column as in a squad-sized patrol. See figure 13-2, which is located on page 13-13).

Night

Night patrols will generally be at least squad-sized and will generally use the same formation as that for day patrols. At night, it may be necessary to close distances between individuals or elements to maintain control. Consideration should be given to the use of night vision devices and thermal weapons sights.

Navigation, Control, and Security Measures

The patrol leader is ultimately responsible for the navigation. The headquarters element normally functions as the base unit during movement. The designated navigator is normally assigned from within the headquarters element. City maps are often inaccurate or outdated; however, when used with aerial photographs and other navigational aids, they can be effective for urban navigation.

Checkpoints and phase lines should be related to major streets (alleys, buildings, bridges) for easy identification. Arrival at checkpoints and crossing of phase

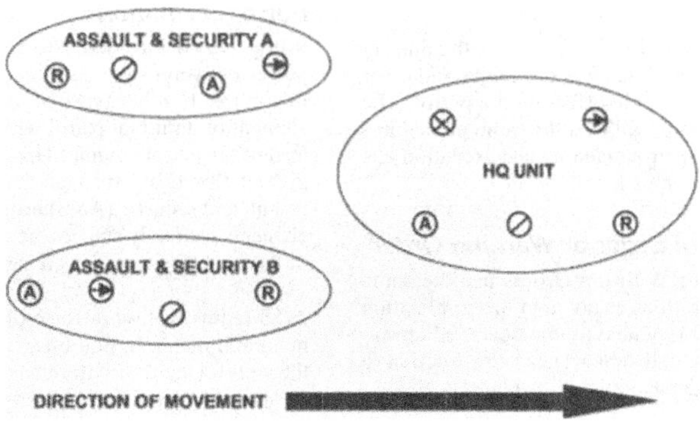

Figure 13-1. Squad-sized Dismounted Urban Patrol Formation.

lines should be relayed to higher headquarters using established brevity or codewords. A detailed patrol overlay is issued to both the reaction force and higher headquarters to keep them advised of routes and control measures used. A terrain model should be maintained at higher headquarters to aid in supervision and control of the patrol.

The use of camouflage should be limited to avoid frightening and confusing the local populace. The often overt nature of urban patrols may negate the need for camouflage. Patrols are sometimes deployed to show force presence and usually move on the urban street in plain view.

During daylight, patrols will routinely vary their rate of movement ranging from short halts to brief periods of double-timing. The British term for this urban patrolling technique is *hard-targeting*, meaning it makes the patrol harder for an enemy to target. Altering the rate of movement is intended to frustrate the enemy's ability to coordinate an attack or ambush against a targeted patrol.

Patrols should use short security halts, with Marines taking up mutually supporting firing positions. Marines must always work in pairs, ensuring mutual support. The last Marine in the element will provide rear security, but stays in his buddy's sight.

Individual Tasks

Individuals may be assigned collateral tasks performed throughout the patrol that may increase the patrol's situational awareness. Individual tasks may include—

- Vehicle spotter: looks for suspicious or known insurgent vehicles.
- Personnel spotter: observes and attempts to recognize previously identified enemy in crowds.
- Talker: attempts to gain information from casual conversations with the local populace. (Talkers are usually subordinate leaders or Marines with foreign language skills.)
- Searcher: conducts physical searches of vehicles and personnel while other patrol members provide cover and security.
- Marksman: engages point targets when the tactical situation does not permit massed or high volume fire. Other patrol members provide security to cover the marksman's engagement.

Departure of Friendly Lines

Urban patrols must vary their departure times to prevent being ambushed while exiting the patrol base.

Individual elements will usually depart exits at staggered times and at different movement rates, especially where sentries cannot provide cover. (This technique is used when the environment adjacent to the patrol base is dangerous.)

Once an element has exited friendly lines, a short halt is conducted in a predetermined, covered initial rally point 50 to 100 meters from the base. This ensures all elements are in position before the patrol continues.

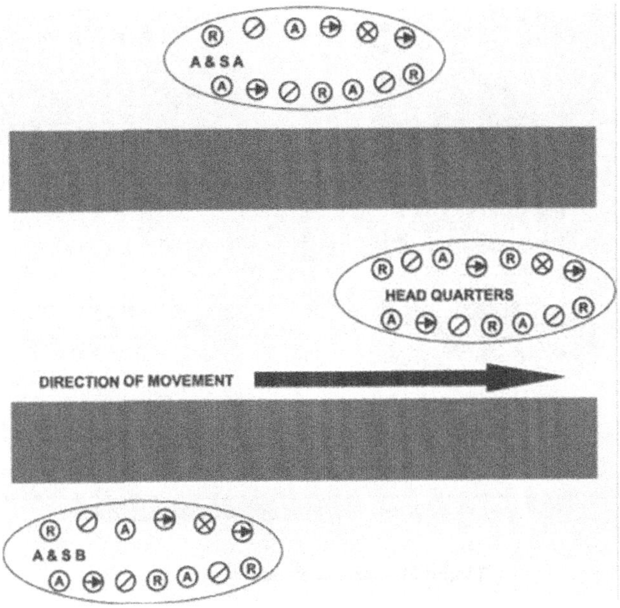

Figure 13-2. Platoon-sized Dismounted Urban Patrol Formations.

Exit points, routes from the base, departure techniques, and locations of IRPs should be varied constantly to avoid setting a pattern. This pattern avoidance may also include using vehicles to insert patrols away from the urban patrol base and employing empty vehicles as part of a deception plan.

Danger Areas

Urban patrols may encounter hundreds of danger areas during a single patrol. The three-dimensional threat requires keen situation awareness by every patrol member. Many danger areas can be dealt with simply by avoidance, while others require an adjustment of patrol formation, movement rate, etc. In the urban environment, places to be treated as danger areas are points that pose a major threat to the patrol, such as local political and religious headquarters, weapons containment areas, roads and routes that canalize movement and direct fire, and any area with a history of repeated contact.

Near and far side rally points are designated and briefed during the issuance of the patrol order. Squad-sized formations may use the A&S teams to provide flank security for the headquarters element and for each other. The headquarters element identifies the danger area and takes up a position on the near side of intersection. Individuals provide all-around security (see fig. 13-3 on page 13-14).

Two Marines (one from each A&S team) are designated to move through the headquarters element and establish respective firing positions on the near side of the danger area covering the patrol's near side flanks. They are followed by a second pair (again, one Marine from each A&S team) that moves across to the far side of the danger area and establishes respective firing positions covering the patrol's far side flanks (see fig. 13-4 on page 13-14).

Once near and far side flank security is established, the headquarters element moves across to the far side of danger area (see fig. 13-5 on page 13-15).

The remaining A&S team members then cross the danger area and join the headquarters element on the far side (see fig. 13-6 on page 13-15).

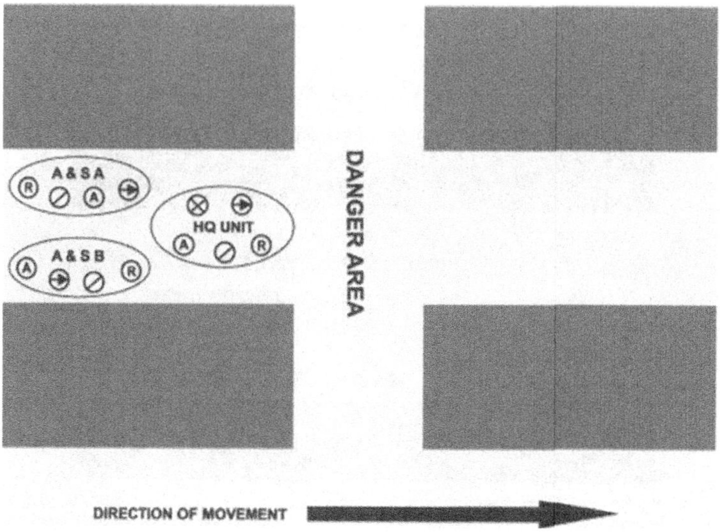

Figure 13-3. Approaching Danger Area.

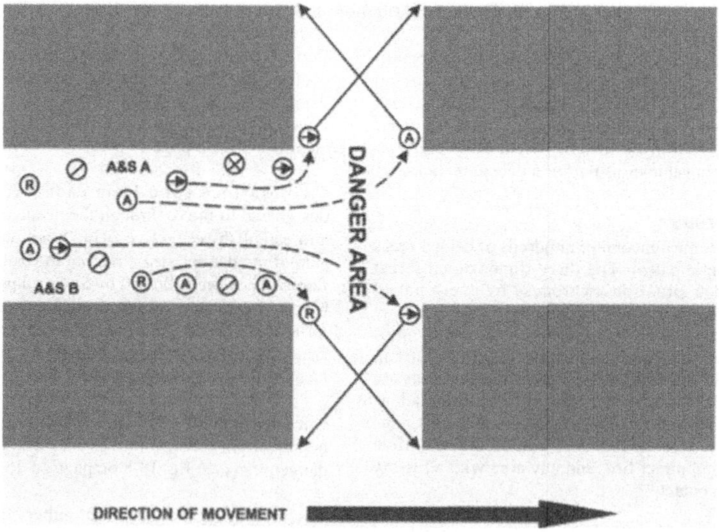

Figure 13-4. Securing Flanks.

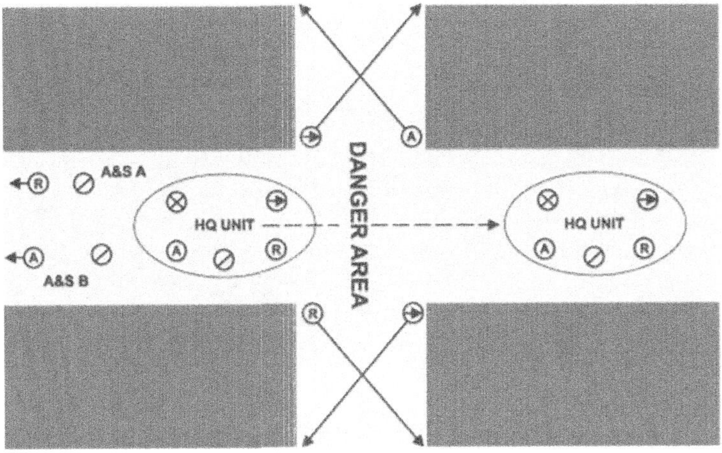

Figure 13-5. HQ Element Moves Through.

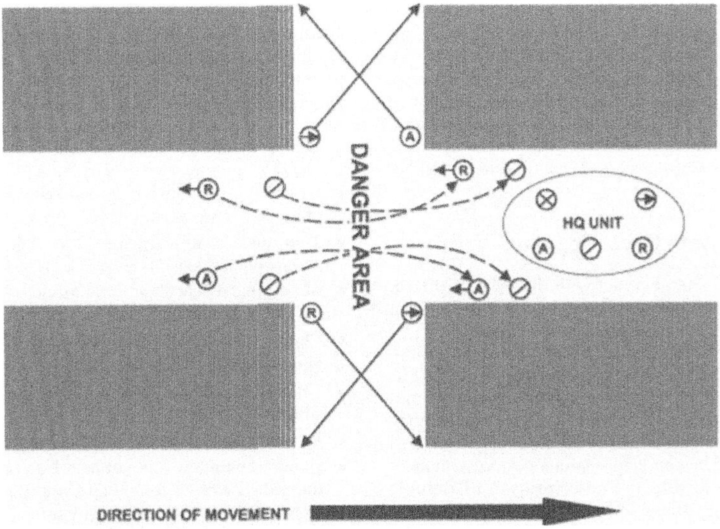

Figure 13-6. A&S Elements Move Through.

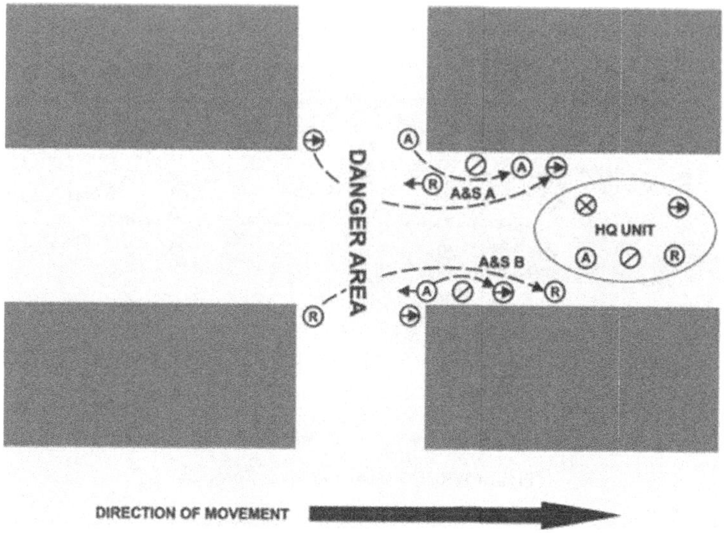

Figure 13-7. Patrol Resumes Movement Along Intended Route.

Once the trail A&S team members cross the danger area, they take up rear security and cover the movements of the near and far side flank security teams as they return to their positions in the patrol formation. The near side security team should collapse back first, followed by their far side counterparts. The patrol then resumes its advance away from the danger area (see fig. 13-7).

Interaction with Local Populace

Urban patrols must interact with the local populace. Patrols are at first a novelty to the civilians but can quickly become an unwelcome intrusion. The movement of the patrol must be fast enough to prevent the enemy from massing their fires upon it, but deliberate enough to ensure adequate security and mutual support. Patrol members must realize that they are usually the only Marines the local populace will encounter and that an inappropriate gesture, comment or act could lead to the deterioration of rapport between U.S. forces and the general population. Marines must remember that the vast majority of the individuals with whom they come in contact will be noncombatants attempting to survive in trying political, economic, and social situations.

Hostile incidents often seem to occur spontaneously, but there are usually indications that can alert Marines to imminent danger. The most obvious are the sudden alteration of normal routines, patterns, and attitudes of the local populace or other unusual activity. Some examples include—

- Observers on rooftops, in windows, etc., who are obviously tracking the patrol.
- The unusual absence of pedestrian traffic and people on porches.
- Stores, markets or street vendors closed suddenly or without explanation.
- Changes in civilian attitude toward patrol members.
- Unknown individuals or vehicles in the patrol area.
- Unfamiliar vehicles parked in the patrol area (possible car bomb).
- Roadblocks.
- Children throwing rocks at patrols to possibly draw the patrol's attention away from a more serious danger, such as a deliberate ambush.
- Vehicles riding unusually low due to overloading (possibly ferrying people, weapons, explosives).
- Agitators trying to provoke an incident with patrol members.

- Absence of the usual stray dogs (dogs are adept at sensing danger and avoiding it).
- Anti-American graffiti suddenly appearing in the patrol area.
- Pictures of enemy leaders and martyrs posted in the patrol area.
- Civilian workers failing to appear at U.S. or friendly bases.
- Normal deliveries and pick-ups conducted late or early without reason.
- Sudden change of civilian sentiment in newspaper articles, radio broadcasts or other media.
- Women and children leaving to live elsewhere.

Immediate Actions Upon Enemy Contact

Reaction to Sniping

Snipings are often executed from a single firing point, but coordinated snipings delievered from multiple points are not uncommon.

Patrol element or team leaders should constantly try to identify likely firing points and anticipate their own reactions to a possible shooting. Normally, snipers in an urban environment have a detailed withdrawal plan. Once a patrol comes into contact with a sniper, the patrol leader must immediately assess the situation and maneuver his patrol accordingly. The patrol's mission, location, size, ROE, and location of the threat often determine whether the patrol will attempt to neutralize the targeted sniper. If the patrol leader decides to kill or capture the sniper, he uses planned and rehearsed immediate actions to maneuver and counter the sniper's assault. The goal is to kill the sniper or cut off his escape and capture him. There are three immediate reactions to neutralizing a sniper: initial contact, immediate follow-up, and subsequent follow-up.

Initial Contact. The initial contact is made when the sniper fires the first shot. The patrol must react immediately and positively to get behind the firing position in order to kill or capture the gunman. The period of contact ends when the gunman is killed or captured, or the patrol element or team leader on the scene ends it. The following technique is the same for both squad- and platoon-sized patrols:

- The element or team in contact attempts to identify the firing position and maneuvers designated marksman into position to return well-aimed and controlled fire. Other members of the patrol take up positions to cover the marksman's engagement.
- The patrol element or team leader in contact sends initial contact report to the patrol leader, who notifies higher headquarters.
- The element or team leader in contact determines appropriate cut-off positions and relays them to flanking elements or teams.
- The patrol element or team leader in contact continues to observe the firing point, but does not enter it due to the possibility of booby traps. Flank elements or teams set up along likely escape routes.
- The incident ends when either the sniper ceases fire or is neutralized.

Immediate Follow-Up. Regardless of the fate of the gunman, isolation of the firing point is necessary to prevent reinforcement and preserve forensic evidence (scent, spent casings, etc.). If not under fire, members of the patrol element or team cordon off the area surrounding the firing point. Flanking elements or teams maintain their positions and prevent civilians from entering the area. The patrol leader moves to link-up with the element or team in contact (if not his own), and makes an estimate of the situation. The patrol leader sends a SPOTREP to higher headquarters.

Subsequent Follow-Up. The aim of the subsequent follow-up is to use follow-on forces to clear the building of remaining resistance or to obtain evidence that can be used to capture the gunman. The patrol leader establishes a position where he can brief arriving units (reaction force commander, S-2 representative, EOD personnel, etc.). Once the arriving units have been briefed, recommendations are made to higher headquarters via radio. No one is allowed into the cordon without the patrol leader's approval.

Reaction to Becoming Decisively Engaged

If a patrol becomes decisively engaged from numerous firing positions, the following immediate action should be taken:

- All patrol members move to available cover and return accurate fire on identified firing points.
- The patrol leader assesses the situation and makes a decision to either request the reaction force or break contact.
- If the reaction force is requested, the patrol will maintain its position until the reaction force arrives. The patrol should use fire and maneuver to gain better tactical positioning and support the arrival of the reaction force. When the reaction force arrives,

its commander may decide to either clear occupied buildings or cover the patrol during its extract.

Reaction to Bomb Threat or Discovery

The use of command-detonated explosive devices is a common ambush tactic employed by a terrorist or insurgent in an urban environment. The appropriate response to a reported threat or an actual discovery is generally involves four steps (known as the four Cs)—

1. The patrol leader **CONFIRMS** the presence of the suspicious item.

2. Without touching or moving anything suspicious, patrol elements or teams **CLEAR** the immediate danger area to a minimum of 100 meters. The area is cleared from the suspected device outward, inform civilians as to the reason for evacuation.

3. A&S teams establish a **CORDON** to secure the cleared area. Avenues of approach are cordoned off to keep people out and to protect EOD or engineer personnel clearing the device. The assistant patrol leader acts as the cordon commander and informs the patrol leader when the cordon is secure. An effective cordon technique is to tape off the area with engineer tape, creating both a physical and psychological boundary.

4. **CONTROL** of the area is maintained throughout the bomb clearing operation by the patrol leader. The patrol leader sends a report to a superior concerning details of the device (if known) and the area affected. The patrol leader coordinates with arriving personnel (EOD, engineers, etc.). The patrol leader maintains communication with the assistant patrol leader and keeps the Marines informed of the progress of the clearing operation.

Reaction to a Bomb Detonation

Bombs may be used by an insurgent as a means of initiating an ambush on mounted or dismounted patrols, in which case the actions for decisive engagement apply. Immediate action in response to an isolated explosion is similar to that used in reaction to a sniping and breaks down into the same three phases:

Initial Contact. The patrol leader attempts to identify the likely initiation point and sends an initial contact report to higher headquarters. If the bomb was command-detonated, the patrol leader sends his A&S teams deep to cut off the bombers' escape routes. Any casualties are moved a minimum of 100 meters from explosion and out of the line of sight to it.

Immediate Follow-Up. The A&S teams may need to maneuver to positions behind the likely initiation point to cut off escape. Once in position, personnel checks are conducted and any suspects are detained. The patrol leader coordinates requests for required support (MEDEVAC, reaction force, etc.)

Subsequent Follow-Up. Due to the possibility of secondary detonations, the four Cs (confirm, clear, cordon, and control) can be conducted as in reaction to a bomb discovery or bomb threat.

Civil Disturbances

Urban patrols must prepare to react to spontaneous aggression by the local populace. In many cases, civil disturbances are organized by the enemy to draw dismounted patrols into a targeted area, or to distract them from enemy activity occurring elsewhere. Civil disturbances are generally divided into two categories: minor aggressive actions, and full-scale rioting.

Minor aggressive actions are activities characterized by rock-throwing or use of devices such as Molotov cocktails and may either be directed at the patrol or take place between different ethnic factions of the population. Minor aggressive actions are normally spontaneous in nature and may have minimal or limited objectives for the insurgents.

Full-scale rioting events are usually in response to another major event or incident that may enflame the populace. Full-scale riots are well-planned and orchestrated, with clear objectives or targets in mind. At times, patrols will need to attempt to maintain control of a civil disturbance situation; however, dismounted and mobile small unit patrols should generally avoid potential flashpoints. Procedures to handle civil disturbances are as follows:

- The patrol leader reports the incident to headquarters and attempts to diffuse the crowd by talking to crowd leaders.
- If the patrol leader determines the size of the disturbance is too large for the force to handle, the patrol should move away from the disturbance to a safer, more remote covered area and occupy positions to observe and report the situation to higher headquarters. To prevent the patrol from being pursued by the crowd, the patrol should move quickly and change direction, often at road junctions, to gain distance from it.

Patrol members should maintain dispersion to create a more difficult target. They should face the crowd at all times to see and avoid any projectiles thrown. Individual self-discipline must be maintained throughout the disturbance. Marines charging into the crowd or throwing objects back at the crowd will only worsen the situation. If pursued or trapped, the patrol leader may consider using riot control agents (combat support, pepper gas, etc.) to disperse or slow a crowd's movement. The use of riot control agents must be authorized under the established ROE. If the patrol leader believes the situation is deteriorating beyond the patrol's ability to control it, the patrol leader should request the reaction force, which may be better equipped to handle a large riot or mob.

Break Contact

As with patrols in rural areas, the patrol leader may be forced to break contact as a result of decisive engagement with the enemy. On the basis of his estimate of the situation, the patrol leader will normally break contact in one of the following ways:

- As a patrol, with elements providing cover for movement as defined by clock direction and distance.
- As individual units/teams taking separate routes out of the area, then linking up at a designated rally point a safe distance away from the engagement.

As in any contact with enemy forces, smoke may be employed to screen movement. Fire support agencies can be utilized to suppress targets; riot control agents can be employed to disrupt enemy movement.

Reentry of Friendly Lines

The reentry of a dismounted patrol into an urban patrol base is no different from that of a patrol conducted in a rural area. The same planning considerations and control methods apply.

Missions Related to Urban Patrolling

House Calls

House calls missions are usually part of a coordinated effort to collect information within the area of operations. They involve obtaining up-to-date information on particular houses and occupants. When possible, local police should accompany patrols to do the actual talking to the occupants. If this is not possible, a technique that may be used by an urban patrol is—

- A&S elements move to provide cover around the target house.
- The headquarters element provides security just outside the house.
- The patrol leader and one Marine for security enter the house, if invited. If not invited, they talk to the occupants on the doorstep. Patience and tact are required in requesting information. An interpreter should be present when language differences exist.

Vehicle Checkpoints

Urban terrorists or insurgents commonly use vehicles to transport personnel, weapons, explosives, and equipment. Civilian vehicles are often used for these purposes, creating the requirement to check as many vehicles as possible. While permanent, fortified checkpoints may be conducted along approaches into an urbanized area, dismounted patrols can be employed to establish hasty vehicle checkpoints to stop vehicles and to keep the enemy off-guard. The two common types of vehicle checkpoints patrols establish are hasty and deliberate.

Hasty checkpoints are deployed anywhere based upon the decision of the patrol leader. Patrols must not set patterns through the frequent use of the same sites.

Deliberate checkpoints are tasked by higher headquarters to achieve a specific purpose. Time and locations are carefully considered to avoid setting patterns.

The general layout for a squad-sized, two-way dismounted checkpoint is depicted in figure 13-8 on page 13-20. The technique is as follows:

- The A&S teams are positioned stealthily in blocking positions on both sides of the road.
- Both the patrol leader and assistant patrol leader act as "talkers" for each direction of traffic (with local police or interrogator or translator Marines acting as interpreters) while a two-man team from the headquarters element physically searches the vehicles.
- Obstacles or parked vehicles may be employed to create a staggered roadblock in center of the checkpoint to slow approaching vehicles.
- The checkpoint location should be sited so that approaching vehicles cannot see it until they have passed a security team, and they have no escape route then available.
- Signs announcing the checkpoint should be displayed a safe distance from the search area for safety to both drivers and Marines.
- Normally higher headquarters will issue criteria that determines which vehicles are searched, but random checks of cars should normally be made as

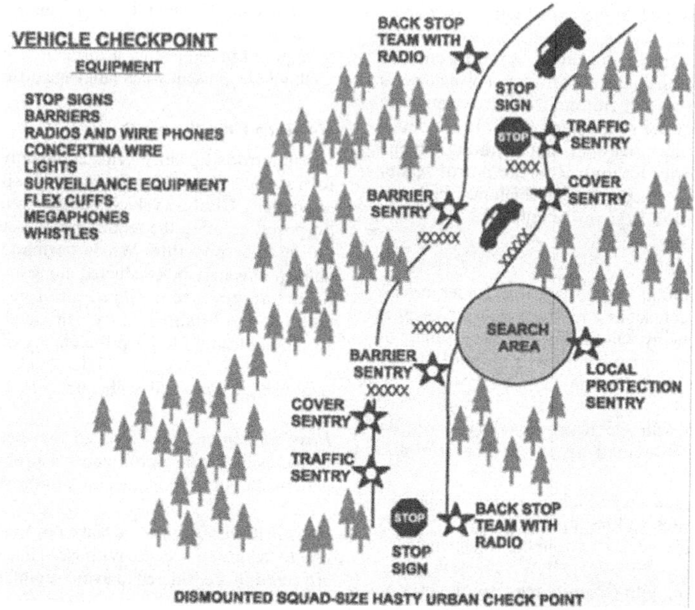

Figure 13-8. Dismounted Squad-sized Hasty Urban Vehicle Checkpoint.

well. While the vehicle is being searched the driver should accompany the searcher around the vehicle.

- Vehicle occupants should be made to exit the vehicle and then searched. Whenever possible, women should be searched by female police or female Marines.
- All patrol members should conduct themselves with courtesy and politeness. If nothing is found, an apology for the inconvenience is recommended.
- A hasty vehicle checkpoint should not be conducted any longer than 30 minutes for security reasons.
- The ROE should dictate whether or not action should be taken against vehicles that fail to stop at the checkpoint. Failure of a vehicle to stop does not automatically give authorization to fire.

Observation Posts

Urban observation posts are established to provide extended security, not only for patrol bases but also for patrols operating within the observation post's sector of observation. Observation posts can be established in conjunction with sniper operations and for providing overwatch for patrols operating within their sector of observation. Observation posts are nor-mally positioned on dominating terrain or in buildings outside the patrol base itself.

Insertion to the observation posts and conduct of observation may be either overt or clandestine in nature. Overt observation posts usually will be hardened positions to increase security. A patrol provides cover while the observation post is being inserted. A clandestine observation post relies on stealth of insertion and occupation for protection. It is normally positioned in abandoned buildings to cover sectors of observation that overt observation posts cannot. Because of their nature, clandestine observation posts are difficult to successfully establish and should not be manned for an extended period of time. Orders establishing observation posts (and patrols) must address the method of extraction as well as actions upon compromise/attack.

Cordon and Search

The cordon and search mission involves isolating a predesignated area by cordoning it off and systematically searching for enemy personnel, weapons, supplies, explosives or communications equipment. While large-scale cordon and search operations are planned and rehearsed in advance and normally entail extensive coordination with local law enforcement agencies, a squad-sized urban patrol may often conduct a cordon and search of a point target—searching one house or building identified by intelligence as a possible weapons cache.

The basic principle of a search of a populated area is to conduct it with limited inconvenience to the population. The populace may be inconvenienced to the point where they will discourage urban guerillas or insurgents from remaining in the area, but not to the point that they will assist the enemy as a result of the search.

Upon receiving intelligence that warrants the searching of a building or a specific tasking from higher headquarters, A&S elements of the patrol move to establish an inner cordon around the target building to seal it off, with the primary intent of preventing movement out of the targeted building.

On order, the designated reaction force deploys to establish an outer cordon, oriented outward some distance from the inner cordon and covering routes leading into the area in order to prevent outside interference/reinforcement. The reaction force maintains a reserve to reinforce either cordon or react to unfolding events (civil disturbance in response to the operation).

Once the cordons have been established, the patrol leader, with the assistance of local police or interpreters, informs the local populace that a building is about to be searched, that a house curfew is in effect (if permitted by higher headquarters), and that all occupants should remain indoors. Occupants of the target house are instructed to gather at a central location to stay out of the way of the search party.

The headquarters element, having linked up with any required assistance (explosive ordnance disposal (EOD), ITT, etc.) now acts as the search party and accompanies local police. A female searcher should be included in the party, if necessary.

Occupants are searched and screened first for possible enemy personnel. Apprehended persons are evacuated as soon as possible.

The head of the household should accompany the search party throughout the operation to be able to counter incriminating evidence and possible accusations of theft and looting against Marines. If possible, a prominent member of the local community should act as a witness.

Buildings are best searched from top to bottom. Ideally, the search is conducted with the assistance of combat engineers using mine detectors to locate hidden arms and ammunition.

If the targeted building is empty or the occupant refuses entry, it may be necessary to forcefully enter the premises to conduct a search of the dwelling. If an unoccupied house containing property is searched, arrangements should be made with the local community to secure it until its occupants return. Unnecessary force and damage to property should be avoided during the search.

Motorized Urban Patrols

The advantages of a motorized urban patrols is their ability to capitalize on the speed, mobility, and protection offered by various vehicles. They may be motorized, mechanized or armored vehicles or a combination. Generally, motorized urban patrols possess greater combat power than dismounted patrols and can cover larger areas faster than dismounted patrols.

The disadvantages of motorized urban patrols is that they are restricted to roads and are vulnerable to ambush by the enemy. They are also restricted in their ability to interact with the local populace.

Motorized patrols are generally organized in the same manner as dismounted patrols (see fig. 13-9). Unit integrity is maintained when assigning personnel to specific vehicles.

The urban patrolling principles apply to motorized patrols in much the same manner as dismounted patrols. Mutual support and depth are achieved by maintaining constant observation between vehicles and coordinating support with any dismounted patrols in the area. All-around security is achieved through the use of constant observation as well as the vehicle's mobility and firepower. Positive communications between units or teams are maintained through vehicle radios. Patrol routes and speeds are varied to promote deception or pattern avoidance.

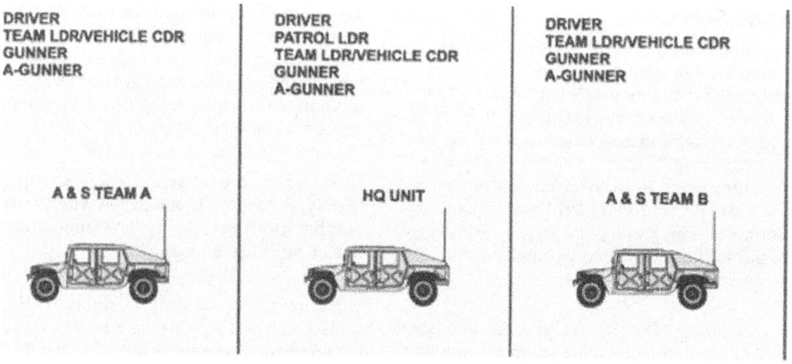

Figure 13-9. Organization of a Squad-sized Motorized Patrol.

The canalizing nature of streets and alleys force vehicular patrols to use a traveling overwatch movement technique to reduce vulnerability to ambushes. All vehicles travel at a moderate rate of speed with the lead vehicle stopping only to investigate potential danger areas. If vehicles must stop in danger areas, designated crew members' will dismount to provide security. The gunner will remain at the ready and in the turret while the driver remains in the driver's seat with the vehicle running.

Vehicles should move at a high rate of speed only when responding to an incident. At all other times, vehicle speed should be between 15 to 20 mph to allow for quick reaction and good observation. Distances between vehicles should be approximately 50 meters (one half to one city block) or such that visual contact and mutual support are ensured. Particular care is taken at major road junctions and other danger areas to ensure individual vehicles do not become isolated.

Vehicles with doors removed generally enhance observation and overall security, yet expose Marines to thrown objects, theft and concealment.

CHAPTER 14. INFORMATION AND REPORTS

It is necessary that patrol leaders and all patrol members be trained in observing and reporting their observations accurately. The leader of a patrol should have all members of the patrol immediately signal or report any information obtained. These reports should not be restricted to information about the enemy, but should also include information about the terrain, such as newly discovered roads, trails, swamps, and streams. The leader includes all information in the report to the officer dispatching the patrol. (Refer to MCRP 2-15.3, *Reconnaissance Reports Guide,* for detailed report formats.)

14001. REPORTING

The officer dispatching the patrol instructs the patrol leader on whether and when messages are to be sent back during the patrol and what communication means to use. Messages may be oral or written. They must be accurate, clear, and complete. Every message should answer the question what, where, and when. For a detailed discussion on reporting, see paragraph 6002.

Verbal Messages

A patrol leader sending a verbal message should make it simple, brief, and avoid using numbers and names. The messenger should accurately repeat the message back to the patrol leader before leaving.

Written Messages

In preparing written messages, the patrol leader must distinguish between fact and opinion. Information about the enemy should include: strength; armament and equipment; actions; location and direction of movement; unit destination, if known; time enemy was observed; and the patrol's location when the observation was made. Use of an overlay or sketch may often simplify the message.

Messengers

A messenger team is given exact instructions as to where to deliver the message and the route to take. Any information obtained along the route should be reported at the time the message is delivered. Messengers must be given all practical assistance. If in danger of capture, the messenger immediately destroys the message.

Radio and Signals

If the patrol is provided with a radio, a definite radio schedule for checking in must be arranged before departure of the patrol. The patrol leader takes every precaution to ensure that codes and copies of messages are not captured by the enemy. If a close reconnaissance of enemy lines is required, the radio should be left in a concealed location at a safe distance from the enemy. Once a report is sent by radio, the patrol should immediately leave the area to avoid the possibility of detection by enemy locating devices. Pyrotechnics (flares, colored smoke, grenades) and air-ground panels may also be used by patrols for reporting information by a prearranged signal.

SALUTE Report

Information must be reported as quickly, accurately, and as completely as possible. An established method to remember how and what to report about the enemy is to use the acronym SALUTE:

Size
Activity
Location
Unit
Time
Equipment

An example of such a report is: "Seven enemy soldiers, unit unknown, traveling SW, crossed road junction on BLACK RIDGE at 211300 August carrying one machine gun and one rocket launcher."

14002. CAPTURED ITEMS

Every patrol should make a practice of searching enemy casualties, prisoners, and installations first for booby traps, then for equipment, papers, maps, messages, orders, diaries, and codes. Search techniques should be practiced often in order to reduce the time exposed to potentially dangerous situations. Items found are collected by the patrol leader and turned in with the patrol report. The items found are marked as to time and place of capture. When possible, captured items should be linked to a specific prisoner who possessed the items or to the place where the items were found. When this is done, the enemy prisoner of war tag and item tag are marked accordingly. The patrol leader must impress upon the members of the patrol the importance of turning in all documents and equipment. Furthermore, the patrol leader must ensure all information gathered by the patrol is quickly disseminated.

14003. PRISONERS

A patrol normally does not capture prisoners unless required by the mission. If prisoners are taken, the "5S and T" rule applies. This memory aid stands for: search, segregate, silence, speed, safeguard, and tag.

Search. Prisoners are body-searched thoroughly for weapons and documents as soon as they have been captured. This search must include the helmet, body armor, and gas mask. These items are left with the prisoner for protection until the patrol is completed. Weapons, equipment, and documents are tagged and immediately sent to the patrol leader.

Segregate. Prisoners are segregated into isolated groups: officers, noncommissioned officers (NCOs), privates, deserters, and civilians. By segregating prisoners, it makes it more difficult for leaders to organize escapes and issue orders to subordinates.

Silence. Silence is essential. Do not allow prisoners to talk to each other.

Speed. Speed is required in getting prisoners to the commander who dispatched the patrol. Timely information secured from prisoners is essential.

Safeguard. Prisoners are safeguarded as they are moved. They are restrained, but not abused. If the patrol will soon reach friendly positions, prisoners are not given cigarettes, food or water until they have been questioned by interrogators. If the patrol will not return to a friendly position for a lengthy time, the prisoners are given food, water and medical aid.

Tag. Equipment and personal effects recovered from prisoners are tagged so that they may later be matched back to specific individuals.

14004. PATROL REPORT

Every patrol leader makes a report when the patrol returns. Unless otherwise directed, the report is made to the person ordering the patrol. If the situation permits, the report is written and supported by overlays and/or sketches. The patrol leader's report should be a complete account of everything of military importance observed or encountered by the patrol while on the assigned mission. It should include the following information:

- Size and composition of patrol.
- Tasks and purpose (mission).
- Time of departure.
- Time of return.
- Routes, out and back (show by sketch, azimuth, trace on map).
- Terrain (general description to include any man-made or natural obstacles and critical terrain features which, if occupied by either enemy or friendly forces, would allow them to control the surrounding area).
- Enemy (size, activity, location, unit, time, equipment).
- Any map corrections (show on map).
- Miscellaneous information not covered elsewhere in report.
- Results of enemy encounters.
- Condition of patrol, including disposition of any dead or wounded.
- Conclusion and recommendations.

14005. PATROL CRITIQUE

After the patrol has rested and eaten, the patrol leader should hold a critique. Constructive criticism is made. It is an excellent time to prepare for future patrols by going over lessons learned as a result of the patrol.

Appendix A. Patrol Warning Order

The warning order is issued as soon as practical with all available information included to assist patrol members in preparation.

1. Situation. Friendly and enemy situation information necessary for initial preparation.

2. Mission. Statement of what the patrol is to accomplish, and the purpose for accomplishing it. When, how, and where will be discussed in the patrol order.

3. Execution

a. Task Organization. General patrol organization and assignment of responsibilities if known; otherwise, promulgate in the patrol order.

b. Tasks. Alert subordinate leaders to patrol tasks (e.g., stream crossing, helicopter rappelling, demolitions) requiring preparation prior to departure.

c. Coordinating Instructions

(1) Time schedule and location for individual preparation, rest, briefings, in-spection, rehearsal, and departure.

(2) Time, place, uniform, and equipment for receiving the patrol order.

(3) Tasks for subordinate leaders to direct and supervise the initial preparation which may include drawing ammunition, rations, and special equipment; conducting immediate action drills or other necessary individual or unit training; meeting and briefing attachment personnel; reconnoitering the area for passage of lines; and coordinating with the necessary unit leaders.

(4) Preliminary guidance to specialists and key individuals regarding their roles and organization within the patrol.

4. Administration and Logistics

a. Individual uniform, equipment, weapons, and prescribed load of rations, water, and ammunition.

b. Crew-served weapons (if required) and guidance regarding distribution of weapons and ammunition during movement.

c. Special equipment requirements (wire cutters, demolitions, radios, flashlights, infrared equipment, mines, binoculars) and their distribution during movement.

d. Restricted or prohibited items.

5. Command and Signal

a. Designation of assistant patrol leader and his role in preparation.

b. Designation of navigators and radio operators (if required).

c. Brief outline of patrol leader's schedule for preparation, and where he can be reached.

Appendix B. Patrol Order

A patrol order follows a warning order. Any of the following subjects that have been addressed in the warning order may be omitted with the exception of the mission statement. A patrol order is more detailed than a 5-paragraph squad order in that a great deal of attention is given to individual duties.

The patrol order also provides orientation information, both enemy and friendly, that impacts the patrol and enemy forces. This information can include, but is not limited to, weather, terrain, visibility, NBC considerations, local population situation, terrain model and/or map orientation, and behavior.

1. Situation

   a. Enemy Forces

      (1) Composition, disposition, and strength are based on size, activity, location, unit, time, and equipment (SALUTE).

      (2) Capabilities and limitations to defend, reinforce, attack, withdraw, and delay (DRAW-D).

      (3) Enemy's most probable course of action.

   b. Friendly Forces

      (1) Mission of next higher unit (task and commander's intent).

      (2) Adjacent unit missions (task and intent). Also identify left, front, right, and rear.

   c. Attachments and Detachments. (date and time effective).

2. Mission. The mission is the task to be accomplished, and its purpose (who, what, where, when, and why). For patrols, specify if the mission or time has priority.

3. Execution

   a. Commander's Intent and Concept of Operations

      (1) Commander's intent.

      (2) The concept of operations tells the where, how, and who and lays out the patrol leader's general scheme of maneuver and fire support plan. It outlines the following:

(a) Task organization of the patrol.

(b) Movement to the objective area, to include navigation method.

(c) Actions in the objective area.

(d) The return movement, to include navigation method.

(e) Use of supporting forces (including illumination, if required).

b. Subordinate Element Missions. Subordinate element missions (task and purpose) are assigned to elements, teams, and individuals, as required.

c. Coordinating Instructions. This paragraph contains instructions common to two or more elements, coordinating details, and control measures applicable to the patrol as a whole. At a minimum, it includes—

(1) Time of assembly in the assembly area.

(2) Time of inspections and rehearsals (if not already conducted).

(3) Time of departure and estimated time of return.

(4) Location of departure and reentry of friendly lines and the actions associated with departure and reentry.

(5) Details on the primary and alternate routes to and from the objective area.

(6) Details on formations and order of movement.

(7) Rally points and actions at rally points.

(8) Final preparation position and actions at this position.

(9) Objective rally point and actions at this point.

(10) Actions at danger areas.

(11) Actions in the event of enemy contact.

(12) Details on actions in the objective area not covered elsewhere.

(13) Estimated time of patrol debriefing upon return.

4. Administration and Logistics

a. Changes/additions to uniform, equipment, and prescribed loads from that given in the warning order.

b. Instructions for handling wounded and prisoners.

5. Command and Signal

a. Command Relationships. Identify key leaders and chain of command.

b. Signal. Challenge and password, arm and hand signals, special signals, and radio frequencies and call signs.

Appendix C. Patrol Evaluation Checklist

This appendix provides a comprehensive checklist of critical patrolling steps, techniques, and procedures to aid unit leaders to critique patrol performance during training. Unit leaders may use the list as they observe the performance of a patrol to provide the patrol leader with a detailed analysis of the performance.

Warning Order	Yes	No
Did the PL ensure all patrol members were present before issuing the warning order?		
Did the PL issue a brief statement on the enemy situation?		
Did the PL issue a brief statement on the friendly situation?		
Did the PL state the mission in a clear, positive manner?		
Did the PL list all members of the patrol including attachments?		
Was the chain of command for the patrol covered fully?		
Were all members of the patrol asigned positions and duties in particular squads and teams?		
Were all the necessary individual duties assigned?		
Did the PL follow established principles in organizing the patrol into squads and teams?		
Was each patrol member assigned a particular weapon to carry on patrol?		
Did the PL list all special equipment needed to accomplish the mission?		
Was required special equipment assigned to the proper element to carry?		
Did the PL select uniform and equiment common to all based on METT-T?		
Coordination with Adjacent Units		
Did the PL coordinate with other patrols operating to the right and left?		
Was the route out and back coordinated?		
Was the time of departure and return cocordinated?		
Were call signs and frequencies coordinated?		
Was a signal for the FPF coordinated so as not to approach friendly lines during this time?		
Coordination With Front Line Units		
Did the coordinator pass the size of the patrol?		
Was the time of departure and return coordinated?		
Did the coordinator give a general area of oeprations for the patrol?		
Did the coordinator ask for information on known or suspected enemy positions and/or obstacles?		
Did the coordinator ask about information on the latest enemy activity?		
Did the coordinator ask for detailed information on friendly fire support avalable and the unit's barrier plan?		
Was the location of the IRP established and coordinated?		
Did the coordinator ask the forward unit to monitor their patrol frequency?		
Was the current challenge and password confirmed?		
Did the coordinator request that all information coordinated be passed on to any relieving unti?		
Patrol Order		
General		
Did the PL check to ensure all patrol members were present before issuing the patrol order?		

	Yes	No
Did the PL issue the patrol order in a forceful, confident manner?		
Did the PL make maximum use of available visual aids in issuing the patrol order (i.e., terrain model, sand table, map board, chalkboard)?		
Did the PL issue the patrol order in correct sequence?		
Did the PL issue the entire patrol order without allowing interruptions by patrol members?		
Did the PL adequately answer all questions asked by patrol members		
Did paragraph 1a (Enemy Situation) include—		
A weather forecast for the period of operation?		
A description of the terrain over which the patrol was to operate?		
Identification or description of enemy units known to bein the area of operations?		
Known locations of enemy units?		
Recent activity of enemy units?		
Stength of enemy units in the operating area?		
Did paragraph 1b (Friendly Situation) include—		
Mission of the next higher unit?		
Locations, missions, and planned actions of units on right and left?		
Fire support available to support the patrol (as per the coordination)?		
Missions and routes of other patrols operating in the immediate area?		
Did paragraph 1c (Commander's Intent) include—		
All views and ides of what the commander wants to be accomplished?		
The final result desired by the commander?		
Did paragraph 1c (Attachments and Detachments) include—		
All attachments to patrol and effective time of attachment?		
All detachments from patrol and effective time of detachment?		
Did paragraph 2 (Mission) include, at a minimum—		
Who was to conduct the patrol?		
What the patrol was tasked to do (i.e., conduct a point recon patrol)?		
Where the action was to take place?		
Why the action was to take place (i.e., the purpose)?		
Did paragraph 3 a (Commander's Intent and Concept of Operations) include—		
A complete concept of operations?		
A detailed description of the mission of elements?		
A detailed description of the mission of teams?		
A detailed description of the duties of specific individuals (i.e., navigator, compass man, pace m an, corpsman, APL)?		
Time of departure and time of return?		

	Yes	No
Type of formations and order of movement to be used?		
Description of the route and alternate routes(s) to include azimuths and distances between checkpoints as a minimum?		
Techniques to be used in the departure from friendly areas?		
Techniques to be used for the reentry into friendly areas?		
Location of the IRP and a tentative ORP, to include grid coordinates and recognizable terrain features?		
Method for designating and passing rally points?		
Actions to be taken at rally points if their use became necessary?		
Actions to be taken in the event of enemy contact?		
Actions to be taken at danger areas?		
Complete, detailed description of actions to be taken at the objective?		
Times and locations for rehearsals and the order of priority for rehearsals?		
Times and lcoations for inspections and the methods of conduct (to include uniform and equipment to be worn and/or carried)?		
Where, when, and by whom the debriefing is to be conducted?		
Did paragraph 4 (Administration and Logistics) include—		
Rations to be carried or reference to the warning order if there is no change?		
Arms and ammunition to be carried or reference to the warning order if there is no change?		
Uniform and equipment to be worn and carried or reference to the warning order if there is no change?		
Method for handling wounded or dead?		
Method for handling prisoners?		
Did paragraph 5a (Signal) include—		
Complete description of all signals, code words, frequencies, and call signs to be used within the patrol?		
Were signals adequate?		
Radio call signs to be used?		
Primary and alternate frequencies to be used?		
Call signs and frequencies to be used with other units in the area of oeprations (if applicable)?		
Required reports to be transmitted to higher headquarters, along with occasions for reporting?		
Code words and brevity codes to be used between the patrol and higher headquarters?		
Current challenge and password to be used in friendly held areas?		
Current challenge and password to be used forward of FEBA?		
Did paragraph 5b (Command) include—		
Chain of command or reference to the warning order if there is not change?		
Location of the PL during all stages of operation?		

	Yes	No
Location of the APL during all stages of operation?		
Inspection Rehearsal		
General		
Did the P L adequately use the allotted inspection time?		
Was the inspection conducted in an orderly manner?		
During the inspection, did the PL check the following:		
Completeness and correctness of uniform?		
Completeness of all equipment necessary to accomplish the mission?		
Operational condition of equipment?		
Did the PL question members to ensure they knew—		
The mission of the patrol?		
The concept of operations?		
Their individual duties and responsibilities?		
Chain of command and succession?		
Duties and responsibilities of key personnel?		
Did the PL utilize the rehearsal area to conform as much as possible to area of operations?		
Were the following major actions rehearsed:		
Actions at the objective?		
Actions at danger areas?		
Actions on enemy contacts?		
Departure and reentry of friendly lines?		
Did the PL critique each action after it had been rehearsed?		
Were interpatrol communications and control measures checked at rehearsal?		
Did the PL maintain control of the patrol during the rehearsal?		
Movement		
Did the PL use proper formations for movement?		
Did the PL adequately control the point team?		
Did the PL check the compass men?		
Was the PL aware of his position at all times?		
Was light and noise discipline enforced?		
Was the rate of movement appropriate for the mission?		
Did the PL make full use of pace?		
Did the PL make full use of count?		
Was proper security maintained during movement?		
Did the PL select and pass proper rallying points?		
Did the PL ensure that all members knew location of rallying points?		

	Yes	No
Did the PL properly use arm and hand signals?		
Did the PL recognize and halt the patrol a safe distance from a danger area?		
If the PL conducted a map check, did he conduct it properly (light discipline, security, etc.)		
Did designated individuals reconnoiter the far side of a danger area properly?		
Were support teams emplaced properly?		
Did the PL use a proper formation in crossing the danger area?		
Were reports made on enemy contact and at checkpoints?		
Actions on Enemy Contact		
Did the PL halt patrol a safe distance from a tentative ORP?		
Did the PL issue an adequate frag order before departing to look for a tentative ORP?		
Did the PL take appropriate personnel to look for the ORP?		
Did the PL select a suitable ORP?		
Did the PL secure the site adequately?		
Did the patrol occupy the ORP as stated in the patrol order or subsequent frag order?		
Did the patrol occupy the ORP in an orderly manner?		
Did the PL maintain control of the patrol during occupation of the ORP?		
Did the PL issue a satisfactory contingency plan to the APL before departing on a leader's recon?		
Did the PL take appropriate personnel on a leader's recon?		
Did the PL maintain or ensure that the patrol leader had communications with the patrol?		
Did the patrol avoid being detected by the enemy during the leader's recon?		
If contact was made, did the PL take appropriate action?		
Was security maintained during the leader's recon?		
Did the PL ensure the objective was kept under surveillance?		
Were appropriate orders given to surveillance teams?		
If a leader recon proved a tentative ORP unsuitable, did the PL move the ORP?		
Was a satisfactory leader's recon conducted?		
Was the ORP move conducted in a satisfactory and orderly manner?		
On return to the ORP, did the PL issue appropriate frag order and allow enough time for dissemination?		
Actions at Objective		
Did the PL issue a frag order for action at the objective?		
Was security emplaced prior to the teams departing from the ORP?		
Did the PL employ the terrain at the objective to the best advantage?		
Was stealth maintained while moving into the objective?		

	Yes	No
Did the Pl exercise positive control of squads, teams, and individuals at the objective?		
Were actions at the objective in accordance with the details outlined in the patrol order?		
Did the PL make use of supporting arms at the objective?		
Was the action at the objective successful?		
Was withdrawal from the objective accomplished quickly and orderly?		
Did units withdraw according to the details reflected in the patrol order?		
Was reorganization at the ORP completed in an expeditious and orderly manner?		
Did the patrol withdraw from the ORP quickly and quietly?		

Appendix D. Ambush Formations

Formations discussed here are identified by names which correspond to the general pattern formed on the ground by deployment of the assault element.

Line Formation

The assault element is deployed generally parallel to the route of movement of the enemy. This positions the assault element parallel to the long axis of the killing zone and subjects the target to heavy flanking fire. The size of the force which can be trapped in the killing zone is limited by the area the assault element can effectively cover. The enemy is trapped in the killing zone by natural obstacles, mines, demolitions, and direct fires. (See fig. D-l.)

A disadvantage of the line formation is the chance that lateral dispersion of the target may be too great for effective coverage. The line formation is appropriate in close terrain that restricts enemy maneuver, and in open terrain where one flank is protected by natural obstacles or can be protected by mines and demolitions. Similar obstacles can be placed between the attack force and the killing zone to provide protection from possible enemy counter attack. When an ambush is deployed in this manner, access lanes are left so that the enemy can be assaulted. (See fig. D-2.)

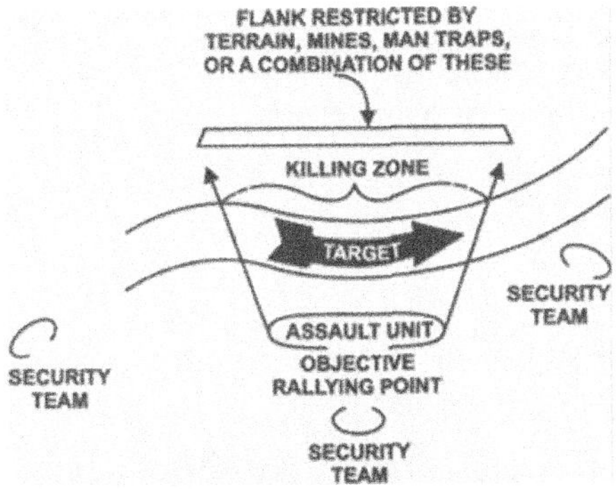

Figure D-1. Line Formation (Harassing or Destruction Ambush).

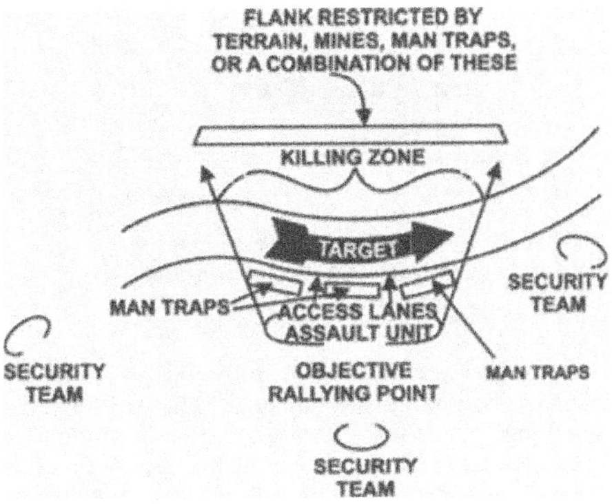

Figure D-2. Line Formation (Access Lanes for Assault of Target).

The main advantage of the line formation is its relative ease of control under all conditions of visibility.

L Formation

The "L" shaped formation is a variation of the line formation. This formation is very flexible because it can be established on a straight stretch of a trail or road (see fig. D-3) or at a sharp bend in a trail or a road (See fig. D-4 on page D-2).The long side of the assault element is parallel to the killing zone and delivers flanking fire. The short side of the attack force is at the

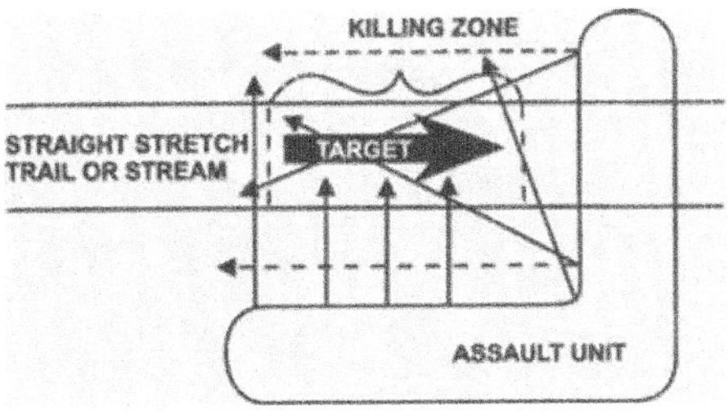

Figure D-3. "L" Formation (Destruction Ambush).

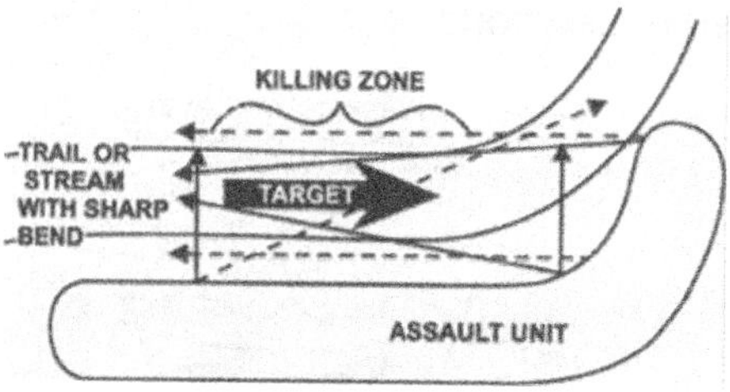

Figure D-4. "L" Formation (Bend of Trail or Stream).

end of, and at right angles to, the killing zone and delivers enfilading fire that interlocks with fire from the long side. When appropriate, fire from the short side can be shifted to parallel the long side if the enemy attempts to assault or escape in the opposite direction. In addition, the short side prevents escape and reinforcement. (See fig. D-5.)

"Z" Formation

The "Z" shaped formation is another variation of the line formation. The assault element is deployed as in the "L" formation, but with an additional side so that the formation resembles a "Z". (See fig. D-6).

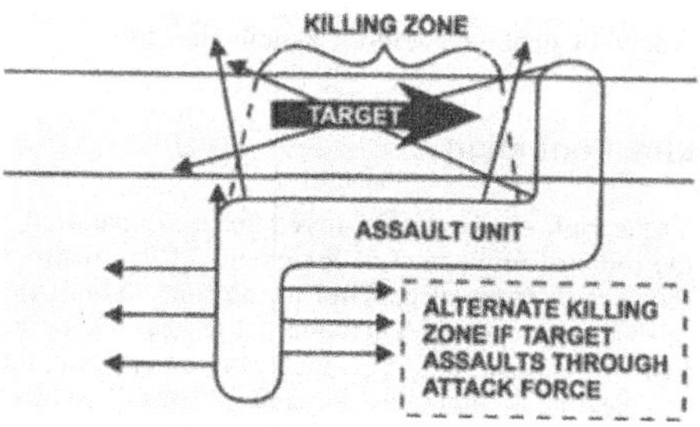

Figure D-6. "Z" Formation.

The additional wing may serve any of the following purposes:

- To engage an enemy force attempting to relieve or reinforce the enemy unit engaged in the kill zone.
- To seal the end of the killing zone.
- To restrict a flank.
- To prevent an envelopment.

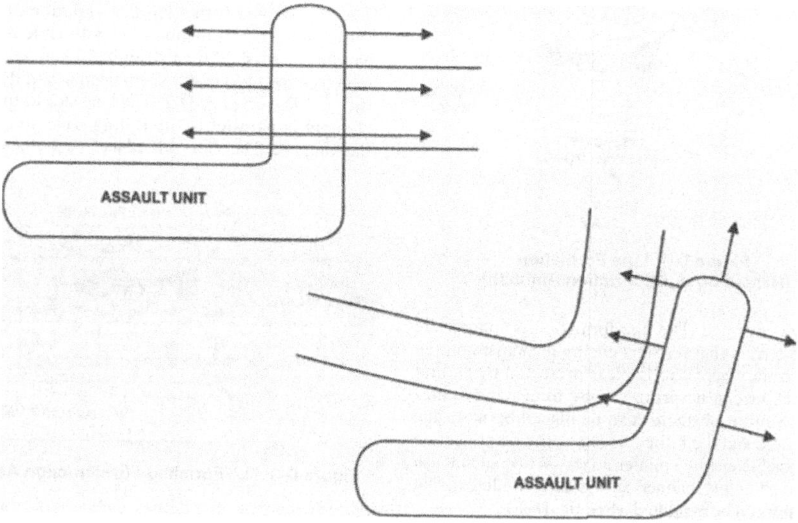

Figure D-5. "L" Formation.

"T" Formation

In the "T" shaped formation, the assault element is deployed across (perpendicular to) the enemy's route of movement so that its position forms the crossing of a "T" at the top. (See fig. D-7.)

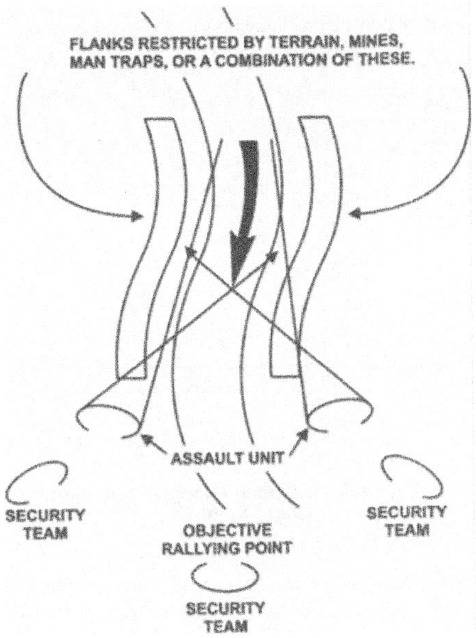

Figure D-7. "T" Formation.

This formation can be used day or night to establish an ambush to interdict movement through open areas that are hard to seal off.

A small force can use the "T" formation to harass, slow, and disorganize a larger force. When the lead elements of the enemy are engaged, they will normally attempt to maneuver right or left to close with the ambush. Mines and other obstacles placed to the flanks of the killing zones slow the enemy's movements and permit the ambush force to deliver heavy fire and withdraw without becoming decisively engaged.

The "T" formation can be used to interdict small groups attempting night movement across open areas. For example, the assault element is deployed along an avenue of approach with every second man facing the opposite direction. The attack of the enemy approaching from either direction requires only that every second man may shift to the opposite side of the formation. Each man fires only to his front and only when the enemy is at a very close range. Attack is by fire only and each man keeps the enemy under fire as long as it remains to his front.

If the enemy attempts to escape in either direction along the killing zone, each Marine takes the enemy under fire as the enemy comes into the Marine's sector of fire. The "T" formation is very effective at halting infiltration. But it has one chief disadvantage; there is a possibility that the ambush will engage a superior force at night while spread out. (See fig. D-8.)

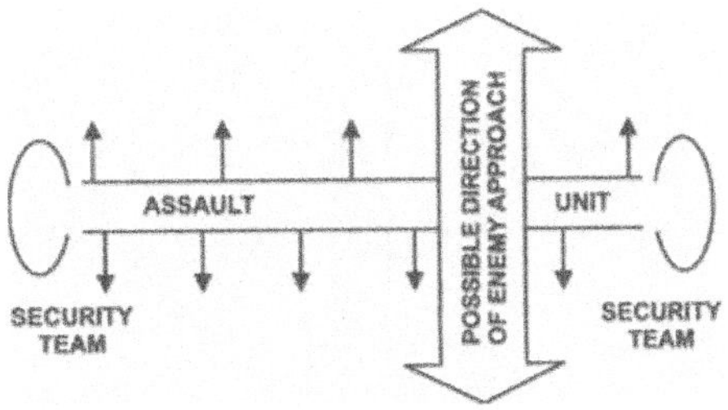

Figure D-8. "T" Formation (Target Approaching from Either Direction).

The "V" Formation

The "V" shaped formation is deployed along both sides of the enemy's route of movement so that it forms a "V"; care is taken to ensure that neither group (within the "V") fires into the other.

This formation subjects the enemy to both enfilading and interlocking fire. The "V" formation is best suited for fairly open terrain but can also be used in close terrain. When established in close terrain, the legs of the "V" close in as the lead element of the enemy force approaches the apex of the "V", and opens fire at a close range.

Here, even more than in open terrain, all movement and fire must be carefully coordinated and controlled to ensure that the fire of one wing does not endanger the other wing. The wider separation of forces makes this formation difficult to control, and there are few sites that favor its use. Its main advantage is that it is difficult for the enemy to detect the ambush until it is well into the killing zone. (See figs. D-9 and D-10.)

Triangle Formation

The triangle is a variation of the "V" formation and can be varied in three ways:

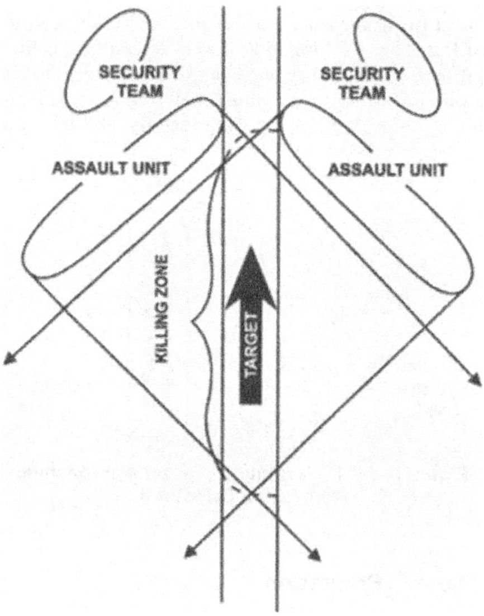

Figure D-9. "V" Formation (Open Terrain).

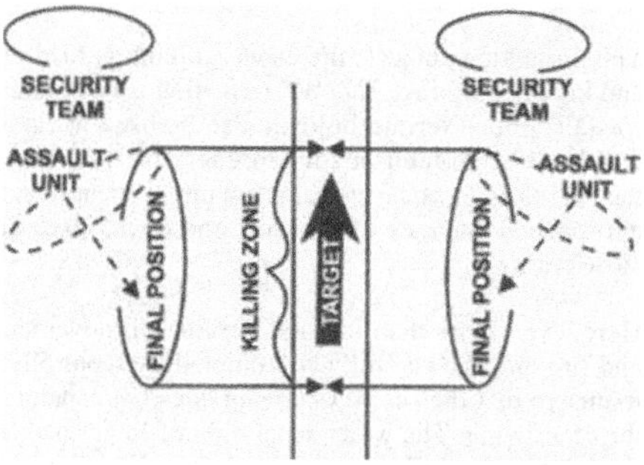

Figure D-10. "V" Formation (Close Terrain).

Closed Triangle Formation

(See fig. D-11.) The assault element is deployed in three teams, positioned so that they form a triangle (or closed "V"). An automatic weapon is placed at each point of the triangle and positioned so that it can be shifted quickly to interlock with either of the others. Men are positioned so that their sectors of fire overlap. Mortars may be positioned inside the triangle.

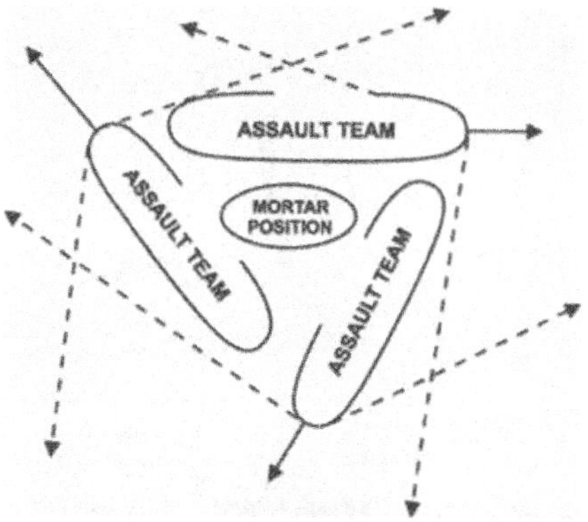

Figure D-11. Closed Triangle Formation (Night Ambush).

When deployed in this manner, the triangle ambush is used to interdict night movement through open areas. When enemy approach is likely to be from any direction, this formation provides all-around security, and security forces are deployed only when they can be positioned so that if detected by an approaching enemy, they will not compromise the ambush. Attack is by fire only, and the enemy is allowed to approach within close range before fire is initiated.

The advantages of the closed triangle formation are ease of control, all-around security, and the enemy can be brought under the fire of at least two automatic weapons, regardless of the direction they approach.

Disadvantages are that it requires a force of platoon size or larger to reduce the danger of being overrun by an unexpectedly large force and that one or more sides of the triangle may come under enfilade fire. The lack of dispersion, particularly at the points, increases the danger from enemy mortar fire.

Open Triangle Harassing Formation

This variation of the triangle ambush enables a small force to harass, slow, and inflict heavy casualties upon a large force without itself being decisively engaged. The assault element is deployed in three teams, positioned so that each team becomes a corner of a triangle containing the killing zone. (See fig. D-12.)

When the enemy enters the killing zone, the team to the enemy's front opens fire on the leading element. When the enemy counterattacks, the group withdraws and the team on the enemy's flank opens fire. When this team is attacked, the team to the opposite flank opens fire. This process is repeated until the enemy is pulled apart. Each team reoccupies its position, if possible, and continues to inflict the maximum damage possible without becoming decisively engaged.

Open Triangle Destruction Formation

The assault element is again deployed in three teams, positioned so that each team is a point of the triangle, 200 to 300 meters apart. The killing zone is the area within the triangle. The enemy is allowed to enter the killing zone; the nearest team attacks by fire. As the enemy attempts to maneuver or withdraw, the other teams open fire. One or more teams, as directed, assault or maneuver to envelop or destroy the enemy. (See fig. D-13 on page D-6.)

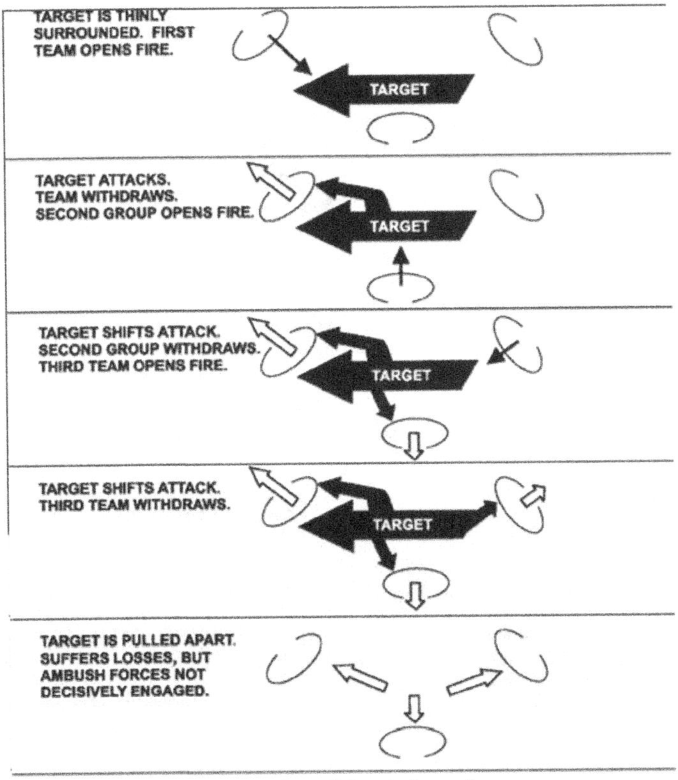

Figure D-12. Open Triangle Formation.

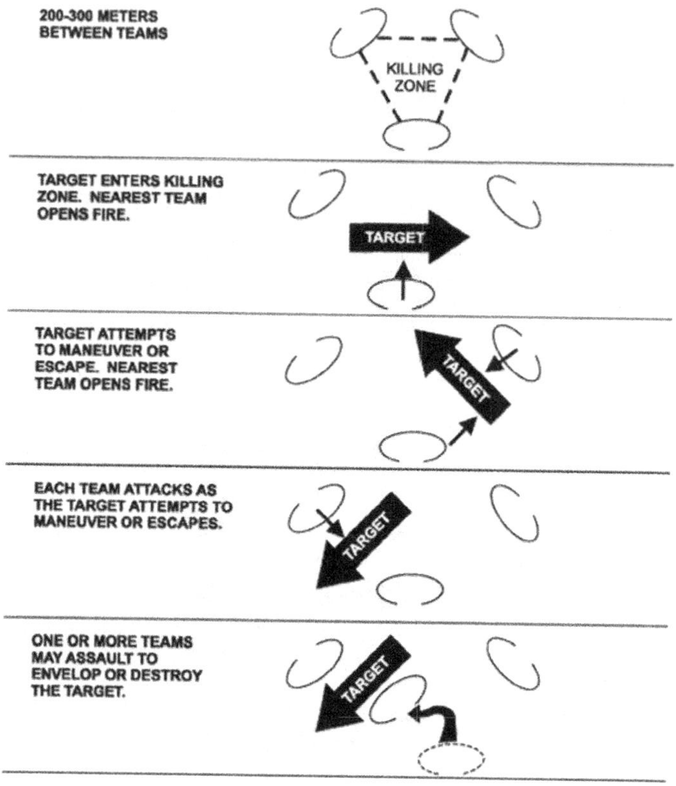

Figure D-13. Open Triangle Formation (Destruction Ambush).

This formation is suitable for platoon-size or larger ambush forces. A smaller force would be in too great a danger of being overrun. Another disadvantage is that control, in assaulting or maneuvering, is very difficult. Very close coordination and control are necessary to ensure that assaulting or maneuvering teams are not fired on by another team. The ambush site must be a fairly level open area that provides concealment around its border for the ambush force.

The Box Formation

The "box" formation is similar in purpose to the open triangle ambush. The assault element is deployed in four teams, positioned so that each team becomes a corner of a square or rectangle containing the killing zone. It can be used for a harassing or destruction ambush in the same manner as the variations of the open triangle formation. (See figs D-14 on page D-7 and D-15 on page D-8.)

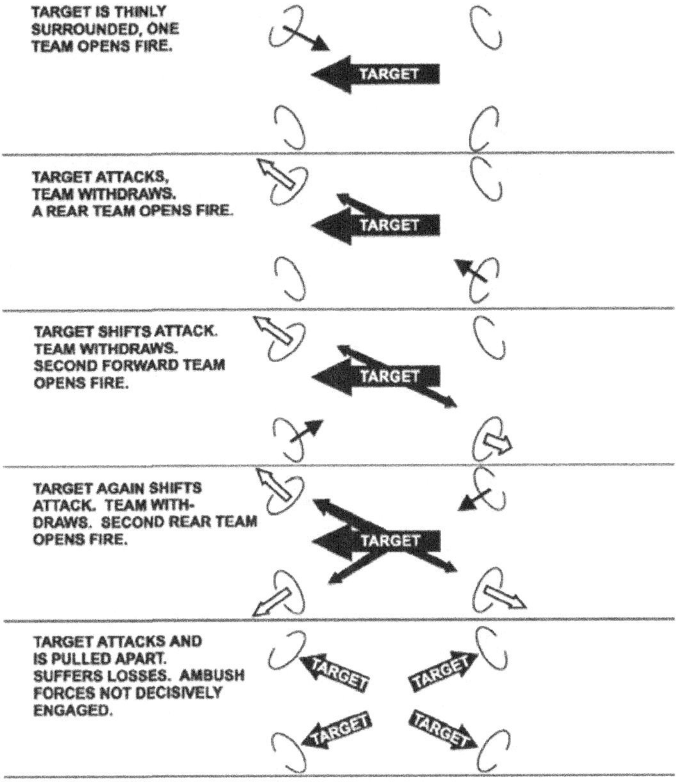

Figure D-14. Box Formation (Harrassing Ambush).

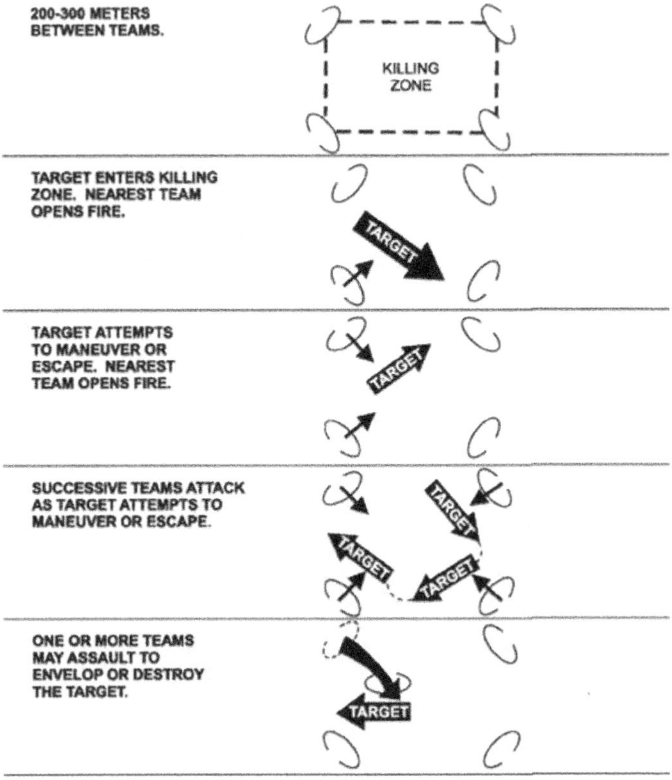

Figure D-15. Box Formation (Destruction Ambush).

APPENDIX E. ACRONYMS

A&S . . . assault and security

CIT . . . counterintelligence team

EOD . . . explosive ordnance disposal

ITT . . . interrogator-translator team

MAGTF . . . Marine air-ground task force
MEDEVAC . . . medical evacuation
METT-T . . . mission, enemy, terrain and weather, troops and support available

NBC . . . nuclear, biological, chemical
NCO . . . noncommissioned officer

OTL . . . overserver-target line

RCA . . . riot control agents
ROE . . . rules of engagement

SMAW . . . shoulder-launched multipurpose assault weapon
SMEAC . . . situation, mission, execution, administration and logistics, and command and signal
SOP . . . standing operating procedures
SPOTREP . . . spot report

TTP . . . tactics, techniques, and procedures

APPENDIX F. REFERENCES

Joint Publication (JP)

1-02 DOD Dictionary of Military and Associated Terms

Fleet Marine Force Manual (FMFM)

6-5 Marine Rifle Squad (proposed MCWP 3-11.2)

Marine Corps Warfighting Publications (MCWP)

3-15.3 Scout Sniping (under development)

3-35.3 Military Operations on Urbanized Terrain (MOUT)

3-41.2 Raids (under development)

Marine Corps Reference Publications (MCRP)

2-15.3A Reconnaissance Patrol Leader's Planning Handbook (under development)

2-15.3B Reconnaissance Reports Guide

3-02C Water Survival Handbook (under development, currently exists as Fleet Marine Force Manual 0-13)

5-12A Operational Terms and Graphics

5-12C Marine Corps Supplement to the DOD Dictionary of Military and Associated Terms

Made in United States
North Haven, CT
04 November 2026

21764783R00378